MW00564657

PRINCIPLES OF LASER MATERIALS PROCESSING

PRINCIPLES OF LASER MATERIALS PROCESSING

Elijah Kannatey-Asibu, Jr.

Published by John Wiley & Sons, Inc., Hoboken, New Jersey

Published simultaneously in Canada

Library of Congress Cataloging-in-Publication Data:

Kannatey-Asibu, E.
Principles of laser materials processing / Elijah Kannatey-Asibu, Jr.
p. cm.
Includes bibliographical references and index.
ISBN 978-0-470-17798-3 (cloth)
1. Lasers–Industrial applications. 2. Materials science. I. Title.

TA1675.K36 2009
621.36'6–dc22 2008044731

Printed in the United States of America

10 9 8 7 6 5 4 3 2 1

To the memory of my parents
Kofi Kannatey and Efuwa Edziiba
And to my children
Bianca, Araba, and Kwame

CONTENTS

PREFACE

Applications of lasers in materials processing have been evolving since the development (first demonstration) of the laser in 1960. The early applications focused on processes such as welding, machining, and heat treatment. Newer processes that have evolved over the years include laser forming, shock peening, micromachining, and nanoprocessing. This book provides a state-of-the-art compilation of material in the major application areas and is designed to provide the background needed by graduate students to prepare them for industry; researchers to initiate a research program in any of these areas; and practicing engineers to update themselves and gain additional insight into the latest developments in this rapidly evolving field.

The book is partitioned into three parts. The first part, Principles of Industrial Lasers (Chapters 1–9), introduces the reader to basic concepts in the characteristics of lasers, design of their components, and beam delivery. It is presented in a simple enough format that an engineering student without any prior knowledge of lasers can fully comprehend it. It helps the reader acquire a basic understanding of how a laser beam is generated, its basic properties, propagation of the beam, and the various types of lasers available and their specific characteristics. Such knowledge is useful to all engineering students, irrespective of their specific interests or area of application. It will enable them to select an appropriate laser for a given application and help them determine how best to utilize the laser. The coverage starts with a discussion of laser generation—basic atomic structure and how it leads to atomic transitions (absorption, spontaneous emission, and stimulated emission). The concepts of population inversion and gain criterion for laser action are introduced. Optical resonators (planar and spherical) are discussed in relation to beam modes (longitudinal and transverse) and stability of optical resonators. Techniques for line and mode selection are outlined. Various pumping techniques that can be used to achieve inversion are then presented, including more recent developments such as diode pumping. The rate equations are then introduced to provide some insight into the conditions necessary for achieving population inversion for both three- and four-level systems. This is followed by a discussion of the broadening mechanisms that are responsible for the spread of a laser's frequency over a finite range. These include both homogeneous (natural and collision broadening) and inhomogeneous (Doppler) broadening. Beam modification mechanisms such as Q-switching and mode locking are presented. After obtaining a fundamental background on laser generation, the characteristics of beams that have a more direct impact on their application are then discussed. These include beam characteristics such as divergence, monochromaticity (with reference to broadening), coherence, polarization, intensity and brightness, frequency stabilization, and focusing. Different types of lasers are then discussed with specific emphasis on high-power lasers used in industrial manufacturing. These include solid-state lasers (Nd:YAG and Nd:glass); gas lasers (neutral atom, ion, metal vapor (copper vapor), and molecular (CO_2 and excimer lasers); dye lasers; and semiconductor (diode) lasers. Finally, beam delivery systems are introduced, discussing concepts such as the Brewster angle, polarization, beam expanders, beam splitters, and transmissive, reflective, and fiber optics.

The second part, Engineering Background (Chapters 10–13), reviews the engineering concepts that are needed to analyze the different processes. Topics that are discussed include thermal analysis and fluid flow, the microstructure that results from the heat effect, solidification of the molten metal for processes that involve melting, and residual stresses that evolve during these processes.

The third part, Laser Materials Processing (Chapters 14–23), provides a more rigorous and detailed coverage of the subject of laser materials processing and discusses the principal application areas such as laser cutting and drilling, welding, surface modification, laser forming, rapid prototyping, and medical and nano applications. Sensors that are normally used for monitoring process quality are also discussed, along with methods for analyzing the sensor outputs. Finally, basic concepts of laser safety are presented. The range of processing parameters associated with each process is outlined. The impact of the basic laser characteristics such as wavelength, divergence, monochromaticity, coherence, polarization, intensity, stability, focusing, and depth of focus, as discussed in Part I, for each process is emphasized.

The material in this book is suitable for a two-course sequence on laser processing. The material in Part I is adequate for an upper-division/first-year graduate course in engineering. Parts II and III can then be used for a follow-up course, or the material in Part I can be skipped if only one course needs to be offered. In either case, Part II can be quickly reviewed, and more time can be spent on Part III.

Two sets of nomenclature are used in this text. There is an overall nomenclature that is reserved for variables that are used throughout the text. In addition, each chapter has its own nomenclature for variables used primarily in that chapter.

The author wishes to express his gratitude to all his colleagues and friends who have provided feedback on the manuscript. Special gratitude goes to all the graduate students who critiqued the course pack on which the book is based and to Mr. Rodney Hill (rodhillgraphics.com) for the skillful illustrations.

University of Michigan
Ann Arbor, MI, USA

ELIJAH KANNATEY-ASIBU JR

PART I
Principles of Industrial Lasers

1 Laser Generation

In this chapter, we outline the basic principles underlying the generation of a laser beam. The term laser is an acronym for light amplification by stimulated emission of radiation, and thus a laser beam, like all other light waves, is a form of electromagnetic radiation. Light may be simply defined as electromagnetic radiation that is visible to the human eye. It has a wavelength range of about $0.37-0.75$ μm, between ultraviolet and infrared radiation, with a median wavelength and frequency of 0.55 μm and 10^{15} Hz, respectively. Lasers, however, may have wavelengths ranging from 0.2 to 500 μm, that is, from X-ray, through ultraviolet and visible, to infrared radiation. Figure 1.1 illustrates the electromagnetic spectrum, which indicates that the visible spectrum is only a minute portion of the entire spectrum. The colors associated with the various wavelengths in the visible range are listed in Table 1.1. The various colors are characterized by specific wavelength ranges. However, white light has the same amplitude over all wavelengths in the visible light region.

In its simplest form, laser generation is the result of energy emission associated with the transition of an electron from a higher to a lower energy level or orbit within an atom. Thus, before proceeding with our discussion on laser generation, we first look at the basic structure of an atom. This is followed by a discussion on atomic transitions and associated absorption, spontaneous emission, and stimulated emission, which form the cornerstone of laser generation. To gain some insight into the timescale associated with transitions, the lifetime or time constant of an excited atom is briefly discussed. The absorption of a beam as it propagates through a medium is then presented to lay the foundation for a discussion on two criteria that are necessary for sustaining laser oscillation, that is, population inversion and threshold gain. Finally, the concept of two-photon absorption is introduced.

1.1 BASIC ATOMIC STRUCTURE

From basic chemistry, we know that an atom consists of a nucleus that is surrounded by electrons and that the nucleus itself is composed of protons and neutrons. The electrons are negatively charged, and the protons positively charged, while the neutrons are electrically neutral. In a simplified description of the atomic structure based on the Bohr model, the electrons are considered to move in circular orbits of

---

Principles of Laser Materials Processing, by Elijah Kannatey-Asibu, Jr.

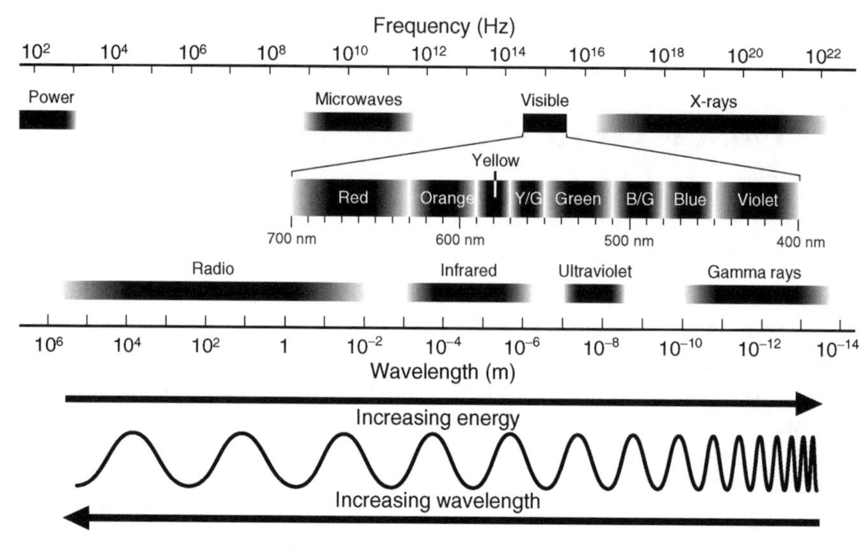

FIGURE 1.1 The electromagnetic spectrum.

specific radii corresponding to discrete energy states, with the nucleus as the center. The atom may then be viewed as consisting of circular shells of electrons, where the lowest shell corresponds to the lowest energy state, with the energy of the higher level shells or energy states being integral multiples, m ($m = 1, 2, 3, \ldots$), of the lowest state (Fig. 1.2). This is a simple description of atomic structure using classical mechanics.

The actual motion of an electron in an atom is best characterized using quantum mechanics rather than classical mechanics due to the uncertainty associated with specifying the position and velocity of the electron, a consequence of the Heisenberg uncertainty principle. Classical mechanics is based on a deterministic description of the motion, that is, assumes precise knowledge of the position and velocity. Quantum mechanics, on the contrary, is based on stochastic theory, which indicates the

TABLE 1.1 Wavelengths Associated with the Visible Spectrum

Wavelength Range (nm)	Color
400–450	Violet
450–480	Blue
480–510	Blue–green
510–550	Green
550–570	Yellow–green
570–590	Yellow
590–630	Orange
630–700	Red

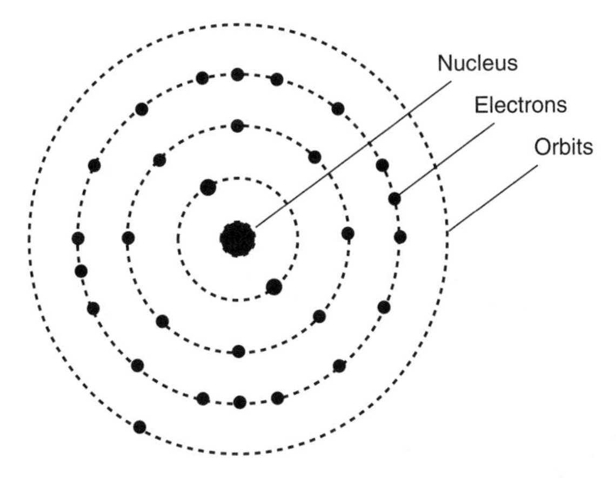

FIGURE 1.2 Schematic of atomic structure (copper).

probability of finding the electron at a certain location rather than a precise specification of the location. This is illustrated in Fig. 1.3. On the basis of this, the motion of electrons in an atom is described by the Schrodinger wave equation, which for the one-dimensional case is given by

$$\frac{d^2\psi}{dx^2} + \frac{8\pi^2 m_e}{{h_p}^2}(E - V_e)\psi = 0 \tag{1.1}$$

where ψ is the wave function of the electron, h_p is the Planck's constant $= 6.625 \times 10^{-34}$ J − s, m_e is the mass of electron $= 9.11 \times 10^{-31}$ kg, E is the total energy of electron (J), V_e is the potential energy of electron (J), and x is the electron position (m).

ψ is a probability function such that $|\psi|^2 dx$ is the probability of the electron being in an interval dx. There are only specific or discrete values of E for which a solution is obtained for the fully three-dimensional form of equation (1.1), indicating that the

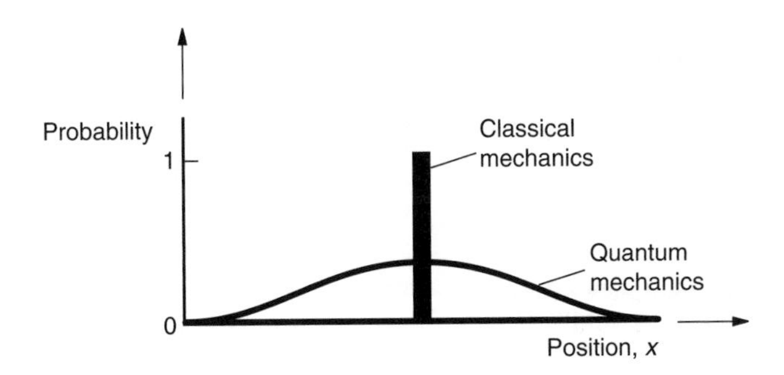

FIGURE 1.3 Classical and quantum mechanics descriptions of electron position in an atom.

electron can only have discrete energy states. This leads to the quantized nature of atoms.

In the more realistic description of the atomic structure given by the Schrodinger equation, the electron orbit is more complex, and not necessarily circular. The motion of the electrons is then described by a set of four quantum numbers indicated by n, l, m, s.

- n: This is the principal quantum number and determines the size of the electron orbit. It takes on integer values $n = 1, 2, 3, 4, \ldots$, with each number corresponding to a specific orbit. These are commonly referred to as the $K, L, M, N, \ldots$ shells. The energy differences between contiguous n values are much greater than those for the other quantum numbers.
- l: This is the quantum number characterizing the orbital angular momentum of the electron and takes on the integer values $l = 0, 1, 2, 3, \ldots, n-1$, which are commonly referred to as the s, p, d, f, ... states. The angular momentum of the s state electrons is zero, giving them a spherical orbit since they move in all directions with the same probability. The other states do not have zero angular momentum, and thus have some directionality to their orbital motion.
- m: This is the magnetic quantum number and indicates the spatial orientation of the angular momentum. It takes on the integer values $-l \leq m \leq l$. It affects the energy of the electron only when an external magnetic field is applied.
- s: This is the quantum number associated with the spin of the electron itself about its own axis and indicates the angular momentum of the electron. It is either $+\frac{1}{2}$ or $-\frac{1}{2}$.

When atoms that have the same n and l values but different m and s values have the same amount of total energy, they are said to be degenerate. The presence of a magnetic field, however, changes the energy of each degenerate level, depending on m and s, thereby removing the degeneracy.

On the basis of the foregoing discussion, we find that each electron in an atom is identified by a set of quantum numbers that uniquely defines its motion, and according to the *Pauli exclusion principle*, no two electrons in an atom can have the same set of quantum numbers. Thus, there cannot be more than one electron in a quantum state. The quantum states of an atom are occupied starting with the lowest level, that is, $n = 1$, and increasing as each level is filled. Within each given level, say $n = 2$, occupation starts with the $l = 0$ or s sublevel. Thus for, say, copper with an atomic number of 29 (i.e., 29 electrons), the atomic structure (Fig. 1.2), will be

$$1s^2 2s^2 2p^6 3s^2 3p^6 3d^{10} 4s$$

where the principal quantum numbers are indicated by the integers, while the superscripts indicate the number of electrons that have the same principal and orbital quantum numbers.

After briefly reviewing the basic structure of the atom, we now turn our attention to atomic transitions or changes in the energy levels of electrons that form the basis for laser generation. The discussion starts with the selection rules that determine what transitions can occur. This is followed by a discussion on how atoms are distributed among different energy levels. The different forms of transition are then presented, along with the Einstein coefficients, which constitute an integral part of the equations governing the transitions.

1.2 ATOMIC TRANSITIONS

Under the right conditions, electrons within an atom can change their orbits. Light or energy is emitted as an electron moves from a higher level or outer orbit to a lower level or inner orbit and is absorbed when the reverse transition takes place. The emission and absorption of light is explained by the fact that light generally consists of photons, which are small bundles or quanta of energy or particles. A photon is able to impart its energy to a single electron, enabling it to overcome the force of attraction restraining it to the atom surface (the work function) and also providing it with initial kinetic energy that enables it to move from a lower to a higher orbit. There is a specific quantum of energy (a photon), ΔE, of specific wavelength or frequency associated with each transition from one orbit or energy level to another and is given by

$$\Delta E = \frac{h_{\mathrm{p}}c}{\lambda} = h_{\mathrm{p}}\nu \tag{1.2}$$

where c is the velocity of light $= 3 \times 10^8$ (exactly $299, 792, 458$) m/s, λ is the wavelength (m), ν is the frequency of transition between the energy levels (Hz), and ΔE is the energy difference between the levels of interest.

Since there are a number of discrete orbits, there are a variety of different transitions possible, and thus many different frequencies that can be emitted.

1.2.1 Selection Rules

As we learned in Section 1.1, the motion of an electron in an atom can only be described using probability theory. In much the same way, the transition of an electron from one energy level to another can only be described using probability theory. The transitions that have a very high probability of occurring are said to be the allowed transitions. Other transitions have a very low probability of occurring (almost nonexistent). These are the forbidden transitions. The selection rules determine which transitions are permitted. One common rule is that, during a transition, there cannot be a change in the total spin of an atom (which is the sum of the individual electron spins). Let us now take a look at the distribution of atoms at various energy levels under equilibrium conditions.

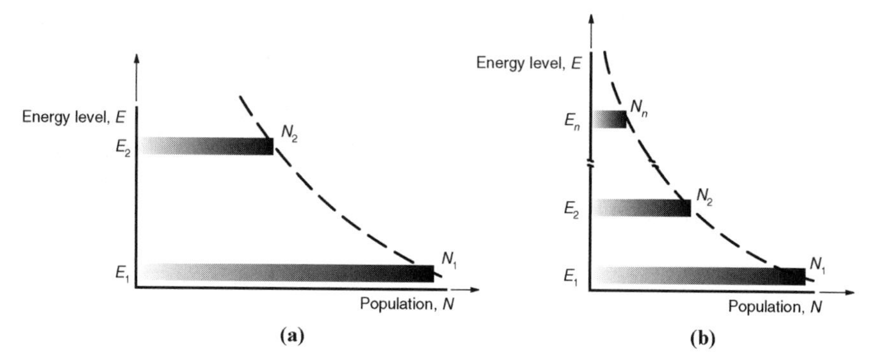

FIGURE 1.4 Schematic of Boltzmann's law. (a) Two-level system. (b) More general case for a multilevel system.

1.2.2 Population Distribution

For simplicity, let us focus our initial discussion on a single frequency, which corresponds to two specific energy levels or orbits, E_1 and E_2, where E_1 is the lower energy level and E_2 is a higher energy level, that is, $E_2 > E_1$. Furthermore, we let the *population* or number of atoms (or molecules or ions) per unit volume at level 1 be N_1 and that at level 2 be N_2. We also assume conditions of nondegeneracy. Degeneracy exists when there is more than one level with the same energy. For example, when atoms with the same values of n and l but different values of m and s have the same total energy.

Now let us consider the distribution of atoms among the energy levels under normal or thermal equilibrium conditions. Under such conditions of thermal equilibrium, the lower energy levels are more highly populated than the higher levels, and the distribution is given by Boltzmann's law that relates N_1 and N_2 as

$$\frac{N_2}{N_1} = e^{-\frac{E_2 - E_1}{k_B T}} = \mathrm{e}^{-\frac{h_p \nu}{k_B T}} \tag{1.3}$$

where k_B is Boltzmann's constant = 1.38×10^{-23} J/K and T is the absolute temperature of the system (K).

This is illustrated in Fig. 1.4a. Figure 1.4b illustrates the equilibrium distribution for the more general case. Boltzmann's law holds for thermal equilibrium conditions, and as such, N_2 will always be less than N_1 under equilibrium conditions. What this means is that the number of atoms with electron configurations corresponding to the excited or higher energy level will be less than those corresponding to the lower energy level. We now look at the various transitions associated with laser generation.

1.2.3 Absorption

Now consider atoms that are in a lower energy state, E_1. Generally, this would be the ground state. When such atoms are *excited* or *stimulated*, that is, they are subjected to some external radiation or photon with the same energy as the energy difference

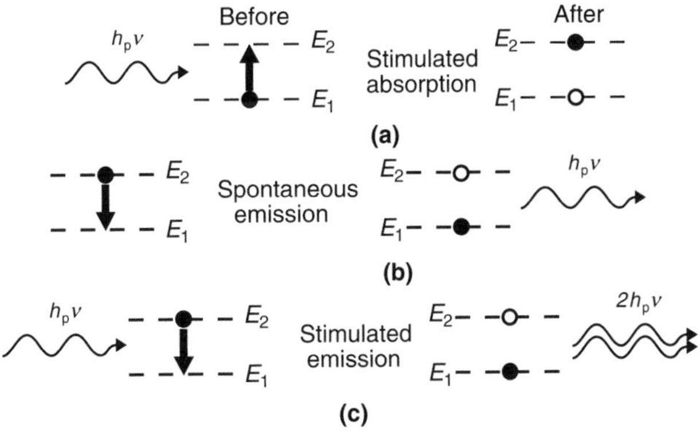

FIGURE 1.5 Schematic of (a) absorption, (b) spontaneous emission, and (c) stimulated emission. (From O'Shea, D. C., Callen, W. R., and Rhodes, W. T., 1977, *Introduction to Lasers and Their Applications*. Reprinted by permission of Pearson Education, Inc.)

between the lower state and a higher state or level, say E_2, the atoms will change their energy level (electrons raised to a higher energy level or molecular vibrational energy increased). If electromagnetic radiation of frequency ν is incident on these atoms, the atoms will absorb the radiation energy and change their energy level to E_2 (Fig. 1.5a) in correspondence with equation (1.2). This process is called *absorption* or more specifically, *stimulated absorption*. The rate at which energy is absorbed by the atoms will be proportional to the number of atoms at the lower energy level and also to the energy density of the incident radiation. Thus,

$$n_{\text{abs}} = B_{12} N_1 e(\nu) \tag{1.4}$$

where B_{12} is a proportionality constant referred to as the Einstein coefficient for stimulated absorption, or stimulated absorption probability per unit time per unit spectral energy density (m^3 Hz/Js), N_1 is the population of level 1 (per m^3), $e(\nu)$ is the energy density (energy per unit volume) at the frequency ν ($\text{J/m}^3\text{Hz}$), and n_{abs} is the absorption rate (number of absorptions per unit volume per unit time).

Once the atom has been excited to a higher energy level, it can make a subsequent transition to a lower energy level, accompanied by the emission of electromagnetic radiation. The emission process can occur in two ways, by *spontaneous emission* and/or *stimulated emission*. Each absorption removes a photon, and each emission creates a photon.

1.2.4 Spontaneous Emission

Spontaneous emission occurs when transition from the excited state to the lower energy level is not stimulated by any incident radiation but occurs more or less naturally (Fig. 1.5b). This happens because the excited atoms want to go back down to their ground state, and if left alone, it is just a matter of time before they do. If the atom was completely stable in its excited state, there would be no spontaneous emission.

The transition between energy levels E_2 and E_1 results in the emission of a photon of energy given by

$$\Delta E = E_2 - E_1 = h_p \nu \tag{1.5}$$

where ν is the frequency of the emitted photon. In spontaneous emission, the rate of emission per unit volume, n_{sp}, to the lower energy level is only proportional to the population, N_2, at the higher energy level and is independent of radiation energy density. Thus, we have

$$n_{sp} = A_e N_2 \tag{1.6}$$

where A_e is the Einstein coefficient for spontaneous emission, or spontaneous emission probability per unit time. The photons emitted by individual atoms under spontaneous emission are independent of each other, and thus there is neither a phase nor directional relationship between them.

1.2.5 Stimulated Emission

If the atom in energy level 2 is subjected to electromagnetic radiation or photon of frequency ν corresponding to the energy difference $\Delta E = E_2 - E_1$ between levels 1 and 2, the photon will stimulate the atom to undergo a transition to the lower energy level. The energy emitted as a result of this transition, which is in the form of an electromagnetic wave or a photon, is the same as the stimulating photon and is superimposed on the incident photon, thereby reinforcing the emitted light (Fig. 1.5c). This results in stimulated emission, where the incident and emitted photons have the same characteristics and are in phase, resulting in a high degree of coherence, and the direction, frequency, and state of polarization of the emitted photon are essentially the same as those of the incident photon. The two photons can generate yet another set, with a resulting avalanche of photons. This is illustrated schematically in Fig. 1.6. The rate of emission per unit volume, n_{st}, in the case of stimulated emission is also proportional to the population at level 2, as well as the energy density, and is given by

$$n_{st} = B_{21} N_2 e(\nu) \tag{1.7}$$

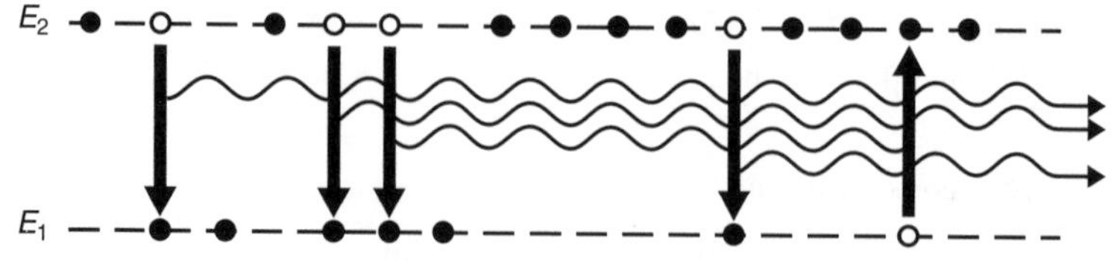

FIGURE 1.6 Illustration of the process of stimulated emission. (From Chryssolouris, G., 1991, *Laser Machining: Theory and Practice*. By permission of Springer Science and Business Media.)

where B_{21} is the Einstein coefficient for stimulated emission, or stimulated emission probability per unit time per unit energy density (m^3 Hz/J s).

1.2.6 Einstein Coefficients: A_e, B_{12}, B_{21}

Under conditions of thermal equilibrium, the rates of upward ($E_1 \rightarrow E_2$) and downward ($E_2 \rightarrow E_1$) transitions must be the same. Thus, we have

$$n_{1\rightarrow 2} = n_{2\rightarrow 1} \tag{1.8}$$

In other words,

$$\text{Stimulated absorption rate} = \text{stimulated emission rate} + \text{spontaneous emission rate} \tag{1.9}$$

or from equations (1.4), (1.6), and (1.7),

$$B_{12}N_1 e(\nu) = B_{21}N_2 e(\nu) + A_e N_2 \tag{1.10}$$

This gives the energy density as

$$e(\nu) = \frac{A_e}{B_{12}\frac{N_1}{N_2} - B_{21}} \tag{1.11}$$

Substituting equation (1.3) gives the energy density as

$$e(\nu) = \frac{A_e}{B_{12}e^{\frac{h_p \nu}{k_B T}} - B_{21}} \tag{1.12}$$

This can be compared with the energy density expression resulting from Planck's law on blackbody radiation, which is given by

$$e(\nu) = \frac{8\pi h_p \nu^3}{c^3}\frac{1}{e^{\frac{h_p \nu}{k_B T}} - 1} \tag{1.13}$$

Since equations (1.12) and (1.13) express the same energy density $e(\nu)$ in two different forms, we find that they can be equivalent only if

$$B_{12} = B_{21} = B$$

and

$$A_e = B\frac{8\pi h_p \nu^3}{c^3} \tag{1.14}$$

The Einstein coefficients B_{12} and B_{21} give the respective probabilities per unit time per unit spectral energy density that a stimulated transition will occur, while A_e is the probability per unit time that a spontaneous transition will occur. The equality of B_{12} and B_{21} indicates that the stimulated absorption and emission have the same probabilities of occurring between the same energy levels. Stimulated and spontaneous emissions, however, are related under equilibrium conditions by

$$\frac{A_e}{B} = (e^{\frac{h_p \nu}{k_B T}} - 1)e(\nu) = \frac{8\pi h_p \nu^3}{c^3} \tag{1.15}$$

Equation (1.15) indicates that for a given temperature, the rate of spontaneous emission is much greater than the rate of stimulated emission at high frequencies, whereas the opposite is true at relatively low frequencies.

Example 1.1

(a) Compare the rates of spontaneous and stimulated emission at room temperature ($T = 300$K) for an atomic transition where the frequency associated with the transition is about 3×10^{10} Hz, which is in the microwave region.

Solution:

From equation (1.15),

$$\frac{h_p \nu}{k_B T} = \frac{6.625 \times 10^{-34}(\text{J s}) \times 3 \times 10^{10}(\text{Hz})}{1.38 \times 10^{-23}(\text{J/K}) \times 300(\text{K})} \approx 5 \times 10^{-3}$$

Thus,

$$\frac{A_e}{B} = (e^{5\times 10^{-3}} - 1)e(\nu) \approx 0$$

This indicates that the stimulated emission rate is much greater than the spontaneous emission rate, and thus amplification is feasible in the microwave range at room temperature.

(b) Repeat Example 1.1a for a transition frequency in the optical region of $\nu = 10^{15}$.

Solution:

$$\frac{h_p \nu}{k_B T} = \frac{6.625 \times 10^{-34}(\text{J s}) \times 10^{15}(\text{Hz})}{1.38 \times 10^{-23}(\text{J/K}) \times 300(\text{K})} \approx 160$$

$$\Rightarrow \frac{A_e}{B} = (e^{160} - 1)e(\nu) \approx \infty$$

indicating that spontaneous emission is then predominant, resulting in incoherent emission from normal light sources. In other words, under conditions of thermal equilibrium, stimulated emission in the optical range is very unlikely.

(c) What will be the wavelength of the line spectrum resulting from the transition of an electron from an energy level of 40×10^{-20} J to a level of 15×10^{-20} J?

Solution:

From equation (1.2), we have

$$\Delta E = \frac{h_p c}{\lambda} = h_p \nu$$

$$\Rightarrow (40 - 15) \times 10^{-20} = \frac{6.625 \times 10^{-34} \times 3 \times 10^8}{\lambda}$$

Therefore,

$$\lambda = 0.792 \times 10^{-6} \text{ m} = 0.792 \ \mu\text{m}$$

1.3 LIFETIME

The time constant or *lifetime*, τ_{sp}, of atoms in an excited state is a measure of how long the atoms stay in that state, or the time period over which spontaneous transition occurs. Strictly speaking, this is how long it takes for the number of atoms in the excited state to reduce to $1/e$ of the initial value. To determine τ_{sp}, we consider the probability p that an atom will leave the excited state in the elemental time interval Δt as

$$p = p_t \Delta t \tag{1.16}$$

where p_t is the probability per unit time that a spontaneous transition will occur. Then if there are N atoms in that state at time t, the change ΔN in the number of atoms in the interval Δt is

$$\Delta N = -N p_t \Delta t$$

The negative sign is due to the fact that there is a reduction in the number of atoms. In the limit, we have

$$\frac{dN}{dt} = -N p_t \tag{1.17}$$

which when integrated gives

$$N = N_0 e^{-p_t t} \tag{1.18}$$

where N_0 is the number of atoms in the excited state at time $t = 0$. Thus for spontaneous emission, the lifetime or time constant is

$$\tau_{sp} = \frac{1}{p_t} \tag{1.19}$$

And since A_e is the probability per unit time that a spontaneous transition will occur,

$$\tau_{sp} = \frac{1}{p_t} = \frac{1}{A_e} \tag{1.20a}$$

or

$$A_e = \frac{1}{\tau_{sp}} \tag{1.20b}$$

We now discuss the absorption of a light beam as it propagates through an absorbing medium.

1.4 OPTICAL ABSORPTION

As a laser beam propagates through an absorbing medium, especially a gaseous medium, absorption by the medium results in the beam intensity diminishing as it propagates. In this section, we analyze the variation of the beam intensity with distance as it propagates. Consider a control volume $\Delta x \Delta y \Delta z$ of material through which a beam of specific frequency, ν, propagates in the x-direction (Fig. 1.7).

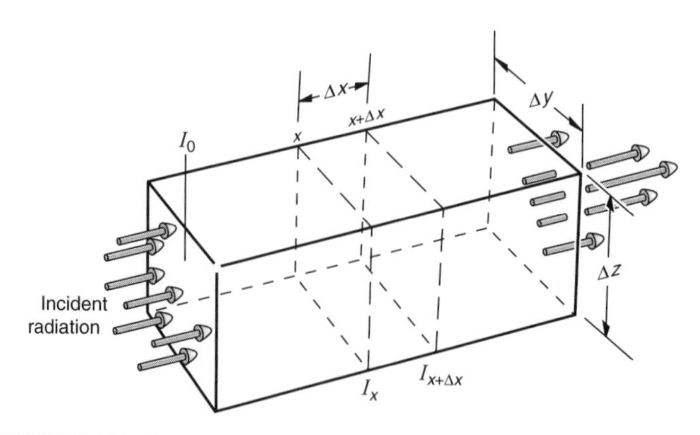

FIGURE 1.7 Propagation of a monochromatic beam in the x-direction.

From equation (1.4), the number of stimulated absorptions per unit time, R_{abs}, in the control volume due to the incident beam is given by

$$R_{abs} = n_{abs} \Delta x \Delta y \Delta z \tag{1.21}$$

And since the energy of each photon is $h_p \nu$, and each photon results in one transition, the energy absorption rate in the control volume $\Delta x \Delta y \Delta z$ is

$$q_{abs} = h_p \nu R_{abs} = n_{abs} h_p \nu \Delta x \Delta y \Delta z \tag{1.22}$$

Likewise, the energy rate of stimulated emission is given by

$$q_{st} = n_{st} h_p \nu \Delta x \Delta y \Delta z \tag{1.23}$$

Since spontaneous emission is transmitted in all directions, its contribution in any specific direction is negligible. We thus consider only the contributions of stimulated emission and absorption. The net rate of energy absorption in the control volume at the frequency ν is then given by

$$q_a = [n_{abs} - n_{st}] h_p \nu \Delta x \Delta y \Delta z \tag{1.24}$$

Let the irradiance or energy intensity of the beam (power per unit area) at a general location x be $I(x)$. Then the flux of energy (energy per unit time) into the element will be

$$q(x) = I(x) \Delta y \Delta z \tag{1.25}$$

and the flux out of the control volume will be

$$q(x + \Delta x) = I(x + \Delta x) \Delta y \Delta z = I(x) \Delta y \Delta z + \frac{\partial I(x)}{\partial x} \Delta x \Delta y \Delta z \tag{1.26}$$

Thus, the net flux of energy out of the control volume, q_e, is

$$q_e = q(x + \Delta x) - q(x) = \frac{\partial I(x)}{\partial x} \Delta x \Delta y \Delta z \tag{1.27}$$

Now under steady-state conditions, the net energy absorbed into the control volume per unit time and that out of it per unit time must be equal in magnitude, but of opposite sign. Thus from equations (1.24) and (1.27), we have

$$\frac{\partial I(x)}{\partial x} \Delta x \Delta y \Delta z = -[n_{abs} - n_{st}] h_p \nu \Delta x \Delta y \Delta z \tag{1.28}$$

Substituting for n_{abs} and n_{st} from equations (1.4), (1.7), (1.14), and (1.20a), we have

$$\frac{\partial I(x)}{\partial x} = -\frac{c^3}{8\pi \nu^2 \tau_{sp}} e(\nu)(N_1 - N_2) \tag{1.29}$$

But the intensity (energy per unit area per unit time) of a light beam propagating in a medium can also be expressed as the product of the propagating speed and energy density (energy per unit volume):

$$I = c_m e(\nu) \tag{1.30}$$

where $c_m = c/n$ is the velocity of the light beam in the medium and n is the refractive index of the medium. Equation (1.29) then becomes, for unidirectional propagation,

$$\frac{dI(x)}{dx} = -\alpha I(x) \tag{1.31}$$

where

$$\alpha = \frac{c^2 n}{8\pi \nu^2 \tau_{sp}}(N_1 - N_2) \tag{1.32}$$

Integration of equation (1.31) results in the following expression for the beam variation in the material:

$$I(x) = I_0 e^{-\alpha x} \tag{1.33}$$

where I_0 is the intensity of the incident beam (W/m^2), and α is the absorption coefficient (m^{-1}).

Equation (1.33) is known as the Beer–Lambert law and indicates that the beam intensity varies exponentially as it propagates into the medium.

Example 1.2 A medium absorbs 1% of the light incident on it over a distance of 1.5 mm into the medium. Determine

(i) The medium's absorption coefficient.
(ii) The length of the medium if it transmits 75% of the light.

Solution:

(i) From equation (1.33), we have

$$I(x) = I_0 e^{-\alpha x}$$

If 1% of the incident light is absorbed over a distance of 1.5 mm, then

$$0.99 \times I_0 = I_0 e^{-\alpha \times 1.5}$$

$$\Rightarrow \log_e 0.99 = -1.5 \times \alpha$$

Therefore,

$$\alpha = 6.7 \times 10^{-3}/\text{mm}$$

(ii) If 75% of the incident light is transmitted, then

$$0.75 \times I_0 = I_0 \mathrm{e}^{-6.7 \times 10^{-3} \times x}$$

$$\Rightarrow \log_e 0.75 = -6.7 \times 10^{-3} \times x$$

Therefore,

$$x = 42.9\,\text{mm}$$

1.5 POPULATION INVERSION

From equation (1.32), it is evident that α is positive if $N_1 > N_2$. The beam intensity then decreases exponentially with distance into the material. Since $N_1 > N_2$ under normal thermal equilibrium conditions, the beam will be attenuated as it propagates through the medium. α is then referred to as the *absorption coefficient* (with units: per unit length) and is positive.

However, if conditions are such that the number of atoms at the higher energy level is greater than those at the lower energy level, that is, $N_2 > N_1$, then α will be negative, in which case the beam intensity will increase exponentially as it propagates through the medium. In other words, the original radiation will be amplified. Equation (1.33) can then be written as

$$I(x) = I_0 \mathrm{e}^{\beta x} \tag{1.34}$$

$$\beta = \frac{c^2 n}{8\pi \nu^2 \tau_{sp}}(N_2 - N_1) \tag{1.35}$$

β is referred to as the *small-signal gain coefficient* and is positive. Such amplification of the original radiation results from stimulated emission that occurs when the right conditions exist.

The condition where $N_2 > N_1$ with more atoms existing at the higher energy level than at the lower energy level is referred to as *population inversion* (Fig. 1.8). From

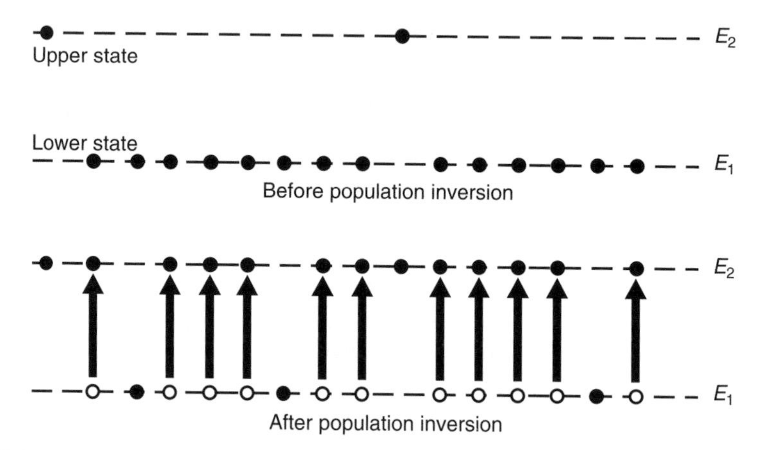

FIGURE 1.8 Population inversion. (From Chryssolouris, G., 1991, *Laser Machining: Theory and Practice*. By permission of Springer Science and Business Media.)

Boltzmann's equation, it is evident that population inversion does not occur under normal thermal equilibrium conditions. For it to be possible, atoms within the laser medium have to be excited or *pumped* to a nonequilibrium state (see Chapter 3). This is done with the fusion of a substantial quantity of energy into the medium using an external source. The process decreases the number of atoms at the lower energy level while increasing the number at the higher level. A material in which population inversion is induced is called an *active medium*.

Unfortunately, even though population inversion is a necessary condition for achieving laser action, it is not a sufficient condition, since a significant number of the excited atoms decay spontaneously to the lower energy level. Thus, there may be no laser action even when a population inversion is achieved. Compensation for the loss due to such decay is accomplished by introducing positive feedback into the system to amplify the laser beam. This is done using *optical resonators*, discussed in greater detail in chapter 2. However, we outline the basic concepts of optical resonators in the following section to gain insight into the other condition necessary for laser action, the threshold gain.

1.6 THRESHOLD GAIN

In an actual laser, the active medium is normally placed between two mirrors and these together constitute the resonator (Fig. 1.9a). Initially, spontaneous emission results in photons being generated in all directions (Fig. 1.9b). However, as stimulated emission becomes significant, an electromagnetic wave travelling along the axis of the resonator oscillates between the two mirrors (Fig. 1.9c). When population inversion exists, the radiation is amplified on each passage through the medium resulting in the buildup of signal intensity (Fig. 1.9d). A useful output beam is obtained by making

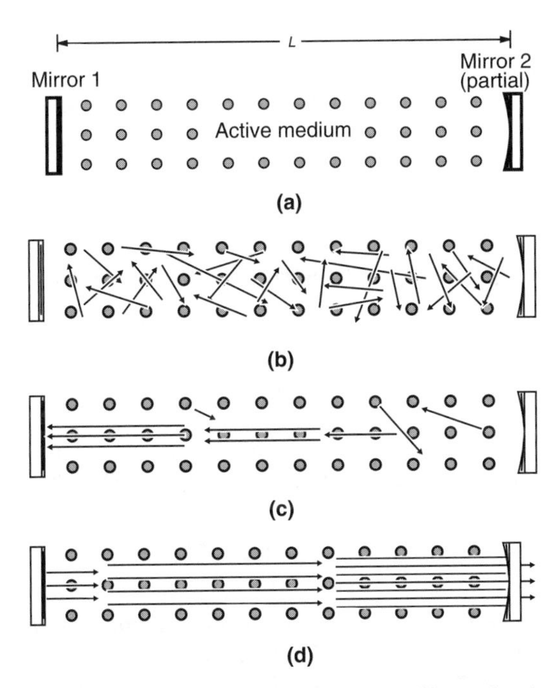

FIGURE 1.9 Illustration of laser amplification. (From Chryssolouris, G., 1991, *Laser Machining: Theory and Practice*. By permission of Springer Science and Business Media.)

one of the mirrors partially transparent. This output beam is the laser that comes out of the system. Such output coupling, along with absorption and scattering by the active medium, absorption by the mirrors, and so on, results in losses in the radiation intensity. These losses may be simply described by equation (1.33). If the losses encountered by the radiation during each passage are greater than the amplification or *gain* of the laser, equation (1.34), then the oscillations cannot continue, and the radiation intensity will eventually die down. Thus for oscillations to be maintained, the gain of the system must at least be equal to the losses in the system. This is accomplished when a *threshold gain*, β_{th}, is reached.

To determine the conditions under which the threshold for achieving laser oscillation is reached, we first note that the amplification factor for the beam during each passage through the medium is determined by the small-signal gain coefficient, β. Likewise, let the factor for losses due to absorption in the material, scattering, and so on in the medium be given by the absorption coefficient α. Equations (1.32) and (1.35) may indicate that α and β are equal and of opposite sign. These are simplifications. In reality, they have different values since α also includes losses due to scattering and other phenomena that were not considered in the development of equation (1.32). Likewise, β.

Now, let the reflectivities or reflection coefficients of the two mirrors be R_1 and R_2. If the length of the active medium is L, and the initial beam intensity as it leaves mirror

2 is I_0, then, on passage through the active medium, its intensity will be amplified by $e^{\beta L}$ as a result of stimulated emission (see equation (1.34)). However, the beam will simultaneously be attenuated by a factor of $e^{-\alpha L}$ because of medium losses (see equation (1.33)). Thus, the beam intensity as it leaves the medium will be

$$I = I_0 e^{(\beta - \alpha)L} \tag{1.36}$$

After reflection from mirror 1, the intensity is further reduced, as a result of reflection losses, to

$$I = I_0 R_1 e^{(\beta - \alpha)L} \tag{1.37}$$

Another transmission through the medium followed by a reflection from mirror 2 results in a beam intensity after one complete passage through the resonator of

$$I = I_0 G_g = I_0 R_1 R_2 e^{2(\beta - \alpha)L} \tag{1.38}$$

where G_g is the round-trip power gain. Oscillation (continued bouncing to and fro within the resonator) can be maintained only if the beam intensity does not diminish after each passage, that is, the amplification or gain must be sufficient to compensate for the energy lost. That means

$$I \geq I_0 \tag{1.39}$$

or

$$G_g = R_1 R_2 e^{2(\beta - \alpha)L} \geq 1 \tag{1.40}$$

This is the other condition for achieving oscillation in a resonator, and the threshold for oscillation is given by the lower bound. Under steady-state conditions, $I = I_0$. The small signal threshold gain is then given by

$$\beta_{th} = \alpha + \frac{1}{2L} \ln\left(\frac{1}{R_1 R_2}\right) \tag{1.41}$$

The second term on the right-hand side of equation (1.41) reflects the losses due to useful output coupling. Substituting for α from equation (1.32) and considering only the threshold condition, we have

$$N_2 - N_1 = -\frac{8\pi \nu^2 \tau_{sp}}{c^2 n}\left(\beta_{th} - \frac{1}{2L} \ln \frac{1}{R_1 R_2}\right) \tag{1.42}$$

That means it is not just sufficient to achieve population inversion for laser action to be initiated, but the population inversion must also attain a certain critical or threshold value given by equation (1.42).

Our discussion thus far has focused on the excitation of an atom from one energy level to another by the absorption of a photon of energy. There are a number of other situations where more than one photon is involved in the excitation, and this is discussed in the next section, focusing on the simpler case of two-photon absorption.

1.7 TWO-PHOTON ABSORPTION

Two-photon absorption is essentially a form of multiphoton absorption, and as such, a nonlinear optical phenomenon. To explain the concepts of nonlinear optics in simple terms, we consider the analogy between electromagnetic phenomena and mechanical oscillation. Excitation of a spring–mass system by an external force will cause the mass to oscillate. For relatively small displacements, the relationship between the force and displacement is linear. However, for large enough displacements, the relationship becomes nonlinear.

In electromagnetic wave theory, the equivalent to the driving force is the electric field that is applied, for example, a laser beam; the mass is equivalent to the electrons; and the displacement is equivalent to the polarization. The relationship between the electric field, E_l, and polarization, P_l, is linear when the strength of the electric field is relatively low, as obtained for ordinary light sources, and is given by

$$P_l(t) = \chi E_l(t) \tag{1.43}$$

where χ is the susceptibility of the dielectric medium and is independent of $E_l(t)$, but is a function of the frequency.

The actual relationship between the electric field and polarization, however, is a power series given by

$$P_l(t) = \chi^{(1)} E_l(t) + \chi^{(2)} E_l(t)^2 + \chi^{(3)} E_l(t)^3 + \cdots \tag{1.44}$$

Here, $\chi^{(m)}$ is a tensor, and $\chi^{(1)} = \chi$. However, $\chi^{(2)}$, $\chi^{(3)}$, and so on are nonlinear susceptibilities and define the degree of nonlinearity. The elements of $\chi^{(m)}$ get smaller and smaller, the higher m gets. Thus for the effects of the higher order terms to be detected, the intensity of the light source has to be very high. This is normally achieved with pulsed lasers.

The nonlinear characteristic of electromagnetic radiation gives rise to several different phenomena. One of these is frequency doubling, which is essentially a $\chi^{(2)}$ effect and enables the frequency of an electromagnetic radiation to be doubled by passing it through a special crystal. For example, the frequency of an Nd:YAG laser

FIGURE 1.10 The two-photon absorption concept.

(infrared wavelength = 1.064 μm) can be doubled to produce a visible green beam of wavelength 0.532 μm after passing through a nonlinear crystal, say β-barium borate. Energy is conserved in the process. In general, higher order frequencies can also be generated.

With this general background, we now turn our attention to the specific phenomenon of two-photon absorption. Normal excitation of an atom or molecule from a lower energy level to a higher level involves absorption of a photon of a specific energy or wavelength defined by the energy levels (Section 1.2.3). Under normal circumstances, the excitation to a specific higher level cannot take place in steps. However, when the atom is exposed to radiation of a high enough intensity, the atom can simultaneously absorb two longer wavelength photons, resulting in the same effect as a single photon of half the wavelength (Fig. 1.10). The process is referred to as two-photon absorption. The combined energy of the two photons enables the atom to be excited to the higher energy level. In essence, each photon provides half the energy of the electronic transition. This occurs because each of the longer wavelength photons can excite the atom to a transient or virtual state that has a lifetime of a few femtoseconds. The virtual state is forbidden for a single-photon transition. Two-photon absorption can take place only if the second photon is absorbed before the virtual state decays. The concerted interaction of the two photons then results in a combined energy that enables the atom to be excited to a level equivalent to what would be induced by a single photon of half the wavelength or twice the energy (see equation (1.2)).

In essence, the absorption cannot occur sequentially. If it occurs one after the other, then the transition is equivalent to two single-photon transitions and not a two-photon transition. Since the two photons have to be absorbed simultaneously, it is essential for the laser source to be ultrashort pulsed and of high peak power or intensity. Femtosecond lasers have thus been found very effective in a number of applications. Rapid, repeated pulsing of the laser provides adequate instantaneous intensity for the process, while maintaining a low average power.

One of the distinguishing features of single-photon and two-photon absorption is the rate at which energy is absorbed by each process. The rate or probability for a two-photon absorption is proportional to the square of the beam intensity (or the fourth power of the electric field amplitude), while the rate for a single-photon absorption is directly proportional to the beam intensity. In general, the rate of n-photon absorption is proportional to the nth power of the photon flux density. To achieve the high intensity necessary for two-photon absorption to be initiated, it is necessary for the beam to be tightly focused. With a tightly focused beam, two-photon absorption is confined to the highly localized focal volume.

Two-photon absorption has found application in such areas as fluorescence microscopy, 3D optical data storage, lithography, photodynamic therapy, and microfabrication.

1.8 SUMMARY

A laser is a form of electromagnetic radiation with wavelengths ranging from X-ray to infrared radiation. It is the result of energy emission associated with the transition of an electron from a higher to a lower energy level or orbit within an atom. This starts with excitation of an atom to a higher energy state as it absorbs a photon. When the excited atom is stimulated by a photon, it also releases another photon as it undergoes a transition to the lower state. This results in stimulated emission, where the incident and emitted photons have the same characteristics and are in phase, resulting in a high degree of coherence. Otherwise the photon is released spontaneously. Under conditions of thermal equilibrium, the rates of upward and downward transitions are the same. Again, under thermal equilibrium conditions, the distribution of atoms at the various energy levels is given by Boltzmann's law, where the number of atoms at the higher energy levels is lower than that at lower energy levels. For a given temperature, the rate of spontaneous emission is much greater than the rate of stimulated emission at high frequencies, while the opposite is true at relatively low frequencies.

As a laser beam propagates through an absorbing medium, especially a fluid medium, absorption by the medium results in the beam intensity diminishing exponentially, according to the Beer–Lambert law, as it propagates. For the beam intensity to increase as it propagates through the active medium, there has to be population inversion, with the number of atoms at the higher energy levels being higher than that at lower energy levels. For it to be possible, atoms within the laser medium have to be excited or pumped to a nonequilibrium state. In addition, it is necessary that the population inversion should attain a certain critical or threshold value.

Under conditions of very high energy density, multiphoton absorption can take place, where more than one photon is involved in exciting an atom to a higher energy level. In addition to the high energy density required, the incident radiation also has to be ultrashort pulsed, since the multiple photons have to be absorbed almost simultaneously. For two-photon absorption specifically, the atom simultaneously absorbs two longer wavelength photons, resulting in the same effect as a single photon of half the wavelength.

After looking at the conditions under which a laser beam is generated, we now look, in the next chapter, at the basic principle of the resonator, the device in which the radiation is generated and amplified, and how that affects the characteristics of the resulting beam.

REFERENCES

Barrett, C. R., Nix, W. D., and Tetelman, A. S., 1973, *The Principles of Engineering Materials*, Prentice Hall, Englewood Cliffs, NJ.

Berlien, H.-P., and Muller, G. J., 2003, *Applied Laser Medicine*, Springer-Verlag, Berlin.

Chryssolouris, G., 1991, *Laser Machining: Theory and Practice*, Springer-Verlag, Berlin.

Goeppert-Mayer, M., 1931, *Annalenden Physik*, Vol. 9, p. 273.

Henderson, A. R., 1997, *A Guide to Laser Safety*, Chapman and Hall, London.

Laud, B. B., 1985, *Lasers and Non-Linear Optics*, Wiley Eastern Limited, New Delhi.

Luxon, J. T., and Parker, D. E., 1985, *Industrial Lasers and Their Applications*, Prentice Hall, Englewood Cliffs, NJ.

O'Shea, D. C., Callen, W. R., and Rhodes, W. T., 1977, *Introduction to Lasers and Their Applications*, Addison-Wesley, Reading, MA.

Shimoda, K., 1986, *Introduction to Laser Physics*, 2nd edition, Springer-Verlag, Berlin.

Svelto, O., 1989, *Principles of Lasers*, 3rd edition, Plenum Press, New York.

Thyagarajan, K., and Ghatak, A. K., 1981, *Lasers, Theory and Applications*, Plenum Press, New York.

Wilson, J., and Hawkes, J. F. B., 1987, *Lasers: Principles and Applications*, Prentice Hall, New York.

APPENDIX 1A

List of symbols used in the chapter.

Symbol	Parameter	Units
A_e	Einstein coefficient for spontaneous emission	/s
B_{12}	Einstein coefficient for stimulated absorption	Sr m^2/J s
B_{21}	Einstein coefficient for stimulated emission	Sr m^2/J s
$B = B_{12} = B_{21}$	Einstein coefficient	Sr m^2/J s
G_g	round-trip power gain	–
n_{abs}	absorption rate (number of absorptions/unit volume/unit time)	/m^3 s
n_{sp}	spontaneous emission rate	/m^3 s
n_{st}	stimulated emission rate	/m^3 s
N_0	population (number of atoms) of excited state at time $t = 0$	/m^3
P_l	electric polarization	C m^2
q_{abs}	energy absorption rate	W
q_{st}	energy rate of stimulated emission	W
R_{abs}	number of stimulated absorptions per unit time	/s
V_e	potential energy of electron	J
x	electron position in X-direction	m
β	small-signal gain coefficient	/m
β_{th}	small-signal threshold gain coefficient	/m
χ	electric susceptibility	–
ψ	wave function of electron	–

PROBLEMS

1.1. Consider the energy levels E_1 and E_2 of a two-level system. Determine the population ratio of the two levels if they are in thermal equilibrium at room temperature, 27°C, and the transition frequency associated with this system is at 10^{15} Hz.

1.2. For the system in Problem 1.1, determine the fraction of atoms that will be in the lower state at room temperature, 27° C, if the transition wavelength associated with the two energy levels is (a) $\lambda = 1060$ nm and (b) $\lambda = 488$ nm.

1.3. If the population ratio, N_2/N_1, for the energy levels in Problem 1.1 under thermal equilibrium conditions at room temperature is $1/e^2$, determine the transition frequency and the type of electromagnetic radiation associated with this transition.

1.4. The oscillating wavelengths of the He–Ne, Nd:YAG, and CO_2 lasers are 0.6328, 1.06, and 10.6 μm, respectively. Determine the corresponding oscillating frequencies. What energy is associated with each transition?

1.5. Determine the stimulated emission probability, B, for a transition of wavelength $\lambda = 250$ nm, if its spontaneous emission rate, A, is 5×10^5/s. For the stimulated emission probability to be 400% that of spontaneous emission, what must be the irradiance at the given wavelength within the cavity at room temperature, 27°C?

1.6. A medium absorbs 0.5% of the light that passes through it for each millimeter of medium length. Determine its absorption coefficient and the percentage of light that will be transmitted if the overall length of the medium is 120 mm.

1.7. (a) Neglecting any losses, determine β for an active medium of length 2 m if the irradiance of light passing through it increases by 100%.

(b) Determine β if a 20% increase in irradiance occurred for the same length of active medium.

1.8. Determine the radiation density emitted by a blackbody at a frequency of 10^{15} Hz, if the temperature is 1200 K, using Planck's theory.

1.9. What should be the net gain coefficient, $\beta - \alpha$, for an active medium of length 0.1 m to result in a round-trip gain of 10%, assuming 100% reflection coefficients for both mirrors?

1.10. If the mirror reflection coefficients for a laser resonator of length 5 m are 98.5% and 60%, and there are no losses, determine the cavity threshold gain.

1.11. Would you expect the absorption coefficient α of a material to increase or decrease with temperature? Explain.

2 Optical Resonators

As discussed in Chapter 1, amplification of the laser beam is more effectively accomplished by using mirrors to reflect or oscillate the beam back and forth several times through the active medium, getting amplified with each passage. This, in a sense, introduces feedback into the system. A device that accomplishes this is an optical resonator, oscillator, or resonant cavity and is the focus of this chapter. The principle of an optical resonator is illustrated schematically in Fig. 1.9. Before discussing the physical characteristics of resonators, we outline the basic concepts of beam propagation in a medium, using Maxwell's equations. This will provide some insight into the modes of oscillation in a cavity. The oscillation characteristics of different types of resonators are then discussed, starting with the simplest form, planar resonators. This leads to the common modes of oscillation, longitudinal and axial modes, and methods of mode selection. This is followed by other forms of resonators, specifically confocal, generalized spherical, and concentric resonators. The chapter ends with a discussion on the stability of optical resonators. This chapter thus provides an understanding of the basic configuration of an optical resonator and the characteristics of the beam oscillation within it.

2.1 STANDING WAVES IN A RECTANGULAR CAVITY

The electromagnetic waves of light have associated with them electric and magnetic fields that oscillate in a direction perpendicular to the general direction of wave propagation (Fig. 2.1). The two waves are in phase, but perpendicular to each other, and can be represented by electric and magnetic vectors, $\mathbf{E}_l$ and $\mathbf{H}$, respectively. The characteristics of the light waves are described by Maxwell's equations that relate the electric field intensity $\mathbf{E}_l$, electric flux density $\mathbf{D} = \epsilon_p \mathbf{E}_l$, magnetic field intensity $\mathbf{H}$, magnetic flux density $\mathbf{B} = \mu_m \mathbf{H}$, charge density $\mathbf{Q}$, and current density $\mathbf{J}$ in electromagnetic theory. ϵ_p and μ_m are the electric permittivity and magnetic permeability, respectively, of the medium. Maxwell's equations are given by

$$\nabla \times \mathbf{E}_l = -\frac{\partial \mathbf{B}}{\partial t} = -\mu_m \frac{\partial \mathbf{H}}{\partial t} \tag{2.1}$$

$$\nabla \times \mathbf{H} = \mathbf{J} + \frac{\partial \mathbf{D}}{\partial t} = \mathbf{J} + \epsilon_p \frac{\partial \mathbf{E}_l}{\partial t} \tag{2.2}$$

Principles of Laser Materials Processing, by Elijah Kannatey-Asibu, Jr.

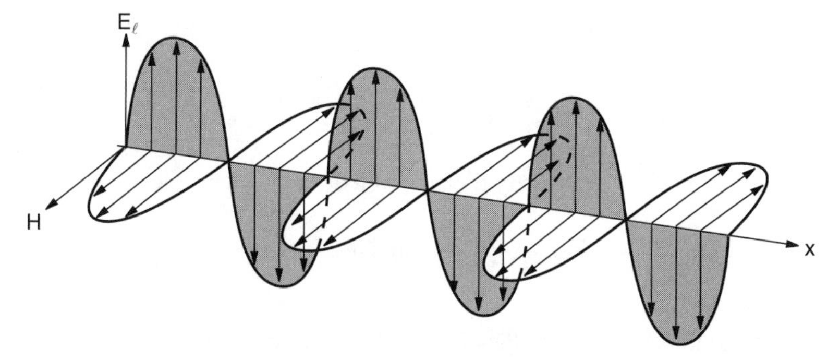

FIGURE 2.1 Electric and magnetic field vectors of an electromagnetic wave.

$$\nabla\cdot\mathbf{D} = \mathbf{Q} \tag{2.3}$$

$$\nabla\cdot\mathbf{B} = 0 \tag{2.4}$$

If we take the curl of equation (2.1), we get

$$\nabla\times(\nabla\times\mathbf{E}_\mathrm{l}) = -\nabla\times\frac{\partial\mathbf{B}}{\partial t} = -\frac{\partial}{\partial t}(\nabla\times\mathbf{B}) = -\mu_\mathrm{m}\frac{\partial}{\partial t}(\nabla\times\mathbf{H}) \tag{2.5}$$

Further expansion of the leftmost term using the identity $\nabla\times(\nabla\times\mathbf{E}_\mathrm{l}) = \nabla(\nabla\cdot\mathbf{E}_\mathrm{l}) - \nabla^2\mathbf{E}_\mathrm{l}$ (see Problem 2.1) reduces the equation to the form

$$\nabla(\nabla\cdot\mathbf{E}_\mathrm{l}) - \nabla^2\mathbf{E}_\mathrm{l} = -\mu_\mathrm{m}\frac{\partial}{\partial t}(\nabla\times\mathbf{H}) \tag{2.6}$$

For good dielectrics, the charge density $\mathbf{Q}$ is zero and thus from the relation $\mathbf{D} = \epsilon_\mathrm{p}\mathbf{E}_\mathrm{l}$, equation (2.3) reduces to

$$\nabla\cdot\mathbf{E}_\mathrm{l} = \frac{\partial E_{\mathrm{l}x}}{\partial x} + \frac{\partial E_{\mathrm{l}y}}{\partial y} + \frac{\partial E_{\mathrm{l}z}}{\partial z} = 0 \tag{2.7}$$

Substituting equation (2.7) into equation (2.6) gives

$$\nabla^2\mathbf{E}_\mathrm{l} = \mu_\mathrm{m}\frac{\partial}{\partial t}(\nabla\times\mathbf{H}) \tag{2.8}$$

Further substituting equation (2.2) into equation (2.8) gives

$$\nabla^2\mathbf{E}_\mathrm{l} = \mu_\mathrm{m}\frac{\partial}{\partial t}\left(\mathbf{J} + \epsilon_\mathrm{p}\frac{\partial\mathbf{E}_\mathrm{l}}{\partial t}\right) \tag{2.9}$$

Or since $\mathbf{J} = \sigma_e \mathbf{E}_l$, where σ_e is the electric conductivity of the medium, we have

$$\nabla^2 \mathbf{E}_l = \mu_m \frac{\partial}{\partial t}\left(\sigma_e \mathbf{E}_l + \epsilon_p \frac{\partial \mathbf{E}_l}{\partial t}\right) \tag{2.10}$$

And since $\sigma_e = 0$ for a perfectly dielectric medium, we have the wave equation

$$\nabla^2 \mathbf{E}_l = \mu_m \epsilon_p \frac{\partial^2 \mathbf{E}_l}{\partial t^2} \tag{2.11}$$

which is Maxwell's equation for the electric intensity vector $\mathbf{E}_l$.

Without loss of generality, we consider one component E_{lx} of the electric vector $\mathbf{E}_l$, resulting in the following Cartesian representation of the three-dimensional form of the wave equation:

$$\frac{\partial^2 E_{lx}}{\partial x^2} + \frac{\partial^2 E_{lx}}{\partial y^2} + \frac{\partial^2 E_{lx}}{\partial z^2} = \frac{1}{c_m{}^2} \frac{\partial^2 E_{lx}}{\partial t^2} \tag{2.12}$$

since the velocity of an electromagnetic wave in the medium is given by $c_m = 1/\sqrt{\mu_m \epsilon_p}$.

Again, without loss of generality, we now consider the medium to be free space. Equation (2.12) can be solved by separation of variables as

$$E_{lx} = X(x)Y(y)Z(z)T(t) \tag{2.13}$$

Now substituting equation (2.13) into equation (2.12) and dividing by $X(x)Y(y)Z(z)T(t)$ gives

$$\frac{1}{X}\frac{d^2 X}{dx^2} + \frac{1}{Y}\frac{d^2 Y}{dy^2} + \frac{1}{Z}\frac{d^2 Z}{dz^2} = \frac{1}{c^2 T}\frac{d^2 T}{dt^2} = -k_w{}^2 \tag{2.14}$$

where c is the velocity of light in free space. If the individual terms of equation (2.14) are expressed as

$$\frac{1}{X}\frac{d^2 X}{dx^2} = -k_x^2 \tag{2.15a}$$

$$\frac{1}{Y}\frac{d^2 Y}{dy^2} = -k_y^2 \tag{2.15b}$$

$$\frac{1}{Z}\frac{d^2 Z}{dz^2} = -k_z^2 \tag{2.15c}$$

we get

$$k_x{}^2 + k_y{}^2 + k_z{}^2 = k_w{}^2 \tag{2.16}$$

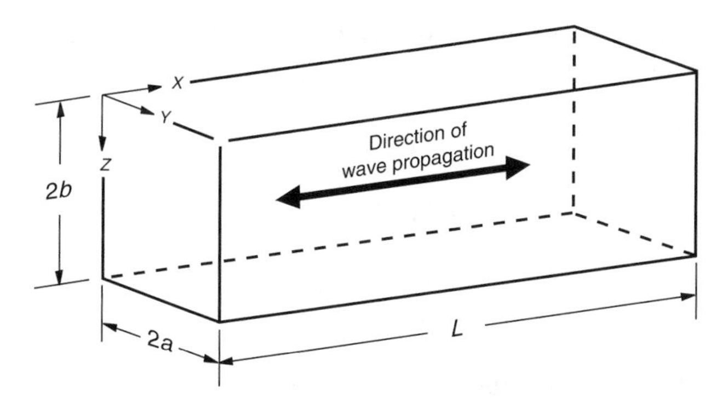

FIGURE 2.2 Rectangular cavity containing a dielectric medium.

where $k_{\mathrm{w}} = 2\pi/\lambda$ is the wavenumber. The general solution for equation (2.14) is

$$E_{1x} = (A_x \cos k_x x + B_x \sin k_x x)(A_y \cos k_y y + B_y \sin k_y y) \times (A_z \cos k_z z + B_z \sin k_z z)(A_t \cos \omega t + B_t \sin \omega t) \tag{2.17}$$

where $\omega = ck_{\mathrm{w}}$.

To determine the coefficients of equation (2.17), the initial and boundary conditions of the system have to be considered. For this, we start with the general case of a rectangular cavity with dimensions as shown in Fig. 2.2, which is filled with a homogeneous and isotropic dielectric medium. Let us further consider the walls or reflecting surfaces of the cavity to be perfectly conducting, which means that the tangential components of the electric field then vanish at the walls, that is,

$$E_1 x = 0 \quad \text{at} \quad \begin{cases} y = 0, \; y = 2a \\ z = 0, z = 2b \end{cases}$$

Or in a more general sense,

$$\mathbf{E_1} \times \underline{\mathbf{n}} = 0$$

at the cavity walls, where $\underline{\mathbf{n}}$ represents a unit vector normal to the cavity walls or reflecting surfaces. Considering the boundary condition at $y = 0$, we have from equation (2.17),

$$0 = A_y(A_x \cos k_x x + B_x \sin k_x x)(A_z \cos k_z z + B_z \sin k_z z)(A_t \cos \omega t + B_t \sin \omega t) \tag{2.18}$$

Thus, $A_y = 0$ since the solution would otherwise be trivial. Similarly, considering the boundary condition at $z = 0$ gives $A_z = 0$. Equation (2.17) then reduces to

$$E_{1x} = E_{01x} \sin k_y y \sin k_z z(A_x \cos k_x x + B_x \sin k_x x)(A_t \cos \omega t + B_t \sin \omega t) \quad (2.19)$$

where $E_{01x} = B_y B_z$. Now considering the other two boundary conditions, we have for the case $y = 2a$,

$$0 = E_{01x} \sin k_y 2a \sin k_z z(A_x \cos k_x x + B_x \sin k_x x)(A_t \cos \omega t + B_t \sin \omega t) \quad (2.20a)$$

And also for the case $z = 2b$,

$$0 = E_{01x} \sin k_y y \sin k_z 2b(A_x \cos k_x x + B_x \sin k_x x)(A_t \cos \omega t + B_t \sin \omega t) \quad (2.20b)$$

Again, for a nontrivial solution, we have

$$\sin k_y 2a = 0$$

$$\sin k_z 2b = 0$$

giving

$$k_y = \frac{m\pi}{2a}, \qquad k_z = \frac{n\pi}{2b}, \qquad m, n = 1, 2, 3, \ldots \quad (2.21)$$

In a similar manner, we obtain the following solutions for the y and z components of the electric vector:

$$E_{1y} = E_{01y} \sin k_x x \sin k_z z(A_y \cos k_y y + B_y \sin k_y y)(A_t \cos \omega t + B_t \sin \omega t) \quad (2.22)$$

$$E_{1z} = E_{01z} \sin k_x x \sin k_y y(A_z \cos k_z z + B_z \sin k_z z)(A_t \cos \omega t + B_t \sin \omega t) \quad (2.23)$$

From which we can again show that

$$k_x = \frac{p\pi}{L}, \qquad p = 1, 2, 3, \ldots \quad (2.24)$$

If we differentiate equation (2.22) with respect to y, we can show that

$$\frac{\partial E_{1y}}{\partial y} = 0 \quad (2.25a)$$

on the planes $x = 0$ and $x = L$. Likewise, we can show that

$$\frac{\partial E_{1z}}{\partial z} = 0 \quad (2.25b)$$

on the same planes. Thus from equation (2.7), we can say that

$$\frac{\partial E_{1x}}{\partial x} = 0 \tag{2.25c}$$

on the same planes. But from equation (2.19), we have

$$\frac{\partial E_{1x}}{\partial x} = E_{01x} k_x \sin k_y y \sin k_z z(-A_x \sin k_x x + B_x \cos k_x x)(A_t \cos \omega t + B_t \sin \omega t) \tag{2.26}$$

Substituting equation (2.25c) into equation (2.26) and again considering the nontrivial solution for the planes $x = 0$ and $x = L$, $B_x = 0$. Thus, the Cartesian components of the electric vector can be expressed as

$$E_{1x} = E_{0x} \cos k_x x \sin k_y y \sin k_z z(A_t \cos \omega t + B_t \sin \omega t) \tag{2.27a}$$

$$E_{1y} = E_{0y} \cos k_y y \sin k_x x \sin k_z z(A_t \cos \omega t + B_t \sin \omega t) \tag{2.27b}$$

$$E_{1z} = E_{0z} \cos k_z z \sin k_x x \sin k_y y(A_t \cos \omega t + B_t \sin \omega t) \tag{2.27c}$$

where E_{0x}, E_{0y}, and E_{0z} are constants. Equations (2.27a)–(2.27c) describe the standing wave patterns of the electric vector in a closed rectangular cavity and represent the modes of oscillation of the cavity. The amplitude of oscillation at any given point of the cavity is time invariant.

From equations (2.16), (2.21), and (2.24), we have

$$\omega^2 = c^2 k_w{}^2 = c^2 (k_x{}^2 + k_y{}^2 + k_z{}^2) \tag{2.28}$$

$$\Rightarrow \omega^2 = c^2 \pi^2 \left(\frac{m^2}{4a^2} + \frac{n^2}{4b^2} + \frac{p^2}{L^2} \right) \tag{2.29a}$$

or

$$\nu^2 = \frac{c^2}{4} \left(\frac{m^2}{4a^2} + \frac{n^2}{4b^2} + \frac{p^2}{L^2} \right) \tag{2.29b}$$

Equation (2.29) represents the permissible oscillating frequencies or waves of the field. The positive integers m, n, p give the number of nodes that the standing wave has along the y, z, and x axes, respectively. Each set of values of m, n, and p represents a well-defined cavity mode with a well-defined resonant frequency.

The number of modes, $M_c(\nu)$, that exist in a closed cavity of volume, V_c, over a frequency interval $\Delta\nu$ can be shown to be given by

$$M_c(\nu) = \frac{8\pi\nu^2 V_c}{c^3} \Delta\nu \tag{2.30}$$

Example 2.1 A laser cavity has a wavelength centered at $\lambda = 1500$ nm. If the cavity volume is 2000 mm^3, determine the number of modes falling in a bandwidth of $\Delta\lambda = 0.2$ nm.

Solution:

The number of modes is given by

$$M_c(\nu) = \frac{8\pi\nu^2 V_c}{c^3}\Delta\nu$$

Now

$$\nu\lambda = c, \quad \text{see equation (9.1a)}$$

where c is the velocity at which electromagnetic radiation propagates in free space.

$$\Rightarrow \nu = \frac{c}{\lambda} = \frac{3 \times 10^8}{1500 \times 10^{-9}} = 2 \times 10^{14} \text{ Hz}$$

Also, from equation (9.2a),

$$\Delta\nu = -\left(\frac{c}{\lambda^2}\right)\Delta\lambda = -\left(\frac{3 \times 10^8}{(1500 \times 10^{-9})^2}\right) \times 0.2 \times 10^{-9} = -2.67 \times 10^{10} \text{ Hz}$$

The negative sign arises since the wavelength increases as the frequency decreases. Now we have

$$M_c(\nu) = \frac{8\pi \times (2 \times 10^{14})^2 \times 2000 \times 10^{-9}}{(3 \times 10^8)^3} \times 2.67 \times 10^{10} = 1.99 \times 10^9$$

As can be seen from the example above, the number of modes is normally very high. An enclosed cavity with this number of modes oscillating would result in a laser beam covering a very wide range of wavelengths. To overcome this problem, an open resonator (see Fig. 2.3), which essentially has the lateral portions of the resonator or cavity removed, is normally used. With such a configuration, the modes traveling along the resonator axis will keep oscillating between the mirrors. However, the modes formed by superposition of waves traveling at an angle to the axis of the resonator will be lost after a few traversals. Thus, it can be said that $m, n << p$, since the modes with high m and n values then experience high diffraction losses, and this considerably reduces the number of modes that oscillate in the laser cavity. The open resonator with its lateral portions removed is essentially a planar resonator.

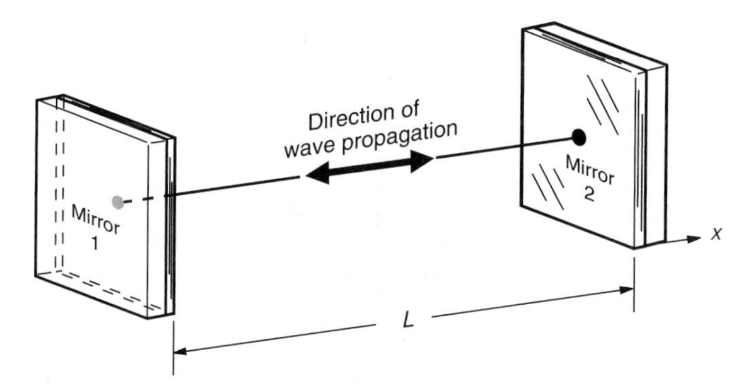

FIGURE 2.3 Planar or Fabry–Perot resonator.

2.2 PLANAR RESONATORS

The planar or Fabry–Perot resonator is one of the simplest forms of optical resonators and consists of two plane and parallel mirrors placed face to face and spaced by a given distance, L (Fig. 2.3). With the lateral sides open, we can make the approximation that $m, n << p$ and use a power series expansion to reduce equation (2.29), for the case where $a = b$, to give the resonant frequencies ω or ν in the cavity, in the form

$$\omega_{mnp} \approx c\pi \left[\frac{p}{L} + \left(\frac{m^2 + n^2}{a^2} \right) \frac{L}{8p} \right] \tag{2.31a}$$

or

$$\nu_{mnp} = \frac{\omega}{2\pi} \approx \frac{c}{2} \left[\frac{p}{L} + \left(\frac{m^2 + n^2}{a^2} \right) \frac{L}{8p} \right] \tag{2.31b}$$

If we ignore the last two terms of equation (2.31a), then the resonant frequencies can be approximated by

$$\omega = \frac{p\pi c}{L} \tag{2.32}$$

or

$$\nu = \frac{1}{2} \frac{pc}{L} \tag{2.33}$$

But the wavenumber k_x is given by

$$k_x = \frac{2\pi}{\lambda} = \frac{p\pi}{L} \tag{2.34}$$

giving the oscillating wavelength as

$$\lambda = \frac{2L}{p} \tag{2.35}$$

Thus, for the resonant or permissible frequencies that oscillate in the cavity, L is an integral multiple (p) of half wavelengths. This latter condition could also have been deduced by considering the fact that the standing wave has to be zero on the two mirrors. And further, by considering the condition that the phase change that a beam undergoes after traveling to one mirror and back must be an integral number multiple of 2π (assuming no phase changes on reflection, and unity refractive index) if a standing wave is to exist, and is given by

$$\phi = \frac{2\pi}{\lambda}(2L) = 2\pi p \tag{2.36}$$

One major drawback of the Fabry–Perot resonator is that it is very difficult to operate in single mode. In addition, the mirror surfaces must be flat to an accuracy of 1% of a wavelength, and alignment of the two mirrors must be precise and normal to the laser axis to an accuracy of a few seconds of an arc. This is because if the mirrors are not precisely aligned, the beam will be reflected off the mirrors after a few reflections back and forth. Spherical mirrors eliminate most of these problems and are thus more widely used in resonators. The advantage with the Fabry–Perot resonator is that since the light beam oscillating between plane-parallel mirrors is not focused, the resonator is able to fully utilize the available active medium for laser action.

Equations (2.27a)–(2.27c) represent the electric field components of the modes of a closed rectangular cavity, where the electromagnetic waves are assumed to be plane. The more rigorous or exact solution for an open cavity based on plane-parallel mirrors is beyond the scope of this chapter.

While we are on the subject of planar resonators, let us address one basic characteristic of laser beams, beam mode, and methodologies for modifying the modes in a laser beam.

2.2.1 Beam Modes

From the discussion in the preceding section, it is evident that the output of a laser may consist of one mode or a superposition of several modes or discrete frequency components. The modes that are sustained depend on the boundary conditions, such as the mirror shape and their separation. Each mode has its own characteristic frequency. There are two broad categories of beam modes:

1. Longitudinal or axial modes
2. Transverse modes.

2.2.1.1 Longitudinal Modes Going back to equation (2.31b), if we consider two modes that have the same values of m and n, but with a unit difference in the p value, the corresponding difference in frequency of oscillation will be

$$\Delta\nu_p = \frac{c}{2L} \tag{2.37}$$

These two contiguous modes will differ only in their field distribution along the longitudinal (or x) axis. Thus, the different modes resulting from the different values of p are referred to as the *longitudinal* or *axial modes.* The various axial modes appear as a single spot in the laser output, even though they are separated at discrete frequencies in the spectral domain.

For illustration, we note that a typical value for the mirror spacing in a laser would be about $L = 1500$ mm. Then,

$$\Delta\nu_p \approx \frac{3 \times 10^{11}}{2 \times 1500} = 100 \text{ MHz}$$

Thus, the separation between the longitudinal modes for this cavity is 100 MHz (Fig. 2.4a). The laser transition linewidth (see Section 5.1) (Fig. 2.4b), is normally much wider, say, 1 GHz for a Doppler broadened medium (see Section 5.2.2 and Table 8.1). As a result, the output of the laser will consist of a number of oscillating discrete frequencies, the axial modes, within the broadened linewidth (Fig. 2.4c).

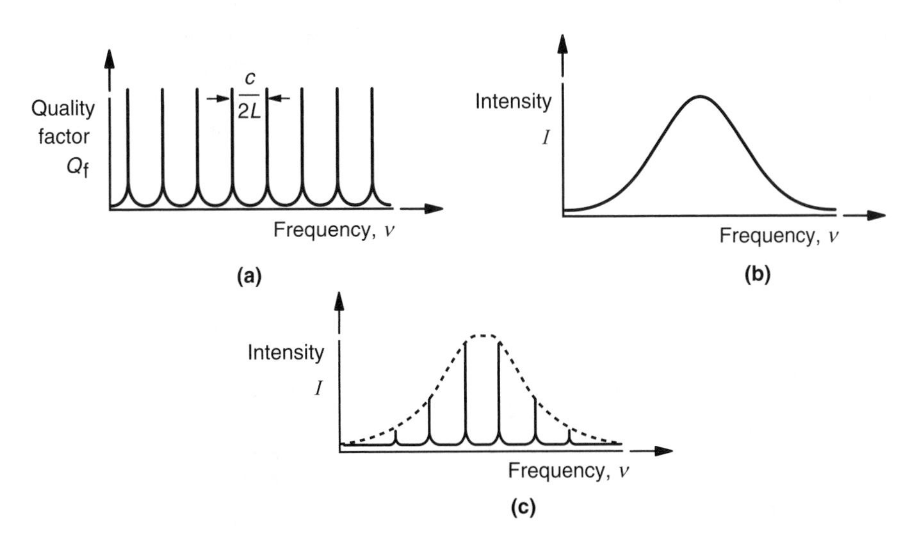

FIGURE 2.4 Longitudinal modes in a laser output. (a) Modes that can exist within the cavity. (b) Overall laser transition linewidth. (c) Output of the laser indicating the modes within it. (From O'Shea, D. C., Callen, W. R., and Rhodes, W. T., 1977, *Introduction to Lasers and Their Applications*, Reprinted by permission of Pearson Education, Inc.)

2.2.1.2 ***Transverse Modes*** When the modes have the same p value but differ in their values for m and n, the resulting field distributions differ in the transverse direction, and are thus referred to as the *transverse modes* of the cavity. In essence, the transverse modes give an indication of the distribution of intensity within the beam cross section. If we consider adjacent transverse modes where $m = n$ but the value of p does not change, and with $\Delta m = 1$, then we have from equation (2.29) (see Problems 2.5 and 2.6),

$$\Delta \nu_m = \frac{cL}{8a^2 p}\left(m + \frac{1}{2}\right) \tag{2.38}$$

or

$$\Delta \nu_m \approx \Delta \nu_p \frac{\lambda L}{8a^2}\left(m + \frac{1}{2}\right) \tag{2.39}$$

And considering the same example used for the axial modes, but with the mirror dimensions $a = b = 10$ mm, and with the beam in the infrared region of wavelength $\lambda \approx \frac{2}{3} \times 10^{-3}$ mm, then since $m \approx 1$, we find that

$$\Delta \nu_m \approx 100 \times \frac{1500 \times \frac{2}{3} \times 10^{-3}}{8 \times 100}\left(1 + \frac{1}{2}\right) \approx 0.2 \text{ MHz}$$

Thus, $\Delta \nu_p >> \Delta \nu_m$, indicating that the separation in frequency between longitudinal modes is much greater than that between transverse modes (Fig. 2.5).

The transverse modes show as a pattern of spots and are often referred to as *transverse electromagnetic* or *TEM* modes. They are characterized by two integers (m, n) that indicate the number of modes in two orthogonal directions and

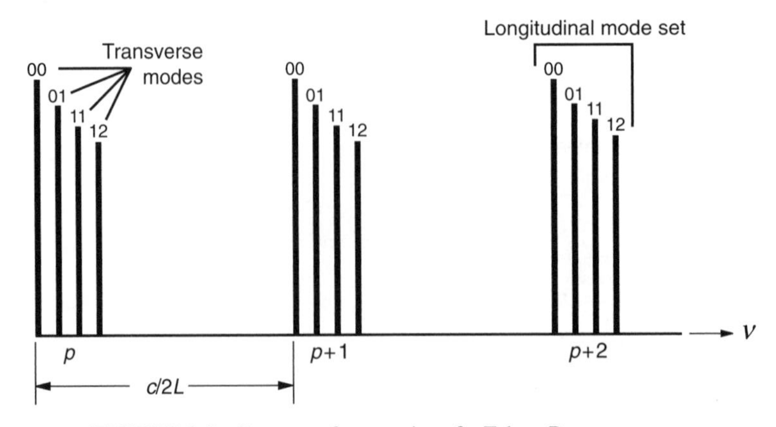

FIGURE 2.5 Resonant frequencies of a Fabry–Perot resonator.

are thus designated by TEM_{mn}. The m and n values are typically small, about 10 or less, and the larger the values, the more complex and spread out are the beam patterns.

The modes may have either rectangular or circular symmetry. For rectangular symmetry, the mode number is one less than the number of illuminated zones or pattern of spots in the corresponding direction. The first number corresponds to the y-axis while the second number corresponds to the z-axis. Some common mode shapes are shown in Fig. 2.6. Even when the mirrors and discharge tube are round, the laser normally oscillates with rectangular symmetry. TEM_{00} is the zero-order mode and ideally has a Gaussian distribution (Fig. 2.7). A laser operating in this mode has the highest spectral purity and degree of coherence and can be focused to the laser's theoretical minimum radius. This mode is more common in lower power lasers and is useful for laser cutting. The distribution of laser intensity across the beam is then given by

$$I(r) = I_{\mathrm{p}} \mathrm{e}^{\left(-\frac{2r^2}{w^2}\right)} \tag{2.40}$$

where $I(r)$ is the beam intensity at any radial position, r (W/cm^2), I_{p} is the peak beam intensity (W/cm^2), r is the radial coordinate (cm), and w is the radius at which beam intensity is $(1/e^2) \times I_{\mathrm{p}}$ or $0.14 \times I_{\mathrm{p}}$. That means 86% of the total energy is contained in a spot of radius w (cm).

A multimode beam, on the contrary, has a more evenly distributed energy (see, for example, Fig. 14.4b). Such higher order modes are more common with the high-power lasers. The higher order modes tend to have greater diffraction losses than the Gaussian (TEM_{00}) mode and have larger spot sizes. Generally, a laser oscillating in

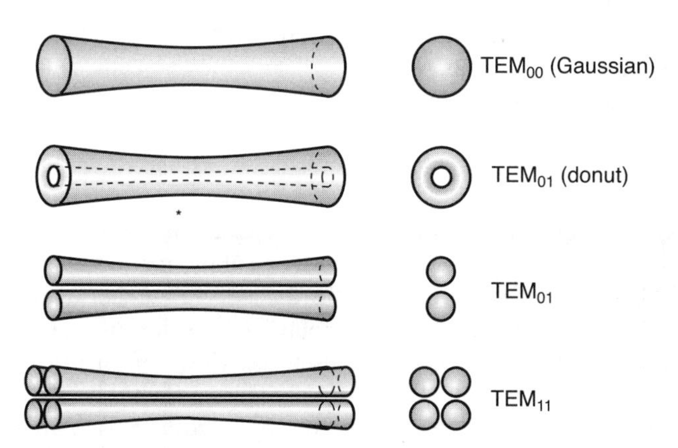

FIGURE 2.6 Mode shapes or schematic of TEM_{mn} mode patterns. (From Chryssolouris, G., 1991, *Laser Machining: Theory and Practice.* By permission of Springer Science and Business Media.)

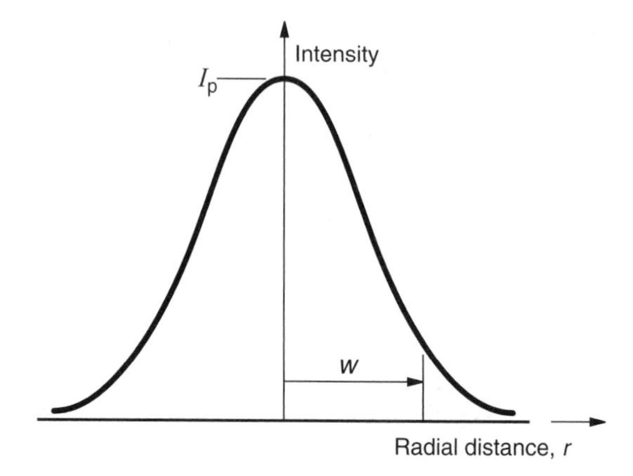

FIGURE 2.7 Gaussian (TEM_{00}) mode distribution.

a higher order transverse mode also generates lower order modes. In some manufacturing processes such as laser cutting, it is desirable to have the Gaussian mode so that the focused beam diameter that corresponds to the highest power density in the beam is achieved. However, the multimode beam is more desirable for laser butt welding applications where the wider beam area reduces restrictions on joint precision requirements.

The Fresnel number, F_r, which gives a measure of the tendency for a laser output to be either in low- or high-order mode, is given by

$$F_r = \frac{{r_a}^2}{\lambda L} \tag{2.41}$$

where r_a is the radius of the laser aperture and L is the cavity length. Low Fresnel numbers result in low-order modes. We now take a look at methods for suppressing undesirable modes that may be present in a cavity.

2.2.2 Line Selection

As will become evident later in our discussion on various ways in which lasers are generated (see Chapter 8), lasers often undergo transition simultaneously on a number of transition lines or wavelengths. Thus, there may be a number of these transition lines present in the output. Since each line contributes to the output power, the power may be relatively high under such circumstances. However, in situations where a higher degree of monochromaticity is desired, the laser should be made to oscillate on only one of the transitions. This can be done by inserting a wavelength dispersive element such as a prism or diffraction grating into the cavity (Fig. 2.8). As the beam propagates through the prism, the individual wavelengths undergo different levels of refraction. The wavelength whose ray is directed normal to the mirror is reflected back

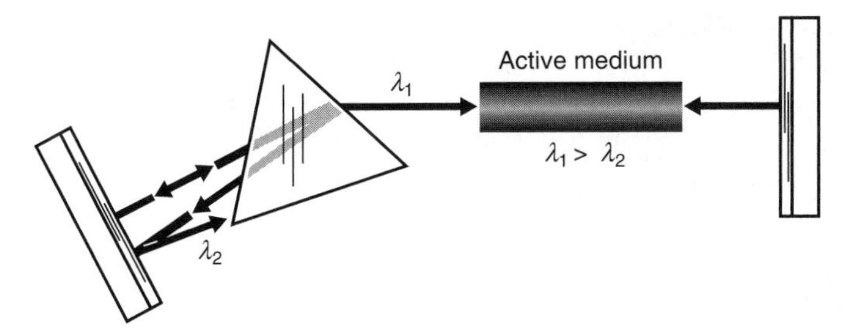

FIGURE 2.8 Wavelength selection.

into the cavity, while the other wavelengths experience losses, resulting in only one wavelength oscillating. The oscillating wavelength can be changed, that is, the laser tuned, by rotating the prism or diffraction grating. Even when the laser is oscillating on a single line, that line itself may be broadened and could be oscillating on a number of modes.

2.2.3 Mode Selection

As explained in Sections 2.1 and 2.2.1, the number of modes that oscillate within a laser cavity can be very large. This has an influence on the properties of the output beam. For example, the directionality of the output beam is reduced as the number of transverse modes increases, and likewise, the spectral purity is reduced as the number of longitudinal modes increases since the beam then contains a large number of discrete frequency components. Although these characteristics may be acceptable for some applications, there are other situations where they may not be appropriate. A case in point is when a laser is used for alignment. This requires low beam divergence, and the lowest divergence is obtained with the TEM_{00} mode. Thus for such applications, it might be necessary to suppress all the higher transverse modes for the laser to operate only in the TEM_{00} mode. Mode selection techniques are used to change the oscillating modes of a laser. In the next two subsections, we consider various methods for selecting transverse and longitudinal modes of a laser.

2.2.3.1 Transverse Mode Selection As will be shown later in Chapter 5, the field distribution of the laser beam is such that the Gaussian or TEM_{00} mode is smaller in size than all the other transverse modes. In fact, the higher the order of the mode, the wider the beam becomes. Furthermore, the energy of the higher order modes tends to be concentrated away from the resonator axis (see, for example, Fig. 2.6). Thus, the higher order modes can be eliminated by placing a diaphragm in the cavity and having it normal to the cavity axis. If the aperture of the diaphragm is made small enough, the higher order modes are severely attenuated, leaving only the TEM_{00} mode. One disadvantage with this technique is that reduction of the aperture also increases losses associated even with the fundamental mode, thereby reducing the overall output power available.

2.2.3.2 Longitudinal Mode Selection Longitudinal mode selection is used when it is desirable to have a highly monochromatic output beam. There are various techniques available for reducing the number of longitudinal modes that oscillate within a laser cavity. These include cavity length variation, the Fabry–Perot etalon, and the Fox–Smith interferometer.

Cavity Length Variation As discussed earlier in Section 2.2.1, the longitudinal modes are spaced apart by

$$\Delta\nu_p = \frac{c}{2L} \tag{2.42}$$

From equation (2.42), it is evident that decreasing the cavity length L increases the spacing between the discrete frequency modes. Thus, if the cavity length is decreased to the point that the frequency spacing is greater than the transition linewidth (see Section 5.1) (Fig. 2.9), then only one mode will oscillate. This technique is effective for cavities where the laser linewidths are relatively small, such as for gas lasers. One drawback of the technique, though, is that decreasing the cavity length decreases the volume of active material available for lasing, which can significantly reduce the output power. For solids and liquids where the laser linewidths are much greater, this technique is usually not appropriate.

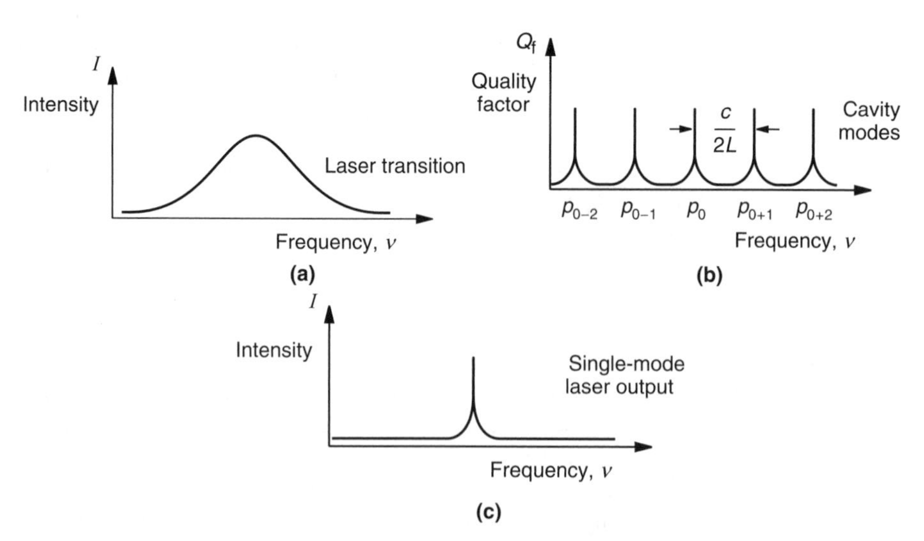

FIGURE 2.9 Longitudinal mode selection by frequency spacing. (a) Overall laser transition linewidth. (b) Modes that can exist within the cavity. (c) Output of the laser, indicating the single mode within it. (From O'Shea, D. C., Callen, W. R., and Rhodes, W. T., 1997, *Introduction to Lasers and Their Applications*. Reprinted by permission of Pearson Education, Inc.)

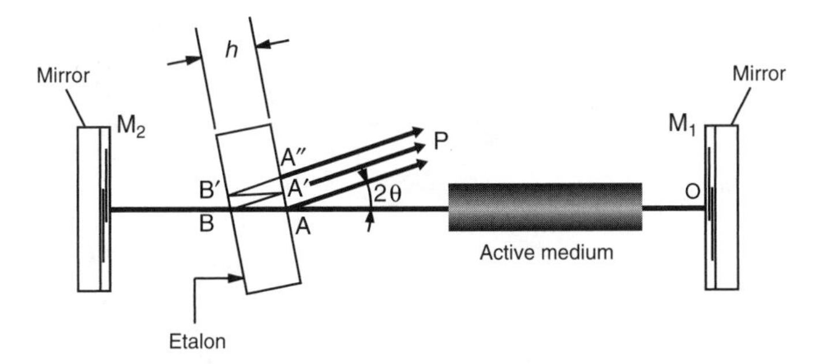

FIGURE 2.10 Fabry–Perot etalon setup.

Fabry–Perot Etalon Another technique involves the use of a Fabry–Perot etalon that may consist of a glass block with two of its faces ground parallel to a high degree of accuracy. This is inserted in the laser cavity as shown in Fig. 2.10. Let the thickness of the block or spacing between the parallel faces be h, and let the normal to the block be inclined at an angle θ to the cavity axis. A beam that is incident on the etalon is partially reflected from the first surface at point A, and the partially transmitted portion is reflected from the second surface at point B and retransmitted through the first surface at point A′. If the initially reflected beam OAP and retransmitted beam OBP, along with those resulting from multiple reflections, destructively interfere for certain modes, then those modes will have low loss. Since the incident beam OA is in the medium with the lower refractive index, it undergoes a 180° (π) phase shift on reflection at A, and it can be shown that OBP undergoes a phase shift, ϕ, of

$$\phi = \frac{2\pi n\nu}{c} 2h \cos\theta' \tag{2.43}$$

where n is the refractive index of the block, ν is the mode frequency, c is the velocity of light in free space, and θ' is the angle of refraction of the beam within the etalon.

For destructive interference, the phase difference between the two rays must be 180°, and thus we have

$$\frac{4\pi n\nu}{c} h \cos\theta' - \pi = (2m - 1)\pi, \quad m = 1, 2, 3, \ldots \tag{2.44}$$

From equation (2.44), we find that the modal frequencies that will be associated with minimum cavity loss when the etalon is inserted in the resonator are given by

$$\nu = \frac{mc}{2nh\cos\theta'} \tag{2.45}$$

with contiguous modes that encounter minimum loss being separated by

$$\Delta\nu = \frac{c}{2nh\cos\theta'} \tag{2.46}$$

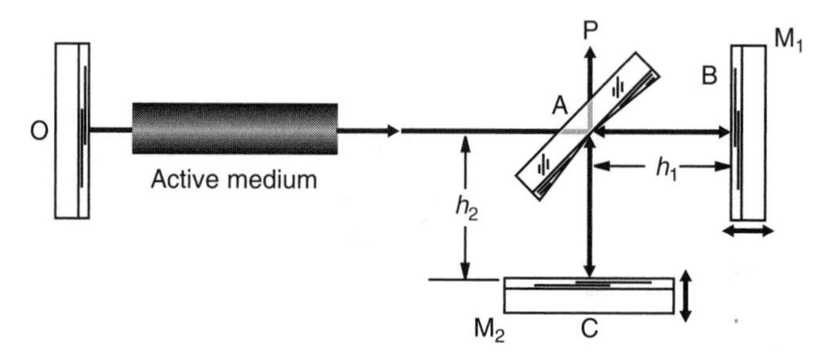

FIGURE 2.11 Fox–Smith interferometer. (From Thyagarajan, K., and Ghatak, A. K., 1981, *Lasers, Theory and Applications*. By permission of Springer Science and Business Media.)

Thus by decreasing the thickness of the etalon, the separation between modes can be increased to the point that only one mode oscillates in the resonator. Furthermore, the orientation θ of the etalon can be adjusted such that only the mode at the center of the linewidth oscillates. In the same vein, the laser can be tuned (oscillating frequency varied) over a narrow frequency range by adjusting θ.

Fox–Smith Interferometer The Fox–Smith interferometer, shown schematically in Fig. 2.11, is another technique that is used for longitudinal mode selection. It consists of two mirrors, M_1 and M_2, along with a beam splitter, A. The beam from O is partially reflected from the splitter as OAP, while the rest of the beam first undergoes 100% reflection at M_1, and then yet another 100% reflection at M_2 as OABACP. Just as in the case of the Fabry–Perot etalon, beam OAP undergoes a 180° phase shift after being reflected. The shift in phase of beam OABACP is

$$2\frac{2\pi}{\lambda}(h_1 + h_2) \tag{2.47}$$

where h_1 and h_2 are the distances of M_1 and M_2, respectively, from the beam splitter. Thus for destructive interference, we have

$$2\frac{2\pi}{\lambda}(h_1 + h_2) - \pi = (2m - 1)\pi, \quad m = 1, 2, 3, \ldots \tag{2.48}$$

resulting in a frequency difference between contiguous modes with low losses of

$$\Delta\nu = \frac{c}{2n(h_1 + h_2)} \tag{2.49}$$

where n is the refractive index of the medium in which the beam propagates. Again, by reducing $h_1 + h_2$, the modal separation can be increased until only one mode oscillates.

We now take a look at the characteristics of other forms of mirror arrangements that could be used for resonant cavities.

2.3 CONFOCAL RESONATORS

A confocal resonator consists of a pair of spherical mirrors that have the same radius, r_m, and are positioned such that the focal points of the two mirrors coincide, that is, $F_1 \equiv F_2$. The center of curvature of each mirror then lies on the surface of the other mirror, so that the distance L separating the mirrors is equal to the radius, r_m, (Fig. 2.12). For such a configuration, if the electric field is described by a scalar quantity, which is a reasonable approximation for a number of applications in optics, then the field distribution, g_{mn}, which represents the variation of the electric field magnitude over the mirrors, can be shown to be given for the Hermite–Gaussian (or rectangular symmetry) mode by

$$g_{mn}(y, z) = H_m\left[y\left(\frac{2\pi}{L\lambda}\right)^{1/2}\right] H_n\left[z\left(\frac{2\pi}{L\lambda}\right)^{1/2}\right] e^{-\frac{\pi}{L\lambda}(y^2+z^2)} \tag{2.50}$$

where H_m and H_n are Hermite polynomials of order m and n, respectively. They are given in general form by

$$H_n(y) = (-1)^n e^{y^2} \frac{\partial^n}{\partial y^n}\left(e^{-y^2}\right) \tag{2.51}$$

Example Hermite polynomials are

$$H_0(y) = 1 \tag{2.52a}$$

$$H_1(y) = 2y \tag{2.52b}$$

$$H_2(y) = 4y^2 - 2 \tag{2.52c}$$

$$H_3(y) = 8y^3 - 12y \tag{2.52d}$$

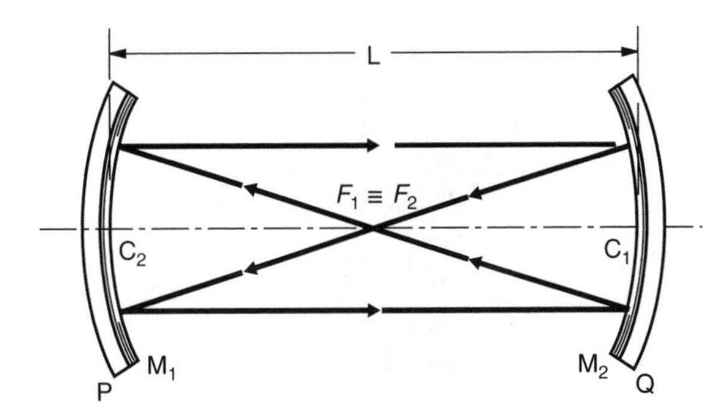

FIGURE 2.12 Confocal resonator. (From Svelto, O., 1989, *Principles of Lasers*, 3rd edition. By permission of Springer Science and Business Media.)

In equation (2.50), m and n are the mode numbers and they determine the transverse electric field amplitude distribution on the mirrors. Let us now consider some specific examples.

(a) For the case where $m = n = 0$, we have the TEM_{00} mode. This is the lowest order mode, and the field distribution is then given by

$$g_{00}(y, z) = e^{-\frac{\pi}{L\lambda}(y^2+z^2)} \tag{2.53}$$

The amplitude of the electric field on the mirror reduces to $1/e$ of the maximum amplitude at the center, at a radial distance w from the center, given (see Problem 2.11) by

$$w_m = \left(\frac{L\lambda}{\pi}\right)^{\frac{1}{2}} \tag{2.54}$$

w_m is characterized as the beam spot size on the mirror. The field has a Gaussian distribution (Fig. 2.13). The size of the focused beam is thus much smaller for a beam in the TEM_{00} mode than for one with the higher order modes.

Equation (2.53) describes the amplitude field distribution on the resonator mirrors. Equation (2.40), however, describes the intensity distribution at any section. The intensity relates to the power, which is proportional to the square of the electric field amplitude.

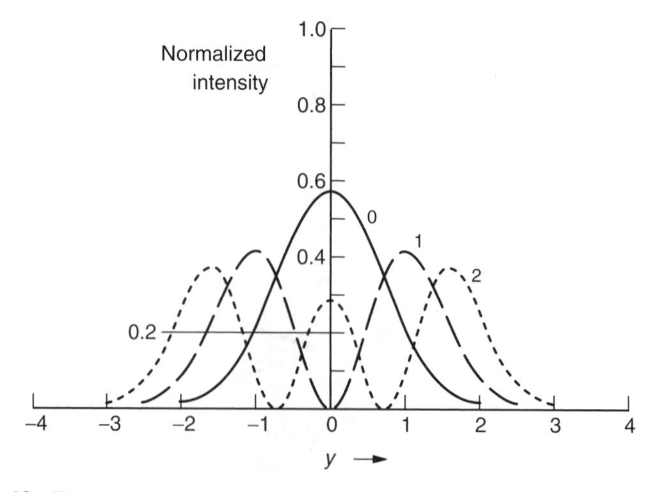

FIGURE 2.13 Transverse distributions in a confocal resonator. (From Thyagarajan, K., and Ghatak, A. K., 1981, *Lasers, Theory and Applications*. By permission of Springer Science and Business Media.)

(b) For $m = 0,\ n = 1$, we have the TEM_{01} mode. The field distribution is then given by

$$g_{01}(y, z) = H_1(z)\mathrm{e}^{-\frac{\pi}{L\lambda}(y^2+z^2)} \tag{2.55}$$

The amplitude variation along the y-axis is then as shown in Fig. 2.13 as order 0, while that along the z-axis is illustrated as order 1 in Fig. 2.13.

(c) For $m = 1,\ n = 1$, we have the TEM_{11} mode. The field distribution is given by

$$g_{11}(y, z) = H_1(y)H_1(z)e^{-\frac{\pi}{L\lambda}(y^2+z^2)} \tag{2.56}$$

and the variation in amplitude along both y and z axes are as shown in Fig. 2.13 as order 1.

Figure 2.13 thus illustrates the distributions for the low- order modes. The resonant frequencies in the confocal resonator cavity can be shown to be given by

$$\nu_{mnp} = \frac{c(2p + m + n + 1)}{4L} \tag{2.57}$$

Modes that have the same value of $(2p + m + n)$ have the same resonant frequency ν_{mnp} and are therefore frequency degenerate. The frequency difference between two contiguous longitudinal modes is given by

$$\Delta\nu_p = \frac{c}{2L} \tag{2.58}$$

while that between contiguous transverse modes is given by

$$\Delta\nu_m = \Delta\nu_n = \frac{c}{4L} \tag{2.59}$$

These are illustrated in Fig. 2.14.

The general field distribution everywhere within the resonant cavity can be shown to be given by

$$g(x, y, z) = \frac{w_\mathrm{w}}{w(x)} H_m\left(\frac{\sqrt{2}y}{w(x)}\right) H_n\left(\frac{\sqrt{2}z}{w(x)}\right) \mathrm{e}^{-\frac{y^2+z^2}{w^2(x)}} \mathrm{e}^{\left\{-i\left[k_\mathrm{w}\frac{y^2+z^2}{2r(x)}+k_\mathrm{w}x-(m+n+1)\phi(x)\right]\right\}} \tag{2.60}$$

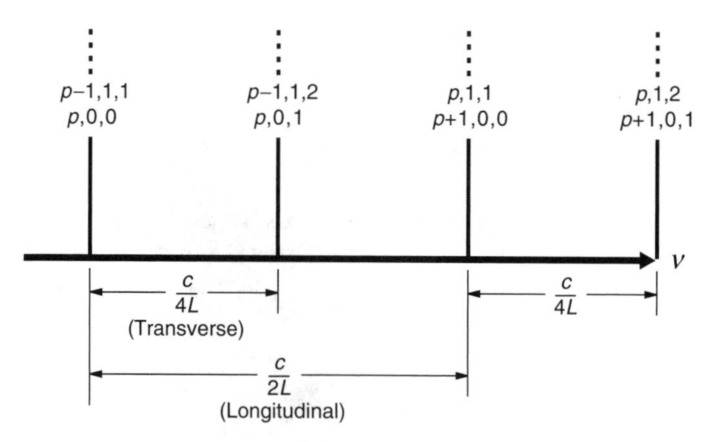

FIGURE 2.14 Oscillating frequencies of a confocal resonator. (From Svelto, O., 1989, *Principles of Lasers*, 3rd edition. By permission of Springer Science and Business Media.)

Where i is the complex variable, $k_{\mathrm{w}} = 2\pi/\lambda$ is the wavenumber, w_{w} is the radius of the beam cross section at the center of the cavity, referred to as the spot size at the beam waist (Fig. 2.15). It is the spot size of the beam, occurring at $x = 0$:

$$w_{\mathrm{w}} = \left(\frac{L\lambda}{2\pi}\right)^{1/2} \tag{2.61}$$

while that at any general location along the cavity axis, $w(x)$ is

$$w(x) = w_{\mathrm{w}}\left[1 + \left(\frac{2x}{L}\right)^2\right]^{1/2} \tag{2.62}$$

This is illustrated schematically in Fig. 2.15 by the solid curve.

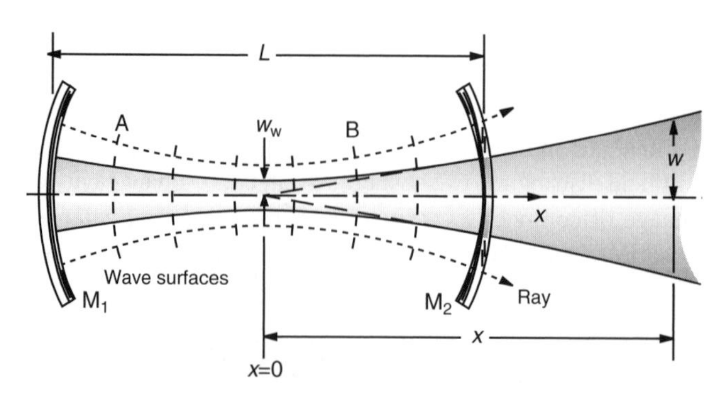

FIGURE 2.15 Beam size and phase for a TEM_{00} mode propagating in a confocal resonator.

$r(x)$ is the radius of curvature of the surface locations of the beam that are in phase and is positive if it is concave toward the resonator, looking from the direction in which the beam is propagating, that is, the center of curvature is to the wavefront's left. It is given by

$$r(x) = x\left[1 + \left(\frac{L}{2x}\right)^2\right] \tag{2.63}$$

Since $x = 0$ at the center of the resonator cavity, the radius of curvature $r(x) = \infty$ at that location and the wavefront is thus a plane.

$\phi(x)$ gives the phase of the beam at any section x and is given by

$$\tan\phi(x) = \frac{2x}{L} \tag{2.64}$$

Surfaces that are in phase are shown as dashed curves in Fig. 2.15. From the phase distribution, the wavefront is found to be spherical, irrespective of the beam mode.

It is evident from Fig. 2.15 that the beam expands as if it is emanating from a point source, with the envelope becoming asymptotic with a line originating from the center of the waist, x_0. In fact, it can be shown (see Problem 2.12) that for very large values of x, equation (2.62) reduces to the form

$$w(x) \approx \frac{\lambda x}{\pi w_w} \quad \text{as} \quad x \to \infty \tag{2.65a}$$

The Rayleigh range (also known as the near-field distance), x_R, is defined as the distance between the beam waist and the point where the beam size is $\sqrt{2}$ times the waist size:

$$x_R = \frac{\pi {w_w}^2}{\lambda} \tag{2.65b}$$

Distances within the Rayleigh range are referred to as the near-field region, while those much greater than the Rayleigh range are referred to as the far-field region. Thus, the beam size varies linearly with distance in the far-field region, that is, for $x >> x_R$.

One advantage of confocal resonators is that they exhibit the lowest diffraction losses for the TEM_{00} mode. One disadvantage, however, is that the waist spot size w_w that is obtained is normally much smaller than the cross-sectional area of the active medium, so that the resonator does not make full use of the available medium or mode volume, that is, the volume of active medium that is filled by radiation, is not 100%. As a result, confocal resonators are seldom used. Next, we look at generalized spherical resonators.

2.4 GENERALIZED SPHERICAL RESONATORS

The generalized resonator consists of mirrors with different radii, r_{m1} and r_{m2}, and spaced a distance L apart. The radius of curvature is considered to be positive if the mirror is concave toward the resonator. Thus for the system shown in Fig. 2.16,

$$r(x_1) = -r_{m1} \tag{2.66a}$$

$$r(x_2) = r_{m2} \tag{2.66b}$$

The analysis of such a system for a Gaussian beam can be simplified using a reverse process by first identifying an equivalent confocal system. To understand this process, we note from Fig. 2.15 and equation (2.63) that the radius of the spherical surface over which the phase of the beam is the same varies with location on the resonator axis. If we can find two such radii where the radius of curvature of the wavefront at either mirror matches the specified mirror radius (r_{m1} or r_{m2}), then the radiation will oscillate in the cavity formed by the two mirrors. The two radii corresponding to r_{m1} and r_{m2} (indicated by A and B in Fig. 2.15) could thus constitute the resonator system. However, to be able to use equation (2.60), we need to find the confocal system that will give the same beam configuration as the one that results from r_{m1} and r_{m2}, and is thus equivalent to the latter system. In other words, we need to find the imaginary mirrors A and B.

Knowing r_{m1}, r_{m2}, and their spacing L, we can substitute into equation (2.63), giving

$$-r_{m1} = x_1 \left[1 + \left(\frac{L_e}{2x_1}\right)^2\right] \tag{2.67a}$$

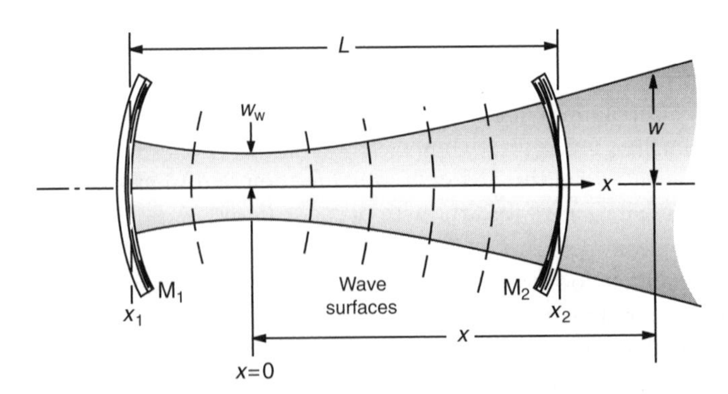

FIGURE 2.16 Generalized spherical resonator. (From Thyagarajan, K., and Ghatak, A. K., 1981, *Lasers, Theory and Applications*. By permission of Springer Science and Business Media.)

and

$$r_{m2} = x_2 \left[1 + \left(\frac{L_e}{2x_2} \right)^2 \right] \tag{2.67b}$$

where L_e is the spacing between the mirrors for the equivalent confocal system. In addition to equations (2.67a) and (2.67b), we have

$$x_2 - x_1 = L \tag{2.68}$$

Thus we can determine x_1, x_2, and L_e, given the generalized system parameters r_{m1}, r_{m2}, and L. Equation (2.60) can then be used to determine the field distribution in the generalized resonator cavity after determining w, w_w, $r(x)$, and $\phi(x)$, where L_e has been substituted for L. The configuration of the generalized spherical resonator is shown in Fig. 2.16. It can be shown that the oscillating frequencies for such a generalized resonator are given by

$$\nu_{mnp} = \frac{c}{2L} \left[p + (m + n + 1) \frac{\cos^{-1}(g_1 g_2)^{1/2}}{\pi} \right] \tag{2.69}$$

where

$$g_1 = 1 - \frac{L}{r_{m1}} \tag{2.69a}$$

$$g_2 = 1 - \frac{L}{r_{m2}} \tag{2.69b}$$

2.5 CONCENTRIC RESONATORS

Concentric resonators are also a special form of generalized spherical resonators, but where the mirrors have the same radii of curvature, r_m, and are located such that the centers of curvature, C_1 and C_2, coincide (Fig. 2.17). The separation distance, L, is then equal to twice the radius of curvature. Concentric resonators have the disadvantage that they are sensitive to alignment of the mirrors. Furthermore, the beam spot size at the center of the resonator is small, and that can make it difficult to generate high power beams, since a relatively small portion of the active medium is used in the amplification process. As a result, they are not frequently used.

Most resonators thus use concave mirrors with large radii of curvature, that is, $r_{mi} >> L$. The spot sizes obtained for such mirrors are relatively large, and the resonators of which they are made are not as sensitive to alignment of the mirrors. Of course, as $r_{mi} \to \infty$, we end up with a plane mirror.

Now that we have obtained some insight into different types of optical resonators, we turn our attention to the conditions under which these resonators are stable.

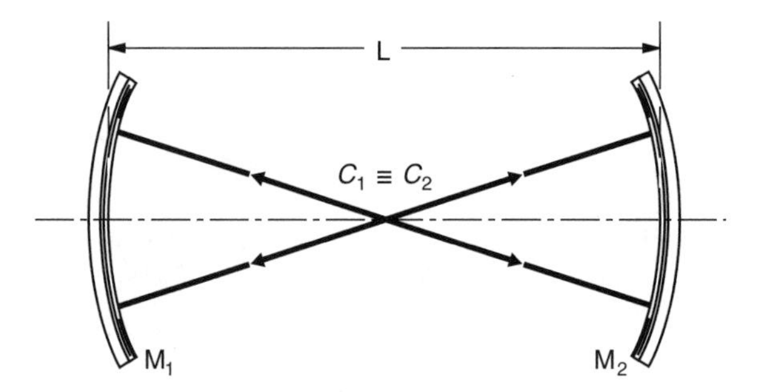

FIGURE 2.17 Concentric resonator. (From Svelto, O., 1989, *Principles of Lasers*, 3rd edition. By permission of Springer Science and Business Media.)

Example 2.2 A laser with mirror radii of $r_{\mathrm{m1}} = 10$ m and $r_{\mathrm{m2}} = 25$ m has a cavity length of 2.5 m. Determine the frequency separation between the TEM_{11} and TEM_{15} modes for the laser output.

Solution:

From equation (2.69), we have

$$\nu_{mnp} = \frac{c}{2L}\left[p + (m+n+1)\frac{\cos^{-1}(g_1 g_2)^{1/2}}{\pi}\right]$$

	TEM_{11}	TEM_{15}
p	p	p
m	1	1
n	1	5

Thus,

$$\Delta\nu = \frac{c}{2L}\left[\Delta n \frac{\cos^{-1}(g_1 g_2)^{1/2}}{\pi}\right]$$

But

$$g_1 = 1 - \frac{L}{r_{m1}} = 1 - \frac{2.5}{10} = 0.75, \qquad g_2 = 1 - \frac{L}{r_{m2}} = 1 - \frac{2.5}{25} = 0.9$$

Therefore,

$$\Delta\nu = \frac{3 \times 10^8}{2 \times 2.5}\left[4 \frac{\cos^{-1}(0.75 \times 0.9)^{1/2}}{\pi}\right] = 26.55 \times 10^8 \text{ Hz} = 2.66 \text{ GHz}$$

2.6 STABILITY OF OPTICAL RESONATORS

A resonator is said to be stable if a ray of light is bounded or stays confined within the cavity after being reflected to and fro indefinitely between the cavity mirrors. An unstable resonator, on the contrary, results in the ray diverging away from the cavity axis after a few traversals. A rigorous analysis of resonator stability can be obtained using wave theory, but a geometrical-optics approximation provides a reasonably accurate and simplified insight into the phenomenon. In the following discussion, we consider the paraxial approximation that only considers rays that have a small angle of orientation with respect to the resonator axis, that is, close to the axis.

Consider a ray propagating in a homogeneous medium along a straight line AB in the $x - y$ plane with the x-axis being coincident with the resonator axis (Fig. 2.18). Furthermore, represent the coordinates of any point along the ray by the height (y) and slope (θ) at that point. It can then be shown that any two points on the ray are related by

$$\begin{bmatrix} y_2 \\ \theta_2 \end{bmatrix} = \begin{bmatrix} 1 & d_c \\ 0 & 1 \end{bmatrix} \begin{bmatrix} y_1 \\ \theta_1 \end{bmatrix} \tag{2.70a}$$

or

$$\mathbf{y_2} = \mathbf{T}\mathbf{y_1} \tag{2.70b}$$

where d_c is the distance separating the two points along the cavity axis, $\mathbf{T} = [\begin{smallmatrix} 1 & d_c \\ 0 & 1 \end{smallmatrix}]$ is the translation matrix, $\mathbf{y_1} = [\begin{smallmatrix} y_1 \\ \theta_1 \end{smallmatrix}]$ the initial point, and $\mathbf{y_2} = [\begin{smallmatrix} y_2 \\ \theta_2 \end{smallmatrix}]$ the final point along the ray.

Further consider the mirror formula, equation (9.30),

$$\frac{1}{u_o} + \frac{1}{v_i} = \frac{1}{f_l} = \frac{2}{r_m} \tag{9.30}$$

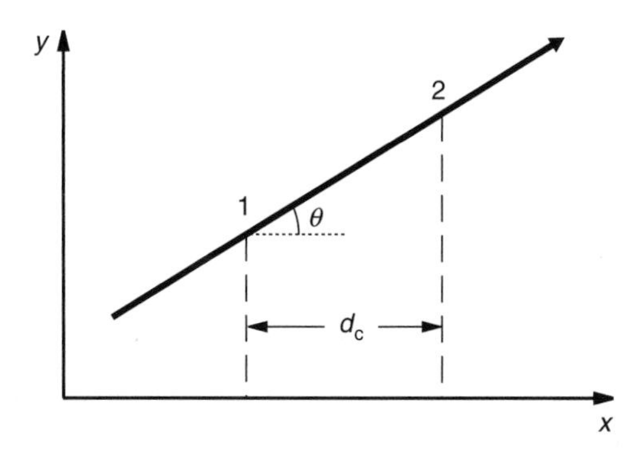

FIGURE 2.18 Beam propagation along a straight line.

where r_m is the mirror radius, u_o is the object distance, v_i is the image distance, and f_l is the focal length.

It can again be shown (see Problem 2.18) that the coordinates, $\mathbf{y_2}$ (made up of the location and direction of the ray at that point), of a ray at point P on a concave mirror, after reflection, are related to the coordinates, $\mathbf{y_1}$, at the same point P on the mirror, of the ray before reflection, see Fig. 2.19, by

$$\begin{bmatrix} y_2 \\ \theta_2 \end{bmatrix} = \begin{bmatrix} 1 & 0 \\ -2/r_m & 1 \end{bmatrix} \begin{bmatrix} y_1 \\ \theta_1 \end{bmatrix} \tag{2.71}$$

If we extend this to the more general case where the points are different, then the relationship between a point P_1 on ray 1 and another point P_2 on ray 2 where the second ray is obtained after the first one is reflected from a concave mirror (Fig. 2.20) is given by

$$\begin{bmatrix} y_2 \\ \theta_2 \end{bmatrix} = \begin{bmatrix} 1 & d_{c2} \\ 0 & 1 \end{bmatrix} \begin{bmatrix} 1 & 0 \\ -2/r_m & 1 \end{bmatrix} \begin{bmatrix} 1 & d_{c1} \\ 0 & 1 \end{bmatrix} \begin{bmatrix} y_1 \\ \theta_1 \end{bmatrix} \tag{2.72}$$

or

$$\begin{bmatrix} y_2 \\ \theta_2 \end{bmatrix} = \mathbf{M} \begin{bmatrix} y_1 \\ \theta_1 \end{bmatrix} \tag{2.73}$$

where

$$\mathbf{M} = \begin{bmatrix} 1 & d_{c2} \\ 0 & 1 \end{bmatrix} \begin{bmatrix} 1 & 0 \\ -2/r_m & 1 \end{bmatrix} \begin{bmatrix} 1 & d_{c1} \\ 0 & 1 \end{bmatrix} = \begin{bmatrix} P & Q \\ S & T \end{bmatrix}$$

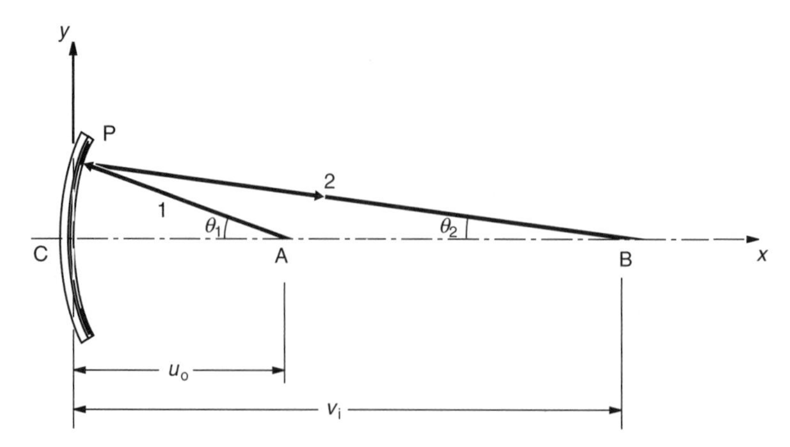

FIGURE 2.19 Beam reflection from a concave mirror. (From Thyagarajan, K., and Ghatak, A. K., 1981, *Lasers, Theory and Applications*. By permission of Springer Science and Business Media.)

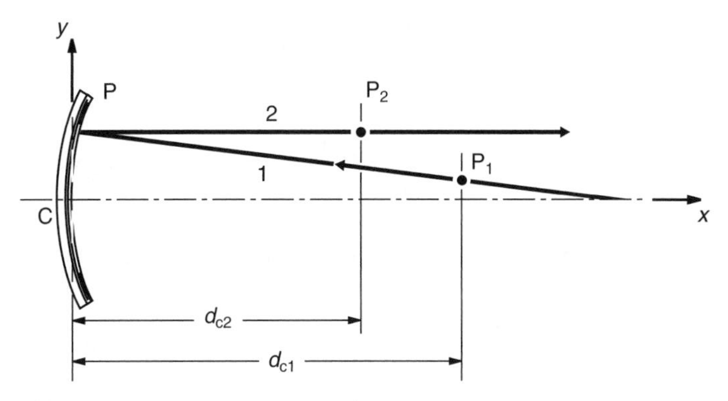

FIGURE 2.20 Relationship between one point on a ray and another point on the corresponding ray after reflection from a concave mirror. (From Thyagarajan, K., and Ghatak, A. K., 1981, *Lasers, Theory and Applications*. By permission of Springer Science and Business Media.)

where d_{c1} is the distance of point P_1 from the mirror, measured along the mirror axis, and d_{c2} is the distance of point P_2 from the mirror, measured along the mirror axis.

For a resonator cavity that consists of two concave mirrors M_1 and M_2 separated by a distance L, if the two points P_1 and P_2 are selected to lie in the same vertical plane located midway between the two mirrors, then the transformation matrix **M** that relates points P_1 and P_2 after reflection from both mirrors (a complete traversal) can be shown to be given (see Problem 2.19) by

$$\mathbf{M} = \begin{bmatrix} P & Q \\ S & T \end{bmatrix} \tag{2.74}$$

where

$$P = 1 - \frac{L}{r_{m1}} - \frac{3L}{r_{m2}} + \frac{2L^2}{r_{m1}r_{m2}}$$

$$Q = -\frac{2}{r_{m1}} - \frac{2}{r_{m2}} + \frac{4L}{r_{m1}r_{m2}}$$

$$S = -\frac{L}{2} - \frac{L^2}{2r_{m1}} + \left(1 - \frac{L}{r_{m2}}\right)\left(\frac{3L}{2} - \frac{L^2}{r_{m1}}\right)$$

$$T = -\frac{L}{r_{m1}} + \left(1 - \frac{L}{r_{m2}}\right)\left(1 - \frac{2L}{r_{m1}}\right)$$

where r_{m1} and r_{m2} are the radii of curvatures of mirrors M_1 and M_2, respectively.

Each time the ray makes a complete traversal starting from the midplane, gets reflected from mirror M_1, and is further reflected from mirror M_2, and finally gets back to the midplane, its coordinates in the plane are modified (or rather multiplied) by the transformation matrix $\mathbf{M}$. Thus after n complete traversals, the relationship between the initial point in the midplane, $\mathbf{y_0} = [\begin{smallmatrix} y_0 \\ \theta_0 \end{smallmatrix}]$, and the final point in the plane, $\mathbf{y_n} = [\begin{smallmatrix} y_n \\ \theta_n \end{smallmatrix}]$, is given by

$$\mathbf{y_n} = \begin{bmatrix} P & Q \\ S & T \end{bmatrix}^n \mathbf{y_0} \tag{2.75}$$

As mentioned earlier, a resonator is stable if a ray of light stays confined within the cavity after being reflected to and fro indefinitely between the cavity mirrors, that is, if $\mathbf{y_n} = [\begin{smallmatrix} y_n \\ \theta_n \end{smallmatrix}]$ does not diverge as $n \to \infty$. That means $\mathbf{M}^n$ should not diverge. It can be shown (see Problem 2.20) that this is true if

$$-1 < \frac{1}{2}(P + T) < 1 \tag{2.76}$$

which reduces to the stability criterion

$$0 < g_1 g_2 < 1 \tag{2.77}$$

that a resonator must satisfy to be stable. This is depicted in graphical form in Fig. 2.21, where the stable zones are shown shaded. Symmetric resonators, that is, those that

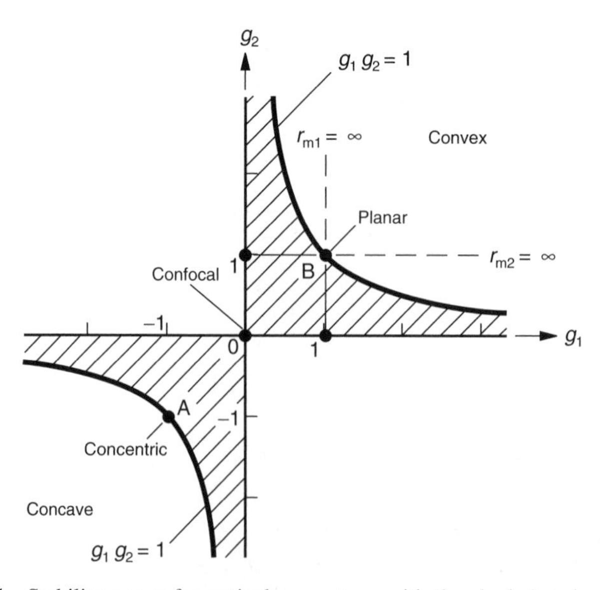

FIGURE 2.21 Stability zones for optical resonators, with the shaded regions representing the stable zones.

have mirrors with equal radii of curvature, lie on line AOB, where point A corresponds to concentric ($g_1 = g_2 = -1$), O to confocal ($g_1 = g_2 = 0$), and B to planar ($g_1 = g_2 = 1$) resonators.

Examples of unstable resonators are shown in Fig. 2.22. The losses are primarily geometrical, as is evident from the figure, and are therefore much greater than the losses for stable resonators, which result primarily from diffraction. However, the portion of the beam that diverges from the cavity as losses is normally tapped as the output coupling, and this results in an annular disc-shaped output. (Some stable modes, $TEM_{01}{}^*$, also do have the donut shape.) Due to the relatively high losses, unstable resonators are appropriate with an active medium that has a relatively high small-gain coefficient, for example, as obtained with the CO_2 laser. Unstable resonators have the advantage that the size of the beam in the cavity is relatively large, thereby making full use of the active medium. Also, the optics used can be all reflective. However, even though the annular shape of the output beam does not appear at the focal point, it makes the resulting peak intensity at the point of focus lower than that of a beam of uniform cross section with the same diameter as the outer diameter of the annular ring. Finally, the unstable resonators tend to be more sensitive to changes within the cavity than do stable resonators. Unstable resonators are more suited for high-power lasers.

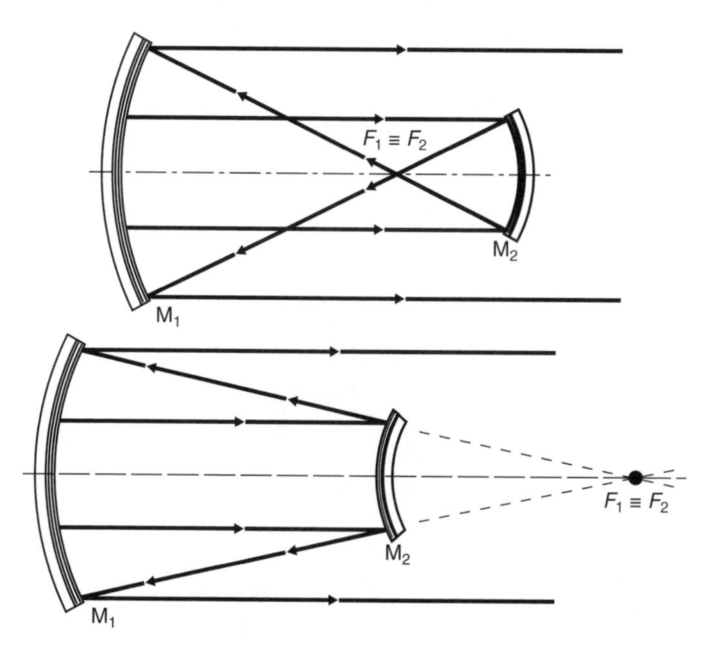

FIGURE 2.22 Different forms of unstable resonators. (From Svelto, O., 1989, *Principles of Lasers*, 3rd edition. By permission of Springer Science and Business Media.)

2.7 SUMMARY

In this chapter, we have discussed the basic concepts of optical resonators, starting with an understanding of how different modes are generated in an idealized rectangular cavity, using Maxwell's equations. The large number of modes that result in a closed cavity can be reduced by using an open cavity, one of the simplest being the planar resonator. The major mode forms that are sustained in such a resonator are the longitudinal (axial) and transverse modes. Contiguous longitudinal modes only differ in their field distribution along the longitudinal axis, while transverse mode field distributions differ in the transverse direction. Undesirable modes can be suppressed using mode selection methods. Higher order transverse modes may be suppressed by placing a diaphragm in the cavity. Longitudinal mode selection methods include cavity length variation, the Fabry–Perot etalon, and the Fox–Smith interferometer. In addition, the monochromaticity of a laser can be enhanced by having the laser oscillate on only one transition line. This can be done by inserting a wavelength dispersive element such as a prism or diffraction grating into the cavity. Other forms of resonators include the confocal, generalized, and concentric resonators.

Finally, we studied the conditions necessary for stability of an optical resonator. A resonator is said to be stable if a ray of light is bounded or stays confined within the cavity after being reflected to and fro indefinitely between the cavity mirrors. An unstable resonator, on the contrary, results in the ray diverging away from the cavity axis after a few traversals.

We are now in a position to study pumping, which is the principal mechanism by which population inversion is achieved in an active medium.

APPENDIX 2A

List of symbols used in the chapter.

Symbol	Parameter	Units
$2a \times 2b \times L$	dimensions of rectangular cavity	m
$\mathbf{B}$	magnetic flux density vector	T
d_c	distance between two points along cavity axis	m
$\mathbf{D}$	electric flux density vector	C/m^2
F_i	focal points	–
g_{mn}	field distribution	–
g_1	$g_1 = 1 - \frac{L}{r_{m1}}$	–
g_2	$g_2 = 1 - \frac{L}{r_{m2}}$	–
h_i	distance of mirror M_i from beam splitter	m
$\mathbf{H}$	magnetic field intensity vector	Wb
H_i	Hermite polynomial of order i	–
$\mathbf{J}$	current density vector	A/m^2

L_e	spacing between mirrors for equivalent confocal system	m
m, n, p	integers	–
M_i	mirror i	–
$\mathbf{Q}$	charge density vector	C/m^3
r_a	radius of laser aperture	m
$r(x)$	radius of surface locations of the beam in phase	m

PROBLEMS

2.1. Show that

$$\nabla\times(\nabla\times\mathbf{E}_l) = \nabla(\nabla\cdot\mathbf{E}_l) - \nabla^2\mathbf{E}_l$$

2.2. Given that in a rectangular cavity,

$$E_{ly} = E_{01y}\sin k_x x \sin k_z z(A_y \cos k_y y + B_y \sin k_y y)(A_t \cos\omega t + B_t \sin\omega t)$$

$$E_{lz} = E_{01z}\sin k_x x \sin k_y y(A_z \cos k_z z + B_z \sin k_z z)(A_t \cos\omega t + B_t \sin\omega t)$$

Show that

$$\frac{\partial E_{lx}}{\partial x} = 0$$

on the planes $x = 0, \quad x = L$.

2.3. A laser cavity has a wavelength centered at $\lambda = 500$ nm. If the cavity volume is 3500 mm^3, determine the number of modes that fall in a bandwidth of $\Delta\lambda = 0.3$ nm.

2.4. Using a power series expansion, derive the following equation for a rectangular cavity where $a = b$:

$$\nu = \frac{\omega}{2\pi} = \frac{c}{2}\left[\frac{p}{L} + \left(\frac{m^2 + n^2}{a^2}\right)\frac{L}{8p}\right]$$

given that

$$\omega^2 = c^2\pi^2\left(\frac{m^2}{4a^2} + \frac{n^2}{4b^2} + \frac{p^2}{L^2}\right)$$

2.5. Given that

$$\omega^2 = c^2\pi^2\left(\frac{m^2}{4a^2} + \frac{n^2}{4b^2} + \frac{p^2}{L^2}\right)$$

and that p is constant, $m = n$, and m changes by one, that is, $\Delta m = 1$, show that the corresponding change in frequency is given by

$$\Delta\nu_m = \frac{cL}{8a^2 p}\left(m + \frac{1}{2}\right)$$

2.6. Given

$$\Delta\nu_m = \frac{cL}{8a^2 p}\left(m + \frac{1}{2}\right)$$

Show that

$$\Delta\nu_m \approx \frac{\Delta\nu_p \lambda L}{8a^2}\left(m + \frac{1}{2}\right)$$

2.7. A laser with one plane mirror and another mirror of radius $r_{m1} = 5$ m has a cavity length of 1.0 m. Determine the frequency separation between the TEM_{1n} and TEM_{3n} modes for the laser output.

2.8. If the mirrors of a Fabry-Perot resonator are placed at a distance of 30 cm apart, determine

(a) The transmitted wavelengths that will be immediately next to the 0.633 μm wavelength.

(b) The m value for that wavelength.

(c) The wavelength of the next integer mode transmission.

(d) The frequency spacing between two contiguous transverse mode transmissions.

2.9. Given that for a Fox–Smith interferometer we have

$$2\frac{2\pi}{\lambda}(h_1 + h_2) - \pi = (2q - 1)\pi, \quad q = 1, 2, 3, \ldots$$

Show that

$$\Delta\nu = \frac{c}{2n(h_1 + h_2)}$$

2.10. For a confocal resonator, show that the beam spot size at any general location along the cavity axis is also given by

$$w(x) = w_{\mathrm{w}} \left[1 + \left(\frac{\lambda x}{\pi w_{\mathrm{w}}{}^2} \right)^2 \right]^{1/2}$$

2.11. The Gaussian field distribution for a confocal resonator is given by

$$g_{00}(y, z) = \mathrm{e}^{-\frac{\pi}{L\lambda}(y^2+z^2)}$$

Show that the beam spot size of the electric field amplitude on the mirrors is

$$w_{\mathrm{m}} = \left(\frac{L\lambda}{\pi} \right)^{\frac{1}{2}}$$

2.12. Given that

$$w(x) = w_{\mathrm{w}} \left[1 + \left(\frac{2x}{L} \right)^2 \right]^{1/2}$$

Show that the following relationship holds:

$$w(x) \approx \frac{\lambda x}{\pi w_{\mathrm{w}}} \qquad \text{as} \quad x \to \infty$$

2.13. For the confocal resonator, show that the waves at the mirror surfaces are in phase, that is, the mirrors are equiphase surfaces.

2.14. A CO_2 laser with output wavelength 10.6 μm has a cavity of length 1500 mm and a confocal mirror system.

(a) What is the beam spot size at the resonator center?

(b) What will be the spot size on the mirrors?

(c) What will be the frequency difference between two contiguous longitudinal modes?

2.15. Given that the beam size from a confocal resonator is given by

$$w_{\mathrm{w}} = \left(\frac{L\lambda}{2\pi} \right)^{1/2}$$

and

$$w(x) = w_{\mathrm{w}} \left[1 + \left(\frac{2x}{L} \right)^2 \right]^{1/2}$$

Show that the beam divergence may be approximated by

$$\theta \approx \frac{\lambda}{\pi w_{\mathrm{w}}}$$

2.16. Given that for a generalized resonator,

$$-r_{\mathrm{m1}} = x_1\left[1 + \left(\frac{L_{\mathrm{e}}}{2x_1}\right)^2\right]$$

$$r_{\mathrm{m2}} = x_2\left[1 + \left(\frac{L_{\mathrm{e}}}{2x_2}\right)^2\right]$$

and

$$x_2 - x_1 = L$$

Show that

(a) The mirror locations are given by

$$x_1 = -\frac{L(1 - g_1)g_2}{g_1 + g_2 - 2g_1g_2}$$

$$x_2 = -\frac{L(1 - g_2)g_1}{g_1 + g_2 - 2g_1g_2}$$

(b) The spot sizes at the two mirrors are given by

$$w(x_1)^2 = \frac{\lambda L}{\pi}\left[\frac{g_2}{g_1(1 - g_1g_2)}\right]^{1/2}$$

$$w(x_2)^2 = \frac{\lambda L}{\pi}\left[\frac{g_1}{g_2(1 - g_1g_2)}\right]^{1/2}$$

(c) The spot size at the waist is

$$w_{\mathrm{w}}{}^4 = \left(\frac{\lambda L}{\pi}\right)^2\left[\frac{(1 - g_1g_2)g_1g_2}{(g_1 + g_2 - 2g_1g_2)^2}\right]$$

2.17. If a He–Ne laser that has mirror radii $r_{\mathrm{m1}} = 15$ m, $r_{\mathrm{m2}} = 12$ m, and a cavity length of 1 m operates in the TEM_{00} mode, calculate

(a) the location of the waist

(b) the spot size at the waist

(c) the spot size at the mirror r_{m1}.

2.18. Show that the coordinates (made up of the location, y_2, and direction, θ_2, of the ray at that point) of the point $\mathbf{y_2}$ on a concave mirror of radius, r_m, after reflection are related to the same point on the mirror before reflection, $\mathbf{y_1}$, see Fig. 2.19, by

$$\begin{bmatrix} y_2 \\ \theta_2 \end{bmatrix} = \begin{bmatrix} 1 & 0 \\ -2/r_{\mathrm{m}} & 1 \end{bmatrix} \begin{bmatrix} y_1 \\ \theta_1 \end{bmatrix}$$

2.19. The relationship between a point $\mathrm{P_1}$ on ray 1 and another point $\mathrm{P_2}$ on ray 2 where the second ray is obtained after the first one is reflected from a concave mirror is given by

$$\begin{bmatrix} y_2 \\ \theta_2 \end{bmatrix} = \mathbf{M} \begin{bmatrix} y_1 \\ \theta_1 \end{bmatrix}$$

where

$$\mathbf{M} = \begin{bmatrix} 1 & d_2 \\ 0 & 1 \end{bmatrix} \begin{bmatrix} 1 & 0 \\ -2/r_{\mathrm{m}} & 1 \end{bmatrix} \begin{bmatrix} 1 & d_1 \\ 0 & 1 \end{bmatrix} = \begin{bmatrix} P & Q \\ S & T \end{bmatrix}$$

Show that for a resonator cavity that consists of two concave mirrors $\mathrm{M_1}$ and $\mathrm{M_2}$ separated by a distance L, if the two points $\mathrm{P_1}$ and $\mathrm{P_2}$ are selected to lie in the same vertical plane located midway between the two mirrors, then the transformation matrix $\mathbf{M}$ that relates points $\mathrm{P_1}$ and $\mathrm{P_2}$ after reflection from both mirrors (a complete traversal) is given by

$$\mathbf{M} = \begin{bmatrix} P & Q \\ S & T \end{bmatrix}$$

where

$$P = 1 - \frac{L}{r_{\mathrm{m1}}} - \frac{3L}{r_{\mathrm{m2}}} + \frac{2L^2}{r_{\mathrm{m1}} r_{\mathrm{m2}}}$$

$$Q = -\frac{2}{r_{\mathrm{m1}}} - \frac{2}{r_{\mathrm{m2}}} + \frac{4L}{r_{\mathrm{m1}} r_{\mathrm{m2}}}$$

$$S = -\frac{L}{2} - \frac{L^2}{2r_{\mathrm{m1}}} + \left(1 - \frac{L}{r_{\mathrm{m2}}}\right)\left(\frac{3L}{2} - \frac{L^2}{r_{\mathrm{m1}}}\right)$$

$$T = -\frac{L}{r_{\mathrm{m1}}} + \left(1 - \frac{L}{r_{\mathrm{m2}}}\right)\left(1 - \frac{2L}{r_{\mathrm{m1}}}\right)$$

where r_{m1} and r_{m2} are the radii of curvatures of mirrors $\mathrm{M_1}$ and $\mathrm{M_2}$, respectively.

2.20. Consider a resonator cavity that consists of two concave mirrors $\mathrm{M_1}$ and $\mathrm{M_2}$ separated by a distance L. If two points $\mathrm{P_1}$ and $\mathrm{P_2}$ are selected to lie in the same vertical plane located midway between the two mirrors, then the transformation matrix **M** that relates points P_1 and P_2 is given by

$$\mathbf{M} = \begin{bmatrix} 1 - \frac{L}{r_{m1}} - \frac{3L}{r_{m2}} + \frac{2L^2}{r_{m1}r_{m2}} & -\frac{2}{r_{m1}} - \frac{2}{r_{m2}} + \frac{4L}{r_{m1}r_{m2}} \\ -\frac{L}{2} - \frac{L^2}{2r_{m1}} + \left(1 - \frac{L}{r_{m2}}\right)\left(\frac{3L}{2} - \frac{L^2}{r_{m1}}\right) & -\frac{L}{r_{m1}} + \left(1 - \frac{L}{r_{m2}}\right)\left(1 - \frac{2L}{r_{m1}}\right) \end{bmatrix}$$

Show that the criterion for stability of the resonator is given by

$$-1 < \frac{1}{2}(P + T) < 1$$

Further show that it implies

$$0 \leq g_1 g_2 \leq 1$$

2.21. The mirrors of a He–Ne laser resonator have radii of curvature $r_{m1} = 1.5$ m and $r_{m2} = 2.0$ m. Determine the spacing or range of spacing between the mirrors for the resonator to be

(a) stable

(b) marginally stable

(c) unstable.

2.22. Consider a cylindrical tube of radius r and length L.

(a) If a Gaussian beam propagates through the cylinder, with its beam waist at the tube center, obtain an expression for the beam waist when the transmittance is maximum, assuming the power that is incident at the entrance of the cylinder can be transmitted.

(b) If the cylinder diameter is 5 mm and the length is 100 mm, determine the fraction of incident He–Ne laser Gaussian beam that will be transmitted.

2.23. If the life of a cavity is τ_c, show that the time constant of the resulting laser light amplitude is $2\tau_c$.

2.24. Why does the Gaussian beam exhibit the smallest possible beam divergence?

2.25. A laser beam would normally contain a number of axial modes. What mechanism could be used in the generation of the beam to produce only a single axial mode?

2.26. Show that the Rayleigh range, x_R, is given by

$$x_R = \frac{\pi w_w^{\,2}}{x}$$

3 Laser Pumping

Pumping is the process by which atoms are raised from a lower to a higher energy level to achieve a population inversion. There are various techniques for laser pumping, and the two that are commonly used for industrial lasers are

1. Optical pumping
2. Electrical pumping.

In this chapter, we outline the basic principles of these pumping techniques. For each technique, we start with a discussion on the common methods used in pumping and outline their advantages and disadvantages.

3.1 OPTICAL PUMPING

This form of pumping involves the excitation of the active medium using an intense source of light. Two types of light sources are normally used, and these are

1. An arc lamp or a flash lamp
2. Another laser, typically a diode laser (see Chapter 8).

3.1.1 Arc or Flash Lamp Pumping

Arc and flash lamps are more commonly used with solid-state or liquid lasers (see Chapter 8). This is because the output of the lamp normally has a relatively broad bandwidth, and thus a good portion of the source energy may be lost. Thus, for it to be more efficiently absorbed, the active medium also has to be one with a broad-linewidth. Solid and liquid active media tend to have broad energy bands due to the line-broadening mechanisms (see Chapter 5) associated with them. Thus, the solid and liquid media can more efficiently absorb the energy from the lamp source.

One form of lamp-based optical pumping is illustrated schematically in Fig. 3.1, where a linear pumping source (cylindrical in shape) is placed along one principal axis (focus, F_1) of an elliptical cavity (reflector) and the laser rod (active medium) along the other axis, F_2. Efficient coupling between the source and rod is achieved

Principles of Laser Materials Processing, by Elijah Kannatey-Asibu, Jr.

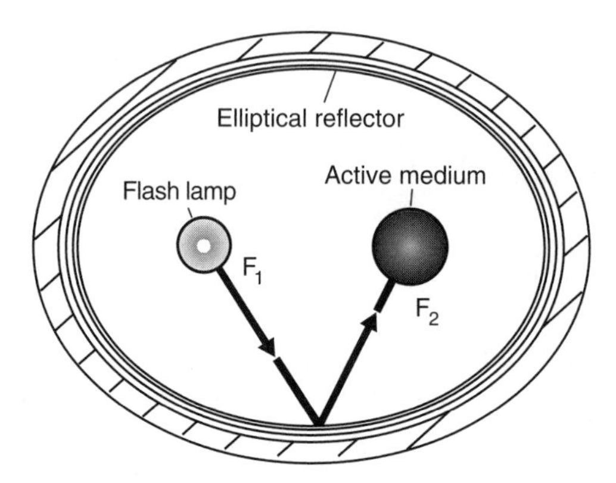

FIGURE 3.1 Schematic of an optical pumping system using an elliptical cavity.

when both are of the same length and diameter. Light from either axis (focus) of a specular (mirror-like) elliptical cavity in a plane orthogonal to the cylinder axis is ideally reflected in such a way that it passes through the second axis. Thus, a significant portion of the light from the source is directed, after reflection, toward the laser rod. The reflector may be either specular or diffuse. Diffuse reflectors tend to reflect light in all directions. Specular reflectors result in a higher pumping efficiency, but less uniform pumping of the active material, compared to diffuse reflectors.

Another type of configuration involves a lamp source that is in the form of a helix and is placed within a cylindrical container that has a specular interior surface. The active material, which is in the form of a rod, is placed at the center of the helix (Fig. 3.2.). Light from the lamp source propagates either directly to the rod or after reflection from the interior of the cylindrical container.

Pulsed lasers are normally pumped using medium pressure (500–1500 Torr [mmHg]) xenon or krypton lamps. An intense flash of light is applied for a period of the order of microseconds. Continuous wave lasers on the other hand are based on a

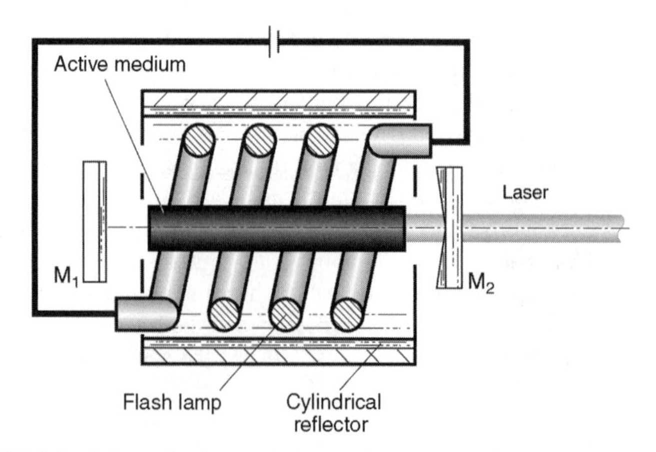

FIGURE 3.2 Schematic of an optical pumping system using a helical flash lamp.

high-pressure (4000–8000 Torr) krypton or tungsten–iodine lamp that is continuously applied. Typical lamp life is about 800 h.

3.1.2 Diode Laser Pumping

The efficiency of optical pumping is further enhanced when another laser is used for pumping. Since the pumping laser infuses energy at a specific wavelength, very little of its energy is wasted. However, for this to be true, the output wavelength range of the pumping diode laser must be nearly or the same as the absorption bandwidth of the active medium being pumped. Even though any laser can essentially be used, diode lasers (see Chapter 8) have been found very convenient and developed specifically for pumping other lasers, usually solid-state lasers. They are used mainly because

1. They have a relatively high efficiency, which results in a reduced electrical power consumption for a given desired output power.
2. They can generate high output power when stacked.
3. They are available in many wavelengths.
4. They require reduced maintenance. The working life of diode-pumped lasers is generally more than 12,000 h, while that for flash lamp lasers is in the range of 600 – 1,000 h.
5. They are much smaller in size, making them more portable.

Two common forms of pumping configuration are normally used:

1. Longitudinal or end pumping
2. Transverse or side pumping.

3.1.2.1 Longitudinal Pumping The longitudinal or end pumping technique is illustrated in Fig. 3.3. The beam being used for pumping is coupled to the laser medium from one end of the laser rod, along the resonator axis. A lens arrangement is used to concentrate the beam to a small and circular spot. The pumping beam can also be delivered using a fiber optic cable.

3.1.2.2 Transverse Pumping With transverse pumping, the laser diodes are radially arranged around the active medium (Fig. 3.4.) Several diode pumped rods may

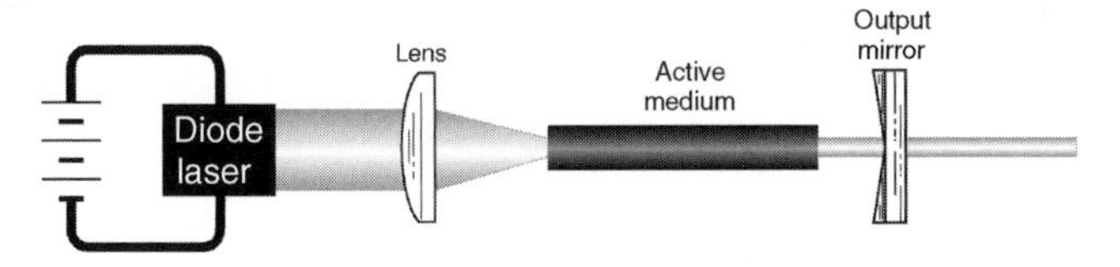

FIGURE 3.3 Longitudinal pumping using a diode laser.

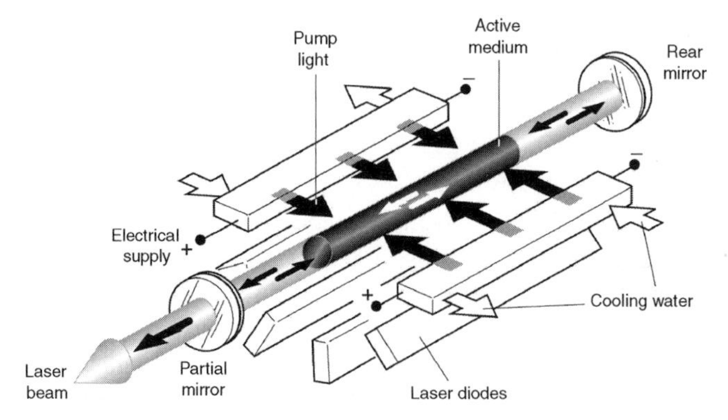

FIGURE 3.4 Transverse pumping using a diode laser. (By permission of ROFIN-SINAR Laser GmbH.)

be arranged in series to achieve high output powers in the kilowatt range. As in the case of longitudinal pumping, the beam can also be delivered using fiber optic cables.

3.1.3 Pumping Efficiency

The efficiency of the pumping process is determined by a number of factors, including

1. The conversion of electrical energy into optical energy by the pumping source in the frequency range of interest.
2. The transmission of the optical energy from the source into the rod (determined primarily by the design of the pumping configuration). The helical configuration is always lower in efficiency than linear pumping. However, the helical configuration results in a more uniform pumping process, making it attractive for high-power systems.
3. The absorption by the active medium.
4. The actual number of atoms resulting in lasing action as compared to the number of atoms excited by the optical energy absorbed by the rod.

The efficiency of a lamp-based system is normally in the range of 1–3%. This is the fraction of the input electrical energy that is converted to useful output. The 97–99% of energy that is not used is removed in the form of heat using a chiller. Using a diode laser for pumping increases the efficiency to the range of 30–40%.

3.1.4 Energy Distribution in the Active Medium

The distribution of pumped energy within the active rod depends on a number of factors, including

1. The relative dimensions of the source and rod (active medium).
2. The refractive index, n, of the active medium relative to the surrounding medium.
3. The absorption coefficient of the active medium at the pumping frequency.

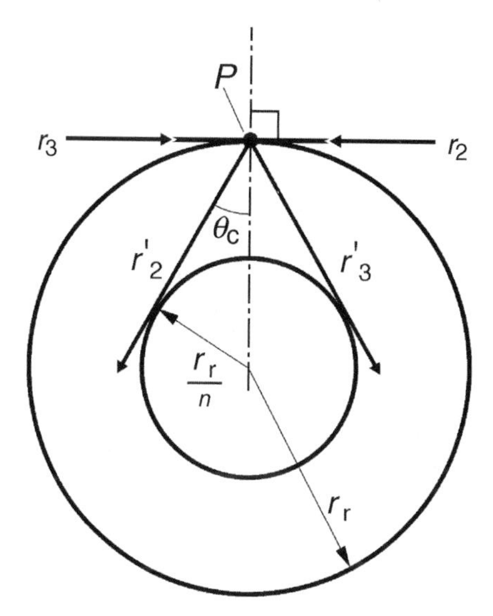

FIGURE 3.5 Light ray propagation in a cylindrical rod due to refraction. (From Svelto, O., 1989, *Principles of Lasers*, 3rd edition. By permission of Springer Science and Business Media.)

When the source radius is greater than or equal to the rod radius, r_r, the light entering the rod tends to be more concentrated in the central core of the rod, in a region of radius r_r/n, where the light is uniformly distributed (Fig. 3.5). This is better understood by considering the possible ways in which a ray of light can enter the rod, assuming the surface to be smooth. The largest angle of incidence that a ray can have will be 90°, equivalent to tangential rays, r_2 and r_3, at a point P on the surface. These rays will be refracted at the critical angle, θ_c:

$$\theta_c = \sin^{-1}(1/n)$$

This results in r_2' and r_3'. Any incident rays will be refracted to fall between r_2' and r_3', since these can only be incident between r_2 and r_3. The set of refracted critical rays r_2' and r_3' that result from points of incidence around the circumference of the rod will form an inner circle of radius r_r/n within which all incident light will be concentrated. This core of the active rod will thus be subjected to more pumping than the outer portion. This is illustrated in Fig. 3.6, which shows the variation of the density of pumped energy in the rod with radius, for different absorption coefficients, α. The dashed line is the ideal case for $\alpha r_r = 0$. However, since light enters the rod from any direction, the actual distribution is as shown with a solid curve for $\alpha r_r = 0$. The relative energy density values are higher for this case than for the cases where $\alpha r_r \neq 0$ because in the latter case part of the energy is absorbed as the beam propagates through the rod in the radial direction, r. Thus, the amount of energy that reaches the central portion is reduced.

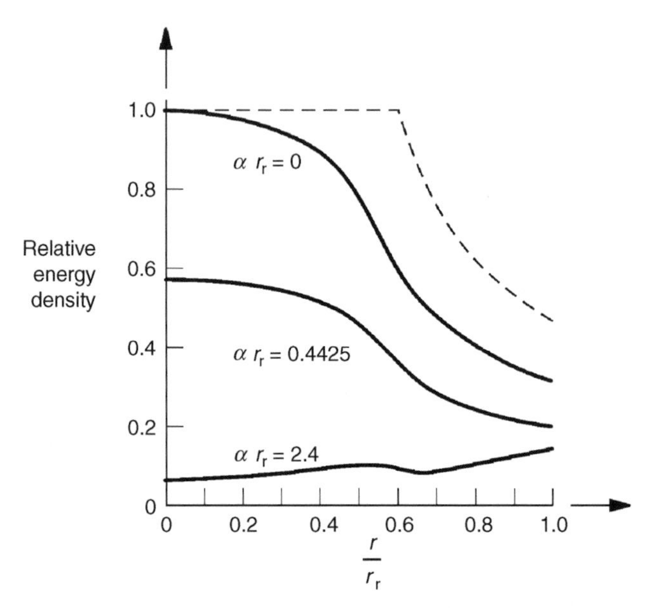

FIGURE 3.6 Energy density distribution in an active rod. (From Svelto, O., 1989, *Principles of Lasers*, 3rd edition. By permission of Springer Science and Business Media.)

It is evident from the graph that the energy density distribution within the rod in the radial direction tends to be more uniform as the absorption coefficient increases, that is, as the beam gets attenuated with distance into the rod. The nonuniformity in energy distribution within the rod for lower α values can be eliminated by cladding the rod with a transparent material whose refractive index is the same as that of the rod. The outer radius of the cladded rod is then made equal to nr_r. This concentrates the light energy density within the rod, where it is also more uniformly distributed. An alternative means of providing a more uniform energy distribution within the rod is to have the rod surface slightly rough so that the source light is diffused as it enters the rod.

3.2 ELECTRICAL PUMPING

Electrical pumping is more commonly used with gas and semiconductor lasers. Whereas semiconductor lasers can be efficiently pumped using either optical or electrical pumping, the latter is normally preferred since it is more convenient to apply. However, the narrow absorption bands of gas lasers make them inappropriate for optical pumping, and are thus normally pumped electrically. This is because optical pumping is a broad band energy source in most cases, and only a small fraction of the available energy would be absorbed by the narrow band gas, making the process inefficient.

The pumping process involves the creation of an electrical discharge by passing electrical energy, of the order of 1–2 kV at a low current of about 50 mA, through the gas medium, in the case of gas lasers (Fig. 3.7). Electrons are generated and their kinetic energy is increased as they are accelerated by the electric field. The resulting electrical medium is referred to as a discharge. A much higher voltage is required to initiate the discharge than to maintain it. This is because a significant amount of energy is required to overcome the ionization potential and ionize a gaseous medium.

When the electrons are accelerated by the electric field, they are able to excite atoms, ions, or molecules with which they collide, from the ground state to a higher level by imparting their kinetic energy to the atoms. Other excitation mechanisms include dissociation of molecules, recombination of ions and electrons, resonance absorption, and so on. The ground state refers to the situation when all the electrons are in their normal orbits or quantum states. For an inert gas, for example, this would be the closed shell.

When the gaseous medium consists of only one component, the electron energy is transferred directly to its atoms. When there are two or more components in the medium, collision between an electron and the atom of one component, say P, will excite the latter to the higher energy state. When this atom collides with an atom of component Q, the latter may also become excited while the P atom falls back to the lower energy level. This is most useful when component P is metastable in its excited state, so that it can act as an energy storage source for exciting component Q.

The electrical energy may be applied as a dc (direct current) or rf (radio frequency alternating) electric field. In dc-generated discharges, the electrodes are contained within the discharge tube or container, and over a period of time, the electrodes may become contaminated. Changing the electrodes requires access to the interior of the discharge tube, a process that may in itself introduce additional contamination. However, with rf-generated discharges, the electric field is generated from outside the discharge tube, and thus the electrodes can be changed without accessing the interior of the tube.

Excitation frequencies for rf-excited lasers are typically of the order of 30 MHz, and thus they tend to have a fast response to a change in power, as compared to dc-excited lasers. This characteristic makes it easier to modulate rf-excited lasers, resulting in very high pulse rates, up to 15 kHz. In addition, higher levels of electrical

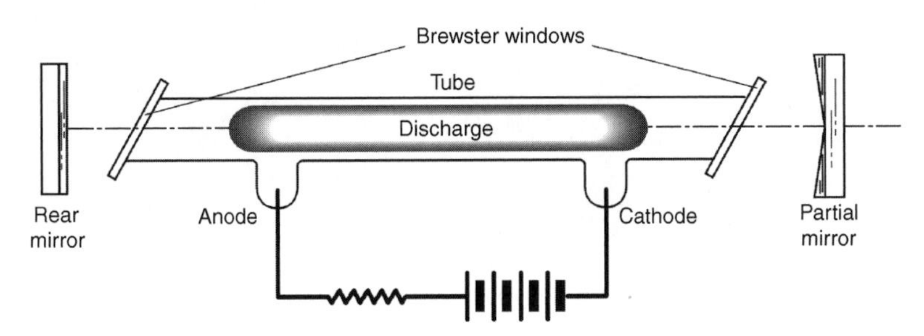

FIGURE 3.7 Schematic of an electrical pumping system.

power densities can be achieved, along with greater homogeneity of the plasma, in rf-excited lasers, resulting in higher power densities and higher discharge stability and homogeneity. This further improves beam quality and also enables more compact generator configuration to be achieved for the same power output. Radio frequency excitation does not require the use of permanent anodes and cathodes. Thus, cathode problems that result from gas reactions are eliminated. Furthermore, a more stable discharge may be obtained by placing a nondissipative element such as a dielectric slab in series with the discharge circuit. However, rf-excited lasers tend to have lower overall efficiencies compared to dc-excited lasers.

3.3 SUMMARY

In this chapter, we have discussed the basic techniques used in pumping active media to achieve population inversion. Our focus has been on optical and electrical pumping, which are more commonly used for industrial lasers.

Optical pumping can be done using either lamp-based or diode laser-based systems. Arc and flash lamps are normally used with solid-state or liquid lasers since the lamp output usually has a relatively broad bandwidth and is thus more efficient when used with active media that have broad linewidths. The lamp source configuration may be linear or helical. Configurations that are used when pumping with diode lasers are longitudinal (end) and transverse (side) pumping. Lamp-based systems are generally about 1–3% efficient, while the efficiency of diode laser-based systems range between 30 and 40%.

Electrical pumping is normally done using radio frequency or direct current systems and is more commonly used with gas and semiconductor lasers. Even though semiconductor lasers can be pumped using either optical or electrical methods, gas lasers have narrow absorption bands, and thus can be efficiently pumped only electrically.

We now discuss, in the next chapter, the rate equations that determine the time variation of the populations of individual levels when subjected to pumping.

4 Rate Equations

The rate equations provide a simplified approach to understanding temporal variations in the populations of different energy levels during pumping. They also enable the conditions under which population inversion can be achieved to be investigated. The pumping rate required to maintain continuous wave (steady-state) operation and the population inversion that exists can also be obtained. A more rigorous analysis would involve the semiclassical approach where the laser beam is analyzed using classical electromagnetic wave theory based on Maxwell's equations, while the atomic behavior is quantized, or the fully quantum approach where both atoms and the light field are analyzed using quantum mechanics. Either approach is beyond the scope of this text. Before discussing the rate equations for configurations for which population inversion is feasible, that is, the three- and four-level laser systems, we first look at the simpler two-level system and try to understand why it is not normally used.

4.1 TWO-LEVEL SYSTEM

As indicated in equation (1.3), the distribution of the state of atoms under thermal equilibrium at any given temperature is given by Boltzmann's law that may be expressed in general form as

$$\frac{N_j}{N_i} = \mathrm{e}^{-\frac{\Delta E_{ij}}{k_B T}} \tag{4.1}$$

where N_i and N_j are number of atoms per unit volume in the lower and higher energy states, respectively, ΔE_{ij} is the energy difference between the two states, k_B is Boltzmann's constant, and T is the absolute temperature.

Thus, the number of atoms in the higher energy states is lower than that in the lower states (see Fig. 1.4). However, the number of atoms at the higher energy levels increase with increasing temperature. For simplicity, let us consider only two energy levels (Fig. 1.4a) with the populations per unit volume of the lower and upper levels being N_1 and N_2, respectively. At relatively low temperatures, that is, $k_B T << \Delta E$, practically no atoms exist in the higher energy state. However, for the situation where the temperature is relatively high, that is, $k_B T >> \Delta E$, the populations of the two energy levels are almost equal, that is, $N_1 \approx N_2$. As the temperature increases, N_2 approaches N_1 asymptotically, but can never exceed it under thermal equilibrium conditions. The only conditions under which N_2 can become greater than N_1, that is, a population inversion can be achieved, is through pumping, which would result in a nonequilibrium distribution.

Still considering the two levels, let the total population per unit volume be N, and the lifetime of the upper state, τ_{sp}. Further let the higher energy level, 2, be pumped to increase its population by infusion of radiation of energy density $e(\nu)$. Then the rate of change of the population of the higher energy level is (using equations (1.4), (1.6), and (1.7)) given by

$$\frac{dN_2}{dt} = N_1 B_{12} e(\nu) - N_2 B_{21} e(\nu) - A_e N_2 \tag{4.2}$$

Or from equation (1.20),

$$\frac{dN_2}{dt} = (N_1 - N_2) B_{12} e(\nu) - \frac{N_2}{\tau_{sp}} \tag{4.3}$$

The first term on the right-hand side of equation (4.2) is the rate of increase of atoms at level 2 as a result of excitation from level 1; the second term the rate of loss due to stimulated emission from level 2; and the third term the rate of loss due to spontaneous emission. We also have

$$N_1 + N_2 = N \tag{4.4}$$

Now let

$$\Delta N = N_1 - N_2 \tag{4.5}$$

Thus,

$$N_1 = \frac{1}{2}(N + \Delta N) \tag{4.6}$$

$$N_2 = \frac{1}{2}(N - \Delta N) \tag{4.7}$$

Equation (4.3) can then be rewritten as

$$\frac{d}{dt}(\Delta N) = \frac{N}{\tau_{sp}} - \Delta N \left[2B_{12} e(\nu) + \frac{1}{\tau_{sp}} \right] \tag{4.8}$$

Under steady-state conditions, the rate of change of the population is zero, that is, $d(\Delta N)/dt = 0$. Thus, we have

$$\Delta N = \frac{N}{1 + 2\tau_{sp} B_{12} e(\nu)} \tag{4.9}$$

Equation (4.9) indicates that the population difference decreases as the optical energy density, $e(\nu)$, increases. The energy density depends on the number of photons in the cavity mode. As absorption increases and $\tau_{sp} B_{12} e(\nu) \to \infty$, $\Delta N \to 0$, or in the limit,

$$N_1 = N_2 = \frac{N}{2} \tag{4.10}$$

Thus, for a two-level system, absorption occurs only insofar as there are more atoms in the lower energy state than in the higher energy state. As the atoms are pumped to the point where the two energy levels are equally populated, absorption stops. Thus, population inversion is difficult to achieve in a two-level system.

4.2 THREE-LEVEL SYSTEM

We now consider a three energy level system (Fig. 4.1), with populations per unit volume given by N_1, N_2, and N_3. The lowest energy level is 1, and 3 is the highest, with transition occurring only between these three levels. Atoms are normally pumped from the lowest energy level, 1, to the highest level, 3. From there, they undergo rapid decay to level 2, and radiative transition that results in lasing action occurs between

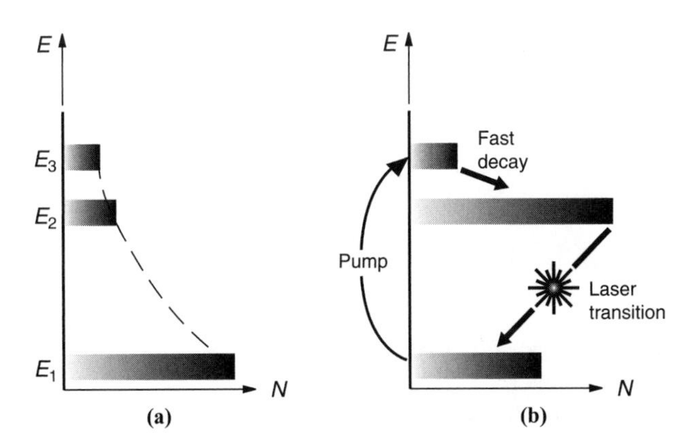

FIGURE 4.1 Schematic of a three-level system. (a) No pumping. (b) Intense pumping. (From O'Shea, D. C., Callen, W. R., and Rhodes, W. T., 1977, *Introduction to Lasers and Their Applications*. Reprinted by permission of Pearson Education, Inc.)

levels 2 and 1. The rate of change of the population of E_3 is

$$\frac{dN_3}{dt} = W_p N_1 + W_{13} N_1 + W_{23} N_2 - W_p N_3 - W_{31} N_3 - W_{32} N_3 \tag{4.11}$$

or

$$\frac{dN_3}{dt} = (W_p + W_{13}) N_1 + W_{23} N_2 - (W_p + W_{32} + W_{31}) N_3 \tag{4.12}$$

where $W_p = B_{13} e(\nu)$ is the pump rate or probability per unit time of induced transition caused by the exciting radiation in the frequency range ν_{13}, $e(\nu)$ is the pump beam energy density, and W_{ij} is the probability per unit time of the atom being thermally excited between levels i and j.

Generally, for spontaneous emission,

$$W_{ij} = A_{ij} + S_{ij} \tag{4.13}$$

where A_{ij} is the Einstein coefficient associated with spontaneous transition and S_{ij} results from nonradiative decay processes such as transferring energy to surrounding atoms in the form of kinetic, vibrational, or other form of energy.

The first term in equation (4.11) is the rate of increase of atoms at level 3 due to pumping; the second term is the rate of increase due to thermal excitation; the third term is due to thermal excitation from level 2 to level 3; the fourth term is due to loss from level 3 to level 1 as a result of pumping; and the fifth and sixth terms are due to loss from level 3 to levels 1 and 2, respectively, as a result of thermal excitation. If we assume that $k_B T << \Delta E_{ij}$, then from equation (4.1) we can say that

$$W_{ij} << W_{ji} \tag{4.14}$$

where i refers to the lower energy level and j the higher energy level, or as an example, $W_{12} << W_{21}$. Equation (4.12) then reduces to the form

$$\frac{dN_3}{dt} = W_p (N_1 - N_3) - (W_{32} + W_{31}) N_3 \tag{4.15}$$

Similarly, we have, for N_2,

$$\frac{dN_2}{dt} = W_s N_1 + W_{12} N_1 - W_s N_2 - W_{21} N_2 - W_{23} N_2 + W_{32} N_3 \tag{4.16}$$

which reduces to

$$\frac{dN_2}{dt} = W_s (N_1 - N_2) - W_{21} N_2 + W_{32} N_3 \tag{4.17}$$

where $W_s = B_{12}e(\nu)$ is the probability per unit time of induced transition caused by the lasing action in the frequency range ν_{12}.

Finally, we have

$$\frac{dN_1}{dt} = -W_pN_1 - W_sN_1 - W_{12}N_1 - W_{13}N_1 + W_sN_2 + W_{21}N_2 + W_pN_3 + W_{31}N_3 \tag{4.18}$$

which also reduces to the form

$$\frac{dN_1}{dt} = W_p(N_3 - N_1) + W_s(N_2 - N_1) + W_{21}N_2 + W_{31}N_3 \tag{4.19}$$

If the total number of atoms per unit volume in the system is N, then

$$N = N_1 + N_2 + N_3 = \text{constant} \tag{4.20}$$

Equations (4.15), (4.17), and (4.19) constitute the rate equations for the three-level laser system and represent the time-dependent behavior of the laser. To simplify the analysis, we consider the steady-state conditions, $dN_i/dt = 0$, and when the pumping rate is constant. Thus we can set the left-hand side of each of these equations to zero, which gives

$$N_1 = \left(\frac{W_p + W_{31} + W_{32}}{2W_p + W_{31} + W_{32}}\right)(N - N_2) \tag{4.21}$$

$$N_2 = \left[\frac{W_s}{W_s + W_{21}} + \frac{W_pW_{32}}{(W_s + W_{21})(W_p + W_{31} + W_{32})}\right]N_1 \tag{4.22}$$

From which we get

$$\frac{N_2 - N_1}{N} = \frac{W_p(W_{32} - W_{21}) - W_{21}(W_{31} + W_{32})}{W_s(3W_p + 2W_{31} + 2W_{32}) + W_p(2W_{21} + W_{32}) + W_{21}(W_{31} + W_{32})} \tag{4.23}$$

Equation (4.23) gives the steady-state population inversion for a given pumping rate W_p and describes the continuous wave (CW) behavior of the laser. Under these conditions, the losses are exactly equal to the gain that results from both stimulated and spontaneous emissions. It is evident from equation (4.23) that a population inversion between energy levels 1 and 2 can be achieved for a three-level system, that is, $N_2 > N_1$, only if

$$W_{32} > W_{21} \tag{4.24}$$

However, this is only a necessary condition for laser oscillations to occur, but not a sufficient one, since the population inversion has to be high enough to overcome excess cavity losses. In other words, a threshold level of the population inversion must also be achieved. Now from equations (1.20b) and (4.13), we have the approximation

$$W_{ij} \propto \frac{1}{\tau_{ij}} \tag{4.25}$$

where τ_{ij} is the lifetime for transition from level i to level j.

Thus, the lifetime for transition from level 3 to level 2 must be smaller than the lifetime for transition from level 2 to level 1, for population inversion to be possible in a three-level system. As a result, for the system to be effective, level 2 has to be metastable, while there has to be rapid decay from level 3 to level 2 (Fig. 4.1). Level 3 is then practically empty, while population buildup occurs on level 2, resulting in population inversion between levels 2 and 1.

Lasing transition occurs between levels 2 and 1, and since level 3 is normally not a laser level, it can have a broad band, and that makes it possible to pump the system effectively using an optical source with a broad bandwidth. Now level 1 is both the lower laser level and the lowest energy level in the system. Thus, most of the atoms are in this energy level under thermal equilibrium, and for population inversion to occur, it is essential that the pumping must be high enough to lift at least half of the atoms from level 1 into level 2.

It must be noted that the achievement of population inversion also depends on the pumping rate W_p. The critical pumping rate, W_{pc}, necessary to achieve equality in population between levels 2 and 1 is obtained by setting N_2 equal to N_1, in equation (4.23), giving

$$W_{pc} = \frac{W_{21}(W_{31} + W_{32})}{W_{32} - W_{21}} \tag{4.26}$$

However, the threshold pumping rate, W_{pt}, necessary to maintain laser oscillation can be obtained by first determining the threshold population inversion using equation (1.42) and combining with equation (4.23) above. A typical example of a three-level laser is the ruby laser (Chapter 8).

4.3 FOUR-LEVEL SYSTEM

For the four-level laser (Fig. 4.2), let us denote the levels of interest as 0, 1, 2, and 3, so that the lasing levels correspond to those of the three-level system. The atoms are pumped from level 0 (ground level) to level 3, while lasing occurs between levels 2 and 1. The rate equations for this case are then given, starting with level 3, by

$$\frac{dN_3}{dt} = W_p(N_0 - N_3) - (W_{32} + W_{31} + W_{30})N_3 \tag{4.27}$$

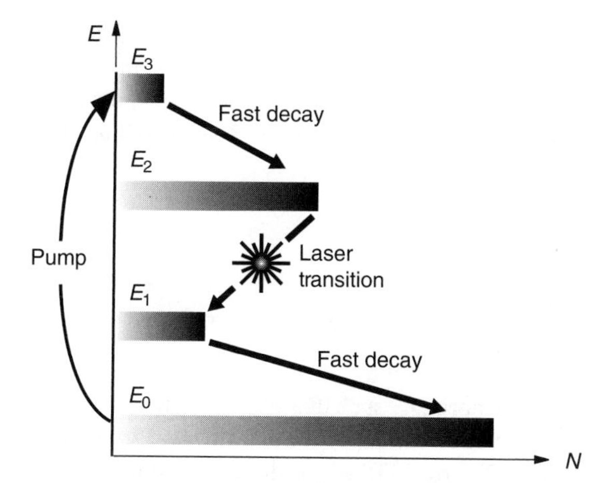

FIGURE 4.2 Schematic of a four-level system. (From O'Shea, D. C., Callen, W. R., and Rhodes, W. T., 1977, *Introduction to Lasers and Their Applications*. Reprinted by permission of Pearson Education, Inc.)

or

$$\frac{\mathrm{d}N_3}{\mathrm{d}t} = W_\mathrm{p}(N_0 - N_3) - W_3 N_3 \tag{4.28}$$

where $W_3 = W_{32} + W_{31} + W_{30}$, and the number of atoms that are thermally excited from the lower to the upper levels have been neglected. The first term in the equation indicates the net rate of excitation of atoms from the ground to the highest level by pumping, while the second term refers to the net rate of decay from the highest level to the lower levels.

Similarly, the rate of change of level 2 is

$$\frac{\mathrm{d}N_2}{\mathrm{d}t} = W_\mathrm{s}(N_1 - N_2) - W_{20}N_2 - W_{21}N_2 + W_{32}N_3 \tag{4.29}$$

or

$$\frac{\mathrm{d}N_2}{\mathrm{d}t} = W_\mathrm{s}(N_1 - N_2) - W_2 N_2 + W_{32}N_3 \tag{4.30}$$

where $W_2 = W_{20} + W_{21}$. For level 1, we have

$$\frac{\mathrm{d}N_1}{\mathrm{d}t} = W_\mathrm{s}(N_2 - N_1) - W_{10}N_1 + W_{21}N_2 + W_{31}N_3 \tag{4.31}$$

Finally, we have

$$\frac{\mathrm{d}N_0}{\mathrm{d}t} = W_\mathrm{p}(N_3 - N_0) + W_{10}N_1 + W_{20}N_2 + W_{30}N_3 \tag{4.32}$$

and

$$N = N_0 + N_1 + N_2 + N_3 = \text{constant} \tag{4.33}$$

Four-level systems are normally such that $W_{32} >> (W_{31}$ or $W_{30})$ since most of the spontaneous transition from level 3 occurs directly to level 2. Likewise, $W_{21} >> W_{20}$. Thus, from equations (4.28) to (4.33), we get the population difference between levels N_2 and N_1, under steady-state conditions as

$$\frac{N_2 - N_1}{N} = \frac{(W_{10} - W_{21})W_{32}}{W_{32}(W_{10}+W_{21}) + W_3W_{31} + \frac{2W_\mathrm{p}+W_3}{W_\mathrm{p}}W_2W_{10} + W_\mathrm{s}\left\{2W_{32} + \left[\frac{2W_\mathrm{p}+W_3}{W_\mathrm{p}}\right]W_{10}\right\}} \tag{4.34}$$

It is also evident from equation (4.34) that a population inversion between energy levels 1 and 2 can be achieved, that is $N_2 > N_1$, only if

$$W_{10} > W_{21} \tag{4.35}$$

In other words, under steady-state or CW operation, the lifetime for transition from level 1 to level 0 must be smaller than the lifetime for transition from level 2 to level 1, for population inversion to be possible in a four-level system. Again, this is only a necessary condition, and a threshold level of the population inversion must also be achieved. Just as in the case of the three-level laser, the threshold pumping rate, W_{pt}, necessary to maintain laser oscillation can be obtained by first determining the threshold population inversion using equation (1.42), and combining with equation (4.34) above.

Since lasing action occurs between levels 2 and 1, level 2 is essentially in a metastable state and thus has a relatively long lifetime, but there is rapid transition from level 3 to level 2, and also from level 1 to level 0; that is, levels 3 and 1 have relatively short lifetimes. Thus, $W_{32} >> W_{21}$, and $W_{10} >> W_{21}$, further reducing equation (4.34) to the form

$$\frac{N_2 - N_1}{N} = \frac{W_{10}W_{32}}{W_{32}W_{10}+W_3W_{31}+\frac{2W_p+W_3}{W_p}W_2W_{10}+W_s\left\{2W_{32}+\left[\frac{2W_p+W_3}{W_p}\right]W_{10}\right\}} \tag{4.36}$$

which, when compared with equation (4.23) for the three-level system, shows that population inversion can be obtained in a four-level system much more easily, and with very little pumping power. This is because the achievement of population inversion in a four-level system is independent of pumping power, whereas that for a three-level system depends on the pumping power. It must be noted, however, that the amount of inversion depends on the pumping rate, as equation (4.36) indicates.

For illustration, the critical pumping rate, W_{pc}, necessary for achieving population inversion in a three-level laser (ruby) can be obtained by substituting the appropriate values in equation (4.26), and would typically be about 330 per second. For comparison with the four-level laser (Nd:YAG), we rather obtain the threshold pumping rate, W_{pt}, for the latter case, which is easier, by first determining the threshold population inversion for laser oscillation and then using equation (4.36), or a simplification of it, to determine the threshold pumping rate, and would be about 0.3 per second. Corresponding pumping power at the active medium can also be estimated to be about 1100 and 5 W/cm^3, for the three-level and four-level lasers, respectively.

Just as in the case of the three-level system, level 3 in the four-level system can have a broad band since it is not a laser level, and thus it is possible to pump the system effectively using an optical source with a broad bandwidth. Furthermore, the energy difference between levels 1 and 0 has to be much greater than $k_B T$, that is, $\Delta E_{01} >> k_B T$ or level 1 has to be much higher than level 0, so that the population of level 1 is essentially negligible at ordinary temperatures. Otherwise, level 1 would be populated by atoms from level 0 through thermal excitation, thereby reducing the population inversion and the efficiency of the system.

It can be shown, for both three-level and four-level systems, that once the threshold population inversion is reached for steady-state or CW operation, the population inversion stays constant, no matter how far the pumping rate exceeds the threshold pumping rate. This is because under steady-state conditions, the gain must always equal the losses that are constant. Thus, the gain must also be constant, and since it is linearly proportional to the population inversion, the latter must also be constant at threshold under steady-state conditions.

4.4 SUMMARY

The distribution of the state of atoms under thermal equilibrium at any given temperature is given by Boltzmann's law, which shows that the number of atoms in the higher energy states is lower than that in the lower states under equilibrium conditions. This can be reversed by pumping to achieve population inversion. However, population inversion is difficult to achieve in a two-level system.

For a three-level system, the lifetime for transition from level 3 to level 2 must be smaller than the lifetime for transition from level 2 to level 1, for population inversion to be possible. In other words, level 2 has to be metastable, while there has to be rapid decay from level 3 to level 2. For laser oscillation to be possible, it is also necessary that a threshold population inversion be achieved.

Population inversion in a four-level system is achieved when the lifetime for transition from level 1 to level 0 is smaller than the lifetime for transition from level 2 to level 1. Thus, level 2 is essentially in a metastable state, while levels 3 and 1 have relatively short lifetimes.

The achievement of population inversion in a four-level system is independent of pumping power, while that for a three-level system depends on the pumping power. Thus, population inversion can be obtained in a four-level system much more easily, and with very little pumping power.

With this background, we discuss, in the next chapter, the mechanisms that result in broadening of laser transitions, as a result of which laser outputs are not exactly monochromatic.

APPENDIX 4A

List of symbols used in the chapter.

Symbol	Parameter	Units
n_{sp}	spontaneous emission rate	$/m^3s$
S_{ij}	probability per unit time of nonradiative decay between levels i and j	/s
W_{ij}	probability per unit time of the atom being thermally excited between levels i and j	/s
W_{pc}	critical pumping rate	/s
W_{pt}	threshold pumping rate	/s
W_s	probability per unit time of induced transition caused by laser action	/s
τ_{ij}	lifetime for transition from level i to level j	s

PROBLEMS

4.1. Given

$$\frac{N_2 - N_1}{N} = \frac{W_p(W_{32} - W_{21}) - W_{21}(W_{31} + W_{32})}{W_s(3W_p + 2W_{31} + 2W_{32}) + W_p(2W_{21} + W_{32}) + W_{21}(W_{31} + W_{32})}$$

Derive an expression for the critical pumping rate, W_{pc}. Show all the appropriate steps.

4.2. Given the following equations for a three-level laser:

$$\frac{dN_3}{dt} = W_p(N_1 - N_3) - (W_{32} + W_{31})N_3$$

$$\frac{dN_2}{dt} = W_s(N_1 - N_2) - W_{21}N_2 + W_{32}N_3$$

$$\frac{dN_1}{dt} = W_p(N_3 - N_1) + W_s(N_2 - N_1) + W_{21}N_2 + W_{31}N_3$$

$$N = N_1 + N_2 + N_3 = \text{constant}$$

Show that

$$N_1 = \left(\frac{W_p + W_{31} + W_{32}}{2W_p + W_{31} + W_{32}}\right)(N - N_2)$$

$$N_2 = \left[\frac{W_s}{W_s + W_{21}} + \frac{W_p W_{32}}{(W_s + W_{21})(W_p + W_{31} + W_{32})}\right] N_1$$

$$\frac{N_2 - N_1}{N} = \frac{W_p(W_{32} - W_{21}) - W_{21}(W_{31} + W_{32})}{W_s(3W_p + 2W_{31} + 2W_{32}) + W_p(2W_{21} + W_{32}) + W_{21}(W_{31} + W_{32})}$$

4.3. Given the following equations for a four-level laser:

$$\frac{dN_3}{dt} = W_p(N_0 - N_3) - W_3 N_3$$

$$\frac{dN_2}{dt} = W_s(N_1 - N_2) - W_2 N_2 + W_{32}N_3$$

$$\frac{dN_1}{dt} = W_s(N_2 - N_1) - W_{10}N_1 + W_{21}N_2 + W_{31}N_3$$

$$\frac{dN_0}{dt} = W_p(N_3 - N_0) + W_{10}N_1 + W_{20}N_2 + W_{30}N_3$$

$$N = N_0 + N_1 + N_2 + N_3 = \text{constant}$$

Show that

$$\frac{N_2 - N_1}{N}$$

$$= \frac{(W_{10} - W_{21})W_{32}}{W_{32}(W_{10} + W_{21}) + W_2 W_{31} + \frac{2W_p + W_3}{W_p} W_2 W_{10} + W_s\left\{2W_{32} + \left[\frac{2W_p + W_3}{W_p}\right] W_{10}\right\}}$$

4.4. Obtain an expression for the threshold pumping rate, W_{pt}, for a four-level laser, given

$$\frac{N_2 - N_1}{N} = \frac{(W_{10} - W_{21})W_{32}}{W_{32}(W_{10} + W_{21}) + W_3 W_{31} + \frac{2W_p + W_3}{W_p} W_2 W_{10} + W_s \left\{ 2W_{32} + \left[\frac{2W_p + W_3}{W_p} \right] W_{10} \right\}}$$

5 Broadening Mechanisms

Our discussion thus far has been based on the idealized assumption that there exists a single frequency or spectral line, the resonant frequency, at which lasing occurs. However, in reality, the transition frequency is not exactly monochromatic, that is, it is spread over a finite frequency range. In this chapter, the principal mechanisms that cause broadening of the transition frequency and their corresponding line-shape functions will be discussed. Broadening mechanisms and line-shape functions are important because they determine the performance characteristics of the laser. For example, the line-shape function determines how many axial modes oscillate in the laser, the threshold gain, and so on.

We begin our discussion by looking at the line-shape function and its relation to broadening mechanisms. This is then followed by a discussion on the two general categories of broadening: homogeneous and inhomogeneous broadening, and the different mechanisms that fall under each category. Finally, we compare the basic characteristics of the individual mechanisms. Due to the need to use the Fourier transform, we shall, for convenience, use the angular frequency, ω, for the frequency in this chapter.

5.1 LINE-SHAPE FUNCTION

The finite frequency range over which the transition frequency is spread is called the *atomic linewidth* and is centered about a central or resonant frequency, ω_0, in accordance with the uncertainty in the transition energy levels. The strength of the interaction between the atom and incident radiation is thus a function of frequency, and is known as the *line-shape function*, designated $g(\omega, \omega_0)$. The function $g(\omega, \omega_0)\mathrm{d}\omega$ represents the probability that the transition frequency lies between ω and $\omega + \mathrm{d}\omega$, and is centered at the resonant frequency ω_0.

Thus, in the normalized form, the line-shape function is defined such that

$$\int_{-\infty}^{\infty} g(\omega, \omega_0)\mathrm{d}\omega = 1 \tag{5.1}$$

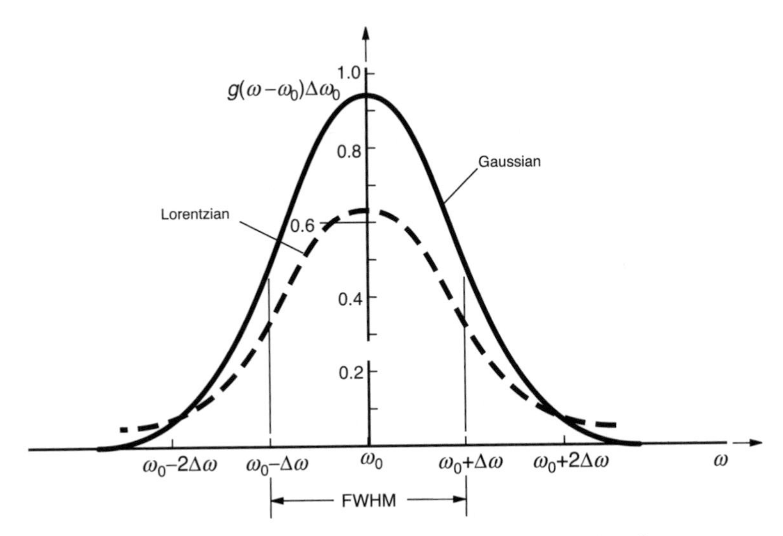

FIGURE 5.1 Lorentzian and Gaussian line-shape functions.

It is characterized using the width or frequency range $\Delta\omega_0 = 2\Delta\omega = 2(\omega - \omega_0)$ at the point of the curve where its amplitude is half the maximum amplitude. This is referred to as the full width at half maximum (FWHM) (Fig. 5.1).

With this background on the line-shape function, we now discuss the principal mechanisms that contribute to broadening of the transition (oscillating) frequency.

5.2 LINE-BROADENING MECHANISMS

Broadening mechanisms can be grouped under two main categories:

1. Homogeneous broadening
2. Inhomogeneous broadening.

5.2.1 Homogeneous Broadening

In homogeneous broadening, the mechanism has the same effect on each atom, and thus the response of each atom is broadened in an identical manner. Each atom in the ensemble has the same resonant frequency and atomic line shape. Homogeneous broadening mechanisms include natural or intrinsic broadening, and collision broadening.

5.2.1.1 Natural Broadening Natural broadening is the basic form of broadening that would exist even if there were no other form of broadening. It is associated with the finite lifetime of the excited state due to spontaneous emission and can

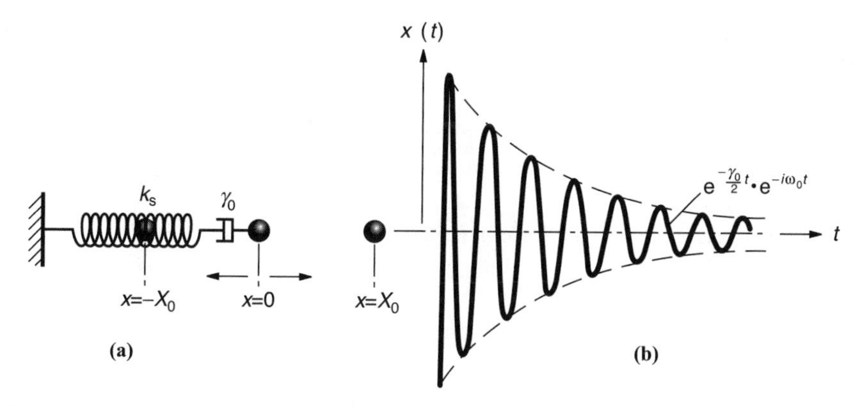

FIGURE 5.2 Classical electron oscillator of an electron orbiting a nucleus. (a) Spring–mass–dashpot analogy. (b) System response.

only be rigorously analyzed using quantum mechanics. However, we adopt a simpler approach to analyze this form of broadening by assuming that the emission process can be modeled as the classical electron oscillator (Fig. 5.2). As the electron oscillates and radiates, it loses energy. Thus, in a sense, the motion can be said to be damped. The governing equation for the motion, assuming a unit mass, is then given by

$$\ddot{x} + \gamma_0 \dot{x} + k_s x = 0 \tag{5.2}$$

where x is the displacement of the electron from an equilibrium position, γ_0 is the damping coefficient, and k_s is the spring constant.

Solution of equation (5.2) results in the following electric field associated with such an electron oscillator:

$$x(t) = X_0 \mathrm{e}^{-\frac{1}{2\tau_{sp}}t} \mathrm{e}^{-i\omega_0 t} \tag{5.3}$$

where X_0 is the amplitude of oscillation, $\tau_{sp} = \frac{1}{\gamma_0}$ is the spontaneous emission lifetime, and $\omega_0 = \sqrt{k_s}$ is the central or natural frequency of oscillation.

Taking the Fourier transform of equation (5.3), we get the amplitude spectrum of the electric field associated with the spontaneous emission as

$$x(\omega) = \frac{1}{2\pi} \int_0^\infty X_0 \mathrm{e}^{-\frac{1}{2\tau_{sp}}t} \mathrm{e}^{-i\omega_0 t} e^{i\omega t} \mathrm{d}t \tag{5.4}$$

or

$$x(\omega) = \frac{1}{2\pi} \frac{X_0}{\frac{1}{2\tau_{sp}} - i(\omega - \omega_0)} \tag{5.5}$$

The corresponding power spectrum, which indicates the intensity distribution, is then given by

$$I(\omega) \propto |x(\omega)|^2 = \frac{{\tau_{sp}}^2}{\pi^2} \frac{{X_0}^2}{1 + 4{\tau_{sp}}^2(\omega - \omega_0)^2} \tag{5.6}$$

which when normalized to the form in equation (5.1) becomes

$$g(\omega, \omega_0) = \frac{2\tau_{sp}}{\pi} \frac{1}{1 + 4{\tau_{sp}}^2(\omega - \omega_0)^2} \tag{5.7}$$

This spectral distribution, the line-shape function, is said to be Lorentzian (Fig. 5.1). The maximum value of the function occurs at $\omega = \omega_0$ and is $2\tau_{sp}/\pi$. The full width at half maximum is then given by

$$\Delta\omega_0 = 2\Delta\omega = \frac{1}{\tau_{sp}} \tag{5.8}$$

Natural broadening sets a lower limit on the atomic linewidth and because of its relatively small magnitude tends to be predominated by other broadening mechanisms.

5.2.1.2 Collision Broadening Collision broadening is more predominant in gaseous media and results primarily from collision between atoms. From equation (5.3), we note that when damping is neglected, the radiation emitted by an atom undergoing emission can be approximately represented by the following equation:

$$x(t) = X_0 e^{i(-\omega_0 t + \phi)} \tag{5.9}$$

where ϕ is the phase of the radiated wave at time $t = 0$. However, since atoms in a gaseous medium are continuously moving in all directions, such motion results in random collisions between them and other atoms, molecules, container walls, and so on, especially at high temperatures. The phase ϕ of the radiated wave changes abruptly or discontinuously if the atom suffers a collision while undergoing emission (Fig. 5.3). Such random changes in phase cause broadening of the atomic linewidth.

Equation (5.9) above holds for the period $t_0 \le t \le t_0 + t_c$ between collisions, that is,

$$x(t) = X_0 e^{i(-\omega_0 t + \phi)}, \quad t_0 \le t \le t_0 + t_c \tag{5.10}$$

where t_c is the time between collisions. During this time, ϕ is constant, but changes randomly at each collision. The line-shape function of the emission associated with such collisions is thus given by the power spectrum of the waveform shown in Fig. 5.3,

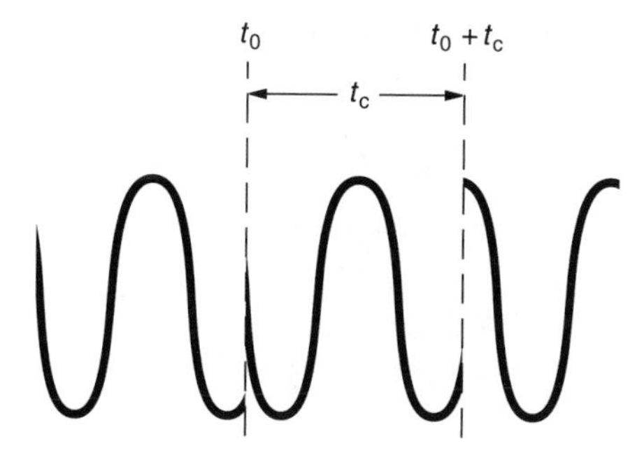

FIGURE 5.3 Phase changes associated with radiation emission resulting from atomic collision.

which is obtained by first taking the Fourier transform of equation (5.10):

$$x(\omega) = \frac{1}{2\pi} \int_{t_0}^{t_0+t_c} X_0 \mathrm{e}^{-i\omega_0 t + i\phi} \mathrm{e}^{i\omega t} \mathrm{d}t \tag{5.11}$$

$$\Rightarrow x(\omega) = \frac{1}{2\pi} X_0 \mathrm{e}^{[i(\omega-\omega_0)t_0 + i\phi]} \frac{\mathrm{e}^{[i(\omega-\omega_0)t_c]} - 1}{i(\omega - \omega_0)} \tag{5.12}$$

where the limits of integration for the Fourier transform are, in this case, between t_0 and $t_0 + t_c$. The resulting power spectrum is given by

$$I(\omega) \propto |x(\omega)|^2 = \frac{{X_0}^2}{\pi^2} \frac{\sin^2 [(\omega - \omega_0)\frac{t_c}{2}]}{(\omega - \omega_0)^2} \tag{5.13}$$

Equation (5.12) is based on the assumption that t_c is constant. In reality, t_c varies with each collision. Thus in determining the spectral density, we have to consider the probability $p_t(t_c)\mathrm{d}t_c$ that an atom undergoes collision in the interval $\mathrm{d}t_c$, that is, between t_c and $t_c + \mathrm{d}t_c$. The probability per unit time, $p_t(t_c)$, is well developed in gas dynamics theory and is defined such that

$$\int_0^\infty p_t(t_c)\mathrm{d}t_c = 1 \tag{5.14}$$

It is given by

$$p_t(t_c) = \frac{\mathrm{e}^{-\frac{t_c}{t_{ca}}}}{t_{ca}} \tag{5.15}$$

where t_{ca} is the average time between collisions. The power spectrum of the radiation due to the random collisions is then given by

$$I(\omega) \propto \int_0^\infty \frac{X_0{}^2}{\pi^2} \frac{\sin^2[(\omega-\omega_0)\frac{t_c}{2}]}{(\omega-\omega_0)^2} p_t(t_c)\mathrm{d}t_c \tag{5.16}$$

$$= \int_0^\infty \frac{X_0{}^2}{\pi^2} \frac{\sin^2[(\omega-\omega_0)\frac{t_c}{2}]}{(\omega-\omega_0)^2} \frac{e^{-\frac{t_c}{t_{ca}}}}{t_{ca}} \mathrm{d}t_c \tag{5.16a}$$

$$= \frac{1}{2}\left(\frac{t_{ca}{}^2}{\pi^2}\right) \frac{X_0{}^2}{1 + t_{ca}{}^2(\omega-\omega_0)^2} \tag{5.17}$$

When the spectral density function is normalized, we get the corresponding lineshape function, which is also Lorentzian, as

$$g(\omega, \omega_0) = \frac{1}{\pi} \frac{t_{ca}}{1 + (\omega-\omega_0)^2 t_{ca}{}^2} \tag{5.18}$$

The maximum value of the function at $\omega = \omega_0$ is t_{ca}/π. The corresponding full width at half maximum is then given by

$$\Delta\omega_0 = 2\Delta\omega = \frac{2}{t_{ca}} \tag{5.19}$$

An estimate of the average time between collisions of an atom or molecule, t_{ca}, can be obtained by considering the mean free path and the average thermal velocity and is given by

$$t_{ca} = \frac{\text{mean free path}}{\text{average thermal velocity}} = \frac{(m_m k_B T)^{\frac{1}{2}}}{16\pi^{\frac{1}{2}} P_g r_a{}^2} \tag{5.20}$$

where k_B is Boltzmann's constant, m_m is the atomic or molecular mass, P_g is the gas pressure, r_a is the radius of the atom or molecule, and T is the absolute temperature.

For example, for neon gas that is at room temperature and under a pressure of 0.5 Torr, we have

$$t_{ca} \approx 0.5 \times 10^{-6}\ \text{s}.$$

The corresponding FWHM is then

$$\Delta\omega_0 = 2\Delta\omega = \frac{2}{0.5 \times 10^{-6}} = 4 \times 10^6\ \text{rad/s}$$

or

$$\Delta\nu_0 = 0.64 \text{ MHz}$$

Similar effects arise in solids as a result of the interaction of atoms with the lattice.

5.2.2 Inhomogeneous Broadening

In inhomogeneous broadening, the resonant frequency of each atom is different or spread out, resulting in a band of frequencies for the entire ensemble, even though the lines of individual atoms are not necessarily broadened. Thus, each atom has a different resonant frequency for the same transition. Examples include Doppler broadening and broadening resulting from crystalline defects.

The most common form of inhomogeneous broadening in gas lasers is Doppler broadening. Doppler broadening is essentially a result of the *Doppler effect*, which is the phenomenon whereby an observer who is moving at a velocity $\mathbf{u}$ detects a frequency ω_0 (centerline frequency) that is different from the frequency ω detected by an observer who is in a stationary frame. A simple illustration of this is the sound of a moving car as detected by an observer standing by the roadside. Even though the frequency of the sound is the same to the driver (observer in the car, ω_0), the sound heard or detected by the standing observer (ω) has a higher frequency as the car approaches and a lower frequency as the car recedes from the stationary observer. This is because the number of oscillations that reach the stationary observer per unit time is increased or decreased depending on whether the car is approaching or receding.

Gaseous atoms, as we know, are in constant motion at rather high velocities. Now let the radiation emitted by an atom while in motion have a frequency ω_0 (as measured in the moving coordinate system or as seen by the atom). Further, let the absolute velocity of the atom (as measured relative to the stationary coordinate system) be u. If the component of this velocity in the direction of wave propagation is u_x, then the frequency ω of the emitted radiation as observed in the stationary coordinate system will be related to ω_0 by

$$\omega_0 = \omega\left(1 \pm \frac{u_x}{c}\right) \tag{5.21}$$

The positive or negative sign depends on whether the velocity is in the opposite or same direction as the emitted wave. Thus, even though the transition or centerline frequency ω_0 might be constant, with reference to the stationary frame, each atom would be observed to be emitting radiation at a different resonant frequency depending on its direction, and that is what results in broadening of the emission line shape.

To determine the line-shape function, we first consider the probability $p(u_x)\mathrm{d}u_x$ that an atom has a velocity component between u_x and $u_x + \mathrm{d}u_x$. This is given by the

Maxwell–Boltzmann velocity distribution:

$$p(u_x)\mathrm{d}u_x = \left(\frac{m_m}{2\pi k_B T}\right)^{\frac{1}{2}} \mathrm{e}^{\left(-\frac{m_m u_x{}^2}{2k_B T}\right)} \mathrm{d}u_x \tag{5.22}$$

If we substitute for ω from equation (5.21) and consider only the difference in velocities, then we have the probability $p(\omega)\mathrm{d}\omega$ that the emitted radiation will be in the frequency range ω to $\omega + \mathrm{d}\omega$ being given by

$$p(\omega)d\omega = \frac{c}{\omega_0}\left(\frac{m_\mathrm{m}}{2\pi k_\mathrm{B} T}\right)^{\frac{1}{2}} \mathrm{e}^{\left[-\frac{m_\mathrm{m}c^2}{2k_B T}\left(\frac{\omega-\omega_0}{\omega_0}\right)^2\right]} \mathrm{d}\omega \tag{5.23}$$

Now considering that the intensity at any frequency is proportional to the probability at that frequency, that is,

$$p(\omega)\mathrm{d}\omega = g(\omega)\mathrm{d}\omega \tag{5.24}$$

we have

$$g(\omega, \omega_0) = \frac{c}{\omega_0}\left(\frac{m_\mathrm{m}}{2\pi k_\mathrm{B} T}\right)^{\frac{1}{2}} \mathrm{e}^{\left[-\frac{m_\mathrm{m}c^2}{2k_\mathrm{B} T}\left(\frac{\omega-\omega_0}{\omega_0}\right)^2\right]} \tag{5.25}$$

This is the line-shape function for Doppler broadening and has a Gaussian distribution (Fig. 5.1). It has a maximum at $\omega = \omega_0$ given by $c/\omega_0\,(m_\mathrm{m}/2\pi k_\mathrm{B} T)^{1/2}$, and the FWHM is given by

$$\Delta\omega_0 = 2\Delta\omega = 2\omega_0\left(\frac{2k_\mathrm{B} T \ln 2}{m_\mathrm{m}c^2}\right)^{1/2} \tag{5.26}$$

Equation (5.25) can be simplified by substituting the expression for the full width at half maximum, equation (5.26), giving

$$g(\omega, \omega_0) = \frac{2}{\Delta\omega_0}\left(\frac{\ln 2}{\pi}\right)^{\frac{1}{2}} \mathrm{e}^{\left[-4\ln 2\left(\frac{\omega-\omega_0}{\Delta\omega_0}\right)^2\right]} \tag{5.27}$$

5.3 COMPARISON OF INDIVIDUAL MECHANISMS

To get an idea of the relative impacts of each of these broadening mechanisms, we consider a He–Ne laser with a transition at $\lambda = 632.8$ nm, and which is at room temperature, 300 K and pressure of 0.5 Torr. For natural broadening, $\tau_\mathrm{sp} \approx 10^{-8}$ s.

Thus,

$$\Delta \nu_{0\text{nat}} = 2\Delta \nu_{\text{nat}} = \frac{1}{2\pi \tau_{\text{sp}}} \approx 20 \text{ MHz}$$

For collision broadening, $t_{\text{ca}} \approx 0.5 \times 10^{-6}$ s. Thus,

$$\Delta \nu_{0\text{col}} = 2\Delta \nu_{\text{col}} = \frac{1}{\pi t_{\text{ca}}} \approx 0.64 \text{ MHz}$$

For Doppler broadening,

$$\Delta \nu_{0\text{Dop}} = 2\Delta \nu_{\text{Dop}} = 2\nu_0 \left(\frac{2k_{\text{B}}T \ln 2}{m_{\text{m}}c^2} \right)^{1/2} \approx 1.7 \text{ GHz}$$

The example shown indicates that Doppler broadening is the predominant form of broadening, with collision broadening being relatively negligible. This is the case for the He–Ne laser where the cavity pressures are relatively low. However, at high enough pressures, such as that with the CO_2 laser, collision broadening becomes significant.

When two or more mechanisms are equally significant, the overall line-shape function is the convolution of the individual functions. If both mechanisms have a Lorentzian function, then the width $\Delta\omega_0$ of the resulting function is

$$\Delta\omega_0 = \Delta\omega_{01} + \Delta\omega_{02} \tag{5.28}$$

where $\Delta\omega_{01}$ and $\Delta\omega_{02}$ are the widths of the individual functions. For two Gaussian functions, we have

$$\Delta\omega_0 = ({\Delta\omega_{01}}^2 + {\Delta\omega_{02}}^2)^{1/2} \tag{5.29}$$

For comparison, the two line-shape functions, Lorentzian (for homogeneous broadening) and Gaussian (for inhomogeneous broadening), are shown superimposed in Fig. 5.1 in dimensionless plots such that they have the same FWHM. It is evident from the figure that the Gaussian curve is much sharper and has a higher peak than the Lorentzian curve. The peak of either curve occurs at $\omega = \omega_0$. The Gaussian curve peak is

$$g(\omega_0) = \frac{0.939}{\Delta\omega_0} \tag{5.30}$$

while that for the Lorentzian curve is

$$g(\omega_0) = \frac{0.637}{\Delta\omega_0} \tag{5.31}$$

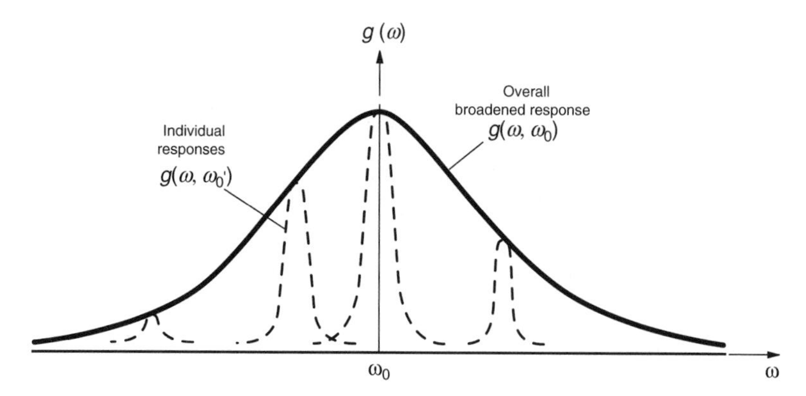

FIGURE 5.4 Line-shape function of an inhomogeneously broadened medium. (From O'Shea, D. C., Callen, W. R., and Rhodes, W. T., 1977, *Introduction to Lasers and Their Applications*. Reprinted by permission of Pearson Education, Inc.)

In homogeneous broadening, all the atoms have the same line-shape function $g(\omega, \omega_0)$ with the same central frequency, ω_0. For inhomogeneous broadening, the overall line-shape function $g(\omega, \omega_0)$ is the convolution of the functions $g(\omega, \omega_0')$ of individual atoms (Fig. 5.4).

5.4 SUMMARY

We started this chapter with a basic discussion of the line-shape function, which describes the strength of the interaction between an atom undergoing transition and the radiation that is incident on it. This is the result of the finite frequency range over which a transition occurs, which is referred to as the atomic linewidth. The concept of the full width at half maximum was then introduced. This is the frequency range ($\Delta\omega_0$) which corresponds to the point on the line-shape function curve at which the amplitude is half the maximum amplitude.

This was followed by a discussion of the line-broadening mechanisms, which can be grouped into two main categories, homogeneous and inhomogeneous broadening. Under homogeneous broadening, we have natural broadening and collision broadening as two of the most common mechanisms. Natural broadening is associated with the finite lifetime of the excited state as a result of spontaneous emission and is the basic form that would exist even if there were no other form of broadening. Collision broadening results primarily from collision between atoms and is therefore more predominant in gaseous media. In homogeneous broadening, the mechanism has the same effect on each atom, and thus the response of each atom is broadened in an identical manner, with each atom in the ensemble having the same resonant frequency and atomic line shape. The entire system is thus broadened in the same way.

Inhomogeneous mechanisms include Doppler broadening, which is a result of the Doppler effect, the phenomenon whereby an observer moving at a velocity u detects a frequency ω_0 (centerline frequency) that is different from the frequency ω detected by an observer who is in a stationary frame. In inhomogeneous broadening, the resonant frequency of each atom is different or spread out, resulting in a band of frequencies for the entire ensemble, even though the lines of individual atoms are not necessarily broadened. Thus, each atom has a different resonant frequency for the same transition.

Homogeneous broadening mechanisms result in a Lorentzian line-shape function, while inhomogeneous mechanisms result in a Gaussian line-shape function.

APPENDIX 5A

List of symbols used in the chapter.

Symbol	Parameter	Units
r_a	Bohr radius of an atom	m
$I(\omega)$	power spectrum of intensity	W/m^2
k_s	spring constant	N/m
m_m	mass of atom or molecule	mg
P_g	gas pressure	N/m^2
t_c	time between collisions	s
t_{ca}	average time between collisions	s
x	displacement of electron from equilibrium position in x-direction	m
X_0	amplitude of oscillation	m
γ_0	damping coefficient	/s

PROBLEMS

5.1. Given

$$\ddot{x} + \gamma_0 \dot{x} + k_s x = 0$$

Show that for $\gamma_0 \ll k_s$,

$$x(t) = X_0 e^{-\frac{1}{2\tau_{sp}}t} e^{-i\omega_0 t}$$

5.2. Given

$$x(t) = X_0 e^{-\frac{1}{2\tau_{sp}}t} e^{-i\omega_0 t}$$

Show that the spectrum of spontaneous emission is

$$x(\omega) = \frac{X_0}{\frac{1}{2\tau_{sp}} - i(\omega - \omega_0)}$$

Further, show that the normalized form of the power spectrum is

$$g(\omega, \omega_0) = \frac{2\tau_{sp}}{\pi} \frac{1}{1 + 4\tau_{sp}{}^2(\omega - \omega_0)^2}$$

5.3. Show that the full width at half maximum of natural broadening is given by

$$\Delta\omega_0 = 2\Delta\omega = \frac{1}{\tau_{sp}}$$

5.4. Derive the power spectrum for collision broadening.

5.5. Given that the power spectrum of the radiation due to random collisions is given by

$$f(\omega) = \int_0^\infty \frac{X_0{}^2}{\pi^2} \frac{\sin^2 [(\omega - \omega_0)\frac{t_c}{2}]}{(\omega - \omega_0)^2} \frac{e^{-\frac{t_c}{t_{ca}}}}{t_{ca}} dt_c$$

Show that the corresponding line-shape function is

$$g(\omega, \omega_0) = \frac{1}{\pi} \frac{t_{ca}}{1 + (\omega - \omega_0)^2 t_{ca}{}^2}$$

5.6. Given

$$p(u_x)\mathrm{d}u_x = \left(\frac{m_m}{2\pi k_B T}\right)^{\frac{1}{2}} e^{\left(-\frac{m_m u_x{}^2}{2k_B T}\right)} du_x$$

Show that the line-shape function for Doppler broadening is

$$g(\omega, \omega_0) = \frac{c}{\omega_0} \left(\frac{m_m}{2\pi k_B T}\right)^{\frac{1}{2}} e^{\left[-\frac{m_m c^2}{2k_B T}\left(\frac{\omega - \omega_0}{\omega_0}\right)^2\right]}$$

5.7. Show that the full width at half maximum for Doppler broadening is

$$\Delta\omega_0 = 2\Delta\omega = 2\omega_0 \left(\frac{2k_B T \ln 2}{m_m c^2}\right)^{1/2}$$

5.8. Determine the number of oscillations that an atom will undergo over its lifetime, τ_{sp}, of 1.5 μs, if the transition frequency is 10^{13} Hz.

5.9. Determine the frequency that corresponds to the transition wavelength ($\lambda =$ 632.8 nm) of the helium–neon laser. Further, determine the value of p of the laser mode closest to the line center if the separation between the mirrors of the laser cavity is 60 cm. How many longitudinal modes will be contained in the linewidth of the laser if the gain curve has a linewidth of 1.5 GHz?

5.10. The transition wavelength of a CO_2 molecule at a temperature of 400 K is $\lambda = 10.6\ \mu m$. Determine its Doppler linewidth.

5.11. Determine the pressure at which the Doppler and collision broadening mechanisms for a CO_2 laser contribute equally to the linewidth if collision broadening is approximately 6 MHz/Torr.

6 Beam Modification

Some of the basic characteristics of a pulsed beam, such as the peak power, pulse duration (pulse width), and pulse repetition rate (pulse frequency), can be changed during the beam generation to enable certain desired beam properties to be achieved. For example, a higher peak power may be required to initiate cutting or welding in some materials, with the power level that is needed being significantly reduced once the material melts. In this chapter, the techniques used for varying the beam characteristics are discussed. We start with a discussion on the measure of energy losses in the laser cavity, that is, quality factor (Q_f), which constitutes the framework for controlling the peak power and pulse duration. This is followed by a discussion on Q-switching, which is a method that is used in producing pulsed laser outputs of relatively short duration, and different methods for achieving Q-switching. The theory of Q-switching is then presented. Next, we talk about mode locking, which is used to achieve pulsed outputs of even higher power and shorter duration. Again, different mode-locking methods are presented. Finally, the concepts of laser spiking and Lamb dip are discussed.

6.1 QUALITY FACTOR

The Q-factor or quality factor, Q_f, is a measure of the losses or energy dissipation in any mode of the laser cavity and is defined as

$$Q_f = \frac{\omega_0 \times \text{energy stored in the mode}}{\text{energy lost or dissipated in the mode per unit time}} \tag{6.1}$$

$$= \frac{2\pi \times \text{energy stored in the mode}}{\text{energy lost or dissipated per cycle}} = \frac{\omega_0 \times Q}{q_l} \tag{6.2}$$

where Q is the energy stored in the mode at any instant in time, q_l is the power loss in the mode, and ω_0 is the oscillation or transition frequency of that mode.

A cavity whose Q-factor is high stores energy well, whereas the opposite is true for a low Q_f cavity. Equation (6.2) can be rewritten as

$$q_l = \frac{\omega_0 \times Q}{Q_f} \tag{6.3}$$

Principles of Laser Materials Processing, by Elijah Kannatey-Asibu, Jr.

or

$$\frac{\mathrm{d}Q}{\mathrm{d}t} = -\frac{\omega_0 \times Q}{Q_\mathrm{f}} \tag{6.4}$$

where the negative sign reflects a loss or reduction in energy with time. The solution of equation (6.4) is

$$Q(t) = Q_0 \mathrm{e}^{-\frac{\omega_0}{Q_\mathrm{f}} t} \tag{6.5}$$

where Q_0 is the initial energy at time $t = 0$.

The time constant of the system described by equation (6.5), which is equivalent to the cavity lifetime, τ_c, is thus given by

$$\tau_\mathrm{c} = \frac{Q_\mathrm{f}}{\omega_0} \tag{6.6}$$

The corresponding electric field amplitude of the mode can thus be described by

$$E_1(t) = E_0 \mathrm{e}^{-\frac{t}{2\tau_\mathrm{c}}} \mathrm{e}^{i\omega_0 t} \tag{6.7.a}$$

or

$$E_1(t) = E_0 \mathrm{e}^{\left(i\omega_0 - \frac{\omega_0}{2Q_\mathrm{f}}\right) t} \tag{6.7.b}$$

since the time constant of the field amplitude is twice that of the field intensity. E_0 in equation (6.7b) is the initial electric field amplitude at time $t = 0$. The first exponential in equation (6.7a) expresses the decay of the signal intensity with time.

Taking the Fourier transform of equation (6.7b) gives the field spectral density as

$$E_1(\omega) = \frac{1}{2\pi} \int_0^\infty E_1(t) \mathrm{e}^{-i\omega t} \mathrm{d}t \tag{6.8}$$

which reduces to

$$E_1(\omega) = \frac{1}{2\pi} \frac{E_0}{\frac{\omega_0}{2Q_\mathrm{f}} + i(\omega - \omega_0)} \tag{6.9}$$

The spectral density of the signal intensity then becomes

$$I(\omega) = |E_1(\omega)|^2 = \frac{1}{4\pi^2} \frac{{E_0}^2}{\left(\frac{\omega_0}{2Q_\mathrm{f}}\right)^2 + (\omega - \omega_0)^2} \tag{6.10}$$

Equation (6.10) describes a Lorentzian distribution with a full width at half maximum or linewidth $\Delta\omega_0$ that is given by

$$\Delta\omega_0 = \frac{\omega_0}{Q_f} \tag{6.11}$$

Equation (6.11) indicates that a cavity mode with a high Q-factor (low losses) tends to have a smaller linewidth.

Now let us develop an explicit expression for the quality factor. In doing so, we consider the primary losses in the cavity to be due to absorption in the active medium and at the mirrors. We then use equation (1.38) that expresses the amplitude of the beam energy after one complete passage through the cavity, but this time without considering amplification due to stimulated emission. Thus, $\beta = 0$. Now, if the cross-sectional area of the cavity is A, then the beam energy, Q, can be expressed as $Q = A \times I \times t$, where I is the beam intensity. The beam intensity I in equation (1.38) can then be replaced with the beam energy Q, giving

$$Q = Q_0 R_1 R_2 e^{-2\alpha L} \tag{6.12}$$

If the time t in equation (6.5) is taken as the time to complete one passage through the cavity, then equations (6.5) and (6.12) both express the energy available in the cavity after one cycle or round trip of the beam through the cavity, and thus should be the same. Equating them gives

$$R_1 R_2 e^{-2\alpha L} = e^{-\frac{\omega_0}{Q_f} t} \tag{6.13}$$

From which we get

$$Q_f = \frac{\omega_0 t}{2\alpha L - \ln R_1 R_2} \tag{6.14}$$

But the time to complete one passage through the active medium is given by

$$t = \frac{2L}{c_m} \tag{6.15}$$

where c_m is the velocity of light in the active medium. Substituting into equation (6.14) gives the quality factor as

$$Q_f = \frac{2\omega_0 L}{c_m(2\alpha L - \ln R_1 R_2)} \tag{6.16}$$

6.2 Q-SWITCHING

The output of some pulsed solid-state lasers such as the ruby laser usually consists of a number of random spikes, each of about a microsecond duration, with the individual spikes spaced apart by about 1 μs, and with peak powers of the order of kilowatts. The entire pulse duration may be about 1 ms. Q-switching is a technique that is used to produce laser outputs of higher power (of the order of megawatts) and shorter duration (of the order of nanoseconds). It must be noted, however, that even though the output power is increased, the total energy content of the pulse is not, and may even be less.

To understand the principle behind Q-switching, we consider a laser cavity in which a shutter is placed in front of one of the mirrors. When the shutter is closed, it prevents light energy from reaching the second mirror, and thereby being reflected back into the cavity. In other words, most of the light energy is lost and oscillation cannot take place. The Q-value of the cavity is then very low (Fig. 6.1a-1), since the losses are high (Fig. 6.1a-2). As pumping of the laser continues (Fig. 6.1a-3), the population inversion keeps building up (Fig. 6.1a-4), far in excess of the threshold value, without any oscillation taking place. When a significantly high value of population inversion has been achieved, the shutter is suddenly opened to reduce the losses. At this point, the gain of the laser (due to the high population inversion) is much greater than the losses. The high energy accumulated as a result of the large difference between the instantaneous and threshold population inversions is then released as an intense beam

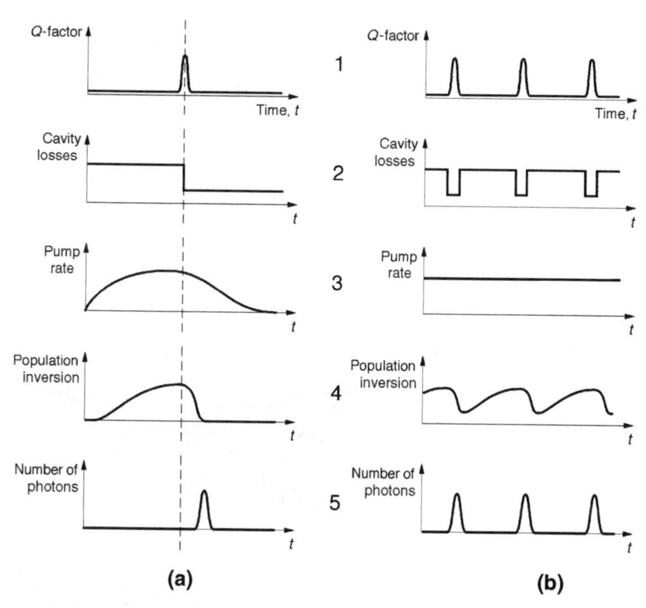

FIGURE 6.1 Time evolution of the Q-switching process. (a) Single pulse. (b) Continuous pulsing. (From Svelto, O., 1989, *Principles of Lasers*, 3rd edition. By permission of Springer Science and Business Media.)

of short duration (Fig. 6.1a-5). Opening of the shutter to reduce the losses increases the Q-value of the cavity. Hence, the name Q-switching.

Q-switching can either result in a single pulse, in which case the pump rate is also pulsed (Fig. 6.1a), or it can be repetitively pulsed, in which case pumping is continuous (Fig. 6.1b). The necessary conditions for a laser to be Q-switchable are

1. The lifetime, τ_u, of the upper level has to be longer than the cavity buildup time, t_c, that is,

$$\tau_u > t_c$$

This enables the upper level to store the extra energy pumped into it.

2. The pumping duration, t_p, has to be longer than the cavity buildup time, t_c. Preferably, it has to be at least as long as the lifetime of the upper level, τ_u, that is,

$$t_p > t_c \text{ and } t_p \geq \tau_u$$

3. The initial cavity loss must be high enough during pumping to prevent oscillation during that period.
4. The change in Q_f value must be sudden.

There are various techniques available for Q-switching, and these are outlined in the following sections.

6.2.1 Mechanical Shutters

With mechanical shutters, one of the cavity mirrors, usually the total reflector, is rotated about an axis perpendicular to the resonator axis (Fig. 6.2). Just before the two mirrors become parallel, the flash lamp is triggered to initiate pumping action.

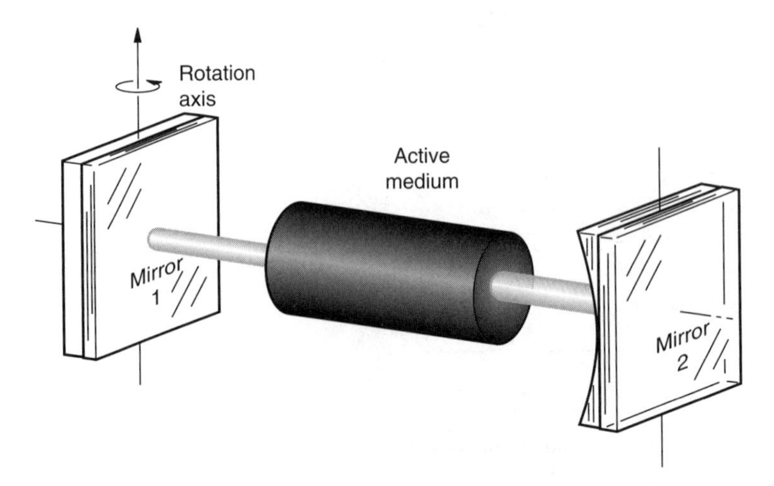

FIGURE 6.2 Mirror rotation for mechanical shutter action.

However, since the mirrors are then not aligned (closed shutter case), the cavity losses are too high to sustain oscillation. The population inversion is thus increased beyond the threshold value. The flash lamp trigger is timed in such a way that the population inversion will be a maximum just when the two mirrors are parallel. The energy stored in the form of high population inversion is then released as a high-power pulse when the mirrors become aligned (open shutter case).

The mechanical shutter technique is robust and independent of wavelength. However, the normal mirror rotational speeds of up to 30,000 revolutions per minute (RPM) result in a slow switching action, which in turn results in a relatively lower peak power. Furthermore, the high rotational speeds may result in mechanical vibration that can influence the beam stability. Also, obtaining ideal beam alignment is a significant problem. The vibrational problem may be minimized by using a quadrilateral- or hexagonal- shaped mirror assembly (Fig. 6.3). If one considers a laser with an upper level lifetime of 3 ms, which is Q-switched using a hexagonal mirror, the mirror assembly would have to rotate one-sixth of a turn every 3 ms, or a full turn in 18 ms. That corresponds to a rotational speed of 3333 RPM, which is quite high, even though it is much lower than would be the case for a single-sided mirror. As a result, it is difficult to use mechanical shutters for Q-switching, especially for lasers of very short upper level lifetime, such as argon ion lasers.

The typical pulse duration obtained when using mechanical shutters is of the order of 400 ns. Increasing the rotational speed would increase the peak power and reduce the pulse duration. However, that would also increase noise and vibrational problems. Mechanical shutters are good for both single-pulsed and repetitive Q-switching operations. However, they are not extensively used anymore because of the mechanical problems associated with them.

6.2.2 Electro-Optic Shutters

The principle of electro-optic shutters is illustrated in Figs 6.4 and 6.5. A polarizer is placed between the active medium and one of the mirrors. An electro-optic cell is

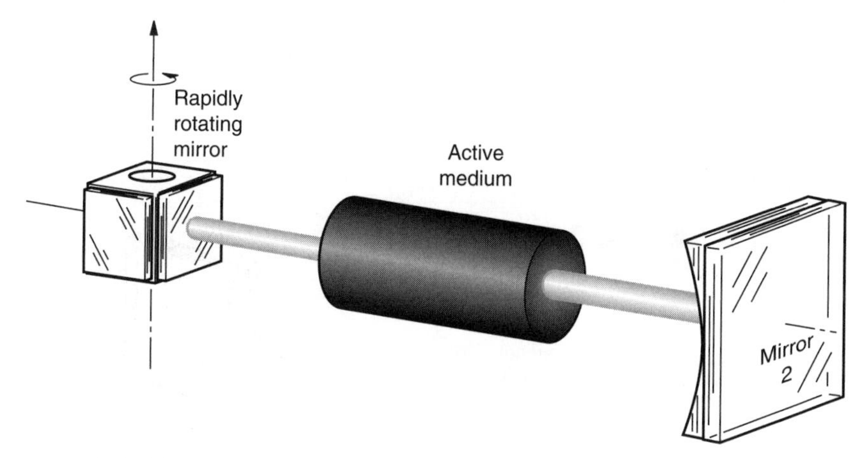

FIGURE 6.3 Quadrilateral mirror assembly for mechanical Q-switching.

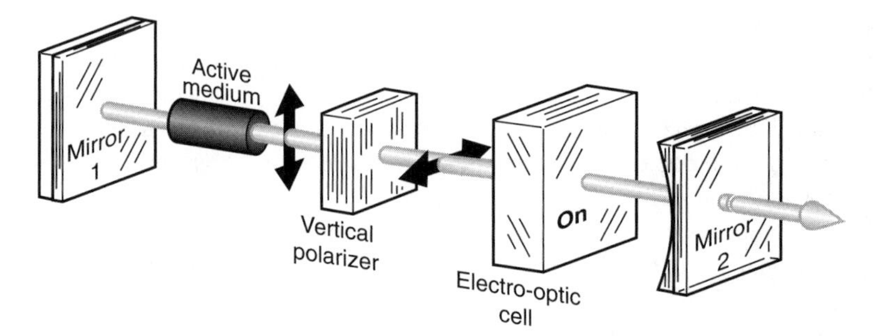

FIGURE 6.4 Electro-optic shutter with the source dc voltage on.

then placed between the polarizer and the mirror closest to it. The electro-optic cell becomes birefringent when a dc voltage is applied to it, and the extent of induced birefringence depends on the applied voltage. Birefringence is the phenomenon whereby light entering a medium is refracted in two slightly different directions to form two rays (see Section 9.3). For switching action, the cell is oriented such that the induced birefringence lies in a plane normal to the resonator axis. Light from the active medium is first linearly polarized as it passes through the polarizer. The linearly polarized light then becomes circularly polarized by the cell. After reflection from mirror 2, the light is converted back to a linearly polarized light by the cell, but this time, with the plane of polarization orthogonal to the original direction. It is therefore blocked by the polarizer on the return trip (Fig. 6.4). In this state, the system acts as a closed shutter, not permitting any light back into the active medium, and that results in high cavity losses and low Q-value. When the dc voltage is removed from the cell, the birefringence vanishes and light from the active medium is reflected back without change in polarization (Fig. 6.5), thereby being retransmitted, resulting in an open shutter. Peak power outputs obtained by this method are of the order of 100 MW, with pulse duration of the order of 10 ns.

The electro-optic cell is called a Pockels cell when the induced birefringence is proportional to the applied voltage. The electric field is applied along the optic axis

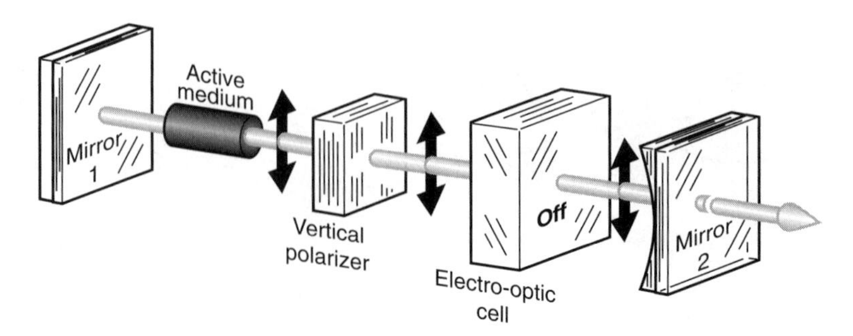

FIGURE 6.5 Electro-optic shutter with the source dc voltage off.

of the crystal and in the laser beam direction. The cell is a Kerr cell when the degree of birefringence that is induced is proportional to the square of the applied voltage. The field is applied in a direction transverse to the beam. The voltage required for a Pockels cell is of the order of 1 kV, while that for a Kerr cell is of the order of 10 kV.

Electro-optic shutters are usually not as robust as mechanical shutters, but are much faster. They are normally useful for single-pulse operation.

6.2.3 Acousto-Optic Shutters

In acousto-optic modulation, an optical material such as ordinary glass is positioned between the active medium and the output mirror (Fig. 6.6). A beam of ultrasonic waves applied to the optical material (using, for example, a piezoelectric transducer) induces local strains in the material, resulting in spatial variations in its refractive index. This simulates a phase grating effect in the optical material, with a period and amplitude that depend on those of the ultrasonic wave. If the ultrasonic waves are made to propagate in a direction normal to the direction of the incident laser beam, the resulting phase grating effect causes a portion of the laser beam to be diffracted out of the cavity, thereby inducing cavity losses. If the losses are high enough, then oscillation cannot be sustained and the system then acts as a closed shutter. Removal of the transducer voltage removes the phase grating effect, thereby acting as an open shutter to permit most of the laser beam to be transmitted with minimal losses. Acousto-optic modulators are normally used when pumping is continuous to produce repeated pulsing, such as is done with neodymium and argon ion lasers.

6.2.4 Passive Shutters

Passive Q-switching does not involve the use of either electronic circuits or mechanical devices, but is based on saturable dyes or absorbers, that is, materials whose absorptivity decrease with increasing beam intensity (Fig. 6.7), and is thus relatively simple to implement. Normally, the material used absorbs energy at the appropriate laser wavelength. A cell containing the dye material is placed between the active medium and one of the mirrors. In the initial stages of pumping, most of the incident

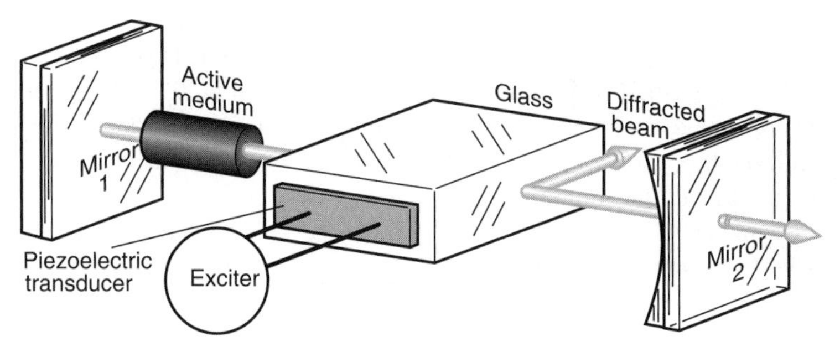

FIGURE 6.6 Acousto-optic modulation.

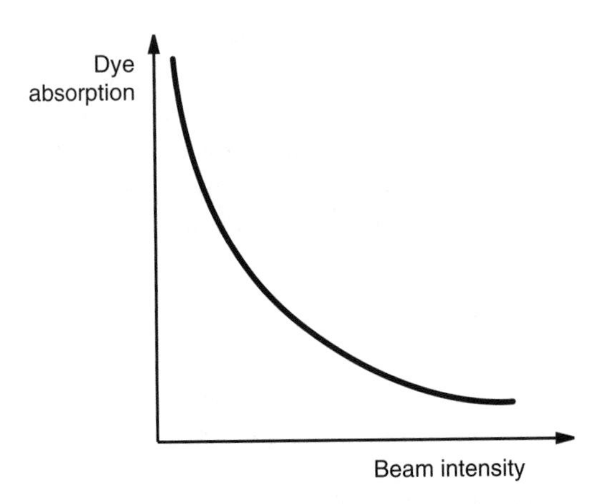

FIGURE 6.7 Absorption characteristics of a saturable absorber.

laser beam is absorbed by the dye due to the relatively low intensity of the beam. The population inversion thus increases with pumping, increasing the light intensity inside the cavity. At a certain level of intensity, the dye begins to bleach, that is, its transmittance is significantly increased. The increased transmittance further enhances the rate of increase of power level within the cavity, which in turn increases the rate of bleaching. The accelerated bleaching results in the dye becoming almost completely transparent when the population inversion is still very high. When the dye is almost transparent (that means the shutter is open) oscillation then begins, causing a high-power pulse to be generated. Passive shutters are normally used for single-pulsed operation.

6.3 Q-SWITCHING THEORY

The dynamic characteristics of the Q-switching mechanism can be evaluated using the appropriate population rate equations and the corresponding photon rate equation for the system of interest. However, for simplicity, we consider the two-level system, but with pumping being directed at higher levels, such as occurs in a four-level system. The population of the lower level 1 is also neglected; that is, level 1 has such a high decay rate to the lower levels that it is essentially unpopulated. This would be typical of a four-level system. The laser is further assumed to oscillate on only one cavity mode. Even though the number of cavity modes is usually very high, only a few modes with the highest gain will normally oscillate since most of the modes tend to have high losses and thus high threshold pumping rate.

Let the number of photons in the cavity mode be q_0. The energy density, $e(\nu)$, is then related to the number of photons by

$$e(\nu) = \frac{q_0 h_p \nu}{V_c} \tag{6.17}$$

where V_c is the cavity volume. The population inversion for the two-level system is given by $N = N_2 - N_1$, and if we assume that $N_1 \approx 0$, we can write $N = N_2$. Then using equation (4.2), and further neglecting the contribution due to spontaneous emission, leaves only the second term on the right-hand side of equation (4.2) that results from stimulated emission:

$$\frac{dN_2}{dt} = -N_2 B_{21} e(\nu) \tag{6.18a}$$

or substituting for $e(\nu)$ from equation (6.17) gives

$$\frac{dN}{dt} = -\frac{C_1}{V_c} q_0 N \tag{6.18b}$$

where $C_1 = B_{21} h_p \nu$ is the rate of stimulated emission per photon per mode in the cavity volume, V_c.

For photon generation in the cavity, we again neglect the contribution of spontaneous transitions. The rate of change of the number of photons in the cavity mode can then be expressed as

$$\frac{dq_0}{dt} = N_2 B_{21} e(\nu) V_c - \frac{q_0}{\tau_c} \tag{6.19a}$$

or

$$\frac{dq_0}{dt} = C_1 q_0 N - \frac{q_0}{\tau_c} \tag{6.19b}$$

where τ_c is the cavity lifetime, or time constant, which is the time in which the energy stored in the mode (the intensity of the light) decays to $1/e$ of its initial value. At the same time, τ_c, the light amplitude will reduce to $1/\sqrt{e}$ of its initial value. The time constant of the light amplitude, that is, the time in which the amplitude becomes equal to $1/e$ of its initial value is $2\tau_c$.

The first term of equation (6.19b) accounts for an increase in the number of photons as a result of stimulated emissions from level 2 into the cavity mode, and the second term accounts for cavity losses.

Equations (6.18b) and (6.19b) constitute a set of equations that describe the dynamic behavior of the photons in the cavity mode, as well as the number of atoms at the upper lasing level. Solution of these equations requires the appropriate initial conditions. Typically, these would be $N_2(0) = N_{in}$ and $q_0(0) = q_{0i}$, where q_{0i} is a rather small number of photons that exist in the initial stages of pumping as a result of spontaneous emission. It is this small number of photons that initiate laser oscillation. Even these simplified equations are nonlinear in q_0 and N, and the dynamic behavior of the population inversion and number of photons or pulse power can be obtained only numerically. This behavior is illustrated qualitatively in Fig. 6.8 and further discussed in Section 6.5.

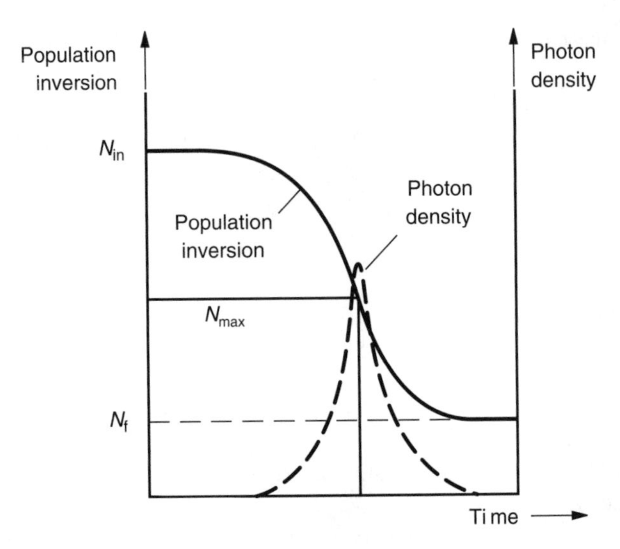

FIGURE 6.8 Temporal variation of population inversion and number of photons. (From Laud, B. B., 1985, *Lasers and Non-Linear Optics*, John Wiley & Sons, Inc. Reprinted by permission.)

The population inversion in the cavity when the number of photons is maximum can be obtained by setting the left-hand side of equation (6.19b) to zero ($\mathrm{d}q_0/\mathrm{d}t = 0$), giving

$$N_{\max} = \frac{1}{C_1 \tau_c} \tag{6.20}$$

Using the chain rule in connection with equations (6.18b) and (6.19b), we have

$$\frac{\mathrm{d}q_0}{\mathrm{d}N} = V_c \left(\frac{1}{C_1 \tau_c N} - 1 \right) \tag{6.21}$$

which when integrated gives

$$\int_{q_{0i}}^{q_0} \mathrm{d}q_0 = V_c \int_{N_{in}}^{N} \left(\frac{1}{C_1 \tau_c N} - 1 \right) \mathrm{d}N \tag{6.22}$$

Thus, we have

$$q_0 = \frac{V_c}{C_1 \tau_c} \ln \left(\frac{N}{N_{in}} \right) - V(N - N_{in}) \tag{6.23}$$

where N_{in} is the initial population inversion, but the initial number of photons, q_{0i}, has been taken as zero. Setting $N = N_{\max}$, where $N_{\max}$ is given in equation (6.20) gives the maximum number of photons as

$$q_{0\max} = \frac{V_c}{C_1 \tau_c} \ln\left(\frac{N_{\max}}{N_{in}}\right) - V_c(N_{\max} - N_{in}) \tag{6.24}$$

Multiplying by $h_p\nu$ to get the maximum energy and further dividing by τ_c gives the peak output power, q_p, as

$$q_p = \frac{V_c h_p \nu f_q}{C_1 {\tau_c}^2} \ln\left(\frac{N_{\max}}{N_{in}}\right) - \frac{V_c h_p \nu f_q}{\tau_c}(N_{\max} - N_{in}) \tag{6.25}$$

where f_q accounts for the fraction of the power that is coupled out of the cavity. The total energy output during the pulse is approximated by

$$Q_{out} = \frac{1}{2}(N_{in} - N_f) V_c h_p \nu f_q \tag{6.26}$$

where N_f is the population inversion at the end of the pulse. Equation (6.26) is essentially a product of the average power generated during the pulse and the cavity lifetime. The factor of 2 in the denominator results from the fact that each emission of a photon results in a twofold change in the population inversion.

6.4 MODE-LOCKING

Mode-locking is used to generate pulses of higher peak power (of the order of gigawatts or more) than can be achieved by Q-switching and of very short duration (of the order of picoseconds or even femtoseconds). A laser cavity generally sustains a large number of oscillating modes (see Chapter 2). The oscillation of each of these modes is normally independent of other modes. Mode-locking is achieved by combining a number of distinct longitudinal modes of a laser in phase, with each mode having a slightly different frequency.

In the ideal case, the various longitudinal modes in a Fabry–Perot resonator are equally spaced (Figs 2.4 and 6.9), with a frequency separation given by

$$\delta\omega = \pi \frac{c}{L} \tag{6.27}$$

where L is the cavity length.

However, nonlinear dispersion of the active medium results in the various modes having random phases, ϕ_m, with the frequency differences between adjacent modes also being random. The result is a number of irregular spikes in the laser output. The total electric field, $E_l(t)$, of the corresponding electromagnetic wave is a superposition

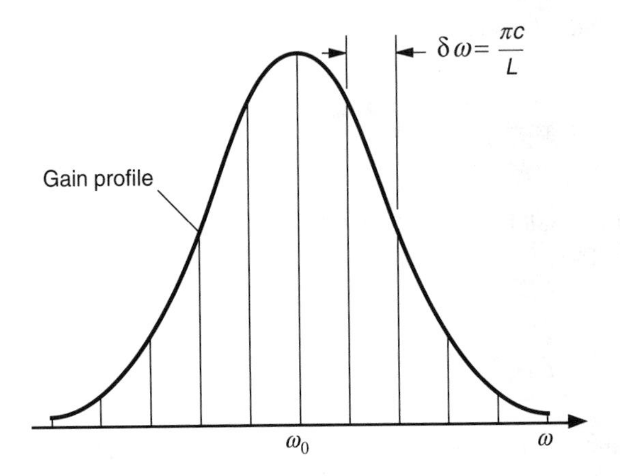

FIGURE 6.9 An active medium's gain profile (see Fig. 2.4).

of the fields of all modes and can generally be expressed as

$$E_l(t) = \sum_{m=1}^{M_c} E_{0m} e^{i(\omega_m t + \phi_m)} \tag{6.28}$$

where E_{0m} = amplitude of oscillation of mode m, ω_m = frequency of mode m, ϕ_m = phase of mode m, and M_c = number of oscillating modes.

For simplicity, let us consider the case where the various modes are of equal amplitude, E_0. The overall peak beam intensity is then a summation of the intensities of the individual modes:

$$I_p = M_c {E_0}^2 \tag{6.29}$$

Let us now consider the case where the individual modes are locked in phase (forced to maintain fixed phase relationships) such that in addition to having the same amplitude, they also oscillate with equal relative phase:

$$\phi_m = \phi \tag{6.30}$$

If the separation between modal frequencies, $\delta\omega$, is also equal, then we have

$$\omega_m = \omega + m\delta\omega \tag{6.31}$$

where ω is the frequency of the lowest mode.

The electric field can then be expressed as

$$E_l(t) = E_0 \sum_{m=1}^{M_c} e^{i[(\omega + m\delta\omega)t + \phi]} \tag{6.32}$$

or

$$E_l(t) = E_0 e^{i(\omega t + \phi)} \sum_{m=1}^{M_c} e^{im\delta\omega t} \tag{6.33}$$

which when expanded becomes

$$E_l(t) = E_0 e^{i(\omega t + \phi)} \left[e^{i\delta\omega t} + e^{i2\delta\omega t} + e^{i3\delta\omega t} + \cdots + e^{iM_c\delta\omega t} \right] \tag{6.34}$$

The overall beam intensity, which is proportional to $|E_l(t)|^2$, is then given, for simplicity, as

$$I(t) = E_0{}^2 \left[e^{i\delta\omega t} + e^{i2\delta\omega t} + e^{i3\delta\omega t} + \cdots + e^{iM_c\delta\omega t} \right]^2 \tag{6.35}$$

which forms a geometric series that reduces to the form

$$I(t) = E_0{}^2 \frac{\sin^2\left(\frac{M_c\delta\omega t}{2}\right)}{\sin^2\left(\frac{\delta\omega t}{2}\right)} \tag{6.36}$$

The peak intensity, I_p, is obtained by considering small values of time when $t \approx 0$ and thus $\sin\theta \approx \theta$, giving

$$I_p = M_c{}^2 E_0{}^2 \tag{6.37}$$

This is the peak intensity of the mode-locked laser that is thus M_c times that of the nonmode-locked laser. For a typical solid-state laser where the number of oscillating modes may be very large, of the order of 10^4, this can be very significant. The average power, however, is unaffected by mode-locking. The peak intensity occurs at times when $\sin \delta\omega t/2 = 0$:

$$\frac{\delta\omega t}{2} = m\pi, \qquad m = 1, 2, 3, \ldots \tag{6.38}$$

or

$$t = \frac{2m\pi}{\delta\omega} = \frac{2mL}{c} \tag{6.39}$$

The separation between contiguous pulses is then given by

$$\Delta t = \frac{2L}{c} \tag{6.40}$$

which is the time taken for a round trip of the cavity. The mode-locked laser may thus be considered as constituting a pulse that propagates back and forth in the cavity. It can be shown from equation (6.36) that the pulse duration, $\Delta\tau_p$ (full width at half maximum), of the beam intensity is given (see Problem 6.3) by

$$\Delta\tau_p \approx \frac{2\pi}{M_c \delta\omega} = \frac{2\pi}{\Delta\omega_0} \tag{6.41}$$

Thus, the larger the oscillating bandwidth, $\Delta\omega_0$, the shorter the output pulse duration. For gas lasers, the pulse duration is typically of the order of nanoseconds, whereas for solid-state lasers, the pulse duration may be of the order of picoseconds or less.

The mode-locking phenomenon can be more easily conceptualized by considering a simple case involving the superposition of three pure sinusoidal waveforms whose frequencies are equally spaced with a frequency difference of $\delta\omega = \pi c/L$. Each of the waveforms would constitute a mode in the cavity. For the case pertaining to a normal laser where the individual modes have different phases at time $t = 0$, the resulting output power, which is an addition to or superposition of the three waveforms, is shown in Fig. 6.10, where the overall intensity or output is found to be relatively uniform and low in magnitude. However, when the individual phases at time $t = 0$ are the same for the three waveforms, the resulting output is a series of pulses spaced apart by $\Delta t = 2L/c$, (Fig. 6.11).

We now discuss some of the common techniques used for mode locking. These may be categorized as being either active or passive.

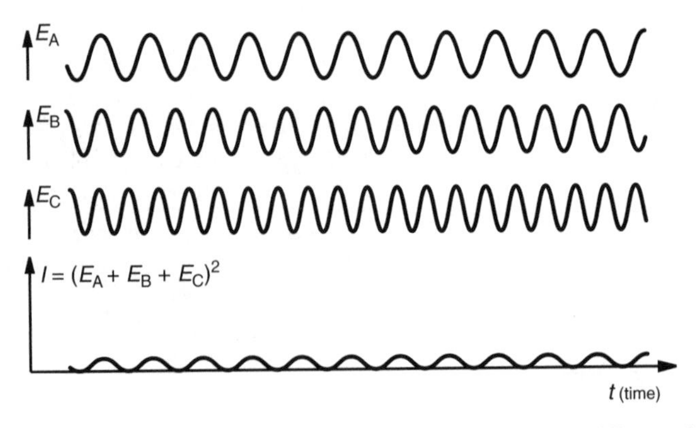

FIGURE 6.10 Superposition of three equally spaced waveforms with different phases. (From O'Shea, D. C., Callen, W. R., and Rhodes, W. T., 1997, *Introduction to Lasers and Their Applications* Reprinted by permission of Pearson Education, Inc.)

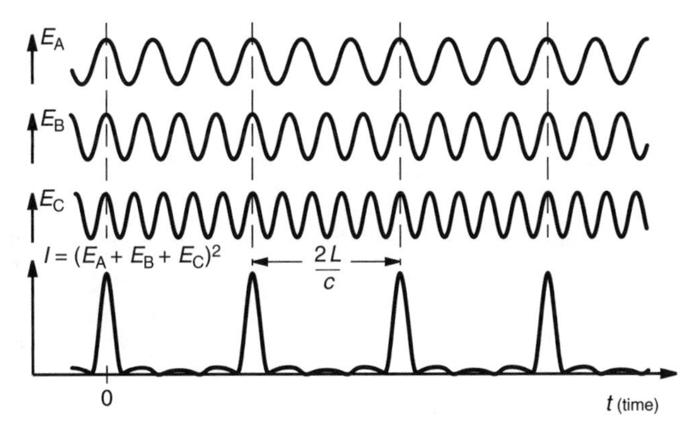

FIGURE 6.11 Superposition of three equally spaced waveforms of the same phase. (From O'Shea, D. C., Callen, W. R., and Rhodes, W. T., 1977, *Introduction to Lasers and Their Applications*. Reprinted by permission of Pearson Education, Inc.)

6.4.1 Active Mode Locking

In active mode locking, an external signal is used to modulate or vary cavity losses or quality factor, $Q_f(t)$, at a frequency equal to the intermodal separation (Fig. 6.12). This can be done using either electro-optic or acousto-optic shutters. We consider the simple case where the shutter is open very briefly, $\delta\tau$, after every period of $\Delta t = 2L/c$ and closed the rest of the time $(\Delta t - \delta\tau)$ during that period. A pulse of light of exactly the same length as the time $\delta\tau$ that the shutter stays open, and which arrives exactly when the shutter is open, will then be unaffected by the shutter. Any portion of the signal that arrives when the shutter is closed will be eliminated. A single pulse that propagates back and forth thus results.

In more rigorous terms, we note that on starting the laser, the mode closest to the center frequency, ω_0, will be the first to start oscillating since it has the highest gain. However, with losses in the cavity modulated at a frequency $\delta\omega$, the amplitude of this

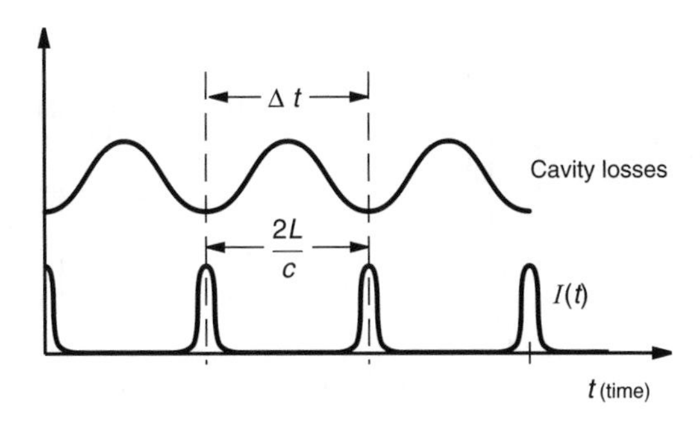

FIGURE 6.12 Amplitude modulation of cavity losses. (From Svelto, O., 1989, *Principles of Lasers*, 3rd edition, By permission of Springer Science and Business Media.)

mode will also vary at the frequency $\delta\omega$. The electric field is then described by

$$E_1(t) = (E_0 + E_{01}\cos\delta\omega t)\cos\omega_0 t$$
$$= E_0\cos\omega_0 t + \frac{1}{2}E_{01}\cos(\omega_0 + \delta\omega)t + \frac{1}{2}E_{01}\cos(\omega_0 - \delta\omega)t \quad (6.42)$$

where E_0 and E_{01} are amplitudes of oscillation. The right-hand side of equation (6.42) shows that the modulated central mode ω_0 is a superposition of modes of frequencies ω_0, $\omega_0 + \delta\omega$, and $\omega_0 - \delta\omega$. The modes corresponding to these frequencies are thus in perfect phase with the central mode. Since the amplitudes of these other modes are also modulated, they, in a similar manner, induce adjacent modes $\omega_0 + 2\delta\omega$ and $\omega_0 - 2\delta\omega$ into a perfect phase relationship. Consequently, all oscillating modes are forced to have the same phase, resulting in mode-locking.

Mode-locking induced by cavity losses as described is said to be amplitude modulated. It can also be induced by variation in the refractive index of the cavity, which results in variations in optical path length. It is then said to be frequency modulated.

6.4.2 Passive Mode-Locking

Passive mode-locking involves the use of nonlinear optical material (saturable absorber) (Fig. 6.13). Let us consider a series of random pulses that constitute the initial beam that is not mode-locked. If these waves are incident on the saturable absorber, the low-intensity pulses will be attenuated more while the higher intensity pulses tend to bleach the dye and thus are more easily transmitted. Eventually, the more intense pulses are amplified at the expense of the less intense ones. For this to be effective, the response or recovery time of the absorber must be much shorter than the pulse duration. The use of a saturable absorber results in simultaneous mode-locking and Q-switching. Thus the output will generally consist of a series of mode-locked pulses of very short duration that are contained in an envelope of longer duration.

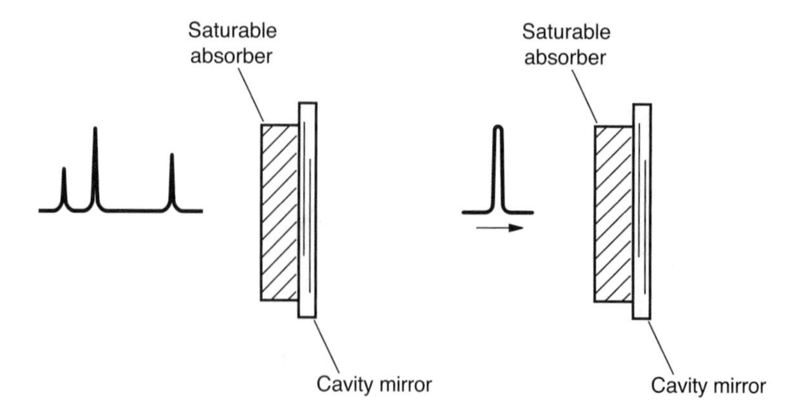

FIGURE 6.13 Passive mode-locking using a saturable absorber. (From Svelto, O., 1989, *Principles of Lasers*, 3rd edition. By permission of Springer Science and Business Media.)

6.5 LASER SPIKING

The transient behavior of a laser can be obtained by solving the appropriate set of rate equations corresponding to the type of laser of interest. General closed-form analytical solution of the equations, including the photon rate equation, is often not feasible since they are nonlinear. A numerical solution of the rate equations for, say, a three-level laser oscillating in a single mode produces the behavior shown in Fig. 6.14 for the case where the system is subjected to a step input in pump rate, that is, when the pump rate is suddenly changed from zero to a fixed value at time $t = 0$:

$$W_{\mathrm{p}}(t) = \begin{cases} 0 & \text{for} \quad t < 0 \\ W_{\mathrm{p}} & \text{for} \quad t \geq 0 \end{cases}$$

The solution shows an oscillation of the laser output power, represented by the photon number, $q_0(t)$, and the population inversion, $N(t)$, about their steady state values (q_{0s} and N_s), until they finally reach steady state. The oscillation results from the fact that when the pump is suddenly activated (step input), the population inversion increases until the threshold inversion is reached. At that point, laser oscillation begins and thus the number of photons also begins to increase. However, it takes time for the photon number to reach its threshold value, and by that time the population inversion is much higher than its threshold since the pumping continues to build up the population. Thus, there is a delay between the two. The high population inversion causes the photon number to increase beyond its threshold, resulting in an increased rate of stimulated emission that reduces the population inversion below its threshold value since it depletes the higher level population much faster than the pumping action can restore it. In fact, the photon number reaches its peak value at the instant that the population inversion threshold is reached. When the population inversion gets below

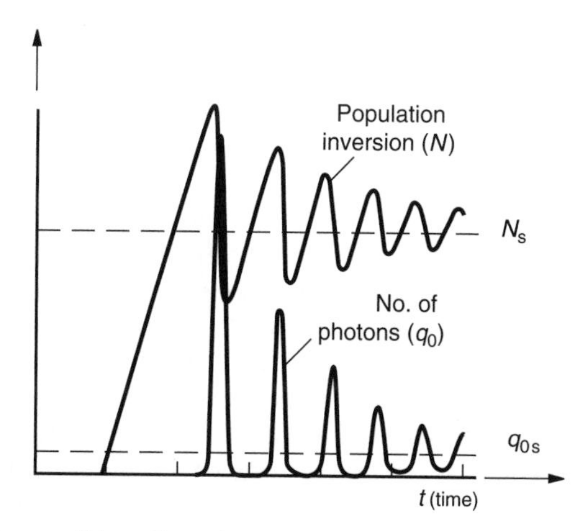

FIGURE 6.14 Laser spiking. (From Svelto, O., 1989, *Principles of Lasers*, 3rd edition. By permission of Springer Science and Business Media.)

its threshold, laser action dies out. The pumping action now drives the population inversion back up, repeating the entire cycle.

The periodic variation in the photon number results in a series of bursts or spikes in the laser output that eventually decays to the steady-state value for continuous wave operation, or until the flash light intensity falls to such a level that it is unable to sustain the population inversion for pulsed operation. The separation between individual spikes tends to be regular and is of the order of microseconds, with the spikes themselves also being of the order of microseconds.

6.6 LAMB DIP

The Lamb dip is a phenomenon that occurs primarily in gas lasers when oscillating in a single mode, and when inhomogeneous broadening (primarily Doppler broadening) is more significant than homogeneous broadening. Let us now consider the situation where the oscillating mode, ω, does not coincide with the centerline frequency, ω_0. Note that each mode is essentially a standing wave and thus consists of two waves that travel in opposite directions between the mirrors. The waves that travel from left to right ($+x$ direction) will interact with those atoms that have a velocity component u_x along the resonator axis such that

$$\omega = \omega_0 \left(1 + \frac{u_x}{c}\right) \tag{6.43}$$

Likewise, the waves that travel from right to left ($-x$ direction) will interact with those atoms with a velocity component $-u_x$ along the resonator axis such that

$$\omega' = \omega_0 \left(1 - \frac{u_x}{c}\right) \tag{6.44}$$

This is essentially due to the Doppler effect. The mode thus interacts with two sets of atoms whose x-component velocities are of the same magnitude but moving in opposite directions.

If the mode frequency is changed to change the mode, two different sets of atoms will interact with the new mode since the velocity magnitude that satisfies equations (6.43) and (6.44) would have changed for the new frequency. However, when the tuning is such that the mode frequency coincides with the centerline frequency, then u_x is zero. That means the mode only interacts with those atoms that have zero velocity component along the x-axis. Thus, instead of two sets of atoms, only one set interacts with the mode at this frequency, and thus the output power must statistically be less than that at the other frequencies. A plot of the laser output power versus the mode frequency for a fixed pumping rate, obtained by varying the cavity length over a half wavelength, will therefore show a dip at the centerline frequency (Fig. 6.15). This is the Lamb dip and is used as a basis for maintaining a stable laser frequency (see Section 7.5).

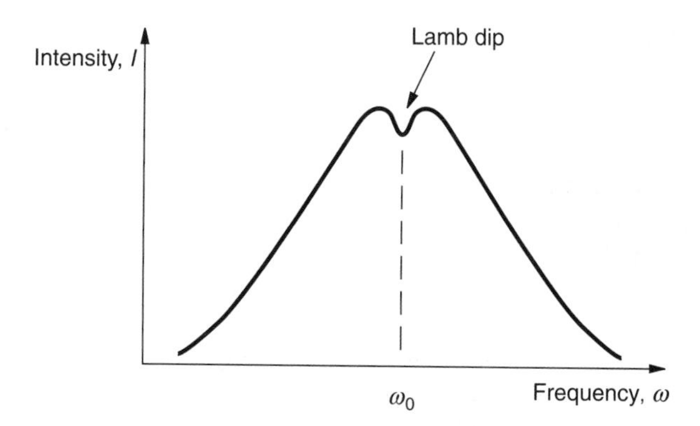

FIGURE 6.15 Lamb dip.

6.7 SUMMARY

The Q-factor of a laser cavity is a measure of the losses or energy dissipation in any mode of the laser cavity. A cavity with a high Q-factor has low losses, and thus has a smaller linewidth.

Q-switching is a technique that is used to produce pulsed laser outputs of higher power but shorter duration. It involves using a shutter to control the Q-factor by preventing oscillation in the cavity until a high population inversion is achieved, at which time the shutter is opened, resulting in an intense beam of short duration being released. Q-switching can be achieved using mechanical, electro-optic, acousto-optic, or dye (passive) shutters. Mechanical shutters are limited by noise and vibrational problems, since higher rotational speeds are required to reduce the pulse duration, which is thus limited to about 400 ns. Electro-optic shutters enable higher peak powers to be obtained, with pulses as short as 10 ns in duration, but are normally limited to a single-pulse operation, as are passive shutters. Acousto-optic shutters, however, can generate repeated pulsing.

Pulses with higher peak powers and shorter duration than can be obtained by Q-switching are achieved through mode-locking, by combining a number of distinct longitudinal modes of a laser in phase. Mode-locking is achieved by using electro-optic or acousto-optic shutters to modulate or vary losses in the cavity at a frequency equal to the intermodal separation. Dyes are also used, but these generally result in simultaneous Q-switching and mode-locking.

The series of spikes that are normally observed in a laser output result from oscillation of the number of photons and population inversion about their steady-state values. The Lamb dip is a phenomenon where the variation of the laser output with frequency shows a dip at the oscillating frequency, ω_0. It occurs primarily in gas lasers when oscillating in a single mode, and when inhomogeneous broadening (primarily Doppler broadening) is more significant than homogeneous broadening.

APPENDIX 6A

List of symbols used in the chapter.

Symbol	Parameter	Units
C_1	rate of stimulated emission per photon per mode	sr m^2/s
E_{0m}	amplitude of oscillation of mode m	V/m
f_q	fraction of power coupled out of cavity	–
N_f	population inversion at the end of pulse	$/m^3$
N_i	population (number of atoms) of energy level i	$/m^3$
N_{in}	initial population inversion	$/m^3$
N_{max}	population inversion corresponding to maximum number of photons in cavity	$/m^3$
N_s	steady-state population inversion	$/m^3$
q_l	power loss in a mode	W
q_0	number of photons in a cavity mode	–
q_{0i}	initial number of photons in a cavity mode	–
q_{0s}	steady state number of photons	–
q_p	peak output power	W
Δt	time separation between contiguous pulses	s
t_c	cavity buildup time	s
t_p	pumping duration	s
u_x	x-velocity component of atom	m/s
Q	energy stored in a mode	J
Q_0	initial energy stored in a mode	J
Q_{out}	total energy output during a pulse	J
ω_m	frequency of mode m	rad/s
$\delta\omega$	frequency separation between longitudinal modes	rad/s
ϕ_m	phase of mode m	°
$\delta\tau$	shutter opening time	s
τ_u	upper level lifetime	s

PROBLEMS

6.1. Given the rate equations for population inversion and the number of photons in a two-level system as

$$\frac{dN}{dt} = -\frac{C_1}{V} q_0 N$$

and

$$\frac{dq_0}{dt} = C_1 q_0 N - \frac{q}{\tau_c}$$

Obtain the variation of the population inversion and number of photons with time if $C_1 = 10.5 \times 10^{-6}$ sr m^2/s, V =150 mm^3, and $\tau_c = 15$ ns.

6.2. Given

$$E_1(t) = E_0 e^{\left(i\omega_0 - \frac{\omega_0}{2Q_f}\right)t}$$

Show that the spectral density of the signal intensity is

$$I(\omega) = |E_1(\omega)|^2 = \frac{{E_0}^2}{(\omega - \omega_0)^2 + \left(\frac{\omega_0}{2Q_f}\right)^2}$$

6.3. Knowing that

$$I(t) = {E_0}^2 \frac{\sin^2\left(\frac{M_c \delta\omega t}{2}\right)}{\sin^2\left(\frac{\delta\omega t}{2}\right)}$$

Show that the pulse width (full width at half maximum) is given by

$$\Delta\tau_p \approx \frac{2\pi}{M_c \delta\omega} = \frac{2\pi}{\Delta\omega_0}$$

6.4. (a) Would you expect the Q-factor of a cavity to be high or low when a shutter in front of one of the mirrors is closed? Explain.

(b) Why is it that a mode-locked laser is considered to act as a pulse that propagates back and forth in the cavity?

6.5. If a Q-switched Nd:YAG laser (wavelength = 1.06 μm) generates a rectangular pulse with an energy content of 10 J and of duration 20 ps., what will be the peak output power?

6.6. If the laser in Problem 6.5 is to have an output pulse energy of 7.5 J, determine the fraction of the total population that will be needed for the population inversion if the active medium contains 10^{20} Nd^{3+} ions. Assume that each ion in the upper state undergoes only one transition while the pulse is being generated.

6.7. If an Nd:YAG laser has 500 longitudinal modes in a cavity of length 1500 mm, what will be the width of pulses generated by mode-locking? Determine the time spacing between the pulses.

6.8. For the Nd:YAG laser in problem 6.7, determine the cavity Q-factor if the reflectances of the resonator mirrors are 100% and 80%.

6.9. If 200 of the axial modes in the laser in Problem 6.8 are mode-locked, determine the spacing between pulses, time between pulses, and frequency bandwidth.

7 Beam Characteristics

The properties of a laser beam that have significant influence on materials processing include divergence, monochromaticity, coherence, brightness, stability, size, and mode. We shall discuss each of these characteristics in the following sections. We start with a discussion on beam divergence, where an expression is provided for calculating the divergence, assuming it to be primarily due to diffraction. This is followed by monochromaticity, in light of the broadening phenomena presented in Chapter 5. Expressions are provided for the theoretical minimum oscillating bandwidth for both continuous wave and pulsed lasers. Next, the concept of beam coherence is discussed, considering both space-and time-dependent coherence. These phenomena are then explained using Young's experiment and Michelson's interferometer. Finally, the intensity and brightness, frequency stabilization based on the Lamb dip phenomenon, beam size, beam, focusing, and radiation pressure are discussed in that order.

7.1 BEAM DIVERGENCE

Beam divergence may be defined as the characteristic property of a light beam that results in an increase in its cross-sectional area with distance from the source. Light from an ordinary source is uniformly radiated in all directions, and as a result, its divergence is high, (Fig.7.1a). Its intensity is inversely proportional to the square of the distance from the source. A laser, on the other hand, is such that it is only light that is almost parallel with the resonator axis, and which is close to the axis that comes out, resulting in relatively low divergence, that is, almost parallel output beam (Fig.7.1b). The wavefront of the output beam is thus almost plane. If the beam is considered to have perfect spatial coherence (see Section 7.3), then the resulting divergence is due only to diffraction (the divergence is then said to be diffraction-limited).

Divergence of the diffraction-limited beam is better understood by considering a monochromatic beam of wavelength λ that consists of plane waves that are incident on a small aperture (Fig.7.2a). If a screen is placed at a reasonably large distance behind the aperture, diffraction fringes consisting of a series of alternating dark and bright circles or lines will be seen on the screen depending on whether the aperture is circular or rectangular. The fringes result from interference of waves from different sections of the aperture, a consequence of Huygens' principle. In simple terms, Huygens'

Principles of Laser Materials Processing, by Elijah Kannatey-Asibu, Jr.

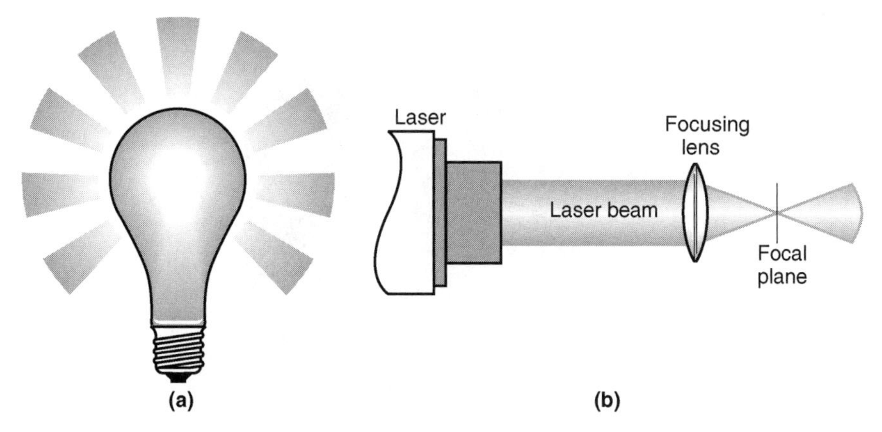

FIGURE 7.1 (a) Divergent light from an ordinary light source (b) Light from a laser beam showing minimal divergence.

principle states that each point on a wave front is the source of elementary waves that spread out from these points. The intensity distribution on the screen for a circular aperture is illustrated in Fig. 7.2b. The central bright fringe will contain 84% of the incident beam energy. The angle, $2\theta_d$, subtended by the edges of this bright region with the center of the aperture is used as a measure of the beam's divergence.

For a circular aperture of diameter D, the divergence can be expressed, from diffraction theory, by the semiangle θ_d, (Fig.7.2a) as

$$\theta_d = \sin^{-1}\left(\frac{K_1\lambda}{D}\right) \approx \frac{K_1\lambda}{D} \tag{7.1}$$

where K_1 is a constant ≈ 1.22.

The divergence may be expressed in another form by considering the variation of the beam size $w(x)$ with distance x from its waist (Fig.2.16). Using equations (2.61) and (2.62) for a confocal resonator, it can be shown that at distances much greater than the Rayleigh range (Section 2.3), that is $x >> x_R$, the beam size from a resonator made up of two curved mirrors may be approximated as

$$w(x) \approx \frac{\lambda x}{\pi w_w} \tag{7.2}$$

From this, the divergence semiangle for a confocal resonator may be obtained as

$$\theta_d \approx \frac{w(x)}{x} \approx \frac{\lambda}{\pi w_w} \tag{7.3}$$

where w_w is the radius of the beam at its waist (Fig. 2.16). Equation (7.3) indicates that increasing the size of the beam waist reduces the divergence.

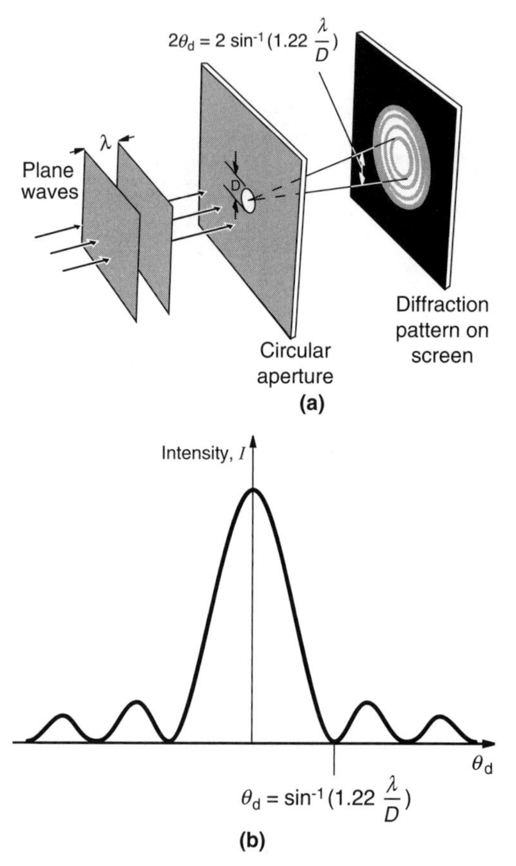

FIGURE 7.2 (a) Beam divergence from a circular aperture. (b) Intensity distribution on the screen. (From O'Shea, D. C., Callen, W. R., and Rhodes, W. T., *Introduction to Lasers and Their Applications*, 1977. Reprinted by permission of Pearson Education, Inc.)

The low divergence and thus almost parallel nature of a laser beam enables it to be focused to a very small radius, since a parallel beam with a plane wave can ideally be focused to a point. Single-mode lasers generally experience less divergence than multimode lasers. Also, a Gaussian beam, TEM_{00} or fundamental mode, exhibits the smallest possible beam divergence, that is, provides true diffraction-limited beam divergence. This is because the high intensity region of the laser beam for this mode is farther from the geometrical boundaries of the walls of the resonant cavity and aperture edges. A typical value for the divergence, θ_d, may be 0.05°. However, for semiconductor lasers, the divergence can be much higher, by about two orders of magnitude (see Table 8.1).

For some manufacturing processes such as laser cutting and microfabrication, it is essential to have a highly focused beam. Thus, low beam divergence is important for such processes. It may not be as important for high-power laser welding, and even less important for surface heat treatment or laser forming, where a wider beam area is more desirable.

7.2 MONOCHROMATICITY

A light beam is said to be monochromatic when it is composed of only a single wavelength. Unlike conventional sources whose emission extends continuously over a broad band, lasers are generally considered to be monochromatic (Fig. 7.3). However, in reality, due to the broadening mechanisms discussed in Chapter 5, the wavelength of a light source, lasers included, is never exactly monochromatic but covers a frequency bandwidth that depends on the nature of the source and the broadening mechanisms involved. For lasers, the oscillating bandwidth can be very narrow, and may range from $\Delta\nu = 500$ Hz ($\Delta\lambda = 10^{-11}$ μm at $\lambda = 0.6$ μm, (see Section 9.1)) for a high-quality stable gas laser to greater than 1 GHz ($\Delta\lambda = 10^{-6}$ μm) for a solid-state laser. In comparison, the bandwidth for a conventional monochromatic source such as the sodium lamp may be of the order of several gigahertz.

Theoretically, the minimum oscillating linewidth (or fluctuation of the laser frequency) $\delta\nu$ of a laser oscillating in a single mode under steady state is determined by spontaneous emission and can be shown to be given by

$$\delta\nu = \frac{4\pi h_p \nu_o (\Delta\nu)^2}{q} \tag{7.4}$$

where q is the power in the oscillating field, h_p is Planck's constant, $\Delta\nu$ is the half-width of the resonance at half maximum intensity, ν_o is the oscillation frequency, and $\delta\nu$ is the minimum oscillating linewidth.

Equation (7.4) indicates that an increase in the laser power results in a decrease in the theoretical linewidth. This is because an increase in power is due to an increase in the ratio of stimulated to spontaneous emissions.

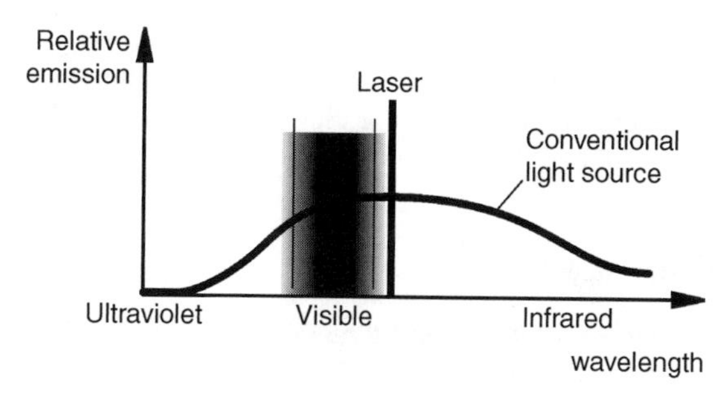

FIGURE 7.3 Comparison of the spectral characteristics of a conventional light source and a laser source.

For a pulsed laser beam, the minimum linewidth is given by the inverse of the pulse duration, $\Delta\tau_p$ (cf. equation (6.41)):

$$\delta\nu = \frac{1}{\Delta\tau_p} \tag{7.5}$$

Thus, the shorter the pulse duration, the wider the linewidth. The oscillating linewidth is also wider for a laser that oscillates on many modes.

Monochromaticity is important in laser processing of materials, because it determines the extent to which a laser beam can be focused. Since the radius at the point of focus of a light beam depends on the wavelength, a monochromatic beam can be more sharply focused than a beam with a broad bandwidth. This may not be significant in laser welding or heat treatment, but in laser cutting or microfabrication, the process efficiency depends significantly on how well the beam is focused.

7.3 BEAM COHERENCE

Coherence is the term used to describe the situation where a fixed-phase relationship exists between two waves or between two points of the same wave (Fig. 7.4a). Coherence is important for interference-based measurements. The coherence phenomenon is both space and time dependent, and thus one normally has to consider spatial coherence and temporal coherence.

Spatial coherence refers to the phenomenon whereby the phase difference between two points on a wave front of an electromagnetic field remains constant with time (Figs 7.4a and 7.4b). It often describes the phase relationship between two beams of light, even though a single beam may also be spatially incoherent. Temporal coherence, on the other hand, refers to the situation where the phase difference between the wave front of an electromagnetic field at a given point P at a time t and that at the same point P and time $t + \tau_0$ remains constant with time. The electromagnetic wave is then temporally coherent over the period τ_0. This is illustrated in Fig. 7.4b, which shows an electromagnetic wave whose phase changes at intervals of τ_0. Temporal coherence often describes the phase relationship associated with a single beam of light. Even though a beam may have both spatial and temporal coherence, the two do not necessarily coexist. In other words, a wave that is spatially coherent may be temporally incoherent, and vice versa. For clarity, an incoherent beam is illustrated in Fig. 7.4c.

7.3.1 Spatial Coherence

The concept of spatial coherence is best understood by considering Young's experiment. A light source S is placed in front of a narrow slit A_o contained in a screen A. We set up a coordinate system as shown in Fig. 7.5, with origin O at the center of the slit.

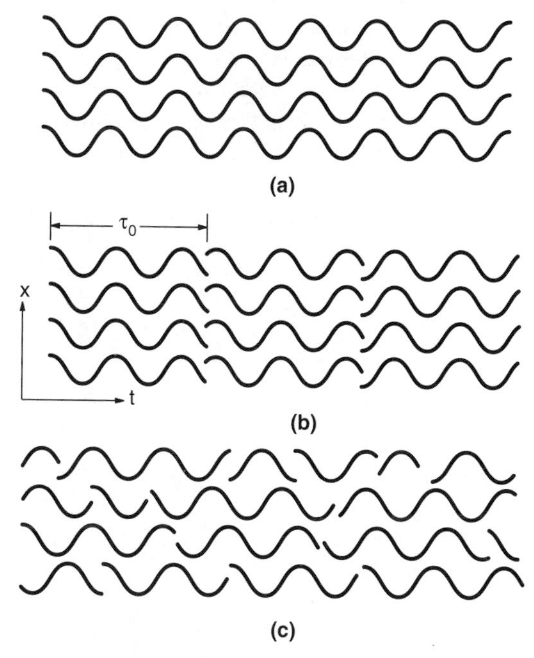

FIGURE 7.4 Coherence. (a) Perfect coherence. (b) Spatially coherent beam with only partial temporal coherence with coherence time τ_0 (c) Almost completely incoherent beam. (From Wilson, J., and Hawkes, J. F. B., *Lasers, Principles and Applications*, 1987. Reprinted by permission of Pearson Education, Inc., Harlow, Essex, UK.)

A second screen B containing two parallel and narrow slits B_1 and B_2 is placed at a distance x_1 from A such that the y-axis is exactly halfway between the two slits B_1 and B_2 that are parallel to the y-axis and are located at a distance $\pm z_b$ from the x-axis. A third screen C is now placed at a distance x_2 from B such that the x-axis passes through it at D. The three screens are all parallel to the y–z plane.

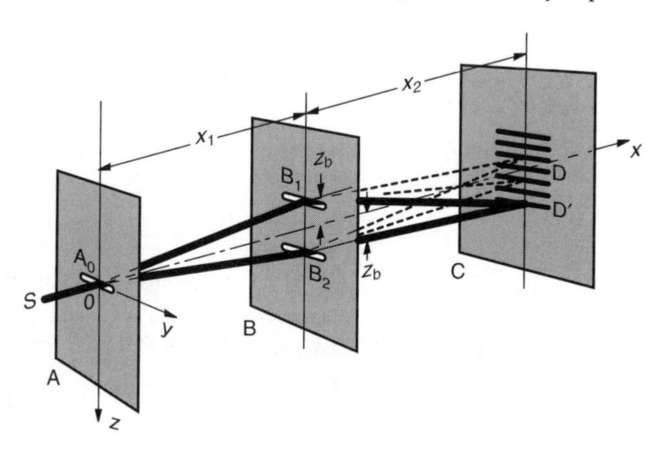

FIGURE 7.5 Young's experiment.

Light emanating from the slit A_o and passing through the slits B_1 and B_2 will form fine interference fringes (alternate bright and dark lines) on the screen C. To understand the formation of the fringes, we consider a general point $D'(x_1 + x_2, z)$ located very close to D on screen C. Light at this point is a superposition of light from two sources: in one case, starting from O through B_1 to D', and in the other case, starting from O through B_2 to D'.

The optical path length (distance traveled by the ray) from O through B_1 to D' is given by

$$l_1 = |\mathbf{OB_1}| + |\mathbf{B_1D'}| = \sqrt{x_1{}^2 + z_b{}^2} + \sqrt{x_2{}^2 + (z - z_b)^2} \tag{7.6}$$

while that from O through B_2 to D' is

$$l_2 = |\mathbf{OB_2}| + |\mathbf{B_2D'}| = \sqrt{x_1{}^2 + z_b{}^2} + \sqrt{x_2{}^2 + (z + z_b)^2} \tag{7.7}$$

The optical path length difference l between the two rays is then given by

$$l = l_2 - l_1 = \sqrt{x_2{}^2 + (z + z_b)^2} - \sqrt{x_2{}^2 + (z - z_b)^2} \tag{7.8}$$

For $x_2 \gg z, z_b$, equation (7.8) can be approximated by

$$l = \frac{2zz_b}{x_2} \tag{7.9}$$

Since the two light sources are from the same origin O, they will reinforce each other if the path difference at $D'(x_1 + x_2, z)$ is an integral number of wavelengths, that is, if

$$l = \frac{2zz_b}{x_2} = m\lambda \tag{7.10}$$

and they will nullify each other if the path difference is an integral number of wavelengths and a half, that is, if

$$l = \frac{2zz_b}{x_2} = (m + 1/2)\lambda \tag{7.11}$$

Thus as D' varies with distance from D, the light intensity will show alternate bright and dark fringes. In this particular example, since B_1 and B_2 are symmetrically placed with respect to OD, a bright fringe will be formed at D. From equation (7.10), the position of the mth bright fringe can be obtained as

$$z(m) = \frac{mx_2\lambda}{2z_b}, \quad m = 0, \pm 1, \pm 2, \cdots \tag{7.12}$$

The spacing between contiguous fringes will be

$$z(m+1) - z(m) = \frac{x_2\lambda}{2z_b} \tag{7.13}$$

Let us now consider another slit in screen A that is parallel to the one at O and located at O′, a distance d_s from O. We assume that there is no phase relationship between the light waves from this slit and those from O. If O′ is very close to O, then the fringes formed by O′ will be such that there will be a bright fringe in the neighbourhood of D. If the distance d_s between O and O′ is now gradually increased by moving O′ in the positive z direction, its fringes will shift relative to those due to O. At some critical separation d_c, the dark fringes of O′ will be superimposed on the bright fringes of O, eliminating the entire interference fringe pattern on the screen. d_c can be shown to be given by

$$d_c = \frac{\lambda x_1}{4z_b} \tag{7.14}$$

If O and O′ are considered to be part of a relatively wide single slit of width d_s, then it can be said that good fringes will be obtained if

$$d_s \ll \frac{\lambda x_1}{z_b} \tag{7.15}$$

or

$$z_b \ll \frac{\lambda x_1}{d_s} \tag{7.16}$$

Thus, for a wide enough slit, no fringes will be observed for an ordinary light source, and this is because the waves emitted from different parts of the source are incoherent. A laser beam, on the other hand, will produce interference fringes even for a wide slit since it has high spatial coherence, and thus light from any two points on the beam can interfere.

The spatial coherence of a laser beam is determined by the number of transverse modes present. A single-mode laser having only one longitudinal and transverse mode is fully spatially coherent.

7.3.2 Temporal Coherence

The concept of temporal coherence is also best explained using the Michelson interferometer (Fig. 7.6). A parallel beam of near-monochromatic light from a source S is split into two using a beam splitter A, such that part of the beam is reflected and the other part transmitted. Each of the two beams is reflected back from a plane mirror. One of the mirrors, M_1, is fixed, while the other, M_2, is movable along the direction of the beam. Each of the reflected beams is again partly reflected and partly transmitted

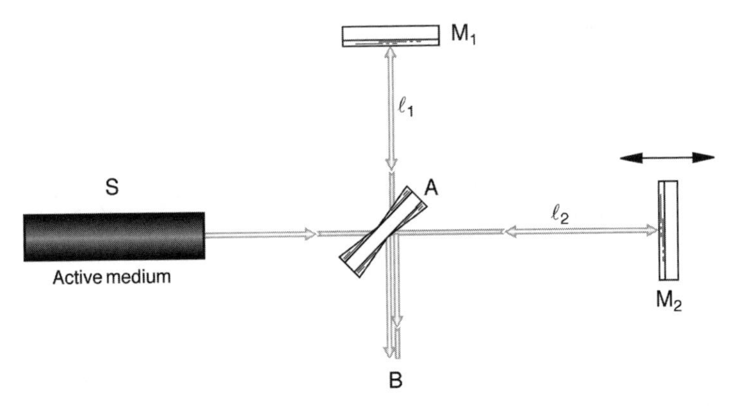

FIGURE 7.6 Michelson interferometer.

at A such that they are both directed onto a detector at B where they are superimposed to form interference fringes. If the distance from A to M_1 is the same as that from A to M_2, then the optical path difference between the reflected beams is zero. Good interference fringes are then observed at B. On gradually moving the mirror M_2 away from A, the interference fringes become less and less distinct, until they completely disappear when the optical path difference is a few centimeters (Fig. 7.7a).

The reduction in fringe visibility with increasing optical path difference is due to the fact that the light waves generated by the source are not of an infinitely long duration, but rather a series of wave trains of average duration say, τ_{ct}. The phases of the individual wave trains are random, especially if the time period between the transitions is greater than τ_{ct}. This is because there are several excited atoms, with each one emitting a wave train as it undergoes a transition to a lower energy level, and that wavetrain is uncorrelated with the transition of other atoms. Despite the randomness of the wave trains, a light source appears continuously bright because the wave trains are being constantly produced and superimposed on each other. The wave train resulting from each transition may be approximated as a damped oscillation (Fig. 7.7b) with a time constant for the light intensity of τ_{ct}. In other words, the intensity reduces to $1/e$ of its original intensity in time τ_{ct}, which is also a measure of the lifetime of the excited state of the atom.

The duration of a wave train, τ_{ct}, is referred to as the coherence time, and its length $c\tau_{ct}$ is the coherence length. The finite duration or length of the wave trains of an ordinary light source is the reason why the fringes produced by the Michelson interferometer reduce in quality as the optical path difference between the two beams increases. When the optical path difference is less than the coherence length, light waves are being superimposed, which were essentially split from the same wave train, and are thus in phase. Good or clear interference fringes are obtained between light waves which are coherent or in phase. When the optical path difference is greater than the coherence length, different light waves are being superimposed, which were generated at times separated by an interval greater than the coherence time, and are thus uncorrelated. Just for illustration, we find that for a light source with

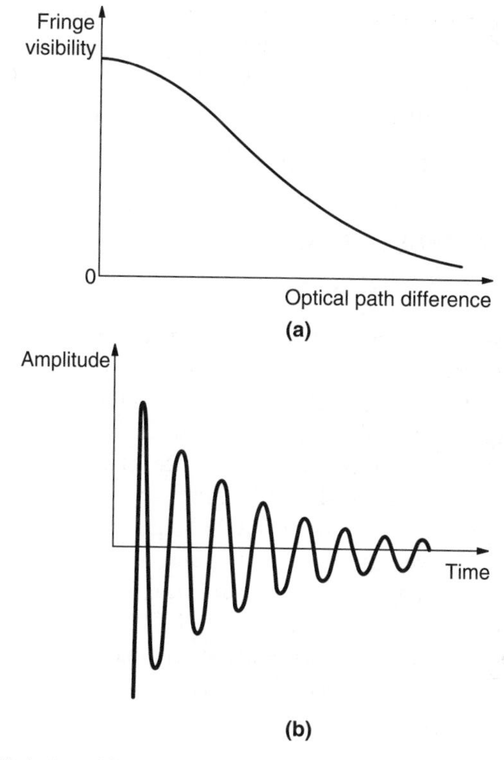

FIGURE 7.7 (a) Variation of fringe visibility with optical path difference. (b) Approximation of the wave train associated with an atomic transition to a lower energy level. (From Shimoda, K., *Introduction to Laser Physics*, 2nd edition, 1986. By permission of Springer Science and Business Media.)

$\tau_{ct} = 10^{-8}$ s, the coherence length is about 3 m, while for $\tau_{ct} = 10^{-10}$s, the corresponding coherence length is about 0.03 m.

The linewidth $\Delta\nu_o$ associated with a light wave having a coherence time of τ_{ct} can be shown to be given by

$$\Delta\nu_o = \frac{1}{\tau_{ct}} \tag{7.17}$$

The linewidth associated with the emission from an ordinary source with a coherence time of about $\tau_{ct} = 10^{-10}$ s, will thus be 10^{10} Hz. However, due to broadening, the actual bandwidths can even be much broader. Since the frequency of light waves emitted by individual atoms is spread over a certain range, their respective fringes are not uniformly spaced, and that reduces the fringe contrast. On the other hand, the oscillating linewidth for a high-quality stable gas laser may be as small as 500 Hz. The corresponding coherence time and length are 2×10^{-3} s and 6×10^5 m (600 km), respectively. Thus, the interference properties of a laser are very good and can be used for producing interference fringes even when the optical path difference is large.

In general, the presence of several modes affects both the spatial and temporal coherence. Continuous wave gas lasers often have a much higher degree of coherence than pulsed lasers.

7.4 INTENSITY AND BRIGHTNESS

The high directionality or low divergence of a laser beam tends to concentrate all the available energy into a very small region, resulting in very high intensity (power per unit area) of the laser beam. On the contrary, light from an ordinary bulb is radiated uniformly in all directions. Thus, at any distance from the source, only a small fraction of the emitted energy is available over a given area, resulting in low intensity of the light. As a result, a laser beam of power 1 mW (typical of a He–Ne laser) would appear more intense than a 100 W light bulb. Furthermore, the intensity of the ordinary light source decreases rapidly with distance from the source.

The brightness of a power source at a point is defined as the power emitted per unit surface area per unit solid angle (angular cone). Brightness is a basic characteristic of the source radiation and cannot be increased. The brightness of different light sources may be more easily compared by specifying the brightness temperature, T_B, which is defined as that temperature at which a blackbody would emit the same radiant flux as the given body and may be expressed as

$$T_B = \frac{q}{k_B \Delta \nu_o} \tag{7.18}$$

where q is the power of the source, $\Delta \nu_o$ is the linewidth, and k_B is Boltzmann's constant.

For a 1 W laser of linewidth 1 MHz, the corresponding brightness temperature is of the order of 10^{19}K. For comparison, we note that the brightness temperature of an incandescent lamp is about 10^4K.

7.5 FREQUENCY STABILIZATION

The stability of pulsed lasers is usually determined by the repeatability of the output amplitude and duration. Thus, for the output of a pulsed laser to be stable, the energy supplied to the flash tubes or laser electrodes has to be the same for each pulse. Variations in the output pulse may also result from sources such as

1. Temperature change of the active medium
2. Cavity expansion
3. Internal heating and associated deterioration of discharge tubes

4. Electrode material sputtering
5. Gas leaks in gas laser cavities.

For CW lasers, temperature variations and flexibility in the tube structure lead to instability of the cavity length, thereby resulting in variation of the center frequency with time. This can be minimized or eliminated by using rigid tube construction incorporating vibration isolation to ensure mechanical stability or by maintaining precise control over the ambient temperature. Another approach to stabilizing the frequency involves feedback control based on the Lamb dip phenomenon. This is done by making one of the mirrors movable along the resonator axis. Fine motion can be achieved if the mirror is mounted on a piezoelectric transducer in such a way that varying the transducer excitation voltage in turn varies the mirror location, and thereby the cavity length. Variation in the output frequency resulting from any of the disturbances mentioned, such as the cavity length variation, will cause a difference in the irradiance of two modes that are initially located symmetrically with respect to the Lamb dip. The difference in the irradiance of the two modes can be used as an output in a feedback loop to correct for the frequency shift by sending an input to the transducer through its excitation voltage to change the cavity length.

7.6 BEAM SIZE

As indicated in Chapter 2, the beam radius, w, or spot size is usually taken as the distance at which the electric field amplitude reduces to $1/e$ (0.368) the center (peak) amplitude. It may also be defined as the distance at which the intensity becomes $1/e^2$ (0.135) the center (peak) intensity. The output laser intensity is proportional to $|E_1(x, t)|^2$, the electric field amplitude. About 87.5% of the total beam energy is contained in the region of radius w. The $1/e^2$ peak intensity definition is most accurate for near-Gaussian beams, while the 87.5% definition is more useful for cylindrically symmetrical beams. The two definitions are equivalent for a Gaussian beam.

For a single-mode beam with a Gaussian profile, the central beam intensity (power per unit area), I, and radiant exposure or fluence (energy per unit area), F_q, may be obtained from the beam radius w specified at the $1/e$ points as

$$I = \frac{q}{\pi w^2} \tag{7.19a}$$

and

$$F_q = \frac{Q}{\pi w^2} \tag{7.19b}$$

where q is the power output (usually for CW lasers) and Q is the energy output (usually for pulsed lasers).

If the beam radius is based on $1/e^2$, then equations (7.19a) and (7.19b) give the average intensity and fluence, respectively. The radius at $1/e^2$ is the product of $\sqrt{2}$ and the radius at $1/e$.

7.7 FOCUSING

The power density of a beam of given power at any section along the beam depends on its radius at that section. Of critical importance is the power density at the point of focusing. For an almost parallel Gaussian beam of wavelength λ incident on a lens, the radius w_f at the focus is given, Fig. 7.8, by

$$w_f = \frac{\lambda}{\pi\theta} \approx \frac{1}{\pi}\frac{f_l\lambda}{w_0} \tag{7.20a}$$

where f_l is the lens focal length, w_0 is the beam radius at the lens, and θ is the angular aperture of the lens, or the angle subtended by the lens at the focus.

The minimum spot diameter, d_{min}, that can be obtained at the focal point is given by

$$d_{min} = 2w_{min} = \frac{1.22\lambda}{\text{NA}} \approx \lambda \tag{7.20b}$$

where NA $= n\sin(\theta/2)$ is the numerical aperture and n is the refractive index of the medium in which the lens is working.

For traditional welding and cutting applications, the beam diameter at the focal point is about 250 μm. However, with appropriate optics, this can be reduced to the order of the beam's wavelength.

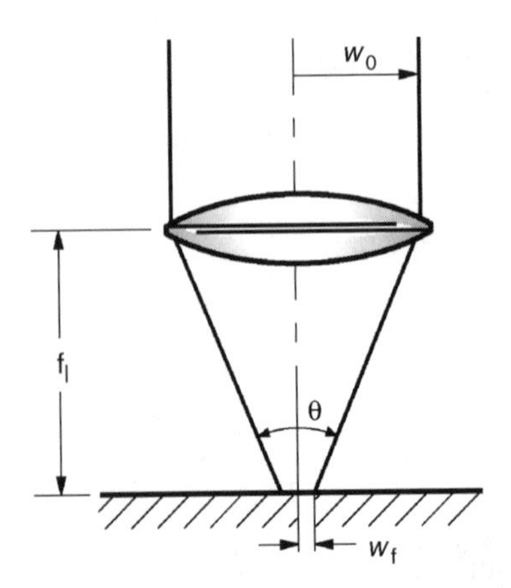

FIGURE 7.8 Beam focusing.

For a general mode distribution, the radius at the focus may be expressed as

$$w_{\rm f} = K_{\rm m}\frac{1}{\pi}\frac{{\rm f}_{\rm l}\lambda}{w_0} \tag{7.21}$$

where K_m represents the energy mode distribution. A typical value for K_m would be 6.

It is evident from equation (7.20) that the smaller the wavelength of the laser beam, the smaller the focused beam size, and thus the greater the power density produced by a beam of given power. As a result, for a given beam power and mode, the power density produced with an Nd:YAG laser will be two orders of magnitude greater than that produced using a CO_2 laser.

7.8 RADIATION PRESSURE

When radiation is absorbed by a surface, the transfer of momentum from photons to the surface results in a certain amount of pressure, P, being induced on the surface and is given by

$$P = \frac{F}{A} = \frac{1}{A}\frac{{\rm d}M_{\rm o}}{{\rm d}t} = \frac{I}{c} \tag{7.22}$$

where F is the force, A is the area of incidence, M_o is the momentum, c is the speed of light, and I is the intensity.

This can be expressed in simpler form as

$$P = \frac{q(1+R_{\rm s})}{Ac} \tag{7.23}$$

where R_s is the reflectivity of the surface, and q the incident power. For a 2 MW pulse focused on an area of 10^{-3}cm^2, the resulting pressure is about 1atm.

7.9 SUMMARY

In this chapter, we have discussed the basic characteristics of a laser beam. We started with beam divergence that, for a diffraction-limited beam, is proportional to the beam wavelength and inversely proportional to the diameter of the aperture from which the beam is emerging. Whereas ordinary light is uniformly radiated in all directions, the divergence of a laser beam may be as low as $0.05°$. This property makes the laser beam very useful for laser cutting and microfabrication. Next, we learnt that no light source is exactly monochromatic. However, lasers are as close to being monochromatic as one can get, with the linewidth of a high-quality stable gas laser being as low as $\Delta\nu_o = 500$ Hz, while that of a conventional monochromatic source may be several gigahertz. Monochromaticity is important in materials processing because it determines the extent to which a laser beam can be focused. Following that, we learnt about beam coherence, considering both spatial (space-dependent) and temporal (time-dependent) coherence. Whereas coherence is very important for interference-based measurements, it is also important for some laser processing applications such as cutting and microfabrication, since it again determines how well the beam can be focused.

The dependence of the beam intensity and brightness on the directionality and monochromaticity were then discussed. This was followed by a discussion of the factors that influence the stability of both pulsed and CW lasers and steps that might be taken to maintain a stable output, including feedback control based on the Lamb dip phenomenon. Finally, beam size, beam focusing, and radiation pressure were presented, where we learnt that the size of the focused beam is proportional to the wavelength of the incident beam and inversely proportional to the size of the beam that is incident on the lens.

We have thus far covered all basic aspects of beam generation and characteristics, including the concepts of population inversion, gain, laser oscillation, resonant cavity, pumping, the rate equations, broadening mechanisms, quality factor, Q-switching, mode-locking, divergence, monochromaticity, and coherence. In the next chapter, we take a look at some of the principal types of lasers, especially those used as industrial lasers.

REFERENCES

Laser Institute of America, 2007, *American National Standard for Safe Use of Lasers*, ANSI Z136.1-2007, Orlando, FL.

Henderson, R. A., 1997, *A Guide to Laser Safety*, Chapman & Hall, London, UK.

Rofin-Sinar, *Technical Note—CO_2 Laser Welding*, Rofin Sinar, Plymouth MI.

APPENDIX 7A

List of symbols used in the chapter

Symbol	Parameter	Units
d_c	critical separation between slits	m
d_s	separation betwen slits	m
d_{min}	minimum focused beam diameter	μm
D	aperture diameter	m
K_1	constant	—
K_m	energy mode distribution	—
R_s	reflection coefficient of surface	—
T_B	brightness temperature	K
w_{min}	minimum focused beam radius	μm
Q	laser energy output	m

x_1	x-coordinate of plane B	m
x_2	distance between planes B and C	m
z_b	z-coordinate of slits in plane B	m
$\Delta\nu$	half-width of the resonance at half maximum intensity	Hz
τ_{ct}	coherence time	s
τ_0	coherence period	s
θ_d	divergence semiangle	°

PROBLEMS

7.1. Given that the optical path difference between two rays is

$$l = l_2 - l_1 = \sqrt{x_2{}^2 + y + y_b^2} - \sqrt{x_2{}^2 + y - y_b^2}$$

show that

$$l \approx \frac{2yy_b}{x_2}$$

7.2. Show that the critical distance, d_c, at which the interference fringes of two narrow slits that are close to each other will nullify each other is given by

$$d_c = \frac{\lambda x_1}{4y_b}$$

7.3. (a) Show that for a frequency bandwidth of

$$\Delta\nu = 500 \text{ Hz}$$

the corresponding wavelength bandwidth is

$$\Delta\lambda = 10^{-11} \ \mu m \text{ at } \lambda = 0.6 \ \mu\text{m}$$

(b) Determine the coherence length of a laser that has a transition wavelength of $\lambda = 0.63$ μm and a bandwidth of 1.5 nm. If a coherence length of 2.5 m is desired, what should the bandwidth be?

7.4. Given that the beam size from a confocal resonator is given by

$$w(x) = w_w \left[1 + \left(\frac{2x}{L}\right)^2\right]^{1/2} \tag{7.24}$$

and

$$w_{\mathrm{W}} = \left(\frac{L\lambda}{2\pi}\right)^{1/2} \tag{7.25}$$

show that the beam divergence may be approximated by

$$\theta \approx \frac{\lambda}{\pi w_{\mathrm{W}}} \tag{7.26}$$

7.5. (a) Given the fact that the bandwidth of a good He–Ne laser is about 500 Hz, while that of an ordinary light source is about 10 GHz, explain why one would result in good interference fringes while the other would not.

(b) What will be the coherence time and length of an Nd:YAG laser of linewidth 195 GHz?

7.6. (a) Determine the divergence of an Nd:YAG laser beam of wavelength 1.06 μm and diameter 5 mm if its divergence is equivalent to that of a plane monochromatic wave that is incident on a circular aperture of the same diameter.

(b) Determine the diameter that a diffraction-limited red He–Ne laser beam will have on venus, which is at a distance of about 48×10^6km from the earth, after going through the 500 mm aperture of a telescope.

7.7. What is the distance between the slits B_1 and B_2 if screen B is located at a distance of 1,500 mm from screen C in Young's experiment, and the interference fringes observed have a period of 2 mm for a He–Ne laser of wavelength 0.633 μm?

7.8. Make a list of applications (one in manufacturing and one general) for which the following laser properties are useful:

(a) Spatial coherence

(b) Temporal coherence

(c) Monochromaticity

(d) Divergence

(e) Gaussian mode

(f) Multimode.

8 Types of Lasers

There are various types of lasers. Each one has different characteristics that depend, to a large extent, on the active medium used for laser action. In this chapter, we first discuss the major types of lasers, based primarily on the active medium. In addition, we also discuss recent developments in laser technology, especially with regard to industrial lasers, specifically those used for materials processing. The principal laser categories include the following.

1. Solid-state lasers
2. Gas lasers
3. Liquid dye lasers
4. Semiconductor (diode) lasers
5. Free electron lasers.

Even though semiconductor lasers are also based on solid material, they are considered separately since the pumping and lasing mechanisms are different. In each category, a number of lasers have been developed using different active media. The different laser types are discussed in the order listed above in the following sections. The basic concepts related to each general type are discussed using one or two examples for illustration. The basic characteristics of some major industrial lasers are summarized in Appendix 8B, and the wavelength range of common lasers are summarized in Appendix 8C.

The notations used to represent the electron configurations in other laser texts are usually complex. In this book, we use an unconventional but very simple representation, which starts with the letter E with a subscript that designates the energy level, for example, E_x, indicating energy level x.

In addition to the major laser categories, we shall discuss new developments in industrial laser technology. Ultrafast (utrashort pulse or femtosecond), slab, disk, and fiber lasers are considered separately under this category. Ultrafast lasers (which are also based on solid-state laser technology) use special techniques, in addition to mode-locking, to achieve the ultrashort pulse duration. Slab and disk lasers are also given special consideration even though they both use solid-state active media. Slab lasers can also use gas active media, and may be optically or electrically pumped.

Principles of Laser Materials Processing, by Elijah Kannatey-Asibu, Jr.

Fiber lasers are also based on solid-state laser technology and can generate ultrashort pulse outputs.

8.1 SOLID-STATE LASERS

Solid-state lasers normally use an insulating crystal or glass as the host lattice. In it is embedded the active medium, which is either a dopant or an impurity in the host material. The crystal host material does not participate directly in the lasing action. The dopant is the component that participates directly in laser action. It is normally a transition metal or rare earth element, and it substitutes for some of the atoms in the host material, rather than being an interstitial impurity. For the solid-state lasers that we consider under this section, the active medium is shaped in the form of either a rod or a slab. Since the rod form is the traditional shape for the active medium, we use it as illustration in most of our discussion. The slab configuration is discussed at the end of this section. The rod is normally cylindrical in shape, with ends that are ground and polished to be plane and parallel. Its dimensions depend on the active medium being used. The ends of the rod may either be silvered (one completely and the other partially), in which case the rod constitutes the optical cavity, or the rod may be placed between two external mirrors. Pumping is commonly done using a flash lamp, and the configuration involves either a cylindrical cavity with elliptical cross-section or a helical flashtube (see Chapter 3, Fig. 3.1). The rod temperature may be controlled by circulating air or liquid around it. Otherwise, the heat generated can change the cavity dimensions and consequently, the cavity modes.

Most solid-state lasers generate pulsed beams, even though some generate continuous wave outputs. Their coherence lengths are thus relatively short, making them unsuitable for a number of interference-based applications. However, the significant amount of energy available in each pulse makes them attractive to applications that require such energy bursts. These include resistor trimming, initiation of thermonuclear fusion, and spectroscopic research. Common types of solid-state lasers include the ruby, Nd:YAG, and Nd:Glass lasers. These are further discussed in the following sections.

8.1.1 The Ruby Laser

The significance of the ruby laser is that it was the first laser to be successfully made and operated. The rod is a single crystal of ruby, which consists of crystalline aluminum oxide (Al_2O_3) that is doped with chromium. The chromium constitutes about 0.05% of the rod by weight and replaces some of the aluminum ions. The resulting material is pinkish in color. The aluminum oxide (sapphire) is the host lattice while the chromium ions, Cr^{3+}, constitute the active medium. The size of the ruby rod normally ranges from 0.5 to 1.0 cm in diameter and from 5 to 20 cm in length.

The energy levels involved in the lasing action are those of the chromium ions, and these are illustrated in Fig. 8.1. This is a three-level system. The ground level (level 1)

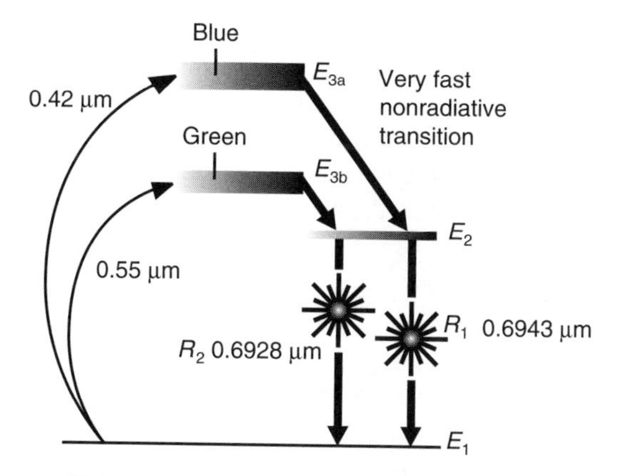

FIGURE 8.1 Energy levels of the ruby laser.

is indicated as E_1. The pump level (level 3), indicated by E_3, has two bands, E_{3a} that has a mean wavelength of 0.42μm that corresponds to blue and E_{3b} that has a mean wavelength of 0.55μm corresponding to green. The upper lasing level (level 2), E_2, constitutes two subbands, E_{2a} with a wavelength of 0.6943μm (commonly referred to as R_1) and E_{2b} with a wavelength of 0.6927μm (commonly referred to as R_2). The R_1 transition usually dominates. Atoms that are pumped to level 3 where the lifetime is about 10^{-4}ms. decay rapidly through nonradiative transition to level 2, which is a metastable state, with a lifetime of about 3 ms. The lasing transition normally occurs between E_{2a} and E_1 (the R_1 transition), even though the R_2 transition can also occur under special circumstances. The transition is homogeneously broadened due to interaction between the chromium ions and lattice phonons, with a FWHM of about 330 GHz.

Pumping may be done using a xenon flash lamp operated at a pressure of about 500 Torr, which provides a pulsed white light of the order of megawatts with a period of about a millisecond and a pulse rate of one per second. The pumping action excites chromium ions from the ground state (level 1) to level 3 due to the absorption of radiation in the blue and green wavelength range, when the threshold light intensity is achieved. The excited ions undergo rapid transition to the metastable state corresponding to the upper lasing level. Due to the relatively long lifetime of this level (level 2), its population keeps increasing as pumping continues, until a population inversion is eventually achieved between levels 2 and 1. Lasing action is then triggered by the initial small spontaneous emission that naturally results from atoms in the higher energy states. The photons from such spontaneous emission are radiated in all directions. Those corresponding to the 2–1 transition and that are directed parallel to the cavity axis then stimulate the chromium ions in the upper laser level to radiate. Lasing action ceases when the lamp stops operating. The net result is an output pulse from the laser. Even during a single pulse of about a millisecond, a number of sharp peaks that result from spiking (Chapter 6) can be observed.

The output of a ruby laser may be about 10–50 MW with a pulse duration of about 10–20 ns when Q-switched. Since ruby lasers are three-level lasers, high-threshold energy is needed to excite at least half of the Cr^{3+} ions to achieve population inversion. High levels of energy are thus required to operate them, resulting in extensive heating of the laser rod. This, coupled with the low thermal conductivity of ruby, makes them difficult to operate in continuous mode. As a result, they are not widely used anymore.

8.1.2 Neodymium Lasers

Neodymium (Nd) lasers have either a crystal or glass material as the host lattice, and this is doped with neodymium ions, Nd^{3+}, that constitutes the active medium. The energy levels are thus those of the neodymium ions and are illustrated in Fig. 8.2. These are four-level lasers. The ground level (level 0) is the E_0 energy level. Again, the pump level (level 3) has two bands, with E_{3a} having a mean wavelength of 0.73μm, while E_{3b} has a mean wavelength of 0.8μm. The upper lasing level (level 2) is E_2 with a lifetime of about 0.23 ms., and from there, transition occurs to the lower lasing level (level 1), E_1. This transition is also primarily homogeneously broadened with a FWHM of about 195 GHz, and a transition wavelength of 1.06μm in the infrared range. Atoms are pumped from level 0 to level 3, from which they undergo rapid nonradiative decay to level 2, which is a metastable state. The energy difference between level 1 and the ground state, level 0, is such that $h_p\nu \gg k_B T$ at room temperature, so that level 1 is practically empty under thermal equilibrium conditions. Furthermore, transition from level 1 to level 0 is very fast, and occurs by nonradiative processes. Any atoms that make a transition to level 2 then essentially result in a population inversion. Neodymium lasers thus have a much lower inversion threshold requirements than ruby lasers. The traditional pumping schemes are similar to those outlined for the ruby laser. There is a variety of host materials that are used for neodymium lasers, but the most common ones are YAG and glass.

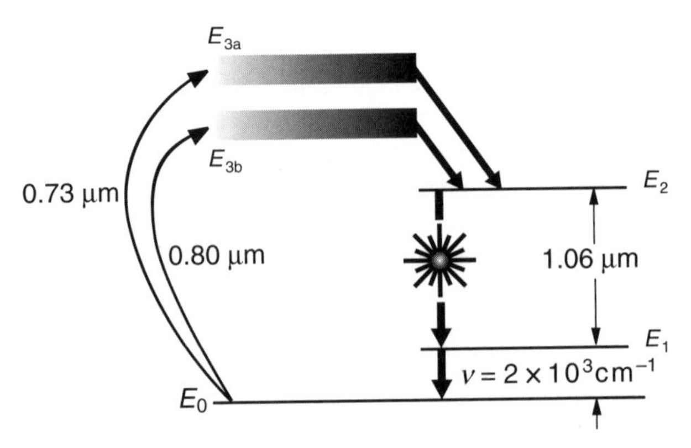

FIGURE 8.2 Energy levels of neodymium lasers.

8.1.2.1 The Nd:YAG Laser The host lattice for this laser is a crystal of yttrium aluminum garnet (YAG) with the chemical composition $Y_3\ Al_5\ O_{12}$, where the Nd^{3+} ions (about 0.1–2%) substitute for some of the Y^{3+} ions. Operation of the Nd:YAG laser can be either in the continuous wave (CW) or in pulsed mode, depending on whether pumping is continuous or intermittent. A xenon flash lamp may be used at medium pressure (500–1500 Torr) or a krypton lamp at high pressure (4–6 atm), and the rod size is similar to that of the ruby laser. The power output in CW mode varies from 150 W to 6 kW, and that in Q-switched pulsed mode is of the order of 50 MW with a pulse duration of about 20 ps, at a repetition rate of 1–100 Hz. The high-power (up to 2 kW) CW laser may be achieved by having three Nd:YAG rods in line as a single oscillator, with each rod being arc lamp pumped. The very high powers (up to 6 kW) are obtained by pumping with a diode laser (Figs 3.3 and 3.4). The efficiency of both pulsed and CW Nd:YAG lasers typically ranges between 1 and 3%. Applications of Nd:YAG lasers include laser surgery and materials processing, for example, welding, cutting, drilling, and surface modification.

8.1.2.2 The Nd:Glass Laser The host lattice for the Nd:Glass laser is glass such as silicate, phosphate, or fluoride glass, with 1–5% Nd^{3+}. It has the advantage that the rod size can be much larger (up to 1 m in length and about 50 mm in diameter) than that for Nd:YAG since glass can be easily made with high-quality or optical homogeneity (free of residual stresses) to large sizes because of its lower melting temperature. With power output per unit volume being comparable to that of Nd:YAG, relatively larger power outputs can be obtained with this type of laser. Since glass has an amorphous structure, it has short range order. Thus, the environment and therefore the field of each ion is different from that of any other ion. This variation of ion environments in the glass matrix results in additional inhomogeneous broadening, and thus a much broader bandwidth. The Nd:Glass laser therefore normally operates in multimode. Very short pulse periods, of the order of 5 ps, with high-output powers, can thus be obtained when the output is mode-locked.

One major disadvantage of Nd:Glass lasers is the low thermal conductivity of glass, which is about an order of magnitude smaller than that of Nd:YAG. This limits their usefulness for CW operation, and are thus only used for low-rate pulsed outputs, with a typical output pulse rate of 1 Hz. The relatively high-peak pulse outputs (up to 20 TW) of Nd:Glass lasers enable them to be used in laser-induced nuclear fusion reactions. However, the low pulse rates limit their applicability in manufacturing to such operations as drilling and spot welding, and in the cases where they are acceptable, are able to produce superior quality holes at depth-to-diameter ratios up to 50:1, much higher than what can be achieved with the Nd:YAG laser.

8.2 GAS LASERS

Gas lasers are among the most common form in the laser industry. The power levels range from several kilowatts (carbon dioxide (CO_2) lasers) to milliwatts (helium–neon (He–Ne) lasers). They can be operated in either the continuous mode or pulsed mode, with output frequencies ranging from ultraviolet to infrared.

As the name implies, gas lasers use a gaseous medium as the active medium. Common examples are the He-Ne and CO_2 lasers. The broadening mechanisms in gas lasers are not as strong as those in solids. Thus, the resulting linewidths, determined primarily by Doppler broadening, are relatively small. This is because collision broadening is relatively small due to the low pressures normally used in gas lasers. As a result, the energy levels are relatively narrow, and thus a sharp emission line is essential for excitation. Optical pumping, with its broad emission spectrum, is therefore not suitable for pumping gas lasers, since it would result in inefficient pumping. Electrical pumping is thus the most common means of exciting the active medium in gas lasers. Pumping is also done by chemical means, with an electron beam, or by gas-dynamic expansion.

When classified by the active medium, one can identify four principal types of gas lasers:

1. Neutral atom lasers
2. Ion lasers
3. Metal vapor lasers
4. Molecular lasers.

8.2.1 Neutral Atom Lasers

The most common example of neutral atom lasers (using inert gases as the active medium) is the He–Ne laser, which was also the first laser to generate CW output. The active medium in this case consists of 1 part neon to 10 parts helium. This is a four-level laser, and Fig. 8.3 illustrates the energy level schemes of He and Ne.The helium atoms are more easily or efficiently excited to the higher levels by electron collision than are neon atoms. Thus, the electrons that are accelerated by the passage

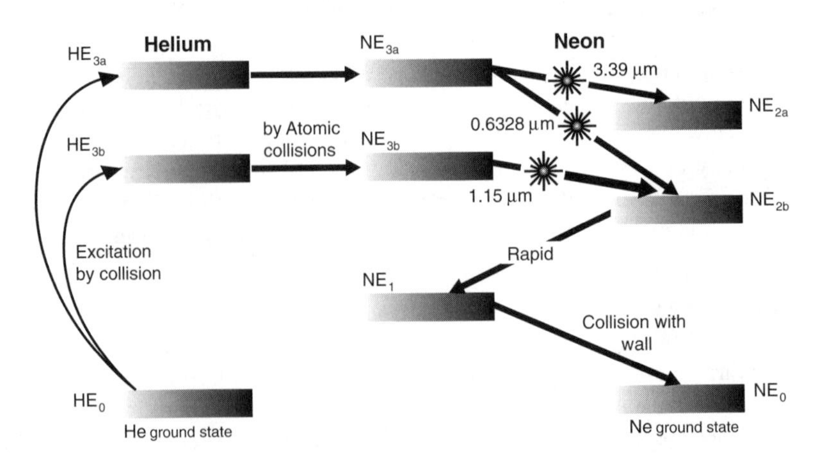

FIGURE 8.3 Energy levels of the He–Ne system.

of a discharge through the mixture excite the helium atoms to the metastable higher energy levels denoted by HE_{3a} and HE_{3b}. These energy levels of helium happen to coincide with some of the excited states of neon, the NE_{3a} and NE_{3b} levels. The excited helium atoms are thus able to transfer energy to neon atoms, which are in the ground state when collision takes place between the two. This results in de-excitation of the helium atoms involved to the ground state while the neon atoms are excited to the higher energy level. This process of energy transfer through collision is referred to as *resonant collision* or *resonant energy transfer,* and is so called because the corresponding energy levels coincide. The process increases the population of the NE_{3a} and NE_{3b} levels of neon relative to the lower NE_{2a} and NE_{2b} levels, resulting in a population inversion between the two sets of levels.

Lasing action occurs between energy levels of the neon atoms. The role of helium is thus to assist in the pumping process. The metastable state of the excited helium atoms (staying at the high level for a relatively long time) makes the energy transfer process more efficient. The decay time of the upper lasing levels of neon (about 100 ns) is about an order of magnitude greater than that of the lower lasing levels (about 10 ns), thereby satisfying a necessary condition for CW operation (see Section 4.3).

Even though there are a large number of possible transitions between the sublevels of the Ne laser transition states, the principal or strongest ones are shown in Fig. 8.3, with transition at wavelengths of 0.633, 1.15, and 3.39 μm. The last two wavelengths fall in the infrared region, while the first one results in the red light, and is the most common mode in which the laser is used. The desired wavelengths can be selected using techniques discussed in Chapter 2.

After transition, the atoms in the lower lasing level relax spontaneously to the NE_1 level which is metastable. Thus, if the temperature of the neon gas is high enough, electrons from the NE_1 state may jump back to the NE_{2a} and NE_{2b} levels (Boltzman's law). This phenomenon is known as radiation trapping, and will cause the population inversion between the upper lasing level and the NE_{2a} and NE_{2b} levels to decrease, resulting in quenching.

The three main factors that affect radiation trapping and the population inversion of He–Ne lasers are current density, total and partial gas pressures of He and Ne, and diameter of the discharge tube. With a higher current density in the discharge tube, the temperature of the gas becomes higher. Therefore, there is an optimum current density that provides maximum pumping power without quenching laser action.

The gas pressures of He and Ne influence the population of the NE_1 state.If the population of the NE_1 state is increased due to a pressure change, the chances of radiation trapping are increased. The optimum pressures have been found to be about 1 mmHg (1 Torr) for helium and about 0.1 mmHg for neon. Finally, in order to reduce the population of the NE_1 level and thus reduce the chances for radiation trapping, the NE_1 neon atoms must collide with the discharge tube walls to de-energize. If the inner diameter of the discharge tube wall is too large, there will be less chance of collisions with the wall. Thus, the tube wall must be made small enough to ensure a high-enough probability of wall collisions. In practice, tube diameters range from about 1 to 6 mm. Beyond that range, the output power is reduced because the population inversion is lessened due to a fewer number of wall collisions by atoms in the NE_1 state.

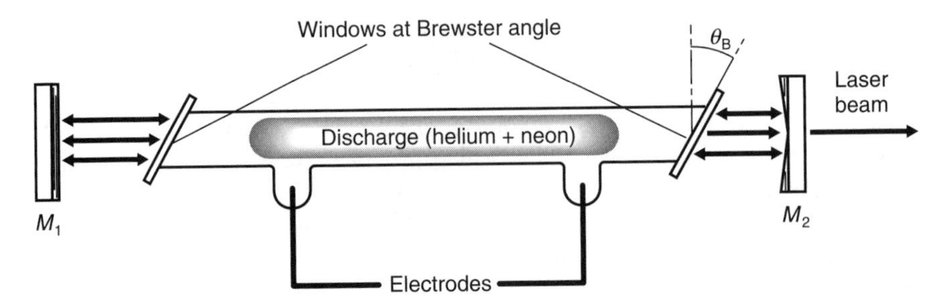

FIGURE 8.4 Schematic of a He–Ne laser (external mirrors).

Figure 8.4 is a schematic of the He–Ne laser. The discharge tube is typically 20–80 cm in length. At the ends are the mirrors, which may be either internal (sealed inside the discharge tube) or external to the sealed tube (Fig. 8.4). The internal system has the disadvantage that it needs to be replaced periodically due to erosion by the discharge. On the contrary, with the external system, the windows at the ends of the tube reflect part of the beam away, resulting in losses. This problem is mitigated by positioning the windows at the Brewster angle θ_B (see Section 9.4), which is given by

$$\tan \theta_B = n \tag{8.1}$$

where n is the refractive index of the window material.

The output beam of the He–Ne laser has a bandwidth of about 1.6 GHz due primarily to Doppler broadening. Thus, there may be several axial modes operating simultaneously. A single axial mode operation can be obtained by reducing the cavity length to about 10–15 cm. This enables high stability to be achieved by controlling the cavity length (see Section 7.5).

The outputs of He–Ne lasers typically range from 0.5 to 50 mW, with an input power requirement of 5–10 W. The overall efficiency is thus of the order of 0.02 %. There are a number of other neutral atom gas lasers, mostly inert gases, essentially based on the same principle as the He–Ne laser.

Common applications for He–Ne lasers include position sensing, character or bar-code reading, alignment, displacement measurement by interferometry, holography, and video disk memories.

8.2.2 Ion Lasers

Ion gas lasers are generally four-level lasers, and the active medium is an ionized inert gas, with a typical operating pressure of about 1 Torr. There are a variety of ion lasers, including argon, krypton, xenon, and mercury ion lasers. These have essentially the same design. We shall use the argon laser for illustration.

The energy level scheme of the argon ion (Ar^+) laser is shown in Fig. 8.5. Excitation to the higher laser level, E_3, can occur in three ways:

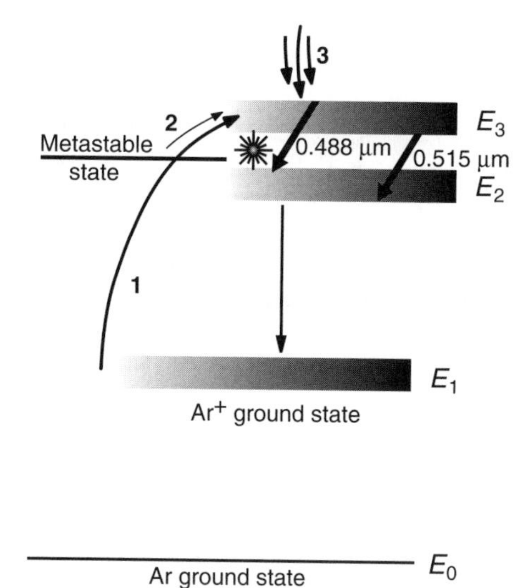

FIGURE 8.5 Energy levels of the Ar^+ system.

1. Through collision of (Ar^+) ions in their ground state with electrons;
2. Through collision of (Ar^+) ions in the metastable state with electrons;
3. Through radiative cascading from higher levels (to the E_3 level).

Laser transition occurs between the E_3 (higher) and E_2 (lower) levels, with the lifetime of the E_3 level being about 10^{-8}, which is an order of magnitude greater than that of the E_2 level (about 10^{-9}), thereby satisfying the condition for CW operation.

Unfortunately, the pumping process involves two stages instead of one. In the first stage, the neutral atom is ionized by colliding with an electron in the discharge tube. The ion may then be excited to a higher energy level by another collision with an electron, in the second stage. Since pumping requires two stages, the pumping process, and therefore the ion laser in general, is inefficient. The efficiency is approximately 0.1%. Current densities of around 1000 A/cm^2 are therefore required to maintain the threshold population inversion in ion lasers.

The need for such high-current densities complicates the design of ion lasers. To begin with, the tube diameter must remain relatively small to achieve such high-current densities. In addition, the high-current densities necessary for laser operation raise the temperature of the ions in the discharge tube to very high levels (of the order of 3000 K). This has the effect of broadening the Doppler linewidth to about 3.5 GHz (see equation (5.26)). Furthermore, with the ions at such high temperatures, collisions with the tube walls would normally cause significant damage. This damage might be reduced by increasing the diameter of the tube wall, but that requires higher pumping currents for the same power output. Thus, discharge tubes must be made of ceramic materials to withstand collisions with the hot ions. Wall damage may be

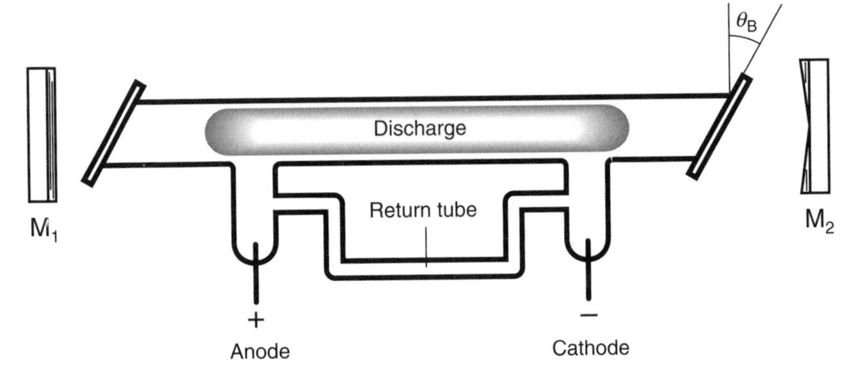

FIGURE 8.6 Schematic of an Ar^+ laser.

further reduced by applying a static magnetic field along the tube axis in the discharge zone. This is done by wrapping a current-carrying coil around the tube. It forces the electrons to undergo a spiral motion along the tube axis, thereby confining them to the tube center, and reducing contact with the walls. This solution also has the benefit of increasing the electron density along the tube center, which increases the pumping rate and thus the output power.

In addition to raising the temperature of the discharge tube, the high-current density causes the ions to migrate toward the cathode. Since the ions have a lower mobility than electrons, there is a tendency for the ions to accumulate at the cathode. This might result in the discharge being extinguished due to a loss of active medium. To prevent this from happening, a return path is provided between the anode and the cathode (Fig. 8.6). The return tube is made longer than the discharge tube to prevent the discharge from passing through the return tube (path of least resistance). Overall, the two-stage pumping scheme of ion lasers significantly complicates their design and restricts their performance. Yet, ion lasers have been put to practical use.

Output of the Ar^+ ion laser occurs at a number of wavelengths, with the strongest being the 0.4881 μm (blue) and 0.515 μm (green) wavelengths. Power outputs typically range from about 1 to over 20 W CW, but can also be lower than 1 W. Ar^+ ion lasers are used in laser printers, in surgery, for pumping dye lasers, and in spectroscopy. Other ion lasers include the krypton (Kr^+) ion laser with an output wavelength of 0.6471 μm and a CW output of 5-6 mW, and the xenon (Xe^+) ion laser with an output wavelength of 0.5395–0.995 μm, and a pulsed peak power output of about 200 W.

8.2.3 Metal Vapor Lasers

The design of a metal vapor laser is similar to that of the He–Ne laser, since both use He to assist the pumping process. However, the metal vapor laser employs a metal vapor instead of a second inert gas. The metal vapor is generated by a metal component that is contained in a small reservoir close to the anode, where it is heated to produce the vapor in the discharge tube. The ionized vapor atoms migrate toward the cathode where the vapor eventually condenses. This process is referred to as cataphoresis. An adequate supply of metal is therefore necessary for the life of the laser (about 5000 h). It may also result in coating of a laser mirror with the metal, causing the laser beam to be quenched. The metal vapor lasers are used in facsimiles, spectroscopy, and photochemical experiments. Typical CW output powers for these lasers fall between 50 and 100 mW, filling the range between He–Ne and Ar^+ ion lasers.

Several different tube configurations have been designed to prevent the output windows from being coated with metal. One example is where the tube is designed with a second anode between the output window and the cathode. The second anode maintains a discharge that keeps the metal from approaching the output window (Fig. 8.7.). There are a variety of metal vapor lasers, but the most common ones are the helium–cadmium (He–Cd) and helium–selenium (He–Se) lasers. In the following paragraphs, we discuss the He–Cd laser as an example of a metal vapor laser.

Figure 8.8 illustrates the energy level scheme of the He–Cd system. The energy levels of the excited He states, HE_{3a} and HE_{3b}, are much higher than those of the excited Cd^+ ions ($(Cd^+)^*$—combined ionization and excitation energies (CE_3)). Collision between the excited He atoms (He^*) and Cd atoms in their ground state result in transfer of energy to the Cd atoms causing their ionization and excitation to the higher energy level:

$$He^* + Cd \rightarrow He + (Cd^+)^* + e^- \tag{8.2}$$

This mode of energy transfer is called Penning ionization, and is not resonant. Unlike the case of ion lasers, the excitation or pumping process involves a single stage. As a result, the pump rate is proportional to the current density. The excess energy is transformed into kinetic energy of the electron that is ejected. The excited helium states are metastable, thereby making the energy transfer efficient.

Even though both the CE_2 and CE_3 states of the Cd^+ ion can be excited, the CE_3 states are more easily excited. Moreover, they have a much longer lifetime (10^{-7}s) than the CE_2 states (10^{-9} s). Population inversion is thus easily achieved between

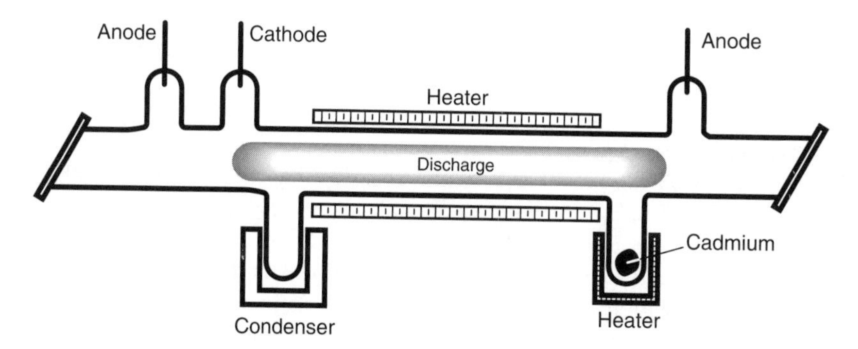

FIGURE 8.7 Metal vapor laser with two anodes. (From Duley, W. W.,1983, *Laser Processing and Analysis of Materials*. By permission of Springer Science and Business Media.)

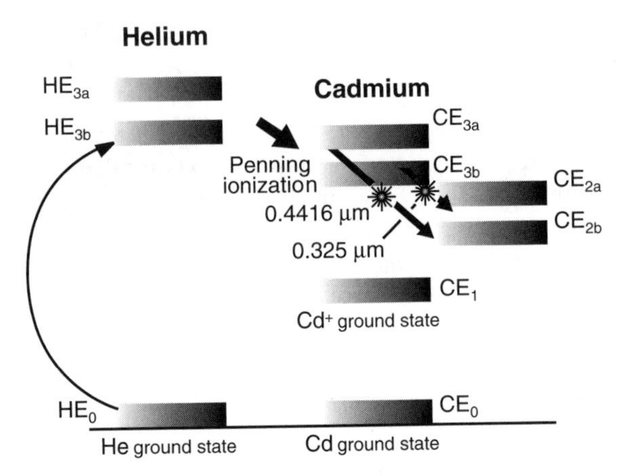

FIGURE 8.8 Energy levels of the He–Cd system.

the CE_2 and CE_3 states. Laser transition occurs on the $CE_{3b} \rightarrow CE_{2a}$ line with a wavelength of 0.325μm in the ultraviolet range, and the $CE_{3a} \rightarrow CE_{2b}$ line with a wavelength of 0.4416μm in the blue range. Efficiencies may be up to 0.02 %, while the oscillation bandwidths are about 5 GHz.

8.2.4 Molecular Gas Lasers

Unlike the neutral atom and ion gas lasers, molecular gas lasers derive their transitional energy from a molecule. The overall energy of a molecule is normally composed of four principal components:

1. Electronic energy, which results from electron motion about each nucleus.
2. Vibrational energy, which results from vibrations of the constituent atoms about an equilibrium position.
3. Rotational energy, which results from rotation of the molecule as a whole about an axis.
4. Translational energy, which results from linear or curvilinear motion of the molecule as a whole.

Each electronic energy level (Fig. 8.9a), consists of a number of vibrational levels (Fig. 8.9b) and each vibrational level in turn consists of a number of rotational levels (Fig. 8.9c). In Fig. 8.9, E_1 refers to the ground level, and E_2 refers to the first excited state (cf. Fig. 8.19) for the case where the atoms of the molecule are held fixed at a nuclear separation distance r. The electronic, vibrational, and rotational energies are quantized, and transitions between the various levels can produce laser oscillations. Translational energy, on the contrary, is not quantized, and is thus not useful for laser oscillations. There are three common transitions associated with the quantized energies. These are normally used to categorize the lasers as

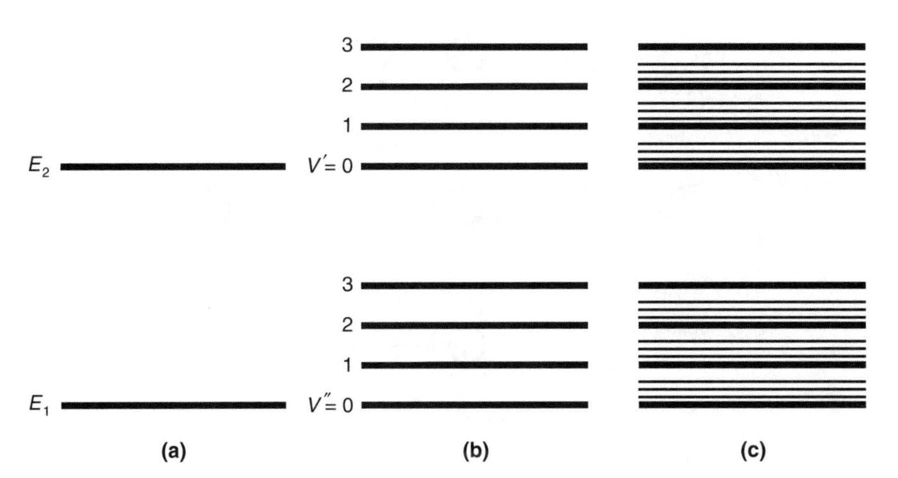

FIGURE 8.9 Energy levels of a molecule. (a) Electronic. (b) Vibrational. (c) Rotational.

1. Vibrational–rotational.
2. Vibronic (vibrational–electronic).
3. Pure rotational.

A fourth type of molecular laser, which falls in a category of its own, but some of which may classify as a vibronic laser, is the excimer laser. In the following sections, we shall discuss the vibrational–rotational, vibronic, and excimer lasers, since these are more easily produced commercially. Since laser action is relatively difficult to achieve in rotational lasers, these are not further discussed.

8.2.4.1 Vibrational–Rotational Lasers In vibrational-rotational lasers, oscillation is achieved as a result of transitions that occur between vibrational levels (rotational level of one vibrational state and the rotational level of another vibrational state) of a given electronic state, normally the ground state. The wavelength of the output beam is usually in the range 5–300μm (middle to far-infrared).

The CO_2 laser is the most common example. It has the advantage of relatively high efficiencies, about 10–30 %. Common applications of the CO_2 laser include materials processing, communications, spectroscopy, and surgery.

Figure 8.10 shows the following three fundamental vibration modes of the CO_2 molecule:

1. Asymmetric stretching.
2. Bending.
3. Symmetric stretching.

The energy level or oscillation behavior of the system is described using the number of energy quanta in each vibrational mode. For example, (pqr) indicates that there are p quanta in the symmetric mode, q quanta in the bending mode, and r quanta in the asymmetric mode.

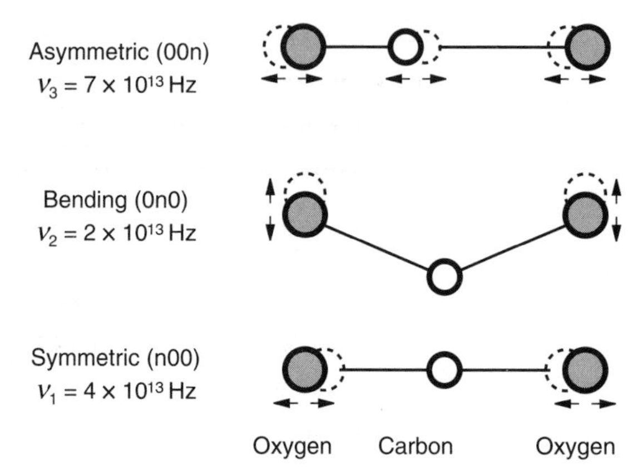

FIGURE 8.10 Vibrational modes of the CO_2 molecule.

The active medium for the CO_2 laser is a mixture of CO_2, N_2, and He. The CO_2 molecule provides the transitions that generate the laser beam. N_2 helps to increase the population of the upper lasing level, while He helps to depopulate the lower lasing level, thereby enhancing the achievement of population inversion. Since N_2 is a diatomic molecule, it has only one vibrational mode, unlike the CO_2 molecule. The contributions of the individual gases are now further discussed.

The energy level scheme of the electronic ground state for this type of laser is shown in Fig. 8.11. The (001) level of the CO_2 molecule, which is the upper lasing level, is pumped by the following:

1. Direct electron collision.
2. Resonant energy transfer.
3. Radiationless transition from upper excited levels.

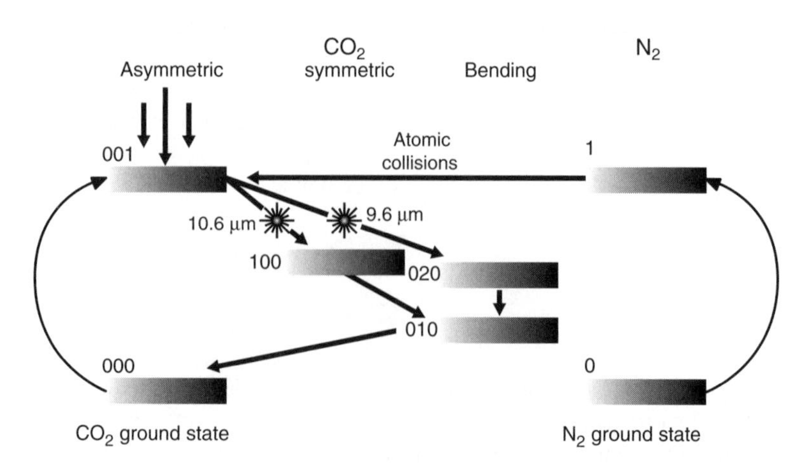

FIGURE 8.11 Energy level scheme of the CO_2 laser.

Collision between electrons accelerated in the discharge and the CO_2 molecule results in the molecule being excited preferentially to the (001) energy level. This process can be represented as

$$e^- + CO_2(000) \rightarrow e^- + CO_2(001) \tag{8.3}$$

With resonant energy transfer, electron collision with an N_2 molecule first results in the latter being excited from the ground level (0) to the upper level (1). This is a very efficient process, and since the upper level is a metastable state and has about the same energy level as the (001) level of CO_2, resonant transfer of energy can occur easily between the (1) level of N_2 and the (001) level of the CO_2 molecule. Finally, higher excited levels ($00r$) of CO_2 decay rapidly to the metastable (001) state. As a result of these various processes, pumping of the (001) level is very efficient, contributing to the relatively high efficiency (up to 30 %) of the CO_2 laser.

The upper lasing level is the (001) level, and the lower levels are the (100) and (020) levels. Transition between the (001) and (100) levels results in oscillation at the wavelength 10.6μm, while that between the (001) and (020) results in oscillation at the wavelength 9.6μm, with the transition at 10.6 μm having a much greater likelihood (about 20 times) of occurring than the 9.6μm transition, since it has a higher gain. The 10.6μm output is thus more common. Frequency-selection techniques can be used to suppress the 10.6μm output if the 9.6μm output is desired.

In reality, each of the vibrational levels mentioned, say (001), is composed of several rotational levels that have a population distribution given by the Boltzmann distribution. Thus ideally, all the rotational levels would be oscillating. However, only the rotational level with highest gain oscillates since the thermalization rate of the rotational levels (about $10^7\, s^{-1}\, (\text{mmHg})^{-1}$) is much greater than the rate at which the population of the oscillating rotational level decreases. Thermalization refers to redistribution of the population of various levels to achieve equilibrium or steady-state distribution.

The lifetimes of the various levels are influenced by the partial pressures of the gas mixture. For a typical mixture with partial pressures of 1.5mmHg for CO_2, 1.5 mmHg for N_2, and 12 mmHg for He, the lifetime of the upper lasing level is found to be about 0.4 ms. The energy difference between the lower lasing levels (100) and (020) is very small, much less than $k_B T$, and thus the (100) $\rightarrow$ (020) decay is very fast. Likewise, the (100) $\rightarrow$ (010) and (020) $\rightarrow$ (010) decays are very fast. Thermal equilibrium is thus reached between the three levels (100), (020), and (010) in a relatively short period. As a result, if the decay from the (010) level to the ground state (000) is not fast enough, then there would be an accumulation of atoms in all three lower levels since they are in thermal equilibrium. This would lead to a quenching of population inversion. The decay of the (010) level to the ground state is accelerated or enhanced by the presence of He in the gas mixture. For the partial pressures mentioned, the corresponding lifetime of the (010), (100), and (020) levels is about 20 μs, therefore satisfying the condition for laser oscillation.

The presence of He in the mixture serves another purpose. To prevent repopulation of the (010) level and thus lower lasing levels by thermal excitation, the energy difference between (010) and the ground state has to be much greater than $k_B T$. This occurs if the temperature of the CO_2 gas is kept relatively low. Since He has a high thermal conductivity, it can conduct heat away to the tube walls effectively, thereby keeping the CO_2 gas at a low temperature.

The long lifetime of the upper lasing level makes CO_2 lasers highly suited to Q-switching. However, the average output power that results from the Q-switched laser of this type is only about 5% of the power available from one that is operated in the CW mode. This results from the fact that the pulse period is of the same order of magnitude as the time taken for thermalization of the various rotational levels of the vibrational lasing level. Thus only a small fraction of the rotational levels are able to contribute to the lasing action.

There are various designs for CO_2 lasers, including the following:

1. Sealed tube lasers
2. Longitudinal-flow lasers
3. Transverse-flow lasers
4. Transversely excited atmospheric (TEA) lasers
5. Gas-dynamic lasers.

For the first four types, excitation may be by either radio frequency (rf) or direct current (dc) (Section 3.2).

Sealed Tube Lasers In sealed tube lasers, the gas mixture or active medium is contained in an enclosed chamber (the discharge tube). Therefore, there is no fresh supply of gases or active medium to the laser. In the case of CO_2 lasers, dissociation products (simpler chemical constituents of larger compounds) that are undesirable for laser operation may form in the tube. For example, carbon monoxide (CO) forms from CO_2:

$$2CO_2 \rightarrow 2CO + O_2 \tag{8.4}$$

The formation of carbon monoxide is undesirable in CO_2 lasers, because it absorbs light at 10.6 μm (the main transition wavelength of a CO_2 laser). Thus, if a large amount of dissociation products exist within the discharge tube, the laser may be quenched.

To prevent quenching of laser action due to accumulation of dissociation products, a catalyst is used to regenerate CO_2 from CO. There are a number of ways this can be done:

1. A small amount (about 1 %) of water vapor (H_2O) is added to the gas mixture;
2. A Ni cathode that is heated to about 300°C is used as the catalyst.

With these catalysts, the laser can be used for over 10, 000 h. Typical output powers for sealed tube lasers are of the order of 60 W/m. The output beam of the sealed tube CO_2 laser normally operates in the low-order mode. Therefore, it has been used in

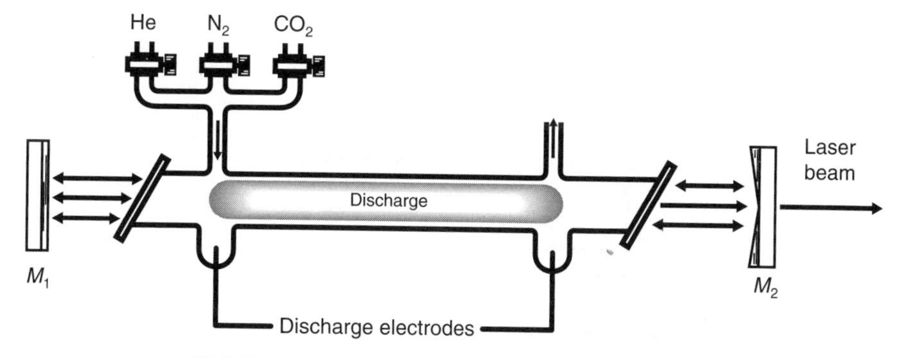

FIGURE 8.12 Schematic of the axial flow CO_2 laser.

microsurgery and micromachining where high accuracy of the low-order mode beam is essential. Machining applications include drilling and cutting of thin metal sheets (about 0.5–1.5 mm in steels), and non–metals.

Axial Flow Lasers Figure 8.12 shows the configuration of a typical axial (longitudinal) flow CO_2 laser, where the gas mixture is made to flow continuously through the tube. Continuous flow of the gas mixture is used mainly to enable the dissociation products of the discharge, especially CO, to be removed, thereby preventing contamination. The cavity mirrors can be either internal or external. For the external system, the windows at each end are configured at the Brewster angle. The tube walls are water cooled to remove heat that is dissipated in the discharge. The pressure of the gas in the tube is normally kept around 15 mmHg. Broadening of the output laser is then primarily due to Doppler broadening, with a linewidth of about 50 MHz, even though collision broadening is also significant. Because of the relatively small linewidth, the system can easily be operated in a single longitudinal mode by confining the cavity length to less than 1 m (see Section 2.2.3.2). For mechanical and thermal stability, the tubes are mounted in an invar rod frame to reduce expansion during operation. There are two types of axial flow lasers:

1. Slow axial flow lasers.
2. Fast axial flow lasers.

SLOW AXIAL FLOW LASERS In this type of laser, the gas mixture flows at a relatively slow speed (about 20 L/min), with the prime objective of removing the dissociation products. Heat is removed by conduction from the tube center to the walls. The output power is typically about 50–60 W/m, or 50–500 W overall, and does not depend on the tube diameter which is about 25 mm. The higher output powers are obtained by increasing the optical path length of the beam in the resonator, using folding mirrors to redirect the beam along different paths (Fig. 8.13). The lower power levels are used in laser surgery, while the higher powers are more appropriate for scribing, resistor trimming, welding of thin sheets, and cutting nonmetals.

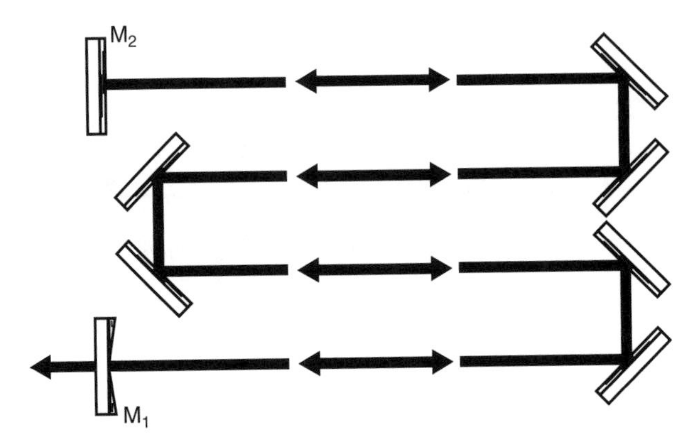

FIGURE 8.13 One possible resonator configuration.

FAST AXIAL FLOW LASERS The output power of axial flow lasers is limited by the heat generation within the laser, and the manner in which it is extracted. Since the efficiency of the laser is up to 30 %, over 70 % of the energy input is dissipated as heat that, in slow axial flow lasers, is removed by diffusion from the tube center to the walls. Removal of the dissipated heat can be made more efficient by flowing the gas mixture through the tube at much higher speeds of about 300–500 m/s to carry away the heat by convection. The gas is then recycled through a heat exchanger. This enables the power to be greatly increased to about 600 W/m, or 500 W to 6 kW in overall power output and units as large as 20 kW have been built. This is a *fast axial flow laser*. Axial flow lasers are extensively used in materials processing, for example, welding and cutting. They tend to have better beam characteristics than transverse flow lasers described in the following section.

TRANSVERSE FLOW LASERS In transverse flow lasers, the gas flow, electrical discharge or current flow, and resonator axis are mutually perpendicular to one another (Fig. 8.14a and b). This is because the heat generated within the discharge tube can be more efficiently extracted by flowing the gas at a rate of about 60 m/s in a direction normal to the resonator axis. This results in a reduced operating temperature, and under such circumstances, the laser can be operated at higher pressures. But the output power for CO_2 lasers increases with the pressure. Thus, for a given discharge current density, higher output powers are achieved by increasing the total pressure from 15 to 100 mmHg. However, the electric field required to maintain the discharge at the higher pressures also increases to very high levels of about 500 kV/m. For standard axial flow lasers of cavity length 1 m or more, the amount of voltage required would be impractical. To circumvent this, the discharge is applied normal to the resonator axis so that the distance between the electrodes is as small as possible, of the order of 10 mm. This is also sometimes referred to as a transversely excited (TE) laser.

There are two principal ways in which the discharge is sustained in transverse flow lasers.

1. The discharge is said to be self-sustained when the gas is ionized by the discharge. To initiate the discharge, a high electric field needs to be applied. However, the electric field necessary to maintain the discharge is relatively low due to the short distance between the electrodes, but the current is quite high. The output power is about 600 W/m or up to 9 kW overall power. The beam is often folded or reflected back and forth a number of times between the mirrors to increase the power output (Figs 8.13 and 8.14a).
2. The nonself-sustained discharge involves the use of an electron beam to ionize the gas, such that the excitation voltage that is applied is just optimum for attaining population inversion, but itself not adequate for sustaining the discharge. Power outputs obtained with this mode of excitation may be about 10 kW/m or up to 15 kW overall power.

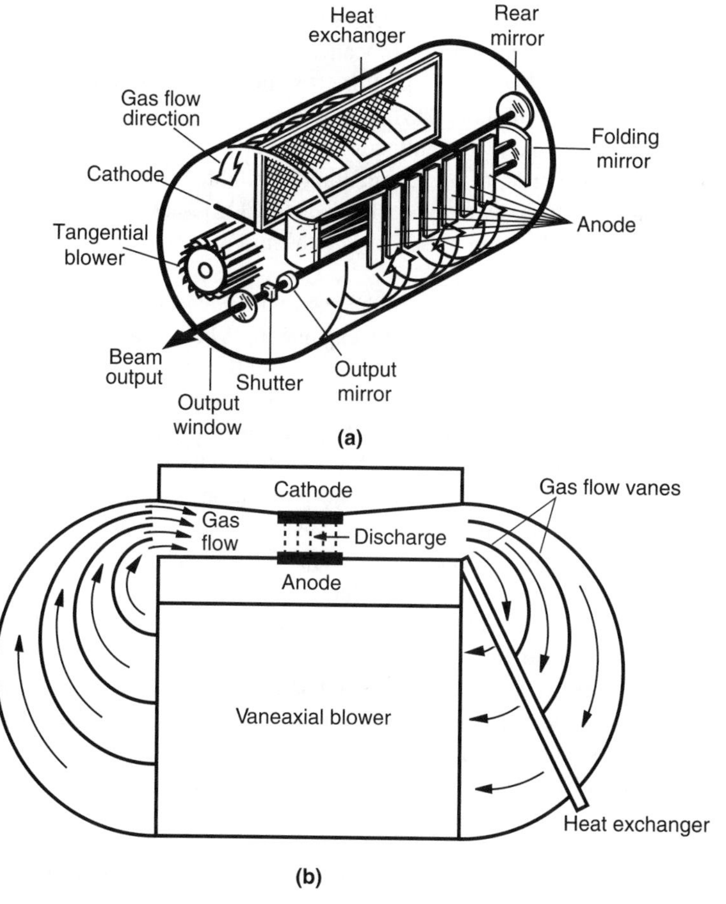

FIGURE 8.14 (a) Three-dimensional view of a transverse flow CO_2 laser. (b) Simplified two-dimensional sectional view of a transverse flow CO_2 laser. (From Svelto, O., 1989, *Principles of Lasers*, 3rd edition. By permission of Springer Science and Business Media.)

The high powers attainable from transverse flow lasers make them highly suitable for materials processing applications such as welding, surface heat treatment, and cladding.

Transversely Excited Atmospheric Pressure Lasers Since the output power of the CO_2 laser increases with operating pressure, the power output of the transversely excited laser is significantly increased if it is made to operate at atmospheric pressure. This is the TEA pressure laser. Unfortunately, in the CW mode, glow discharge instabilities develop, causing arcing in the discharge, when the laser is operated at atmospheric pressure. The instability is prevented by pulsing the electric field that is applied to the electrodes. For a short enough pulsing period, of the order of a microsecond, there is no time for the instability to fully develop, resulting in a pulsed output beam with a peak power of up to 20 TW. The potential for arc formation may be further reduced by preionizing the gas mixture just before applying the voltage pulse. This may involve the use of a trigger electrode that applies a high-voltage trigger pulse to ionize the mixture in a localized area, or it may involve the use of a pulsed electron beam. When the gas mixture is sealed, low pulse rates of about 1 Hz are obtained, while rates of about 1 kHz are obtained when it is transversely flowed.

The linewidths, due mainly to collision broadening at atmospheric pressure, are relatively broad, of the order of 4 GHz, making it suitable for achieving even shorter pulse duration (about 1 ns) by mode-locking. TEA CO_2 lasers are often used in laser fusion experiments, and also in pulsed laser marking.

Gas-Dynamic Lasers The principle of the gas-dynamic laser is based on the change in state of a gas as it undergoes adiabatic expansion. When a gas mixture is initially kept at a high temperature and pressure (about 1400 K and 17 atm, respectively) the population distribution given by the Boltzmann distribution law at that temperature and pressure is such that the population of the upper lasing level (001) is relatively high, about 10% that of the ground level. However, that of the lower lasing level, (100), is still higher, about 25%, so that there is no population inversion. If, however, the mixture is now made to undergo adiabatic expansion through a nozzle, the temperature of the gas is significantly reduced downstream of the nozzle. This redistributes the populations to the equilibrium level equivalent to the corresponding temperature at any downstream location. Even though both the upper and lower lasing level populations will be reduced, the shorter lifetime of the lower lasing level will result in its population being reduced sooner (at a location closer to the nozzle) than that of the upper lasing level. Consequently, there will be a significant region downstream of the nozzle where a population inversion will exist (Fig. 8.15a).

There are two conditions that need to be satisfied for a population inversion to be possible:

1. The period over which there is a reduction in pressure and temperature during expansion has to be smaller than the lifetime of the upper lasing level. This requires expansion to very high downstream velocities.
2. The period has to be longer than the lifetime of the lower lasing level.

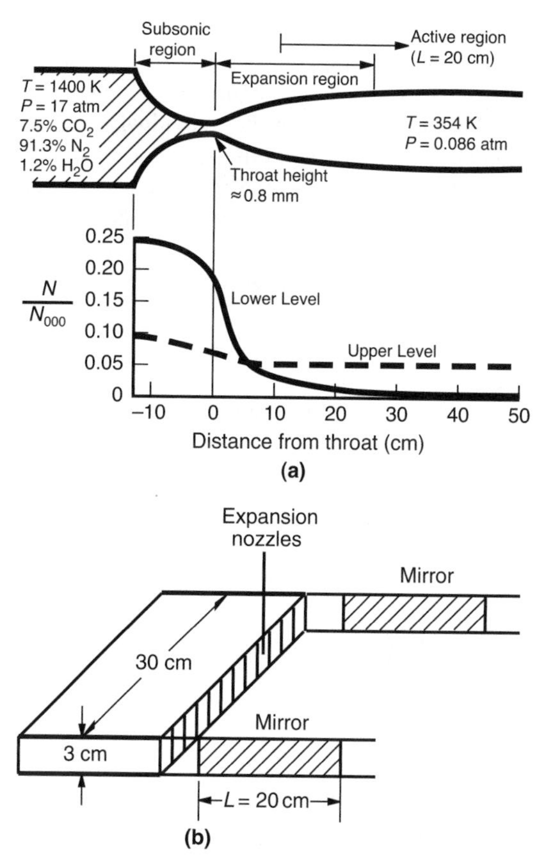

FIGURE 8.15 (a) Gas-dynamic CO_2 laser principle. (b) Schematic of a gas-dynamic CO_2 laser. (From Svelto, O., 1989, *Principles of Lasers*, 3rd edition. By permission of Springer Science and Business Media.)

The lifetime of the lower lasing level can be further reduced by the addition of small amounts of water vapor at the nozzle exit. This increases the population inversion. The cavity mirrors are mounted in the downstream region where most of the population inversion occurs (Fig. 8.15b). Power outputs of up to 80 kW have been produced, and due to the high energy involved, can only be used for short periods in the CW mode. A disadvantage of this type of laser is the size and noise associated with the lasing action.

*8.2.4.2 **Vibronic Lasers*** As the name implies, oscillation is achieved in vibronic lasers as a result of transitions between different electronic states (vibrational level of one electronic state and the vibrational level of another electronic state). A common example is the nitrogen (N_2) laser with an output wavelength of 0.3371 μm in the ultraviolet range. Another type of vibronic laser is the hydrogen (H_2) laser with output wavelengths of 0.160 and 0.116μm in the vacuum ultraviolet range. At such low wavelengths, the threshold power increases rapidly with decreasing wavelength, making it difficult to achieve laser oscillation.

8.2.4.3 Excimer Lasers Excimer lasers are based on the transition of an excited diatomic molecule to a lower energy state where it dissociates into single atoms. The potential energy of a diatomic molecule (dimer), M_2, depends on the separation distance, r, between the nuclei centers of the individual atoms, and is of the form shown in Fig. 8.16a. The overall curve is the resultant of repulsive and attractive energy curves. The resultant curve, that is, the potential energy well, has a minimum, Q_o, which occurs at the equilibrium separation distance, r_o, between the atoms. Thus, each electronic energy level has a potential energy well, as shown in Fig. 8.16b.

In some molecules, the ground-state potential energy curve does not have a minimum, and consists only of the repulsive component, whereas the excited state does have a minimum (Fig. 8.16c). Since the ground state does not have an equilibrium separation, the molecule does not exist in this state, as the constituent atoms repel each other in this state. However, the individual atoms, M, may exist in the ground state. However, since the excited state of the molecule does have a minimum in the energy curve, the molecule can exist in the excited state $M_2{}^*$. A molecule that exists under such conditions is referred to as an *excimer* (excited dimer). Common

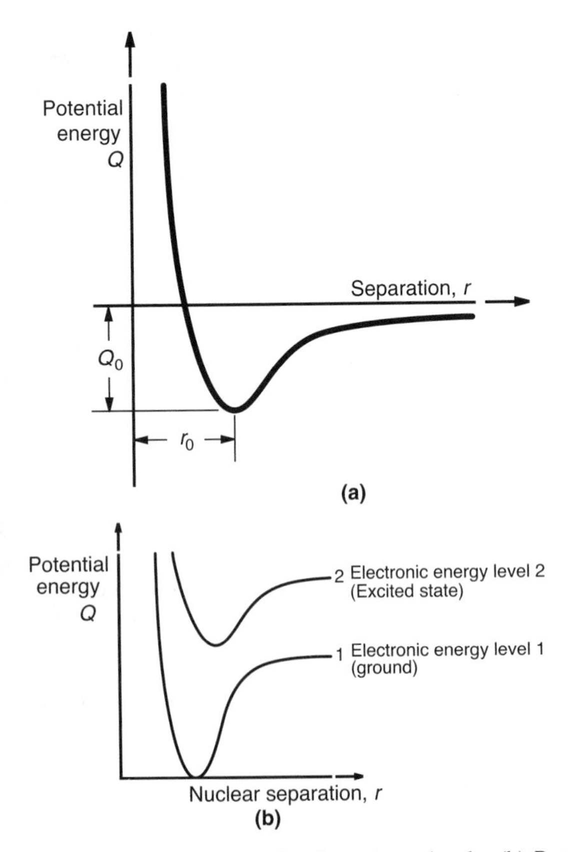

FIGURE 8.16 (a) Potential energy well of a diatomic molecule. (b) Potential energy for various electronic energy levels. (c) Energy levels of an excimer laser.

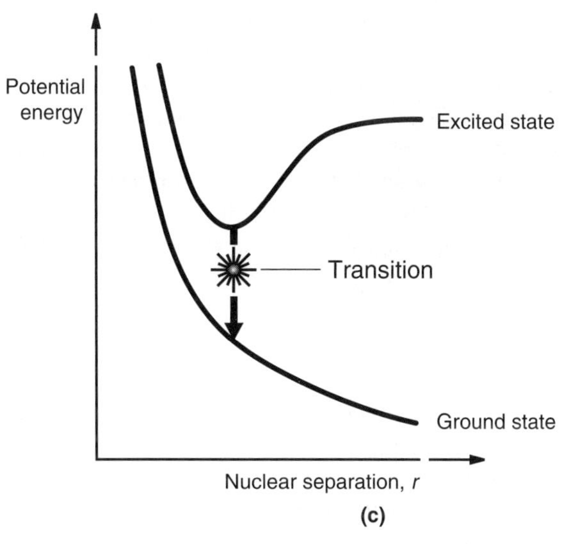

FIGURE 8.16 *(Continued)*

examples are the noble gases ($Ar_2{}^*$, $Kr_2{}^*$, $Xe_2{}^*$) or their oxides, (e.g., ArO^*), or halides (e.g., ArF^*). The noble gas halide excimer lasers tend to have better beam properties than the pure noble gas excimer lasers.

For such a material, the excimer corresponds to the upper lasing level, while the free atoms represent the lower lasing level. Transition occurs between different electronic states. Thus when the molecule undergoes a transition from the upper (excimer state) to the lower lasing level (ground state), it dissociates into individual atoms. As a result, the ground state of the molecule is always empty, making it easier to achieve population inversion. The transition is broadband since there are no well defined vibrational–rotational transitions. The resulting output laser beam is thus tunable over a range of wavelengths.

Pumping may involve subjecting the noble gas to a pressure of about 10 atm and using an electron beam or an electrical discharge to excite it to the higher level, where the atoms combine to form the excited $M_2{}^*$ excimer. For the $Xe_2{}^*$ excimer, transition occurs at a wavelength of 0.172μm with a linewidth of $\Delta\lambda = 0.015$ μm, over which the laser beam is tunable, while the output wavelength of the $Ar_2{}^*$ excimer laser is about 0.125μm. The noble gas halide excimer lasers also generate beams in the ultraviolet range. Examples include ArF, 0.193μm; KrF, 0.248μm; XeCl, 0.308μm; and XeF, 0.351μm. The linewidth of the ArF excimer laser, for example, is about $\Delta\lambda = 500$ pm. This can be reduced to about $\Delta\lambda = 1$ pm[1] using wavelength dispersive optical elements in the laser cavity.

The output beam is normally pulsed with a pulse duration of about 10 ns, average output power of up to 200 W, and a pulse rate of up to 1 kHz. The efficiency of the

[1] http://www.excimers.com/Narrowband.htm

laser is about 1 %. Excimer lasers are useful in applications such as isotope separation where the ultraviolet output is very useful. XeCl excimer lasers have been used in laser-assisted chemical vapor deposition for semiconductor manufacture. Generally, excimer lasers are useful for removal processes such as photochemical reactions, or more specifically, photoablation, and are typically used for polymers, ceramics, and glass. One application involves making holes in the nozzles of ink jet components for printers. Other applications include wire stripping and burr removal. In recent years, they have also been gaining increasing importance in materials processing.

8.3 DYE LASERS

The dye laser is a form of liquid laser where the active medium is an organic dye, which is solid that is dissolved in a solvent such as water, ethyl alcohol, or methyl alcohol. Various dyes are used, including scintillator dyes with wavelengths less than 0.4 μm; coumarin dyes with wavelengths in the range 0.4–0.5 μm; xanthene dyes (e.g., rhodamine 6 G) from 0.5–0.7 μm; and polymethine dyes in the range 0.7–1.0 μm.

The energy level scheme of a typical dye laser is illustrated in Fig. 8.17. There are two types of states in the energy level structure of lasing dyes. One is the singlet state, indicated by S_i, where the total spin of the electrons in each level is zero. The other is the triplet state T_i with a total spin of 1. Each electronic energy level of either state consists of a number of vibrational levels, with each vibrational level in turn consisting of a number of rotational levels. The energy difference between the vibrational levels is of the order of 1500/cm, while that between the rotational levels is

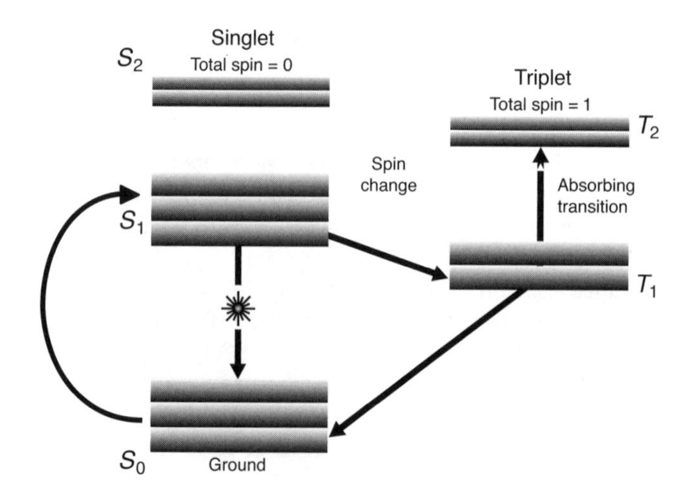

FIGURE 8.17 Energy levels of a dye laser.

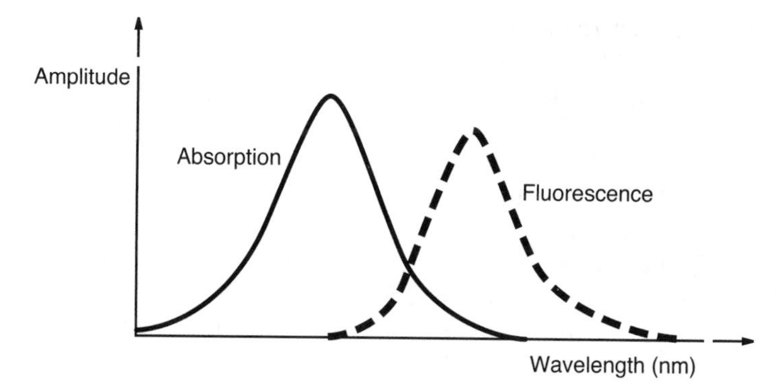

FIGURE 8.18 Fluorescence in a dye laser.

about 15/cm, which is relatively small.[2] The rotational levels thus essentially overlap. This overlap, when coupled with line broadening, results in electronic energy levels that are broadband or almost continuous from one vibrational level to another.

Pumping results in excitation of the dye molecule from the ground level S_0 to a vibrational level in the next electronic level, S_1. Radiative transitions between S and T levels, that is, singlet–triplet transitions, are forbidden, since selection rules indicate that the total spin of an atom or molecule cannot change during a transition (see Section 1.2.1). Thus, pumping can only excite the molecule to the S_1 level or higher. Due to the continuous nature of the electronic energy levels, absorption can occur over a broadband. After excitation to a vibrational level of S_1, the molecule undergoes a nonradiative transition to the lowest vibrational level of S_1. The lifetime for this transition is of the order of 1 ps. In reality, it is thermalization of the S_1 level that occurs. This results in the lowest vibrational level of the S_1 state being effectively populated, and laser transition then occurs between this level and any of the vibrational levels of S_0, from which the molecule undergoes another very fast nonradiative transition, also with a lifetime of about 1 ps, to the lowest vibrational level of S_0. The latter nonradiative transition also effectively depopulates the lower lasing level. By comparison, the lifetime of the laser transition from S_1 to S_0 is of the order of 1 ns. These are the reasons why lasing action occurs from what might appear to be a two-level system. Since absorption occurs between the lowest S_0 vibrational level and any S_1, while lasing occurs between the lowest S_1 level and any S_0 level, the absorption and emission wavelengths are usually different, with the latter being longer (Fig. 8.18). This is termed fluorescence, where light is absorbed at one frequency and re-emitted at another frequency. Since lasing action can result from transition between a number of vibrational levels, the resulting emission spectrum is very broad.

Even though the triplet T_i states are not directly involved in lasing action, they do have significant influence on the process, essentially limiting laser action. This is

[2] When frequency is expressed in units of cm^{-1}, the actual frequency is obtained by multiplying the number by the velocity of light in vacuum (3×10^{10} cm/s). The corresponding energy is obtained by multiplying the frequency by Planck's constant (see equation (1.2)).

due to the fact that intermolecular collisions can result in a transition from state S_1 to T_1. This is referred to as intersystem crossing, and reduces the population of the upper lasing level S_1. Furthermore, transition from T_1 to S_0, which is referred to as phosphorescence, is forbidden radiatively, but may also occur by collision. However, if its lifetime is longer than the lifetime for intersystem crossing, then accumulation of molecules in level T_1 will result. Now transition from T_1 to T_2 is allowed, and occurs in a wavelength range that corresponds to the lasing transition wavelength range, S_1 to S_0. The absorption associated with the $T_1 \rightarrow T_2$ transition results in a loss in laser gain, and can quench laser action after pumping has progressed for some time. Thus, pumping sources with very fast rise times, short relative to the time it takes for molecules to accumulate in level T_1 to an appreciable degree, are necessary for good lasing action. The lifetime of the triplet state T_1 can be reduced from about 10^{-3} s. to about 10^{-7} s. by the addition of oxygen to the dye solution. This has the effect of reducing the accumulation of molecules in state T_1, thus minimizing the quenching effect of the T_1 level. Either short-pulsed lasers, for example, the N_2 laser, or flashlamps with very fast rise times are used as pumping sources.

The broad output of each dye enables the resulting laser to be tuned over a wide frequency range, about 50 nm wavelength range. This can be done using wavelength-selecting devices such as a diffraction grating unit or a dispersive prism. The use of a large variety of dyes, which are automatically interchanged further enables tuning to be extended over an even broader range of frequencies from near-ultraviolet, through near-infrared. However, the dyes have a tendency to degrade, and often require a complex liquid-handling system.

The CW output power of dye lasers is of the order of 10–100 mW, while the pulsed outputs may average about 100 W with peak powers of about 1 kW. The broad oscillation bandwidth enables the output of dye lasers to be mode locked to produce very short pulses of the order of picoseconds. The tunable characteristics of dye lasers make them very useful for spectroscopy, pollution detection, photochemical processing, and isotope separation.

Tunable solid-state lasers are easier to operate, and also generate output powers in comparison with dye lasers. Furthermore, they do not involve the use of toxic materials that may pose significant disposal problems. On the contrary, dye lasers can be tuned over a wider range of the electromagnetic spectrum, and are also capable of generating much shorter pulses.

8.4 SEMICONDUCTOR (DIODE) LASERS

Semiconductor lasers, also known as diode lasers, are based on the generation of photons when electrons in the conduction band of an appropriate semiconductor material recombine with holes in its valence band. The discussion in this section starts with a background on semiconductor materials in general. This is followed by the basic principles of semiconductor lasers, and finally the different types of semiconductor lasers.

8.4.1 Semiconductor Background

As discussed in Sections 1.1 and 1.2, electrons in an atom can only occupy specific quantum states, and the distribution of electrons in the atom depends on the temperature. In semiconductors, it is more appropriate to consider the energy levels of the entire solid rather than individual atoms. This is because the available energy levels for electrons in these solids can be grouped into two broad bands, where the energy levels within each band are so closely spaced that they essentially overlap. These are the following:

1. The valence band where the energy levels are almost completely filled by electrons such that the electrons are not freely mobile in the solid.
2. The conduction band where the energy levels are either empty or partially occupied by electrons that are able to move freely within the solid (Fig. 8.19).

The energy separation between these two bands is referred to as the bandgap or forbidden band, and cannot be occupied by electrons in the solid.

The energy levels in the valence and conduction bands are described by Fermi–Dirac statistics where the probability, $p(E)$, of occupation of an energy level is given by

$$p(E) = \frac{1}{1 + e^{\left(\frac{E-F_e}{k_B T}\right)}} \tag{8.5}$$

where E is the energy of the level of interest, F_e is the energy of the Fermi level, k_B is Boltzmann's constant, and T is the absolute temperature.

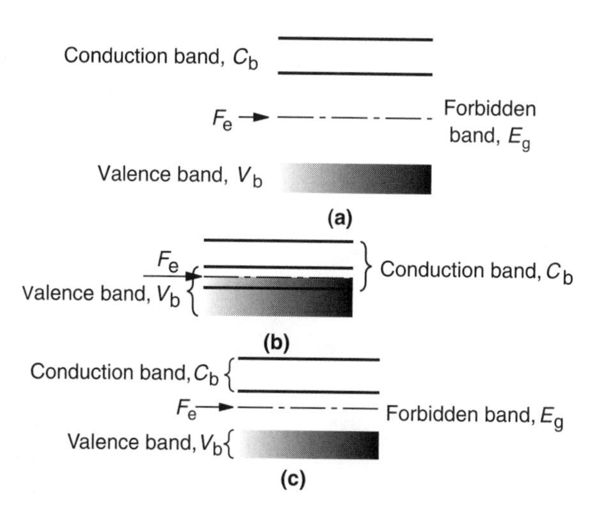

FIGURE 8.19 Energy bands of (a) insulators, (b) metals, and (c) semiconductors. (From O'Shea, D. C., Callen, W. R., and Rhodes, W. T., 1977 *Introduction to Lasers and Their Applications*. Reprinted by permission of Pearson Education, Inc.)

The Fermi energy defines the energy level where the probability that an electron will be in that state (i.e., the probability of occupation) is 0.5. From this equation it is found that at absolute zero,

$$p(E) = \begin{cases} 0 & \text{for} \quad E > F_e, \\ 1 & \text{for} \quad E < F_e. \end{cases} \tag{8.6}$$

At absolute zero temperature, the electrons occupy the lowest possible energy levels allowed, with all lower energy levels being completely filled before the next level is occupied. Under these conditions, the valence band is completely occupied while the conduction band is empty, and the boundary between them is the Fermi energy level. The material is then an insulator (Fig. 8.19a). This is because conduction of current results from flow of electrons from one level in a band to another level in a band, when the material is subjected to an electric field. With the valence band being completely occupied, then, the electron has no place to go since all levels that it could occupy are already filled, and the energy required to cross the forbidden gap, E_g and reach the conduction band is much greater than the available thermal energy, that is, $E_g >> k_B T$. However, in good electrical conductors such as metals, either the conduction band is filled to some extent or there is an overlap between it and the valence band (Fig. 8.19b). There are then empty levels available for electrons to move into when the material is subjected to a voltage source. The material is classified as a semiconductor if the forbidden gap is small enough that electrons can be thermally excited from the valence to the conduction band, that is, $E_g \approx k_B T$ (Fig. 8.19c). The material's conductivity then lies in between that of a good conductor and an insulator.

When an electron is raised from the valence band into the conduction band, a *hole* is created in the valence band (Fig. 8.20). This has essentially the same characteristics as an electron, except that it has an opposite, that is, positive, charge. The hole also moves when subjected to an electric field. In fact, when a voltage is applied to the material, holes will flow toward the negative terminal, while electrons flow toward the positive terminal.

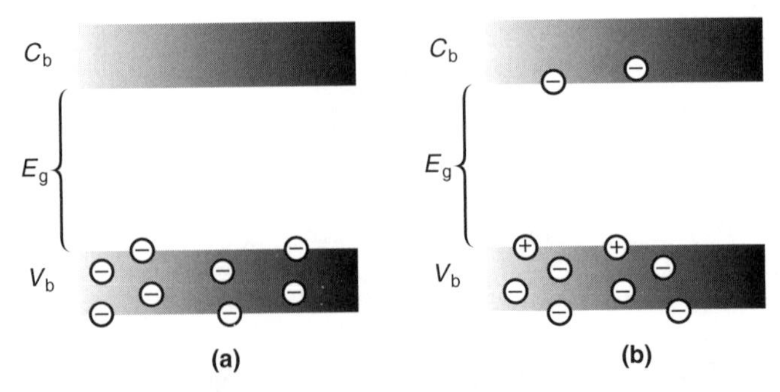

FIGURE 8.20 Creation of holes in a valence band. (a) Filled valence band. (b) Partially occupied conduction band with holes in valence band.

A pure semiconductor that contains no impurities is referred to as an *intrinsic semiconductor*. Such a semiconductor has an equal number of conduction electrons and holes since the holes are created by the excitation of electrons to the conduction band. The number of holes in the valence band, and electrons in the conduction band (which are available for electrical conduction), increases with temperature since occupation of higher energy levels increases with temperature, according to Boltzmann's law. Thus, the conductivity of such materials increases with temperature (the opposite is true of metals). Examples of such class of materials are silicon and germanium, which are group IV elements of the periodic table, and each has four electrons in the outermost orbits of their atoms. Each atom shares its four outermost electrons with four other surrounding atoms to form covalent bonds (Fig. 8.21), and since each bond involves two electrons, the outermost shell now has eight electrons altogether and is thus completely filled. It thus behaves as an insulator at absolute zero, with the conductivity increasing with temperature.

If now a group V element, say arsenic, which has five valence electrons, is added in small amounts as impurities to the germanium solid (a process called doping), four of its electrons will form stable bonds with the surrounding germanium atoms leaving the fifth atom not tightly bound and thus essentially free and mobile in the solid (Fig. 8.22a). When a semiconductor is doped in this manner, it is said to be n-type, where the "n" indicates negative since there is an excess of electrons available for conduction. Arsenic is then a donor element. On the contrary, if the original solid is doped with a group III element, say gallium, which has three valence electrons, then there is a deficiency in bonding electrons, resulting in the creation of holes (Fig. 8.22b). The resulting semiconductor is said to be p-type. The gallium atom is then an

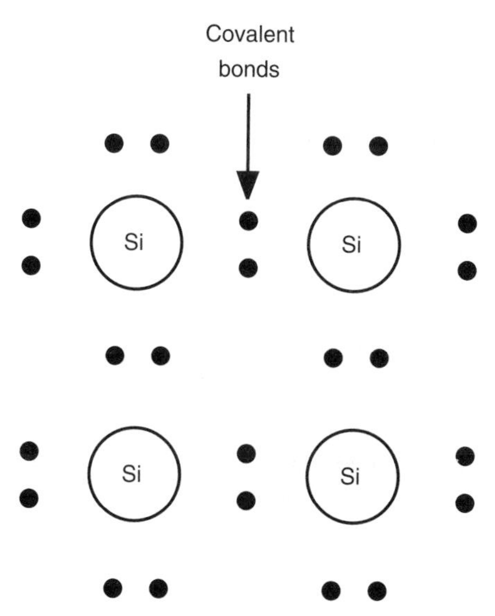

FIGURE 8.21 Covalent bonding in an intrinsic semiconductor.

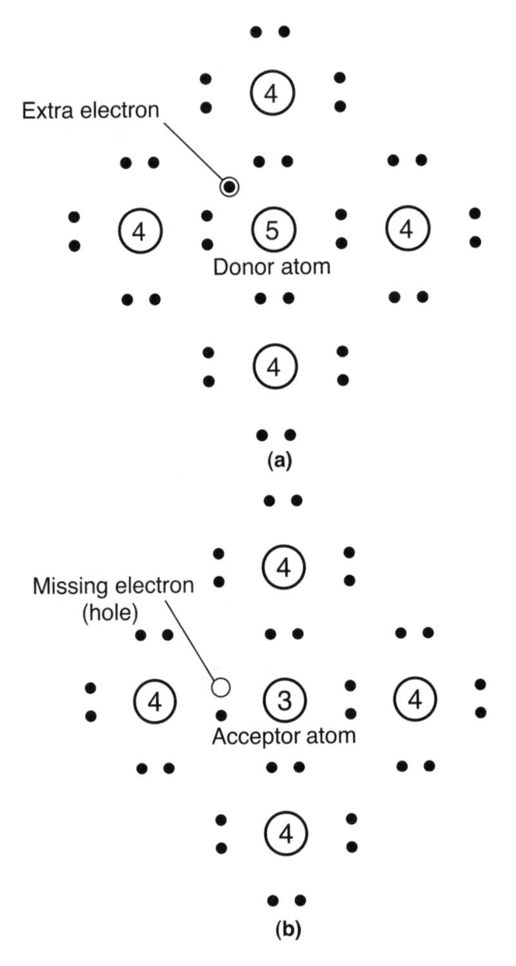

FIGURE 8.22 (a) n-Type semiconductor. (b) p-Type semiconductor.

acceptor atom. It essentially takes an electron from the valence band, leaving holes behind.

8.4.2 Semiconductor Lasers

To understand the principle of operation of a semiconductor laser, first consider an intrinsic semiconductor at absolute zero temperature (Fig. 8.23a), where the valence band (V_b) is completely filled (indicated by the shaded region). If electrons are now excited from the valence band into the conduction band (C_b), by say, pumping, these electrons rapidly drop down to the lowest states in the conduction band with a lifetime of about 0.1 ps. Likewise, the electrons in the valence band are also redistributed to fill the lowest levels that are empty, resulting in holes being created at the top of

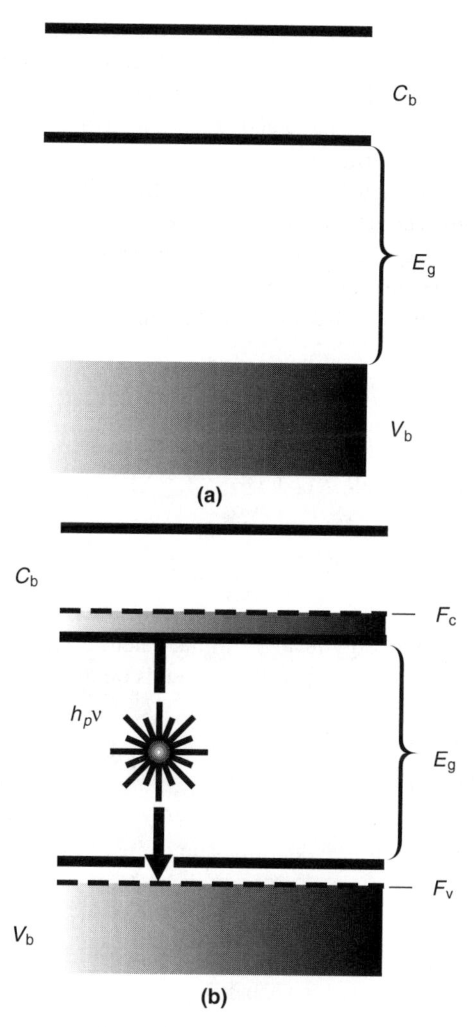

FIGURE 8.23 Basic concepts in semiconductor laser operation. (a) A completely filled valence band. (b) A population inversion resulting from a partially filled conduction band. (From Svelto, O., 1989, *Principles of Lasers*, 3rd edition. By permission of Springer Science and Business Media.)

the valence band (Fig. 8.23b). Under these circumstances, the semiconductor as a whole is not in thermal equilibrium, even though the redistribution process results in localized equilibrium in each band. Within each band, there is a quasi-Fermi level that separates the fully occupied and empty levels in that band. These are indicated by F_c and F_v for the conduction and valence bands, respectively, and their values depend on the number of electrons pumped to the conduction band. When the semiconductor as a whole is in equilibrium, $F_c = F_v$.

With the bottom of the conduction band being occupied by electrons while the top of the valence band contains holes, a population inversion then exists between the conduction and valence bands (Fig. 8.23b). When the electrons in the conduction band drop down into the valence band and recombine with the holes, either a photon (light wave) or a phonon (thermal wave) will be emitted depending on the material. In materials such as silicon and germanium where the radiation is emitted in the form of heat (phonons), no laser action can be obtained. Laser action can be achieved in materials like gallium arsenide where the radiation is in the form of light (photons).

The probability of occupation of any energy level, E, of each band is given by a relationship similar to equation (8.5), with that for the conduction band, $p_c(E)$, being

$$p_c(E) = \frac{1}{1 + e^{\left(\frac{E - F_c}{k_B T}\right)}} \tag{8.7}$$

while that for the valence band, $p_v(E)$, is

$$p_v(E) = \frac{1}{1 + e^{\left(\frac{E - F_v}{k_B T}\right)}} \tag{8.8}$$

During pumping from energy level E_v in the valence band to energy level E_c in the conduction band, the rate of absorption depends on the following:

1. The density of incident radiation, $e(\nu)$
2. The transition coefficient, B
3. The probability that the valence band energy level is occupied, $p_v(E_v)$
4. The probability that the conduction band energy level is empty, $1 - p_c(E_c)$,

It is therefore given by

$$\frac{dN_1}{dt} = k_c B p_v(E_v)[1 - p_c(E_c)]e(\nu) \tag{8.9}$$

where k_c is a constant. Likewise, the rate of stimulated emission is given by

$$\frac{dN_2}{dt} = k_c B p_c(E_c)[1 - p_v(E_v)]e(\nu) \tag{8.10}$$

Here, N_1 and N_2 are the number of photons absorbed and emitted, respectively. Laser amplification will prevail only when $\frac{dN_2}{dt} > \frac{dN_1}{dt}$, that is,

$$p_c(E_c)[1 - p_v(E_v)] > p_v(E_v)[1 - p_c(E_c)] \tag{8.11}$$

Substituting for $p_c(E)$ and $p_v(E)$ from equations (8.7) and (8.8) gives

$$F_c - F_v > E_c - E_v = h_p\nu \tag{8.12}$$

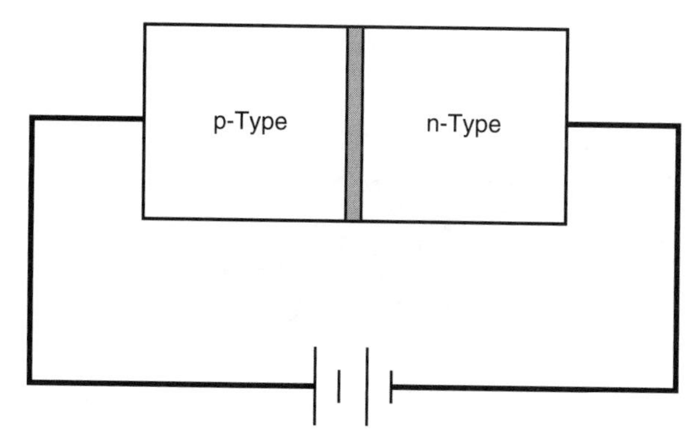

FIGURE 8.24 p–n Junction diode (forward bias).

where ν is the emitted photon frequency, while E_c and E_v are the upper and lower laser levels, corresponding to energy levels in the conduction and valence bands, respectively.

When an electron undergoes a transition from an energy level E_c in the conduction band to an energy level E_v in the valence band, the energy associated with the transition is also greater than the energy gap, E_g. Thus,

$$F_c - F_v > h_p\nu = \frac{h_p c}{\lambda} > E_g \tag{8.13}$$

Laser action is more easily achieved in a semiconductor when using a p-n junction diode (Fig. 8.24), which is obtained by joining an n-type to a p-type semiconductor. The energy levels for a diode are shown in Fig. 8.25. The energy levels of the conduction and valence bands are different on the two sides of the junction, with those for the n-type being at a lower level than those for the p-type, Fig. 8.25a. Population inversion is achieved if the diode is heavily doped with acceptors and donors. Under such circumstances, the Fermi level F_n for the n-type lies in the conduction band, while that for the p-type F_p lies in the valence band (Fig. 8.25a). A portion of the conduction band in the n-type region is then occupied by electrons up to F_n, and in the p-type region, holes are added by the acceptors down to F_p.

When there is no voltage applied across the junction, excess electrons flow from the n-type to the p-type, setting up an electric field that opposes and thus stops the flow. The Fermi levels on the two sides of the junction, F_n and F_p, are then at the same level or have the same energy (Fig. 8.25a). The energy levels E_c and E_v in Fig. 8.25 indicate the lower level of the conduction band and upper level of the valence band, respectively.

If the junction is reverse biased by connecting the n-type portion of the junction to a positive electrode and the p-type portion to a negative electrode, the electrons flow toward the anode while the holes flow toward the cathode, until there are no more free electrons available. Current flow then stops. On the contrary, if the positive electrode

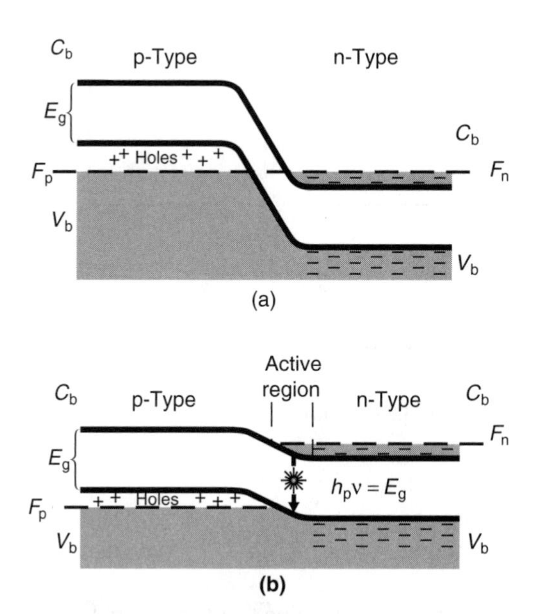

FIGURE 8.25 Energy levels of a heavily doped p–n junction diode. (a) Zero bias voltage. (b) Forward bias. (From O'Shea, D. C., Callen, W. R., and Rhodes, W. T., 1977, *Introduction to Lasers and Their Applications*. Reprinted by permission of Pearson Education, Inc.)

is connected to the p-type side while the negative electrode is connected to the n-type side, forward biasing, (Figs 8.24 and 8.25b), electrons flow into the junction as they move toward the anode. At the junction, they may recombine with holes, which are attracted in the opposite direction, to generate radiation. Current flow continues as the n-type region is supplied with more electrons while more holes are created in the p-type side (as it is depleted of electrons) by the voltage source. This is the basic principle of a diode, conducting current in one direction, and preventing current flow in the other direction.

The flow of current due to forward biasing injects electrons from the conduction band of the n-type and holes from the valence band of the p-type materials into a narrow region in the immediate neighborhood of the p–n junction, called the depletion layer, which is only about 0.1 μm in thickness. That results in the creation of a population inversion in that region. Recombination of electrons and holes in this region generates radiation of energy equal to the bandgap energy. The recombination is essentially equivalent to the de-excitation of an electron from the conduction band to the valence band. If the resulting emission is not stimulated, then the device is a light-emitting diode (LED). If, however, the applied current is high enough to exceed a certain threshold value, then a population inversion is achieved which results in stimulated emission and laser action. Above the threshold current, the beam intensity increases rapidly with increasing current, becomes more monochromatic, and its divergence is significantly reduced.

Normally, laser oscillation involves using two mirrors to form a cavity. In semiconductor lasers, this is achieved by cleavage of the material along natural crystalline planes to obtain highly reflecting and parallel surfaces at two ends of the crystal which are also perpendicular to the junction. The other two surfaces are left rough to suppress oscillation in undesired directions. Furthermore, since semiconductors have a very high index of refraction, about 3.6, the inherent reflectivity at the crystal–air interface is very high, about 35 %, making it unnecessary to use reflecting coatings.

Semiconductor lasers have the advantage of being relatively small in size and easy to fabricate by mass production, thereby being low cost. Typical dimensions are 1 mm $\times$ 1 mm $\times$ 200μm, with the thickness of the active region, where most of the oscillation is concentrated, being about 1 μm (Fig. 8.26). However, due to diffraction, the beam itself is much wider, about 40μm, extending away from the interface into the p- and n- regions. The extension of the beam into the regions away from the interface results in high cavity losses since absorption is much greater than emission as one moves away from the interface. These losses result in high threshold current densities being required for laser oscillation.

The threshold current increases with temperature and this is because $p_{\mathrm{v}}(E_v)$ $[1 - p_{\mathrm{c}}(E_c)]$ increases, and $p_{\mathrm{c}}(E_c)[1 - p_{\mathrm{v}}(E_v)]$ decreases as temperature increases, and from equation (8.11), that means the oscillation gain decreases with increasing temperature. The relatively narrow width of the beam, about 40μm, as it comes out of the optical cavity, results in high beam divergence, compared to other lasers, as it is diffracted at the cavity aperture. This may range from 5° to 15°. Also, since transition occurs between two broad bands, its monochromaticity is not as good as in other lasers. The direct conversion of electrical energy into optical energy, however, results in very high efficiencies for semiconductor lasers, up to about 50 %. The output

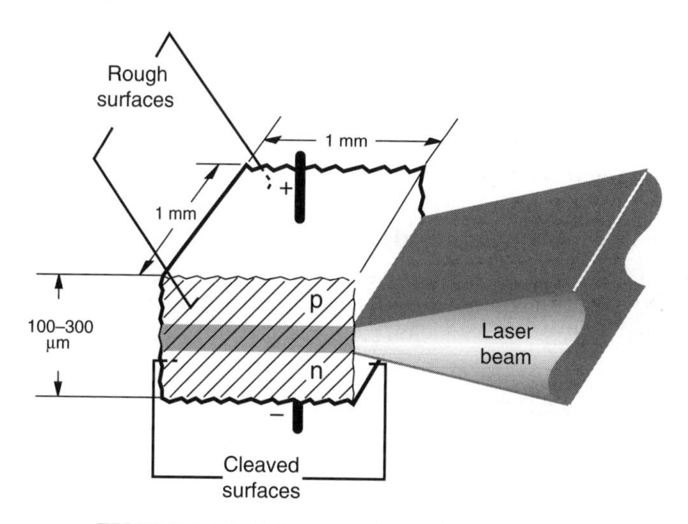

FIGURE 8.26 Schematic of a semiconductor laser.

wavelengths range from 0.7 to 30μm in the infrared region. Semiconductor lasers are particularly suited to optical fiber communications.

8.4.3 Semiconductor Laser Types

There are several forms of semiconductor lasers. These include the following:

1. Homojunction lasers.
2. Heterojunction lasers.
3. Quantum well lasers.
4. Separate confinement heterostructure lasers.
5. Distributed feedback lasers.
6. Vertical-cavity surface-emitting lasers (VCSELs).

In the following sections, we shall provide highlights of the first three types that are listed above.

8.4.3.1 Homojunction Lasers Homojunction lasers are made out of a single semiconductor material such as gallium arsenide, GaAs. In such lasers, the high concentration of electrons and holes in the active region increases the refractive index of the material in that region. This results in the emitted light being internally reflected and thus confined to that region. However, the amount that is lost out of the region is significant enough to require high threshold current densities for oscillation. For example, the threshold current density at room temperature for GaAs is about 10^5 A/cm^2. Thus, homojunction lasers require cryogenic temperature environment for CW operation. At room temperatures, they can only be operated in pulsed mode. At liquid nitrogen temperature of about 77K, the GaAs laser can be operated under CW conditions. The output wavelength at room temperature is about 0.9μm, decreasing to 0.84μm at 77K, and can thus be tuned by varying the temperature.

8.4.3.2 Heterojunction Lasers Heterojunction lasers are made out of two or more materials. This significantly reduces the threshold current density to the extent that for some material combinations, CW operation can be achieved at room temperature. This is because the different materials used have different refractive indices and bandgaps. These differences cause radiation generated in the junction region to be reflected back into the region, resulting in higher efficiency. The difference in bandgap results in the carriers being confined to the junction region, and this reduces the threshold current density.

Heterojunction lasers can be one of two forms:

1. Single heterojunction.
2. Double heterojunction.

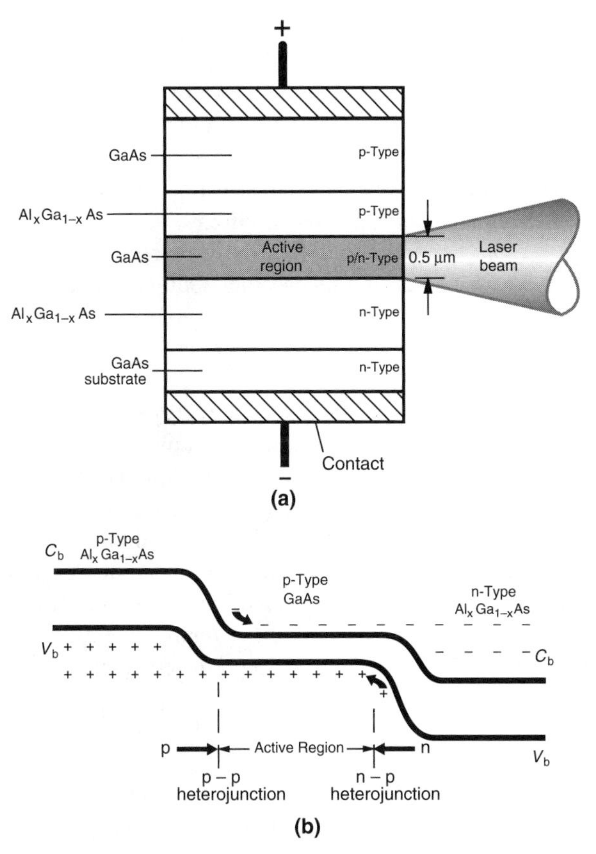

FIGURE 8.27 (a) Schematic of a double heterojunction laser. (b) Energy level scheme of a double heterojunction laser. (From Luxon, J. T., and Parker, D. E., 1985, *Industrial Lasers and Their Applications*, Prentice Hall, Englewood Cliffs, NJ.)

In a single heterojunction, the junction is formed between say, a p-type $Al_xGa_{1-x}As$ semiconductor and an n-type GaAs semiconductor, where x is a fraction. A typical double heterojunction has an active region of p-type GaAs which is a thin layer of about 0.1-0.5 μm, sandwiched between an n-type and a p-type $Al_xGa_{1-x}As$ semiconductor (Fig. 8.27a). There are thus two junctions between different materials. This reduces the threshold current density to about 10^3 A/cm^2, enabling room temperature CW operation to be achieved. This is due to a number of reasons.

(1) The difference in refractive indices of the materials (being about 3.4 for $Al_xGa_{1-x}As$ and about 3.6 for GaAs) confines the laser to the GaAs active region where there is high gain, thereby reducing losses due to absorption.

(2) The energy gap in the $Al_xGa_{1-x}As$ semiconductor is greater than that in the GaAs semiconductor (Fig. 8.27b). This acts as an energy barrier that reflects both holes and electrons back into the active region where they are confined, increasing their concentration and thus, gain in that region.

Since the output wavelength depends on the fraction x of Al in the semiconductor, the oscillation wavelength can be tuned by varying x. Peak pulse powers of 10–20 W can be obtained at room temperature with a pulse repetition rate of about 10^4 Hz, and a pulse duration of the order of a picosecond. Output powers of 5–100 mW can be achieved at room temperature under CW operation, but 5–10 mW is typical.

8.4.3.3 Quantum Well Lasers If the potential energy well of Fig. 8.16a becomes so narrow that it confines the motion of electrons or holes to two dimensions (planar motion), then we have a quantum well. Confining motion to one dimension results in a quantum wire, and a quantum dot is obtained when motion is restricted in all three dimensions. The structure of the quantum well laser is similar to the heterojunction laser illustrated in Fig. 8.27a, except that with the quantum well laser, the middle layer is made thin enough to confine electron motion in this region to two dimensions. This improves the efficiency of the laser. Further improvements in efficiency can be obtained with further restrictions in motion to produce a quantum wire, and further yet, a quantum dot.

8.4.4 Low-Power Diode Lasers

The CW power output of diode lasers is comparable to that of helium-neon lasers. Thus, the two types of lasers are competitive in the 5–10 mW power range. The characteristics of common industrial lasers are summarized in Appendix 8A. Semiconductor lasers are much smaller, and their power requirements are lower (due to the higher efficiency) compared to He–Ne lasers. On the contrary, the output beam quality is much better for He–Ne lasers. Also, the 0.6328 μm output of He–Ne lasers is more sensitive to the human eye. However, they are more expensive than semiconductor lasers.

There have been significant improvements in the output powers of diode lasers in recent years. This is achieved by using stacks of diode lasers. In the next section, we present a brief background on high-power diode lasers.

8.4.5 High-Power Diode Lasers

The basic unit of high power diode lasers is an individual diode laser with an output power of up to 100 mW, (Fig. 8.26). The power that is necessary for materials processing is acquired by combining the individual diode lasers in a linear setup to form an array. A number of arrays are further combined to form a laser bar of approximate dimensions 10 mm × 0.6 mm × 0.1 mm, and which is mounted on a

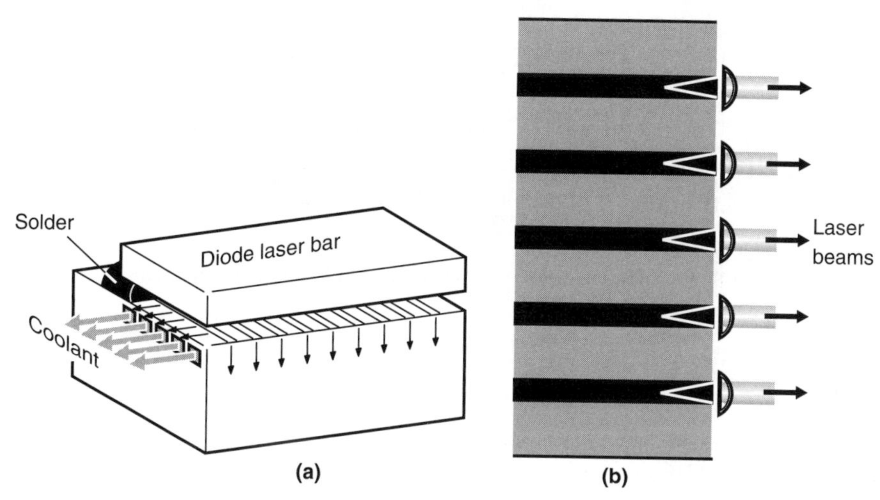

FIGURE 8.28 High-power diode laser construction. (a) Diode laser bar. (b) Stacked laser.

heat sink or cooling device (Fig. 8.28a). The output power of each bar is about 30–50 W. The individual bars are then stacked one on top of the other to form the laser diode stack (Fig. 8.28b). The number of bars used determines the overall power output of the stack. Combining stacks enables output powers of up to 5 kW to be achieved.

The lifetime of the laser diode stack is influenced by its operating temperature. More efficient operation is achieved at lower temperatures. Different methods have been developed for cooling the stack. One involves forcing a coolant through microchannels etched into a silicon manifold layer, with each channel being approximately 25μm in width and 150μm in depth. Others involve the use of microfins or porous materials.

The beam from a diode laser is strongly divergent in the direction of the diode junction. Beam shaping therefore generally starts with collimating the beam in that direction. As an example, the divergence angle of an InGaAs-based diode laser along the fast axis (see Section 9.5) is about 45–60°, while that along the slow axis is about 12°. Microlenses may be mounted directly in front of the diodes to collimate the fast axis and reduce the divergence angle in that direction to about 0.6°. Since the emitting aperture of the laser is rather large, about $10 \times 15\,\text{mm}^2$, and the divergence differs in different directions, cylindrical lenses are necessary to collimate the slow axis divergence, before using an ordinary lens to focus it down to about $8.56 \times 0.62\ \text{mm}^2$ (Fig. 8.29), which would be desirable for say, heat treatment.

The advantages of high-power diode lasers compared to the conventional CO_2 and Nd:YAG lasers include the following:

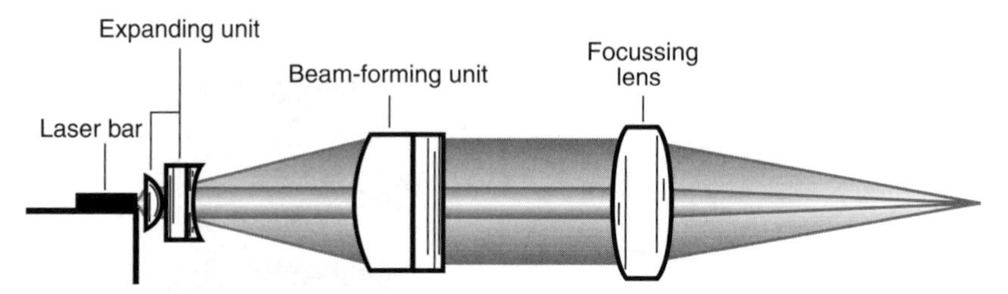

FIGURE 8.29 Setup of a stacked diode laser with beam transformation optics.

- Compact size.
- High efficiency (over 30 %).
- Almost maintenance-free operation.
- Fiber optic beam guiding capability.
- Higher absorption at metallic surfaces.

The disadvantages include the following:

- Low power.
- Poor beam quality.

8.4.6 Applications of High-Power Diode Lasers

Early applications of diode lasers were limited to pumping of solid-state lasers, where the diode laser substitutes for the lamp. This increases efficiency, beam quality, lifetime, and reliability of the solid-state laser. Other applications in materials processing were in areas such as surface treatment, soldering small components, and welding plastics. With increased output powers of up to about 5 kW, diode lasers have also found application in laser welding, but are mainly limited to conduction mode welding of thin sheets, due to the relatively low power densities of up to $3\text{–}4 \times 10^5$ W/cm^2.

Since the output of the diode laser is usually either rectangular or square in shape, special optics are required to transform the beam from the linear emission characteristics of the laser bars to a round (or square) shape for coupling into an optical fiber.

8.5 FREE ELECTRON LASER

Compared to the other lasers that we have discussed thus far, with free electron lasers, energy levels of atoms or molecules are not involved in the laser radiation. Even though they are not commercially available, we briefly introduce them here because of their potential as high-power, high-efficient lasers of excellent beam quality, and with the capability of being tuned over a wide range of wavelengths, from ultraviolet through the infrared.

The principle of operation is based on electrons that are accelerated to speeds approaching that of light, and become coupled by Coulomb interaction to a beam of photons traveling in the same direction. Part of the electron-beam energy can be transferred to the photon beam under suitable circumstances. To achieve this, a magnetic arrangement is used to induce a transverse oscillatory (wiggly) motion to the electron beam (Fig. 8.30). Thus, the electron beam develops a velocity component perpendicular to the direction of travel, which can then couple with the electric field of the light beam that is naturally transverse to the direction of travel. The wavelength of the radiation emitted by the electrons under these conditions is given by

$$\lambda = \frac{\lambda_{\mathrm{m}}}{2\gamma_{\mathrm{e}}^2}\left[1 + k_{\mathrm{m}}^2\right] \tag{8.14}$$

where, λ_{m} is the oscillatory period of the magnet, γ_{e} is the ratio of the electron energy to its energy at rest, and k_{m} is a constant, the undulator parameter, which is usually less than 1.

The corresponding linewidth (FWHM) of the radiation, $\Delta\nu_0$, may be approximated by

$$\frac{\Delta\nu_0}{\nu} = \frac{1}{2n_{\mathrm{m}}}$$

where n_{m} is the number of periods in the magnetic assembly.

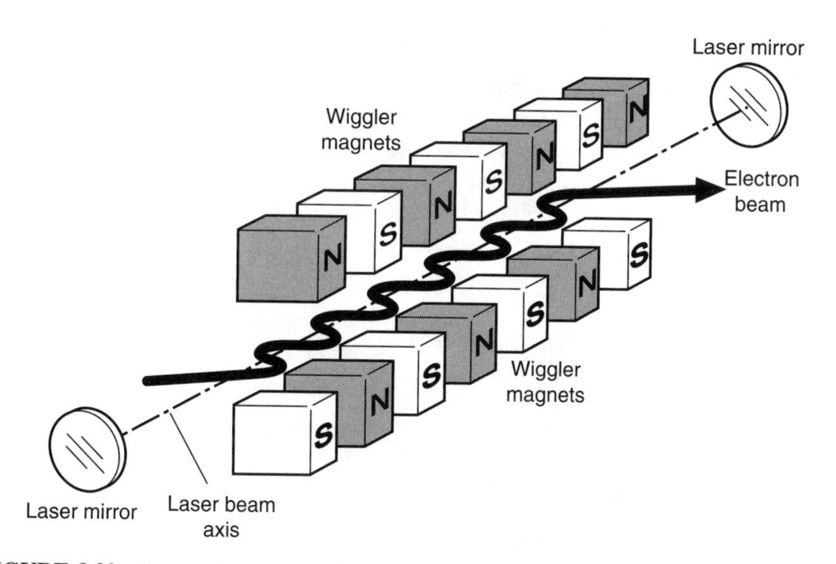

FIGURE 8.30 Schematic of a free electron laser. (From Wilson, J., and Hawkes, J. F. B., 1987, *Lasers, Principles and Applications*. Reprinted by permission of Pearson Education, Inc., Essex, UK.)

Free electron lasers have the advantage that they can be tuned to specific wavelengths, and over a very wide range. They are also capable of producing intense laser pulses with extreme precision, and since there is no active medium in the laser cavity, high power levels can be achieved without overheating the cavity. However, they are rarely used because they currently require large electron-beam accelerators to operate. Also, they have very low efficiencies (of the order of 0.001–0.01 %). However, since they have great potential, efforts are being made to increase their efficiency.

8.6 NEW DEVELOPMENTS IN INDUSTRIAL LASER TECHNOLOGY

There have been significant advances in laser technology in recent years. In this section, we discuss the principal developments in laser technology that are useful for materials processing. We focus on the following:

1. Slab lasers.
2. Disk lasers.
3. Ultrafast lasers.
4. Fiber lasers.

8.6.1 Slab Lasers

The active medium for solid-state lasers is traditionally shaped in the form of a cylindrical rod, with pumping and cooling occurring in the radial direction (Figs 3.1 and 3.2), while the beam propagates axially. However, with this configuration, a radial temperature gradient is induced in the rod. This leads to thermal focusing effects that cause the beam to be distorted at high laser power levels. This further makes it difficult to obtain a low order mode of operation or high beam quality at such power levels. This phenomenon is referred to as *thermal lensing*.

The use of a slab geometry for the active medium enables the beam quality to be significantly improved. The slab configuration is shown schematically in Fig. 8.31.

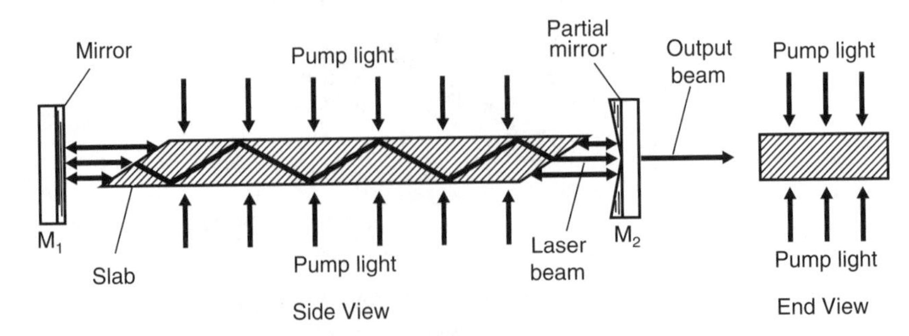

FIGURE 8.31 General schematic of a slab laser.

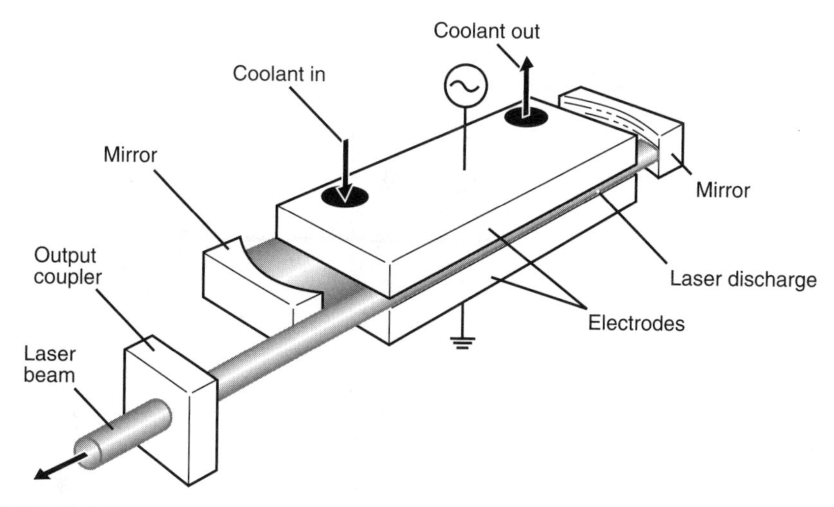

FIGURE 8.32 Schematic of a CO_2 slab laser. (By permission of ROFIN-SINAR Laser GmbH.)

Both pumping and cooling occur in a direction normal to the flat faces of the slab, and is thus one dimensional, resulting in only one-dimensional temperature gradient. The beam is subjected to total internal reflection, thus encountering the entire array of temperature variation in the medium. This minimizes thermal lensing effects and improves beam quality.

Slab lasers can have either solid-state or gas-active media, and Fig. 8.32 illustrates a CO_2 slab laser. Here, the active medium is excited using two parallel rf-electrodes, which are water cooled. The system is thus said to be diffusion cooled. As a result, the conventional gas circulation systems involving roots blowers or turbines are not required.

8.6.2 Disk Lasers

Disk lasers are also solid-state lasers and were developed to circumvent essentially the same problems for which slab lasers were developed—thermal lensing at high power levels. The principle of the disk laser is illustrated in Fig. 8.33. The active medium, which is shaped in the form of a thin disk, is pumped using a diode laser. Parabolic mirrors are used to direct the diode laser output to one face of the disk. Excess heat that is generated in the disk is removed by a heat sink that is attached to the other face of the disk, resulting in efficient cooling. Since the resulting thermal gradients are parallel to the laser beam axis, the thermal lensing effect is minimized. Beam quality is thus significantly improved, and is about four times better than that of a comparable diode pumped rod laser. For example, a beam quality of $M^2 < 1.5$ can be obtained at a laser power of about 1 kW. Multikilowatt output laser powers, up to about 4 kW can be obtained by connecting disks in series.

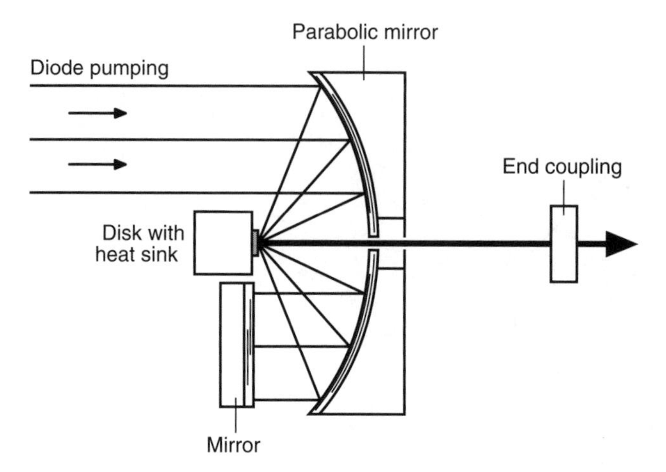

FIGURE 8.33 Schematic of a disk laser.

The higher beam quality enables a smaller focused beam size to be achieved. This results in a much higher power density, which enables higher processing speeds to be achieved. For example, cutting speeds of about 60 m/min have been achieved in cutting 0.2 mm thick aluminum sheets. Welding speeds of over 50 m/min have also been achieved in 0.2 mm thick stainless steel sheets.

Advantages of the thin disk laser system may thus be summarized as:

1. Reduced thermal lensing effect.
2. More efficient cooling of the laser.
3. High beam quality, and thus high power density.
4. Higher laser efficiency.

8.6.3 Ultrafast (Femtosecond) Lasers

The laser systems discussed thus far are capable of generating either continuous wave or pulsed beams with pulse duration of up to hundreds of picosecond duration by either Q-switching or mode-locking. Shorter laser pulses of pulse duration down to femtoseconds are obtained using ultrafast (or ultrashort pulse) lasers. These enable much higher peak intensities to be obtained. For example, a peak intensity of 10^{15} W/cm^2 can be obtained from a 100 fs (10^{-13} s) laser pulse with a pulse energy of 0.33 mJ when focused to a 20 μm diameter. To achieve the same peak intensity with a 10 ns laser pulse, the energy content would have to be 100 J.

Different methods have been developed for generating ultrashort pulse beams. In this section, we discuss one of the basic principles, chirped-pulse amplification. As indicated in Section 6.4, the spectrum of a laser beam is inversely proportional to its pulse duration, equation (6.41). Thus, a laser gain medium with a broad, continuous emission spectrum is necessary for generating an ultrashort pulse. The emission spectrum of traditional media such as CO_2 and Nd:YAG is such that they cannot support pulse duration down to the femtosecond range. Nd:YAG lasers, for example can only generate pulse duration down to 30 ps. Typical materials that are used as active media for ultrafast lasers include Ti:sapphire (6 fs), Nd:glass (100 fs), Yb:glass, Yb:YAG, Cr:YAG, and dyes.

The generation of ultrashort pulses starts with mode-locking (Section 6.4). This generally results in a pulse repetition rate of about 100 MHz, with a pulse energy of the order of nanojoules. For a number of applications, it is necessary to amplify the pulse to the microjoule and millijoule range. However, due to the high intensity of the pulses, they cannot be directly amplified since this would damage the optical components. The technique of chirped-pulse amplification (CPA) is one approach that is used to overcome this limitation. The CPA principle is illustrated in Fig. 8.34. The peak power of the pulse generated by the oscillator is first reduced by stretching the pulse duration by a factor of 10^3 to 10^4 to make it longer. The stretched pulse is then amplified, and then finally recompressed back to its short pulse state using a compressor. To avoid damage to the compressor, the laser beam diameter is expanded before being sent to the compressor.

Pulse stretching is normally achieved using a pair of optical gratings which are not parallel, or by using optical fibers. For pulses of relatively narrow bandwidths, a fiber optic cable is used. This first broadens the pulse bandwidth by about 50 times, and then stretches the widened pulse, usually about three to five times. Optical gratings are used for pulses that have large bandwidths. The principle of the latter stretching process is illustrated in Fig. 8.35a. The width of a pulse that has a broadband spectrum is changed as a result of the dispersive properties of the gratings. In the original pulse, all the frequency components travel together. However, the gratings disperses the different frequency components into different directions, with each one having a different optical path length. And since the frequency component with the shorter optical path length emerges earlier than the one with the longer optical path length, the pulse becomes stretched in time.

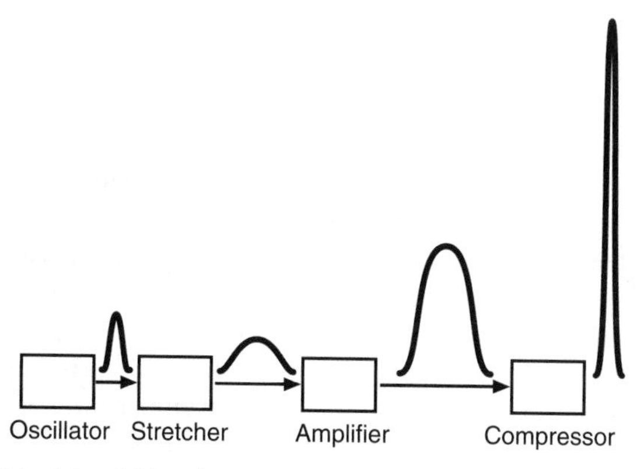

FIGURE 8.34 Principle of chirped-pulse amplification for ultrashort pulse generation. (From Liu, X., Du, D., and Mourou, G., 1997, *IEEE Journal of Quantum Electronics*.)

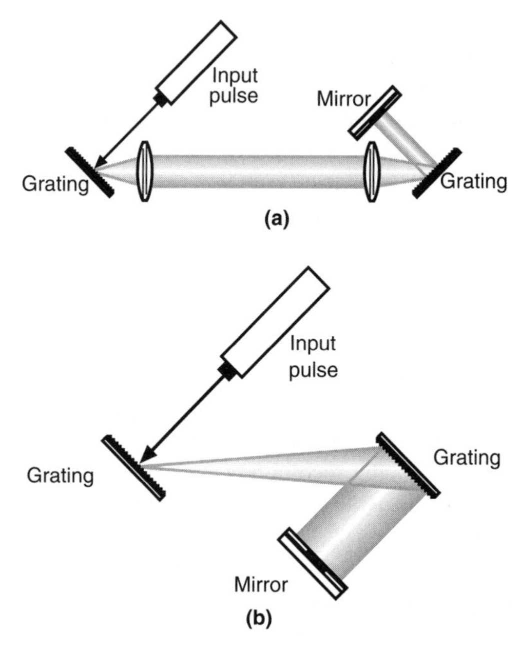

FIGURE 8.35 Chirped-pulse amplification. (a) Pulse stretching using an antiparallel grating pair. (b) Pulse compression using a parallel grating pair. (From Liu, X., Du, D., and Mourou, G., 1997, *IEEE Journal of Quantum Electronics*.)

For compression, the gratings are arranged to be parallel (Fig. 8.35b). The separation between the gratings determines the amount of compression that is achieved.

A typical ultrafast laser that has a Ti:sapphire active medium operates at a wavelength of 800 nm with a pulse duration of 100 fs, and generates pulses up to 1 mJ at a pulse repetition rate of 1 kHz, resulting in an average power of about 1 W.

8.6.4 Fiber Lasers

As the name implies, a fiber laser uses an optical fiber (Section 9.9.2) with a core that is doped with a rare earth element as the active medium. Common elements used for doping include erbium (Er), neodymium (Nd), and ytterbium (Yb). Pumping is normally done using a diode laser (Section 3.1.2), either longitudinally or transversely.

In the simplest case, the laser cavity consists of the doped optical fiber, with the ends cleaved and butted with dielectric mirrors (multiple thin layers of different transparent optical materials) (Fig. 8.36). In practice, however, fiber Bragg gratings are used for the end mirrors since these are easier to mass produce, and fiber-coupled laser diodes are used for pumping. In a fiber Bragg grating, the effective refractive index in the core of the optical fiber is varied (periodically or aperiodically).

The output power of fiber lasers is typically up to hundreds of watts, but some lasers have been produced in the kilowatt power range. Developments in the higher power

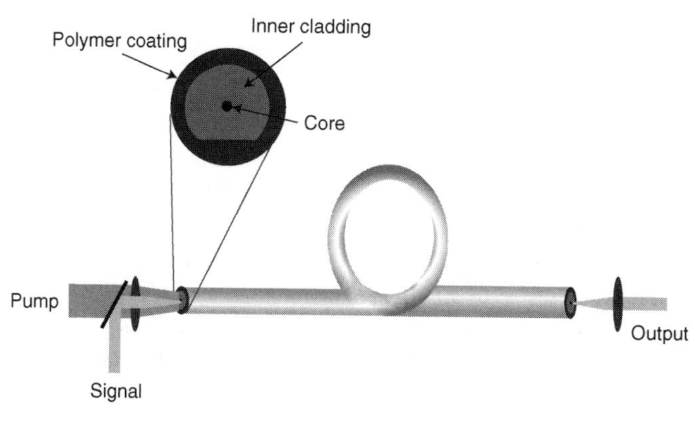

FIGURE 8.36 Schematic of a fiber laser. (From *Encyclopedia of Laser Physics and Technology. With permission of RP Photonics Consulting GmbH.)*

range are on the rise, and these are normally achieved by using double-clad optical fibers where there is an additional outer cladding (usually a polymer) which restricts the pump beam to the inner cladding and the core. The fiber core can be either single-mode or multimode. Single-mode fibers enable diffraction-limited output beams to be generated, but result in relatively lower output powers, while multimode fibers are able to produce much higher output beams, but of relatively lower quality. The cross-sectional area of the inner cladding is typically about 100–1000 times that of the core, enabling higher power diode lasers to be used for pumping. To ensure that all modes of the inner cladding are coupled to the core, the optical fiber is either coiled or the cladding is made noncircular in shape (Fig. 8.36).

Advantages of fiber lasers include the following:

1. Compact designs, since the fibers can be packaged in a coiled configuration. In fact the term "briefcase laser" has been used to refer to these types of lasers.
2. High beam quality (diffraction limited) that results from the use of single-mode fiber cores.
3. Ability to generate ultrashort pulse laser beams and/or wide tunable range, due to the broad emission spectrum of the glass gain medium.
4. Relatively high output efficiency of about 50 % compared to about 10–30 % for CO_2 and 2 % for Nd:YAG lasers.

Disadvantages of fiber lasers include the following:

1. Difficulty in aligning the pump laser output to single-mode core optical fibers.
2. Potential for fiber damage at high powers.
3. Need to use long cavity lengths since the pump absorption per unit length is limited.

8.7 SUMMARY

In this chapter, we have discussed some of the major types of lasers, focusing on those that are either currently used, or have the potential to be used for manufacturing applications. The discussion started with solid-state lasers, with a focus on the ruby laser (which is a three-level laser) and Nd:YAG laser (which is four-level, and is extensively used in manufacturing applications). The Nd:YAG laser is capable of transmission using a fiber optic cable, and can be used in either the pulsed or CW mode, with CW power levels up to 6 kW. The Nd:Glass laser is able to deliver higher output peak power pulses than the Nd:YAG laser. However, it has a more limited use in manufacturing because of its inability to deliver CW output power for any reasonable period of time.

The discussion on solid-state lasers was followed by gas lasers, where the focus was on neutral atom (He–Ne), ion (argon ion), metal vapor (He–Cd), and molecular (CO_2 and excimer) lasers. The low power CW output of the He–Ne laser makes it suitable for position sensing and alignment, among other applications. The CW power output of the argon ion laser is higher than that of the He–Ne laser, but much lower than can be obtained from the CO_2 laser. Metal vapor lasers were discussed for completeness, even though they are not used in manufacturing. The CO_2 laser, which is the other laser that is extensively used in processing, is much more efficient than the Nd:YAG laser, and generates much higher CW output powers, as well as higher beam quality. However, the longer wavelength output does not lend itself to fiber optic cable transmission. The CO_2 lasers that are most extensively used in industry are the fast axial flow and transverse flow lasers, which produce the highest power outputs. Other forms are the sealed tube, transversely excited atmospheric, and gas-dynamic lasers. The excimer laser, which normally generates pulsed beams in the ultraviolet wavelength range, is beginning to see increasing use in manufacturing.

Dye lasers, which are used primarily in spectroscopy, photochemical processing, etc., were also discussed for the sake of completeness. They were followed by diode lasers, which constitute the fastest growing laser type in materials processing. Even though the power output of an individual diode laser is relatively small, stacking of the individual lasers enables high output powers suitable for materials processing to be achieved. The near infrared wavelength of the diode laser makes it suitable for transmission using fiber optic cables. The major drawback of this type of laser is the high beam divergence, which requires special optics for collimating and focusing the beam. Even then, it is limited to processes that require a relatively wide beam area, such as surface modification, and conduction mode welding.

Following diode lasers, free electron lasers were briefly mentioned, to put things in perspective for this type of laser which holds promise for the future, but is not fully commercialized yet. Next, slab lasers were discussed. These can have either solid state or gas as the active medium. However, the pumping configuration enables higher quality output beams to be obtained compared to the beams generated using traditional pumping techniques. This was followed by a discussion on disk lasers, which produce even better beam quality because of reduced thermal lensing effects. Ultrafast lasers were then discussed. Even though these are also solid-state lasers, they were considered separately because of the special characteristics imparted by the ultrashort pulse duration in materials processing, enabling materials to be ablated or removed with minimal thermal damage to the surrounding material. Finally, fiber lasers, where the active medium is a doped optical fiber, were discussed. These have the advantage of compactness, coupled with high beam quality, and the potential for very high output powers.

REFERENCES

Abil'siitov, G. A., and Velikhov, E. P., 1984, Application of CO_2 lasers in mechanical engineering technology in the USSR, *Optics and Laser Technology*, Vol. 16, No. 1, pp. 30–36.

Barrett, C. R., Nix, W. D., and Tetelman, A. S., 1973, *The Principles of Engineering Materials*, Prentice Hall, Englewood Cliffs, N.J.

Duley, W. W., 1982, *Laser Processing and Analysis of Materials*, Plenum Press, New York.

Harry, J. E., 1974, *Industrial Lasers and Their Applications*, McGraw Hill, London.

Henderson, R. A., 1997, *A Guide to Laser Safety*, Chapman & Hall, London, UK.

Koechner, W., 1995, *Solid-State Laser Engineering*, 4th edition, Springer, Berlin, Germany.

Laud, B. B., 1985, *Lasers and Non-Linear Optics*, Wiley Eastern Limited, New Delhi.

Liu, X., Du, D., and Mourou, G., 1997, Laser ablation and micromachining with ultrashort laser pulses, *IEEE Journal of Quantum Electronics*, Vol. 33, No. 10, pp. 1706–1716.

Loosen, P., Treusch, G., Haas, C. R., Gardenier, U., Weck, M., Sinnhoff, V., Kasperowski, St., and Esche, R., 1995, High-power diode-lasers and their direct industrial applications, *Proceedings of SPIE*, Vol. 2382, pp. 78–88.

Luxon, J. T., and Parker, D. E., 1985, *Industrial Lasers and Their Applications*, Prentice Hall, Englewood Cliffs, N.J.

Mann, K., and Morris, T., 2004, Disk lasers enable application advancements, *Photonics Spectra*, Vol. 38, No. 1, pp. 106–110.

Marabella, L. J., 1994, Future solid state lasers, *Automotive Laser Applications Workshop*, Dearborn, MI, 1994.

O'Shea, D. C., Callen, W. R., and Rhodes, W. T., 1977, *Introduction to Lasers and Their Applications*, Addison-Wesley, Reading, MA.

Rofin-Sinar, *Technical Note:* CO_2 *Laser Welding*, Rofin Sinar, Plymouth, MI.

Shimoda, K., 1986, *Introduction to Laser Physics*, 2nd edition, Springer-Verlag, Berlin.

Solarz, R., et al., 1994, The future of diode pumped solid state lasers and their applicability to the automotive industry, *Automotive Laser Applications Workshop*, Dearborn, MI.

Svelto, O., 1982, *Principles of Lasers*, Plenum Press, New York.

Thyagarajan, K., and Ghatak, A. K., 1981, *Lasers, Theory and Applications*, Plenum Press, New York.

Tiffany, W. B., 1985, Drilling, marking, and application for industrial Nd:YAG lasers, *SPIE, Applications of High Powered Lasers*, Vol. 527, pp. 28–36.

Walsh, C. A., Bhadeshia, H. K. D. H., Lau, A., Matthias, B., Oesterlein, R., and Drechsel, J., 2003, Characteristics of high-power diode-laser welds for industrial assembly, *Journal of Laser Applications*, Vol. 15, No. 2, pp. 68–76.

Wilson, J., and Hawkes, J. F. B., 1987, *Lasers Principles and Applications*, Prentice Hall International Series in Optoelectronics, New York.

APPENDIX 8A

List of symbols used in the chapter.

Symbol	Parameter	Units
C_b	conduction band	–
$(Cd^+)^*$	excited cadmium ion	–
e^-	electron	–
E_c	energy level in conduction band	J
E_g	energy of forbidden gap	J
E_v	energy level in valence band	J
F_c	energy of Fermi level in conduction band	J
F_n	Fermi level for n-type semiconductor	J
F_p	Fermi level for p-type semiconductor	J
F_v	energy of Fermi level in valence band	J
He^*	excited helium atom	–
k_c	constant	$/m^3 - Hz^2$
k_m	constant	–
n_m	number of periods in magnetic assembly	–
p_c	probability of occupation of conduction band	–
p_v	probability of occupation of valence band	–
V_b	valence band	–
γ_e	ratio of electron energy to its energy at rest	–
λ_m	oscillatory period of magnet	m
$\Delta\nu_0$	linewidth of radiation	Hz

APPENDIX 8B

Comparison of Common Industrial Lasers

Characteristics	CO_2	Nd:YAG	Excimer	Diode	He–Ne
Wavelength (μm)	10.6	1.06	0.125–0.351	0.6–10	0.633
Divergence (Degree)	0.1–0.2	0.06–1.15	0.06–0.3	5–30	0.03–0.06
FWHM (GHz)	0.1–4	up to 195	8–4000	1875	1.4
Output power, CW (W)	up to 30×10^3	up to 8×10^3	5–200	1–100 $\times 10^{-3}$	(0.5–100) $\times 10^{-3}$

Peak power (MW)	0.024	50	10–100	10^{-5} to 10^{-6}	–
Overall efficiency (%)	5–30	1–3	1–4	30–50	0.02
Focused beam intensity (W/cm^2)	10^{6-8}	10^{5-9}	–	10^{3-5}	–
Laser head size ($10^{-3}\,m^3$)	10^3	10^2	–	1	–
Price (2009 $/W)	35–120	100–120	1000	60–90	–
Maintenance periods (H)	2000	200 (Lamps)	–	Maintenance free	–
Pulsed or CW	CW/pulsed	pulsed/CW	pulsed/CW	CW/pulsed	CW
Pulse rate (Hz)	1–5000	1–100	up to 1000	12×10^6	–
Pulse duration (sec)	10^{-4}	10^{-3} to 10^{-8}	10^{-9}	10^{-12}	–

APPENDIX 8C

Wavelength Range of Common Lasers (Table 1.1, Henderson, 1997)

Type	**Wavelength** (μm)
Krypton fluoride excimer	0.249
Organic dye	0.3–1.0 (tunable)
Xenon chloride excimer	0.308
Helium–cadmium	0.325, 0.442
Krypton ion	0.335–0.800
Nitrogen	0.337
Argon ion	0.450–0.530 (0.488 and 0.515 strongest)
Copper vapor	0.510, 0.578
Helium–neon	0.543, 0.633, 1.15
Gold vapor	0.628
Titanium sapphire	0.660–1.06 (tunable)
Semiconductor (GaInP family)	0.670-0.680
Ruby	0.694
Alexandrite	0.700–0.830 (tunable)
Semiconductor (GaAlAs family)	0.750–0.900
Nd:YAG	1.064
Semiconductor (InGaAsP family)	1.3–1.6
Carbon monoxide	5.0–6.0
Carbon dioxide	9.0–11.0 (main line 10.6)
Free electron lasers	Any wavelength

PROBLEMS

8.1. (a) List three ways in which solid-state lasers may be pumped.

(b) Explain why it is necessary to have temperature control of the rod in solid-state lasers.

8.2. (a) What is a typical gas ratio by volume for a He–Ne laser? (e.g. He:Ne - XX:YY)

(b) What is a typical total gas pressure in a He–Ne laser?

(c) How many different transitions normally occur in a He–Ne laser?

(d) What types of mirrors would typically be used in a He–Ne laser cavity?

(e) Would you prefer the mirrors in a He–Ne laser to be internal or external to the discharge tube? Why?

8.3. How would you expect the gain of an Ar^+ ion laser to vary with the internal diameter of the discharge tube? Explain.

8.4. (a) What is the purpose of Nitrogen in the CO_2 laser?

(b) What is the purpose of Helium in the CO_2 laser?

(c) What is the purpose of CO_2 in the CO_2 laser?

(d) In CO_2 lasers, transition occurs at 10.6 μm and at 9.6 μm. What method would be used to ensure that only the 10.6 μm wavelength beam is available at the output?

(e) Which would you recommend for butt welding of relatively thick steel, slow axial flow or transverse flow CO_2 laser? Explain.

8.5. (a) What type of CO_2 laser would you recommend for cutting relatively thin materials?

(b) What type of CO_2 laser would you recommend for butt-welding two thin pieces?

(c) Everything else being the same, which laser type would you recommend for cutting a given material, CO_2 or Nd:YAG? Explain.

8.6. Give two reasons why excimer lasers are a serious candidate for materials processing.

8.7. (a) Explain the basis of fluorescence in dye lasers. Show why the output beam wavelength is longer than the wavelength of the pumping source.

(b) What is the level structure of dye lasers?

8.8. Show why $p_v(E)[1 - p_c(E)]$ increases, and $p_c(E)[1 - p_v(E)]$ decreases as temperature increases.

8.9. (a) Relative to most lasers, the diode laser is highly efficient. Why?

(b) Give two reasons why double heterojunction diode lasers require a lower threshold current density than homojunction diode lasers.

(c) Show that the divergence of a semiconductor laser should be about 5–15°.

8.10. Determine the power efficiency of a semiconductor laser which is operated in pulsed mode if the power input to the laser is 20 W , and the input current pulse is rectangular in shape. The peak output power is 7.5 W, with a duration of 0.5 ps, and the pulse repetition rate is 10 kHz. Further determine the duty cycle of the laser. The duty cycle is the ratio of the time that the laser is on to the total time.

9 Beam Delivery

Beam delivery is the means by which the laser beam is delivered or propagates from the laser generator to the target of interest. This normally involves the use of mirrors, lenses, or fiber-optic cables, or any combination of these elements. Since light is part of the electromagnetic spectrum, the manner in which the beam propagates is determined, to a large extent, by the characteristics of the electromagnetic spectrum. Thus, we start our discussion in this chapter with a brief introduction to the electromagnetic spectrum. This is followed by a brief review of reflection and refraction, where the concept of the Brewster angle is introduced. Next, birefringence is discussed, followed by polarization, mirrors and lenses, expanders, and beam splitters. A general discussion on beam delivery systems then follows. Fiber-optic systems are discussed as a major component of the chapter.

9.1 THE ELECTROMAGNETIC SPECTRUM

The electromagnetic spectrum, of which visible light is part, exhibits both wave-like and particle-like properties. The wave-like properties are analyzed using Maxwell's equations (see Chapter 2 for a brief introduction). The particle-like properties are based on quantum mechanics. The radiations of the electromagnetic spectrum are characterized by their wavelength λ and frequency ν, which are related by

$$\nu\lambda = c \tag{9.1a}$$

where c is the velocity at which electromagnetic radiation propagates in free space and is given (see also Section 2.1) by

$$c = \frac{1}{\sqrt{\mu_{\mathrm{m0}}\epsilon_{\mathrm{p0}}}} \approx 3 \times 10^8 \text{ m/s} \tag{9.1b}$$

The velocity, c_{m}, at which an electromagnetic wave propagates in any medium, is given by

$$c_{\mathrm{m}} = \frac{1}{\sqrt{\mu_{\mathrm{m}}\epsilon_{\mathrm{p}}}} \tag{9.1c}$$

Principles of Laser Materials Processing, by Elijah Kannatey-Asibu, Jr.

where μ_{m0} and ϵ_{p0} are the magnetic permeability and electric permittivity, respectively, in free space and μ_m and ϵ_p are the magnetic permeability and electric permittivity, respectively, in the medium.

c and c_m are related by the refractive index of the medium, n, as

$$n = \frac{c}{c_m} = \sqrt{\frac{\mu_m \epsilon_p}{\mu_{m0}\epsilon_{p0}}} = c\sqrt{\mu_m \epsilon_p} \tag{9.1d}$$

From equation (9.1a), it can be shown (see Problem 9.11) that for small $\Delta\lambda$,

$$\Delta\nu = -\left(\frac{c}{\lambda^2}\right)\Delta\lambda \tag{9.2a}$$

or

$$\Delta\lambda = -\left(\frac{c}{\nu^2}\right)\Delta\nu \tag{9.2b}$$

The negative sign arises since the wavelength increases as the frequency decreases. Let us now look at some basic characteristics of electromagnetic radiation that are associated with their propagation.

9.2 REFLECTION AND REFRACTION

Even though beam propagation is often illustrated using rays, we start this section with a brief discussion of waves and how they relate to rays. When we consider the wave nature of light propagating from a source, the series of points on different radial lines from the source, which have the same amplitude at any time, is referred to as the wave front. For an optically homogeneous and isotropic medium, the wave front is perpendicular to the direction of the wave propagation, and the waves from a point source have a spherical wave front and radial rays. Very far from the source, these wave fronts may be considered planar. A ray is a line perpendicular to the wave front, and is therefore in the direction of wave propagation.

9.2.1 Reflection

When a beam of light is incident on an opaque surface, specular reflection occurs if the surface is smooth enough such that the wavelength of the incident beam is much greater than the roughness of the surface. Under such circumstances, the beam

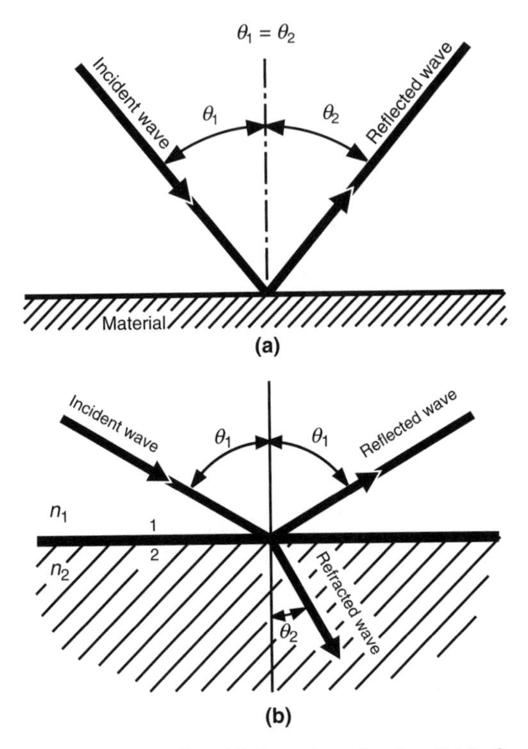

FIGURE 9.1 Reflection and refraction. (a) Specular reflection. (b) Refraction at an interface.

is reflected in such a way that the angle of incidence, θ_1, is equal to the angle of reflection, θ_2 (Fig. 9.1a), that is,

$$\theta_1 = \theta_2 \tag{9.3}$$

The incident ray, reflected ray, and the normal to the surface all lie in the same plane, the *plane of incidence.* The intensity of the reflected beam, I_r, is related to that of the incident beam, I_0, by the following relation:

$$I_r = RI_0 \tag{9.4}$$

where R is the coefficient of reflection or reflectivity of the surface.

Diffuse reflection, where the incident beam is randomly reflected in all directions, occurs when the wavelength is relatively smaller than the roughness of the surface.

9.2.2 Refraction

When light enters a medium, its velocity in the medium differs from, and is always less than its velocity in free space. This results in the phenomenon of refraction. If the material on which the light beam is incident is transparent, part of the beam is reflected and the other part is refracted, with the refracted ray also lying in the plane of incidence (Fig. 9.1b). The *index of refraction, n*, is the ratio of the light velocity in free space to its velocity in the medium, that is,

$$n = \frac{\text{velocity of light in free space}}{\text{velocity in the medium}} = \frac{c}{c_{\mathrm{m}}} \tag{9.5}$$

where c_{m} is the velocity in the medium. Typical values for n range from about 1.0003 for gases to about 1.8 for solids. n for a given material also depends on the wavelength of the light incident on it.

Let the original beam be incident from medium 1, and let it be refracted as it enters medium 2. The incident ray is refracted in such a way that the angle of refraction, θ_2, is related to the angle of incidence, θ_1 by the law of refraction or Snell's law:

$$\frac{\sin\theta_1}{\sin\theta_2} = \frac{n_2}{n_1} = n_{21} \tag{9.6a}$$

where n_1 is the refractive index of medium 1, n_2 is the refractive index of medium 2, n_{21} is the refractive index of medium 2 relative to medium 1.

If $n_1 > n_2$, in other words, if the original beam is incident from the medium with the higher refractive index, then there is a critical angle of incidence, θ_c, above which total internal reflection occurs, and there is no refraction of the incident beam. Then from Snell's law, we have

$$\frac{\sin\theta_{\mathrm{c}}}{\sin 90^o} = \frac{n_2}{n_1} = n_{21} \tag{9.6b}$$

$$\Rightarrow \sin\theta_{\mathrm{c}} = \frac{n_2}{n_1} = n_{21} \tag{9.6c}$$

There cannot be total internal reflection when the original beam is incident from the medium with the lower refractive index.

9.3 BIREFRINGENCE

Birefringence is the phenomenon whereby a light beam that is incident on a material is refracted into two rays, the ordinary ray and the extraordinary ray. Materials that exhibit this characteristic are said to be birefringent (optically anisotropic). The two rays are plane-polarized (see next section) in orthogonal directions. The ordinary ray obeys Snell's law and has the same velocity in any direction in the material. The extraordinary ray, on the contrary, has different velocities in different directions and does not follow Snell's law. Both the ordinary and extraordinary rays have the same velocity along the optic axis of the material. Since the two rays generally have different velocities, the birefringent material normally has two different refractive indices. An example of such a material is the calcite crystal.

9.4 BREWSTER ANGLE

In Fig. 2.1, the electric and magnetic vectors of the light wave are shown as lying in specific planes. Light having this characteristic where the electric vector is restricted to a specific plane is said to be *linearly* or *plane polarized.* The electric vector for a plane polarized beam is illustrated schematically in Fig. 9.2a, while that for an unpolarized beam is illustrated in Fig. 9.2b.

The electric vector associated with a beam that is incident on a surface can be resolved into two components, one lying in, or parallel to the plane of incidence, *p-component,* and the other normal to the plane of incidence, *s-component.* The intensities of the two components of the incident beam are equal if the original beam is unpolarized. However, the intensities of the components of the reflected beam are not necessarily equal and depend on the reflection coefficients for the two components of

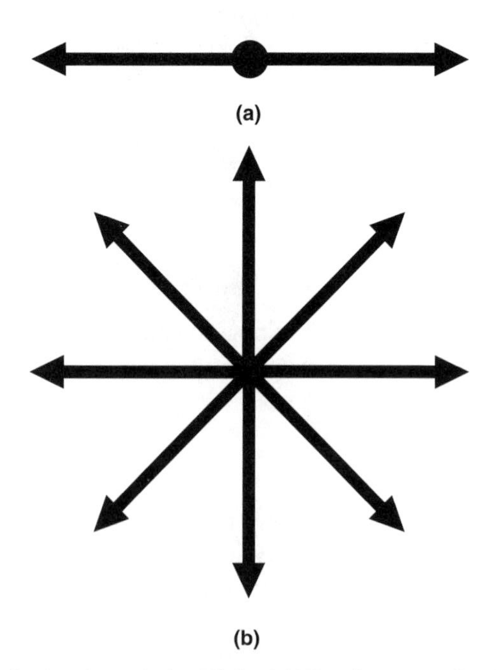

FIGURE 9.2 Polarized and unpolarized light. (a) Electric vector of a plane-polarized light. (b) Electric vector of an unpolarized light.

the reflected beam, which are different. These are given by

$$R_p = \frac{\tan^2(\theta_1 - \theta_2)}{\tan^2(\theta_1 + \theta_2)} \tag{9.7a}$$

$$R_s = \frac{\sin^2(\theta_1 - \theta_2)}{\sin^2(\theta_1 + \theta_2)} \tag{9.7b}$$

where R_p is the reflection coefficient of the reflected ray that is plane polarized in the plane of incidence, and R_s is the normal component. θ_1 and θ_2 are the respective incident and refracted angles. R_p and R_s are the intensity or power reflection coefficients and are also known as *reflectances*. The corresponding amplitude reflection coefficients are given by

$$R_{pa} = \frac{\tan(\theta_1 - \theta_2)}{\tan(\theta_1 + \theta_2)} \tag{9.8a}$$

$$R_{sa} = \frac{\sin(\theta_1 - \theta_2)}{\sin(\theta_1 + \theta_2)} \tag{9.8b}$$

Variation of the reflection coefficients with the angle of incidence is shown in Fig. 9.3a. From equation (9.7a), we observe that when $\theta_1 + \theta_2 = 90°$, the reflection coefficient, R_p, for the p-component in the plane of incidence is zero since $\tan(\theta_1 + \theta_2)$ is then infinity. The reflected beam is then plane polarized normal to the plane of incidence (Fig. 9.3b). This occurs when the reflected and refracted rays are perpendicular to each other. The angle of incidence at which this occurs is referred to as the *polarizing angle* or *Brewster angle*, θ_B. Since $\theta_1 + \theta_2 = 90°$, we have $\sin\theta_2 = \cos\theta_1$, and thus from Snell's law, we have

$$\tan\theta_B = \frac{n_2}{n_1} = n_{21} \tag{9.9}$$

When the original beam is incident from the medium with the lower refractive index, that is, medium 1, the beam undergoes a 180° (or $\lambda/2$ linear) phase shift during reflection. However, there is no phase shift if the light is incident from the higher refractive index medium.

The preceding discussion applies primarily to transparent materials such as glass. A general behavior of metals in response to an incident beam is illustrated in Fig. 9.3c. It is evident that the reflection coefficients for metals are relatively large. The reflection coefficient for the s-component (normal) increases monotonously with increasing angle of incidence. However, that of the p-component (parallel) first decreases with increasing angle of incidence until (depending on the material and wavelength of the incident beam), about 85°, when it begins to increase to 100% reflection at 90° incidence. Except for 0° incidence and 90° incidence, the reflection coefficient for the parallel component is always lower than that of the normal component.

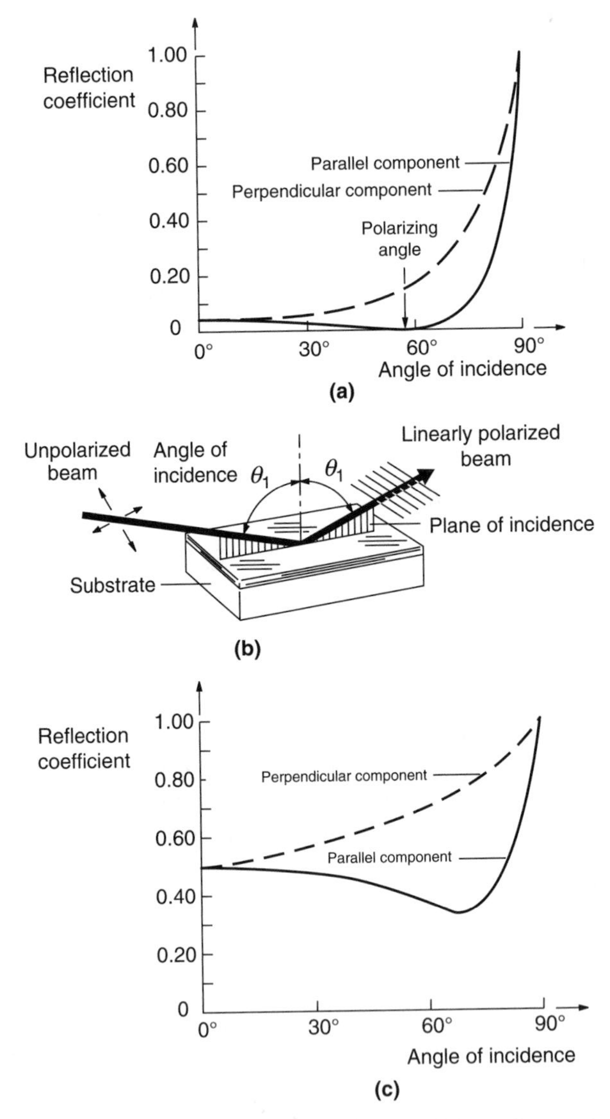

FIGURE 9.3 Reflection coefficients (a) Reflection coefficients for an air–glass interface ($n = 1.5$). (b) Reflection of an unpolarized beam at the polarizing angle. and (c) Reflection coefficients for an air–gold interface. (From Dally, J. W., and Riley, W. F., *Experimental Stress Analysis,* 1991, McGraw-Hill, Inc., New York.)

9.5 POLARIZATION

In discussing polarization, we first consider a simplified one-dimensional propagation of the electric vector of an electromagnetic plane wave

$$E_1(x, t) = E_0 \cos \frac{2\pi}{\lambda}(x - ct) = E_0 \cos(k_w x - \omega t) \tag{9.10}$$

where E_l is the magnitude of the electromagnetic wave (electric field vector), E_0 is the amplitude of the electromagnetic wave, $k_w = \frac{2\pi}{\lambda}$ is the wavenumber, and ω is the angular frequency.

This is illustrated schematically in Fig. 9.4a. If we consider measurements that are made at a fixed position along the beam, then equation (9.10) can be expressed as

$$E_l(t) = E_0 \cos(\phi - \omega t) \tag{9.11}$$

where ϕ is the phase angle of the wave at the position of interest.

Two different waves, 1 and 2, of the same frequency (Fig. 9.4b) can be represented as

$$E_{l1}(t) = E_{01} \cos(\phi_1 - \omega t) \tag{9.12}$$

and

$$E_{l2}(t) = E_{02} \cos(\phi_2 - \omega t) \tag{9.13}$$

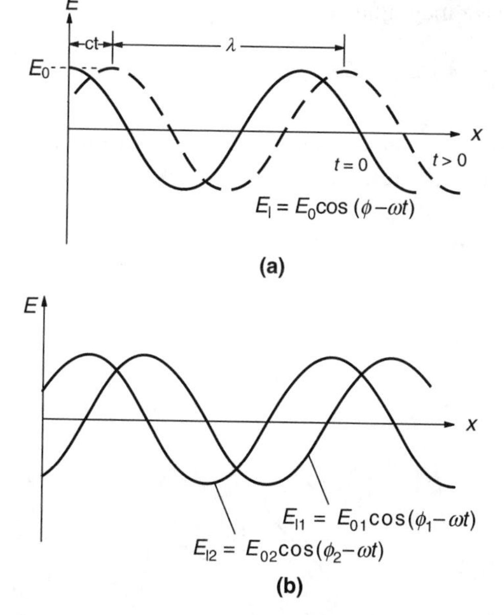

FIGURE 9.4 One-dimensional electromagnetic wave propagation. (a) A single wave at different times. (b) Two different waves of the same frequency and a phase difference. (From Dally, J. W., and Riley, W. F., *Experimental Stress Analysis,* 1991, McGraw-Hill, Inc., New York.)

If these two waves are plane polarized in the same plane and are superimposed on each other, then the resulting light vector $E_l(t)$ is given by

$$E_l(t) = E_{l1}(t) + E_{l2}(t) \tag{9.14}$$

which can be shown to reduce to the form

$$E_l(t) = E_0 \cos(\phi - \omega t) \tag{9.15}$$

where

$$E_0{}^2 = E_{01}{}^2 + E_{02}{}^2 + 2E_{01}E_{02}\cos(\phi_2 - \phi_1) \tag{9.16}$$

and

$$\tan\phi = \frac{E_{01}\sin\phi_1 + E_{02}\sin\phi_2}{E_{01}\cos\phi_1 + E_{02}\cos\phi_2} \tag{9.17}$$

Thus, the frequency of the resulting light vector is the same as that of the original light vectors. However, the amplitudes and phase angles are different.

Now let us consider the situation where the two plane waves $E_{ly}(t)$ and $E_{lz}(t)$ lie in different planes which are orthogonal to each other (Fig. 9.5). The resulting light vector will then have a magnitude of

$$E_l(t)^2 = E_{ly}(t)^2 + E_{lz}(t)^2 \tag{9.18}$$

If we go back to the original equation for the wave motion, we can show that at any moment in time, the resultant electric vector along the x-axis or direction of propagation describes an elliptical helix (Fig. 9.6a). If we now look at a projection of this resultant vector on a plane perpendicular to the x-axis, the equation describing

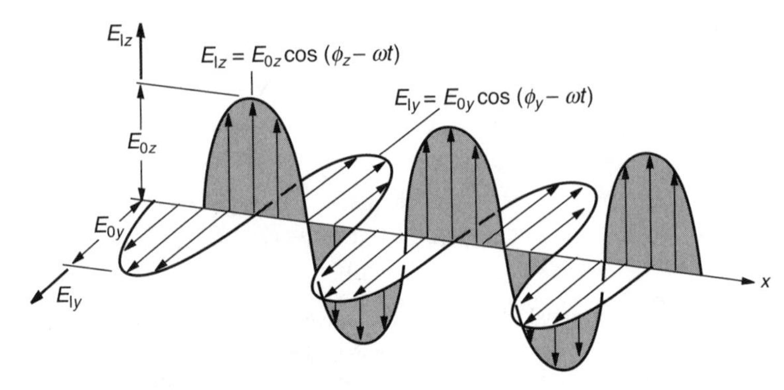

FIGURE 9.5 Plane-polarized electromagnetic waves in two orthogonal planes.

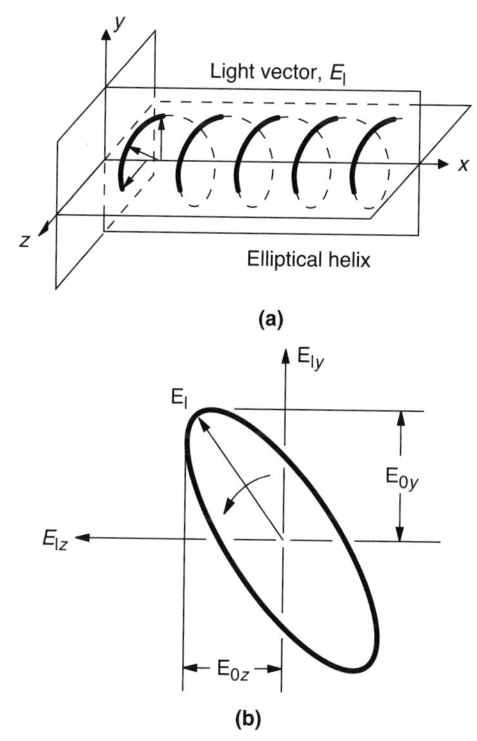

FIGURE 9.6 (a) Instantaneous configuration of the resultant electric vector of two plane polarized electromagnetic waves in two orthogonal planes. (b) Normal projection of the resultant electric vector of two plane polarized electromagnetic waves in two orthogonal planes. (From Dally, J. W., and Riley, W. F., *Experimental Stress Analysis,* 1991, McGraw-Hill, Inc., New York.)

the outermost trace of the projected image can be obtained by eliminating time from equations (9.12) and (9.13), giving

$$\left[\frac{E_{1y}(t)}{E_{0y}}\right]^2 - 2\frac{E_{1y}(t)E_{1z}(t)}{E_{0y}E_{0z}}\cos(\phi_z - \phi_y) + \left[\frac{E_{1z}(t)}{E_{0z}}\right]^2 = \sin^2(\phi_z - \phi_y) \qquad (9.19)$$

which is the equation of an ellipse (Fig. 9.6b). Thus the electromagnetic radiation obtained by superimposing two beams that are plane polarized in two orthogonal planes is said to be *elliptically polarized.* The projected electric vector rotates clockwise in the projected plane as time varies, and the resulting helix is a right circular helix if the phase difference is such that

$$\phi_z - \phi_y = \frac{2m-1}{2}\pi \quad m = 0, 1, 2, 3, \cdots \qquad (9.20)$$

and rotates in the counterclockwise direction, producing a left circular helix if

$$\phi_z - \phi_y = \frac{2m+1}{2}\pi \quad m = 0, 1, 2, 3, \ldots \tag{9.21}$$

If $E_{0y} = E_{0z} = E_0$, that is, the amplitudes of the two light vectors are equal, and the phase difference between them is 90° or a quarter of a wavelength, then equation (9.19) becomes

$$E_{ly}(t)^2 + E_{lz}(t)^2 = E_0{}^2 \tag{9.22}$$

This is the equation of a circle. The resulting light is then said to be *circularly polarized.*

Plane polarized light may be obtained by using a special optical element (plane polarizer) such as a Polaroid sheet, or by positioning a reflector such that the light beam is incident on it at the Brewster angle. Furthermore, if we compare equations (9.7a) and (9.7b) for R_p and R_s, we find that R_s is always greater than R_p. Thus repeated reflections from mirrors will eventually result in light that is plane polarized normal to the plane of incidence.

A lightwave that is incident on a transmissive plane polarizer is resolved into two components, one normal to the axis of polarization, and the other parallel to the axis of polarization (Fig. 9.7). The parallel component is transmitted, while the normal component may be absorbed. If the original light wave $E_l(t)$ is plane polarized and oriented at an angle θ to the polarization axis, then the transmitted, $E_t(t)$, and absorbed, $E_a(t)$, light vectors will be given by

$$E_t(t) = E_l(t)\cos\theta = E_0 \cos\omega t \cos\theta \tag{9.23}$$

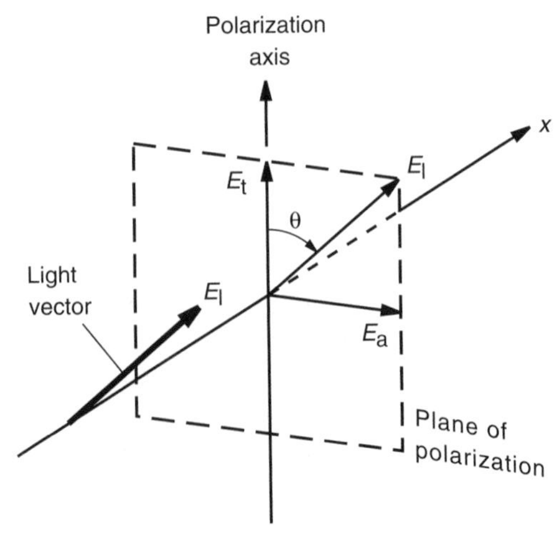

FIGURE 9.7 Transmission of a beam by a plane polarizer.

and

$$E_a(t) = E_l(t) \sin\theta = E_0 \cos\omega t \sin\theta \tag{9.24}$$

where we have neglected the initial phase ϕ of the light wave for simplicity. $E_t(t)$ will be along the axis of polarization (Fig. 9.7).

Consider a birefringent material that is formed into the shape of a cuboid or rectangular block such that the optic axis of the crystal is normal to two of the sides or faces of the block. Now any light that enters any of the six crystal faces in a direction perpendicular to that face is not refracted, and thus continues along the same line or direction through the crystal. If the light is along the optic axis, then the ordinary and extraordinary rays emerge in phase since they then propagate with the same velocity. However, for the other crystal faces where the normal to the crystal face is also normal to the optic axis, since the ordinary and extraordinary rays have different velocities, they fall out of phase. Further, since the two rays are plane polarized in orthogonal directions, the resulting beam that comes out of the crystal is elliptically polarized. Such a crystal or optical element that changes the polarization of a light beam is called a *wave plate* or *phase retarder.*

For a plate of birefringent material, since the velocity of the extraordinary ray will be different along different directions in this plate, it will be a maximum along a certain direction. This is called the *fast axis* of the plate. The direction normal to this, the *slow axis,* will have the minimum velocity. A plane polarized lightwave incident on the plate will have the transmitted portion resolved into two components along the fast and slow axes (Fig. 9.8). If the fast axis makes an angle α_f with the plane of polarization, then the two components will be

$$E_{tf}(t) = E_t(t) \cos\alpha_f = E_0 \cos\omega t \cos\theta \cos\alpha_f \tag{9.25}$$

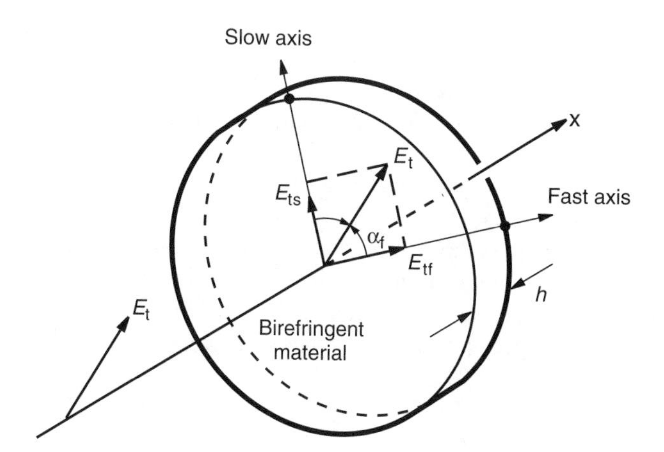

FIGURE 9.8 Plane-polarized light vector propagating into a birefringent material.

and

$$E_{ts}(t) = E_t(t) \sin \alpha_f = E_0 \cos \omega t \cos \theta \sin \alpha_f \tag{9.26}$$

where the subscript t refers to the transmitted ray, f the fast axis, and s the slow axis. Since the two beam components have different velocities, they emerge with a phase difference ϕ, between them, and this can be shown to be given by

$$\phi = \frac{2\pi h(n_s - n_f)}{\lambda} \tag{9.27}$$

where n_f and n_s are the refractive indices along the fast and slow axes, respectively, and h is the plate thickness. The two components coming out of the plate can then be expressed as

$$E_{tf}(t) = E_{0\theta} \cos \alpha_f \cos \omega t \tag{9.28}$$

and

$$E_{ts}(t) = E_{0\theta} \sin \alpha_f \cos(\omega t - \phi) \tag{9.29}$$

where $E_{0\theta} = E_0 \cos \theta$. The polarization of the beam that comes out of the plate depends on the phase shift ϕ and the angle α_f.

If the phase shift ϕ is set to be 90°, then the plate is said to be a quarter wave plate. For this situation, if α_f is set to be 45°, then the emerging beam is circularly polarized. If, on the contrary, α_f is set to any other angle than $m\pi/4$, then an elliptically polarized beam is obtained.

9.6 MIRRORS AND LENSES

For completeness, we briefly review in this section, the basic concepts of image formation by lenses and mirrors. The formation of an image by a spherical mirror is illustrated, for a concave mirror as in Fig. 9.9. The object is located at O, a distance u_o from the mirror vertex, while the image is located at I, a distance v_i from the vertex. Paraxial rays, that is, those that are parallel and close to the mirror axis are reflected in such a way that they pass through the focal point, F, of the mirror. F is located at a distance f_l, the focal length, from the mirror vertex, and is halfway between the center of curvature, C, and the vertex. Thus the focal length is half the mirror radius. For paraxial rays, the object distance, u_o, image distance, v_i, and focal length, f_l, are related by

$$\frac{1}{u_o} + \frac{1}{v_i} = \frac{1}{f_l} \tag{9.30}$$

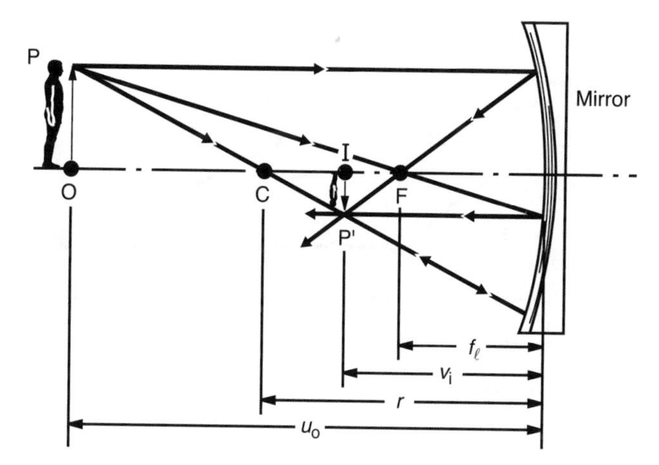

FIGURE 9.9 Spherical mirror image formation illustrating the three most common rays used in locating the image.

This is valid for concave, convex, and plane mirrors, provided the appropriate sign convention is used. The object and image distances are positive for real objects and images, and negative for a virtual image. A real image results if the reflected rays pass through the image points, and a virtual image results if the rays are reflected such that they appear to be coming from an image. That would be the case for a convex mirror. The magnification, m_g, which is the ratio of the image to object size, can be shown to be given by

$$m_g = -\frac{v_i}{u_o} \tag{9.31}$$

The negative sign results since the image is inverted.

For thin lenses, where the lens thickness is considered negligible relative to the other distances, the image formation process is illustrated in Fig. 9.10. Equations (9.30) and (9.31) also apply to the relationship between magnification, object distance, image distance, and focal length. For a lens system, the image is formed by the refracted light, as opposed to the reflected light for a mirror. Refraction is assumed to occur at the principal plane, which is essentially the plane of symmetry for a symmetrical lens. Again, the sign convention is such that distances are positive for real images and negative for virtual images, and the same holds for both concave and convex lenses.

Common lens types used for laser processing are the meniscus and plano-convex lenses. The meniscus lens results in a smaller focal spot size, but is more expensive than the plano-convex lens. For normal applications, a thin lens may be used, but for high-pressure applications, say, 10–25 bar as may occur when nitrogen is used as the assist gas for cutting, then a thick or high-pressure lens is required. Protective windows may be used to cover and protect the lens in hazardous environments when

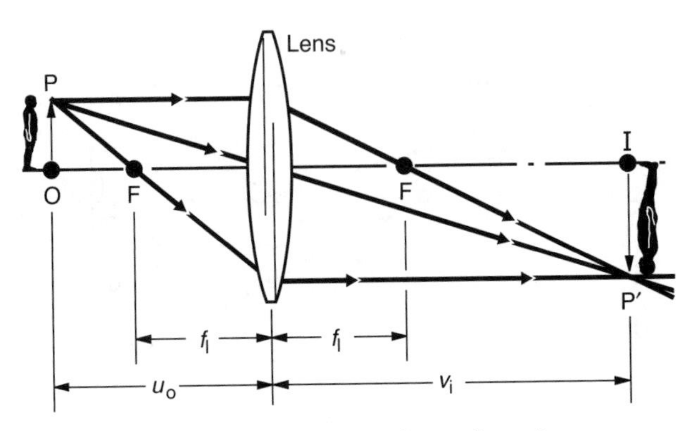

FIGURE 9.10 Thin lens image formation.

say, extensive spatter is produced. However, care must be taken in using such components since they are normally made from potassium chloride or sodium chloride, which are hygroscopic and may absorb moisture from the atmosphere. When this happens, the absorption of the beam energy increases, reducing the power available at the workpiece. Even worse, the increased absorption may damage the window and consequently, the lens.

As mentioned earlier, the lens and mirror equations are valid for paraxial systems or situations where the object rays make a small angle with the axis of the system. Image distortion may result if these conditions do not hold. A common defect source is what is referred to as spherical aberration, which arises when the parallel rays are not confined to the immediate neighborhood of the axis. In such a situation, rays farther away from the axis are focused to a point closer to the vertex than those closer to the axis. Thus the parallel rays are not focused at a single point. Astigmatism results when the rays from the object make a relatively wide angle with the lens or mirror. Chromatic aberration is due to the variation of refractive index with wavelength of light, resulting in different wavelengths in a given beam being focused at different points. Other forms of defect associated with image formation include coma and curvature of field.

9.7 BEAM EXPANDERS

Beam expanders are normally used for changing the size of the output beam. From equation (7.20a), it is evident that the size of the focused beam is inversely proportional to the original beam size:

$$w_f = \frac{1}{\pi} \frac{f_l \lambda}{w_0} \tag{7.20a}$$

Thus a more intensely focused beam can be obtained from a laser beam of the same power if its diameter as it enters the focusing lens is increased. This could be achieved using a beam expander, which essentially consists of a series of mirrors (Fig. 9.11a), or lenses (Fig. 9.11b). The transmissive or lens beam expanders are mounted in such a way that they have the same focal point (Fig. 9.11b). The size of the expanded beam, w_2, is determined by the product of the original beam size, w_1, and the ratio of the focal length of the two lenses:

$$w_2 = w_1 \frac{f_{l2}}{f_{l1}} \tag{9.32}$$

where w_1 is the size of the incoming beam, w_2 is the size of the expanded beam, f_{l1} is the focal length of the lens on which the original beam is incident, f_{l2} is the focal length of the output lens.

Increasing the beam size also has the advantage of reducing the diffraction spreading associated with it (see Section 7.1).

For the lens system, another configuration is based on a nonfocusing arrangement. This type has the advantage of being more compact and suitable for high-power applications where the focused beam could present problems. The focusing arrangement, on the contrary, enables the original beam to be focused at a pinhole where higher-order diffraction fringes are filtered or blocked out. This results in a beam whose intensity distribution is more uniform because of the noise reduction.

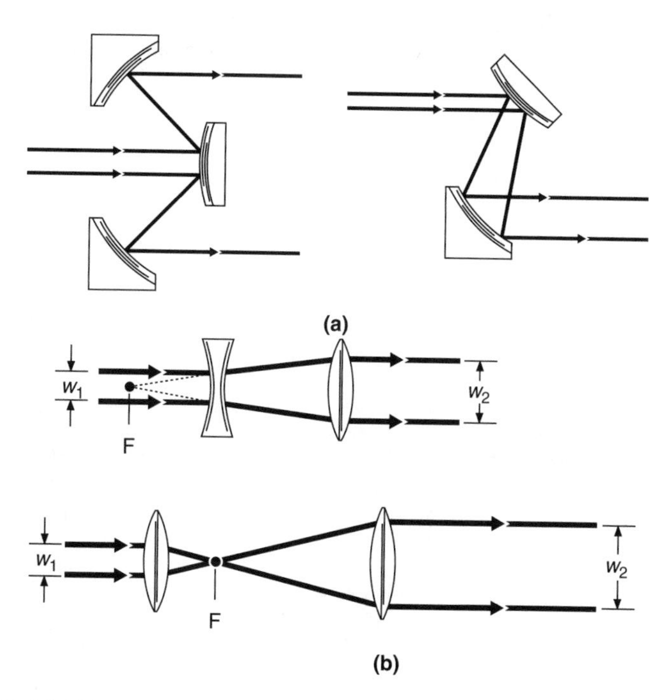

FIGURE 9.11 Examples of beam expanders (a) mirrors and (b) lenses.

9.8 BEAM SPLITTERS

Beam splitters are optical elements that are used to divide an incoming beam into two, and through cascading, into any number of parts. One simple method is based on reflection, using a wedge-shaped reflector, Fig. 9.12, with the edge pointing directly into the laser beam path to reflect the incoming beam into two parts. For the split beams to come out parallel, the wedge angle has to be 90°. The power ratio of the split beams is determined by the position of the wedge edge relative to the axis of the incoming beam. The characteristics of the split beams depend on the stability and spatial mode (distribution) of the original beam. The quality of the beams also depends on the dimensional accuracy of the wedge and smoothness of the reflector, and a clean environment is essential for high beam quality.

Another technique for splitting a laser beam makes use of a partially transmissive and partially reflective optical element (Fig. 9.13). When the laser beam is incident on such a material, part of the beam is reflected, while the remaining portion is transmitted, separating the beam into two parts. Such an optical element is normally obtained by coating a transparent material such as a ZnSe block with a reflective material such as inconel. The amount of coating determines the ratio of transmitted to reflected beam power. The latter can thus be varied for a given application by using an element with a graded or varied coating density on the surface and changing the location at which the beam is incident on the surface of the optical element.

Other techniques for splitting laser beams are based on the principles of the Fabry–Perot interferometer or a combined linear polarizer and electro-optic cell.

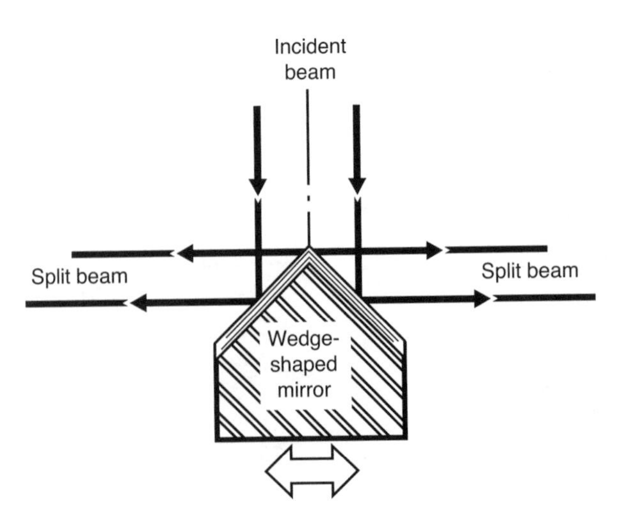

FIGURE 9.12 Wedge-shaped beam splitter. (From Chryssolouris, G., *Laser Machining: Theory and Practice*, 1991. By permission of Springer Science and Business Media.)

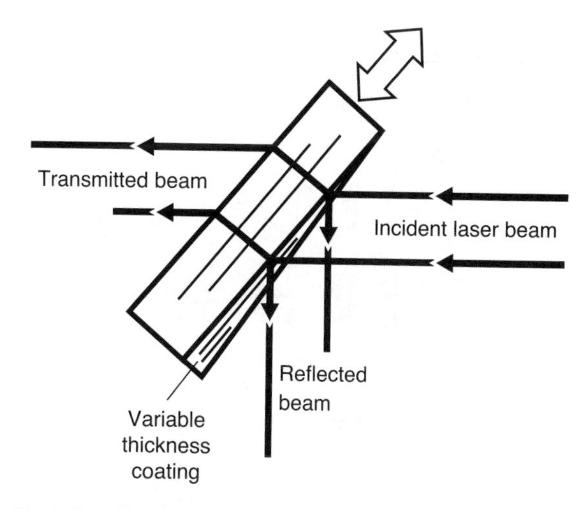

FIGURE 9.13 Partially reflective, partially transmissive beam splitter. (From Chryssolouris, G., *Laser Machining: Theory and Practice*, 1991. By permission of Springer Science and Business Media.)

9.9 BEAM DELIVERY SYSTEMS

A beam delivery system is essentially a means for directing the laser beam from the generator to the point of application. The two principal types are

1. Conventional beam delivery.
2. Fiber optic beam delivery.

9.9.1 Conventional Systems

Conventional beam delivery systems are conceptually simple, as shown schematically in Fig. 9.14a, typically consisting of three main components:

1. The beam bending assembly, which is usually reflective, that is, consists of mirrors.
2. The focusing assembly, which may be transmissive or reflective.
3. Interconnecting beam guard tubes.

The beam bender assembly usually contains mirrors oriented at 45°, which reflect the unfocused beam at right angles to the original direction. Common materials used for the mirrors include copper or molybdenum that may be bare or coated with silicon. The advantage of bare molybdenum is that it is durable and easily cleaned, making it suitable for industrial applications. However, it absorbs about 1–4% of the incident light. The coated version, on the contrary, absorbs only about 0.1% of the light, but cannot be restored by cleaning.

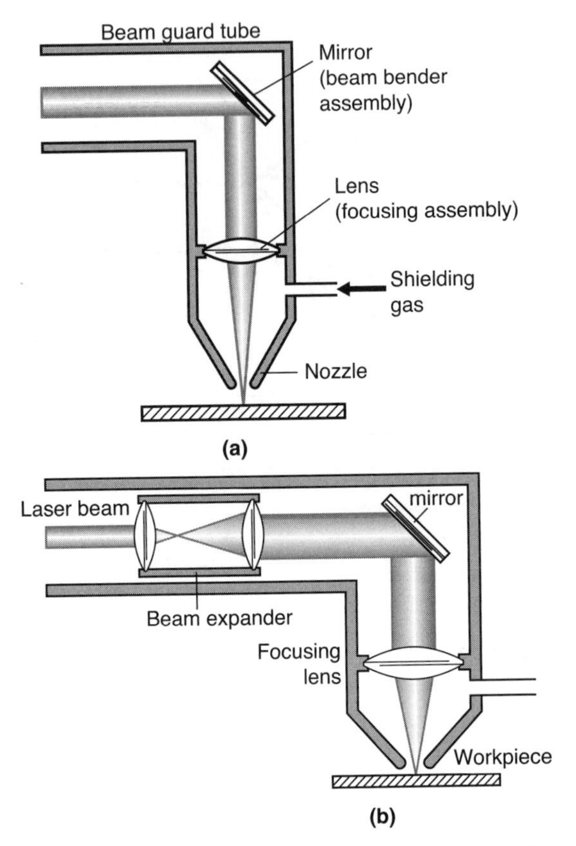

FIGURE 9.14 (a) Simplified laser beam delivery system. (b) Beam delivery system with an expander.

The lens is contained in the focusing assembly, which is normally made adjustable so that the distance between the focal point and nozzle tip can be varied to suit the application. It also has an inlet for shielding or cutting assist gas flow. For high-power applications, both the beam bender and focusing assembly have built-in circulating systems for air or water-cooling to prevent overheating of the optics, which may cause thermal lensing or distortion of the laser beam, and thereby change the focusing characteristics of the lens. Common lens materials include ZnSe, KCl, GaAs, and CdTe. ZnSe is more commonly used with CO_2 lasers.

A number of beam benders can be used depending on the application. The workstation is sometimes placed a distance of about 6 m or more from the output coupler to permit the beam to be fully developed in the far-field mode, and also to minimize the impact of back reflection into the laser generator. Immediately behind the output coupler (aperture), the resulting wave front is not planar. This is due to diffraction effects that result from Huygen's principle. A planar wave front is fully developed after a distance of about five Rayleigh ranges (see Sections 2.3 and 7.1).

Where necessary, a beam expander may be inserted in the beam path to provide the desired beam size (Fig. 9.14b). For high-power lasers in the infrared range, it is normal practice to provide beam guards made from aluminum tubes for the beam along its path from the generator to the bender for safety since the beam is invisible. The beam guard also serves to keep dust and other contaminants such as oil and mist off the delivery system. Internal cleanliness is assured by flowing clean dry air or nitrogen through the optical path in a direction away from the mirrors and lenses, at a pressure of about 30 *psi*.

Fiber optic systems provide a very flexible means of beam delivery. However, currently available systems are heavily absorbed by the longer wavelength CO_2 lasers, and are thus suitable primarily for the shorter wavelength lasers, for example, the Nd:YAG laser. The basic principles of fiber optic systems will be discussed next.

9.9.2 Fiber Optic Systems

Optical fiber systems are normally used in transmitting laser beams or electromagnetic energy from one location to another. This application is quite common in optical communications where the power levels are generally low. For materials processing, where high-power industrial lasers are typically used, usage of optical fibers for transmitting the laser beam is limited to the shorter wavelength lasers. The longer wavelength CO_2 laser results in heating of the cable. Optical fibers are thus not used for high-power CO_2 laser beam transmission. The use of fiber optic transmission makes beam delivery more flexible, compared to the conventional mirrors and lenses.

The basic structure of an optical fiber system is shown schematically in Fig. 9.15. The principal components include:

Transmitter—It converts an electrical signal into an optical signal. This includes a light source.

Transmission Line—It conducts the optical signal, and this comprises the optical fibers.

Receiver—It converts the optical signal back to the original electrical form. The receiver normally consists of photodetectors and signal amplifiers.

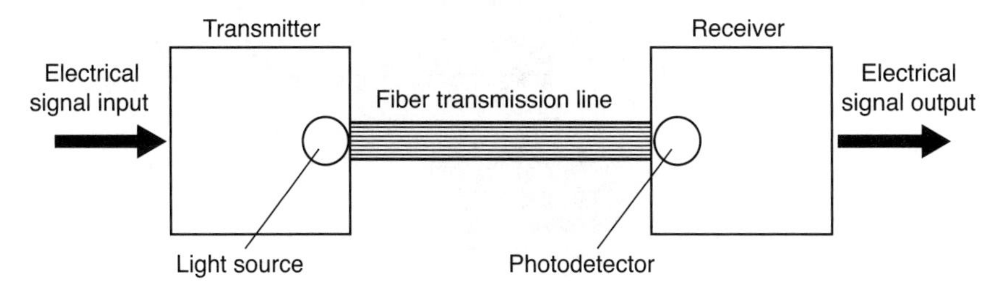

FIGURE 9.15 Schematic of a fiber optic system.

Our focus in this section will be on the transmission line since fiber optic cables are used in laser processing primarily to transmit a beam from the source to the delivery point without any electrical conversions. First, we look at the characteristics of optical fibers in general.

9.9.2.1 Optical Fiber Characteristics Optical fibers can be characterized on the basis of either their physical or optical properties. On the physical side, optical fibers consist of glassy fibers of diameter about 125 μm each, with lengths up to the order of kilometers. Inorganic oxide glasses of high silica content are often used, and the resulting fibers are very flexible. For the optical characteristics, the major limitations of optical fibers are the attenuation and bandwidth dispersion. Attenuation results in dissipation of optical energy during transmission and is caused mainly by absorption and scattering. The bandwidth determines the limits of frequencies that can be transmitted.

Optical fibers are normally operated in the temperature range from −55 to 125°C, but may extend to a range from −250 to 500°C. Due to the dielectric nature of glass and the small size of fibers (i.e., much smaller than the wavelength of radio and microwaves), they are immune to interferences from such signals. Thus fibers can pass through certain regions of high electromagnetic fields without interference. Also, waveguide covers prevent interference from other optical signals.

9.9.2.2 Waveguide Structure The optical fiber waveguide is the thread-like medium through which the light propagates. It is generally circular in section and has a long cylindrical structure. The basic structure has two components (Fig. 9.16):

Core—This is the inner portion through which light is transmitted and is made of transparent glass (or plastic) of relatively high refractive index. Core diameters

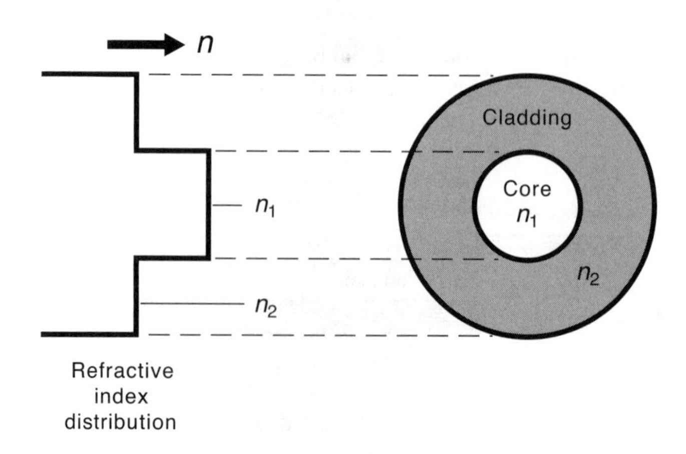

FIGURE 9.16 Basic structure of a fiber optic cable.

typically vary between 4 and 12 μm for single-mode fibers (see Section 9.9.2.4.2), and between 50 and 200 μm for multimode fibers (see Section 9.9.2.4.1).

Cladding—This is the concentric outer layer with a relatively lower index of refraction. It reduces scattering loss from discontinuities at the core surface; adds mechanical strength to the fiber; and protects the core from absorbing surface contaminants. The cladding may also be a glass or plastic material.

The difference in refractive indices of the core (n_1) and cladding (n_2) is often expressed in the normalized form, Δ, as

$$\Delta = \frac{n_1 - n_2}{n_1} \tag{9.33}$$

Typical values for Δ would be 0.2% for single-mode and 1% for multimode fibers. Since the core has a higher index than the cladding, total internal reflection occurs, enabling the waves to stay within the fiber. The optical fiber waveguide is itself contained in another coaxial layer that supports the fiber structure.

9.9.2.3 Background Consider the simple system shown in Fig. 9.17, with a core of radius, r_c. Let the light incident on the fiber end make an angle θ with the waveguide axis. If θ is less than θ_a, the *angle of acceptance,* total internal reflection occurs because the refracted light is then incident at the core-cladding interface at an angle greater than θ_c, the critical angle. The light then stays within the core. For angles greater than θ_a, no internal reflection will occur and the light will propagate into the cladding. The incident rays that propagate within the core are more commonly defined using the *numerical aperture, NA.* It is a measure of the light gathering or collecting power of an optical fiber and is defined as

$$\text{NA} = \sin\theta_a \tag{9.34}$$

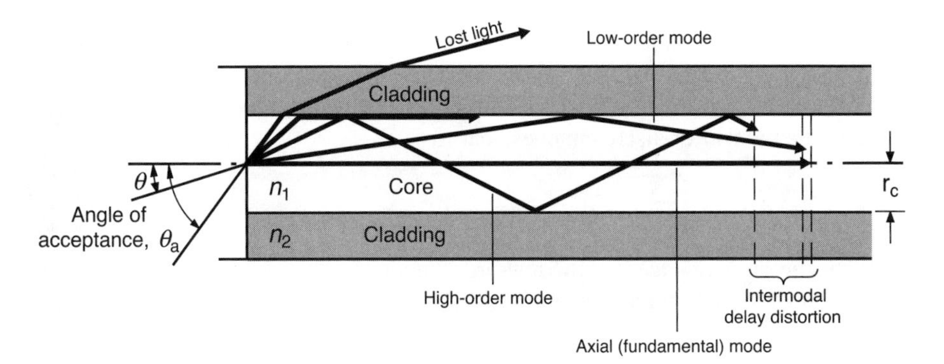

FIGURE 9.17 Propagation of a light ray in an optical fiber.

For step-index fibers, this is given by

$$\mathrm{NA} = \sqrt{n_1{}^2 - n_2{}^2}$$

The amount of light that can enter and stay within the fiber increases as the numerical aperture increases. However, the bandwidth that can be transmitted decreases as the NA increases.

Only specific quantized wave modes are propagated, and the mode volume parameter or normalized frequency, ν_n, of the modes guided by the fiber is defined as

$$\nu_\mathrm{n} = k_\mathrm{w} r_\mathrm{c} \sqrt{n_1{}^2 - n_2{}^2} = k_\mathrm{w} r_\mathrm{c} \mathrm{NA} \tag{9.35}$$

where $k_\mathrm{w} = 2\pi/\lambda$ is the wavenumber and r_c is the core radius.

The total number of modes, M_c, that can exist in a step-index fiber is approximated by

$$M_\mathrm{c} = \frac{4\nu_\mathrm{n}{}^2}{\pi^2} \approx \frac{\nu_\mathrm{n}{}^2}{2} \tag{9.36}$$

9.9.2.4 Fiber Types Three common types of fibers normally used are

(1) Step-index multimode fibers.
(2) Single-mode fibers.
(3) Graded-index multimode fibers.

Step-Index Multimode Fibers A step-index fiber is one with a uniform core refractive index, n_1, which abruptly changes at the core/cladding interface to the refractive index of the cladding, n_2 (Fig. 9.16).

$$n_1 > n_2 \tag{9.37}$$

The core may be silica, with a refractive index of about 1.46. The step-index multimode fiber usually has a core cross-sectional area that is large enough for transmitting a significant amount of energy. Thus the normalized frequency is high enough to support a number of discrete modes, that is,

$$\nu_\mathrm{n} \geq 2.405$$

It has the disadvantage that the low-order modes (with angle of incidence at the fiber end θ very small) travel a shorter ray path through the guide than do higher order modes, that is, those with θ close to θ_a, Fig. 9.17. This causes *intermodal dispersion* (see Section 9.9.2.5) where rays starting at the same time become out of phase. It restricts the fiber bandwidth or the frequency range over which it is useful. For this and other reasons, the step-index multimode fiber is only suitable for short distance applications.

Single-Mode Fibers Single-mode fibers are another form of step-index fibers. However, the core is much smaller. Also, the difference in refractive index between the core and cladding is much smaller. As the normalized frequency, ν_n, is reduced, so does the number of propagated modes. When $\nu_n < 2.405$, all the higher order modes are cutoff, and only the fundamental mode propagates. This reduction in ν_n is achieved either by a reduction in core diameter or a reduction in the index difference (equation 9.35). All modes that are incident at an angle are therefore cutoff and propagation occurs only along the fiber axis. This eliminates intermodal dispersion. A much greater bandwidth is thus obtained. However, due to the minute core size, installation is a lot more difficult, especially with respect to alignment during joining. To further minimize dispersion, the single-mode fiber requires a laser beam source. A single-mode fiber is illustrated schematically in Fig. 9.18.

Graded-Index Fibers In graded-index fibers, the refractive index of the core varies (is graded) from a maximum, n_1, at the center of the core to the value n_2, of the cladding (Fig. 9.19), according to the relationship

$$n_r = \begin{cases} n_1\sqrt{\left[1 - 2\Delta\left(\frac{r}{r_c}\right)^{k_1}\right]} & 0 \le r \le r_c \\ n_1\sqrt{(1 - 2\Delta)} & r > r_c \end{cases} \tag{9.38}$$

where k_1 is a constant, n_r is the core refractive index at radius r.

$k_1 = \infty$ for step-index fibers. For $k_1 = 2$, light from a point source experiences periodic focusing in the fiber (Fig. 9.20). For modes with larger entrance angles θ,

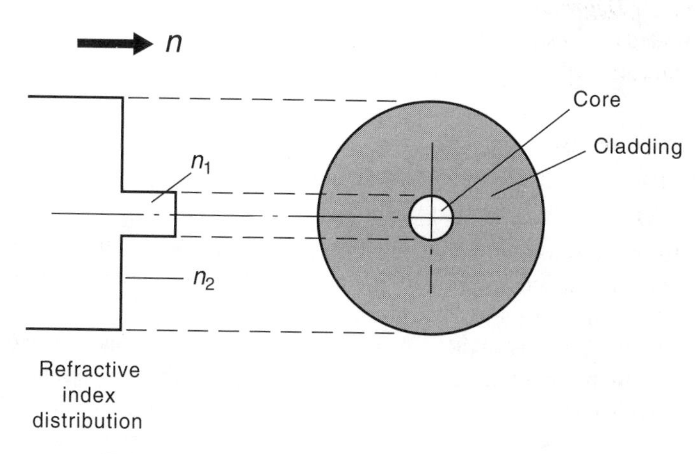

FIGURE 9.18 Cross section of a single-mode fiber.

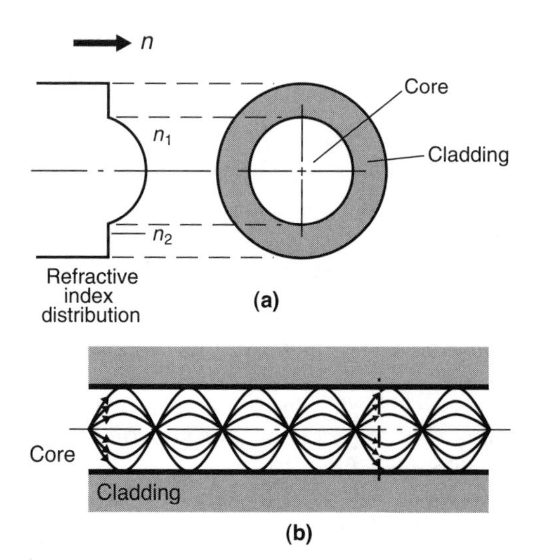

FIGURE 9.19 (a) Cross section of a multimode graded-index fiber. (b) Periodic focusing in a multimode graded-index fiber. (From Cherin, A. H., *An Introduction to Optical Fibers,* 1983. With permission of The McGraw-Hill Companies.)

the travel path is longer. However, they also travel in a region where the refractive index is lower, and thus the wave velocity is greater. Thus, they reach a given point about the same time as the lower order modes that travel mainly in the slower regions. This reduces the intermodal dispersion. Thus the bandwidth increases accordingly. The graded fiber index is thus good for large bandwidth and medium distance applications.

9.9.2.5 Beam Degradation As it propagates through a fiber, the light beam becomes degraded as a result of several factors that affect the beam characteristics. The two principal forms of degradation are

1. Dispersion or distortion.
2. Optical losses.

Before further discussing the degradation phenomena, we outline other ways in which optical fibers are characterized. The characteristics of an optical fiber can also be described in terms of its bandwidth or its impulse response in the time domain (rms pulse duration). Either of these characterizations can be obtained by exciting the fiber with an impulse input and observing the output pulse or its spectrum. The fiber characteristics are generally specified as either a bandwidth-length product, with units of megahertz kilometers, or a pulse dispersion denoted in nanoseconds per meter. We now take a look at the different forms of degradation.

Dispersion Dispersion or distortion of the light beam results from various causes. The principal contributing mechanisms are

1. Chromatic (material) dispersion.
2. Waveguide dispersion.
3. Modal dispersion.

To understand the phenomenon of dispersion, let us consider the concept of group velocity or group delay associated with the energy that is transmitted through the fiber. From equation (9.10), we have the following one-dimensional equation for the electric vector of a monochromatic electromagnetic plane wave propagating in a dielectric medium:

$$E_1(x, t) = E_0 \cos(k_w x - \omega t) \tag{9.39}$$

By considering the phase of the wave to be constant, that is, $k_w x - \omega t = \text{constant}$, we obtain the phase velocity (also known as wave velocity), u_w, as

$$u_w = \frac{dx}{dt} = \frac{\omega}{k_w} \tag{9.40}$$

The phase velocity is also given, equation (9.1d), by

$$u_w = \frac{c}{n} \tag{9.41}$$

where n is the refractive index of the medium and c is light velocity in free space. Thus,

$$u_w = \frac{c}{n} = \frac{\omega}{k_w} \tag{9.42}$$

For a nonmonochromatic source, the phase velocity in a dispersive medium may be determined using the center wavelength of the source. However, it is customary to use the group velocity, u_g, which in the limit, is defined as

$$u_g = \frac{dx}{dt} = \frac{d\omega}{dk_w} \tag{9.43}$$

The group velocity is the velocity of the wave envelope and is the velocity at which information that is modulated on the wave will propagate. Generally, the group and phase velocities of a wave in a dispersive medium will be different. The inverse of the group velocity is referred to as the group delay per unit length, χ_g:

$$\chi_g = \frac{1}{u_g} = \frac{dk_w}{d\omega} \tag{9.44}$$

Now from equation (9.42), this can also be expressed as

$$\chi_g = \frac{dk_w}{d\omega} = \frac{1}{c}\frac{d}{d\omega}(n\omega) \tag{9.45}$$

$$= \frac{1}{c}\left(n + \omega\frac{dn}{d\omega}\right) \tag{9.46}$$

The expression in brackets on the right hand side of equation (9.46) is referred to as the group index, n_g:

$$n_g = n + \omega\frac{dn}{d\omega} \tag{9.47a}$$

Equation (9.47a) can also be expressed (see Problem 9.6) as

$$n_g = n - \lambda\frac{dn}{d\lambda} \tag{9.47}$$

The group velocity and group delay can now be expressed as

$$u_g = \frac{c}{n_g} \tag{9.48}$$

and

$$\chi_g = \frac{n_g}{c} \tag{9.49}$$

CHROMATIC DISPERSION Chromatic dispersion is also sometimes referred to as material dispersion and results from the nonlinear variation of wavelength with the refractive index of a material, that is, change in n_1 and n_2 with λ. A dielectric medium whose refractive index is a function of wavelength is said to be a dispersive dielectric medium. As discussed in Chapters 5 and 7, even laser beams are not exactly monochromatic and generally cover a range of wavelengths. Thus, if the refractive index–wavelength relationship is not linear over that range, the different wavelengths will propagate with different velocities within the fiber, resulting in distortion of the optical signal.

Now consider a nonmonochromatic source with a wavelength band $\Delta\lambda$ that spreads between λ_1 and λ_2 and is centered at λ_0. If it propagates energy over a single-mode fiber of length L_f, the difference in arrival times of energies propagated by the wavelengths λ_1 and λ_2, the delay distortion or pulse broadening, $\Delta\tau_{cd}$, can be expressed as

$$\Delta\tau_{cd} = L_f[\chi_g(\lambda_1) - \chi_g(\lambda_2)] = \frac{L_f}{c}n_g(\lambda_1) - \frac{L_f}{c}n_g(\lambda_2) \tag{9.50a}$$

$$= -\frac{L_f}{c}\frac{dn_g}{d\lambda}\Delta\lambda \tag{9.50}$$

But from equation (9.47),

$$\frac{dn_g}{d\lambda} = \frac{dn}{d\lambda} - \frac{dn}{d\lambda} - \lambda\frac{d^2n}{d\lambda^2} \tag{9.51a}$$

$$= -\lambda\frac{d^2n}{d\lambda^2} \tag{9.51}$$

Substituting equation (9.51) into (9.50) gives the pulse broadening due to chromatic dispersion in a single-mode fiber as

$$\Delta\tau_{cd} = \frac{L_f}{c}\lambda\frac{d^2n}{d\lambda^2}\Delta\lambda \tag{9.52a}$$

or

$$\Delta\tau_{cd} = L_f\left(\frac{\lambda}{c}\frac{d^2n}{d\lambda^2}\right)\Delta\lambda$$

$$= L_f D_d \Delta\lambda \tag{9.52b}$$

where D_d is the dispersion coefficient, with units ps/(km nm). This is shown schematically in Fig. 9.20.

Example 9.1 Consider a beam of wavelength 1.0 μm propagating through a fiber of length 5 km, and whose dispersion coefficient is as shown in Fig. 9.20. If the spectral

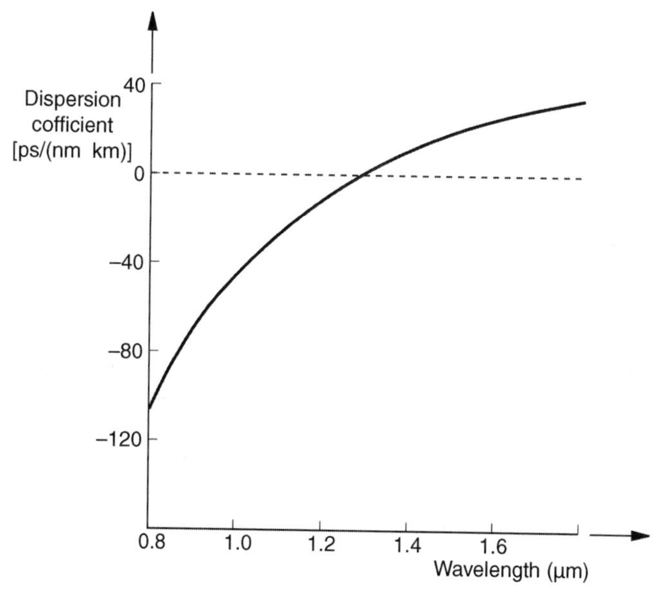

FIGURE 9.20 Dispersion coefficient as a function of wavelength. (From Keiser, G., *Optical Fiber Communications,* 1991. With permission of The McGraw-Hill Companies.)

width, $\Delta\lambda$, of the beam is 0.050 μm, determine the pulse broadening due to chromatic dispersion.

Solution:

From Fig. 9.20, the absolute value of the dispersion coefficient is obtained as 42 ps/(kmnm).

Thus the pulse broadening is

$$\Delta\tau_{\mathrm{cd}} = L_{\mathrm{f}} D_{\mathrm{d}} \Delta\lambda = 5 \times 42 \times 50$$
$$= 10.5 \text{ ns}$$

The effect of chromatic dispersion can be minimized by appropriate selection of beam wavelength to fall in the region where $\mathrm{d}^2 n/\mathrm{d}\lambda^2 \approx 0$. This is normally in the range 1.2–1.4 μm for fused silica. The impact of chromatic dispersion in multimode fibers is generally negligible. However, it is the more dominant form of distortion, and thus limits the bandwidth in single-mode fibers.

WAVEGUIDE DISPERSION Waveguide dispersion results from the fact that the propagating characteristics of a mode depend on the beam wavelength. Longer wavelengths have a longer path since they reflect at more oblique angles with the cladding. Analysis of the delay distortion for waveguide dispersion is beyond the scope of this book. It can be shown, however, that the pulse broadening, $\Delta\tau_{\mathrm{wd}}$, due to waveguide dispersion may be expressed as

$$\Delta\tau_{\mathrm{wd}} = \frac{L_{\mathrm{f}}}{c}\frac{\Delta\lambda}{\lambda}(n_2 - n_1) D_{\mathrm{w}}(\nu_{\mathrm{n}}) \tag{9.53}$$

where $D_{\mathrm{w}}(\nu_{\mathrm{n}})$ is a dimensionless dispersion coefficient and L_{f} = fiber length.

The pulse broadening due to waveguide dispersion is relatively small compared to chromatic dispersion.

INTERMODAL DISPERSION Intermodal dispersion results from differences in the distances propagated by the different modes sustained by the fiber (Fig. 9.17). If all the rays shown in the figure start out at the same instant, the bouncing rays reach the end of the fiber at a later time than the axial ray. The temporal delay (dispersion) in the arrival times of the rays causes delay distortion or change in the spectrum of the original input beam. The derivation of the expression for the delay distortion for this case is also beyond the scope of this book. However, for a step-index multimode fiber, the group delay, χ_{g}, can be expressed as

$$\chi_{\mathrm{g}} = \frac{L_{\mathrm{f}}}{c}\frac{\mathrm{d}(n_2 k_{\mathrm{w0}})}{\mathrm{d}k_{\mathrm{w0}}} + \frac{L_{\mathrm{f}}}{c}\frac{\nu_{\mathrm{n}}}{k_{\mathrm{w0}}}\frac{\mathrm{d}(n_2 k_{\mathrm{w0}} b_{\mathrm{c}} \Delta)}{\mathrm{d}\nu_{\mathrm{n}}} \tag{9.54}$$

where $k_{w0} = \frac{2\pi}{\lambda_0}$ is the wavenumber in free space, $\nu_n, n_2, \Delta,$ and r_c are as defined in Sections 9.9.2.2 and 9.9.2.3, and b_c is a normalized propagation constant defined as

$$b_c = \frac{(\gamma_c r_c)^2}{\nu_n^2} \tag{9.55}$$

where γ_c is the rate of decay of the electric field in the cladding.

The first term on the right-hand side of equation (9.54) corresponds to the group delay due to chromatic dispersion in the step-index multimode (cf. equation (9.45)), and is independent of a particular mode.

The second term corresponds to the modal waveguide delay. It is different for every mode. When a laser pulse enters the fiber, it is shared by many guided modes in the fiber. The pulse thus splits up into a number of pulses. Each of these exits the fiber at a different time since the delays of the modes are different. The difference in arrival times of the mode with the largest waveguide group delay and that with the least delay can be derived, through rigorous analysis (which is beyond the scope of this book) to be

$$\Delta\tau_m = \frac{L_f}{c}(n_1 - n_2)\left(1 - \frac{\pi}{\nu_n}\right) \tag{9.56a}$$

Equation (9.56a) expresses the difference in the arrival times of the leading and trailing edge of the output pulse of the fiber. A simpler approximation of the pulse broadening due to intermodal distortion may be based on the difference between the travel time of the slowest mode (corresponding to the highest order mode) and that of the fastest (corresponding to the fundamental mode) (Fig. 9.17). Using ray tracing, this is obtained as

$$\Delta\tau_m = t_{max} - t_{min} = \frac{L_f n_1 \Delta}{c} \tag{9.56b}$$

where t_{max} is the travel time of the slowest mode (one with the longest ray congruence path) and t_{min} is the travel time of the fastest mode (one with the shortest ray congruence path).

Example 9.2 Determine the pulse broadening for a 5 km fiber with core and cladding refractive indices of $n_1 = 1.48$ and $n_2 = 1.46$, respectively.

Solution:

The approximate value of the pulse broadening is obtained as

$$\Delta\tau_m = \frac{L_f n_1 \Delta}{c} = \frac{5000 \times 1.48}{3 \times 10^8}\left(\frac{1.48 - 1.46}{1.48}\right) = 333 \text{ ns}$$

Thus, the time delay between the axial ray and the ray that just enters the fiber at the angle of acceptance is about 333 ns. This corresponds to a modal dispersion-limited bandwidth of approximately 3.0 MHz.

It is evident that modal dispersion will generally not be a problem with single-mode fibers, since single-mode fibers only sustain a single mode of the beam. However, for step-index multimode fibers, modal dispersion is much greater than chromatic dispersion. It is the dominant form of distortion, and is the primary mechanism that limits step-index multimode fibers to low bandwidth (less than 100 MHz km) applications.

Optical Losses There are several factors that contribute to optical losses associated with beam transmission by a fiber optic system. These can be categorized as

1. Input-coupling losses.
2. Connector/splice losses.
3. Fiber losses.
4. Output-coupling losses.

The most significant of these losses, especially in regard to fiber optic beam delivery for manufacturing applications, are the input-coupling and fiber losses. We shall thus provide further discussion only on these two sources of optical losses.

INPUT-COUPLING LOSSES Input-coupling losses occur at the interface between the beam source and the fiber. These include

1. Mismatch between the source's emitting area and the fiber core area, which contributes to input-coupling losses.
2. Effect of the numerical aperture, which defines the light-gathering ability of the fibers themselves.
3. The packing fraction, which is associated with fiber-bundle cables, where several fibers are grouped together and illuminated by one light source. This loss is given by the ratio of the collective core areas of the fibers to the total bundle cross-sectional area.

FIBER LOSSES Losses in the fiber result in attenuation of the beam as it propagates through the fiber. The attenuation is expressed as the ratio of input power to output power (in decibels) per unit length of fiber:

$$\alpha_l = \frac{1}{L_f} 10 \log_{10} \left(\frac{q_{in}}{q_{out}} \right) \tag{9.57}$$

where α_l = attenuation, q_{in} = input power, and q_{out} = output power.

Fiber losses may be categorized into intrinsic and extrinsic losses.

Intrinsic Fiber Losses Intrinsic fiber losses are associated with the basic microstructural characteristics of the material used in making the fiber. They can generally be attributed to three main causes:

1. *Ultraviolet Absorption*: This results from excitation of electrons of the glass material from the valence to the conduction band by photons of ultraviolet wavelength and is thus electronic in origin. The absorption is wavelength-dependent, and may be expressed as

$$\alpha_{\mathrm{uv}} = K_{\mathrm{uv}} e^{\frac{a_{\mathrm{uv}}}{\lambda}} \qquad \mathrm{dB/km} \tag{9.58}$$

   where K_{uv} is a constant, which is a function of the dopant in the glass, a_{uv} is a constant (with units of microns), and the wavelength λ is also expressed in microns. The absorption peaks at approximately 0.14 μm wavelength for fused silica.
2. *Infrared Absorption*: This is due to the lattice vibrational modes of the fused silica and dopants. Depending on the type of dopant, the absorption peak for this type of loss may fall between 7 and 11 μm wavelengths. The power loss due to infrared absorption increases exponentially with wavelength and is expressed as

$$\alpha_{\mathrm{ir}} = K_{\mathrm{ir}} e^{\frac{-a_{\mathrm{ir}}}{\lambda}} \qquad \mathrm{dB/km} \tag{9.59}$$

   where K_{ir} and a_{ir} are constants, and the wavelength λ is expressed in microns.
3. *Rayleigh Scattering*: This results from inhomogeneities (such as density and composition variations) within the fiber due to the amorphous nature of glass. The microscopic density and composition variations cause local variations of the refractive index, which are of dimensions much smaller than a wavelength, and thus act as scattering centers. The Rayleigh scattering loss is strongly wavelength dependent and is given by

$$\alpha_{\mathrm{Rs}} = K_{\mathrm{Rs}}/\lambda^4 \qquad \mathrm{dB/km} \tag{9.60}$$

   where K_{Rs} is the Rayleigh scattering coefficient, with units (dB/km) $\mu\mathrm{m}^4$.

The general dependence of the intrinsic losses on wavelength is illustrated in Fig. 9.21 that shows an increase in the ultraviolet and Rayleigh scattering losses as the wavelength decreases, while the infrared losses increase with the increasing wavelength. The end effect is that the total intrinsic loss curve has a minimum, usually in the near-infrared region between 0.6 and 1.6 μm, depending on the fiber material and dopants.

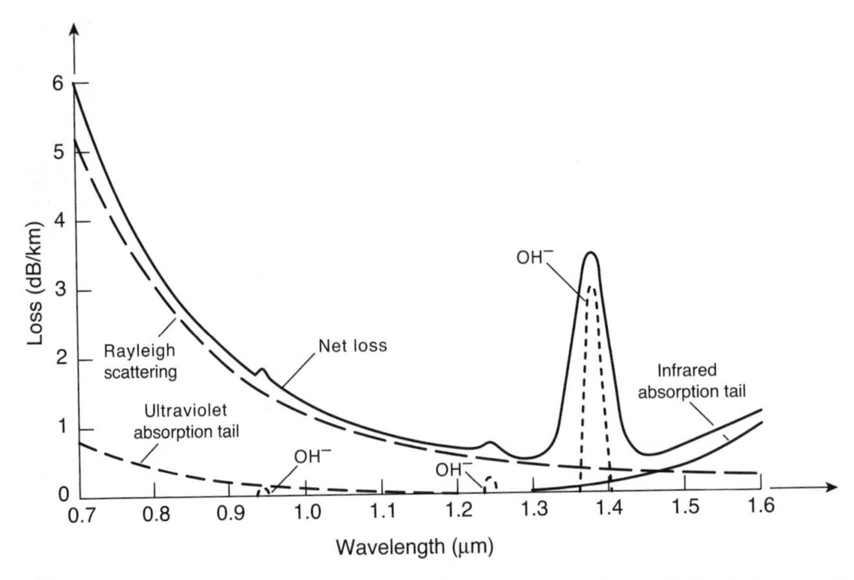

FIGURE 9.21 Sources of beam loss within a glass fiber. (From Cherin, A. H., *An Introduction to Optical Fibers,* 1983. With permission of The McGraw-Hill Companies.)

Extrinsic Fiber Losses Extrinsic fiber losses include impurity and structural imperfection losses, macrobending losses, and microbending losses.

1. *Impurity and Structural Imperfection Losses*: These result from imperfections associated with the fabrication process, such as impurities within the glass and structural imperfections, and can be minimized by improvements to the fabrication process. An example of an impurity loss is illustrated by the peak at about 1.375 μm of the loss curve in Fig. 9.21, caused by the hydroxyl group, OH^-.
2. *Macrobending Losses*: These are associated with axial bends of relatively large radius, of the order of 1 mm or more, in the fiber. This type of loss increases with increasing wavelength and decreasing bend radius. The loss is also mode dependent, with higher order modes being more severely attenuated by bending. Generally, bend radii of the order of a few centimeters should be avoided.
3. *Microbending Losses*: These are caused by random deflections of the fiber along its axis. The deflections are of the order of a few microns and can cause scattering losses.

9.9.2.6 Application of Optical Fibers in High-Power Laser Systems From the viewpoint of manufacturing applications, the principal effects that beam degradation (dispersion and attenuation) have on a laser beam as it propagates through an optical fiber are changes in the beam mode and intensity. Thus a beam that enters the fiber with a Gaussian mode may be incident at the workpiece with a significantly different mode form, and with the intensity significantly reduced. It is also obvious from

Fig. 9.21 that there is a certain wavelength range (typically in the near-infrared range) over which beam attenuation is minimized. Thus it is more feasible to transmit the 1.06 μm wavelength Nd:YAG laser beam using fiberoptic cables than it is to transmit the 10.6 μm wavelength CO_2 laser beam, or the much shorter wavelengths in the ultraviolet region.

Since most high-power Nd:YAG lasers have beam quality of M^2 between 50 and 100, fiber sizes that are used typically range between 800 and 1000 μm in diameter. Smaller fiber sizes can only be used when beam quality significantly improves. This is because the beam size needs to be smaller than the fiber core diameter. Otherwise, power losses and heating may create significant problems.

Based on the preceding discussions, following are a few points to be noted in using fiber optic delivery systems in high-power laser applications.

- Shorter fibers tend to have less effect on the beam characteristics, whereas a relatively long fiber may result in an output beam whose characteristics are significantly different from that of the incident beam.
- Increasing the number of bends in the fiber cable or the tightness of the bends tends to generate more modes and thus change the beam profile or quality.
- Smaller fibers limit the number of modes and thereby increase the beam quality.
- Graded index fibers tend to produce higher quality beams.
- Smaller fibers and graded index fibers are more appropriate for processing applications that require high-power densities such as laser cutting.
- For applications such as heat treatment that require a widely distributed beam (top-hat profiles), step-index fibers are preferable.
- Fiber optic connectors are required at both the input end to launch the beam into the cable and at the output end for delivering the beam to the workpiece.
- The output optics system may also include a pair of lenses to collimate and focus the beam.

9.10 SUMMARY

In electromagnetic wave theory, the product of the wavelength and frequency of electromagnetic radiation is equal to the velocity at which the radiation propagates in free space. When electromagnetic radiation or light beam enters a birefringent material, it is refracted into two rays, the ordinary and extraordinary rays. The ordinary ray obeys Snell's law and has the same velocity in any direction in the material, while the extraordinary ray has different velocities in different directions and does not follow Snell's law.

A light beam is said to be linearly or plane polarized when the electric vector is restricted to a specific plane. When a light beam is incident on a surface, the electric vector associated with the beam can be resolved into two components, the p-component (lying in, or parallel to the plane of incidence) and the s-component (normal to the plane of incidence). The angle of incidence at which the reflected beam becomes plane polarized normal to the plane of incidence is called the Brewster angle.

If two waves that have the same frequency and are plane polarized in the same plane are superimposed on each other, the frequency of the resulting light vector is the same as that of the original light vectors. However, the amplitudes and phase angles are different. Now if the two plane waves lie in different planes which are orthogonal to each other, the resulting light vector will be elliptically polarized. Further, if the amplitudes of the two light vectors are equal, and the phase difference between them is 90°, the resulting light is circularly polarized.

Beam expanders are usually used to increase the size of a beam before focusing, and this reduces the size of the focused beam, increasing the power density at the focal point. This can be achieved using transmissive or reflective optics. On the contrary beam splitters are used to split a light beam into two or more beams and can be done using either a wedge-shaped reflector or a partially reflective and partially transmissive optical element.

In this chapter, we have discussed the two major ways in which a beam can be delivered from the laser generator to the workpiece:

1. Using a conventional combination of transmissive and reflective optics (lenses and mirrors)
2. Using a fiber optic cable (which will usually also include some lenses).

Major components of a conventional beam delivery system include the beam bending assembly; focusing assembly; interconnecting beam guard tubes; and different types of mirrors and lenses.

Beam delivery can also be achieved using fiber optic systems, even though these may not be applicable to all types of high-power lasers. The basic structure consists of the transmitter; transmission line (of interest in processing); and receiver. The waveguide, in which the light beam propagates, consists of two principal components—core and cladding. The core is the inner portion through which light is transmitted, and the cladding is the concentric outer layer with a relatively lower index of refraction. The waveguide is characterized by the numerical aperture, normalized frequency, and angle of acceptance. The major fiber types are the step-index multimode, single-mode, and graded-index multimode fibers. Fiber optic systems are normally limited by beam degradation, and the major forms of degradation are dispersion or distortion (including chromatic or material dispersion, waveguide dispersion, and modal dispersion), and optical losses such as input-coupling losses, connector/splice losses, fiber losses, and output-coupling losses.

REFERENCES

Buck, J. A., 1995, *Fundamentals of Optical Fibers,* John Wiley & Sons, New York.

Cherin, A. H., 1983, *An Introduction to Optical Fibers,* McGraw-Hill, New York.

Chryssolouris, G., 1991, *Laser Machining, Theory and Practice*, Springer-Verlag, Berlin.

Dally, J. W., and Riley, W. F., 1991, *Experimental Stress Analysis*, 3rd edition, McGraw Hill, New York.

Gnanamuthu, D.S., and Shankar, V.S., 1985, Laser heat treatment of iron-base alloys, *SPIE, Applications of High Powered Lasers*, Vol. 572, pp. 56–72.

Keiser, G., 1991, *Optical Fiber Communications,* McGraw-Hill, New York.

Killen, H. B., 1991, *Fiber Optic Communications,* Prentice Hall, Englewood Cliffs, NJ.

Kleekamp, C., and Metcalf, B., 1978, *Designer's Guide to Fiber Optics—Part 1,* Cahners Publishing Company, Boston, MA.

Lachs, G., 1998, *Fiber Optic Communications,* McGraw-Hill Telecommunications, New York.

Lacy, E. A., 1982, *Fiber Optics,* Prentice Hall, Englewood Cliffs, NJ.

Leong, K. H., and Hunter, B. V., 1996, High-power fiberoptic laser beam delivery moves forward, *Industrial Laser Review*, Vol. 11, pp. 7–12.

Luxon, J. T., and Parker, D. E., 1985, *Industrial Lasers and Their Applications*, Prentice Hall, Englewood Cliffs, NJ.

Ungar, S., 1990, *Fiber Optics, Theory and Applications*, John Wiley & Sons, New York.

Wilson, J., and Hawkes, J. F. B., 1987, *Lasers—Principles and Applications*, Prentice Hall International Series in Optoelectronics, New York.

APPENDIX 9A

List of symbols used in the chapter

Symbol	Parameter	Units
b_c	normalized propagation constant	–
D_d	dispersion coefficient	ps/(km nm)
D_w	dimensionless dispersion coefficient	–
E_a	amplitude of absorbed light	V/m
E_t	amplitude of transmitted light	V/m
E_{tf}	amplitude of transmitted light along fast axis	V/m
E_{ts}	amplitude of transmitted light along slow axis	V/m
k_1	constant	–
$k_{wo} = \frac{2\pi}{\lambda_o}$	wavenumber in free space	/m
K_{ij}	constant	dB/km
K_{Rs}	constant	$(dB/km) - \mu m^4$
L_f	fiber length	m
m_g	magnification	–
n_1	index of refraction of core	–
n_2	index of refraction of cladding	–
n_f	index of refraction along fast axis	–
n_g	group index of refraction	–
n_i	index of refraction of medium *i*	–

n_r	core index of refraction at radius r	–
n_s	index of refraction along slow axis	–
n_{ij}	index of refraction of medium i relative to medium j	–
q_{in}	input power	W (Watts)
q_{out}	output power	W (Watts)
r_c	core radius	m
R_p	reflection coefficient of ray in plane of incidence	–
R_s	reflection coefficient of ray normal to plane of incidence	–
t_{max}	travel time of the slowest mode	s (seconds)
t_{min}	travel time of the fastest mode	s (seconds)
u_g	group velocity	m/s
u_w	wave velocity	m/s
α_f	angle between fast axis and plane of polarization	°
α_l	attenuation	/m
α_{ir}	attenuation due to infrared absorption	/m
α_{Rs}	attenuation due to Rayleigh scattering	/m
α_{uv}	attenuation due to ultraviolet absorption	/m
γ_c	rate of decay of electric field in cladding	/m
λ_o	wavelength in free space	m
χ_g	group delay	s/m
ϵ_{p0}	electric permittivity in free space	farad/m (F/m)
μ_{m0}	magnetic permeability in free space	henry/m (H/m)
ν_n	normalized frequency	–
Δ	$\frac{n_1 - n_2}{n_1}$	–
$\Delta\tau_m$	difference in arrival times of mode with largest waveguide group delay and that with least delay	s
$\Delta\tau_{cd}$	pulse broadening or delay distortion	s
$\Delta\tau_{wd}$	pulse broadening due to waveguide dispersion	s
θ_1	angle of incidence	°
θ_2	angle of refraction	°
θ_a	angle of acceptance	°
θ_c	critical angle of incidence	°

PROBLEMS

9.1. Given two plane-polarized waves that are polarized in orthogonal planes and are superimposed on each other,

$$E_{1y}(t) = E_{0y}\cos(\phi_y - \omega t)$$

and

$$E_{1z}(t) = E_{0z}\cos(\phi_z - \omega t)$$

show that

$$\left[\frac{E_{1y}(t)}{E_{0y}}\right]^2 - 2\frac{E_{1y}(t)E_{1z}(t)}{E_{0y}E_{0z}}\cos(\phi_z - \phi_y) + \left[\frac{E_{1z}(t)}{E_{0z}}\right]^2 = \sin^2(\phi_z - \phi_y)$$

9.2. Given two plane-polarized waves which are polarized in the same plane and are superimposed on each other,

$$E_{11}(t) = E_{01}\cos(\phi_1 - \omega t)$$

and

$$E_{12}(t) = E_{02}\cos(\phi_2 - \omega t)$$

show that

$$E_1(t) = E_0\cos(\phi - \omega t)$$

where

$$E_0{}^2 = E_{01}{}^2 + E_{02}{}^2 + 2E_{01}E_{02}\cos(\phi_2 - \phi_1)$$

and

$$\tan\phi = \frac{E_{01}\sin\phi_1 + E_{02}\sin\phi_2}{E_{01}\cos\phi_1 + E_{02}\cos\phi_2}$$

9.3. Show that for a plate of birefringent material, the phase difference, ϕ, between the transmitted ray along the fast and slow axes, is given by

$$\phi = \frac{2\pi h(n_s - n_f)}{\lambda}$$

9.4. Given that

$$\text{NA} = \sin\theta_a$$

And assuming that the refractive index for air is 1, show that for step-index fibers, NA may also be expressed as

$$\text{NA} = \sqrt{n_1{}^2 - n_2{}^2}$$

9.5. Determine the core radius and refractive index of a step-index multimode fiber that has a normalized frequency of 75, and a numerical aperture of 0.25. Assume the wavelength of the propagating medium is 1.06 μm, and that the cladding refractive index is 1.52.

9.6. Using equation (9.47a), obtain an expression for the group index, n_g, in terms of the wavelength λ.

9.7. Show that the normalized frequency, ν_n, can also be expressed as

$$\nu_n = k_w r_c n_1 \sqrt{2\Delta - \Delta^2}$$

9.8. (a) The refractive index of glass is 1.52 and that of water is 1.33. If light travels from glass into water, determine the resulting Brewster and critical angles.

(b) Consider the He–Ne laser shown in Fig. 8.4. What should the angle θ_B be if the light reflected from the windows are to be totally polarized perpendicular to the plane of incidence? Assume the windows are made of glass of refractive index 1.52.

9.9. The core of a step-index fiber optic cable which is in air has a refractive index of 1.62, while the cladding has an index of 1.52. What is the entrance cone angle for light that enters the fiber? What will be the corresponding numerical aperture? Determine the modal dispersion time delay per kilometer, for the fiber.

9.10. A perfect polarizer is placed in the path of a plane-polarized beam such that its axis is horizontal. If the plane of polarization of the beam is vertical, determine the fraction of light that will be transmitted by the polarizer. If another polarizer, which is similar to the first one is placed before the first one such that its axis is at 45° to the horizontal, what will be the fraction of light transmitted?

9.11. (a) Given

$$\nu\lambda = c$$

Show that for small $\Delta\lambda$,

$$\Delta\nu = -\left(\frac{c}{\lambda^2}\right)\Delta\lambda.$$

(b) Given a frequency bandwidth of $\Delta\nu = 5500$ Hz, determine the corresponding wavelength bandwidth at a wavelength of $\lambda = 0.8$ μm.

9.12. A multimode waveguide has a core refractive index of 1.5, a core radius of 30 μm, and a Δ of 1.5%. Determine the numerical aperture and normalized frequency for a laser beam of wavelength $\lambda = 1.06$ μm that enters the waveguide.

9.13. For the setup in Problem 9.12, determine how the number of modes in the waveguide will vary as

(a) The core refractive index
   (i) increases
   (ii) decreases

(b) The incident wavelength
   (i) increases
   (ii) decreases.

9.14. A single mode step-index waveguide has a core refractive index of 1.52, a cladding refractive index of 1.50, a normalized frequency of 2.0, and an input signal linewidth of $\Delta\lambda = 0.002\ \mu\text{m}$. If the input signal wavelength is $\lambda = 1.06\ \mu\text{m}$, determine the delay distortion per kilometer of fiber waveguide.

9.15. (a) An Nd:YAG laser beam is delivered to a workpiece through an optical fiber cable of length 5 m. If the power that is measured at the workpiece is 500 W, determine the power output of the laser, assuming the only losses are those incurred in the fiber cable.

Hint: Use Fig. 9.21.

(b) What would be the loss in dB/km if 95% of the beam was delivered?

PART II
Engineering Background

10 Heat And Fluid Flow

The heat and the fluid flow that occur during laser processing influence the microstructures (through the grain structure and phases that are formed), residual stresses (through the thermal stresses that result from differential strains), and distortions that evolve during the process. These in turn affect the mechanical properties and thus the quality of the process. The discussion in this chapter is in three sections. In the first section, simplified lumped parameter energy balance equations are presented that enable quick estimates to be made of energy requirements for a given process. In the second section, only heat flow in the solid part of the workpiece is considered. The analysis in this section is applicable, as a first approximation, to all the processes discussed and is more accurate in the solid part of the workpiece. To enable closed form solutions to be obtained, the heat source is initially modeled as either a point or line source. Analytical expressions are then obtained for the temperature distribution for the quasistationary state, together with the cooling rates and peak temperatures. A Gaussian distribution that is more representative of the heat source is then considered. This is followed by the two-temperature model, which is appropriate for ultrashort pulse laser beams in metals. In the third section, flow in the molten pool for processes that involve melting (such as welding, cladding, surface melting, and cutting) is discussed. In the second and third sections, the basic governing equations are first presented. Then, where feasible, approximate closed form analytical solutions are presented that enable a quick analysis of the process to be made.

10.1 ENERGY BALANCE DURING PROCESSING

During processing, part of the beam that is incident on the workpiece is reflected away, and the remaining part is absorbed by the workpiece. The absorbed energy, Q_a, is what is used in processing and may involve

1. Heating up the solid to the melting temperature, $Q_1 = m_a c_p \Delta T_m$
2. Melting, which requires the latent heat of melting, $Q_2 = m_a L_m$
3. Heating up the molten material to the vaporization temperature, $Q_3 = m_a c_p \Delta T_v$
4. Vaporization, which requires the latent heat of vaporization, $Q_4 = m_a L_v$
5. Energy lost by conduction, convection, and radiation, Q_l.

where c_p is the specific heat (J/kg K), L_m is the latent heat of melting per unit mass (J/kg), L_v is the latent heat of vaporization per unit mass (J/kg), m_a is the mass (kg), T_m is the melting temperature (°C), T_v is the vaporization temperature (°C), $\Delta T_m = T_m - T_0$, $\Delta T_v = T_v - T_m$.

Thus, an energy balance gives

$$Q_a = m_a c_p \Delta T_m + m_a L_m + m_a c_p \Delta T_v + m_a L_v + Q_l \quad (10.1a)$$

Assuming all the absorbed energy is used in processing, and neglecting any losses, we have the following energy balance for the process:

$$Q_a = m_a c_p \Delta T_m + m_a L_m + m_a c_p \Delta T_v + m_a L_v \quad (10.1b)$$

or

$$Q_a = \rho V (c_p \Delta T_m + L_m + c_p \Delta T_v + L_v) \quad (10.1c)$$

where ρ = density (kg/m^3) and V = volume (m^3).

Since part of the absorbed energy is conducted into the surrounding material, it experiences a temperature rise, with subsequent cooling. Conduction into the workpiece is discussed in the next section.

10.2 HEAT FLOW IN THE WORKPIECE

The principal thermal factors that determine the microstructure of the workpiece are the temperature distribution, the peak temperatures, and the cooling rates that are experienced during the process. Once an expression for the temperature distribution is obtained, it can be used to derive the peak temperature and cooling rate expressions. The following analysis of the heat flow in the solid workpiece is therefore discussed along these lines.

10.2.1 Temperature Distribution

The discussion on the temperature distribution starts with the governing equation for conduction heat flow. This is based on the assumption that heat conduction through the workpiece is usually much greater[1] than any heat exchange with the surroundings by natural convection or radiation. It is further assumed that the workpiece surfaces are adiabatic; that is, there is no heat loss or gain by either convection or radiation.

With these assumptions, the governing equation for heat flow in a solid with a coordinate system fixed at a stationary origin relative to the solid is given

[1] For materials with very low thermal conductivity, this assumption can lead to significant error and thus might need to be reconsidered.

(see Appendix 10B) by

$$\rho c_p \frac{dT}{dt} = \frac{\partial}{\partial x}\left(k\frac{\partial T}{\partial x}\right) + \frac{\partial}{\partial y}\left(k\frac{\partial T}{\partial y}\right) + \frac{\partial}{\partial z}\left(k\frac{\partial T}{\partial z}\right) + q_s \tag{10.2a}$$

$$\Rightarrow \rho c_p \frac{dT}{dt} = k\left(\frac{\partial^2 T}{\partial x^2}+\frac{\partial^2 T}{\partial y^2}+\frac{\partial^2 T}{\partial z^2}\right) + \frac{\partial k}{\partial T}\left[\left(\frac{\partial T}{\partial x}\right)^2 + \left(\frac{\partial T}{\partial y}\right)^2 + \left(\frac{\partial T}{\partial z}\right)^2\right] + q_s \tag{10.2b}$$

or in a more compact form,

$$\rho c_p \frac{dT}{dt} = \nabla \cdot (k\nabla T) + q_s \tag{10.2c}$$

where k is the thermal conductivity (W/mm K), q_s is the rate of local internal energy generated per unit volume (W/mm^3), t is the time (s), and T is the temperature (°C).

Equation (10.2) is nonlinear, and thus it is difficult to solve in closed form. To make the problem more tractable analytically, the following simplifications are made:

1. The equation is linearized by assuming that the material's physical coefficients such as thermal conductivity are independent of temperature. Even though the thermal conductivity for plain carbon steel, for example, may vary from about 65 W/m K at 0°C to about 30 W/m K at 1200°C, using an average value (about 50 W/m K) provides a reasonable approximation and enables a closed form solution to be obtained. Thus $\partial k/\partial T = 0$.
2. The internal heat generation is neglected. This means $q_s = 0$. This assumption is reasonable for a number of applications, especially when one compares the external heat sources associated with some laser processes with any heat that might be generated within the material. However, this is not necessarily true of oxygen-assisted laser cutting where the exothermic reaction can be considerable.
3. The workpiece material is assumed to be homogeneous and isotropic. This is a reasonable approximation for many noncomposite materials that have not been work-hardened.

With these simplifications, equation (10.2) reduces to the form

$$\frac{dT}{dt} = \kappa\left(\frac{\partial^2 T}{\partial x^2} + \frac{\partial^2 T}{\partial y^2} + \frac{\partial^2 T}{\partial z^2}\right) \tag{10.3a}$$

or

$$\frac{dT}{dt} = \kappa\nabla^2 T \tag{10.3b}$$

where $\kappa = k/\rho c_p$ is the thermal diffusivity (mm^2/s).

Solution of equation (10.3) requires six boundary conditions, two in each direction, and an initial condition.

One general approach to solving equation (10.3a) is to use separation of variables. Just as was the case for Maxwell's equation for the electric vector E_l discussed in Chapter 2, equation (10.3a) can also be solved using separation of variables by first expressing $T(x, y, z, t)$ as

$$T = X(x)Y(y)Z(z)\tau(t) \tag{10.4}$$

However, before further discussing the solution of equation (10.3), we first consider possible forms of the heat flow problem in laser processing to see if we can further simplify the analysis. The heat source used in laser processing is usually one of the two possible configurations:

1. A stationary source, for example, as used in laser drilling or spot welding, and such processes may be directly analyzed using the preceding equations.
2. A moving source, for example, as used in laser cutting, welding, or heat treatment.

Analysis of the moving heat source case is facilitated by using a coordinate system that is attached to the heat source. Let us, therefore, consider a coordinate system moving with the heat source along the x-axis, as shown in Fig. 10.1. The corresponding governing equation is obtained by a coordinate transformation from the plate to the heat source (see Appendix 10C), with x being replaced by ξ, y by y', z by z', and t by t', that is,

$$\xi = x - u_x t, \quad y' = y, \quad z' = z, \quad t' = t$$

where u_x is the traverse velocity of the heat source in the x-direction (mm/s).

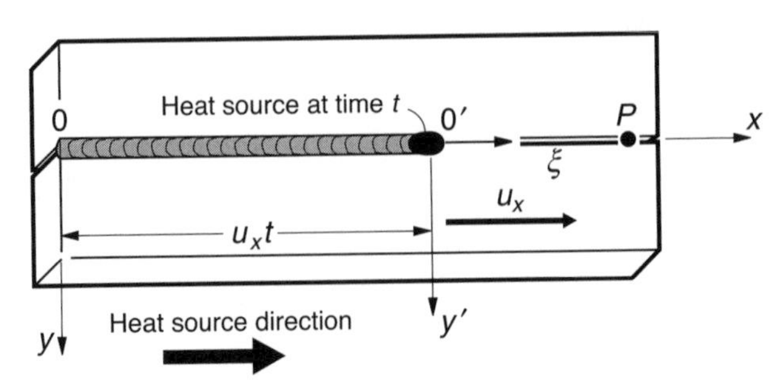

FIGURE 10.1 Schematic of moving coordinate system associated with laser processing.

The transformation gives the governing equation (10.3) in terms of the moving coordinate system (see Appendix 10C) as

$$\frac{\partial T}{\partial t} = \kappa \left(\frac{\partial^2 T}{\partial \xi^2} + \frac{\partial^2 T}{\partial y^2} + \frac{\partial^2 T}{\partial z^2} \right) + u_x \frac{\partial T}{\partial \xi} \tag{10.5}$$

where ξ, y, z is a coordinate system attached to the moving heat source, with positive ξ in the direction in which the heat source is moving; x, y, z is a coordinate system with origin O and fixed to the workpiece, with positive x in the direction in which the heat source is moving; and $\partial T/\partial t$ is the time rate of change of temperature in the moving coordinate system.

When a uniform prismatic workpiece is processed over a long enough period, experimental results indicate that a state is reached when an observer positioned at the heat source or moving origin detects no change in the temperature distribution around the source. This is the *quasistationary state*, and for this case, the temperature undergoes no change with time with respect to the coordinate system attached to the heat source. The time derivative in equation (10.5) thus vanishes ($\partial T(\xi, y, z, t)/\partial t = 0$) and the governing equation then becomes

$$\frac{\partial^2 T}{\partial \xi^2} + \frac{\partial^2 T}{\partial y^2} + \frac{\partial^2 T}{\partial z^2} = -\frac{u_x}{\kappa} \frac{\partial T}{\partial \xi} \tag{10.6}$$

This is the differential equation for the quasistationary case.

A convenient way of solving an equation of this form is to assume that T is of the form

$$T = T_0 + \exp\left(-\frac{u_x \xi}{2\kappa} \right) \phi(\xi, y, z) \tag{10.7}$$

where T_0 is the initial temperature of the workpiece, $\phi(\xi, y, z)$ is a function that is yet to be determined.

Substituting equation (10.7) into (10.6) results in the following equation:

$$\frac{\partial^2 \phi}{\partial \xi^2} + \frac{\partial^2 \phi}{\partial y^2} + \frac{\partial^2 \phi}{\partial z^2} - \left(\frac{u_x}{2\kappa} \right)^2 \phi = 0 \tag{10.8}$$

Similar differential equations are encountered in analysis of electric wave problems. Thus, in this form, appropriate solutions are more easily obtained, depending on the boundary conditions that are imposed. To facilitate the analysis and obtain a closed form solution of the preceding equation, the following additional simplifying assumptions are made in this section:

1. A point or line heat source is used. This assumption enables a closed form solution to be obtained. A more accurate representation would be that of a

Gaussian distribution. However, for that, we would have to resort to numerical methods.
2. No phase changes occur; that is, the effect of latent heat of fusion is negligible. As a first-order approximation, this is a reasonable assumption, since latent heat is absorbed in front of the heat source and evolved behind it.

The Gaussian heat source is considered separately in Section 10.2.5. Thus, for now, we consider two forms of solutions:

1. One case is that of a thick plate on which a point heat source moves and involves three-dimensional heat flow. This might be the case, for example, in conduction mode welding (see Section 16.3.1).
2. The other case is that of a thin plate with a line heat source that penetrates through the thickness and involves two-dimensional heat flow. Examples would be keyhole welding (Section 16.3.2) or laser cutting (Chapter 15).

To determine whether a plate is thin or thick, the following equation may be used as an initial approximation:

$$\beta_c = h\sqrt{\frac{\rho c_p u_x (T - T_0)}{q}} \tag{10.9}$$

where h is the plate thickness (m) and q is the heat flux (power) input (W).

The plate is considered to be thin when $\beta_c < 0.6$ and thick when $\beta_c > 0.9$. When high accuracy is desired and $0.6 < \beta_c < 0.9$, then it is best to solve the equations numerically. We now consider the closed form solutions for the thick and thin plate cases separately.

10.2.1.1 Thick Plate with Point Heat Source (Three Dimensional) Figure. 10.2 schematically illustrates this configuration. Ideally, the plate is considered to be semi-infinite and the heat source a point. Now let $q =$ heat flux (power) input, and $r = \sqrt{\xi^2 + y^2 + z^2}$, where r is the distance of any general point, $P(\xi, y, z)$, from the moving origin (Fig. 10.1).

However, to make the analysis even simpler, we start by considering the fully infinite case. Since the temperature of the plate remains unchanged far away from the source, the boundary conditions for this case can be constructed as follows:

$$\frac{\partial T}{\partial \xi} \to 0 \quad \text{as} \quad \xi \to \pm\infty \tag{10.10a}$$

$$\frac{\partial T}{\partial y} \to 0 \quad \text{as} \quad y \to \pm\infty \tag{10.10b}$$

$$\frac{\partial T}{\partial z} \to 0 \quad \text{as} \quad z \to \pm\infty \tag{10.10c}$$

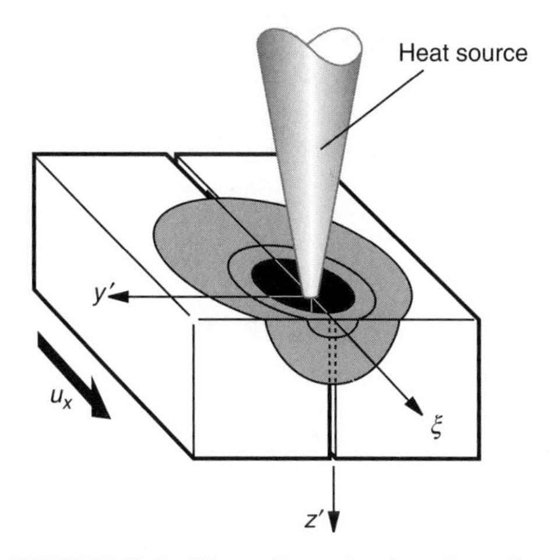

FIGURE 10.2 Three-dimensional configuration.

To account for the heat input, we consider the fact that for the point source, the heat flux through the surface of a sphere of radius r drawn around the source approaches q as $r \to 0$, that is,

$$\lim_{r\to 0} -4\pi r^2 \times k \times \frac{\partial T}{\partial r} = q \tag{10.11}$$

Now we have to obtain the solution for equation (10.8) that satisfies the preceding boundary conditions and accounts for the heat input as given in equation (10.11). Due to the nature of the boundary conditions and the heat input equation, as well as the symmetrical nature of equation (10.8), ϕ depends only on r. Thus, using polar coordinates, equation (10.8) can be reexpressed as

$$\frac{\mathrm{d}^2\phi}{\mathrm{d}r^2} + \frac{2}{r}\frac{\mathrm{d}\phi}{\mathrm{d}r} - \left(\frac{u_x}{2\kappa}\right)^2 \phi = 0 \tag{10.12}$$

But

$$\frac{1}{r}\frac{\mathrm{d}^2(\phi r)}{\mathrm{d}r^2} = \frac{\mathrm{d}^2\phi}{\mathrm{d}r^2} + \frac{2}{r}\frac{\mathrm{d}\phi}{\mathrm{d}r} \tag{10.13}$$

Thus, equation (10.12) becomes

$$\frac{\mathrm{d}^2(\phi r)}{\mathrm{d}r^2} - \left(\frac{u_x}{2\kappa}\right)^2 \phi r = 0 \tag{10.14}$$

An obvious solution of equation (10.14) that also satisfies the boundary conditions is

$$\phi r = k_1 \mathrm{e}^{-\left(\frac{u_x r}{2\kappa}\right)} \tag{10.15a}$$

or

$$\phi = k_1 \frac{\mathrm{e}^{-\left(\frac{u_x r}{2\kappa}\right)}}{r} \tag{10.15b}$$

where k_1 is to be determined.

This solution, equation (10.15), also satisfies the heat input condition, equation (10.11), since

$$\frac{\mathrm{d}\phi}{\mathrm{d}r} \times r^2 \quad \rightarrow \quad \text{constantvalue} \quad \text{as} \quad r \rightarrow 0 \tag{10.15c}$$

Now substituting equation (10.15b) into equation (10.7) gives the expression for the temperature distribution for the infinite case as

$$T - T_0 = \frac{q}{4\pi k r} \exp\left(-\frac{u_x(r+\xi)}{2\kappa}\right) \tag{10.16}$$

The corresponding solution for the semiinfinite case, which is more representative of laser processing of an infinite plate (semiinfinite space), follows from equation (10.16) as

$$T - T_0 = \frac{q}{2\pi k r} \exp\left[-\frac{u_x(r+\xi)}{2\kappa}\right] \tag{10.17}$$

The factor 2 that results in equation (10.17) is due to the fact that we are now considering only one-half of the space. For laser processes that involve melting, equation (10.17) applies only outside the fused zone, that is, for $T < T_{\mathrm{m}}$ (the melting temperature), and is more accurate farther away from the fusion boundary.

10.2.1.2 Thin Plate with Line Heat Source (Two Dimensional) The configuration for this case is illustrated schematically in Fig. 10.3. In this case, heat flow is in two directions ξ (or x)-direction and y-direction. There is no flow in the z-direction. The heat source is considered to be a line that goes through the entire plate thickness uniformly. Thus, heat is input to the system as power per unit thickness. Now let $r = \sqrt{\xi^2 + y^2}$, the radius of a cylinder drawn around the heat source.

Since the heat source is uniform through the thickness, there can be no change in temperature in the thickness direction. Thus, we have

$$\frac{\partial T}{\partial z} = 0 \quad \text{for all } z$$

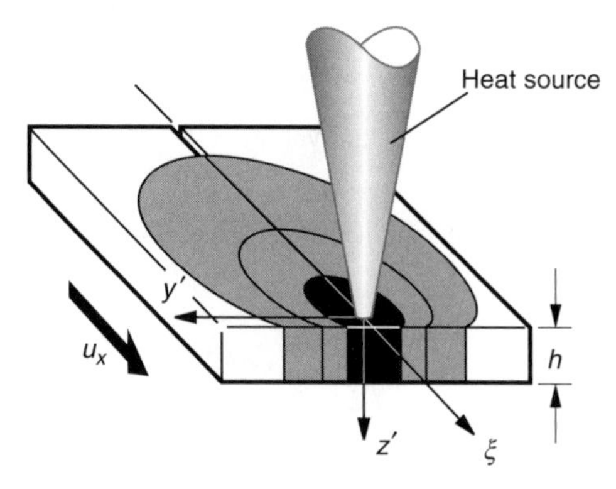

FIGURE 10.3 Two-dimensional configuration.

Equation (10.8) then becomes

$$\frac{\partial^2 \phi}{\partial \xi^2} + \frac{\partial^2 \phi}{\partial y^2} - \left(\frac{u_x}{2\kappa}\right)^2 \phi = 0 \tag{10.18}$$

Again since the temperature of the plate remains unchanged far away from the source, the boundary conditions for this case can be constructed as follows:

$$\frac{\partial T}{\partial \xi} \to 0 \quad \text{as} \quad \xi \to \pm\infty \tag{10.19a}$$

$$\frac{\partial T}{\partial y} \to 0 \quad \text{as} \quad y \to \pm\infty \tag{10.19b}$$

And yet again to account for the heat input, we consider the fact that since this is a line source, the heat flux through the surface of a cylinder drawn around the line source approaches q as $r \to 0$, that is,

$$\lim_{r \to 0} -2\pi r h k \times \frac{\partial T}{\partial r} = q \tag{10.20}$$

Since equation (10.18) is symmetrical in nature, and ϕ depends only on r, it can be expressed in cylindrical coordinates as

$$\frac{\mathrm{d}^2 \phi}{\mathrm{d}r^2} + \frac{1}{r}\frac{\mathrm{d}\phi}{\mathrm{d}r} - \left(\frac{u_x}{2\kappa}\right)^2 \phi = 0 \tag{10.21}$$

The solution to equation (10.21) that satisfies the boundary conditions is the modified Bessel function of the second kind and zero order, $K_0\,(u_x r/2\kappa)$.

It can be shown that

$$K_0\left(\frac{u_x r}{2\kappa}\right) \to \ln r \quad \text{as} \quad r \to 0$$

Thus,

$$\frac{\mathrm{d}K_0}{\mathrm{d}r} \to \text{constant value} \quad \text{as} \quad r \to 0$$

Equation (10.20) is therefore satisfied by $K_0\,(u_x r/2\kappa)$.
Furthermore,

$$K_0\left(\frac{u_x r}{2\kappa}\right) \to \sqrt{\frac{\pi}{\left(\frac{u_x r}{\kappa}\right)}}\mathrm{e}^{-\left(\frac{u_x r}{2\kappa}\right)} \quad \text{as} \quad r \to \infty$$

Thus, the boundary conditions, equations (10.19a) and (10.19b), are satisfied. The solution for the thin plate case can therefore be expressed as

$$T - T_0 = \frac{q}{2\pi kh} \exp\left(-\frac{u_x \xi}{2\kappa}\right) \times K_0\left(\frac{u_x r}{2\kappa}\right) \tag{10.22}$$

where $K_0(\chi)$ is the modified Bessel function of the second kind of order zero, Fig. 10.4 and Appendix 10G and χ is the argument of the modified Bessel function.

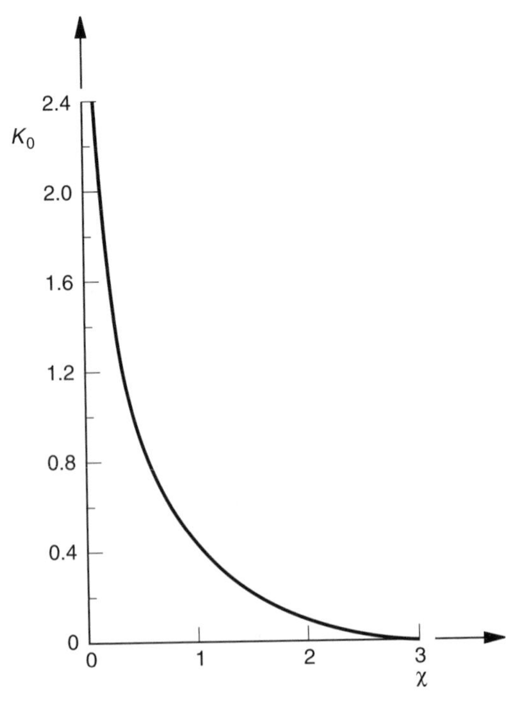

FIGURE 10.4 Plot of the modified Bessel function of the second kind of order zero, $K_0(\chi)$ (see also Appendix 10G). (From Abramowitz, M., and Stegun, I. A., editors, 1964, *Handbook of Mathematical Functions*, NBS, Washington, DC.)

When χ is small,

$$K_0(\chi) \approx -\left(0.5772 + \log \frac{\chi}{2}\right)$$

When χ is large,

$$K_0(\chi) \approx \sqrt{\left(\frac{\pi}{2\chi}\right)} e^{-\chi}$$

Equations (10.17) and (10.22) are also sometimes known as the Rosenthal equations, after the person who first derived them. Samples of the temperature distribution as represented by a family of isotherms drawn around the instantaneous heat source position ($x - y$ plane) are shown in Figs 10.5–10.7. Figure 10.5 shows the effect of thermal conductivity by comparing the isotherms for a relatively low thermal conductivity material (say steel) and a relatively high thermal conductivity material (say aluminum) when other processing conditions are the same. Figure 10.6 shows the effect of speed on the isotherms, for the same input power. Finally, Fig. 10.7 compares the isotherms obtained for a thin plate and a thick plate, when the processing conditions are the same. These graphs were obtained by solving equations (10.17) and (10.22). From these figures and the equations, the following deductions can be made:

1. The temperature gradient ahead of the heat source is much higher than that behind it.
2. Different points along the y-axis in a given section reach their peak temperature at different times. Points farther away have a lower peak temperature, and that is reached at a later time. The locus of points that reach their peak temperatures at the same instant is indicated by curve $n - n$ in Fig. 10.5. The curve bends backward. This is due to the finite time that it takes for heat to flow in materials, which delays the occurrence of the peak temperature at points along the y-axis. Curve $n - n$ also separates points in the plate with rising temperature from those with falling temperature. Its shape depends on both the traverse speed and the thermal diffusivity of the material.
3. A higher thermal conductivity material such as aluminum makes the isotherms more circular, reducing the temperature gradient in front of the heat source, Fig. 10.5.
4. Increasing the traverse speed makes the isotherms more elongated, while also increasing the lag of the locus $n - n$, Fig. 10.6.
5. Increasing the heat input or preheating does not change the shape of the isotherm but increases the size. This widens the fusion zone, as well as the heat-affected zone (HAZ) (see Section 11.1.3).
6. For the same conditions, a thinner plate results in a greater heat affected zone size than a thicker plate, while the thicker plate results in a higher temperature gradient, Fig. 10.7.

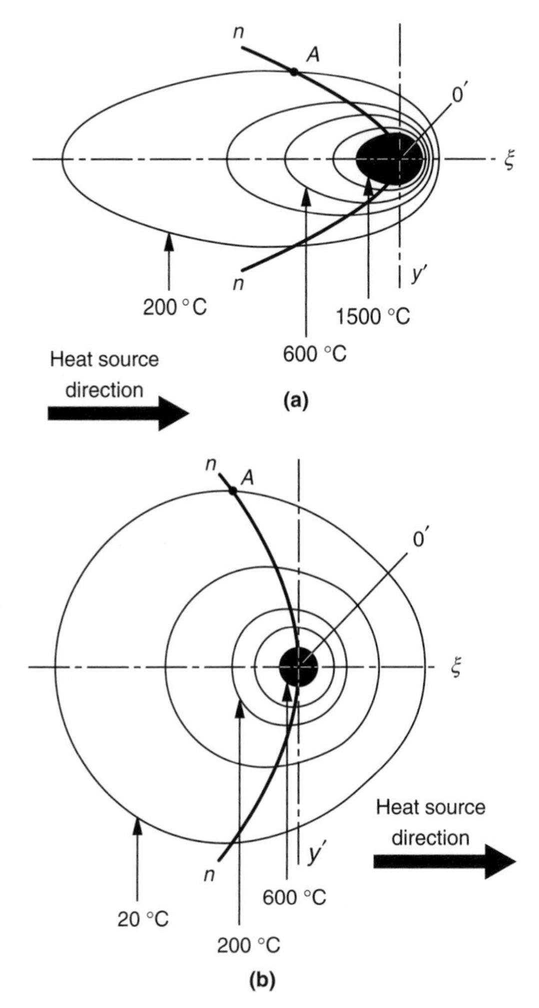

FIGURE 10.5 Temperature distribution in a plate for (a) low thermal conductivity material (low carbon steel) and (b) high thermal conductivity material (aluminum). Other processing conditions are the same. (From Rosenthal, D., 1946, *Transactions of ASME*, Vol. 68, pp. 849–866).

Example 10.1 Two thin steel plates are welded using a laser beam. The following conditions are used:

Power input = 6 kW
Plate thickness = 2.5 mm
Welding speed = 50 mm/s
Initial plate temperature = 25°C
Heat transfer efficiency = 0.7.

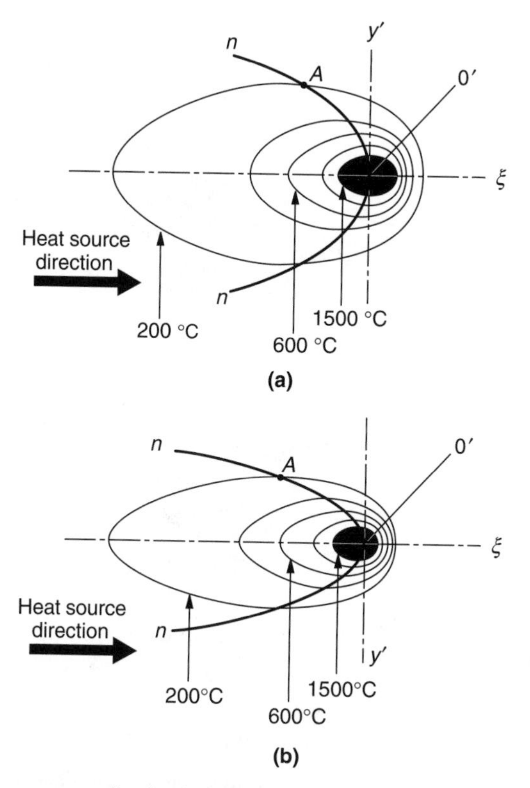

FIGURE 10.6 Temperature distribution as a function of processing speed, other processing conditions being the same. (a) Low speed. (b) High speed (doubled). (From Rosenthal, D., 1946, *Transactions of ASME*, Vol. 68, pp. 849–866).

Calculate the temperature in the plate, 2.5 mm behind the laser beam and 2 mm to one side of it.

Solution:

From Appendices 10D and 10E, the properties of the steel plate are approximated as

Average density, $\rho = 7870$ kg/m^3
Average specific heat, $c_p = 452$ J/kg K
Thermal conductivity, $k = 0.073$ W/mm K.

The distance, r, of the point of interest from the heat source is given by

$$r = \sqrt{\xi^2 + y^2} = \sqrt{(-2.5)^2 + 2^2} = 3.2 \text{ mm}$$

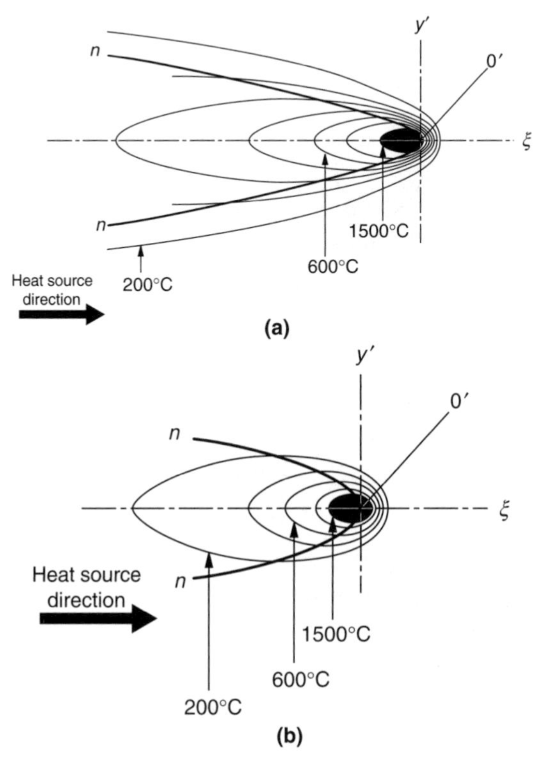

FIGURE 10.7 Temperature distribution as a function of plate thickness, other processing conditions being the same. (a) Thin plate. (b) Thick plate (more than 20 times thicker). (From Rosenthal, D., 1946, *Transactions of ASME*, Vol. 68, pp. 849–866).

The value of ξ is negative since it is behind the heat source and thus in the negative x-direction. The thermal diffusivity, κ, of the material is

$$\kappa = \frac{k}{\rho c_p} = \frac{0.073}{7.87 \times 10^{-6} \times 452} = 20.52 \text{ mm}^2/\text{s}$$

Thus

$$\frac{u_x \xi}{2\kappa} = \frac{50 \times (-2.5)}{2 \times 20.52} = -3.045$$

and

$$\frac{u_x r}{2\kappa} = \frac{50 \times 3.2}{2 \times 20.52} = 3.90$$

giving

$$K_0 \left(\frac{u_x r}{2\kappa} \right) = K_0(3.90) = 0.01248$$

Now the power, q, available at the workpiece is

$$q = 0.7 \times 6000 = 4200\,\mathrm{W}$$

and

$$T - T_0 = \frac{q}{2\pi kh} \exp\left(-\frac{u_x \xi}{2\kappa}\right) \times K_0\left(\frac{u_x r}{2\kappa}\right)$$

Therefore,

$$T = 25 + \frac{4200}{2\pi \times 0.073 \times 2.5} \times \mathrm{e}^{-(-3.045)} \times 0.01248$$
$$\Rightarrow T = 985^\circ\mathrm{C}$$

10.2.2 Peak Temperatures

The peak temperatures experienced throughout the workpiece determine the size of the heat-affected zone. The peak temperature at a given point is experienced by the point shortly after it is passed by the heat source. This is evident from an isotherm (locus of points with the same temperature) of the temperature distribution obtained from either equation (10.17) or equation (10.22) (Fig. 10.5). In other words, as the heat source passes a point, say A, its temperature keeps rising until it gets to a point where the temperature begins falling. By then, the heat source would be at a different location, say point O'.

At any position of the heat source, the isotherms of various temperatures are oval shaped. Higher temperatures have smaller size ovals. The point on any isotherm that is furthest from the x-axis (or line of motion of the heat source) is at its peak temperature at that instant. At these points, the tangent to the isotherm is parallel to the x-axis, and thus the following relation holds:

$$\left(\frac{\partial y}{\partial x}\right)_T = 0 \tag{10.23}$$

Using equations (10.17), (10.22), and (10.23) and considering temperatures in terms of distance from the fusion zone boundary, it can be shown that the peak temperature for a thin plate (line source) is given (see Problem 10.2) by

$$\frac{1}{T_p - T_0} = \frac{\sqrt{2\pi \mathrm{e}}\rho c_p h u_x Y}{q} + \frac{1}{T_\mathrm{m} - T_0} \tag{10.24}$$

while that for a thick plate (point source) is

$$\frac{1}{T_p - T_0} = \frac{2\pi k \kappa e}{q u_x}\left[2 + \left(\frac{u_x Y}{2\kappa}\right)^2\right] + \frac{1}{T_m - T_0} \tag{10.25}$$

where e = natural exponent = 2.71828, T_p is the peak or maximum temperature at a distance Y from the fusion boundary (°C), and Y is the distance from the fusion boundary at the workpiece surface (mm).

Equations (10.24) and (10.25) are applicable to single-pass processes and have to be applied to each pass by itself. They are useful for estimating the heat-affected zone size and also for showing the effect of preheat on the HAZ size. It is evident from the equations that all parameters being constant, preheating increases the size of the HAZ. Also, the size of the HAZ is proportional to the net energy input. Thus, high-intensity processes such as laser welding generally have a smaller HAZ. A high-intensity energy source results in a lower total heat input because the energy used in melting the metal is concentrated in a small region.

In general, the equation (i.e., (10.24) or (10.25)) that gives the higher computed distance from the fusion zone or higher peak temperature at a given location is the more accurate of the two.

Example 10.2 A steel plate is welded using a laser of output power 5 kW, at a speed of 40 mm/s, resulting in a heat transfer efficiency of 75%. The material has a thickness of 4 mm. If the ambient temperature is 30°C, determine

a. The peak temperature at a distance of 1.0 mm from the fusion boundary,
b. the heat-affected zone size.

Solution:

The properties of the steel plate are approximated as

Average density, $\rho = 7870$ kg/m^3
Average specific heat, $c_p = 452$ J/kg K
Thermal conductivity, $k = 0.073$ W/mm K
Melting temperature, $T_m = 1538$°C.

With a heat transfer efficiency of 75%, the net power input available at the workpiece is

$$q = 0.75 \times 5 = 3.75\text{kW} = 3750\text{ W}$$

The thermal diffusivity is given by

$$\kappa = \frac{k}{\rho c_p} = \frac{0.073}{7.87 \times 10^{-6} \times 452} = 20.52\text{ mm}^2/\text{s}$$

We calculate the peak temperature for both the two-and three-dimensional cases, and select the higher value.

1. The peak temperature calculated at a distance of 1.0 mm from the fusion boundary, for the thin plate, is obtained from equation (10.24) as

$$\frac{1}{T_\mathrm{p} - 30} = \frac{\sqrt{2\pi e} \times 7.87 \times 10^{-6} \times 452 \times 4 \times 40 \times 1.0}{3750} + \frac{1}{1538 - 30}$$

giving

$$T_\mathrm{p} = 805°\mathrm{C}$$

For the thick plate, we have from equation (10.25)

$$\frac{1}{T_\mathrm{p} - 30} = \frac{2\pi \times 0.073 \times 20.52 \times 2.71828}{3750 \times 40}\left[2 + \left(\frac{40 \times 1.0}{2 \times 7.35}\right)^2\right] + \frac{1}{1538 - 30}$$

giving

$$T_\mathrm{p} = 888°\mathrm{C}$$

Since the three-dimensional case results in the higher temperature, we consider the plate to be thick.

2. In determining the heat-affected zone size, the temperature at which the material will experience a change in properties will have to be defined. This temperature is used as the peak temperature in calculating the HAZ size. For example, for a plain carbon steel, the critical temperature is the eutectoid temperature. Thus, $T_p = 723°\mathrm{C}$. Since we have determined that the thick plate formulation is more accurate, we now substitute this value in equation (10.25), giving

$$\frac{1}{723 - 30} = \frac{2\pi \times 0.073 \times 20.52 \times 2.71828}{3750 \times 40}\left[2 + \left(\frac{40 \times Y_\mathrm{HAZ}}{2 \times 20.52}\right)^2\right] + \frac{1}{1538 - 30}$$

Thus,

$$Y_\mathrm{HAZ} = 1.65\,\mathrm{mm}$$

10.2.3 Cooling Rates

When a material is heated to a high enough temperature, the rate at which it cools afterwards determine the grain structure and phases that are formed. These in turn affect mechanical properties such as strength and ductility. For example, high cooling rates result in a finer grain structure, which increases the strength but reduces the ductility of the material. Knowledge of cooling rate is most important for materials that are polymorphic in nature, for example, steels. This enables a variety of phases with widely different mechanical properties to be produced. It is of less interest for aluminum, for example, where the cooling rates are always high. In the general case, the cooling rate at any position at any time can be obtained by differentiating equations (10.17) and (10.22) with respect to time.

However, we can make things simpler by observing that the highest cooling rates occur along the line of motion of the heat source, that is, along the centerline ($y = z = 0$ and $r = |\xi|$). We also consider only negative values of ξ, that is, behind the heat source. Substituting into equation (10.17) for the three-dimensional case (thick plate) gives

$$T - T_0 = \frac{q}{2\pi k \xi} \tag{10.26}$$

Differentiating with respect to t gives

$$\frac{dT}{dt} = \frac{\partial T}{\partial t} + \frac{\partial T}{\partial \xi}\frac{\partial \xi}{\partial t} \tag{10.27}$$

However, since the temperature undergoes no change with time with respect to the coordinate system attached to the heat source, the time derivative on the right-hand side of equation (10.27) vanishes ($\partial T(\xi, y, z, t)/\partial t = 0$), reducing equation (10.27) to the form

$$\frac{dT}{dt} = \frac{\partial T}{\partial \xi}\frac{\partial \xi}{\partial t} \tag{10.28}$$

The first term of the product in equation (10.28) is obtained by differentiating equation (10.26) and substituting an equivalent expression for ξ from equation (10.26) as follows:

$$\frac{\partial T}{\partial \xi} = -\frac{q}{2\pi k \xi^2} = -\frac{q}{2\pi k}\left[\frac{2\pi k}{q}(T - T_0)\right]^2 = -\frac{2\pi k}{q}(T - T_0)^2 \tag{10.29}$$

Now, using the chain rule we have

$$\frac{\partial \xi}{\partial t} = \frac{\partial \xi}{\partial x}\frac{\partial x}{\partial t} = u_x$$

The centerline cooling rate then becomes

$$\frac{dT}{dt} = -\frac{2\pi k u_x}{q}(T - T_0)^2 \tag{10.30}$$

Equation (10.30) is applicable to the three-dimensional case, that is, thick materials, and indicates that for such sizable workpieces, the cooling rate is proportional to the square of the temperature rise above the initial temperature.

For thin plates, that is, the two-dimensional case, the centerline cooling rate can be derived in a similar fashion (see Problem 10.5) and is given by

$$\frac{\mathrm{d}T}{\mathrm{d}t} = -2\pi k\rho c_p \left(\frac{u_x h}{q}\right)^2 (T - T_o)^3 \tag{10.31}$$

As indicated in Section 10.2.1, the plate is considered to be thin when $\beta_c < 0.6$, and thick when $\beta_c > 0.9$, (equation (10.9)) or, in general, the equation (i.e., (10.30) or (10.31)) that gives the lower cooling rate is the more accurate of the two (Fig. 10.8). This is especially the case when $0.6 < \beta_c < 0.9$. Even then, the resulting error is such as to put the calculated value on the high side. It must be realized, however, that these methods are only approximate and that where high accuracy is essential, numerical methods ought to be used.

From equations (10.30) and (10.31), the following deductions can be made (see also Fig. 10.9):

(a) Increasing the heat input (q) reduces the cooling rate, while increasing the traverse velocity (u_x) increases the cooling rate.

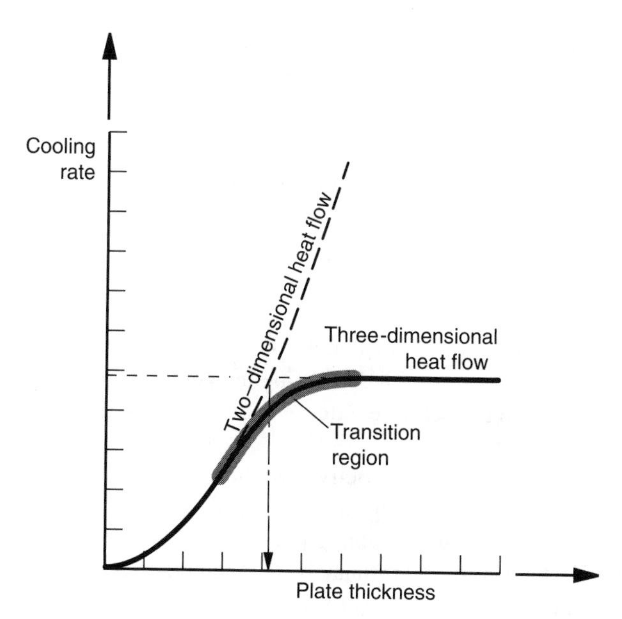

FIGURE 10.8 Representative regions for the cooling rate equations. (From Adams, C. M., 1958, *Welding Journal*, Vol. 37, pp. 210s–215s. By permission of American Welding Society.)

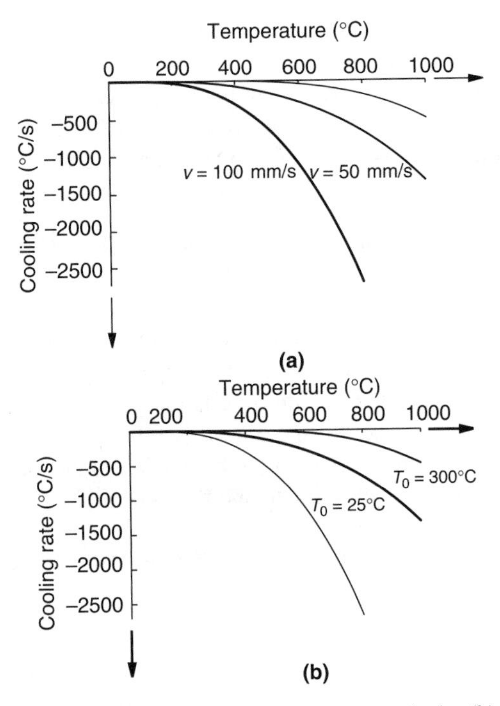

FIGURE 10.9 Dependence of cooling rate on (a) traverse velocity (b) preheat, for the conditions in Example 10.1.

(b) Increasing the initial workpiece temperature (or preheat) (T_0) reduces the cooling rate, and is more effective than increasing the heat input or reducing the traverse velocity.

(c) The cooling rate increases with an increase in plate thickness (h).

(d) A higher conductivity (k) material such as aluminum results in a higher cooling rate.

(e) The cooling rate decreases with an increasing distance (y) from the process centerline.

The last point may not be immediately obvious, unless one considers the fact that the cooling rate decreases with decreasing temperature and that from equations (10.17) and (10.22), the temperature reduces with increasing distance from the weld centerline.

Equations (10.30) and (10.31) strictly give the centerline cooling rates behind a point or line source of heat moving in a straight line at constant velocity on a flat surface and are most accurate for cooling rates at temperatures that are significantly below the melting temperature. Fortunately, the temperatures at which cooling rates are of metallurgical interest, especially for steels, are well below the melting point, and the estimates from these equations are then reasonably accurate. Furthermore, since the centerline cooling rate is only about 10% higher than in the heat-affected zone, these equations also fairly well represent cooling rates in the regions of metallurgical interest.

Example 10.3 Calculate the cooling rate at 500°C in a 3 mm thick aluminum plate clad (see Section 17.3) using a laser with output power of 4.5 kW and speed 60 mm/s. Assume an ambient temperature of 25°C and a heat transfer efficiency of 0.70. Use both the thin and thick plate equations and compare the results.

Solution:

From Appendices 10D and 10E, the properties of the aluminum plate are approximated as

Average density, $\rho = 2700\ \text{kg/m}^3$
Average specific heat, $c_p = 900$ J/kg K
Thermal conductivity, $k = 0.247$ W/mm K.
The actual power available at the workpiece is

$$q = 0.70 \times 4500 = 3150\,\text{W}$$

First, we calculate β_c using equation (10.9) to determine whether the thick or thin plate equation is more appropriate:

$$\begin{aligned}\beta_c &= h\sqrt{\frac{\rho c_p u_x (T - T_0)}{q}} \\ &= 3 \times \sqrt{\frac{2.7 \times 10^{-6} \times 900 \times 60 \times (500 - 25)}{3150}} \\ &= 0.44\end{aligned} \tag{10.9}$$

Since $\beta_c < 0.6$, the plate may be considered thin.
Thus, we have for the thin plate

$$\begin{aligned}\frac{dT}{dt} &= -2\pi \times 0.247 \times 2.7 \times 10^{-6} \times 900 \times \left(\frac{60 \times 3}{3150}\right)^2 \times (500 - 25)^3 \\ &= -1320°\text{C/s}.\end{aligned}$$

For the thick plate solution, we have

$$\begin{aligned}\frac{dT}{dt} = -\frac{2\pi \times 0.247 \times 60}{3150} \times (500 - 25)^2 \\ = -6670°\text{C/s}.\end{aligned}$$

Since the cooling rate value calculated for the thin plate is the lower of the two, it is the more accurate one.

10.2.4 Thermal Cycles

Figure 10.10 shows the variation of temperature with time at three points that are located at different distances from the fusion boundary. This is referred to as the thermal cycle diagram and can be obtained by substituting the relation $\tau = \xi/u_x$ into equation (10.17) or (10.22). This will result in equations of temperature as a function of time. The following deductions can be made from the thermal cycle diagram:

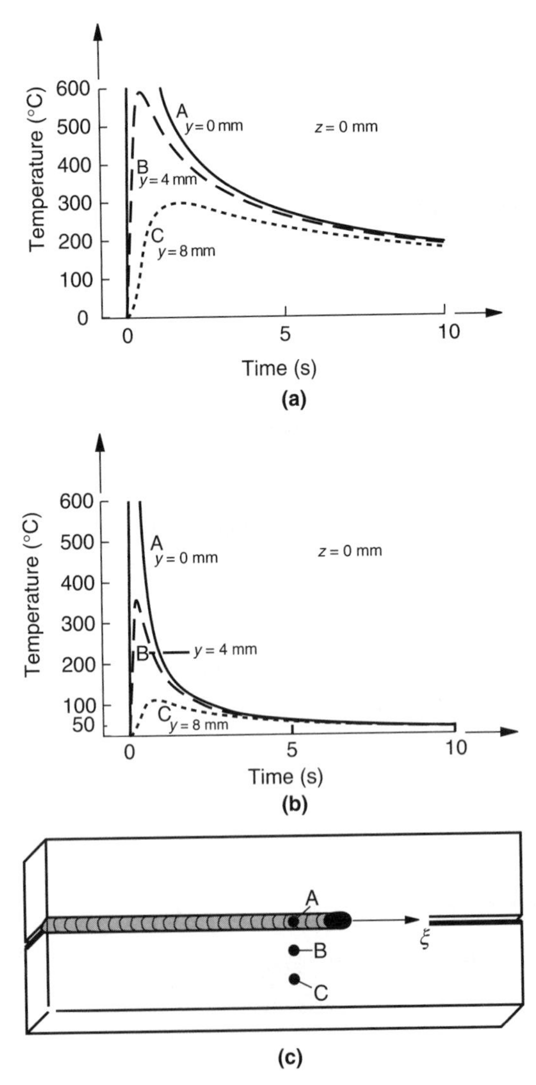

FIGURE 10.10 Thermal cycle diagrams for the conditions shown in Example 10.1. (a) Thin plate. (b) Thick plate. (c) Configuration showing simulation points.

1. The peak temperature decreases rapidly with increasing distance from the centerline.
2. The time required to reach the peak temperature increases with increasing distance from the centerline.
3. The rate of heating and the rate of cooling both decrease with increasing distance from the centerline.

10.2.5 Gaussian Heat Source

The Gaussian distribution provides a better description of most heat sources than the point or line source described in preceding sections. It is especially important for processes such as heat treatment and laser forming, where the heat source is spread over a relatively wide area. However, it does not have a complete closed form solution and requires numerical analysis. The basic elements of the analysis are presented in this section.

In analyzing the heat flow for a Gaussian heat source, the workpiece is assumed to be semi-infinite (three dimensional), that is, $-\infty < x, y < \infty$ and $0 \leq z < \infty$. The heat supply in the material per unit volume and unit time at a point (x', y', z') due to a Gaussian heat distribution for a constant power source q is

$$q_{s1}(x', y', z') = \frac{\alpha q(1-R)}{2\pi w^2} \times \mathrm{e}^{-\left[\frac{x'^2+y'^2}{2w^2}+\alpha z'\right]} \quad \text{for} \quad 0 < z' < \infty \tag{10.32}$$

where w is the Gaussian beam radius, α is the beam power absorption coefficient, which is a property of the base material. It is about 800 m^{-1} for mild steel in keyhole welding (Mazumder and Steen, 1980), R is the reflectivity, which is about 0.3 for steel.

The initial conditions of the problem are given by

$$T(x, y, z, 0) = 0 \quad \text{everywhere in the domain.} \tag{10.33}$$

The boundary conditions are (for $t > 0$)

$$T(x, y, z, t) = 0 \quad \text{for} \quad \begin{matrix} x, y \rightarrow \pm\infty \\ z \rightarrow \infty \end{matrix} \tag{10.34}$$

Convection and radiation heat losses at the top surface of the workpiece are neglected. Thus, the boundary condition at the work surface, $z = 0$, is

$$-k\frac{\partial T(x, y, z, t)}{\partial z} = 0 \quad \text{at} \quad z = 0 \tag{10.35}$$

The last boundary condition can be satisfied by using the imaginary heat source technique. For any real point heat source in the region (x', y', z'), we assume that there is a corresponding imaginary heat source at $(x', y', -z')$ of the same strength as the real source. We also assume an infinite medium instead of a semi-infinite one.

This makes the temperature field symmetric about the $z = 0$ plane; that is, there is no heat transfer across the work surface ($z = 0$). Equation (10.34) then becomes

$$T(x, y, z, t) = 0 \quad \text{for} \quad x, y, z \to \pm\infty \tag{10.36}$$

Considering the real and imaginary heat sources, q_{s1} in equation (10.32) becomes

$$q_{s1}(x', y', z') = \frac{\alpha q(1-R)}{2\pi w^2} \times \mathrm{e}^{-\left[\frac{x'^2+y'^2}{2w^2}+\alpha|z'|\right]} \quad \text{for} \quad -\infty < z' < \infty \tag{10.37}$$

where $|z'|$ is the absolute value of z', so that q_{s1} is symmetric about the $z = 0$ plane.

Now suppose that at time $t = t'$, a unit heat pulse is emitted at (x', y', z') and that an infinite medium moves uniformly past (x', y', z') with velocity $\mathbf{u} = (u_x, 0, 0)$, $u_x < 0$, parallel to the x-axis. The gage temperature at the fixed point (x, y, z) and time t due to the unit heat pulse emitted at (x', y', z') and time t' is

$$G_r(x, y, z, t) = \frac{1}{8\rho c_p[\pi\kappa(t-t')]^{3/2}} \times \mathrm{e}^{-\left[\frac{[x-u_x(t-t')-x']^2+(y-y')^2+(z-z')^2}{4\kappa(t-t')}\right]} \tag{10.38}$$

Thus, for a Gaussian source of power q, and a medium moving with velocity $\mathbf{u}$, the temperature rise over the period $t' = 0$ to t over the body is

$$\begin{aligned} T(x, y, z, t) &= \int_0^t \int_{-\infty}^{\infty} \int_{-\infty}^{\infty} \int_{-\infty}^{\infty} q_{s1}(x', y', z') \\ &\quad \times G_r(x', y', z', t'|x, y, z, t)\mathrm{d}x'\mathrm{d}y'\mathrm{d}z'\mathrm{d}t' \\ &= \int_0^t \int_{-\infty}^{\infty} \int_{-\infty}^{\infty} \int_{-\infty}^{\infty} \frac{\alpha q(1-R)}{2\pi w^2} \times \mathrm{e}^{-\left[\frac{x'^2+y'^2}{2w^2}+\alpha|z'|\right]} \\ &\quad \times \frac{1}{8\rho c_p[\pi\kappa(t-t')]^{3/2}} \\ &\quad \times \mathrm{e}^{-\left[\frac{[x-u_x(t-t')-x']^2+(y-y')^2+(z-z')^2}{4\kappa(t-t')}\right]} \mathrm{d}x'\mathrm{d}y'\mathrm{d}z'\mathrm{d}t' \end{aligned} \tag{10.39}$$

This reduces to

$$\begin{aligned} T(x, y, z, t) &= \int_0^t \frac{\alpha q(1-R)}{4\pi\rho c_p[w^2 + 2\kappa(t-t')]} \mathrm{e}^{\left[\kappa\alpha^2(t-t') - \frac{[x-u_x(t-t')]^2+y^2}{4\kappa(t-t')+2w^2}\right]} \\ &\quad \times \left\{ \mathrm{e}^{\alpha z} \operatorname{erfc}\left(\alpha\sqrt{\kappa(t-t')} + \frac{z}{2\sqrt{\kappa(t-t')}}\right) \right. \\ &\quad \left. + \mathrm{e}^{-\alpha z} \operatorname{erfc}\left(\alpha\sqrt{\kappa(t-t')} - \frac{z}{2\sqrt{\kappa(t-t')}}\right) \right\} \mathrm{d}t' \end{aligned} \tag{10.40}$$

where erfc(x) is the complementary error function, defined by

$$erfc(x) = \frac{2}{\sqrt{\pi}} \int_x^{\infty} e^{-\zeta^2} d\zeta.$$

Equation (10.40) does not have a closed form solution, and thus it is necessary to use numerical integration to obtain the distribution at time t.

10.2.6 The Two-Temperature Model

The two-temperature model to be discussed in this section is more appropriate for describing the interaction between ultrashort pulse beams and metals. The mechanism of laser–material interaction for ultrashort pulse lasers (subpicosecond pulse duration) is different from that for the traditional long-pulse lasers (nanosecond and above). Although the electrons and lattice are essentially in thermal equilibrium at the end of a long pulse, this is not necessarily the case for ultrashort pulse beams. For the latter case, electrons and the metal lattice (phonons) behave as two separate systems. Electrons located within the skin depth first absorb the photon energy. This is then transferred to the metal lattice through electron–phonon collisions. Even though the same process occurs for both long and ultrashort pulse lasers, their separate effects are more evident at short interaction times. As the electrons absorb the photon energy, they get excited to higher energy levels and rapidly thermalize, resulting in a hot free electron gas. The high-temperature electron gas then diffuses into the metal, heating up the metal lattice through electron–phonon collisions. The collision or thermalization time of the electrons is of the order of 20 fs at room temperature. However, the electron–phonon thermal relaxation time is of the order of picoseconds since several collisions are necessary for effective energy exchange between electrons and phonons.

Local thermal equilibrium is thus established between the hot electrons and the lattice during the pulse when the laser pulse duration is longer than the electron–phonon thermal relaxation time. This results in the electrons and lattice having the same temperature by the end on the pulse. The analyses in the preceding sections are then appropriate. These constitute the one-step heating model.

Electrons can normally be heated to very high temperatures within short periods since their heat capacity is relatively low. Thus, for situations where the pulse duration is of the same order of magnitude as the electron–phonon thermal relaxation time or less, the electrons and lattice are no more in local thermal equilibrium. They then have to be considered as separate systems and modeled as a two-step process that constitutes

(a) absorption of photon energy by electrons
(b) lattice heating through electron–phonon coupling.

The corresponding basic governing equations for the two-step process are as follows:

$$c_{\text{vole}}(T_e)\frac{\partial T_e}{\partial t} = \nabla \cdot (k_e \nabla T_e) - \Delta Q_{e\to l} + q_{sl} \tag{10.41}$$

$$c_{\text{voll}}(T_l)\frac{\partial T_l}{\partial t} = \Delta Q_{e\to l} \tag{10.42}$$

where $c_{\text{vol}} = \rho c_p$ is the heat capacity (specific heat per unit volume). The subscripts e and l refer to electron and lattice, respectively, and q_{sl} is the volumetric heat source term given by equation (10.32) for the three-dimensional case.

$\Delta Q_{e\to l}$ is the energy transferred by electrons to the lattice and can be approximated from free electron theory (provided the lattice temperature is not much smaller than the Debye temperature) as

$$\begin{aligned} \Delta Q_{e\to l} &= \frac{\pi^2 m_e n_e {u_s}^2}{6 t_c(T_e) T_e}(T_e - T_l) \\ &= G(T_e - T_l) \end{aligned} \tag{10.43}$$

where m_e is the electron mass, n_e is the electron number density (/m^3), u_s is the speed of sound, and $t_c(T_e)$ is the electron mean free time between collisions at temperature T_e.

This basic form of the two-step model is sometimes referred to as the parabolic two-step model (PTS), while the traditional model is referred to as the parabolic one-step model (POS). Since the expression for G in equation (10.43) can be shown to be only weakly dependent on T_e, it can be further simplified as

$$G = \frac{\pi^4 (n_e u_s k_B)^2}{18k} \tag{10.44}$$

Sample measured values for G are shown in Table 10.1.

This formulation of the governing equations neglects the energy transport by phonons since heat flux is carried mainly by free electrons for metals. The variations of the electron and lattice temperatures with time on an irradiated surface are

TABLE 10.1 Measured Values of the Electron–Phonon Coupling Factor, G, for Some Metals (Units: 10^{16} W/(m^3 K)) (Qiu and Tien, 1992, Table 1)

Metal	G	Metal	G
Copper (Cu)	4.8 ± 0.7	Vanadium (V)	523 ± 37
Silver (Ag)	2.8	Niobium (Nb)	387 ± 36
Gold (Au)	2.8 ± 0.5	Titanium (Ti)	185 ± 16
Chromium (Cr)	42 ± 5	Lead (Pb)	12.4 ± 1.4
Tungsten (W)	26 ± 3		

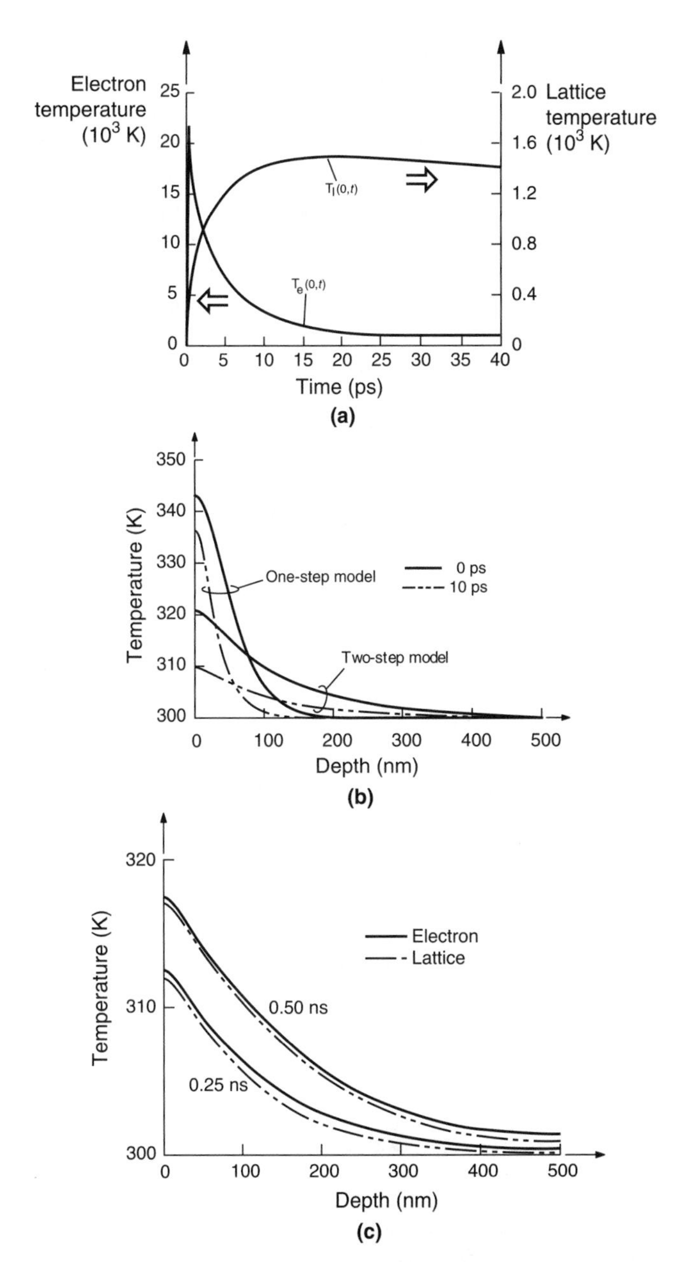

FIGURE 10.11 (a) Electron and lattice temperature variations with time on the surface of a gold film of thickness 200 nm heated with a 100 fs laser pulse of fluence 0.5 J/cm^2. (b) Lattice temperature profiles for a gold film of thickness 0.5μm heated with a 10 ps laser pulse of fluence 10mJ/cm^2. (c) Electron and lattice temperature profiles for a gold film of thickness 1μm heated with a 1 ns laser pulse of fluence 20 mJ/cm^2. (From Chen, J. K., Latham, W. P., and Beraun, J. E., 2005, *Journal of Laser Applications*, Vol. 17, No.1, pp. 63–68; Qiu, T. Q., and Tien, C. L., 1993, *ASME Journal of Heat Transfer*, Vol. 115, pp. 835–841.)

illustrated in Fig. 10.11a. The surface electron temperature is much higher than the lattice temperature in the initial stages of the process as the electrons absorb the incident radiation. However, the two temperatures eventually become equal after a long enough time as thermalization occurs. The lattice temperature profiles obtained using the two-temperature model are compared with those obtained for the corresponding one-temperature or traditional model in Fig. 10.11b for a $10ps$ laser pulse. The traditional model tends to confine the absorbed radiation energy mainly in regions close to the surface, resulting in much higher peak temperatures being predicted, while the two-temperature model predicts a much larger heat-affected zone. Similar prediction for a nanosecond laser pulse, Fig. 10.11c, shows little difference between the two models, indicating that the effects of the microscopic radiation–metal (photon–electron and electron–phonon) interactions are then less important.

The two-temperature model can be extended to account for energy transport in the hyperbolic formulation. This is referred to as the hyperbolic two-step model (HTS) when only energy transport by electrons is considered. When transport in the lattice is also considered, we have the dual hyperbolic two-step model, which is given by the following equations:

$$c_{\mathrm{vole}}(T_{\mathrm{e}})\frac{\partial T_{\mathrm{e}}}{\partial t} = -\nabla \cdot \mathbf{q_e} - G(T_{\mathrm{e}} - T_{\mathrm{l}}) + q_{\mathrm{sl}} \tag{10.45}$$

$$\tau_{\mathrm{Fe}}\frac{\partial q_{\mathrm{e}}}{\partial t} + q_{\mathrm{e}} = -k_{\mathrm{e}}\nabla T_{\mathrm{e}} \tag{10.46}$$

$$c_{\mathrm{voll}}(T_{\mathrm{l}})\frac{\partial T_{\mathrm{l}}}{\partial t} = -\nabla \cdot \mathbf{q_l} + G(T_{\mathrm{e}} - T_{\mathrm{l}}) \tag{10.47}$$

$$\tau_{\mathrm{Fl}}\frac{\partial q_{\mathrm{l}}}{\partial t} + q_{\mathrm{l}} = -k_{\mathrm{l}}\nabla T_{\mathrm{l}} \tag{10.48}$$

where $\mathbf{q_e}$, $\mathbf{q_l}$ is the heat flux vector for the electrons and lattice, respectively, and τ_{F} is the relaxation time evaluated at the Fermi surface. The relaxation time is the mean time for electrons to change their states.

Now the thermalization time, τ_{t}, that is, characteristic time for electrons and phonons to reach equilibrium, can be expressed as

$$\tau_{\mathrm{t}} = \frac{c_{\mathrm{vole0}}(T)}{G} \tag{10.49}$$

where the subscript 0 indicates evaluation at a reference temperature.

The hyperbolic transport effect on the electron temperature is more important when the laser pulse duration is shorter than τ_{F}, and the mechanism of energy deposition becomes important when the pulse duration is shorter than τ_{t}. Values of τ_{Fe} and τ_{t} for some metals are shown in Table 10.2. At room temperature, τ_{Fe} is about an order of magnitude smaller than τ_{t}. Thus, the mechanisms of energy deposition are expected to play much more important roles during short-pulse laser heating of metals than the effects of hyperbolic transport.

TABLE 10.2 Thermalization (τ_t) and Electron Relaxation (τ_{Fe}) Times at 300 K for Some Metals (Units: ps) (Qiu and Tien, 1993, Table 1)

Metal	τ_{Fe}	τ_t
Copper (Cu)	0.03	0.6
Silver (Ag)	0.04	0.6
Gold (Au)	0.04	0.8
Chromium (Cr)	0.003	0.1
Tungsten (W)	0.01	0.2
Vanadium (V)	0.002	0.06
Niobium (Nb)	0.004	0.05
Titanium (Ti)	0.001	0.05
Lead (Pb)	0.005	0.4

The PTS and POS models are special cases of the HTS model. The HTS model becomes the PTS model when the laser pulse duration is much longer than τ_{Fe}, and it becomes the POS model when the pulse duration is much longer than both τ_{Fe} and τ_t.

For one-dimensional analysis, if the spatial distribution of the laser beam is considered uniform, and the temporal shape of the pulse is assumed to be Gaussian with a FWHM (full width at half maximum) pulse duration of τ_p, then q_{s1} in equation (10.41) can be expressed as

$$q_{s1} = \alpha \frac{1}{\tau_p} \frac{Q(1-R)}{\pi w^2} \times e^{\left[-\alpha z - 4\ln 2\left(\frac{t}{\tau_p}\right)^2\right]} \tag{10.50}$$

where Q is the pulse energy, J.

In the next section, we shall address the basic concepts associated with flow in the molten pool. The discussion is useful for processes that involve melting, such as surface melting and welding.

10.3 FLUID FLOW IN MOLTEN POOL

Fluid flow in the molten pool affects the structure and properties of the fused zone, such as the geometry (penetration and bead width), surface smoothness (ripples), porosity, lack of fusion, undercutting, solidification structure, and macrosegregation (mixing). The principal flow factors that determine the process quality include the flow velocities and mode of solidification for the processes that involve melting. In this section, we focus on flow in the molten pool. Solidification is discussed in the next two chapters.

The flow system is illustrated schematically in Fig. 10.12 for conduction mode laser welding (or melting) (see Sections 16.3.1 and 17.2). The primary forces responsible

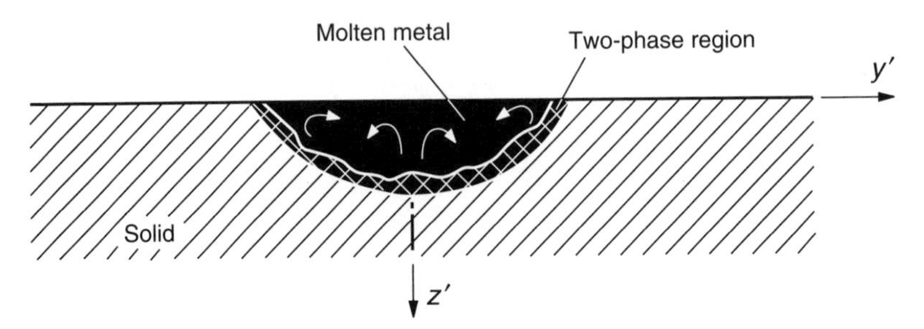

FIGURE 10.12 Configuration of the molten pool.

for flow in the molten pool for this case are the surface tension and buoyancy forces. The equations governing convection and fluid flow in the molten pool are obtained by considering mass conservation in the pool (continuity equation), momentum due to motion of the fluid (Navier–Stokes equations), and conduction in the workpiece. Heat conduction in the workpiece has already been discussed in the preceding section. In the following sections, we discuss the continuity and momentum equations that govern flow in the molten pool. This is followed by a discussion on the surface tension effect for a flat pool surface. Finally, the basic concepts involved in analyzing a pool with a free surface are outlined.

10.3.1 Continuity Equation

Using the same coordinate system as shown in Figs 10.2, 10.3, and 10.12, the general form of the differential equation for conservation of mass in a fluid medium is expressed as

$$\frac{\partial \rho}{\partial t} + \frac{\partial(\rho u_x)}{\partial x} + \frac{\partial(\rho u_y)}{\partial y} + \frac{\partial(\rho u_z)}{\partial z} = 0 \tag{10.51a}$$

or in a more compact form

$$\frac{\partial \rho}{\partial t} + \nabla \cdot (\rho \mathbf{u}) = 0 \tag{10.51b}$$

For a steady flow and incompressible fluid, the density is constant, and thus the equation reduces to the form

$$\frac{\partial u_x}{\partial x} + \frac{\partial u_y}{\partial y} + \frac{\partial u_z}{\partial z} = 0 \tag{10.52a}$$

or

$$\nabla \cdot \mathbf{u} = 0 \tag{10.52b}$$

where $\mathbf{u}$ is velocity vector (m/s) and u_i are velocity components (m/s).

10.3.2 Navier–Stokes Equations

Motion or flow of the fluid in the molten pool is also governed by the Navier–Stokes equations. Assuming incompressible fluid with negligible variations in the viscosity, these become

$$\rho \frac{d\mathbf{u}}{dt} = -\nabla P + \mu \nabla^2 \mathbf{u} + \mathbf{F_b} \tag{10.53a}$$

or

$$\rho \left(\frac{\partial \mathbf{u}}{\partial t} + \mathbf{u} \cdot \nabla \mathbf{u} \right) = -\nabla P + \mu \nabla^2 \mathbf{u} + \mathbf{F_b} \tag{10.53b}$$

which can be expanded to the form

$$\rho \left(\frac{\partial u_x}{\partial t} + u_x \frac{\partial u_x}{\partial x} + u_y \frac{\partial u_x}{\partial y} + u_z \frac{\partial u_x}{\partial z} \right) = -\frac{\partial P}{\partial x} + \mu \left(\frac{\partial^2 u_x}{\partial x^2} + \frac{\partial^2 u_x}{\partial y^2} + \frac{\partial^2 u_x}{\partial z^2} \right) + F_x \tag{10.54a}$$

$$\rho \left(\frac{\partial u_y}{\partial t} + u_x \frac{\partial u_y}{\partial x} + u_y \frac{\partial u_y}{\partial y} + u_z \frac{\partial u_y}{\partial z} \right) = -\frac{\partial P}{\partial y} + \mu \left(\frac{\partial^2 u_y}{\partial x^2} + \frac{\partial^2 u_y}{\partial y^2} + \frac{\partial^2 u_y}{\partial z^2} \right) + F_y \tag{10.54b}$$

$$\rho \left(\frac{\partial u_z}{\partial t} + u_x \frac{\partial u_z}{\partial x} + u_y \frac{\partial u_z}{\partial y} + u_z \frac{\partial u_z}{\partial z} \right) = -\frac{\partial P}{\partial z} + \mu \left(\frac{\partial^2 u_z}{\partial x^2} + \frac{\partial^2 u_z}{\partial y^2} + \frac{\partial^2 u_z}{\partial z^2} \right) + F_z \tag{10.54c}$$

where $\mathbf{F_b}$ is the body force vector with components F_x, F_y, F_z (N/m^3), P is the local fluid pressure (Pa), and μ is the dynamic or absolute viscosity (Pa s).

In laser processing, Maxwell's equations do not play a role, since there are no electromagnetic forces. Thus, the only body force left is the gravitational force in the z-direction. Hence, the body force, using the Boussinesq approximation, for the configuration shown in Fig. 10.12, reduces to

$$F_z = -\rho g \beta_T (T - T_0), \quad F_x = F_y = 0$$

where g is the gravitational acceleration (m/s^2) and β_T is the volumetric thermal coefficient of expansion ($/K$).

Here, the pressure, P, is defined to include the hydrostatic component, $\rho g h$. Equations (10.51)–(10.54), along with equation (10.2), constitute the governing equations for flow in the molten pool when the conduction mode is predominant (i.e., there is no keyhole (see Section 16.3)). They are coupled since the buoyancy term and surface tension forces provide a link between the flow and thermal equations. With the appropriate boundary conditions, they can be solved to obtain the temperature and velocity fields in the molten pool. These equations are more appropriate for processes such as conduction mode welding and surface melting. Fluid flow in laser cutting and keyhole welding will be discussed under the appropriate topics (see Chapters 15 and 16). In the following section, the influence of surface tension on flow in the molten pool is further discussed.

10.3.3 Surface Tension Effect

The surface tension that exists on the molten pool surface induces thermocapillary (surface tension driven) flow on the pool surface. This is because the surface tension force changes with temperature. Moreover, since the temperature varies on the pool surface, a surface tension gradient exists on the surface. This results in fluid being drawn along the surface from the region of lower surface tension to the region of higher surface tension. The flow induced by surface tension is referred to as Marangoni convection. The bead shape that results depends on whether the temperature derivative of the surface tension (surface tension gradient) $d\gamma/dT$ is positive or negative (Fig. 10.13).

The form of the surface tension gradient is determined by surface active elements (i.e., elements that preferentially segregate to the surface of the molten metal) such

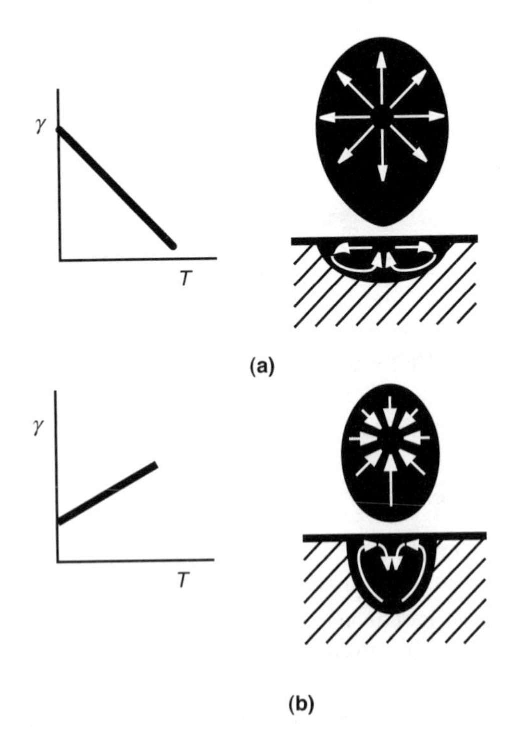

FIGURE 10.13 (a) Flow pattern for a negative gradient. (b) Flow pattern for a positive gradient. (From Heiple, C. R., and Roper, J. R., 1981, *Welding Journal*, Vol. 60, pp. 143s–145s. By permission of American Welding Society.)

as sulfur and oxygen (for iron-based alloys) in the base material. These elements alter both the surface tension and the temperature dependence of the surface tension. Thus, different heats of the same material (compositions within the specified allowable range) are known to produce different shapes of the fusion zone under the same processing conditions. This is because small variations in the sulfur content, for example, result in considerable variations in penetration.

Ordinarily, metals that are either pure or that have insignificant amounts of surface active elements have a negative surface tension gradient and thus induce outward flow of molten metal on the pool surface. Hotter material from the center is thus carried outward toward the periphery at high velocities. As the flow approaches the edge of the molten pool, it is diverted downward and then turns around, moving back to the pool center to complete the recirculation pattern (Fig. 10.13a). This results in a wider but shallower pool. The surface flow normally occurs in a thin layer called the Marangoni layer.

For a molten pool with a positive surface tension gradient, the surface tension forces will increase with temperature, that is, from the periphery or toe (which is relatively cooler) toward the center of the molten pool (which is relatively hotter). The higher surface tension forces at the center will draw or pull the material at the pool surface from the periphery toward the center. This results in surface flow inward (Fig. 10.13b). Since the bead shape is ideally symmetrical, the stream of metal from the two edges of the pool will meet and flow downward at the center, thereby carrying the hottest material from the central portion of the free surface to the bottom of the pool, resulting in deeper penetration.

Sulfur, oxygen, and selenium are among the elements that produce a positive surface tension temperature coefficient. As an illustration, addition of about 50 ppm of sulfur to a steel is adequate to increase the depth of penetration by over 80%, while about 140 ppm of selenium will increase the depth to width ratio by more than 160%.

Addition of surface active elements to the molten pool can be accomplished by doping the filler rod, precoating the joint before processing, or doping the shielding gas.

The surface tension gradient may not be constant as assumed thus far. It may vary with temperature as shown in Fig. 10.14a, being positive at lower temperatures and negative at higher temperatures. This would result in a bifurcated flow pattern (Fig. 10.14b), producing an almost cylindrical pool shape with increased penetration toward the periphery instead of the pool center.

The surface tension effect is specified as boundary shear at the free surface, that is, boundary conditions on the pool surface, since tangential force balance requires that the thermocapillary stresses should be balanced by viscous shear. The equations for the boundary shear (equation (10.55)) can be obtained by considering the fact that for Newtonian flow, the shear stress in the material (right-hand side) is proportional to the resulting shear strain rate (left-hand side). For the simple molten pool shown schematically in Fig. 10.12, we obtain these equations by making the following assumptions:

1. Shear in the shielding gas immediately above the pool surface is negligible
2. The pool surface is flat.

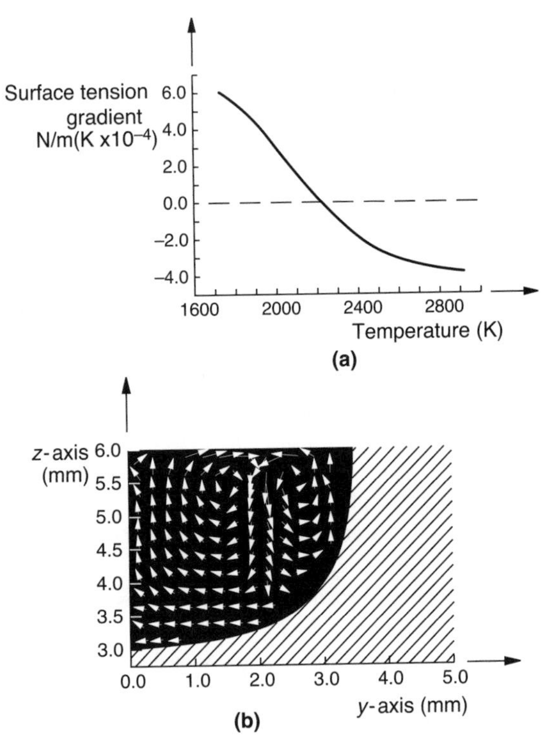

FIGURE 10.14 Effect of variable surface tension gradient on flow pattern in the molten pool. (a) Variation of surface tension gradient with temperature. (b) Flow pattern that results from the variable surface tension gradient. (From Zacharia, T., David, S. A., Vitek, J. M., and Debroy, T., 1989, *Welding Journal,* Vol. 69, pp. 499s–509s, 510s–519s. By Permission of American Welding Society).

Then, we have

$$\mu \frac{\partial u_x}{\partial z} = -\frac{\partial T}{\partial x}\frac{\partial \gamma}{\partial T} \tag{10.55a}$$

$$\mu \frac{\partial u_y}{\partial z} = -\frac{\partial T}{\partial y}\frac{\partial \gamma}{\partial T} \tag{10.55b}$$

$$u_z = 0 \tag{10.55c}$$

where γ is the surface tension of the molten pool material at temperature T (J/m^2).

Surface tension dominated flow will normally result in maximum surface velocity of the order of 10^3 mm/s, while for buoyancy dominated flow, maximum surface velocities are of the order of 10 mm/s. This indicates that the thermocapillary forces are much greater than the buoyancy forces.

When a keyhole is formed, the depth to width ratio is not significantly influenced by the relatively small amounts of surface active elements normally present in steels.

This is because the penetration under such conditions is affected primarily by the magnitude of surface tension rather than by the surface tension gradient.

10.3.4 Free Surface Modeling

In the previous section, the pool surface was assumed to be flat. In reality, the molten pool surface undergoes both spatial and temporal variations during processing. As a result of the surface curvature, the surface tension stresses have both tangential and normal components. The surface stresses may be expressed in the general form:

$$\left[P^{(2)} - P^{(1)} + \frac{\gamma}{r_s}\right] n_i = \left[\mu^{(2)}\left(\frac{\partial u_i^{(2)}}{\partial x_k} + \frac{\partial u_k^{(2)}}{\partial x_i}\right) - \mu^{(1)}\left(\frac{\partial u_i^{(1)}}{\partial x_k} + \frac{\partial u_k^{(1)}}{\partial x_i}\right)\right] n_k - \frac{\partial \gamma}{\partial x_i} \tag{10.56}$$

where $P^{(1,2)}$, $\mu^{(1,2)}$, $u_i^{(1,2)}$, $u_k^{(1,2)}$ are the pressures, dynamic viscosities, and velocity components in phases 1 and 2, respectively, r_s is the radius of curvature of the surface (m) n_i $(i = 1,\ 2,\ 3)$ are the components of the unit vector normal to the surface and directed toward the interior of phase 1.

Summation over a repeated index $(i, k = 1,\ 2,\ 3)$ is assumed.

In two-dimensional problems, the radius of curvature, r_s, is given by

$$\frac{1}{r_s} = \frac{\dfrac{\partial^2 H}{\partial x^2}}{\left[1 + \left(\dfrac{\partial H}{\partial x}\right)^2\right]^{1.5}} \tag{10.57}$$

For three-dimensional problems, $1/r_s$ is replaced by

$$\frac{1}{r_{sx}} + \frac{1}{r_{sy}} = \frac{\dfrac{\partial^2 H}{\partial x^2} + \dfrac{\partial^2 H}{\partial y^2}}{\left[1 + \left(\dfrac{\partial H}{\partial x}\right)^2 + \left(\dfrac{\partial H}{\partial y}\right)^2\right]^{1.5}} \tag{10.58}$$

where H is the height of the surface relative to some reference elevation.

Taking the projections of equation (10.56) in the directions normal and tangential to the interphase boundary, we obtain the curvature and Marangoni effects, respectively. The curvature effects (surface tension effects) are induced from normal stress balance at the free surface and can be represented as

$$P^{(2)} - P^{(1)} + \frac{\gamma}{r_s} = 2\mu^{(2)}\frac{\partial u_n^{(2)}}{\partial n} - 2\mu^{(1)}\frac{\partial u_n^{(1)}}{\partial n} \tag{10.59}$$

where n indicates the direction normal to the interface.

The shear stress also balances the surface tension gradient induced by the temperature variation. This Marangoni effect is expressed as

$$\mu^{(2)}\left(\frac{\partial u_n^{(2)}}{\partial \tau}+\frac{\partial u_\tau^{(2)}}{\partial n}\right)-\mu^{(1)}\left(\frac{\partial u_n^{(1)}}{\partial \tau}+\frac{\partial u_\tau^{(1)}}{\partial n}\right)=\frac{\partial \gamma}{\partial \tau} \tag{10.60}$$

where τ indicates the tangential direction of the free surface.

Because the fluid particle at the free surface must remain attached to the fluid surface, a kinematic condition needs to be satisfied and can be represented as

$$\frac{\partial H}{\partial t}+u_x\frac{\partial H}{\partial x}+u_y\frac{\partial H}{\partial y}=u_z \tag{10.61}$$

10.4 SUMMARY

This chapter addresses basic concepts of heat and fluid flow as they apply to laser processing and is divided into three main sections:

- Energy balance
- Heat flow in a solid workpiece
- Fluid flow in a molten pool.

The analysis on heat flow in a solid workpiece is directly applicable to processes such as laser forming and heat treatment, where no melting occurs. It also provides reasonable approximation for the heat flow in the solid part of the other processes that involve melting. With the heat source being considered as either a point or a line source, the analysis indicates that the temperature gradient ahead of the heat source is much higher than that behind it; increasing the traverse speed elongates the isotherms surrounding the heat source; higher thermal conductivity materials make the isotherms more circular, reducing the temperature gradient in front of the heat source; increasing the heat input or preheating does not change the shape of the isotherm but increases the size, thereby widening the fusion zone and heat-affected zone, and that for the same conditions, a thinner plate results in a greater heat-affected zone size than a thicker plate.

The peak temperatures experienced throughout the workpiece determine the size of the heat-affected zone. The peak temperature at a given point is experienced by the point shortly after it is passed by the heat source. The size of the HAZ increases with both preheating and the net energy input. Thus, the high-intensity laser processes generally have a smaller HAZ.

The cooling rate experienced by a material determines the grain structure and phases that are formed. Increasing the heat input reduces the cooling rate, while increasing the traverse velocity increases the cooling rate. Preheating reduces the cooling rate, while increasing the plate thickness increases the cooling rate. Higher conductivity materials such as aluminum result in a higher cooling rate.

A Gaussian distribution that is more representative of the heat source was then considered. This was followed by the two-temperature model, which is appropriate for ultrashort pulse laser beams in metals. For the latter case, the electron and lattice heat flows are modeled separately, with a coupling term. From this analysis, the surface electron temperature is found to be much higher than the lattice temperature in the initial stages of the process as the electrons absorb the incident radiation. However, the two temperatures become equal as thermalization occurs.

The analysis in the third section on flow in the molten pool is applicable to processes that involve melting such as welding, cladding, surface melting, and cutting. It was noted that a negative surface tension gradient (typical of pure metals) induces outward flow of molten metal on the pool surface, resulting in a wider and shallower pool. A positive surface tension gradient (which occurs when surface active elements such as sulfur are present), on the contrary, induces inward flow of molten metal, resulting in deeper penetration for processes that do not involve a keyhole.

REFERENCES

Abramowitz, M., and Stegun, I. A., 1964, *Handbook of Mathematical Functions*, National Bureau of Standards Applied Mathematics Series 55, Washington, DC.

Adams, C. M., 1958, Cooling rates and peak temperatures in fusion welding, *Welding Journal*, Vol. 37, pp. 210s–215s.

Anisimov, S. I., Kapeliovich, B. L., and Perel'man, T. L., 1974, Electron emission from metal surfaces exposed to ultrashort laser pulses, *Soviet Physics JETP*, Vol. 39, pp. 375–377.

Arpaci, V. S., 1966, *Conduction Heat Transfer*, Addison Wesley Publishing Co., Reading, MA.

Bardes, B. P., editor, 1978a, *Metals Handbook, Properties and Selection: Irons and Steels*, Vol. 1, 9th edition, American Society for Metals, OH.

Bardes, B. P., editor, 1978b, *Metals Handbook, Properties and Selection: Nonferrous Alloys and Pure Metals*, Vol. 2, 9th edition American Society for Metals, OH.

Carslaw, H. S., and Jaeger, J. C., 1959, *Conduction of Heat in Solids*, Clarendon Press, Oxford.

Chan, C., Mazumder, J., and Chen, M. M., 1984, A two-dimensional transient model for convection in laser melted pool, *Metallurgical Transactions A*, Vol. 15A, pp. 2175–2184.

Chande, T., and Mazumder, J., 1984, Estimating effects of processing conditions and variable properties upon pool shape, cooling rates, and absorption coefficient in laser welding, *Journal of Applied Physics*, Vol. 56, pp. 1981–1986.

Chen, J. K., Beraun, J. E., and Tzou, D. Y., 2002, Thermomechanical response of metals heated by ultrashort-pulsed lasers, *Journal of Thermal Stresses*, Vol. 25, pp. 539–558.

Chen, J. K., Latham, W. P., and Beraun, J. E., 2005, The role of electron–phonon coupling in ultrafast laser heating, *Journal of Laser Applications*, Vol. 17, No.1, pp. 63–68.

Cline, H. E., and Anthony, T. R., 1977, Heat treating and melting material with a scanning laser or electron beam, *Journal of Applied Physics*, Vol. 48, pp. 3895–3900.

Heiple, C. R., and Roper, J. R., 1981, Effect of selenium on GTAW fusion zone geometry, *Welding Journal*, Vol. 60, pp. 143s–145s.

Heiple, C. R., Roper, J. R., Stagner, R. T., and Aden, R. J., 1983, Surface active element effects on the shape of GTA, laser, and electron beam welds, *Welding Journal*, Vol. 62, pp. 72s–77s.

Kannatey-Asibu Jr., E., 1991, Thermal aspects of the split-beam laser welding concept, *ASME Journal of Engineering Materials and Technology*, Vol. 113, pp. 215–221.

Kim, Y.-S., McEligot, D. M., and Eagar, T. W., 1991, Analyses of electrode heat transfer in gas metal arc welding, *Welding Journal*, Vol. 70, pp. 20s–31s.

Korn, G. A., and Korn, T. M., 1968, *Mathematical Handbook for Scientists and Engineers*, McGraw-Hill, New York.

Korte, F., Serbin, J., Koch, J., Egbert, A., Fallnich, C., Ostendorf, A., and Chichkov, B. N., 2003, Towards nanostructuring with femtosecond laser pulses, *Journal of Applied Physics A*, Vol. 77, pp. 229–235.

Kou, S., and Wang, Y. H., 1986, Weld pool convection and its effect, *Welding Journal*, Vol. 65, pp. 63s–70s.

Kou, S., 1987, *Welding Metallurgy*, John Wiley and Sons, New York.

Lancaster, J. F., 1965, *Metallurgy of Welding, Brazing, and Soldering*, George Allen & Unwin Ltd, London.

Liu, Y. N., and Kannatey-Asibu Jr., E., 1993, Laser beam welding with simultaneous gaussian laser preheating, *ASME Journal of Heat Transfer*, Vol. 115, pp. 34–41.

Mazumder, J., and Steen, W. M., 1980, Heat transfer model for CW material processing, *Journal of Applied Physics*, Vol. 51, No. 2, pp. 941–947.

Mazumder, J., 1987, An overview of melt dynamics in laser processing, *High Power Lasers*, Vol. 801, pp. 228–241.

Oreper, G. M., and Szekeley, J., 1984, Heat and fluid-flow phenomena in weld pools, *Journal of Fluid Mechanics*, Vol. 147, pp. 53–79.

Paul, A., and Debroy, T., 1988, Free surface flow and heat transfer in conduction mode laser welding, *Metallurgical Transactions B*, Vol. 19B, pp. 851–858.

Qiu, T. Q., and Tien, C. L., 1992, Short-pulse laser heating on metals, *International Journal of Heat and Mass Transfer*, Vol. 35, No.3, pp. 719–726.

Qiu, T. Q., and Tien, C. L., 1993, Heat transfer mechanisms during short-pulse laser heating of metals, *ASME Journal of Heat Transfer*, Vol. 115, pp. 835–841.

Ramanan, N., and Korpela, S. A., 1990, Fluid dynamics of a stationary weld pool, *Metallurgical Transactions A,* Vol. 21A, pp. 45–56.

Rapkin, D. M., 1959, Temperature distribution through the weld pool in the automatic welding of aluminum, *British Welding Journal*, Vol. 6, pp. 132–137.

Raznjevic, K., 1976, *Handbook of Thermodynamic Tables and Charts*, McGraw-Hill Book Company, New York.

Rosenthal, D., 1941, Mathematical theory of heat distribution during welding and cutting, *Welding Journal*, Vol. 20, pp. 220s–234s.

Rosenthal, D., 1946, The theory of moving sources of heat and its application to metal treatments, *Transactions of ASME*, Vol. 68, pp. 849–866.

Russo, A. J., Akau, R. L., and Jellison, J. L., 1990, Thermocapillary flow in pulsed laser beam weld pools, *Welding Journal*, Vol. 69, pp. 23s–29s.

Sanders, D. J., 1984, Temperature Distributions Produced by Scanning Gaussian Laser Beams, *Applied Optics*, Vol. 23, pp. 30–35.

Sherman, F. S., 1990, *Viscous Flow*, McGraw-Hill Publishing Co., New York.

Walsh, D. W., and Savage, W. F., 1985, Technical note: autogenous GTA weldments - bead geometry variations due to minor elements, *Welding Journal,* Vol. 68, pp. 59s–62s.

Wells, A. A., 1952, Heat flow in welding, *Welding Journal*, Vol. 31, pp. 263s–267s.

Wylie, C. R., and Barrett, L. C., 1982, *Advanced Engineering Mathematics*, 5ht edition, McGraw-Hill, Inc., New York.

Yilbas, B. S., 2004, Convergence of electron kinetic, two-temperature, and one-temperature models for laser short-pulse heating, *Journal of Applied Physics A*, Vol. 79, pp. 1775–1782.

Zacharia, T., Eraslan, A. H., and Aidun, D. K., 1988, Modeling of autogenous welding, *Welding Journal,* Vol. 68, pp. 53s–62s.

Zacharia, T., David, S. A., Vitek, J. M., and Debroy, T., 1989, Weld pool development during GTA and laser beam welding of type 304 stainless steel, parts 1 and 2, *Welding Journal,* Vol. 69, pp. 499s–509s, 510s–519s.

APPENDIX 10A

List of symbols used in the chapter.

Symbol	**Parameter**	**Units**
d_e	electron heat penetration depth	m
subscripts e, l	refer to electron and lattice, respectively	–
$\mathbf{F_b}$	body force vector (F_x, F_y, F_z)	N/m^3
G	electron-phonon coupling factor	W/(m^3 K)
H	height of a surface relative to a reference elevation	mm
k_1	constant	m K
q_e	heat flux vector for electrons	W/mm^2
q_l	heat flux vector for the lattice	W/mm^2
q_n	net rate of heat flux per unit volume across element faces	W/mm^3
q_s	rate of local internal energy generated per unit volume	W/mm^3
q_{s1}	heat supply per unit volume by the power input q	W/mm^3
q_v	rate of change of internal energy per unit volume	W/mm^3
$\Delta Q_{e\to l}$	energy transferred by electrons to the lattice	J
T_{es}	electron surface temperature	°C
Y	distance from the fusion boundary at the workpiece surface	mm
$\frac{\partial T}{\partial t}$	rate of change of temperature in moving coordinate system	°C/s
β_c	constant for comparing two-and three-dimensional heat flow	–
$\phi(\xi, y, z)$	analytic function to be determined in relation to temperature	°C
τ	tangential direction of a free surface	–
$\tau_c(T_e)$	electron mean free time between collisions at temperature T_e	s
τ_F	relaxation time evaluated at the Fermi temperature	s
τ_t	thermalization time	s
χ	argument of the Bessel function	

APPENDIX 10B DERIVATION OF EQUATION (10.2A)

In deriving equation (10.2a), use is made of Fourier's law for heat conduction across a surface, which states that

$$q = -kA\frac{\mathrm{d}T}{\mathrm{d}x} \tag{10B.1}$$

where A is the area perpendicular to the flow of heat and q is the heat flux (rate of heat flow) across a surface.

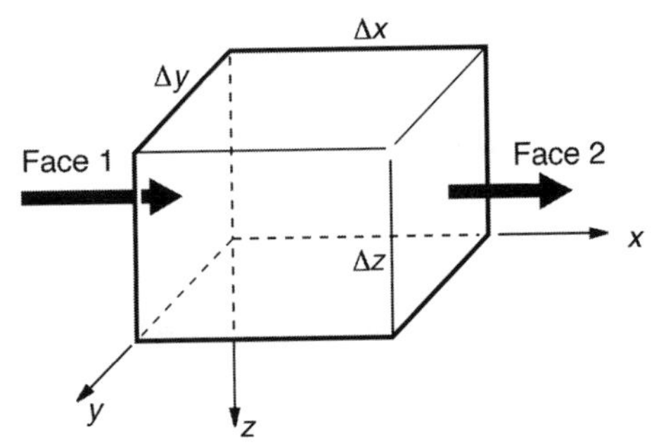

FIGURE 10B.1 Heat flux in a control volume.

To obtain the temperature distribution as a function of time, we consider a control volume as shown in Fig. 10B.1. Considering the principle of energy conservation, the rate of change of internal energy per unit volume (q_v) of the control volume must equal the sum of

1. Net rate of heat flux per unit volume across its faces, q_n, and
2. Any heat sources or sinks within it per unit volume, q_s, such as from chemical reactions or current passing through it (Joule effect), that is,

$$q_\mathrm{v} = q_\mathrm{n} + q_\mathrm{s} \tag{10B.2}$$

The rate of change of internal energy per unit volume q_v is given by:

$$q_\mathrm{v} = \rho c_p \frac{\mathrm{d}T}{\mathrm{d}t} \tag{10B.3}$$

To obtain the net rate of heat flow per unit volume across the faces, q_n, we consider the heat flux across the individual faces. For face 1,

$$q_1 = -\left(k\Delta y\Delta z\frac{\partial T}{\partial x}\right)_x \tag{10B.4a}$$

and across face 2

$$q_2 = -\left(k\Delta y\Delta z\frac{\partial T}{\partial x}\right)_{x+\Delta x} \tag{10B.4b}$$

Thus, the net rate of heat transfer per unit volume in the x-direction into the control volume is

$$\frac{q_1 - q_2}{\Delta x\Delta y\Delta z} = \frac{\left(k\frac{\partial T}{\partial x}\right)_{x+\Delta x} - \left(k\frac{\partial T}{\partial x}\right)_x}{\Delta x} \tag{10B.5}$$

In the limit, as $\Delta x \to 0$, we have

$$\lim_{\Delta x\to 0}\frac{q_1 - q_2}{\Delta x\Delta y\Delta z} = \frac{\partial}{\partial x}\left(k\frac{\partial T}{\partial x}\right) \tag{10B.6}$$

Similar expressions hold for the y and z directions. Thus, considering all three directions, we have

$$q_{\text{n}} = \frac{\partial}{\partial x}\left(k\frac{\partial T}{\partial x}\right) + \frac{\partial}{\partial y}\left(k\frac{\partial T}{\partial y}\right) + \frac{\partial}{\partial z}\left(k\frac{\partial T}{\partial z}\right)$$

Then, from equation (10B.2) we get

$$\rho c_p\frac{\mathrm{d}T}{\mathrm{d}t} = \frac{\partial}{\partial x}\left(k\frac{\partial T}{\partial x}\right) + \frac{\partial}{\partial y}\left(k\frac{\partial T}{\partial y}\right) + \frac{\partial}{\partial z}\left(k\frac{\partial T}{\partial z}\right) + q_{\text{s}} \tag{10B.7a}$$

or in a more compact form

$$\rho c_p\frac{\mathrm{d}T}{\mathrm{d}t} = \nabla\cdot(k\nabla T) + q_{\text{s}} \tag{10B.7b}$$

APPENDIX 10C MOVING HEAT SOURCE

In deriving equation (10.5), let us consider the heat source to be moving along the positive x-axis with a velocity u_x (Fig. 10.1). A new set of coordinate axes is attached to the moving heat source. The new (moving) coordinate system, ξ, y', z', t', is related to the old (fixed) coordinate system, x, y, z, t, as follows:

$$\xi = x - u_x t, \quad y' = y, \quad z' = z, \quad t' = t \tag{10C.1}$$

Then, coordinate transformation gives

$$\frac{\partial T}{\partial x} = \frac{\partial T}{\partial \xi}\frac{\partial \xi}{\partial x} = \frac{\partial T}{\partial \xi}, \quad \frac{\partial^2 T}{\partial x^2} = \frac{\partial^2 T}{\partial \xi^2} \tag{10C.2}$$

$$\frac{\partial T}{\partial y} = \frac{\partial T}{\partial y'}\frac{\partial y'}{\partial y} = \frac{\partial T}{\partial y'}, \quad \frac{\partial^2 T}{\partial y^2} = \frac{\partial^2 T}{\partial y'^2} \tag{10C.3}$$

$$\frac{\partial T}{\partial z} = \frac{\partial T}{\partial z'}\frac{\partial z'}{\partial z} = \frac{\partial T}{\partial z'}, \quad \frac{\partial^2 T}{\partial z^2} = \frac{\partial^2 T}{\partial z'^2} \tag{10C.4}$$

Also,

$$\frac{dT}{dt} = \frac{\partial T}{\partial t'}\frac{dt'}{dt} + \frac{\partial T}{\partial \xi}\frac{d\xi}{dt} = \frac{\partial T}{\partial t'} - u_x\frac{\partial T}{\partial \xi} \tag{10C.3}$$

Substituting into equation (10.2) and noting that $t' = t$ and $\frac{\partial T}{\partial y} = \frac{\partial T}{\partial y'}$, and so on, we get for the moving coordinate system:

$$\frac{\partial T}{\partial t} = \kappa\left(\frac{\partial^2 T}{\partial \xi^2} + \frac{\partial^2 T}{\partial y^2} + \frac{\partial^2 T}{\partial z^2}\right) + u_x\frac{\partial T}{\partial \xi} \tag{10C.4}$$

APPENDIX 10D

Thermal and physical properties of some common materials (Raznjevic, K., 1976; Bardes, 1978a, 1978b)

Material	Density at 20°C (kg/m^3)	Melting Point (°C)	Heat of Fusion (kJ/kg)	Boiling Point (°C)	Heat of Vaporization (kJ/kg)
Aluminum (Al)	2700	660.4	397	2494	10780
Carbon (C)	–	3540	–	4000	5024
Chromium (Cr)	7190	1875	258–283	2680	6168
Copper (Cu)	8930	1085	205	2595	4729
Gold (Au)	19290	1064	62.8	2857	1698.7
Iron (Fe)	7870	1538	247 ±7	2870	7018
Lead (Pb)	11340	327.4	22.98-23.38	1750	945.34
Magnesium (Mg)	1738	650	360-377	1107 ±10	5150–5400
Manganese (Mn)	7300	1244 ±4	267	2095	4090
Molybdenum (Mo)	10220	2610	270	5560	5123
Nickel (Ni)	8900	1453	293	2730	6196
Niobium (Nb)	8570	2468	290	4927	7490
Silicon (Si)	2330	1410	1807.9	3280	10606
Silver (Ag)	10490	961.9	104.2	2163	2630
Tantalum (Ta)	16600	2996	145–174	5427	4160–4270

Tin (Sn)	7168	231.9	59.5	2770	2400
Titanium (Ti)	4507	1668 ±10	440	3260	9830
Tungsten (W)	19254	3410	220 ±36	5700 ±200	4815
Vanadium (V)	6100	1900 ±25	–	–	–
Zinc (Zn)	7130	420	100.9	906	1782
Zirconium (Zr)	6530	1852	25	3700	6520

APPENDIX 10E

Thermal properties of some common materials (Raznjevic, K., 1976; Bardes, 1978a, 1978b)[a]

Material	Coefficient of Linear Expansion at 20–100°C (μm/m K)	Specific Heat at 20°C (J/kg K)	Thermal Conductivity at 20°C (W/m K)
Aluminum (Al)	23.6	900 at 25°C	247 at 25°C
Carbon, Graphite (G)	–	708–934	11.63–174.45
Chromium (Cr)	6.2	459.8	67
Copper (Cu)	16.5	386	398 at 27°C
Gold (Au)	14.2	128 at 25°C	317.9 at 0°C
Iron (Fe)	12 at 100°C	452	73
Lead (Pb)	29.3	128.7 at 25°C	34.8
Magnesium (Mg)	25.2	1025	418
Manganese (Mn)	22.8	477 at 25°C	50.2 at 0°C
Molybdenum (Mo)	5.2	276	142
Nickel (Ni)	13.3	444	82.9 at 100°C
Niobium (Nb)	7.31 at 300°C	270	52.3 at 0°C
Silicon (Si)	2.8–7.9	678	83.7
Silver (Ag)	19	235 at 25°C	428
Tantalum (Ta)	6.5	139.1 at 0°C	54.4
Tin (Sn)	23.1 at 50°C	205 at 10°C	62.8 at 0°C
Titanium (Ti)	10.8 at 100°C	99.3 at 50°C	11.4 at −240°C
Tungsten (W)	4.5 at 100°C	134	151.2 at 100°C
Vanadium (V)	8.3	498	31 at 100°C
Zinc (Zn)	15	382	113 at 25°C
Zirconium (Zr)	5.85	300 at ≈ 100°C	21.1 at 25°C

[a]Thermal properties often vary significantly with temperature. The values given are average and/or approximate. For more precise values at a given temperature, the reader should consult one of the references.

APPENDIX 10F

Reflectance of Some Common Materials (Bardes, 1978a, 1978b)

Material	$\lambda = 0.25\ \mu m$	$\lambda = 1.0\ \mu m$	$\lambda = 10.0\ \mu m$
Aluminum (Al)	86–87	96	97
Chromium (Cr)	67 at $\lambda = 0.3\ \mu m$	63 at $\lambda = 1.0\ \mu m$	88 at $\lambda = 4\ \mu m$
Copper (Cu)	25.9	90.1	98.4
Gold (Au)	40	96	96
Iron (Fe)	–	65 at $\lambda = 1.5\ \mu m$	97 at $\lambda = 15\ \mu m$
Lead (Pb)	62 at $\lambda = 0.589\ \mu m$	–	–
Magnesium (Mg)	72 at $\lambda = 0.5\ \mu m$	74	93 at $\lambda = 9.0\ \mu m$
Molybdenum (Mo)	46 at $\lambda = 0.5\ \mu m$	–	93
Nickel (Ni)	41.3 at $\lambda = 0.3\ \mu m$	–	–
Silicon (Si)	29–39 at $\lambda = 0.13 - 0.2\ \mu m$	–	
Silver (Ag)	≈ 30	≈ 95	≈ 95
Tin (Sn)	80 at $\lambda = 0.546\ \mu m$	–	–
Zinc (Zn)	74.7 at $\lambda = 0.5\ \mu m$	53.3	–

APPENDIX 10G

Table of modified Bessel function of the second kind of order zero (Table F-11, Korn and Korn, 1968)

Bessel Function: $K_0(x)$

x	0	1	2	3	4	5	6	7	8	9
0.0	∞	4.721	4.028	3.624	3.337	3.114	2.933	2.780	2.647	2.531
1	2.427	2.333	2.248	2.170	2.097	2.030	1.967	1.909	1.854	1.802
2	1.753	1.706	1.662	1.620	1.580	1.542	1.505	1.470	1.436	1.404
3	1.372	1.342	1.314	1.286	1.259	1.233	1.208	1.183	1.160	1.137
4	1.115	1.093	1.072	1.052	1.032	1.013	0.9943	9761	9584	9412
5	0.9244	9081	8921	8766	8614	8466	8321	8180	8042	7907
6	7775	7646	7520	7397	7277	7159	7043	6930	6820	6711
7	6605	6501	6399	6300	6202	6106	6012	5920	5829	5740
8	5653	5568	5484	5402	5321	5242	5165	5088	5013	4940
9	4867	4796	4727	4658	4591	4524	4459	4396	4333	4271
1.0	4210	4151	4092	4034	3977	3922	3867	3813	3760	3707
1	3656	3605	3556	3507	3459	3411	3365	3319	3273	3229
2	3185	3142	3100	3058	3017	2976	2936	2897	2858	2820
3	2782	2746	2709	2673	2638	2603	2569	2535	2502	2469
4	2437	2405	2373	2342	2312	2282	2252	2223	2194	2166
5	2138	2111	2083	2057	2030	2004	1979	1953	1928	1904
6	1880	1856	1832	1809	1786	1763	1741	1719	1697	1676
7	1655	1634	1614	1593	1573	1554	1534	1515	1496	1478
8	1459	1441	1423	1406	1388	1371	1354	1337	1321	1305
9	1288	1273	1257	1242	1226	1211	1196	1182	1167	1153
2.0	1139	1125	1111	1098	1084	1071	1058	1045	1033	1020
1	1008	*9956	*9836	*9717	*9600	*9484	*9370	*9257	*9145	*9035
2	0.08927	8820	8714	8609	8506	8404	8304	8204	8106	8010
3	7914	7820	7726	7634	7544	7454	7365	7278	7191	7106
4	7022	6939	6856	6775	6695	6616	6538	6461	6384	6309
5	6235	6161	6089	6017	5946	5877	5808	5739	5672	5606
6	5540	5475	5411	5348	5285	5223	5162	5102	5042	4984
7	4926	4868	4811	4755	4700	4645	4592	4538	4485	4433
8	4382	4331	4281	4231	4182	4134	4086	4039	3992	3946
9	3901	3856	3811	3767	3724	3681	3638	3597	3555	3514
3.0	3474	3434	3395	3356	3317	3279	3241	3204	3168	3131
1	3095	3060	3025	2990	2956	2922	2889	2856	2824	2791
2	2759	2728	2697	2666	2636	2606	2576	2547	2518	2489
3	2461	2433	2405	2378	2351	2325	2298	2272	2246	2221
4	2196	2171	2146	2122	2098	2074	2051	2028	2005	1982
5	1960	1938	1916	1894	1873	1852	1831	1810	1790	1770
6	1750	1730	1711	1692	1673	1654	1635	1617	1599	1581
7	1563	1546	1528	1511	1494	1477	1461	1445	1428	1412
8	1397	1381	1366	1350	1335	1320	1306	1291	1277	1262
9	1248	1234	1221	1207	1194	1180	1167	1154	1141	1129
4.0	1116	1104	1091	1079	1067	1055	1044	1032	1021	1009
1	0.009980	9869	9760	9652	9545	9439	9334	9231	9128	9027
2	8927	8829	8731	8634	8539	8444	8351	8259	8167	8077
3	7988	7900	7813	7726	7641	7557	7473	7391	7309	7229
4	7149	7070	6992	6915	6839	6764	6689	6616	6543	6471
5	6400	6329	6260	6191	6123	6056	5989	5923	5858	5794
6	5730	5668	5605	5544	5483	5423	5363	5305	5246	5189
7	5132	5076	5020	4965	4911	4857	4804	4751	4699	4648
8	4597	4547	4497	4448	4399	4351	4304	4257	4210	4164
9	4119	4074	4030	3986	3942	3899	3857	3814	3773	3732
5.0	3691	3651	3611	3572	3533	3494	3456	3419	3382	3345

Bessel Function: $K_0(x)$

x	0	1	2	3	4	5	6	7	8	9
5.0	0.003691	3651	3611	3572	3533	3494	3456	3419	3382	3345
1	3308	3272	3237	3202	3167	3132	3098	3065	3031	2998
2	2966	2934	2902	2870	2839	2808	2778	2748	2718	2688
3	2659	2630	2602	2574	2546	2518	2491	2464	2437	2411
4	2385	2359	2333	2308	2283	2258	2234	2210	2186	2162
5	2139	2116	2093	2070	2048	2026	2004	1982	1961	1939
6	1918	1898	1877	1857	1837	1817	1798	1778	1759	1740
7	1721	1703	1684	1666	1648	1630	1613	1595	1578	1561
8	1544	1528	1511	1495	1479	1463	1447	1432	1416	1401
9	1386	1371	1356	1342	1327	1313	1299	1285	1271	1258
6.0	1244	1231	1217	1204	1191	1179	1166	1153	1141	1129
1	1117	1105	1093	1081	1070	1058	1047	1035	1024	1013
2	1002	*9918	*9811	*9706	*9602	*9499	*9398	*9297	*9197	*9099
3	0.0009001	8905	8810	8715	8622	8530	8438	8348	8259	8171
4	8083	7997	7911	7827	7743	7660	7578	7497	7417	7338
5	7259	7182	7105	7029	6954	6880	6806	6734	6662	6591
6	6520	6451	6382	6314	6246	6180	6114	6048	5984	5920
7	5857	5795	5733	5672	5611	5551	5492	5434	5376	5318
8	5262	5206	5150	5095	5041	4987	4934	4882	4830	4778
9	4728	4677	4627	4578	4529	4481	4434	4386	4340	4294
7.0	4248	4203	4158	4114	4070	4027	3984	3942	3900	3858
1	3817	3777	3737	3697	3658	3619	3580	3542	3505	3468
2	3431	3394	3358	3323	3287	3253	3218	3184	3150	3117
3	3084	3051	3019	2987	2955	2924	2893	2862	2832	2802
4	2772	2742	2713	2685	2656	2628	2600	2573	2545	2518
5	2492	2465	2439	2413	2388	2363	2338	2313	2288	2264
6	2240	2216	2193	2170	2147	2124	2102	2079	2057	2036
7	2014	1993	1972	1951	1930	1910	1890	1870	1850	1830
8	1811	1792	1773	1754	1736	1717	1699	1681	1664	1646
9	1629	1611	1594	1578	1561	1545	1528	1512	1496	1480
8.0	1465	1449	1434	1419	1404	1389	1374	1360	1346	1331
1	1317	1303	1290	1276	1263	1249	1236	1223	1210	1198
2	1185	1172	1160	1148	1136	1124	1112	1100	1089	1077
3	1066	1055	1043	1032	1022	1011	10002	*9897	*9793	*9690
4	.00009588	9487	9387	9288	9191	9094	8998	8904	8810	8717
5	8626	8535	8445	8356	8269	8182	8096	8011	7926	7843
6	7761	7679	7598	7519	7439	7361	7284	7208	7132	7057
7	6983	6909	6837	6765	6694	6624	6554	6485	6417	6350
8	6283	6217	6152	6088	6024	5961	5898	5836	5775	5714
9	5654	5595	5536	5478	5420	5364	5307	5252	5197	5142
9.0	5088	5035	4982	4930	4878	4827	4776	4726	4677	4628
1	4579	4531	4484	4437	4390	4344	4299	4254	4209	4165
2	4121	4078	4036	3993	3951	3910	3869	3829	3789	3749
3	3710	3671	3632	3594	3557	3519	3483	3446	3410	3374
4	3339	3304	3270	3235	3202	3168	3135	3102	3070	3038
5	3006	2974	2943	2912	2882	2852	2822	2793	2763	2734
6	2706	2678	2650	2622	2595	2567	2541	2514	2488	2462
7	2436	2411	2385	2360	2336	2311	2287	2263	2240	2216
8	2193	2170	2148	2125	2103	2081	2059	2038	2017	1995
9	1975	1954	1934	1913	1894	1874	1854	1835	1816	1797
10.0	1778	1759	1741	1723	1705	1687	1670	1652	1635	1618

PROBLEMS

10.1. The governing equation for conduction heat flow in a solid with a coordinate system fixed at a stationary origin relative to the solid may be expressed as

$$\frac{\mathrm{d}T}{\mathrm{d}t} = \kappa\left(\frac{\partial^2 T}{\partial x^2} + \frac{\partial^2 T}{\partial y^2} + \frac{\partial^2 T}{\partial z^2}\right)$$

Show that for the moving heat source case with a coordinate system moving with the heat source along the x-axis, as shown in Fig. 10.1, the corresponding governing equation obtained by a coordinate transformation from the plate to the heat source, with x replaced by ξ, that is, $\xi = x - u_x t$, is given by

$$\frac{\partial T}{\partial t} = \kappa\left(\frac{\partial^2 T}{\partial \xi^2} + \frac{\partial^2 T}{\partial y^2} + \frac{\partial^2 T}{\partial z^2}\right) + u_x \frac{\partial T}{\partial \xi}$$

10.2. Given that

$$T - T_0 = \frac{q}{2\pi k h} \exp\left(-\frac{u_x \xi}{2\kappa}\right) \times K_0\left(\frac{u_x r}{2\kappa}\right)$$

and that

$$\left(\frac{\partial y}{\partial x}\right)_T = 0$$

and considering temperatures in terms of distance from the fusion zone boundary, show that the peak temperature for a thin plate is given by

$$\frac{1}{T_{\mathrm{p}} - T_0} = \frac{\sqrt{2\pi e}\,\rho c_p h u_x Y}{q} + \frac{1}{T_{\mathrm{m}} - T_o}$$

10.3. Given that

$$T - T_0 = \frac{q}{2\pi k r} \exp\left(-\frac{u_x(r + \xi)}{2\kappa}\right)$$

and that

$$\left(\frac{\partial y}{\partial x}\right)_T = 0$$

and considering temperatures in terms of distance from the fusion zone boundary, show that the peak temperature for a thick plate is given by

$$\frac{1}{T_p - T_o} = \frac{2\pi k\kappa e}{q u_x}\left[2 + \left(\frac{u_x Y}{2\kappa}\right)^2\right] + \frac{1}{T_{\mathrm{m}} - T_0}$$

10.4. The temperature distribution in a semi-infinite plate for a moving point heat source is given as

$$T - T_0 = \frac{q}{2\pi k r}\exp\left[-\frac{u_x(r+\xi)}{2\kappa}\right]$$

Derive an expression for the cooling rate at points of the plate where the cooling rates are highest. Where do these occur?

10.5. The temperature distribution in a thin plate for a moving line heat source is given as

$$T - T_0 = \frac{q}{2\pi k h}\exp\left(-\frac{u_x \xi}{2\kappa}\right) \times K_0\left(\frac{u_x r}{2\kappa}\right)$$

Derive an expression for the cooling rate at points of the plate where the cooling rates are highest.

Hint: For large values of z, the modified Bessel function of the second kind and order zero, $K_0(z)$, can be approximated by

$$K_0(z) = \left(\frac{\pi}{2z}\right)^{1/2} \cdot e^{-z}$$

10.6. A new concept in laser welding involves splitting the laser beam into two as shown in the figure below, with the notion that it would reduce cooling rates. If the heat input of each of the heat sources is q, it can be shown that the temperature distribution at any point for this configuration is given by

$$T - T_0 = \frac{q}{2\pi k}\left[\frac{1}{|\xi|}e^{-\frac{u_x(\xi+|\xi|)}{2\kappa}} + \frac{1}{|\xi'|}e^{-\frac{u_x(\xi'+|\xi'|)}{2\kappa}}\right]$$

And for points behind the second heat source

$$T - T_0 = \frac{q}{2\pi k}\left(\frac{1}{\xi} + \frac{1}{\xi'}\right)$$

$$\xi' = \xi - d$$

$$\xi = x - u_x t$$

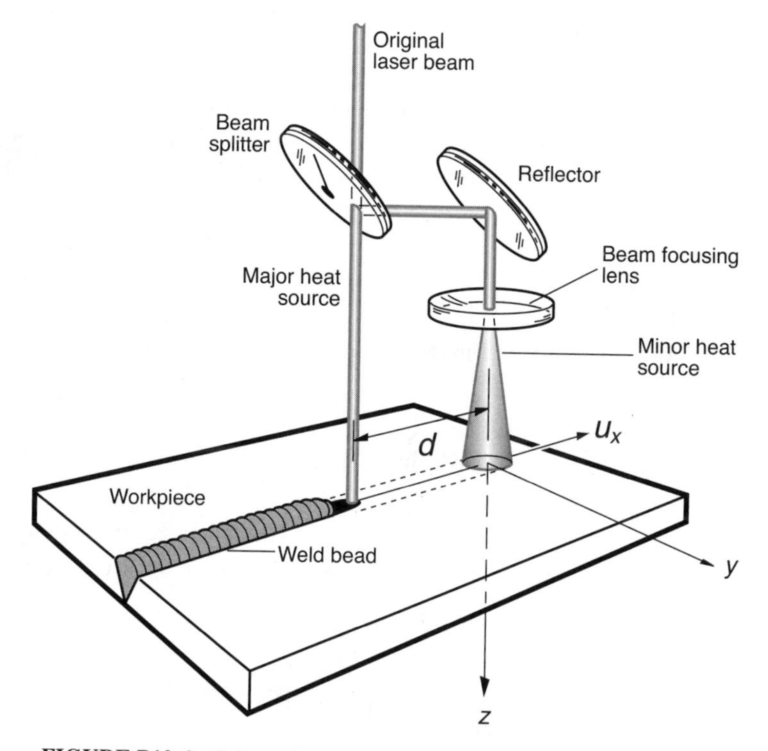

FIGURE P10.6 Schematic of the multiple beam laser welding concept.

On basis of this, show that the cooling rate along the weld centerline and behind the second heat source is given by

$$\frac{\mathrm{d}T}{\mathrm{d}t} = -\left(\frac{2\pi k u_x (T - T_0)^2}{q} - \frac{q u_x}{\pi k \xi \xi'}\right)$$

10.7. Set up the appropriate boundary conditions for the velocity and temperature fields associated with flow in a stationary molten pool with a flat surface.

Hint: Consider the top surface, bottom surface, centerline, side surface, locations far removed from heat source.

10.8. Set up the appropriate boundary conditions for flow in a moving molten pool for:

(a) The top surface of the molten pool
(b) Along the molten pool center plane
(c) Far away from the top surface
(d) Far away from the centerline.

10.9. For the problem in Example 10.2, determine the temperature at a distance of 6 mm directly behind the heat source.

10.10. A 5 kW laser beam is used to remelt silicon in a process to produce material for solar cells. The silicon is 100 mm wide and 2 mm thick. The molten pool passes across the width of the silicon, with a velocity of 4 m/s. The ambient temperature is 293 K and the heat transfer efficiency is 50%.

(a) Calculate the maximum cooling rate in the silicon, after cooling to a temperature 400°C below the melting temperature. At what location is this cooling rate achieved?

(b) At what distance from the fusion zone boundary is a peak temperature of 1580 K achieved?

10.11. A manufacturing engineer is responsible for reducing the cooling rates that occur during laser cladding of a high hardenability material. If the workpiece is determined to be relatively thin,

(a) Identify three primary ways that he or she can reduce the cooling rates.

(b) Which of these three approaches would be most effective? Explain.

10.12. Two wide and thin pieces of cold-worked magnesium plates, 2.5 mm thickness, are to be butt welded using a 5 kW laser.

(a) Calculate the surface temperature at a point 6 mm behind the laser beam and 3 mm to one side.

(b) Estimate the size of the heat-affected zone for this process, assuming recrystallization occurs at a temperature equal to half the melting temperature.

You *may* assume that a centerline cooling rate (*at* 500°C) has been determined to be 3500°C/s, the initial workpiece temperature is 25°C, and the process has a heat transfer efficiency of 75%.

(c) Now assume that the workpieces have been preheated to a temperature of 75°C. Recalculate the surface temperature at the same point (assume that the centerline cooling rate remains unchanged). Does it change? Also, does this result in a larger or smaller HAZ?

11 The Microstructure

The heat flow associated with laser processing does affect the microstructure of the part that is produced. In this chapter, basic microstructural concepts and defect formation, as they relate to laser processing, are discussed. Defects are discussed in this chapter since they are closely tied to the microstructural changes that occur during processing.

The chapter starts with a brief outline of the major microstructural zones that occur in the part during processing, addressing the fusion zone, the zone of partial melting, and the heat-affected zone. Under fusion zone, basic concepts of solidification are introduced, starting with nucleation, where the conditions necessary for the formation of a stable nucleus are outlined. The concepts of homogeneous and heterogeneous nucleation are presented. This is followed by a discussion on the growth of the stable nuclei to form crystals or grains.

The microstructure that results as the molten material solidifies is presented next, outlining the principal types of structure that may result, that is, planar, cellular, or dendritic, and the criterion that determines which structure is most likely to form. Techniques for controlling nucleation and also for grain refinement are then presented. This is followed by a discussion on coring, which results in variation in material properties across the fusion zone.

The zone of partial melting is then discussed, followed by the heat-affected zone that results for pure metals, precipitation-hardening and nonferrous alloys, and steels. Next, common discontinuities that can result during laser processing are discussed. These include porosity, cracking, incomplete penetration, lack of fusion, and undercutting.

The process microstructure is discussed next.

11.1 PROCESS MICROSTRUCTURE

The processes that involve fusion (laser cutting, welding, surface melting, and cladding) have four distinct zones in the region where the heat is applied (Fig. 11.1). These are the

1. Fusion zone
2. Zone of partial melting

Principles of Laser Materials Processing, by Elijah Kannatey-Asibu, Jr.

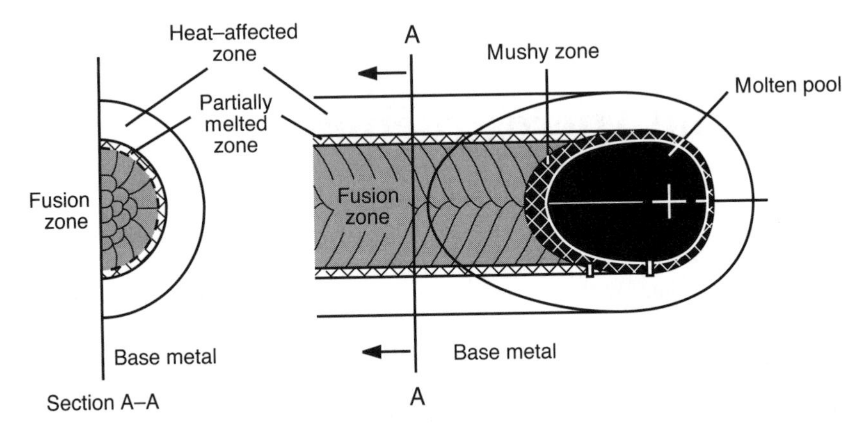

FIGURE 11.1 Major microstructural zones associated with laser processing. (From Kou, S., 1987, *Welding Metallurgy*. Reprinted with permission from John Wiley & Sons, Inc.)

3. Heat-affected zone (HAZ)
4. Base material.

The first three zones will be discussed individually in the following sections. For processes that do not involve fusion (laser heat treatment and laser forming), microstructural changes occur only in the heat-affected zone.

11.1.1 Fusion Zone

The fusion zone is the portion of the material that melts. The discussion on the fusion zone starts with an introduction to the solidification process, focusing on grain nucleation and growth. This is followed by the microstructure of the fusion zone, nucleation and grain refinement in the molten pool, and finally, coring and macrosegregation. Solidification is discussed in greater detail in the next chapter.

11.1.1.1 Initial Solidification Solidification is the process of transformation from the liquid to the solid phase. It generally occurs in two stages:

1. Nucleation, which occurs when a small piece of stable solid (nucleus) forms from the liquid.
2. Growth, which occurs as atoms from the liquid are attached to the nucleus.

Nucleation A nucleus is the smallest particle of the new phase (solid in this case) that can exist in a stable form. Nucleation will generally occur when the part is cooled below an equilibrium temperature, T_e. This is the temperature at which the free energies of the two phases (solid and liquid) are equal. Below it, the solid is the more stable phase since its free energy is lower than that of the liquid phase.

The process of nucleation can occur in two forms:

1. Homogeneous nucleation
2. Heterogeneous nucleation.

In homogeneous nucleation, the nuclei of the second phase form within the bulk or matrix of the original phase, say the liquid, in the case of solidification. This rarely occurs in practice, and for it to be possible, there must be extensive supercooling. In heterogeneous nucleation, the nuclei either form on foreign particles suspended in the liquid or on the walls of the container, and the amount of undercooling necessary for it to occur is much smaller than that required for homogeneous nucleation. In either case, each nucleus constitutes the basis for a crystal or grain, which may subsequently grow.

For homogeneous nucleation, the total free energy change, ΔG, associated with the formation of a spherical solid nucleus in a liquid medium consists of two components:

1. The bulk free energy change
2. The surface energy change.

Thus the total free energy change is given by

$$\Delta G = \frac{4}{3}\pi r^3 \Delta G_v + 4\pi r^2 \gamma_{\mathrm{nl}} \tag{11.1}$$

where r is the radius of the nucleus, ΔG_v denotes the bulk free energy change per unit volume of transformed material, and γ_{nl} stands for surface energy per unit area between the solidified nucleus and liquid (molten metal).

Typical surface energy values for some common metals are shown in Table 11.1.

TABLE 11.1 Surface Energies, γ_{nl}, of Some Common Metals (from Hosford, W., *Physical Metallurgy*, 2005, Taylor & Francis.)

Metal	γ_{nl} J/m^2
Copper (Cu)	0.177
Gold (Au)	0.132
Iron (Fe)	0.204
Lead (Pb)	0.033
Manganese (Mn)	0.206
Nickel (Ni)	0.255
Silver (Ag)	0.126
Tin (Sn)	0.0545

ΔG_v may be expressed as

$$\Delta G_v = \frac{L_{mv} \Delta T_e}{T_e} \tag{11.2}$$

where L_{mv} is the latent heat of fusion per unit volume. The latent heat of fusion is the energy released during the liquid–solid transformation, T_e = equilibrium temperature = T_m (solidification temperature in this case), $\Delta T_e = T_e - T$ = degree of supercooling, and T is the temperature of the molten metal.

Since heat is evolved during solidification, L_{mv} is negative, and thus the bulk free energy change ΔG_v will be negative at all temperatures below the solidification temperature, that is, $T < T_e$, whereas the surface energy change is always positive.

The variation of the two terms in equation (11.1) with the radius of the nucleus is shown in Fig. 11.2. It is evident from Fig. 11.2 that there is a free energy barrier to the nucleation process. In other words, a critical nucleus size, r_c, is necessary for the process to continue. At the critical radius, the system is in metastable equilibrium. A small decrease in the radius at that point will result in the surface energy term dominating, and the nucleus will redissolve, since the total free energy decreases with decreasing radius in the region of the curve below the critical radius. An increase in the radius at the metastable state will result in a continued decrease in free energy with increasing radius as the bulk free energy term dominates. The continued decrease in free energy is a necessary condition for the formation of a stable nucleus. A stable nucleus is then formed.

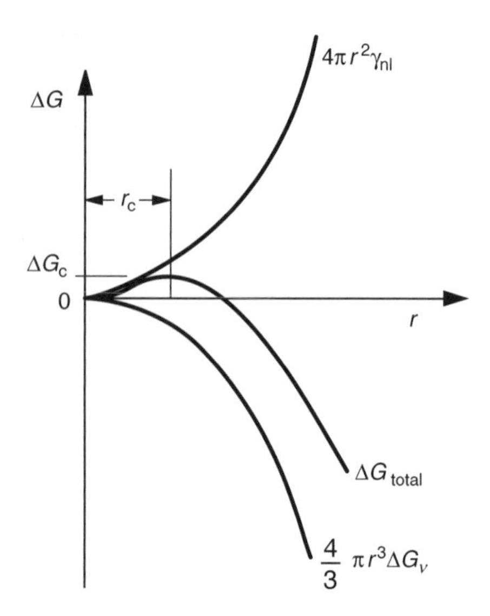

FIGURE 11.2 Free energy change associated with the formation of a spherical nucleus.

To determine the critical radius r_c, at which a nucleus becomes stable, equation (11.1) is differentiated with respect to r and equated to zero, giving

$$r_c = -\frac{2\gamma_{nl}}{\Delta G_v} = -\frac{2\gamma_{nl} T_e}{L_{mv} \Delta T_e} \tag{11.3}$$

and the corresponding critical free energy change, ΔG_c,

$$\Delta G_c = \frac{16\pi\gamma_{nl}{}^3}{3(\Delta G_v)^2} = \frac{16\pi\gamma_{nl}{}^3}{3L_{mv}{}^2} \frac{T_e{}^2}{\Delta T_e{}^2} \tag{11.4}$$

Example 11.1 Two pieces of silver are butt welded together using a laser beam, without any filler metal.

(a) Determine the critical radius necessary for the formation of a stable nucleus if the weld pool was supercooled to a temperature of 724°C.
(b) Determine the critical free energy change necessary for forming the stable nucleus.

Solution:

From appendices 10D and 10E, the properties of silver are approximated as

Average density $\rho = 10490\,\text{kg/m}^3$
Heat of fusion $L_m = 104.2\,\text{kJ/kg}$
Melting temperature $T_m = 962°\text{C} = 1235\,\text{K}$.

(a) The critical radius is given by equation (11.3):

$$\begin{aligned} r_c &= -\frac{2\gamma_{nl}}{\Delta G_v} = -\frac{2\gamma_{nl} T_e}{L_{mv} \Delta T_e} \\ &= -\frac{2 \times 0.126 \times 1235}{(-104.2 \times 10^3) \times 10490 \times (962 - 724)} \\ &= 1.196 \times 10^{-3}\ \mu\text{m} = 1.196\,\text{nm} \end{aligned}$$

(b) The critical free energy change is given by

$$\begin{aligned} \Delta G_c &= \frac{16\pi\gamma_{nl}{}^3}{3L_{mv}{}^2} \frac{T_e{}^2}{\Delta T_e{}^2} \\ &= \frac{16\pi \times 0.126^3}{3 \times (104.2 \times 10^3 \times 10490)^2} \frac{1235^2}{(962 - 724)^2} \\ &= 7.6 \times 10^{-19}\,\text{J}. \end{aligned}$$

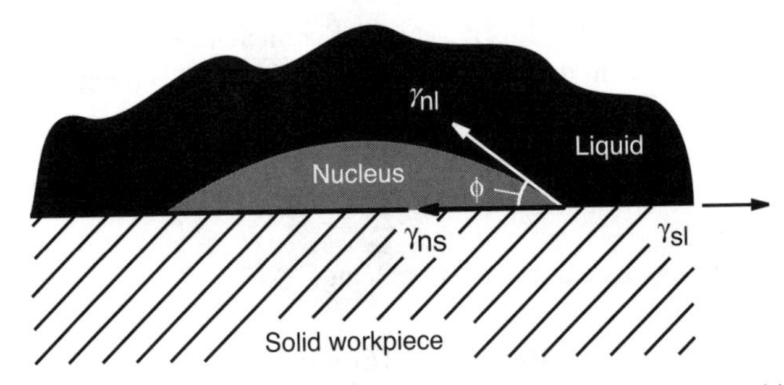

FIGURE 11.3 Heterogeneous nucleation at the base metal–molten pool interface.

In heterogeneous nucleation during cladding, for example, since the nucleation would be on the surface of the unmelted base metal, any nucleus formed would not be completely spherical in shape, but rather can be approximated as a spherical cap (Fig. 11.3). For such a shape, the total free energy change associated with the nucleation would be given by

$$\Delta G = \left[\frac{4}{3}\pi r^3 \Delta G_v + 4\pi r^2 \gamma_{\mathrm{nl}}\right] f(\phi) \tag{11.5}$$

where ϕ is the wetting angle (Fig. 11.3); and

$$f(\phi) = \frac{1}{4}(2 + \cos\phi)(1 - \cos\phi)^2 \tag{11.6}$$

The contact or wetting angle ϕ is obtained by considering the respective surface energies between the fusion zone nucleus and solid base metal, γ_{ns}; the solid base metal and liquid (molten) pool, γ_{sl}; and the fusion zone nucleus and liquid (molten) pool, γ_{nl}. For the spherical cap nucleus shown in Fig. 11.3, the equilibrium of horizontal forces acting on the nucleus gives

$$\gamma_{\mathrm{sl}} = \gamma_{\mathrm{ns}} + \gamma_{\mathrm{nl}} \cos\phi \tag{11.7a}$$

or

$$\cos\phi = \frac{\gamma_{\mathrm{sl}} - \gamma_{\mathrm{ns}}}{\gamma_{\mathrm{nl}}} \tag{11.7b}$$

The corresponding critical free energy change, ΔG_{c}, for this case is given by

$$\Delta G_{\mathrm{c}} = \frac{16\pi \gamma_{\mathrm{nl}}^3}{3(\Delta G_v)^2} f(\phi) = \frac{16\pi \gamma_{\mathrm{nl}}^3}{3L_{mv}^2} \frac{T_{\mathrm{e}}^2}{\Delta T_{\mathrm{e}}^2} f(\phi) \tag{11.8}$$

The number of nuclei per unit volume n_v formed at temperature T may be expressed as

$$n_v = s\mathrm{e}^{\left(-\frac{\Delta G_\mathrm{c}}{k_\mathrm{B}T}\right)} \tag{11.9}$$

where k_B is Boltzmann's constant and s is number of sites available for nuclei formation.

The rate of nucleation $\dot{n}$ depends on two principal parameters:

1. The number of nuclei with sufficient energy to be stable.
2. The ability of atoms to diffuse to the potentially stable nucleus.

It is normally expressed as

$$\dot{n} = \frac{k_\mathrm{B}T}{2\pi h_\mathrm{p} V_\mathrm{m}}\mathrm{e}^{\left(-\frac{\Delta G_\mathrm{c}}{k_\mathrm{B}T}\right)}\mathrm{e}^{\left(-\frac{\Delta E_\mathrm{d}}{k_\mathrm{B}T}\right)} = \frac{k_\mathrm{B}T}{2\pi h_\mathrm{p} V_\mathrm{m}}\mathrm{e}^{\left(-\frac{[\Delta G_\mathrm{c}+\Delta E_\mathrm{d}]}{k_\mathrm{B}T}\right)} \tag{11.10}$$

where h_p is Planck's constant, V_m is volume of a molecule, and ΔE_d is activation energy for diffusion.

The dependence of both the number of nuclei and the nucleation rate on temperature is illustrated in Fig. 11.4a. The diffusion rate is relatively fast at temperatures just below T_e. However, since the undercooling ΔT_e is small, the critical free energy change necessary for nucleation is then large (see equation (11.4)). Thus only a few nuclei attain sufficient energy to reach the critical size. At relatively lower temperatures where ΔT_e is large, the critical free energy change for nucleation is correspondingly small. However, the diffusion rate is then low, and atoms can no longer move about easily to form nuclei. The nucleation rate is then low. A maximum occurs in the nucleation rate at intermediate temperatures where the diffusion rate is relatively high, and the critical free energy change for nucleation is not too large.

The nucleation principles outlined in the preceding paragraphs are also applicable to solid–solid transformations, for example, austenite to pearlite transformation, which may occur in steels after solidification is completed and the fusion zone continues to cool. However, the resulting nucleation rates are generally much smaller in this case for a number of reasons:

1. Much lower diffusion rates in solids.
2. Higher interfacial energies in the solid state, resulting in higher critical free energy change for the formation of a nucleus.
3. Lower bulk free energy change in the solid, which also increases the critical free energy change.

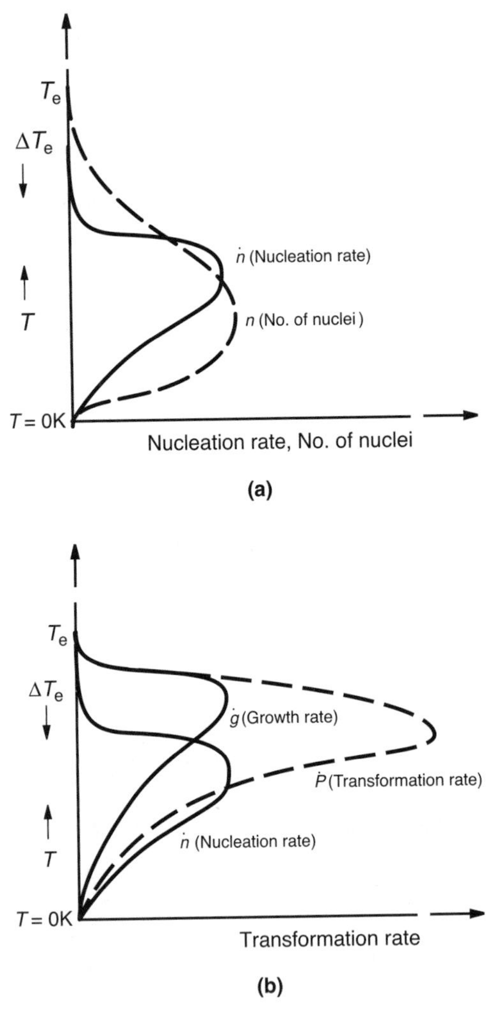

FIGURE 11.4 (a) Dependence of the number of nuclei and nucleation rate on temperature. (b) Dependence of growth and nucleation rates on temperature. (From Barrett, C. R., Nix, W. D., and Tetelman, A. S., 1973, *The Principles of Engineering Materials*. Reprinted by permission of Pearson Education, Inc., Upper Saddle River, NJ.)

4. Since the solid does not flow very easily to accommodate volume changes associated with the formation of a nucleus, there is strain energy associated with the formation of the nucleus. That increases the total free energy change necessary for the nucleus formation, since the strain energy term is positive. This term is absent when there is liquid present since the liquid flows easily to accommodate any deformation or volume change associated with the transformation.

Example 11.2 For the conditions in Example 11.1,

(a) Determine the number of stable nuclei per unit volume. Assume the density of nucleation sites is 2.5×10^{30} per m^3.

(b) Determine the nucleation rate. Assume the radius of a silver atom is 0.144 nm, and the activation energy for diffusion in silver is 44.1 kcal/mol = 184.5 kJ/mol (Shewmon, 1983, p. 66).

Solution:

(a) The number of stable nuclei per unit volume is given by

$$n_v = s\mathrm{e}^{\left(-\frac{\Delta G_c}{k_B T}\right)}$$

Using the results of Example 11.1, we have

$$\begin{aligned} n_v &= 2.5 \times 10^{30} \times \mathrm{e}^{\left(-\frac{7.6\times 10^{-19}}{1.38\times 10^{-23}\times(724+273)}\right)} \\ &= 2.56 \times 10^6 \text{ per m}^3 \end{aligned}$$

(b) The nucleation rate per unit volume is given by

$$\dot{n} = \frac{k_B T}{2\pi h_p V_m}\mathrm{e}^{\left(-\frac{\Delta G_c}{k_B T}\right)}\mathrm{e}^{\left(-\frac{\Delta E_d}{k_B T}\right)}$$

The volume of the silver atom is

$$V_m = \frac{4}{3}\pi r^3 = \frac{4}{3}\pi(0.1444 \times 10^{-9})^3 = 12.61 \times 10^{-30}\text{ m}^3$$

Since the activation energy is given in units of energy per mole, the nucleation rate per unit volume can be written in terms of the gas constant as

$$\dot{n} = \frac{k_B T}{2\pi h_p V_m}\mathrm{e}^{\left(-\frac{\Delta G_c}{k_B T}\right)}\mathrm{e}^{\left(-\frac{\Delta E_d}{R_g T}\right)}$$

where $R_g = 1.98$ cal/mol is the gas constant.

The nucleation rate then becomes

$$\begin{aligned} \dot{n} &= \frac{1.38 \times 10^{-23} \times 997}{2\pi \times 6.63 \times 10^{-34} \times 12.61 \times 10^{-30}}\mathrm{e}^{\left(-\frac{7.6\times 10^{-19}}{1.38\times 10^{-23}\times 997}\right)}\mathrm{e}^{\left(-\frac{44.1\times 10^3}{1.98\times 997}\right)} \\ &= 5.32 \times 10^7 \text{ per (m}^3\text{s)} \end{aligned}$$

Growth The growth of a stable nucleus involves movement of atoms to and attachment to the nucleus. The growth process is thus diffusion controlled. The variation of both nucleation and growth rates with temperature are compared in Fig. 11.4b, where it is seen that the growth rate is greater at temperatures just below the equilibrium transformation temperature, while the nucleation rate is greater at lower temperatures.

Since the base metal is at a lower temperature than the molten pool during laser processing, solidification of the molten metal would ordinarily begin with the formation of solid nuclei at the unmelted surface of the base metal in contact with the molten pool. The nuclei that are formed would then grow into the molten metal, parallel to the direction of heat flow from the molten metal. Such is the case in casting, for example, where nucleation occurs on the mold walls at a temperature just below the melting temperature. However, initial solidification of a molten pool, in welding or surface melting, for example, is usually *epitaxial*. In this case, atoms from the molten metal are deposited directly on the existing crystals of the unmelted base metal, thereby extending the substrate crystalline structure without changing it (Fig. 11.5). This is especially true in autogenous welding and surface melting. For such processes, since the initial grains in the fusion zone grow epitaxially from the substrate grains, the initial crystal structure of the fusion zone is inherited directly from that of the base metal. In other words, the structure and crystallographic orientation of the base metal grains at the fusion zone interface continue into the fusion zone. Thus the nucleation stage is bypassed at the interface.

To understand the driving mechanism for epitaxial growth, the basic kinetics for phase transformation will be considered. The molten metal in autogenous welding or surface melting is normally of the same material as the substrate and is also in intimate contact with it. Thus it completely wets the surface of the base metal, resulting in a zero value for ϕ in equation (11.6). It is thus evident that $f(\phi)$ is also zero, and consequently, so is ΔG_c. In other words, the critical free energy change required for

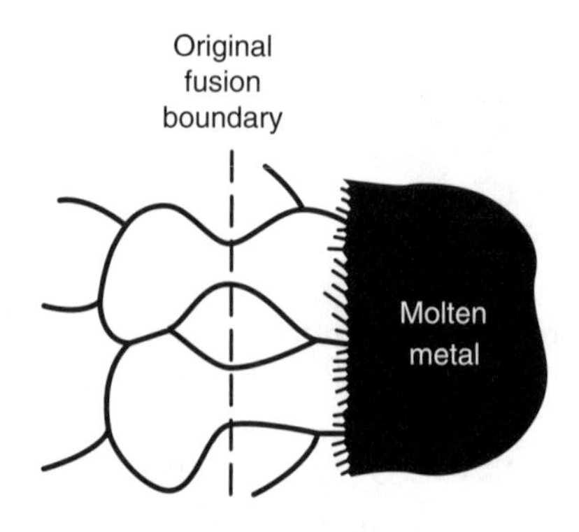

FIGURE 11.5 Epitaxial solidification from the base metal grains at the fusion line.

the formation of a stable nucleus of solid material from the molten metal is zero. Grain growth thus occurs spontaneously from the base metal since there is no resistance to its initiation, that is, the original grains in the base metal in contact with the molten pool continue to grow epitaxially into the pool, without nucleation.

11.1.1.2 Microstructure In this section, we discuss the various microstructures that result from solidification. First, the criterion that determines the microstructure that forms is outlined. The variation of microstructure across the fusion zone is then discussed. This is followed by the factors that determine grain orientation. Finally, the spacing of any dendrite arms that may be formed is discussed.

Structure of the Solidified Metal A pure metal normally solidifies at a constant temperature, with the interface between the solid and liquid media being planar (Figs 11.6a and 11.7a). For an alloy, the temperature may decrease as solidification proceeds, unless a eutectic is formed. The interface in this case can thus be either planar, cellular, or dendritic, with the dendritic structure being further classified as columnar or equiaxed (Figs 11.6a–d and Figs 11.7a–d). Which one prevails is influenced by the solute content of the molten metal and solidification conditions.

In the initial stages of the solidification of alloys, the low-melting elements are rejected by the freezing solid. Thus the enriched liquid directly ahead of the

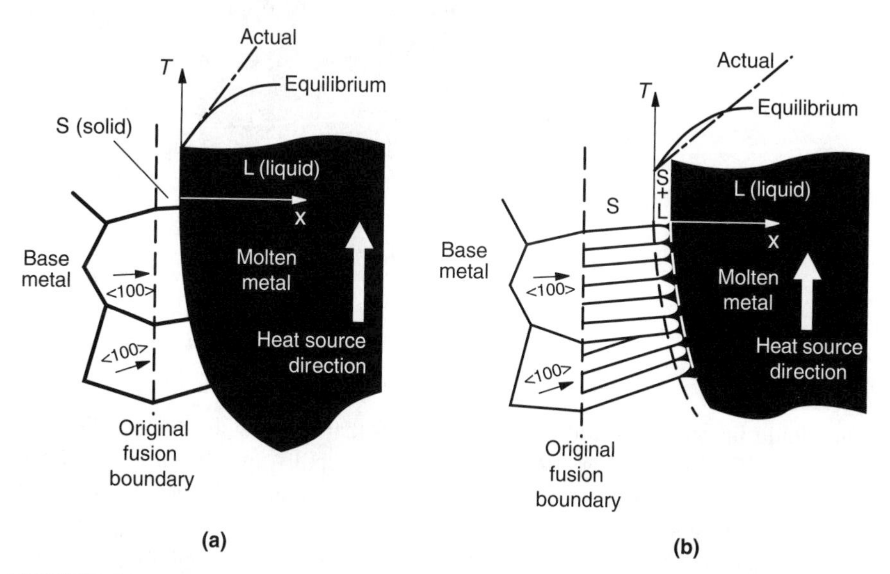

FIGURE 11.6 (a) Planar, (b) cellular, (c) columnar dendritic, and (d) equiaxed dendritic solid/liquid interfaces during solidification. (From Kou, S., 1987, *Welding Metallurgy*. Reprinted with permission from John Wiley & Sons, Inc.)

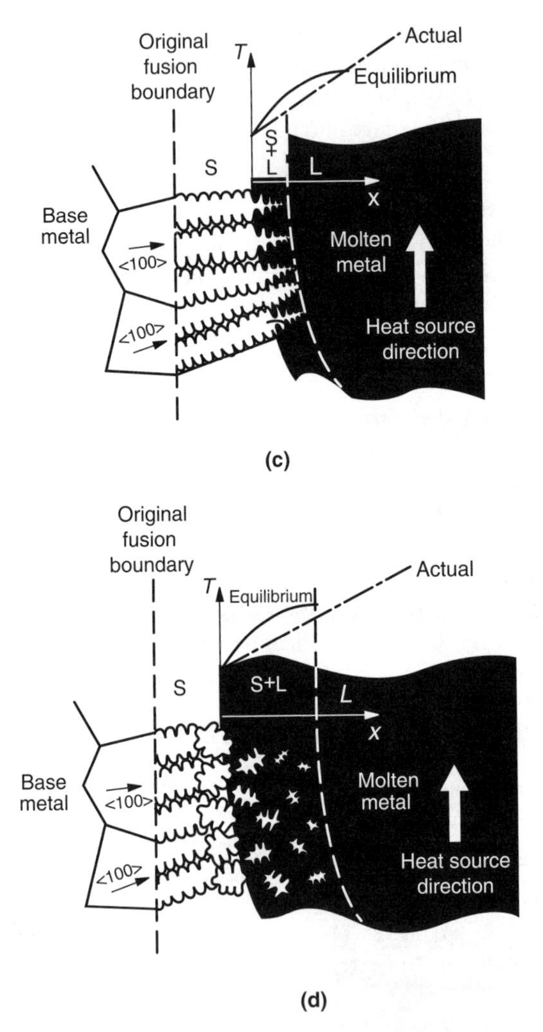

FIGURE 11.6 (*Continued*)

solid–liquid interface has a lower freezing point than the previously solidified material. Since the low-melting atoms cannot diffuse rapidly away from the interface, they tend to retard growth. However the liquid temperature is higher, farther away from the interface, that is, toward the center of the molten metal. Thus portions of the interface protrude or grow rapidly into the liquid, resulting in columns that grow from the initially solidified material, forming a *cellular or columnar* structure. For even lower temperature gradients at the interface, branches can grow from the main columns, resulting in a *dendritic* structure. An individual grain may contain a number of dendrites.

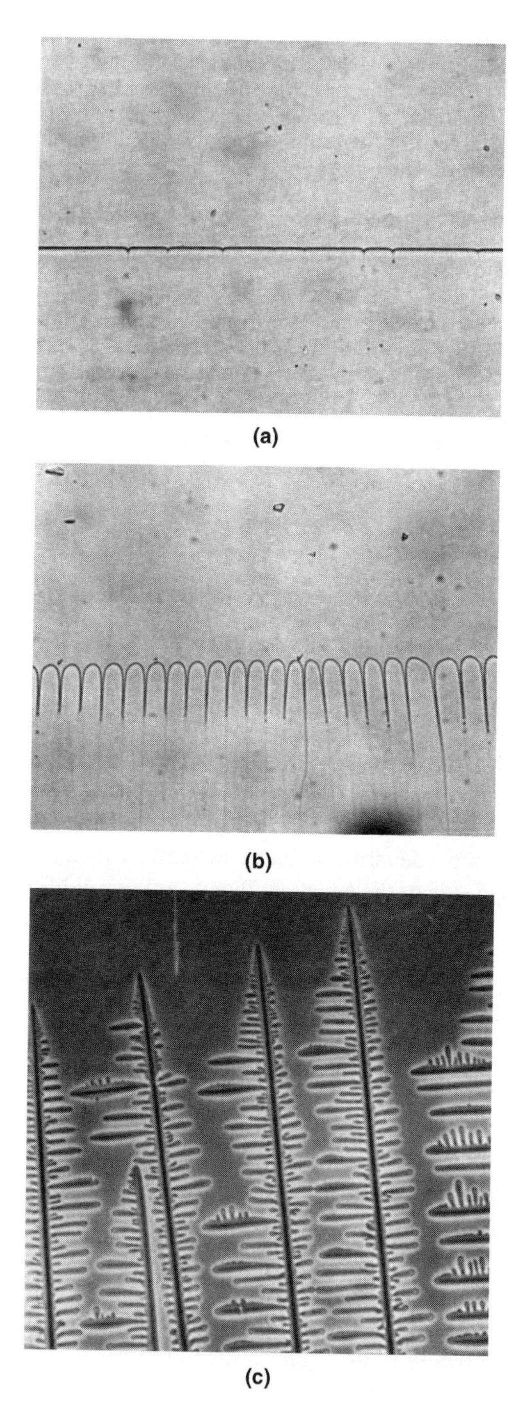

FIGURE 11.7 Pictorial illustrations of (a) planar, (b) cellular, (c) columnar dendritic, and (d) equiaxed dendritic growth. (From Hughel, T. J., and Bolling, G. F. (eds), 1971, *Solidification*, American Society for Metals, Metals Park, OH.)

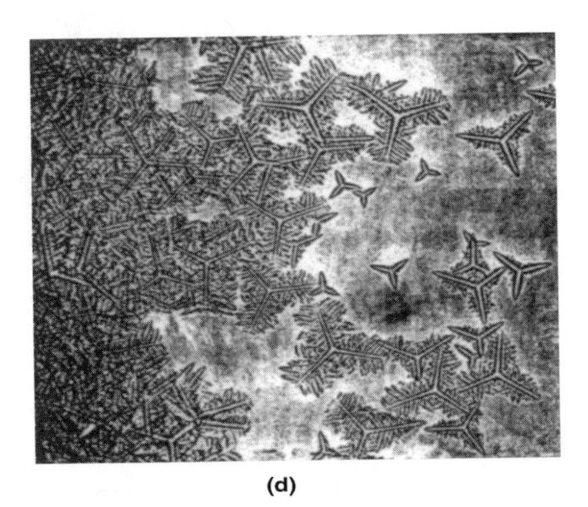

(d)

FIGURE 11.7 (*Continued*)

The planar interface in an alloy breaks down to form either a cellular or dendritic structure if the following condition holds (see Chapter 12 for further details):

$$\frac{G_{\mathrm{Tl}}}{\mathrm{SR}} \leq \frac{\Delta T_f}{D_{\mathrm{l}}} \tag{11.11}$$

where D_{l} is diffusion coefficient of the solute in the liquid, G_{Tl} is temperature gradient of the liquid near the solid–liquid interface, SR stands for solidification rate, $\Delta T_{\mathrm{f}} = T_{\mathrm{l}} - T_{\mathrm{s}}$ = equilibrium freezing range, T_{l} is liquidus temperature, and T_{s} denotes solidus temperature.

The tendency toward a dendritic structure increases with a decreasing value of the ratio on the left-hand side of the equation, that is, as the solidification rate increases, or the temperature gradient of the liquid near the solid–liquid interface decreases. Further reductions in the ratio again result in an increased tendency from a columnar to an equiaxed dendritic structure. This is illustrated schematically in Fig. 11.8. The structure that is determined by the solid–liquid interface exists within the individual grains and is commonly referred to as the substructure or subgrain structure. The formation of a dendritic structure is illustrated schematically in Fig. 11.9.

The product of the average temperature gradient, G_{Tl}, and interface velocity or solidification rate, SR, is a measure of the average cooling rate, CR:

$$\mathrm{CR} = G_{\mathrm{Tl}} \cdot \mathrm{SR} \approx \frac{T_{\mathrm{l}} - T_{\mathrm{s}}}{t_{\mathrm{s}}} \tag{11.12}$$

where t_{s} is local solidification time.

Structural Variation Within Fusion Zone The microstructure or mode of solidification of the fusion zone varies as one moves from the fusion boundary toward the centerline. This results primarily from variations in the solidification rate and temperature gradient across the fusion zone. To further understand this, consider the relationship

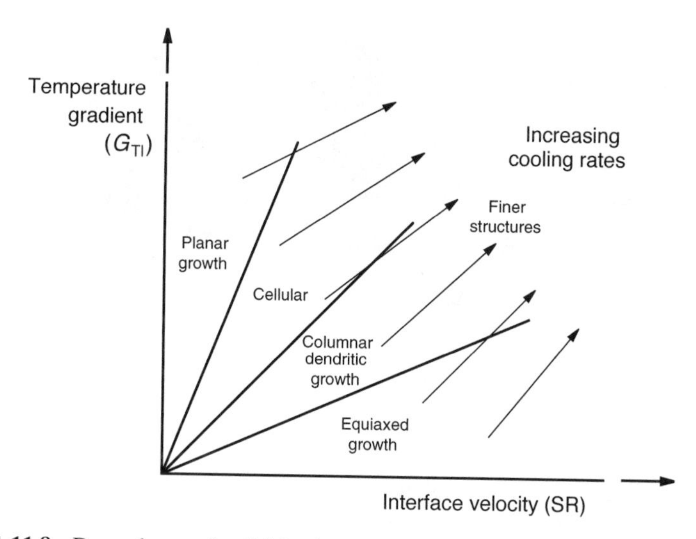

FIGURE 11.8 Dependence of solidification structure on liquid temperature gradient and solidification rate.

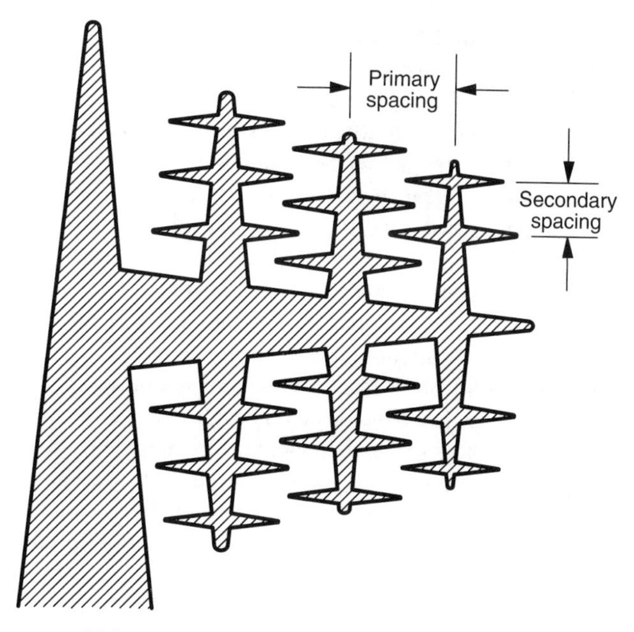

FIGURE 11.9 Schematic of dendrite formation.

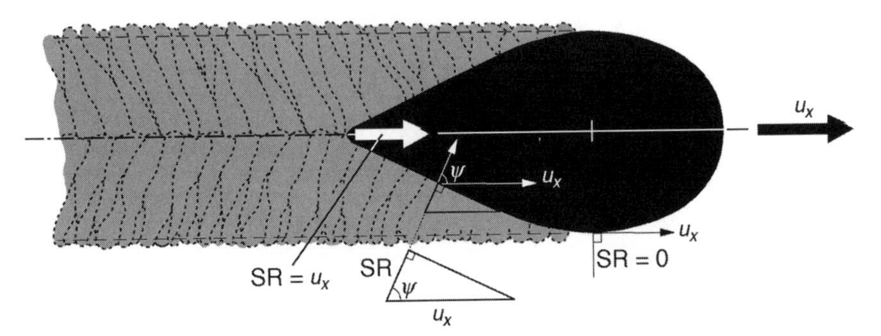

FIGURE 11.10 Variation of solidification rate along the molten pool edge. (From Kou, S., 1987, *Welding Metallurgy*. Reprinted with permission from John Wiley & Sons, Inc.)

between the solidification rate **SR** at any point along the molten pool boundary and the traverse velocity vector $\mathbf{u_x}$, which, as can be inferred from Fig. 11.10, is given by

$$\mathbf{SR} = \mathbf{u_x} \cos \psi \tag{11.13}$$

where ψ is the angle between the traverse velocity vector and the direction of solidification at any point along the molten pool boundary.

Thus the solidification rate SR is lowest (almost zero) along the fusion line where $\psi \approx 90^\circ$ and maximum along the centerline where $\psi \approx 0^\circ$. Also, the elongated nature of the molten pool (resulting from motion) is such that the temperature gradient G_{Tl} is highest along the fusion line and lowest along the centerline. Thus it is evident from equation (11.11) that initial epitaxial growth of the grains along the fusion line is more likely to have a planar front, that is, growth is planar (Figs 11.6a and 11.7a). As one moves inward along the molten pool boundary, the temperature gradient decreases, while the solidification rate increases, with the result that there is an increased tendency toward cellular growth, that is, a transition from planar to cellular growth occurs, resulting in columnar grains (Figs 11.6b and 11.7b). Further reductions in the temperature gradient and increases in the solidification rate toward the centerline may result in subsequent transition to a dendritic structure in the fusion zone (Figs 11.6c and 11.7c). This is especially so along the centerline where the temperature gradient is lowest and the solidification rate is highest. An equiaxed dendritic structure (Figs 11.6d and 11.7d) may thus develop at very high traverse velocities, and it normally occurs in the final stages of solidification, for example, in a weld crater. In considering the processing conditions, increasing the heat input, for example, will normally result in a reduction in temperature gradient and thus an increased tendency toward cellular growth. However, the corresponding reduction in cooling rate will produce a coarser structure.

Grain Orientation The direction of grain growth during solidification is controlled by two factors:

1. The maximum temperature gradient direction.

2. The preferred crystallographic growth direction, which is the $\langle 1\,0\,0 \rangle$ direction for face-centered and body-centered cubic metals and the $\langle 1\,0\,1\,0 \rangle$ direction for hexagonal close-packed metals.

Considering the temperature gradient effect, the grains have a tendency to grow in the direction of steepest temperature gradient. Since, in the case of solidification, this direction is normal to the molten pool boundary, the general direction of grain orientation is also normal to the molten pool boundary. Furthermore, as the shape of the molten pool depends on the traverse speed, so does the direction of grain orientation. Thus, at high traverse velocities, the columnar grains grow straight but are inclined to the fusion boundary or direction of the heat source since they tend to grow in a direction perpendicular to the elongated or teardrop-shaped molten pool boundary as shown in Fig. 11.11a. However, at low traverse velocities, they tend to curve toward the middle of the molten pool, to become perpendicular to the elliptical pool boundary (Fig. 11.11b). In situations where other grains are nucleated in the middle of the molten pool, these grow in the direction of motion. This is because the pool boundary toward the center is normal to the direction of motion at any speed. Thus, for such cases, the elongated shape at high speeds results in a narrow band of grains in the middle (Fig. 11.11c) while the elliptical shape at lower speeds results in a wider band (Figs 11.11d and 11.12).

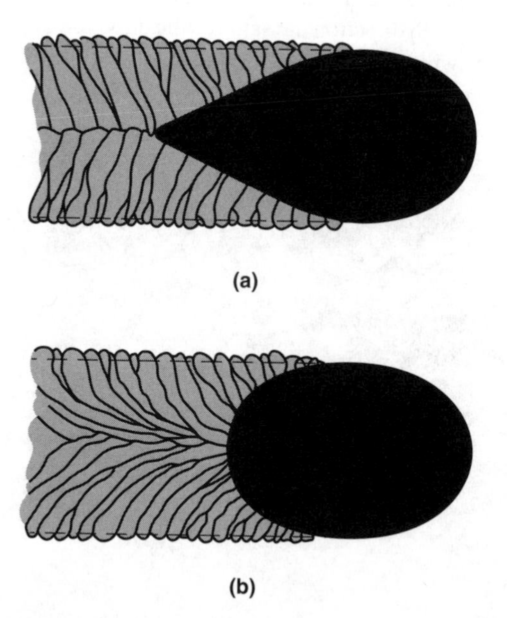

FIGURE 11.11 (a) Grain structure at high traverse velocities. (b) Grain structure at low traverse velocities. (c) Banded grain structure at high traverse velocities. (d) Banded grain structure at low traverse velocities. (From Kou, S., 1987, *Welding Metallurgy*. Reprinted with permission from John Wiley & Sons, Inc.)

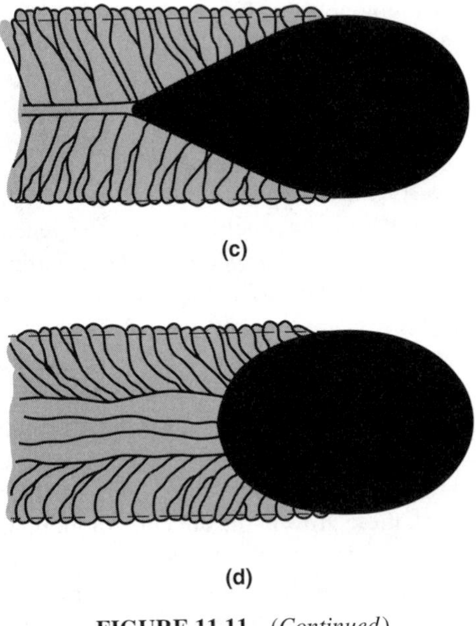

FIGURE 11.11 (*Continued*)

In considering the preferred crystallographic growth direction, it is found that in the initial stages of epitaxial growth, some of the original base metal crystals may have their preferred growth direction the same as the direction of steepest temperature gradient. Such grains will have a greater driving force for growth and are thus more

FIGURE 11.12 Pictorial illustration of a banded grain structure. (From Friedman, P. A., and Kridli, G. T., 2000, *Journal of Materials Engineering and Performance*, Vol. 9, pp. 541–551. By permission of Springer Science and Business Media.)

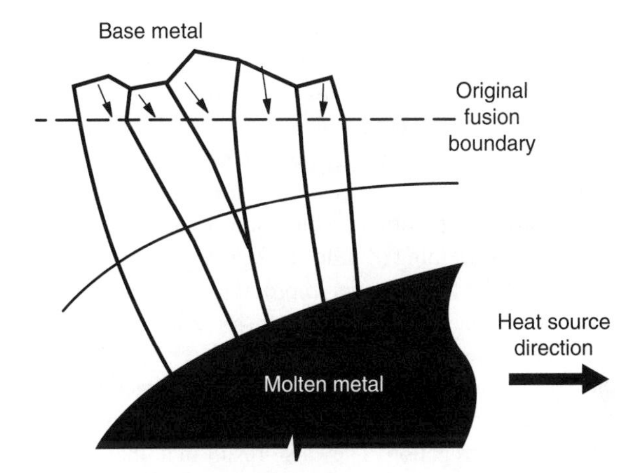

FIGURE 11.13 Preferential growth of crystals that have the two preferred directions of growth being coincident. (From Kou, S., 1987, *Welding Metallurgy*. Reprinted with permission from John Wiley & Sons, Inc.)

likely to grow at the expense of those that have a less favorable orientation (Fig. 11.13). Furthermore, crystals sometimes change orientation during growth to enable the two directions (preferred growth and steepest gradient) to be coincident and thus facilitate growth.

Dendritic Spacing The formation of dendrites result in variation in the composition and properties at different positions within the fusion zone. A faster solidification produces smaller spacing between the dendrite arms, and smaller dendritic spacing often results in greater strength, better ductility, and better toughness. The primary dendrite spacing d_1 (Fig. 11.9) is approximated as

$$d_1 = \left[4\sqrt{2}\frac{r_d k_p \Delta T_f}{G_{Tl}}\left(\frac{C_t}{C_o} - \frac{G_{Tl} D_l}{SR k_p \Delta T_f}\right)\right]^{1/2} \tag{11.14a}$$

The secondary spacing near the dendrite tip d_2 is also approximated as

$$d_2 = t_s{}^{n_d} \tag{11.14b}$$

where C_o is the original alloy composition, C_t is solute concentration at the tip in the liquid, k_p is equilibrium partition coefficient, $n_d = 0.3 - 0.5$, r_d stands for dendrite tip radius, and t_s is solidification time.

11.1.1.3 Nucleation and Grain Refinement in Molten Pool When new grains are nucleated in the molten pool, they tend to form an equiaxed rather than a columnar grain structure. The new grains may be nucleated heterogeneously by fragments of dendrite tips that are broken off by convection within the molten pool, especially in the mushy zone. The mushy zone is the region between the completely liquid and completely solid portions of the material. Both liquid and solid coexist in this region. Grains that are only loosely held together by liquid films in the mushy zone may also be swept away by convection and act as nucleation sites. Finally, new grains may be nucleated by foreign particles in the molten pool. When deliberately applied, the latter technique is called *inoculation* and is used to form a fine equiaxed grain structure. An example is the addition of titanium carbide powder as an inoculant to the molten pool during submerged arc welding of mild steel. A fine grain structure enhances the mechanical properties of the fusion zone, improving such properties as ductility and toughness. It also reduces the tendency of the fusion zone to solidification cracking (see Section 11.2.2.1).

Increasing the cooling rate results in a finer solidified structure (Fig. 11.8), that is, smaller dendritic arm spacing. Thus higher traverse velocities produce finer spacings that in turn result in less segregation. They also result in a finer grain structure.

A structural effect that results in variations in mechanical properties throughout the part is segregation of alloying elements. The basic causes of this phenomenon are discussed in the next section.

11.1.1.4 Coring To understand the basis for alloy segregation, let us consider the solidification of an alloy consisting of two metals such as copper and nickel that are completely soluble in liquid and solid states (Fig. 11.14). Let the starting composition of the alloy be C_0 (40% Cu), and in the liquid phase. As the liquid alloy cools, the first solid forms at temperature T_1, with a composition C_1 that is different from the nominal composition of the alloy C_0 and is richer in solvent. If the system were to be cooled

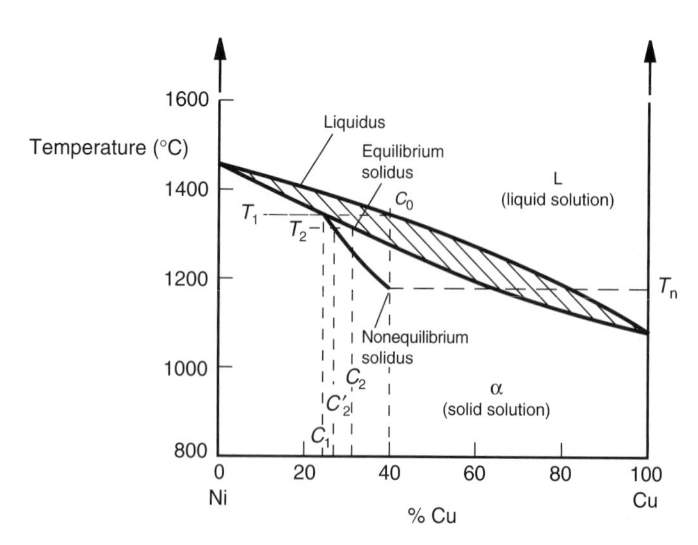

FIGURE 11.14 Nonequilibrium cooling. (From Barrett, C. R., Nix, W. D., and Tetelman, A. S., 1973. *The Principles of Engineering Materials*. Reprinted by permission of Pearson Education, Inc., Upper Saddle River, NJ.)

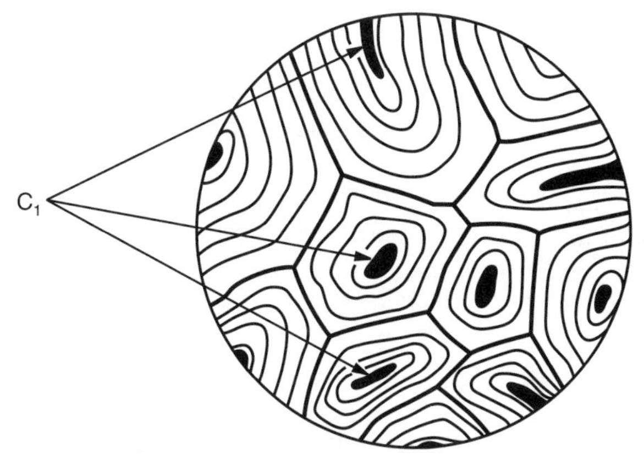

FIGURE 11.15 Coring or macrosegregation. (From Barrett, C. R., Nix, W. D., and Tetelman, A. S., *The Principles of Engineering Materials*, 1973. Reprinted by permission of Pearson Education, Inc., Upper Saddle River, NJ.)

to T_2 under equilibrium conditions, the composition of the entire solid formed at that temperature would be C_2. However, in practice, and especially in the case of laser processing, the cooling rates are too high for equilibrium to be achieved in the solid by diffusion. In other words, cooling and solidification are fast enough that there is not adequate time for diffusion to enable the elements to be uniformly distributed. Thus even though the layer of solid that forms at T_2 will have composition C_2, the composition of the first layer formed will still be essentially C_1, and there will be a composition gradient or difference in composition across the solid formed. In other words, the various sections of the solid will have different compositions (Fig. 11.15). The average composition of the solid that exists at temperature T_2 will thus be $C_2{}'$, which is less than C_2. As solidification continues, the average composition of the solid will continue to be below that of the equilibrium solidus line. This results in a new solidus curve, the nonequilibriumm solidus line, whose position depends on the cooling rate.

A consequence of the nonequilibriumm solidification is that the process is not completed until a temperature T_{n}, which is lower than the equlibrium melting temperature, is reached. In other words, on reheating the material, some portion of it will melt at a temperature T_{n} that is lower than the nominal melting temperature of T_{m}. This segregation of impurities and alloying elements during solidification is called *coring or macrosegregation* and may result in variation in the properties of material across the fusion zone. This would create regions of weakness within the fusion zone. Macrosegregation can be eliminated by homogenization, that is, reheating the material to the temperature T_{n} for some time to permit adequate diffusion to take place.

Though macrosegregation may be significant in the fusion zone, the forces that act on the molten pool generally result in efficient mixing of any filler material and base material, producing a uniform initial composition of molten metal under normal conditions. However, there exists a thin layer of molten metal near the pool boundary where very little mixing, if any, occurs since the molten metal has almost zero velocity in that region. The composition of this layer can be significantly different from that of the bulk molten pool and contributes significantly to macrosegregation.

In conduction mode laser welding, for example, there is no electromagnetic force, and since the surface tension flow effect is only pronounced near the surface, mixing in the bulk of the pool can be poor, enhancing macrosegregation when dissimilar materials are welded. This can be made worse by the higher traverse velocities used.

11.1.2 Zone of Partial Melting

Low-melting constituents of the base material in the region closest to the fusion boundary often melt during processing, resulting in a zone that is partially melted. This region forms a narrow band around the fusion zone (Fig. 11.1). Partial melting is not only most pervasive along grain boundaries, due to a higher concentration of impurities, but may also occur within the grains. It occurs for portions of the base alloy where the temperature falls within the solidus and liquidus range, or for low-melting second-phase particles or inclusions.

Under conditions of partial melting, tensile residual stresses that develop as the part begins to cool can cause separation of material, resulting in hot cracking (see Section 11.2.2.1), which is normally intergranular. Nascent hydrogen has a much greater solubility in molten iron than in solid iron. Thus, parts exposed to hydrogen develop a greater concentration of hydrogen along grain boundaries in the partially melted zone, where hydrogen embrittlement (Section 11.2.2.3) can be severe, resulting in hydrogen-induced cracking. The impact of the partially melted region is reduced in processes where the total heat input is low, such as in laser processing. Reduction in concentration of elements that tend to increase the freezing temperature range of an alloy would also minimize damage in this zone. Finally, the low-melting constituents are normally distributed along the grain boundaries. A reduction in the average grain size, which increases the grain boundary area, would therefore reduce the concentration of the low-melting constituents, thereby minimizing their impact.

11.1.3 Heat-Affected Zone

The HAZ is the portion of the base material that does not melt, but where the temperature rise is high enough to affect the microstructure and mechanical properties of the material (see Fig. 11.1). The microstructure that develops in this region depends on the following:

1. The thermal cycle
2. The type of material (its metallurgical and thermal properties)
3. The state of the material prior to processing (prior thermal and mechanical history).

For this discussion, we group the materials of interest into three main categories:

1. Pure metals
2. Precipitation-hardening and nonferrous alloys
3. Steels.

11.1.3.1 Pure Metals The heat-affected zone microstructure obtained on processing a pure metal such as copper depends on whether the material is originally in the annealed state or is work hardened. An annealed material will only experience grain growth in this area (Fig. 11.16a), that is, there will be no recrystallization since rearrangement of dislocations to form new boundaries is essential for recrystallization. However, a previously work-hardened material will be recrystallized with new grains being nucleated and therefore have a fine grain structure at the outer edge of the heat-affected zone where the temperature is just above the recrystallization temperature (Fig. 11.16b). However, the grain size increases toward the fusion boundary, and a coarse grain structure is almost always inevitable close to the fusion boundary.

The growth kinetics of the grains is expressed as

$$\frac{\mathrm{d}D}{\mathrm{d}t} = K_a \cdot \frac{1}{D} \tag{11.15}$$

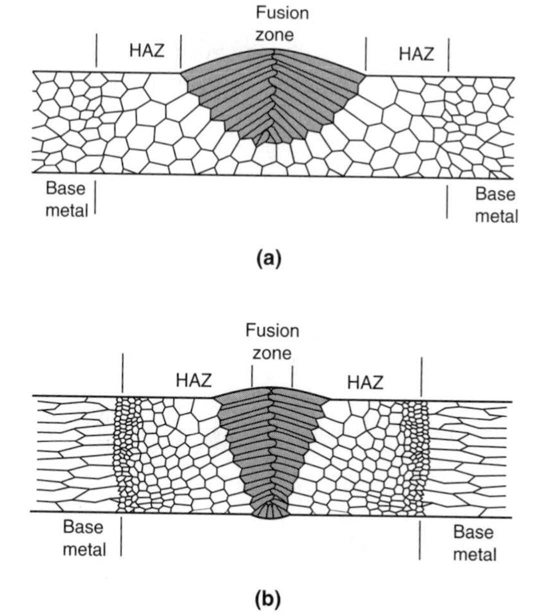

FIGURE 11.16 HAZ microstructure for (a) previously annealed pure metal and (b) previously work-hardened pure metal.

where D is average grain size and K_a denotes rate constant that exhibits Arrhenius temperature behavior.

Thus equation (11.15) can be expressed in a more general form as

$$\frac{dD}{dt} = K_0 \cdot e^{-\frac{\Delta E_d}{k_B T}} \cdot \frac{1}{D} \tag{11.16}$$

where K_0 is a diffusion constant.

For isothermal conditions, integration of equation (11.15) gives

$$D^2 - D_0{}^2 = K_a t \tag{11.17}$$

where D_0 is an initial average grain size.

However, since the temperature at any point in a part that is being processed varies with time, the grain size at any location at any time t is given more appropriately by

$$D^2 = K_0 \int_0^t e^{-\frac{\Delta E_d}{k_B T(t')}} dt' \tag{11.18}$$

where t' is a dummy variable.

The importance of grain structure stems from its influence on the mechanical properties of the component (strength and toughness). A fine grain structure results in an increase in the strength of the material. For a pure material where there are no second-phase particles, inclusions, or other microconstituents, the yield stress of the material, S_y, is related to the average grain size D by the Hall–Petch equation:

$$S_y = \sigma_i + k_y D^{-1/2} \tag{11.19}$$

where σ_i is frictional resistance of the lattice to moving dislocations and k_y is a constant.

This is illustrated in Fig. 11.17 for copper. The dependence of the yield strength on average grain size is more complex for alloys, since other factors also affect the yield strength.

Example 11.3 From Fig. 11.17, obtain the Hall–Petch equation for copper.

Solution:

The Hall–Petch equation is given by

$$S_y = \sigma_i + k_y D^{-1/2}$$

where σ_i is the intercept on the S_y axis and k_y is the slope of the $S_y - D^{-1/2}$ line.

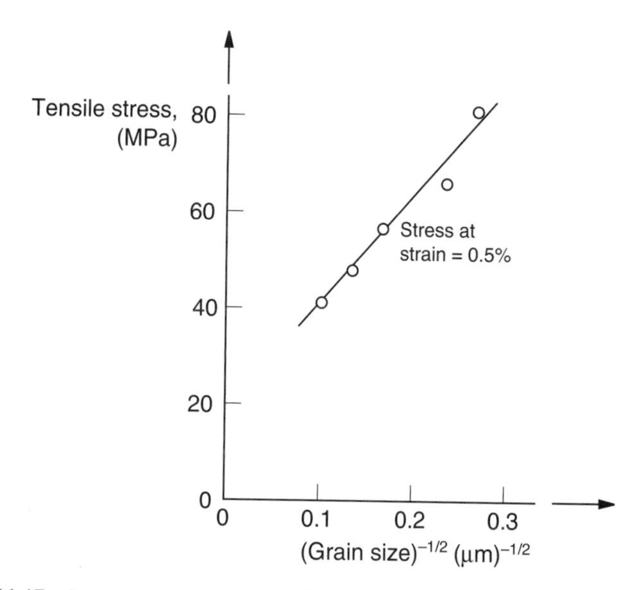

FIGURE 11.17 Stress variation with grain size for copper. (From McLean, D., 1962, *Mechanical Properties of Metals*, John Wiley & Sons, Inc., New York. Courtesy of National Physical Laboratory, UK.)

By extrapolating the line in Fig. 11.17 to the S_y axis, we find the intercept to be about 20 MPa. Thus $\sigma_i = 20$ MPa.

The slope m_g of the line is estimated from the graph to be

$$m_g = \frac{(80 - 20)\,\text{MPa}}{(0.3 - 0)\ \mu\text{m}^{-1/2}} = 200\,\text{MPa}\ \ \mu\text{m}^{1/2}$$

The Hall-Petch equation for this copper is thus given by

$$S_y = 20 + 200D^{-1/2}\,\text{MPa}$$

11.1.3.2 Precipitation-Hardening and Nonferrous Alloys The heat-affected zone in the case of precipitation-hardening materials such as aluminum–copper alloys normally consists of two regions (Fig. 11.18):

(a) There is the inner or solutionized zone, that is, material very close to the fusion zone, where the very high temperatures result in the material transforming into a single-phase solid solution. The new phase results in new grains being nucleated, and these grow as a result of the high temperatures. This weakens the strength of material close to the fusion line, making the strength equivalent to that of an annealed material. Irrespective of the initial condition of the material close to the fusion line, that region will be solution treated and

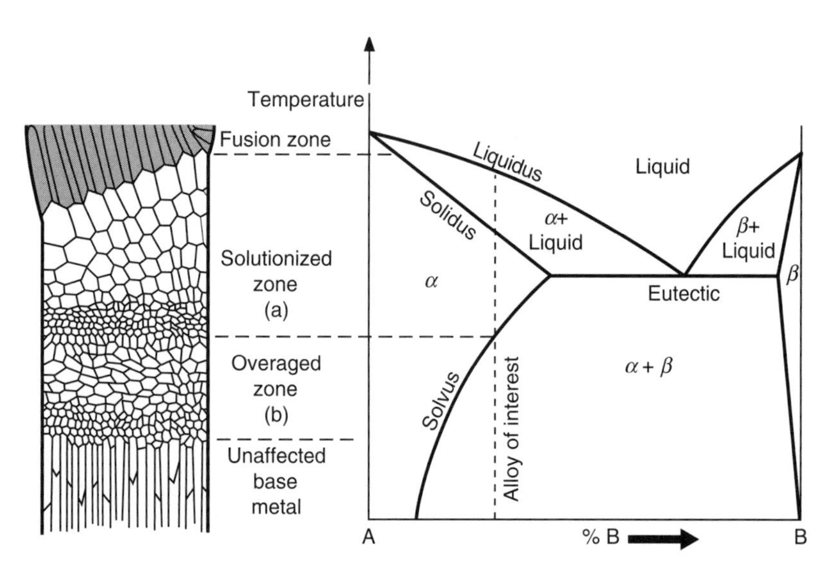

FIGURE 11.18 HAZ microstructure for a previously work-hardened precipitation hardening material. (Copyright American Welding Society, *Welding Handbook*, Vol. 1, 9th edition, 2001. Used with permission.)

quenched and therefore relatively soft, of course, with the coarse grain structure persisting.

(b) Slightly farther away from the fusion boundary, there is the outer or overaged zone where the material will have been heated to a temperature high enough to result in recrystallization with an equiaxed fine grain structure if the material was previously cold worked. Otherwise, grain growth will occur in this zone. Next to this zone, the unaffected base material will have the original structure of the material.

If the material had been previously aged, then it would become solution treated close to the fusion zone, and overaged farther away. To avoid overaging, processing may be done before solution treating and aging. However, this is a difficult approach, due to the inconvenience of heat treating large structures in furnaces and also potential distortion of the structure. Low heat input laser and electron beam processes reduce the overaging effect.

11.1.3.3 Steels In this section, the polymorphic property of steel is discussed in the light of its effect on the heat-affected zone grain structure. This is followed by a discussion on the heat-affected zone microstructure of three major types of steel:

1. Low-carbon steels
2. High-carbon and alloy steels
3. Stainless steels.

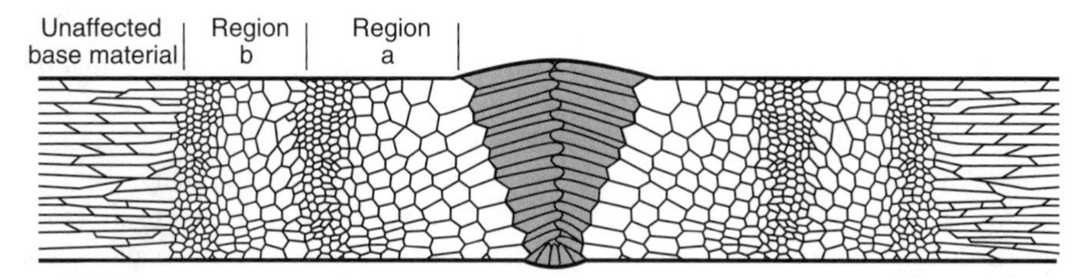

FIGURE 11.19 Schematic of HAZ microstructure for a previously work-hardened steel or precipitation hardening material. (Copyright American Welding Society, *Welding Handbook*, Vol. 1, 9th edition, 2001. Used with permission.)

Polymorphic or allotropic materials such as iron may also exhibit two distinct regions in the heat-affected zone of a previously work-hardened material (Fig. 11.19):

(a) One region is due to recrystallization after transformation to the higher temperature crystal structure, for example, low austenite region. This normally occurs in the temperature range between 723°C and 900°C of the steel, closer to the fusion zone. In this region, the steel becomes either fully or partially austenitized. For the sections that are fully austenitized, new grains are nucleated, resulting in a relatively fine grain structure in the entire section. However, in the regions that are only partially austenitized, it is the newly formed austenite phase that produces a refined grain structure as new grains are nucleated, while the untransformed ferrite phase essentially maintains the original grain structure, with a possible slight growth. Thus, only partial grain refining would occur in this section. In either case, the high temperatures in the immediate vicinity of the fusion zone result in a tendency toward grain growth that may result in weakening in strength of material close to the fusion line.

(b) The other region corresponds to recrystallization of the low-temperature crystal structure, that is, high ferrite region, approximately 700°C depending on the type of steel. An equiaxed fine grain structure may be formed here. This is the outer zone next to the base material. If the steel is not work hardened before being processed, then there will be no recrystallization, and the original grains will simply continue to grow in this region.

Low-Carbon Steels During cooling of polymorphic materials such as steels, additional phase transformation may occur in the solid state. This would often result in a new microstructure being generated that obliterates the microstructure that was generated during solidification. However, the original grain structure may exist in conjunction with new grains that are formed in the new phase.

For low-carbon, low-alloy steels, acicular ferrite may form. This has randomly oriented short ferrite needles, and provides increased resistance to crack propagation, thereby increasing the fracture toughness of the steel. On the contrary, bainite, grain boundary ferrite, and ferrite side plates reduce the fracture toughness of the material.

Bainite and martensite are more likely to form with increasing carbon content and/or cooling rate. Thus processing conditions have to be selected to promote the formation of acicular ferrite.

The coarse grain structure that develops close to the fusion zone may result in a material with strength equivalent to that of an annealed material. Normally, the high cooling rates that occur close to the fusion zone, along with the coarse grain structure in that region, result in the formation of acicular ferrite along grain boundaries, as opposed to blocky ferrite. Even though the low-carbon steels do not readily form martensite, the cooling rates that accompany high intensity (laser or electron beam) processes can produce martensite in the heat-affected zone. The low carbon content tends to reduce the hardness of any martensite that may be formed during cooling. Furthermore, it raises the martensite start temperature such that any martensite formed is tempered during cooling (autotempering), thereby softening it. However, if the cooling rate is fast enough, there is insufficient time for the resulting martensite to be autotempered and it consequently remains as hard martensite. On the contrary, if the cooling rate is very low, the ferrite formed in the initial stages of cooling rejects carbon into the surrounding austenite that may thus subsequently transform into hard martensite since its carbon content is then high. Thus the cooling rates for such steels have to be intermediate.

The portions of the workpiece that are heated to temperatures equivalent to that between the eutectoid temperature and the lower austenite temperature experience partial transformation, with the pearlite phase transforming to austenite that is embedded within the untransformed ferrite matrix. Subsequent cooling of the part may result in the patches of austenite transforming to hard bainite or martensite, embedded in softer ferrite matrix.

High-Carbon and -Alloy Steels For hardenable steels such as high carbon or alloy steels, the relatively high cooling rates in the region closest to the fusion line may transform material in this region to martensite.

Hardenability refers to the ease with which martensite forms. It does not refer to the hardness or strength of the steel. Plain carbon steels, for example, have low hardenability and as such, only very high cooling rates can produce martensite in them. However, even air cooling may result in the formation of martensite in some high-carbon or -alloy steels.

Since different alloying elements affect the hardenability of a steel differently, a simple way of estimating the hardenability or susceptibility of a given steel to cracking is to determine its *carbon equivalent* (CE). This combines the impact of all the alloying elements into a single carbon fraction that provides, in a nutshell, an estimate of the material's hardenability. The carbon equivalent relationship depends on the alloying elements, and there are various forms of it. A commonly used expression is

$$\mathrm{CE} = \mathrm{C} + \frac{\mathrm{Mn}}{6} + \frac{\mathrm{Cr} + \mathrm{Mo} + \mathrm{V}}{5} + \frac{\mathrm{Ni} + \mathrm{Cu}}{15} \tag{11.20}$$

The variables are expressed in percentage by weight. Preheating or postweld heat treatment is often not necessary for materials with a carbon equivalent less than 0.35%. Preheating is normally required when the carbon equivalent is between 0.35 and 0.55%, while for a CE greater than 0.55%, both pre- and postheat treatments are necessary.

Stainless Steels The corrosion resistance of iron increases as its chromium content increases, since the chromium enables a thin, protective surface layer of chromium oxide to be formed on the surface when the material is exposed to oxygen. Stainless steels are essentially iron–chromium alloys that are corrosion-resistant to an extent that depends on the alloying elements of the steel. Common alloying elements include nickel, carbon, and molybdenum, and the types and amounts of theses alloying elements determine the type of the stainless steel. There are three principal types:

(1) Ferritic stainless steels.
(2) Austenitic stainless steels.
(3) Martensitic stainless steels.

Each of these three types behaves differently in the heat-affected zone, and they are briefly described in the following paragraphs.

- *Ferritic Stainless Steels*: This type of stainless steel (Fe–Cr) has a relatively low carbon content (normally less than 0.12% C) and a high chromium content (12–30% Cr). With this alloy content, the material becomes partially or completely ferritic at high temperatures. An example is Type 430. The high chromium content gives it high corrosion resistance, even though it also makes it relatively more expensive.

  Ferritic stainless steels experience severe embrittlement when held in the temperature range of 450–550°C for some time, resulting in a loss in toughness and increase in hardness. The only way to remove the embrittlement involves reheating the material to a higher temperature and quenching through this temperature range. As a result, this type of stainless steel is not appropriate for parts that need to be welded, for example, since the portion of the part that is heated to this temperature range will have poor properties.

  Another problem encountered when ferritic stainless steels are processed with a heat source is the extensive grain growth that occurs, since there is no phase change to produce a refined grain structure, as the material is heated in the solid state. Thus, in this material, grain refinement can only be induced by cold working followed by recrystallization, which is not convenient for parts that are already processed. Finally, ferritic stainless steels that contain more than 16% Cr also encounter sensitization (see the next subsection on austenitic stainless steels) when heated above 925°C. This type of sensitization is not effectively subdued by lowering the carbon content or rapid quenching from above the sensitization temperature, as is commonly done for austenitic stainless steels.

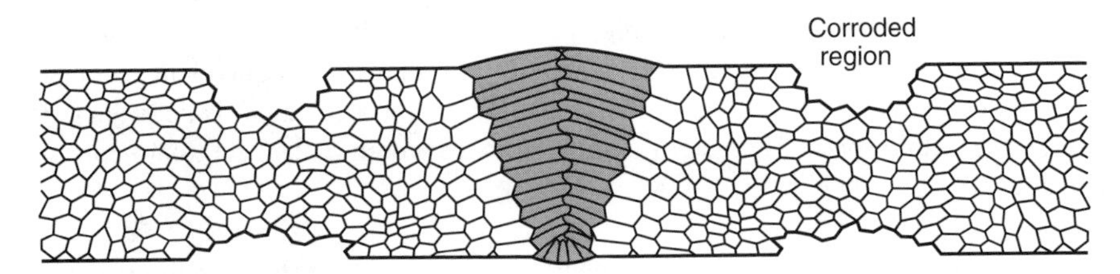

FIGURE 11.20 Corrosion in an austenitic stainless steel after welding, as a result of sensitization.

However, additions of niobium and/or titanium can suppress the sensitization problem.

- *Austenitic Stainless Steels*: Stainless steels with a relatively high chromium content and significant amounts of nickel (Fe–Cr–Ni), typically 18% Cr, 8% Ni, stabilize the austenite phase to the point that it still exists even at room temperature. Thus such alloys cannot be hardened by heat treatment, and are only hardened by cold working. An example is Type 301. When this type of stainless steel is cold worked, the unstable austenite is transformed into martensite, significantly increasing its strength.

  When austenitic stainless steels are heated between 425 and 800°C, carbon tends to react with chromium to form chromium carbide, $Cr_{23}C_6$, and the reaction is most rapid around 650°C. Thus, in the regions of the heat-affected zone that are heated to this temperature range, chromium carbide is precipitated, usually at the grain boundaries. This phenomenon is called *sensitization* and induces corrosion (Fig. 11.20) for two reasons:

  1. An electrolytic cell is formed at the grain boundaries.
  2. The material loses corrosion resistance along grain boundaries due to a significant reduction in the amount of free chromium.

  Sensitization can be prevented by

  1. Reheating the part that has been processed to about 1100°C to redissolve the chromium carbide, and cooling fast to prevent the carbide from precipitating again. However, distortion of the part that sometimes results from quenching often makes this treatment undesirable. Furthermore, if the structure being processed is cumbersome, then the heat treatment process becomes unfeasible.
  2. Reducing the carbon content of the alloy to, say, 0.03% C, so that little chromium is tied up as chromium carbide. An example is Type 304, which is easy to weld.
  3. Adding niobium or titanium that has a greater affinity for carbon than does chromium, thus having all the available carbon tied up leaving the chromium free in solution. Examples are Type 347 that contains niobium

and Type 321 that contains titanium. However, the heat-affected zone of Types 321 and 347 can still suffer attack if subjected to severe corrosive conditions after heating above 1100°C and 1300°C, respectively. This is because the Ti and Nb carbides go into solution when held at those temperatures, resulting in subsequent chromium carbide precipitation along grain boundaries in a narrow zone close to the fusion boundary. The resulting corrosion is referred to as knife-line attack. This type of problem can be eliminated by limiting the carbon content to less than 0.03% C.

Due to the relatively low solubility of sulfur in austenite, austenitic stainless steels tend to have a high susceptibility to hot cracking (see Section 11.2.2.1). This crack sensitivity is minimized by introducing small amounts of δ-ferrite, say 5–10%, to form a duplex structure containing a mixture of ferrite and austenite, since ferrite has a greater solubility for sulfur.

Elements that tend to stabilize the austenite phase include carbon, nickel, manganese, and nitrogen, while those promoting the formation of ferrite are chromium, niobium, aluminum, silicon, and molybdenum.

- *Martensitic Stainless Steels*: Martensitic stainless steels normally have chromium in the range 12–18%, and carbon in the range 0.15–1.20%. An example is Type 410, and most of them form martensite on air cooling from the austenite. They usually encounter severe temper embrittlement when heated in the temperature range 370–600°C. The alloys with a relatively higher chromium content are highly susceptible to hydrogen embrittlement that can result in premature failure.

11.2 DISCONTINUITIES

Discontinuities are normally introduced into a part during processing. As defects (i.e., result in rejectable condition), they are undesirable as they reduce the mechanical properties of the material. They thus influence the safety and performance of the part. There are various types of discontinuities, and the principal ones include coring (see Section 11.1.1.4), porosity, and cracking, which are metallurgical; slag inclusions, incomplete penetration, lack of fusion, and undercutting, which are process and technique related (Table 11.2). It is sometimes also useful to identify discontinuities by geometry, that is, whether they are planar (two-dimensional) or volumetric (three-dimensional). The planar types include cracks, lack of fusion, and incomplete penetration, and these tend to have a much greater damaging effect than do the volumetric types. The volumetric types include porosity and slag inclusions. The closer a given discontinuity is to the surface, the more detrimental it can be to the integrity of the part. Furthermore, if the discontinuity is oriented such that the applied stress tends to open it up, then its damaging effect is greater than one that is oriented otherwise. Temperature, stress state, time, and service environment do have significant influence on defects. Modes of failure include plastic deformation, fatigue, fracture, corrosion, and creep.

TABLE 11.2 Classification of Common Discontinuity Types

DISCONTINUITY CLASSIFICATIONS			
Process and Metallurgical		Geometric	
METALLURGICAL	PROCESS/TECHNIQUE RELATED	PLANAR	VOLUMETRIC
Coring Porosity Cracking	Slag inclusion Incomplete penetration Lack of fusion Undercutting	Cracking Lack of fusion Incomplete penetration	Porosity Slag inclusion

Coring is extensively discussed in Section 11.1.1.4 and will not be further discussed here. In the following setions, the remaining principal discontinuities are discussed.

11.2.1 Porosity

Porosity is the existence of gas pores or pockets within the fusion zone, Fig. 11.21, and may be defined as cavity type discontinuities (voids) formed by gas entrapment during solidification. It basically arises from the absorption of gases into the melt (droplet or pool) and subsequent release of part of the gases by the molten metal during cooling due to reduced solubility in the solid. This is especially the case if the gases exist in amounts greater than their solubility limit in the solid. Surface contamination (e.g., oils, moisture, and paint) is a major source of the gases that cause porosity. Gases may also be released by chemical reactions in the molten metal. Oxygen, hydrogen, and nitrogen are the principal sources of porosity and are absorbed atomically.

The rate of solidification of the molten pool has a pronounced effect on porosity formation. Porosity usually results if the rate is so fast that there is not adequate time for the evolved gases to escape. Porosity formation proceeds by nucleation and growth just as occurs during solidification. The pores become elongated if the solidification

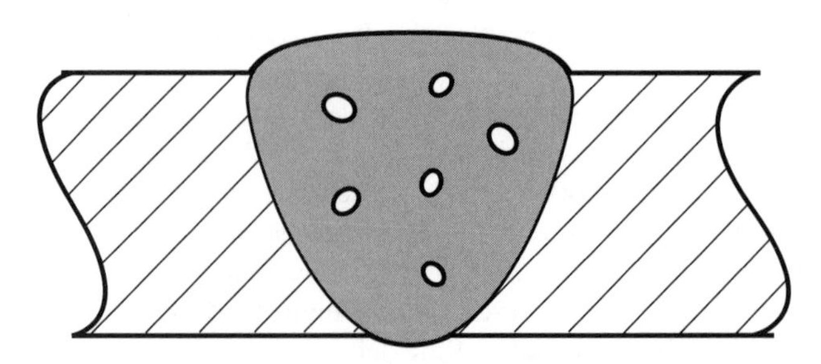

FIGURE 11.21 Porosity in the fusion zone.

and porosity growth rates are about the same. The bubbles that form escape to the surface if the growth rate exceeds the solidification rate, and they are trapped as pores if the reverse is true. Porosity is usually most severe in the initial stages of a process since the conditions are not steady during that period.

Oxygen may be introduced into the molten metal by oxides on filler material or on the base material. It may also be introduced from the atmosphere. The presence of oxygen in the molten pool has two effects. One involves oxidation of the fused metal to produce material with poor properties. The other involves a reaction with carbon in the case of steel to form carbon monoxide that is trapped in the fusion zone. The influence of oxygen may be minimized by adding deoxidants, for example, manganese, silicon, or aluminum to any filler metal used. The oxides so produced are much less damaging than the porous structure that would otherwise result. However, excess deoxidants may reduce the mechanical properties of the fusion zone. The oxides formed when aluminum is used as a deoxidant tend to stay in the fusion zone and give it poor mechanical properties. The addition of silicon and manganese along with aluminum tend to produce oxides that float and are therefore removed with slag.

Hydrogen may, in addition to the porosity, also make the fusion zone and heat-affected zone brittle. It is considered as the principal cause of porosity when welding steels and aluminum. This is because there is a significant difference in the solubility of hydrogen in the molten metal at high temperatures, and its solubility at the melting temperature. The excess hydrogen may be released in the molten pool as it cools. Hydrogen may enter the molten pool from the atmosphere. Moisture absorbed by oxides on the surfaces of filler rod and workpiece and humidity in the air are often the main sources of hydrogen and oxygen. The water vapor dissociates to form hydrogen and oxygen. Thus for highly reactive materials such as aluminum and magnesium, any filler material to be used needs to be protected from the atmosphere until used or cleaned before being used.

Nitrogen may be absorbed from the atmosphere. Its presence results in the formation of nitrides that may make the material brittle.

Brasses are difficult to weld due to volatilization of the zinc alloying element, which may result in porosity and also affect visibility. Copper oxide reacts with hydrogen to release steam, which may induce porosity in the fusion zone unless the copper is deoxidized by adding phosphorus and/or silicon, manganese, titanium, and aluminum.

The most suitable traditional method for detecting porosity after processing is the radiographic technique, followed by the ultrasonic inspection method. Liquid penetrant, magnetic particle, and visual inspection are useful only when the pores appear on the surface. Porosity levels up to about 7% by volume generally do not have any significant influence on the tensile properties of the part. However, the ductility tends to reduce slightly with increasing porosity, the effect being more pronounced as the strength of the material increases. Failure in fatigue is influenced by porosity only when any reinforcement that is present is removed, and surface porosity tends to be more damaging than buried ones. A reinforcement is the portion of the fusion zone that projects above the workpiece surface. More often than not, it results from filler material that is added to the molten pool.

11.2.2 Cracking

In this section, we discuss the basic concepts of cracking that occurs during and after processing. Cracking is the most damaging of all the defects. The principal forms of cracking are

1. Hot cracking.
2. Liquation cracking.
3. Cold cracking.
4. Reheat cracking.
5. Lamellar tearing.

Cracking may occur in either the fusion zone, the heat-affected zone, or both (Figs 11.22 and 11.23). The crack location, in the case of carbon steels, is influenced by the hardenability of both the fusion zone and the base metal. The cracks may be in the form of centerline cracks if the fusion zone and base metal compositions are about the same and the workpiece is not severely restrained. This is because the highest cooling rate occurs along the centerline. On the contrary, in situations where a filler metal is used and its carbon content is lower than that of the base metal, the overall carbon content and thus hardenability of the resulting fusion zone is lower than that of the base metal. The coarse-grained region of the heat-affected zone then tends to be more susceptible to crack formation, resulting in cracks appearing in the heat-affected zone instead.

Cracks could be formed either during the process, that is, soon after solidification (hot cracking) or after the bead has cooled to room temperature (cold cracking) and may continue for several hours, days, or even longer. Hot cracking is more prevalent

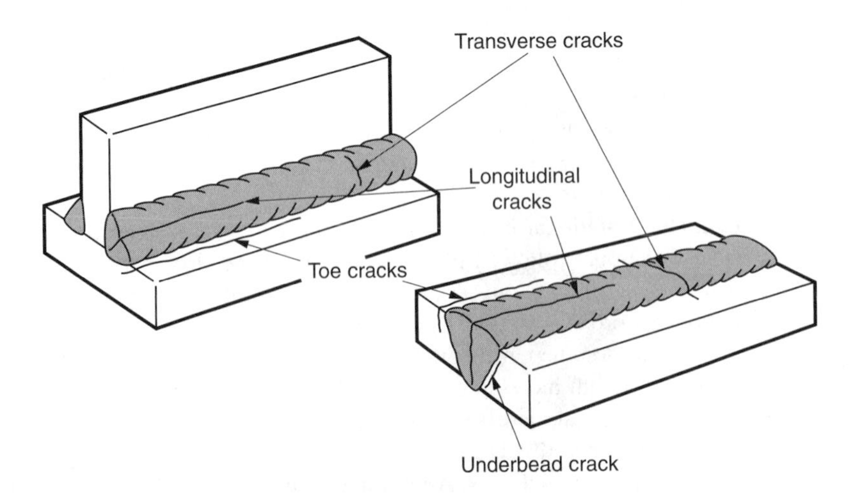

FIGURE 11.22 Schematic illustration of weld cracks. (Copyright American Welding Society, *Welding Handbook*, Vol. 1, 9th ed., 2001. Used with permission.)

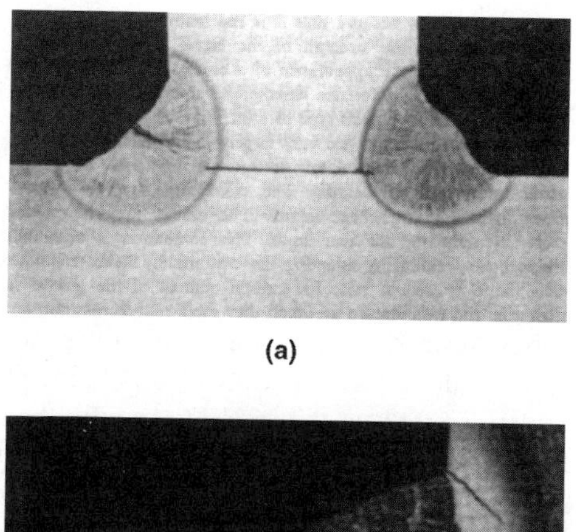

(a)

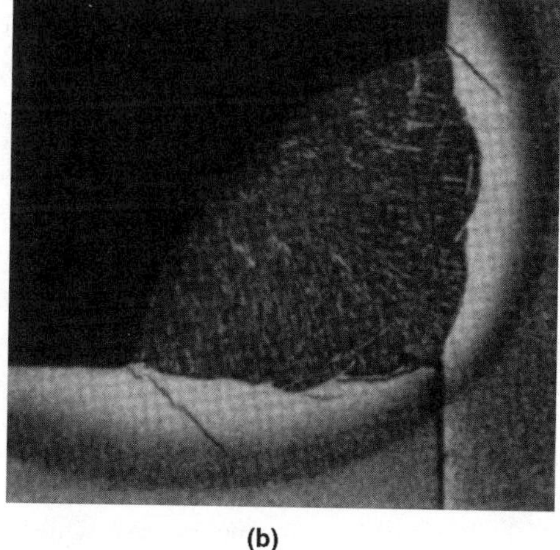

(b)

FIGURE 11.23 Pictorial illustration of weld cracks. (a) Hot crack in a fillet weld. (b) Cold crack in the heat-affected zone of a fillet weld. (From Kou, S., 1987, *Welding Metallurgy*. Reprinted with permission from John Wiley & Sons, Inc.)

in the fusion zone whereas cold cracking is more common in the heat-affected zone. The principal factor inducing cracking is the residual stresses that are set up as the fusion zone contracts during cooling. The stresses are tensile in the fusion zone and gradually change to balancing compressive stresses in the surrounding material. While subjected to residual stresses, the susceptibility of a material to cracking is greatly determined by its ductility in the stressed zone. The ductility is determined by the existence of intergranular liquid films; material phase changes; and introduction of certain gases, especially hydrogen into the part during processing.

Cracks are unfortunately difficult to detect using conventional nondestructive evaluation techniques due to its planar form. Ultrasonic testing is the best method for off-line detection of embedded cracks, but when other forms of discontinuity exist, it becomes difficult to distinguish between them. For real-time or online detection, acoustic emission, audible sound, and infrared/ultraviolet detection have been found to be effective.

11.2.2.1 Hot Cracking Hot cracking usually occurs in the fusion zone either during or immediately after solidification, in the temperature range of about 200–300°C below the melting temperature. It results primarily from low-melting constituents that are capable of wetting the grain boundaries and which form either interdendritic or intergranular liquid films. The segregation of these films result in localized regions of weakness along the grain boundaries and between dendrites that rupture under the tensile residual stresses that develop as the part cools down. This produces interdendritic or intergranular cracks, that is, the cracks occur in-between grains or dendrites, or along their boundaries, instead of occurring within the grains or dendrites themselves. They often develop along the centerline also. The incipient cracks may heal if there is adequate molten metal near the cracks to backfill them. Due to their high-temperature occurrence, hot cracks exposed to the surface can often be identified by an oxide coating that discolors the crack surfaces. Whether a crack is formed or not depends on the type of intergranular films formed, the microstructure (i.e., whether it is coarse or fine), and the residual stresses developed in those regions. Fine equiaxed grains tend to have greater resistance to hot cracking than do coarse columnar grains, since the low-melting constituents are then distributed over a wider surface area and thus are in lower concentrations. Contaminants such as oils that are left on the base metal may also contribute to hot cracking.

An essential condition for cracking to occur is that the stress experienced by the material be greater than its fracture strength. As the molten metal solidifies and individual grains grow, it gets to a point where they impinge on each other, or join together to form a coherent unit, even though there may still be some small amount of molten metal left. At this point, the material attains mechanical strength, even though it then has no ductility and is thus brittle. The temperature at which this occurs is referred to as the coherence temperature. On further cooling, it again gets to a temperature (the nil-ductility temperature) where the solidifying metal begins to gain ductility. The temperature range from the nil-ductility to the coherence temperature is referred to as the brittle range, and it is in this region that the solidifying metal is most liable to crack. Materials with a long brittle range are crack sensitive, while those with a short brittle range tend to be crack resistant (Fig. 11.24).

In steels, nonmetallic alloying elements such as sulfur, phosphorus, and boron that tend to promote the formation of low-melting intergranular films, capable of wetting the grain boundaries, make the material more susceptible to hot cracking. Sulfur forms iron sulfide, FeS, which dissolves in the molten steel and increases the freezing range between the solidus and the liquidus considerably. Furthermore, the surface tension between the molten iron sulfide and the solid grains is low, resulting in wetting of the grain boundaries, that is, the formation of a liquid film. With traces of low-melting phases rich in sulfur still existing in liquid form at relatively low temperatures when significant residual stresses are set up, hot cracking can easily occur. Hot cracking will generally not occur if the sulfur and phosphorus content is less than 0.06%. An increase in the carbon content also increases the cracking tendency, as it increases the freezing range. Nickel has a similar effect, especially in concentrations above 3%. Thus fusion welds are difficult when the nickel content exceeds 4%.

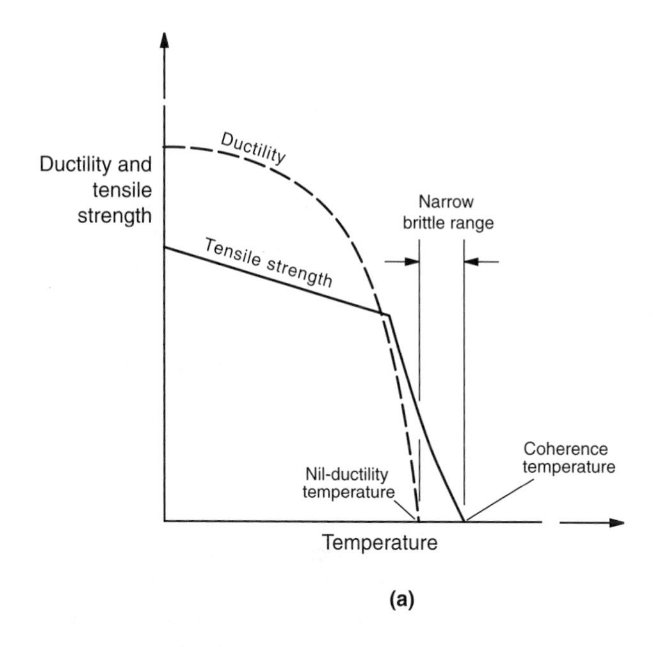

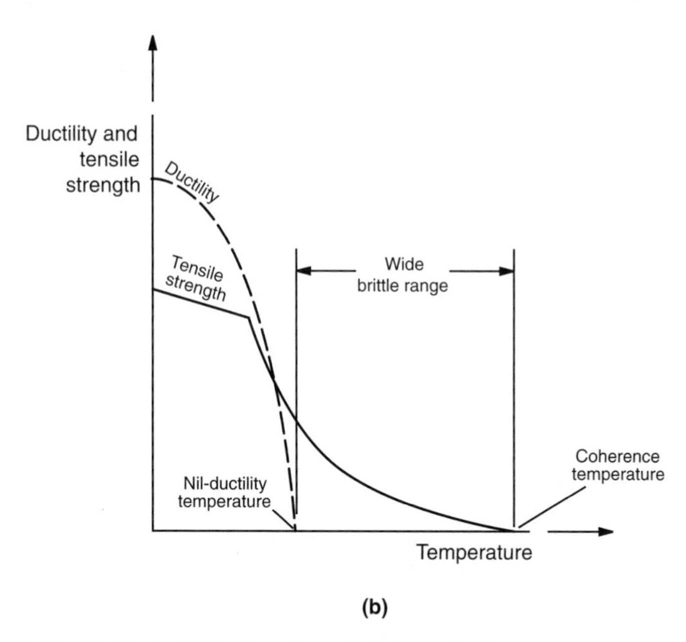

FIGURE 11.24 Mechanical behavior of metals during solidification. (a) Narrow brittle range behavior. (b) Wide brittle range behavior. (From Lancaster, J. F., 1980, *Metallurgy of Welding*, 3rd ed., George Allen & Unwin, London.)

The solubility of sulfur in austenite is lower than in ferrite. There is therefore a greater tendency for the liquid grain boundary film to form and cause cracking in austenitic steels than in ferritic steels since the sulfur is more likely to segregate and form weak regions in the former. Addition of austenite stabilizers such as carbon, nickel, and copper to steels thus makes them more crack susceptible.

Manganese tends to make the sulfides globular and thereby prevents cracking when used in the right amounts. This is because it does not dissolve in molten steel or increase the freezing range and is therefore not dispersed as an intergranular film. As the carbon content of the steel increases, so must the manganese-to-sulfur ratio to prevent hot cracking (Fig. 11.25). However, at high carbon contents, say greater than 0.3% C, higher ratios of manganese-to-sulfur are no longer effective.

Hot cracking in nonferrous alloys may occur for compositions that fall in regions of the phase diagram with a high temperature range between the liquidus and the solidus lines, that is, a long freezing range. Under such circumstances, there is a greater tendency for the formation of low-melting phases.

Thus hot cracking in nonferrous metals results primarily from alloying elements as opposed to the case of steel where it occurs mainly as a result of the presence of low-melting impurities. The sensitivity of the aluminum–silicon alloy (4000 series), for example, to solidification cracking increases with increasing silicon composition, peaking at about 0.5% silicon, and then decreases with further increases in the silicon content. Aluminum–magnesium alloys (5000 series) are also sensitive to hot cracking. In the case of magnesium, zinc (4–6%) and calcium are the alloying elements that result in the highest susceptibility to hot cracking, while manganese, aluminum, and

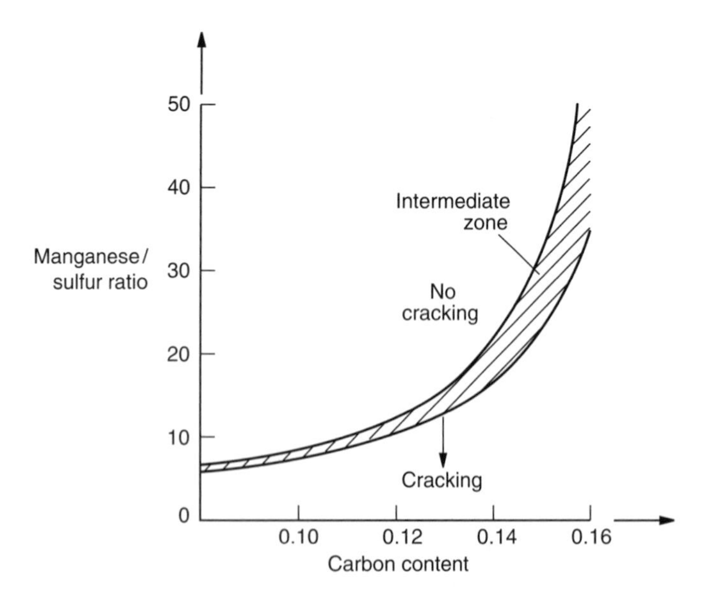

FIGURE 11.25 Dependence of the susceptibility of a steel to cracking on the manganese-to-sulfur ratio and carbon content. (From Lancaster, J. F., 1980, *Metallurgy of Welding*, 3rd ed., George Allen & Unwin, London.)

zirconium alloying elements do not present significant hot cracking problems. The presence of contaminants such as oil and paint on the surface of nickel workpieces are often the source of low-melting constituents that may result in hot cracking or liquation cracking in welding nickel and its alloys. Adequate cleaning of such materials is thus essential for defect-free welding. The high affinity of the reactive metals such as titanium and zirconium to oxygen and nitrogen makes them difficult to weld, since they result in embrittlement of the joint, and may also induce porosity. Thus, with such materials, it is often necessary to introduce shielding gas on both sides of the joint.

The tendency for hot cracking may be minimized by

1. Improved design to minimize constraint effect and thus residual stresses.
2. Using material with low-cracking sensitivity.
3. Adequate cleaning of the workpiece to eliminate surface contaminants such as oil and paint.

11.2.2.2 ***Liquation Cracking*** Sometimes complex carbides or low-melting intermetallic compounds that result from segregation may melt to produce local grain boundary films or regions of localized brittleness in the zone of partial melting during processing. This is especially likely in the temperature range between the solidus and liquidus curves, even though the bulk of material in that region may still be solid. Under sufficiently high residual stresses, cracks may open up in these regions, causing liquation cracking. Liquation cracking may result from sulfur and phosphorus-forming grain boundary liquation in steels or niobium-forming eutectic patches. In the former case, the addition of manganese does not alleviate the problem. The only solution is to limit the amounts of sulfur and phosphorus in the steel to very low amounts. For aluminum, the Al–Mg–Zn alloys or Duralumin are most susceptible to this form of cracking.

Hot cracking involves material that is initially completely molten and is thus common in the fusion zone, whereas liquation cracking involves partial melting, and is most common in the heat-affected zone or zone of partial melting.

11.2.2.3 ***Cold Cracking*** The high cooling rates that occur during processing may result in brittleness in both the fusion zone and portions of the heat-affected zone. The brittleness is even more likely in steels when martensite is formed. These factors, along with the presence of hydrogen, promote crack formation in steels, which may start soon after solidification and continue for some days afterward. This type of crack normally occurs after the material has cooled to relatively low temperatures, and is thus referred to as cold cracking. It is sometimes also referred to as hydrogen cracking or delayed cracking due to the incubation time associated with the development of the crack.

Cold cracking is usually transgranular in nature, but may sometimes also be intergranular. It may appear at the surface of the part, for example, toe cracks, or remain hidden beneath the fusion zone as underbead cracks (Figs. 11.22 and 11.23). It is generally induced by the following:

1. Hardening due to rapid cooling (martensite formation).
2. Residual stresses that are set up in the part.
3. Embrittlement by hydrogen (is enhanced by a moist atmosphere).
4. Existence of hot cracks.

Residual stresses are often introduced in processes that involve a heat source. The formation of cold cracks thus depends primarily on the cooling rate (formation of martensite) and the presence of hydrogen, for steels. Without hydrogen, cold cracking is less likely to occur in steels. Materials with high martensite start and martensite finish temperatures have a lower susceptibility to cold cracking since there is a drastic reduction in the solubility of hydrogen when transformation occurs from austenite to martensite. Also, at the higher temperatures, there is a greater tendency for the supersaturated hydrogen to diffuse out. Furthermore, the higher transformation temperature enables the martensite formed to be tempered to some extent, thereby reducing the associated hardness and residual stresses.

Increasing the hydrogen content of the shielding gas has been shown to directly increase the extent of cold cracking. Hydrogen may be introduced through moisture on electrodes and workpiece or from the atmosphere. Its influence is twofold:

1. Causing porosity
2. Embrittlement.

Even though the presence of hydrogen is almost always essential for cold cracking to occur, a high-carbon, high-alloy steel that is highly restrained can result in cracking in a virtually hydrogen-free atmosphere.

Hydrogen is absorbed by molten droplets as well as the molten pool. A significant portion of the absorbed hydrogen escapes as the molten metal solidifies. However, since the solubility of hydrogen in austenite is high, some of it stays in the austenite that solidifies from the molten metal, while some more diffuses into the adjacent metal that is austenitized. The fusion zone carbon content is usually diluted when filler metal is used. In such cases, the austenite in the fusion zone tends to transform to ferrite and pearlite, which becomes supersaturated with hydrogen since the solubility of hydrogen in ferrite is much lower than in austenite. Hydrogen is thus rejected from the fusion zone into the heat-affected zone that is still austenitic (Fig. 11.26), a process that is enhanced by the high diffusion of hydrogen in ferrite. The lower diffusion of hydrogen in austenite, however, results in entrapment of hydrogen in the heat-affected zone austenite. This may subsequently transform to martensite if the cooling rate is high enough, making the transformed region highly susceptible to cold cracking.

The enhancement of cold cracking by hydrogen results from its embrittling effect on metals, especially ferritic steels. Austenitic stainless steels and nonferrous metals other than copper, however, are not affected to any significant degree by hydrogen, since hydrogen has a relatively high solubility in such materials. The embrittlement of steel by hydrogen may take one of the two forms, permanent damage and reversible embrittlement. The case that results in permanent damage to the material occurs at

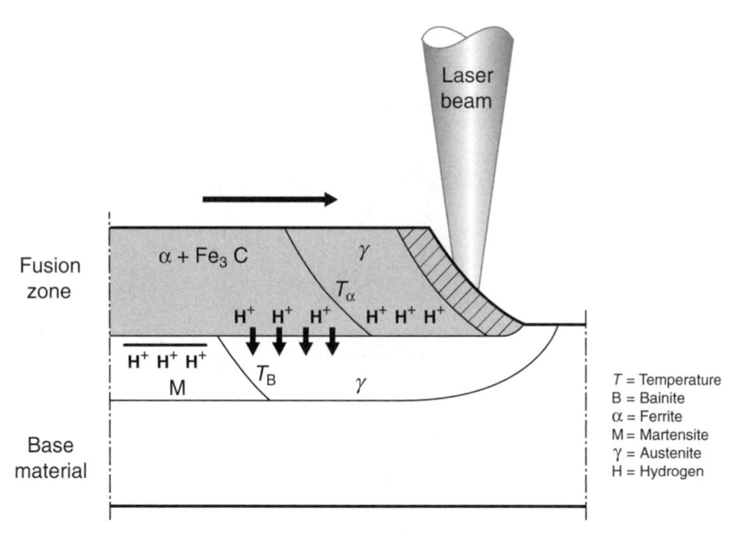

FIGURE 11.26 Diffusion of hydrogen in the fusion zone and HAZ. (From Granjon, H., 1972, *Cracking and Fracture in Welds*, Japan Welding Society, p. IB1.1.)

high temperatures, affecting primarily carbon and low-alloy steels. The damage may be in the form of either cracking or decarburization or both and results from a chemical reaction between hydrogen and carbides. The more common case, which we shall be concerned with in greater detail, occurs at relatively lower temperatures, normally at room temperature, and is reversible, that is, elimination of the gas reverts the steel to its normal toughness.

High strain rates and low temperatures are normally known to enhance brittle behavior. Contrary to such expectations, however, hydrogen embrittlement is enhanced by slow strain rates and high temperatures, that is, fracture ductility decreases with decreasing strain rates and increasing temperatures. Thus the phenomenon of hydrogen embrittlement is expected to be controlled by the diffusion of hydrogen, which depends on temperature. As a result, for a given delay (incubation period), a slow strain rate is more likely to detect embrittlement than a high strain rate. In other words, the mechanism becomes ineffective at high strain rates since the hydrogen then cannot diffuse fast enough to the desired region. Thus, hydrogen has little effect on impact test results, while its effect would be observed in a slow tensile test.

When notched steel workpieces are hydrogenated and subjected to static loading, they are known to exhibit delayed failure over a wide range of applied stress, even though there is very little dependence of the time to failure on stress (Fig. 11.27). Delayed failure occurs when the applied stress does not cause the material to fail immediately, but failure begins to occur after a certain time, even when there is no change in conditions. Failure does not occur below a critical stress level, while above another critical stress, failure occurs without delay. The lower critical stress is found to decrease as the amount of hydrogen dissolved in the steel increases (Fig. 11.28), since the stress required to initiate delayed failure decreases with increasing hydrogen

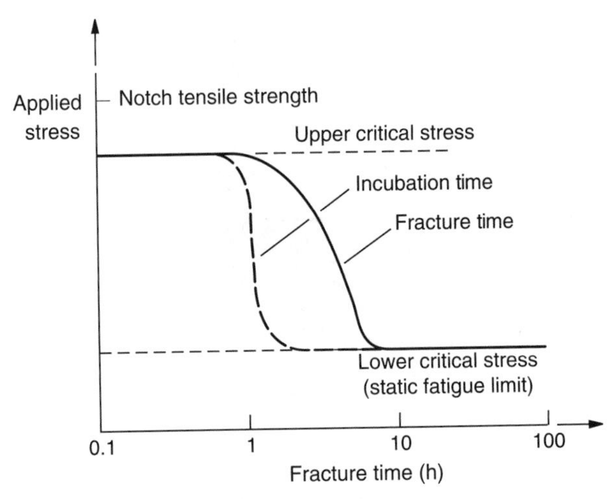

FIGURE 11.27 Failure characteristics of a hydrogenated high-strength steel. (Reprinted with permission of ASM International. All rights reserved. www.asminternational.org.)

concentration. The hydrogen concentration can be reduced by baking the material, the longer the baking time, the lower the resulting hydrogen concentration. The preceding discussion indicates that initiation of a crack is controlled by a critical combination of hydrogen and stress. The lower critical stress also decreases with an increasing material strength, thus making high strength steels more sensitive to hydrogen embrittlement.

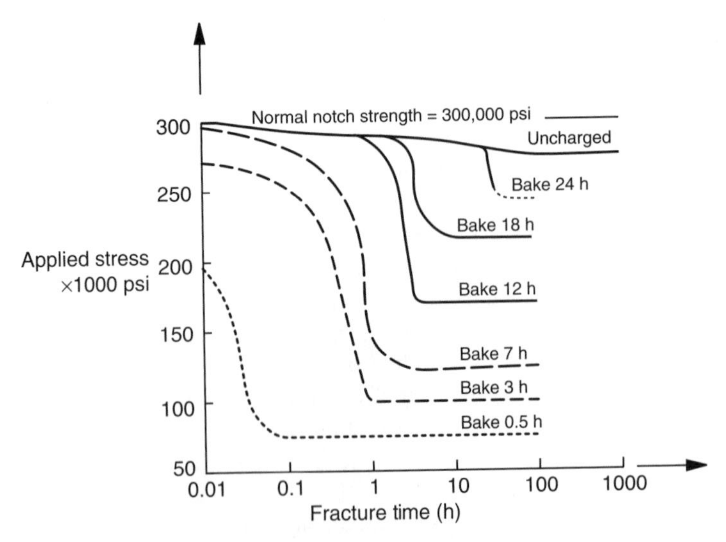

FIGURE 11.28 Static fatigue curves for various hydrogen concentrations. (From Troiano, A. R., 1960, *Transactions of the American Society for Metals*, Vol. 52, pp. 54–80.)

The delay results from the fact that initially the hydrogen may be uniformly distributed throughout the material, and thus the concentration at the crack tip (region of high triaxial stress state) may then be less than the critical value for failure. However, with time, the hydrogen diffuses and concentrates in that region, driven by the stress gradient. A crack is initiated when a critical level of hydrogen accumulates, enough to lower the cohesive strength of the material. Propagation of the crack thus occurs in steps, which occur when hydrogen solute atoms arrive at the crack edge in the right concentrations (Fig. 11.29). The stress influences the process primarily through its ability to produce a critical amount of hydrogen. A critical concentration of hydrogen is developed only when the stress level is greater than the lower limit. The resulting failure process takes place in three stages (Fig. 11.27):

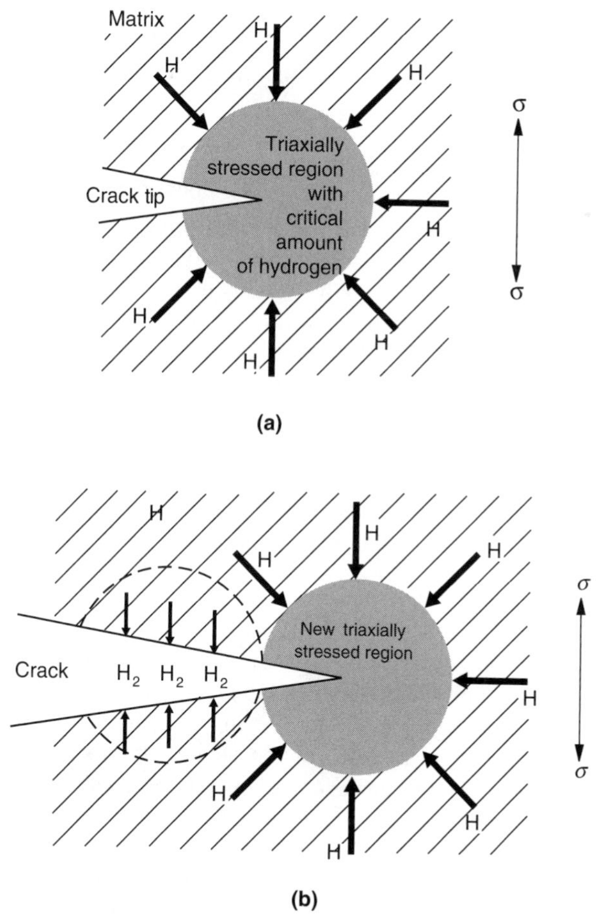

FIGURE 11.29 Hydrogen diffusion to the region of high stress concentration. (*Source: Welding Journal*, American Welding Society.)

1. Delay period or incubation.
2. Slow crack propagation, which is essentially a discontinuous process involving a series of crack initiations or steps.
3. Catastrophic failure.

Despite the fact that the presence of hydrogen can have significant impact on the properties of a steel part, it does not react chemically with iron as do gases such as nitrogen, oxygen, and carbon dioxide. Instead, it is present in the iron as a solute.

Cold cracking may be prevented by

1. Preheating to reduce the cooling rate (see Chapter 10) and thus minimize the formation of martensite in steels (preheating also helps remove moisture).
2. Postweld heat treatment (about 550–650°C for steels) to soften the martensite and help hydrogen diffuse out (also helps remove residual stresses).
3. Improved design to minimize constraint effect and thus residual stresses.
4. Using material with low cracking sensitivity, for example, low carbon steels.

Preheating is about the most commonly used method for preventing cracking when processing high-hardenability materials. Conventional preheating and postweld heat treatment involve either heating the workpiece in a furnace before processing or using an induction or flame heater. For the latter two methods, the heating is often applied locally. In the case of furnace heating, the entire workpiece is heated to the preheat temperature, which is about 250° for a number of steels. These methods are often either too expensive or impractical to use. For laser welding, a more convenient approach may involve using a second laser beam to either preheat the joint or postweld heat treat the fusion zone during the process (see Section 16.7.1.1).

If postweld heat treatment is desired to just prevent cracking, then it may be done at about 200°C for a number of hours to allow hydrogen to diffuse out. An alternative, and perhaps more effective way, is to preheat the specimen and maintain the preheat temperature for a while after the process.

11.2.2.4 Reheat Cracking Elements such as chromium, molybdenum, and vanadium that have a high affinity for carbon tend to precipitate carbides within grains of steels in which they constitute significant alloying elements. This is especially true when the steel is postweld heat treated or reheated to elevated temperature in service. This depletes the grain boundaries of carbon, thus weakening the material along the grain boundaries, which can then rupture under residual stresses. Ferritic creep-resisting and austenitic stainless steels contain significant amounts of these alloying elements and are thus susceptible to this form of cracking, which is more common in the heat-affected zone close to the fusion boundary. The cracks are normally intergranular, following the prior austenite boundaries, and the coarse grained section of the heat-affected zone is more likely to experience reheat cracking due to the higher segregation associated with the coarse grains. It is more common with thick sections of about 20 mm or more due to the higher constraint of the thicker material, and for temperatures exceeding 500°C, since the formation of carbides is more rapid at the higher temperatures.

11.2.2.5 Lamellar Tearing Lamellar tearing results primarily from nonmetallic inclusions that are flattened during prior rolling of the material and occurs primarily in the base metal. This reduces the strength or ductility of the material in the direction normal to the rolling direction, usually the through-thickness direction. Decohesion and tearing at the interface between the nonmetallic inclusion and matrix can then occur under the residual stresses that arise during processing. Joint configurations that result in a fusion boundary that is parallel to the plate surface, such as occurs in corner and tee joints, are thus highly susceptible to lamellar tearing. It is common in the machine tool industry and in offshore oil platforms where the joined plates are usually thick.

11.2.3 Lack of Fusion

When the heat supply is not adequate, it may result in lack of fusion of the molten region, Fig. 11.30, where there is no fusion between the fusion zone and the base material. Lack of fusion may also be caused by flooding of molten metal in front of the heat source. It is more commonly observed in processing pure copper because of the high thermal conductivity that results in most of the heat supplied being lost to the surrounding material but can be minimized or eliminated by preheating the workpiece before processing.

Ultrasonic inspection is most suited for identifying lack of fusion since this type of discontinuity can be either planar or volumetric. Radiographic inspection will detect the volumetric ones but not the more planar type. It is generally considered that up to 10% lack of fusion will not have a significant effect on the tensile properties.

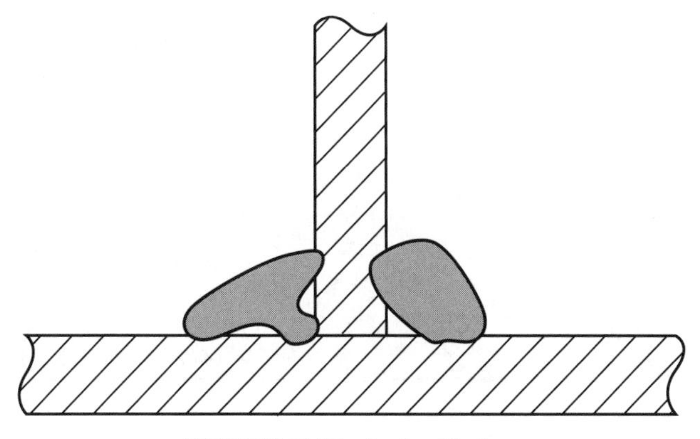

FIGURE 11.30 Lack of fusion.

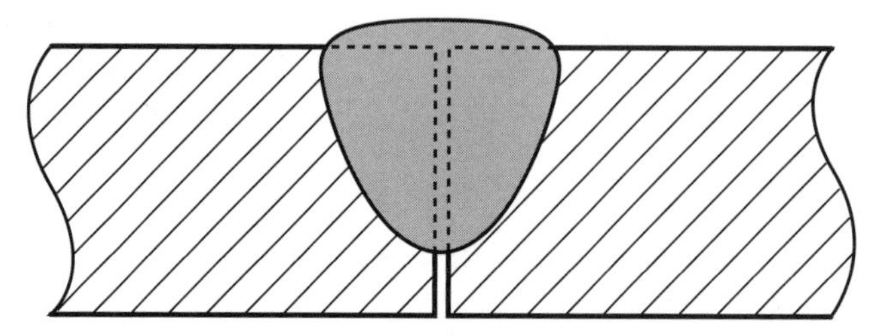

FIGURE 11.31 Incomplete penetration.

However, for a part loaded in fatigue and which has the reinforcement removed and no postweld heat treatment, a 10% reduction in area due to lack of fusion is known to reduce the fatigue strength by about 50%. Lack of fusion is most damaging at a weld face or root.

11.2.4 Incomplete Penetration

Incomplete penetration occurs when the filler and base metals do not fuse at the weld root (Fig. 11.31). The discontinuity is referred to as such only when the penetration is less than the design specification. It is usually caused by low-heat input. The inspection techniques and effect on properties are essentially the same as for lack of fusion. Incomplete penetration reduces the load-carrying capacity of the part.

11.2.5 Undercut

This is the portion of the base metal that is melted but not filled (Fig. 11.32) and forms a groove. It is usually a region of high stress concentration at the weld toe. High processing speed is a common cause of undercutting, since under these conditions, the filler metal being deposited does not have sufficient time to flow to the toe. Conditions that induce a marked convex contour of the fusion zone, such as material with high surface tension, also enhance undercutting. These tend to draw the molten metal in the molten pool upward and to the rear, away from the sides of the bead, with the result that a positive meniscus develops and the molten metal fails to wet the sides

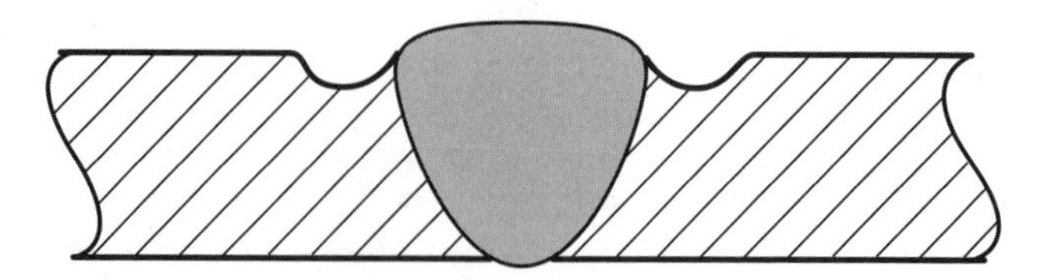

FIGURE 11.32 Undercut.

of the joint. It is generally not harmful unless the part is subjected to severe fatigue loading conditions.

11.3 SUMMARY

In this chapter, we discussed the basic microstructural changes that occur in a material as a result of laser processing, and was broken down into two main sections:

- Process microstructure
- Discontinuity formation.

Under process microstructure, there are four major zones that result after processing: the fusion zone, zone of partial melting, heat-affected zone, and base metal. Solidification in the fusion zone involves nucleation and growth. The concept of epitaxial solidification, which is common with laser processes that result in melting without the addition of filler material, involves solidification without nucleation. In conventional solidification, a critical radius and a critical free energy change are necessary to form a stable nucleus. The two common modes of nucleation are homogeneous and heterogeneous nucleation. After a stable nucleus is formed and begins to grow, the structure that results in the fusion zone may be one of the four types: planar, cellular, dendritic columnar, or dendritic equiaxed. The microstructure varies from a planar to a cellular or dendritic structure as one moves from the fusion boundary toward the centerline, due to changes in solidification rate and temperature gradient. In addition to the general structure, the grain orientation also depends on the direction of maximum temperature gradient and the preferred crystallographic growth direction. A higher cooling rate produces a finer grain structure and also a smaller spacing between the dendrite arms, which in turn results in greater strength, better ductility, and better toughness. A finer grain structure can also be produced by inoculation, a process that involves the introduction of fine particles into the molten pool. Under nonequilibrium cooling conditions, coring or macrosegregation can result in variation in structure in the fusion zone.

The zone of partial melting, which is next to the fusion zone, results primarily from low-melting impurities. Next to it is the heat-affected zone, where no melting occurs, but the microstructure is changed by the heat flow. If the material is previously cold worked, recrystallization occurs in this zone, resulting in a fine grain structure that increases in size toward the fusion boundary. Precipitation hardening alloys and allotropic materials such as steel may exhibit two recrystallized regions in the heat-affected zone. If the material is not cold worked, then the original grains grow in size toward the fusion boundary. Ferritic and martensitic stainless steels may encounter severe embrittlement during processing, while austenitic steels may be subjected to sensitization.

Finally, the principal types of discontinuities that can occur during laser processing were discussed, specifically porosity; cracking (hot cracking, liquation cracking, cold cracking, reheat cracking, and lamellar tearing); lack of fusion; incomplete penetration; and undercutting. Cracking is quite likely in high carbon and alloy steels unless they are preheated before being processed.

REFERENCES

Barrett, C. R., Nix, W. D., and Tetelman, A. S., 1973, *The Principles of Engineering Materials*, Prentice Hall, Englewood Cliffs, NJ.

Easterling, K., 1983, *Introduction to the Physical Metallurgy of Welding*, Butterworths, London.

Felbeck, D. K., and Atkins, A. G., 1996, *Strength and Fracture of Engineering Solids*, 2nd edition, Prentice Hall, Englewood Cliffs, NJ.

Friedman, P. A., and Kridli, G. T., 2000, Microstructural and mechanical investigation of aluminum tailored-welded blanks, *Journal of Materials and Performance*, Vol. 9, pp. 541–551.

Granjon, H., 1972, *Cracking and Fracture in Welds*, Japan Welding Society, p. IB1.1.

Guy, A. G., 1971, *Introduction to Materials Science*, McGraw-Hill, New York.

Hughel, T. J., and Bolling, G. F., editors, 1971, *Solidification*, American Society for Metals, Metals Park, OH.

Hosford, W., *Physical Metallurgy*, 2005, Taylor & Francis, Boca Raton, FL.

Jenney, C. L., and O'Brien, A., editors, 2001, *Welding Handbook*, 9th ed., American Welding Society, Miami, FL.

Kou, S., 1987, *Welding Metallurgy*, John Wiley & Sons, New York.

Lancaster, J. F., 1980, *Metallurgy of Welding*, 3rd edition, George Allen & Unwin, London.

Linnert, G. E., 1968, *Welding Metallurgy*, American Welding Society, Miami, FL.

Lundin, C. D., 1984, Fundamentals of weld discontinuities and their significance, *WRC Bulletin*, Vol. 295, pp. 1–33.

Mallett, M. W., and Rieppel, P. J., 1946, Arc atmospheres and underbead cracking, *Welding Journal*, Vol. 25, pp. 748s–759s.

Masubuchi, K., 1980, Weld cracking and joint restraint, *Analysis of Welded Structures*, Pergamon Press, Oxford, UK, pp. 518–576.

McLean, D., 1962, *Mechanical Properties of Metals*, John Wiley & Sons, Inc., New York.

Petch, H. J., 1952, Delayed fracture of metals under static load, *Nature*, Vol. 169, pp. 84–843.

Savage, W. F., Nippes, E. F., and Szekeres, E. S., 1976, Hydrogen induced cold cracking in a low alloy steel, *Welding Journal*, Vol. 55, pp. 276s–283s.

Shewmon, P. G., 1983, *Diffusion in Solids*, McGraw-Hill, New York.

Trivedi, R. and Somboonsuk, K., 1984, Constrained dendritic growth and spacing, *Materials Science and Engineering*, Vol. 65, pp. 65–74.

Troiano, A. R., 1960, The Role of hydrogen and other interstitials in the mechanical behavior of metals, *Transactions of the American Society for Metals*, Vol. 52, pp. 54–80.

van Vlack, L. H., 1975, *Elements of Materials Science and Engineering*, Addison-Wesley, Reading, MA.

Voldrich, C. B., 1947, Cold cracking in the heat-affected zone, *Welding Journal*, Vol. 26, pp. 152s–169s.

APPENDIX 11A

List of symbols used in the chapter.

Symbol	Parameter	Units
d_1	primary dendrite spacing	m
d_2	secondary dendrite spacing	m
ΔE_d	activation energy for diffusion	J
ΔG_c	critical free energy change necessary to form stable nuclei	J
ΔG_v	bulk free energy change per unit volume	J/m^3
k_p	equilibrium partition coefficient (see Appendix 12A)	–
k_y	Hall–Petch equation constant	$N/m^{3/2}$
K_a	rate constant that exhibits Arrhenius temperature behavior	m^2/s
K_0	rate constant for diffusion	m^2/s
$n_d = 0.3 - 0.5$	dendritic constant	–
$\dot{n}$	rate of nucleation	$/m^3$ s
n_v	number of nuclei per unit volume	$/m^3$
r	radius of a nucleus	m
r_c	critical nucleus radius	m
r_d	dendrite tip radius	m
s	number of sites available for nuclei formation	–
t'	dummy variable of integration	s
t_s	local solidification time	s
V_m	volume of a molecule	m^3
γ_{nl}	surface energy per unit area between the nucleus and liquid	J/m^2
γ_{ns}	surface energy per unit area between the nucleus and solid base metal	J/m^2
γ_{sl}	surface energy per unit area between the solid base metal and liquid	J/m^2
ϕ	wetting angle	°
ψ	angle between traverse velocity and direction of solidification	°
σ_i	frictional resistance of the lattice to moving dislocations	N/m^2
σ_y	yield strength of a material	N/m^2

PROBLEMS

11.1. Most steels derive their strength from quenching, but other materials (including most aluminum alloys) are strengthened by precipitation hardening. During the aging process, the strength of the part varies with time as shown in the figure below.

If a precipitation hardening material is welded, show the relative strengths in the three regions of the weldment, A (weld bead); B (heat-affected zone); and C (base material), in the figure if the material was ideally aged before welding.

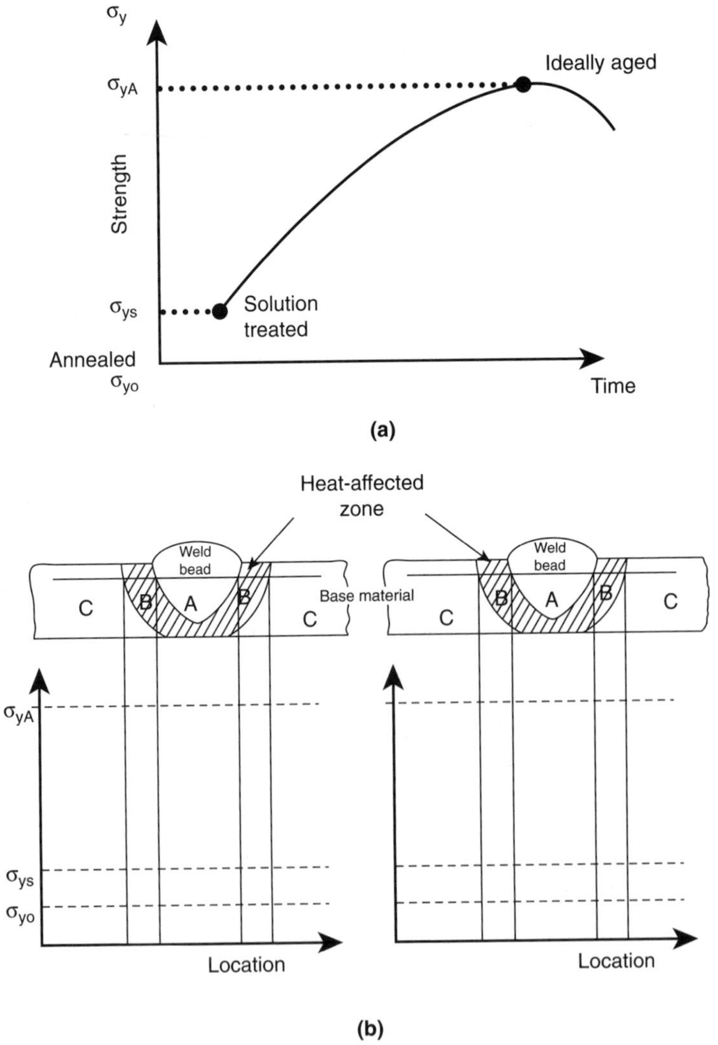

FIGURE P11.1 Aging curve and strength variation in a processed zone.

(a) Immediately after processing
(b) After subsequent artificial aging.

11.2. Given that the total free energy change associated with the formation of a spherical cap solid nucleus is given by

$$\Delta G = \left[\frac{4}{3}\pi r^3 \Delta G_v + 4\pi r^2 \gamma_{sl}\right] f(\phi)$$

show that the critical free energy change necessary for the formation of a stable nucleus is

$$\Delta G_c = \frac{16\pi {\gamma_{sl}}^3}{3(\Delta G_v)^2} f(\phi)$$

11.3. When a material that transforms from one phase to another at an equilibrium temperature is cooled to that temperature, transformation will begin. At that temperature, the change in free energy is zero. If it is cooled below the equilibrium temperature, the change in free energy is negative and there is a greater driving force for the transformation to proceed. The lower the temperature, the greater will be the driving force. Thus one would expect the shape of the top part of a T-T-T diagram (see Fig. P11.3) (above the nose) to be logical since the lower the temperature, the sooner will the transformation begin.

Why then does the transformation take longer at lower temperatures below the nose of the T-T-T diagram?

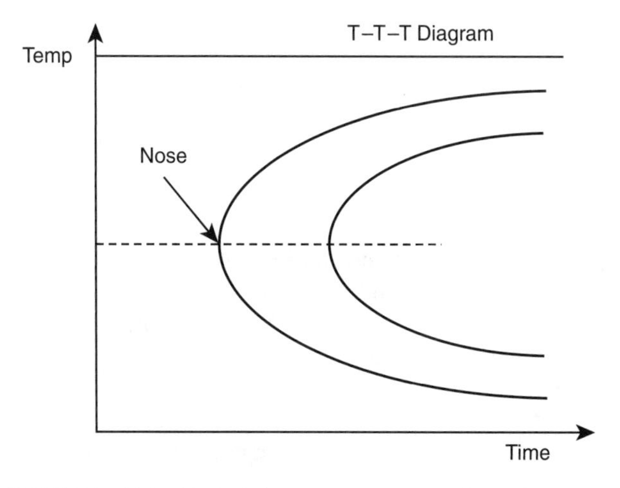

FIGURE P11.3 Typical time temperature transformation curve.

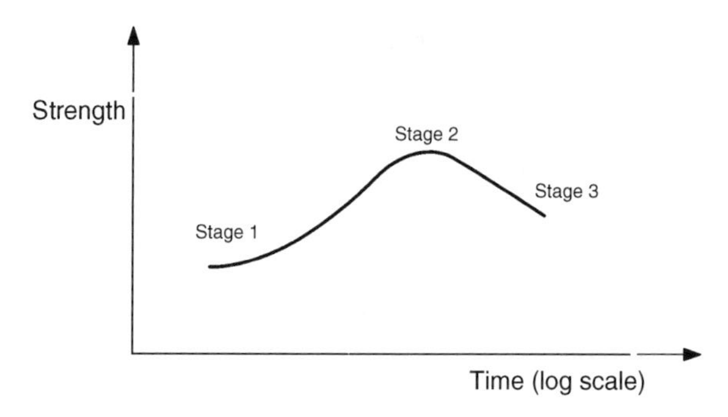

FIGURE P11.4 Variation of strength with time during aging.

11.4. The variation of strength with time during aging at 200°C of an aluminum alloy that was solution treated at 450°C and then quenched to room temperature is shown in Fig. P11.4. Sketch the curve for aging at 250°C on the same graph.

(a) Name two main differences between the two curves.
(b) Why are the two curves different?
(c) Sketch the aging curve for a pure aluminum piece if it were subjected to the same treatments.
(d) Is it different? Explain your solution.

11.5. (a) Determine the heat of fusion per atom for copper if its atomic weight is 0.06354 kg/mol.
(b) Determine the free energy change necessary to form a critical nucleus at a temperature that corresponds to 100°C of supercooling.
(c) Determine the free energy change necessary if the supercooling is 250°C.
(d) Comment on the probability of observing homogeneous nucleation in both (b) and (c).

11.6. The variation of yield stress with grain diameter for different alloys are shown in Fig. P11.6. Obtain the Hall–Petch equation for Fe–N from information given in the graph.

11.7. Two plates of silver are welded together using a laser beam. If the welding conditions are such that solidification of the molten pool starts with homogeneous nucleation, determine the critical radius that will result in the formation of a stable nucleus at a temperature that is 250°C below the melting temperature.

11.8. Determine the critical free energy change necessary for forming the stable nucleus in Problem 11.7.

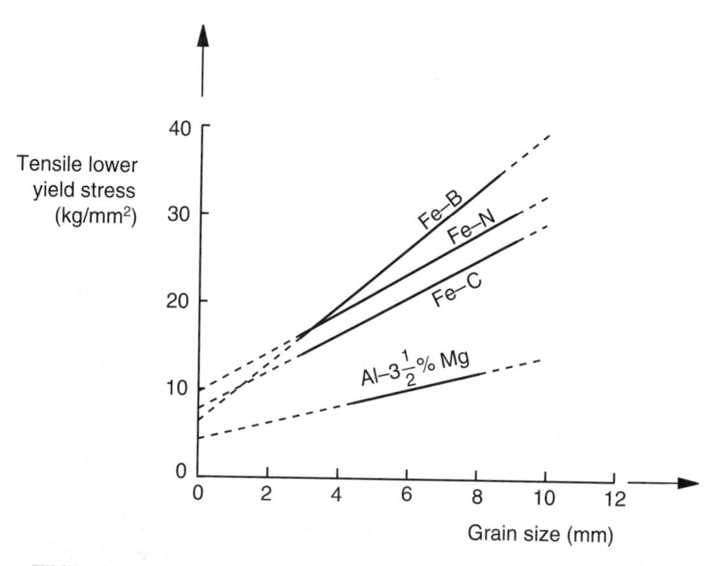

FIGURE P11.6 Stress variation with grain size for different alloys.

11.9. Determine the number of stable nuclei per unit volume for the conditions in Problem 11.7 if the density of nucleation sites is 6×10^{48} per m^3.

11.10. Calculate the nucleation rate for the conditions in Problem 11.7. Assume the radius of a silver atom is 0.1444 nm (van Vlack, p. 95) and the activation energy for self-diffusion in silver is 44.1 kcal/mol (Shewmon, 1983, Table 2.2, p. 66).

11.11. (a) A part made of nickel has its surface modified by surface melting (see Chapter 17) using a laser. Assuming solidification of the molten material occurs primarily by homogeneous nucleation, plot the variation of the critical nucleus size and critical free energy necessary for nucleation with the degree of supercooling.

(b) Now, if instead of just surface melting, the surface is modified by cladding, and the filler metal used results in a wetting angle of 30°, again plot the variation of the critical nucleus size and critical free energy necessary for nucleation with the degree of supercooling.

(c) How does supercooling affect the nucleation process? In what way does the use of a filler material affect the process?

12 Solidification

Solidification is important in laser processing since the behavior of the molten pool during solidification determines, to a great extent, grain shape and size in the fusion zone, microstructure, inclusions, porosity, and cracks, and ultimately, the mechanical properties of the part. In Chapter 11, we briefly outlined the basic concepts of solidification, focusing on nucleation. The criterion that determines which structure prevails in the solidified material was also introduced. Our focus in this chapter will be on growth, on the assumption that nucleation has already taken place.

For processes where local equilibrium is maintained at the solid–liquid interface, conventional solidification theory is directly applicable. However, for those processes that result in rapid solidification, such theories may not be directly applicable. For completeness, we shall discuss both types of solidification systems, starting with conventional solidification processes. Under conventional solidification, we start with the simpler case of solidification without flow, where we first address solidification of a pure metal, followed by that of a binary alloy. As part of our discussion on binary alloy solidification, we further discuss the conditions that lead to the formation of different types of microstructures (planar, cellular, dendritic) and also the mushy zone (where solid and liquid coexist). Then we provide some insight into the basic concepts of solidification with flow. Finally, we outline the specific issues associated with rapid solidification.

12.1 SOLIDIFICATION WITHOUT FLOW

12.1.1 Solidification of a Pure Metal

We start the discussion by considering the simplest case, the classical Stefan problem. This involves a one-dimensional analysis of the solidification of either a pure metal or a binary melt of eutectic composition. Let us consider a semi-infinite liquid material of initial uniform temperature, T_0. Now let the base of the material be suddenly placed in contact with a chilled surface of constant temperature, T_c at time $t = 0$, such that $T_c < T_m$, where T_m is the melting temperature of the liquid material. We set up a coordinate system as shown in Fig. 12.1 with origin at the chilled surface, $x = 0$. Since we are considering a pure metal or eutectic, the composition of the solid that forms and that of the liquid are identical, so composition

Principles of Laser Materials Processing, by Elijah Kannatey-Asibu, Jr.

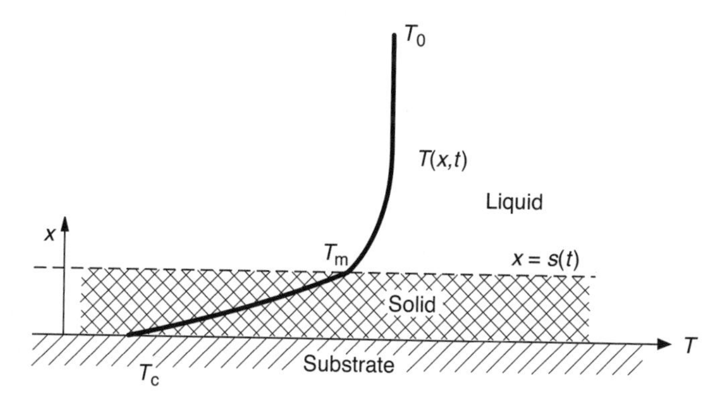

FIGURE 12.1 Temperature field for one-dimensional solidification of a pure material. (From Huppert, H. E., 1990, *Journal of Fluid Mechanics*, Vol. 212, pp. 209–240. Permission of Cambridge University Press.)

does not influence solidification. We further assume that there is no flow in the liquid and neglect any density differences that might exist between the solid and liquid phases.

The problem is one of determining the temperature field, $T(x, t)$, in the solid and liquid phases and also the motion of the solid–liquid interface, $s(t)$. Since this is a pure metal, the interface is a distinct front that separates the solid and liquid phases. Determination of the temperature field and interface requires solving the one-dimensional conduction equation in the two phases:

$$\rho_s c_{ps} \frac{\mathrm{d}T}{\mathrm{d}t} = \frac{\partial}{\partial x}\left(k_s \frac{\partial T}{\partial x}\right) \quad (0 < x \leq s(t)) \tag{12.1a}$$

$$\rho_l c_{pl} \frac{\mathrm{d}T}{\mathrm{d}t} = \frac{\partial}{\partial x}\left(k_l \frac{\partial T}{\partial x}\right) \quad (x \geq s(t)) \tag{12.1b}$$

with the initial and boundary conditions

$$T = T_0 \quad (t = 0) \tag{12.2a}$$

$$T = T_c \quad (x = 0) \tag{12.2b}$$

$$T \to T_0 \quad (x \to \infty) \tag{12.2c}$$

where the subscripts s and l indicate solid and liquid, respectively, and T_c is the temperature of the chilled surface.

Another boundary condition is given by the need to have conservation of heat flux (energy balance) at the interface between the liquid and solid phases:

$$L_{\mathrm{mv}} u_x = k_{\mathrm{s}} G_{\mathrm{Ts}} - k_{\mathrm{l}} G_{\mathrm{Tl}} \tag{12.2d}$$

where $G_{\mathrm{Ts}} = \mathrm{d}T/\mathrm{d}x|_{s-}$ is temperature gradient in the solid at the interface, $G_{\mathrm{Tl}} = \mathrm{d}T/\mathrm{d}x|_{s+}$ is temperature gradient in the liquid at the interface, L_{mv} is the latent heat of solidification (melting) per unit volume of solid, and $u_x = \dot{s} = \mathrm{SR}$ is the interface velocity.

Equation (12.2d) implies that the net difference between the heat fluxes in the solid and liquid phases at the interface is the rate of release of the latent heat of solidification. Very few analytical solutions exist for equations (12.1)–(12.2), and they are normally solved using numerical methods. The temperature field that results from their solution is of the form shown in Fig. 12.1.

The discussion thus far has focused on a pure material where solidification occurs at constant temperature, and the evolution of latent heat occurs as a step change. Most materials used in practice, however, are alloys. The solidification process for such materials takes place over a temperature range, between the liquidus and solidus (or eutectic) temperatures. The latent heat that evolves then has a functional relationship with temperature. In the next section, we discuss the solidification of alloys, focusing on the simpler but yet very common type, a binary alloy.

12.1.2 Solidification of a Binary Alloy

In this section, we first discuss the simple case where the solidification interface is planar. This will be followed by a detailed discussion of the conditions that lead to interface instability. Finally, the mushy zone that exists between the completely solid and completely liquid interfaces will be discussed.

12.1.2.1 Temperature and Concentration Variation in a Solidifying Alloy Let us consider a setup identical to that of the pure metal solidification, except that the material is now an alloy of two elements. We make the following assumptions:

1. The composition of the solid does not change.
2. The solid–liquid interface is flat and distinct; that is, there is no mushy zone.
3. There is no convection in the liquid.
4. Transport occurs only by molecular diffusion and is one dimensional.
5. Local thermodynamic equilibrium exists. That means the concentration of solute in the solid and liquid sides of the interface is given by the equilibrium phase diagram.

The governing heat conduction equations are the same as for a pure metal, equations (12.1a) and (12.1b), with the thermal boundary conditions, equations (12.2a)–(12.2c), and the thermal conservation condition, equation (12.2d).

As solidification proceeds, solute is rejected from the interface into the liquid. Thus, in addition to the energy equations (12.1a) and (12.1b), we must also have a solute transport equation that describes the solute concentration in the liquid and is given by

$$\frac{dC}{dt} = \frac{\partial}{\partial x}\left(D_l \frac{\partial C}{\partial x}\right) \quad (x \geq s(t)) \tag{12.3}$$

where C is the solute concentration (%) and D_l is the diffusion coefficient of the solute in the liquid (m^2/s).

Considering conservation of solute at the interface, we have the following boundary condition:

$$u_x C + D_l \left.\frac{dC}{dx}\right|_{s+} = 0 \quad (x = s(t)+) \tag{12.4}$$

The other boundary and initial conditions are

$$C \to C_0 \quad (x \to \infty)$$

$$C = C_0 \quad (t = 0)$$

where C_0 is the initial liquid composition.

Since solidification occurs under local thermodynamic equilibrium, the solute concentration at the interface and solidification or melting temperature are coupled by a simplified relationship of the form

$$T = -m_g C + b \tag{12.5}$$

where m_g is the absolute value of the slope of the liquidus curve, which is approximated as a straight line.

Numerical solution of equations (12.3)–(12.5) along with (12.1)–(12.2) gives temperature and concentration profiles of the form shown in Fig. 12.2.

For the ideal situation described in the preceding paragraphs, the temperature gradient is in the direction of crystal growth. The solid–liquid interface is thus planar and moves with a steady velocity, u_x. The process is then one dimensional, and the solute and temperature fields vary only in the growth direction. If the liquid state is quiescent (i.e., at rest), the concentration and temperature extending ahead of the interface are exponential (see equation (12B.9) and Fig. 12B.4a).

As briefly discussed in Section "Structure of the Solidified Metal," the interface that develops during solidification of noneutectic alloy compositions is not always flat (planar) as assumed but may become cellular or dendritic as a result of instabilities associated with perturbations at the interface.

12.1.2.2 Interface Stability Theories There are two classes of theory that are used in analyzing *interface instability*, or the conditions under which deviations from the

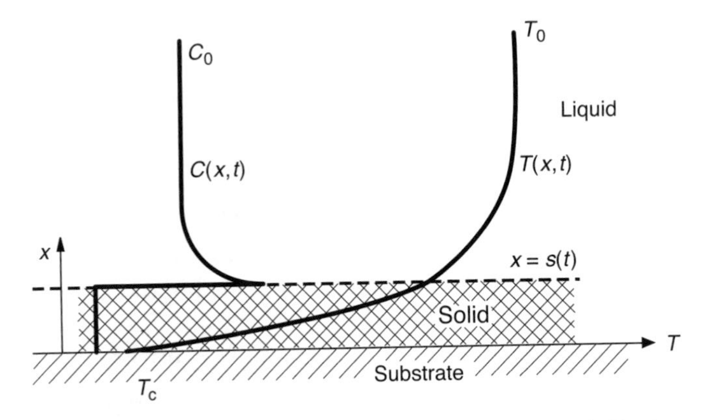

FIGURE 12.2 Schematic temperature field for one-dimensional solidification of a binary alloy. (From Huppert, H. E., 1990, *Journal of Fluid Mechanics*, Vol. 212, pp. 209–240. Permission of Cambridge University Press.)

planar interface will occur. These are

1. The constitutional supercooling theory.
2. The morphological stability theory.

The basic principles of each approach are outlined in the following sections.

Constitutional Supercooling Theory The constitutional supercooling theory has traditionally been used as the basis for a stability criterion for solidification processes that occur at low interface velocities and is based on thermodynamics principles. The rationale is to determine whether the liquid ahead of a moving interface is undercooled with respect to its composition. Such undercooling may result from rejection of solute by the growing solid and tends to make the interface unstable, resulting in the formation of cells or dendrites that could protrude into the liquid. The planar interface may be shown to be stable if

$$k_s G_{Ts} + k_l G_{Tl} > m_g G_c (k_s + k_l) f(A, k_p) \tag{12.6}$$

where $A = {k_p}^2 T_m \gamma_e u_x / m_g (k_p - 1) C_0 D_l L_{mv}$, $f(A, k_p) \approx 1 - A$, $A \leq 1$, $G_c = u_x C_0 (k_p - 1)/D_l k_p$ is the solute concentration gradient at the interface in the unperturbed liquid, neglecting diffusion in the solid, k_p is the equilibrium partition coefficient (see Appendix 12B), T_m is the interface melting temperature in the absence of solute and curvature, and γ_e is the interface surface energy.

Typically, solidification velocities are such that $A << 1$, in which case $f(A, k_p) \approx 1$. The stability criterion then reduces to the form

$$k_s G_{Ts} + k_l G_{Tl} > m_g G_c (k_s + k_l) \tag{12.7}$$

or

$$k_s G_{Ts} + k_l G_{Tl} > m_g \frac{u_x C_0 (k_p - 1)}{D_l k_p}(k_s + k_l) \tag{12.8}$$

which can be simplified by assuming $G_{Ts} \approx G_{Tl}$, giving

$$\frac{G_{Tl}}{u_x} > \frac{m_g C_0}{D_l} \frac{(k_p - 1)}{k_p} \tag{12.9}$$

or

$$G_{Tl} > \frac{m_g u_x C_0}{D_l} \frac{(k_p - 1)}{k_p} \tag{12.10}$$

On the contrary, at high solidification velocities, $f(A, k_p) = 0$, since A cannot exceed 1. The stability criterion then becomes

$$k_s G_{Ts} + k_l G_{Tl} > 0 \tag{12.11}$$

The constitutional supercooling criterion, represented by equation (12.9) or (12.10), is illustrated schematically in Fig. 11.8 as the line between the planar and cellular regions. In the planar region, the actual temperature gradient in the liquid is greater than the equilibrium temperature as obtained from the right-hand side of equation (12.10), and thus planar growth is morphologically stable. This morphology does not result in a segregated structure. In the cellular region, the solute gradient effect dominates, and the actual temperature gradient is less than that given by the right-hand side of equation (12.10). This results in chance protuberances becoming pronounced, leading to cellular growth that also results in microsegregation. At yet higher interface velocities relative to the temperature gradient, or higher solute concentrations, growth becomes dendritic.

A simplified derivation that results in a slightly simpler version of the constitutional supercooling criterion is given in Appendix 12B.

The constitutional supercooling theory is a limiting case of the more general theory of morphological stability and is adequate for a number of conventional solidification processes where local equilibrium conditions hold, and the solidification rates are relatively low. However, since it neglects surface tension effects and latent heat evolution, both of which have a stabilizing effect on the interface, it is generally not adequate for modeling rapid solidification processes. In fact, equation (12.9) predicts a less stable interface as the interface velocity increases, which contradicts experimental results that show stable flat interface morphologies at high interface velocities. For such cases, the morphological stability theory, which incorporates surface energy effects, is more appropriate.

Morphological Stability Theory The morphological stability theory considers the time dependence of a general sinusoidal perturbation of the interface with respect to solute, heat diffusion fields, and surface tension. The morphological stability and constitutional supercooling theories predict the same outcome for a significant number of conventional solidification processes. However, the morphological stability theory predicts a more stable interface during rapid solidification due to the stabilizing influence of surface tension and latent heat evolution.

Let us assume a perturbation at the solidification front of the form

$$x = \delta e^{\omega_t t + i\omega_y y + i\omega_z z} \tag{12.12}$$

where δ is the amplitude of perturbation at time $t = 0$, ω_t is the temporal frequency, and ω_y and ω_z are spatial frequencies.

Using a linear time-dependent stability analysis, it can be shown that the interface is stable when the real part of ω_t is negative for all perturbations. The value of ω_t can be obtained by solving the steady-state heat flow and diffusion equations with appropriate boundary conditions. In other words, the interface is stable with respect to infinitesimal perturbations when

$$m_g G_c \frac{\omega^* - \frac{u_x}{D_l}}{\omega^* - \frac{(1-k_p)u_x}{D_l}} - \frac{T_m \gamma_e \omega^2}{L_{mv}} - \left(\frac{k_s G_{Ts} + k_l G_{Tl}}{k_s + k_l}\right) < 0 \tag{12.13}$$

where

$$\omega^* = \frac{u_x}{2D_l} + \left[\left(\frac{u_x}{2D_l}\right)^2 + \omega^2\right]^{1/2}$$

$$\omega^2 = {\omega_y}^2 + {\omega_z}^2$$

The first term of equation (12.13) is due to the solute concentration, the second due to capillary effects, and the third due to thermal effects. The interface is stable when the inequality is satisfied. The solute term thus tends to make the interface unstable, while the capillary term tends to prevent the growth of a perturbation, and since the thermal term may be either positive or negative, it can be either stabilizing or destabilizing, depending on whether the numerator is negative or positive, that is, if G_{Ts} is sufficiently positive.

It can be shown that the morphological stability criterion, equation (12.13), reduces to the constitutional supercooling criterion, equation (12.9), in the limiting case when the interface velocity is very low and when capillary effects are neglected. The relationship between the constitutional and morphological stability criteria is illustrated in Fig. 12.3. It shows the critical concentration of copper below which interface stability occurs for different interface velocities in directional solidification of an Al–Cu alloy.

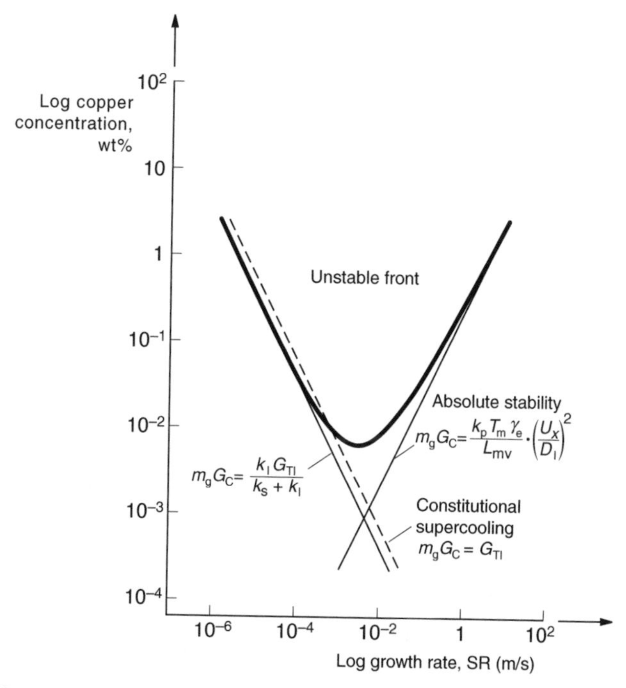

FIGURE 12.3 Relationship between morphological and constitutional supercooling stability criteria for Al–Cu alloys ($G_{\mathrm{Tl}} = 2 \times 10^4$ K/m). (From Mehrabian, R., 1982, *Metals Review*, Vol. 27, pp. 185–208.)

At high interface velocities, when $\omega << u_x/D_\mathrm{l}$, and for a stabilizing thermal field, the morphological stability criterion can be shown to reduce to

$$m_\mathrm{g} G_\mathrm{c} \left(\frac{D_\mathrm{l}{}^2}{u_x{}^2 k_\mathrm{p}} \right) \omega^2 - \frac{T_\mathrm{m} \gamma_\mathrm{e} \omega^2}{L_\mathrm{mv}} < 0 \tag{12.14}$$

for a stable interface. This is referred to as the absolute stability criterion, Fig. 12.3, and is essentially independent of G_{Tl}.

12.1.2.3 Mushy Zone As opposed to the solidification of pure materials, binary alloys do not exhibit a distinct front separating the solid and liquid phases. During solidification of actual alloys, three regions exist in the overall solidification domain. These include the solid and liquid phases, and in-between them is the mushy zone. The latter is a region of finite size where solid and liquid coexist, and is bounded by the liquidus and solidus (or eutectic) curves. For simplicity, we assume that the mushy zone exists between the interface and the base surface (Fig. 12.4). We further assume that the volume fraction, f_s, of the solid in the mushy zone is constant. The

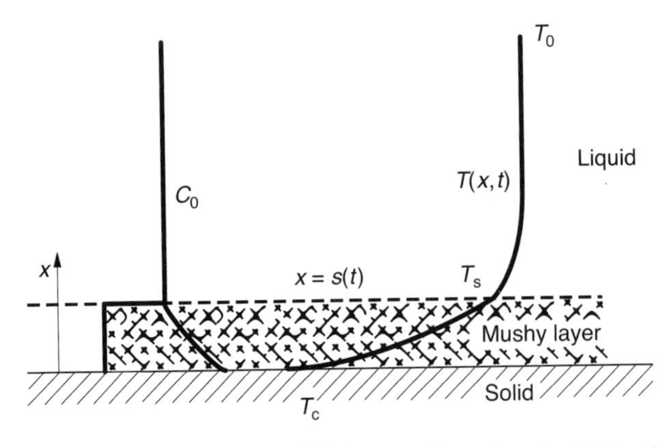

FIGURE 12.4 Schematic temperature field for one-dimensional solidification of a binary alloy with a mushy zone. (From Huppert, H. E., 1990, *Journal of Fluid Mechanics*, Vol. 212, pp. 209–240. Permission of Cambridge University Press.)

governing equations for the temperature field are then given by

$$\bar{\rho}\bar{c}_p \frac{\mathrm{d}T}{\mathrm{d}t} = \frac{\partial}{\partial x}\left(\bar{k}\frac{\partial T}{\partial x}\right) \quad (0 < x \leq s(t)) \tag{12.15}$$

$$\rho_l c_{pl} \frac{\mathrm{d}T}{\mathrm{d}t} = \frac{\partial}{\partial x}\left(k_l\frac{\partial T}{\partial x}\right) \quad (x \geq s(t)) \tag{12.16}$$

$$L_{mv} f_s u_x = \bar{k} G_{Ts} - k_l G_{Tl} \tag{12.17}$$

where the overbars indicate properties in the mushy region and are based on averages weighted by the volume fraction. For example,

$$\bar{k} = f_s k_s + f_l k_l$$

and is an appropriate approximation for a random mixture of solid and liquid. Here, f is the volume fraction.

The solid in the mushy region is assumed to be of constant concentration, while the interstitial liquid concentration corresponds to that of the liquidus temperature:

$$T = -m_g C + b \quad (0 < x \leq s(t)) \tag{12.18}$$

The concentration of the liquid phase is also assumed to be equal to its original concentration, since diffusion effects are considered to be negligible:

$$C = C_0 \quad (x \geq s(t)) \tag{12.19}$$

For global conservation of solute (i.e., the total amount of solute must be conserved), we have

$$(1 - f_s) \int_0^{s(t)} C(x, t) \mathrm{d}x = s(t) C_0 \tag{12.20}$$

The temperature and concentration profiles for this case are shown in Fig. 12.4.

When the volume fraction is variable, the governing equations in the mushy zone become

$$\bar{\rho} \bar{c}_p \frac{\partial T}{\partial t} = \frac{\partial}{\partial x} \left(\bar{k} \frac{\partial T}{\partial x} \right) + L_{\mathrm{mv}} \frac{\partial f_s}{\partial t} \tag{12.21a}$$

$$(1 - f_s) \frac{\partial C}{\partial t} = \frac{\partial}{\partial x} \left(D_l (1 - f_s) \frac{\partial C}{\partial x} \right) + C \frac{\partial f_s}{\partial t} \tag{12.21b}$$

The second term on the right-hand side of equation (12.21a) represents latent heat release into the mushy zone, while that of equation (12.21b) represents the release of solute into the interstitial fluid. The boundary condition for the spatial derivative of the solid volume fraction is given by

$$\frac{\partial T}{\partial x} = -m_g \frac{\partial C}{\partial x} \quad (x = s(t)+) \tag{12.22}$$

which corresponds to equilibrium saturation in the liquid just ahead of the growing dendrites.

Since there are two phases, a natural approach to modeling such a system is to develop separate governing equations for each phase (solid and liquid in this case) and then couple the two sets of equations using boundary conditions at the interface. This is the approach that has thus far been used in this section and is commonly referred to as the multiple domain method. Incorporation of the volume fraction of each phase into the governing equations enables the mushy zone to be analyzed. However, the interfacial geometry between the solid and liquid phases is quite complex, and it is also an unknown function of space and time for which a solution needs to be obtained. The two-phase approach is thus very difficult to use for all but the simplest configurations such as pure materials. To circumvent this problem and simplify the analysis, one-phase or continuum models can be developed from the two-phase governing equations, resulting in the entire domain being treated as a single region governed by one set of conservation equations that are valid for the liquid and solid regions, as well as the mushy zone. Thus, with the continuum models, it is not necessary to track the interface, and there is no need for boundary conditions internal to the solidifying medium. They are therefore suitable for analyzing alloys. The formulation for the continuum models is based on volume averaging techniques, where the microscopic equations valid for each phase are integrated over a small volume element. Classical mixture theory is also used.

12.2 SOLIDIFICATION WITH FLOW

The discussion in the preceding section focused on solidification in a liquid medium where no flow occurs. In a number of applications such as cladding, significant flow occurs in the liquid medium as it solidifies. Such flow may be due to either natural convection (from the solidification process itself) or forced convection (induced by external factors). In either case, the resulting motion in the melt does affect the solidification process. Natural convection results from buoyancy forces that are caused by density variations within the fluid. Such variations in density may be induced by either temperature or compositional (solutal) gradients that occur during the process. Compositional gradients result from solute rejection into the liquid. Thus, we can have thermal or solutal buoyancy, or both.

In such situations where flow occurs in the solidification medium, the governing equations are

1. Mass
2. Momentum
3. Energy
4. Solute concentration equations.

For simplicity, we make the following assumptions:

1. Flow is Newtonian and laminar.
2. The properties of each phase are homogeneous and isotropic.
3. Local thermodynamic equilibrium exists at the solid–liquid interface, that is, at the interface, $T = T_s = T_l$ and $C_s = k_p C_l$.
4. Negligible field disturbances such as dispersion fluxes and supercooling of the liquid.
5. Saturated mixture conditions exist, that is, $f_s + f_l = 1$. That means there cannot be void formation. Void formation occurs when $\rho_s > \rho_l$.
6. Solid phase is free of internal stresses and does not deform. Internal stresses develop when $\rho_s < \rho_l$.
7. Negligible solute diffusion in the solid.

We also note that there are two principal driving forces that influence flow in the molten pool during laser processing, and these are

1. Buoyancy forces that are accounted for in the momentum equations to be discussed shortly.
2. Shear and pressure that act on the pool surface as a result of surface tension gradient and surface curvature. These are modeled as boundary conditions of the governing equations.

Since three-dimensional analysis can be computationally intensive, we simplify things by considering only two-dimensional analysis. The governing equations for the one-phase or continuum model are then given by

Conservation of mass—The general form of the governing equation for mass conservation is

$$\frac{\partial \rho}{\partial t} + \nabla \cdot (\rho \mathbf{u}) = 0 \tag{12.23}$$

where $\mathbf{u}$ is the velocity vector.

Momentum Equations—The momentum equations are modified by a concentration term and for two-dimensional analysis, become

$$\frac{\partial}{\partial t}(\rho u_x) + \nabla \cdot (\rho \mathbf{u} u_x) = -\frac{\partial P}{\partial x} + \nabla \cdot (\mu \nabla u_x) + F_x \tag{12.24a}$$

$$\frac{\partial}{\partial t}(\rho u_z) + \nabla \cdot (\rho \mathbf{u} u_z) = -\frac{\partial P}{\partial z} + \nabla \cdot (\mu \nabla u_z) + F_z - \rho g[\beta_{\mathrm{T}}(T - T_0) + \beta_c(C_{\mathrm{l}} - C_0)] \tag{12.24b}$$

where β_{T} = thermal volume expansion coefficient (/K), β_{C} is the solutal volume expansion coefficient (/%), g is the acceleration due to gravity (m/s^2), P is the pressure (Pa (N/m^2)), F_x, F_z is the external body forces in the x- and z-directions, respectively (N/m^3), μ is the viscosity (Pa s), u_x, u_z are the magnitudes of velocities in the x- and z-directions, respectively (m/s), and T_0 and C_0 are the reference (usually eutectic) temperature and concentration, respectively.

The last term on the right-hand side of equation (12.24b) is the force due to thermal and solutal buoyancy.

Enthalpy (energy) conservation—For convenience, the energy equation is expressed in terms of enthalpy, h_{e}:

$$\frac{\partial}{\partial t}(\rho h_{\mathrm{e}}) + \nabla \cdot (\rho \mathbf{u} h_{\mathrm{e}}) = \nabla \cdot \left(\frac{k}{c_{\mathrm{ps}}} \nabla h_{\mathrm{e}}\right) + S_{\mathrm{h}} \tag{12.25}$$

Solute concentration—The equation for solute concentration is expressed as

$$\frac{\partial}{\partial t}(\rho C) + \nabla \cdot (\rho \mathbf{u} C) = \nabla \cdot (D \nabla C) + S_{\mathrm{c}} \tag{12.26}$$

where S_{h}, S_{c} are enthalpy and concentration source terms, respectively.

In this formulation, the enthalpy, h_{e}, is defined as

$$h_{\mathrm{e}} = \int_0^T c_{\mathrm{ps}} \mathrm{d}T \tag{12.27a}$$

Thus,

$$h_s = h_e \tag{12.27b}$$

$$h_l = h_e + \delta h_e \tag{12.27c}$$

$$\delta h_e = \int_0^T (c_{pl} - c_{ps})\mathrm{d}T + L_m \tag{12.27d}$$

where c_{pl} and c_{ps} are the specific heats of the liquid and solid, respectively (J/kg K), h_l and h_s are the enthalpies of the liquid and solid, respectively (J/kg), and L_m is the latent heat of fusion per unit mass (J/kg).

The exact form of the one-phase model that is used depends on the nature of the mushy zone. Thus, at this point, we consider two limiting cases:

1. Mushy fluid.
2. Columnar dendritic structure.

12.2.1 Mushy Fluid

This type of mushy zone is similar to granular solid particles that are dispersed within a liquid medium (Fig. 12.5). It is more commonly found in the equiaxed zone of some metals and also in amorphous materials such as glasses and waxes. Both the solid and liquid phases then have the same velocity, and as such, this model is sometimes referred to as the no relative motion model:

$$\mathbf{u} = \mathbf{u}_l = \mathbf{u}_s \tag{12.28}$$

where $\mathbf{u}_l$ and $\mathbf{u}_s$ are the liquid and solid velocities, respectively.

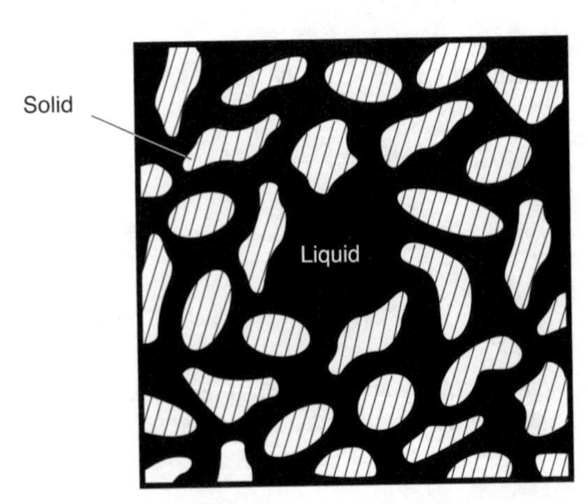

FIGURE 12.5 Mushy zone with solid particles dispersed within the liquid medium.

Since the volume fraction of the solid varies in the mushy zone, an effective viscosity, $\bar{\mu}$, is used at any location as

$$\bar{\mu} = \mu_s f_s + \mu_l f_l \tag{12.29}$$

where f_s and f_l are the solid and liquid volume fractions, respectively, and are related to the respective mass fractions, m_s and m_l, by

$$\bar{\rho} m_s = \rho_s f_s \quad \text{and} \quad \bar{\rho} m_l = \rho_l f_l \tag{12.29a}$$

where ρ_l and ρ_s are the liquid and solid densities, respectively, and $\bar{\rho}$ is the overall mixture density and is given by

$$\bar{\rho} = \rho_s f_s + \rho_l f_l \tag{12.29b}$$

Since there is no flow in the fully solid region, the velocity there has to be zero. In other words, the system velocity has to approach zero as the local liquid fraction approaches zero. This is normally achieved by specifying a very large solid viscosity, μ_s, say of the order of 10^8, such that as the solid volume fraction tends to unity, the effective viscosity $\bar{\mu}$ dominates the momentum equations and eventually reduces the velocity $\mathbf{u}$ to zero when fully solidified. The momentum equations for this case are given by

$$\frac{\partial}{\partial t}(\bar{\rho} u_x) + \nabla \cdot (\bar{\rho} \mathbf{u} u_x) = -\frac{\partial P}{\partial x} + \nabla \cdot (\bar{\mu} \nabla u_x) \tag{12.30a}$$

$$\frac{\partial}{\partial t}(\bar{\rho} u_z) + \nabla \cdot (\bar{\rho} \mathbf{u} u_z) = -\frac{\partial P}{\partial z} + \nabla \cdot (\bar{\mu} \nabla u_z) - \bar{\rho} g[\beta_T (T - T_0) + \beta_C (C_l - C_0)] \tag{12.30b}$$

The corresponding energy or enthalpy equation is given by

$$\frac{\partial}{\partial t}(\bar{\rho} h_e) + \nabla \cdot (\bar{\rho} \mathbf{u} h_e) = \nabla \cdot \left(\frac{k}{c_{ps}} \nabla h_e \right) + \nabla \cdot \left(\frac{k}{c_{ps}} \nabla (h_s - h_e) \right) \tag{12.31}$$

The terms on the right-hand side of the equation represent the net Fourier heat flux. The solute concentration is given by

$$\frac{\partial}{\partial t}(\bar{\rho} C) + \nabla \cdot (\bar{\rho} \mathbf{u} C) = \nabla \cdot (\rho D \nabla C) + \nabla [\bar{\rho} D \nabla (C_l - C)] \tag{12.32}$$

12.2.2 Columnar Dendritic Structure

The columnar dendritic structure is such that the solid and liquid phases are distinct, Fig. 12.6, with the solid having a specified velocity that may or may not be zero and is definitely different from the liquid velocity. The model that results from this formulation is sometimes referred to as the relative motion model. Laser spot welding is an example of the case with zero solid velocity, while laser seam welding is for the case with nonzero solid velocity.

This structure is essentially similar to a porous medium, and as such, flow in the mushy zone is considered to be governed by Darcy's law. This effect is accounted for by modifying the momentum equations to account for flow in the porous medium and requires the following assumptions:

1. The solid is free of internal stresses.
2. Viscous stresses resulting from local density gradients are small.
3. The density in each phase is constant.

With these assumptions, the momentum equations are given by

$$\frac{\partial}{\partial t}(\rho u_x) + \nabla \cdot (\rho \mathbf{u} u_x) = -\frac{\partial P}{\partial x} + \nabla \cdot \left(\mu_l \frac{\rho}{\rho_l} \nabla u_x \right) - \frac{\mu_l}{K_p} \frac{\rho}{\rho_l}(u_x - u_{xs}) \quad (12.33a)$$

$$\frac{\partial}{\partial t}(\rho u_z) + \nabla \cdot (\rho \mathbf{u} u_z) = -\frac{\partial P}{\partial z} + \nabla \cdot \left(\mu_l \frac{\rho}{\rho_l} \nabla u_z \right)$$

$$-\frac{\mu_l}{K_p} \frac{\rho}{\rho_l}(u_z - u_{zs}) - \rho g[\beta_T(T - T_0) + \beta_C(C_l - C_0)] \quad (12.33b)$$

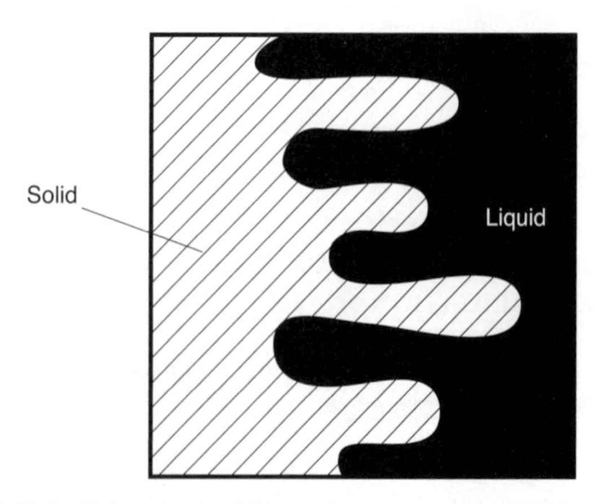

FIGURE 12.6 Columnar dendritic mushy zone at the solidifying interface.

where K_p represents permeability and is assumed to be isotropic. It depends on the cooling rate, temperature gradient, and primary dendrite arm spacing during the phase change. u_{is} represents solid velocity and is set to zero for static phase change systems, or to a constant value for continuous phase change systems.

There are a number of models available for the permeability, K_p. A common isotropic formulation is based on the Carman–Kozeny equation:

$$K_p = K_0 \left[\frac{f_l{}^3}{(1 - f_l)^2} \right] \tag{12.34}$$

where K_0 is a constant that depends on the mushy zone morphology. The Carman–Kozeny equation is generally valid for laminar flow and for liquid volume fractions less than 0.5.

The energy or enthalpy equation is also given by

$$\frac{\partial}{\partial t}(\rho h_e) + \nabla \cdot (\rho \mathbf{u} h_e) = \nabla \cdot \left(\frac{k}{c_{ps}} \nabla h_e \right) + \nabla \cdot \left(\frac{k}{c_{ps}} \nabla (h_s - h_e) \right) - \nabla \cdot [\rho (h_l - h_e)(\mathbf{u} - \mathbf{u_s})] \tag{12.35}$$

The first two terms on the right-hand side of the equation represent the net Fourier heat flux, while the last term represents the energy flux associated with relative phase motion.

The solute concentration equation is given by

$$\frac{\partial}{\partial t}(\rho C) + \nabla \cdot (\rho \mathbf{u} C) = \nabla \cdot (\rho D \nabla C) + \nabla [\rho D \nabla (C_l - C)] - \nabla \cdot [\rho (C_l - C)(\mathbf{u} - \mathbf{u_s})] \tag{12.36}$$

The first two terms on the right-hand side of the equation represent the net diffusive flux, while the last term represents the species flux associated with relative phase motion.

12.3 RAPID SOLIDIFICATION

Rapid solidification results when the rate at which the molten material cools is so high that its temperature falls significantly below its equilibrium freezing temperature, resulting in extensive undercooling before solidification begins. The process then occurs under nonequilibrium conditions with a high solidification velocity, and the interface temperature becomes a function of time and process parameters. The assumption of local thermodynamic equilibrium is then invalid.

Some of the benefits of rapid solidification include

1. Fine grain structure and solidification substructure (dendrite and cell spacing)
2. Reduced microsegregation
3. Extended solubility of alloying elements
4. Metastable crystalline and amorphous phases—rapid solidification is the basis for obtaining surface modification of materials using a laser beam.

There are two extreme heat flow conditions that are associated with rapid solidification. In one case, the molten material is superheated, and so the heat of fusion is withdrawn through the solid, and the interface velocity is controlled primarily by the rate of external heat extraction. This would be the case for surface melting and melt spinning, for example, where the interface velocity is limited by the rate of heat extraction to the substrate and the material thickness.

In the other case, the molten material is supercooled and thus acts as an effective heat sink, quickly removing any released latent heat. Crystal growth kinetics then become the main limiting factor. However, the latent heat that is released at the interface as solidification proceeds may result in reheating of the molten material to a temperature close to the equilibrium solidification temperature, and thereby reducing the growth rate of the freezing front (interface velocity). This is referred to as recalescence and may result in microsegregation, reduced supersaturation, and phase segregation. The interface velocity for this case is determined primarily by the extent of supercooling, and the heat loss to the surroundings does not significantly affect the solidification process in the initial stages, but may become significant as solidification proceeds.

These are illustrated in the enthalpy–temperature diagram (Fig. 12.7). Equilibrium solidification essentially follows the isothermal path. Here, the latent heat that is released is used to maintain the specimen's temperature at the equilibrium value until solidification is completed.

The molten material is said to be supercooled when it is cooled below the melting temperature without nucleation taking place. The supercooled temperature is such that when solidification finally starts, the recalescence that results from the release of latent heat is not enough to raise the melt's temperature even to the solidus temperature. The boundary for the supercooled process is given by the isoenthalpic path. The more common situation is given by the general case where the molten material is supercooled and when solidification begins, the recalescence resulting from the rapid release of latent heat raises the melt temperature close to the equilibrium value. Subsequent solidification then takes place under near-equilibrium conditions.

The conditions under which local thermodynamic equilibrium are valid may be approximated by considering the diffusive velocity, u_D, of the solute in the liquid,

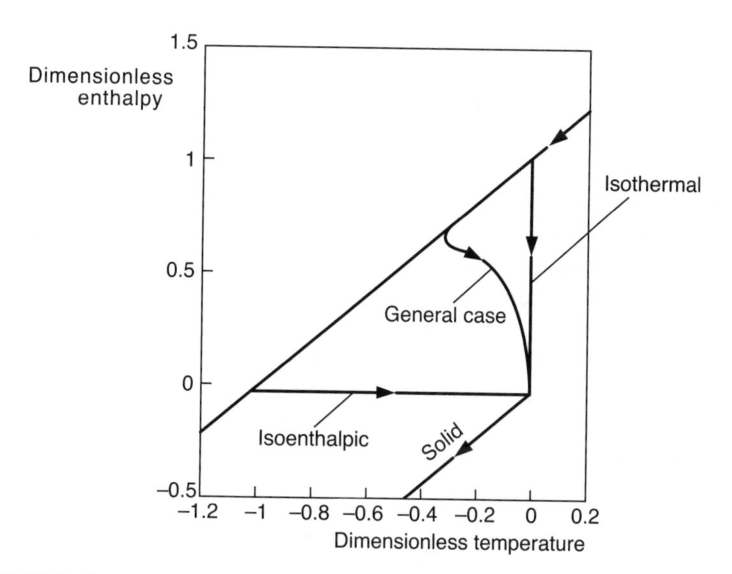

FIGURE 12.7 Enthalpy–temperature diagram. (From Mehrabian, R., 1982, *Metals Review*, Vol. 27, pp. 185–208.)

which may be defined as

$$u_{\mathrm{D}} = \frac{D_{\mathrm{i}}}{l_{\mathrm{a}}} \tag{12.37}$$

where l_{a} is the interatomic spacing and D_{i} is the coefficient of interdiffusion across the interface.

As the interface velocity approaches the diffusive velocity, the solidification front begins overtaking the solute atoms, at which point local thermodynamic equilibrium is no more valid at the interface. Complete solute trapping results as the interface velocity exceeds the diffusive velocity. Generally, as conditions depart from local equilibrium, interface stability is enhanced.

12.4 SUMMARY

The emphasis in this chapter has been on the growth component of the solidification process, having covered the nucleation process in Chapter 11. For a pure metal without flow, the solidification interface is planar. That of an alloy may break down to form a cellular, columnar dendritic, or equiaxed dendritic structure as a result of instability associated with the process. The instability of the interface can be analyzed using either the constitutional supercooling theory or the morphological stability theory. Although the stability criterion based on the constitutional supercooling theory depends only on the temperature gradients in the liquid and solid at the interface, that of the morphological stability theory also takes into account the solute concentration as well as capillary effects. It is shown that the solute term tends to make the interface unstable, while the capillary term tends to prevent the growth of a perturbation, and since the thermal term may be either positive or negative, it can be either stabilizing or destabilizing. For analysis in the mushy zone, similar governing equations are used, except that they are modified to account for the solid volume fraction.

When solidification occurs in conditions involving flow, the analysis also includes mass and momentum conservation equations. The analysis is simplified by considering the mushy zone as either a mushy fluid (which is similar to granular solid particles that are dispersed within a liquid medium) or a columnar dendritic structure (where the solid and liquid phases are distinct, with the solid having a specified velocity that may or may not be zero and is different from the liquid velocity).

With rapid solidification, which is the more likely scenario in laser processes that involve a large mass that acts as a heat sink, there are two extreme heat flow conditions:

1. One case where the molten material is superheated, which is also the more likely case for laser processing.
2. Another case where the molten material is supercooled.

Rapid solidification results in fine grain, dendrite, and cell structures; reduced microsegregation; extended solubility of alloying elements; and metastable crystalline and amorphous phases.

REFERENCES

Beckermann, C., and Viskanta, R., 1988a, Double-diffusive convection during dendritic solidification of a binary mixture, *PCH PhysicoChemical Hydrodynamic*, Vol. 10, No. 2, pp. 195–213.

Beckermann, C., and Viskanta, R., 1988b, Natural convection solid/liquid phase change in porous media, *International Journal of Heat Mass Transfer*, Vol. 31, No. 1, pp. 35–46.

Bennon, W. D., and Incropera, F. P., 1987a, A continuum model for momentum, heat and species transport in binary solid–liquid phase change systems—I. Model formulation, *International Journal of Heat Mass Transfer*, Vol. 30, No. 10, pp. 2161–2170.

Bennon, W. D., and Incropera, F. P., 1987b, A continuum model for momentum, heat and species transport in binary solid–liquid phase change systems—II. Application to solidification in a rectangular cavity, *International Journal of Heat Mass Transfer*, Vol. 30, No. 10, pp. 2171–2187.

Combeau, H., Roch, F., Poitrault, I., Chevrier, J., and Lesoult, G., 1990, Numerical study of heat and mass transfer during solidification of steel ingots, *Heat Transfer*, pp. 79–90.

Glicksman, M. E., Coriell, S. R., and McFadden, G. B., 1986, Interaction of flows with the crystal–melt interface, *Annual Reviews in Fluid Mechanics*, Vol. 18, pp. 307–335.

Huang, S.-C., and Glicksman, M. E., 1981, Fundamentals of dendritic solidification—I. Steady-state tip growth, *Acta Metallurgica*, Vol. 29, pp. 701–715.

Huppert, H. E., 1990, The fluid mechanics of solidification, *Journal of Fluid Mechanics*, Vol. 212, pp. 209–240.

Kou, S., 1987, *Welding Metallurgy*, John Wiley and Sons, New York.

Kurz, W., and Fisher, D. J., 1992, *Fundamentals of Solidification*, 3rd edition, Trans Tech Publications, Switzerland.

Mazumder, J., 1987, An overview of melt dynamics in laser processing, *High Power Lasers*, Vol. 801, pp. 228–241.

Mehrabian, R., 1982, Rapid Solidification, *Metals Review*, Vol. 27, pp. 185–208.

Mullins, W. W., and Sekerka, R. F., 1964, Stability of a planar interface during solidification of a dilute binary alloy, *Journal of Applied Physics*, Vol. 35, pp. 444–451.

Oldenburg, C. M., and Spera, F. J., 1992, Hybrid model for solidification and convection, *Numerical Heat Transfer B*, Vol. 21, pp. 217–229.

Smith, T. J., and Hoadley, A. F. A., 1987, Recent developments in modeling metal flow and solidification, *Modelling the Flow and Solidification of Metals*, Smith, T. J., editor, pp. 277–303.

Szekely, J., and Jassal, A. S., 1978, An experimental and analytical study of the solidification of a binary dendritic system, *Metallurgical Transactions B*, Vol. 9B, pp. 389–398.

Tiller, W. A., Jackson, K. A., Rutter, J. W., and Chalmers, B., 1953, The redistribution of solute atoms during the solidification of metals, *Acta Metallurgica*, Vol. 1, pp. 428–437.

Trivedi, R., and Somboonsuk, K., 1984, Constrained dendritic growth and spacing, *Materials Science and Engineering*, Vol. 65, pp. 65–74.

Tseng, A. A., Zou, J., Wang, H. P., and Hoole, S. R. H., 1992, Numerical modeling of macro and micro behaviors of materials in processing: a review, *Journal of Computational Physics*, Vol. 102, pp. 1–17.

Voller, V. R., and Prakash, C., 1987, A fixed grid numerical modeling methodology for convection-diffusion mushy region phase-change problems, *International Journal of Heat and Mass Transfer*, Vol. 30, No. 8, pp. 1709–1719.

Voller, V. R., Brent, A. D., and Prakash, C., 1989, The Modelling of Heat, Mass and Solute Transport in Solidificatiion Systems, *International Journal of Heat and Mass Transfer*, Vol. 32, No. 9, pp. 1719–1731.

APPENDIX 12A

List of symbols used in the chapter.

Symbol	Parameter	Units
C	alloy Composition or solute concentration	%
C_l	liquid composition or concentration at the interface	%
C_0	initial (reference) liquid composition	%
C_s	solid composition or concentration at the interface	%
D_i	diffusion coefficient of solute across the interface	m^2/s
G_c	solute concentration gradient at the solid–liquid interface	%/m
h_e	enthalpy	J
k_p	equilibrium partition coefficient (see Appendix 12B)	–

K_p	permeability	m^2
l	subscript for liquid	–
l_a	interatomic spacing	m
m_g	absolute value of slope of liquidus curve	°C
s	subscript for solid	–
$s(t)$	position of solid liquid interface	m
$\dot{s}$	solidification rate or interface velocity	m/s
S_c	concentration source term	kg/m^3s
S_h	enthalpy source term	$W\ kg/m^3$
T_c	constant temperature of a chilled surface	°C
u_D	diffusive velocity of solute in liquid	m/s
δ	amplitude of perturbation at time $t = 0$	m
ω_t	temporal frequency	Hz
$\omega_{y,z}$	spatial frequencies	Hz

APPENDIX 12B CRITERION FOR SOLIDIFICATION OF AN ALLOY

The condition under which a planar interface in an alloy breaks down to form a either cellular or dendritic structure was discussed in Section "Structure of the Solidified Metal", and was expressed as

$$\frac{G_{Tl}}{SR} \leq \frac{\Delta T}{D_l} \tag{11.11}$$

Basic theories for interface stability were further discussed in Sections "Constitutional Supercooling Theory" and "Morphological Stability Theory." In this section, a simplified approach is used to develop the criterion expressed by equation (11.11).

Solidification of a molten metal would ideally occur under equilibrium conditions, where there is adequate time for complete diffusion to occur in the solid state, with perfect mixing occurring in the liquid state. However, in practice, the cooling rates are such that diffusion in the solid is not complete. Furthermore, mixing in the liquid is not always perfect. As a result, solidification often occurs under nonequilibrium conditions. In the following analysis, it is assumed for simplicity that there is no diffusion in the solid, that mixing in the liquid occurs primarily by diffusion, that is, convection in the liquid is negligible, and that local equilibrium exists at the solid/liquid interface.

Consider an alloy of initial composition C_0 in the liquid state in a long cylindrical container, where solidification occurs with the solid–liquid interface advancing from left to right (Fig. 12B.1). The first solid that forms at temperature T_1, as discussed in Section 11.1.1.4, has the composition $C_s = C_1$ (Fig. 11.14), which is less than the nominal composition of the alloy C_0 and is thus richer in the solvent. That means solute atoms are rejected into the liquid, increasing the solute concentration in the liquid. Thus, at the interface between the solid and the liquid, the equilibrium concentration

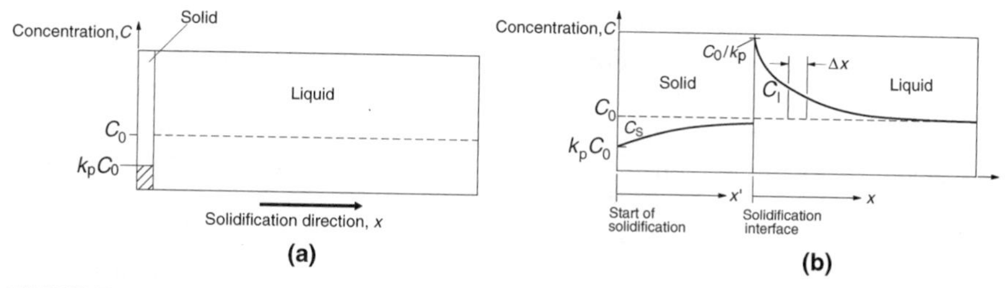

FIGURE 12B.1 One-dimensional solidification. (a) Concentration distribution in the initial stages of solidification. (b) Concentration distribution in the intermediate stages of solidification (From Easterling, K., *Introduction to the Physical Metallurgy of Welding*. Copyright Elsevier, 1983.)

of solute in the solid, C_s, is different from the equilibrium concentration of solute in the liquid adjacent to it, C_l. A partition coefficient, $k_p = C_s/C_l$, is defined as the ratio of solute concentrations in the solid and liquid that are in equilibrium at a given temperature; C_s and C_l are the respective compositions of the solid and liquid at the interface.

In the initial stages of solidification, the solute concentration in the liquid is essentially C_0, that is, $C_l = C_0$, and thus the initial solid that forms has the concentration k_pC_0, Fig. 12B.1a, while the last liquid to solidify has the concentration C_0/k_p. These are illustrated in Fig. 12B.2. As solidification proceeds, the solute concentration in the liquid will increase because of the solute rejected from the interface. As a result, the solute concentration in the solid will also increase until a steady-state condition

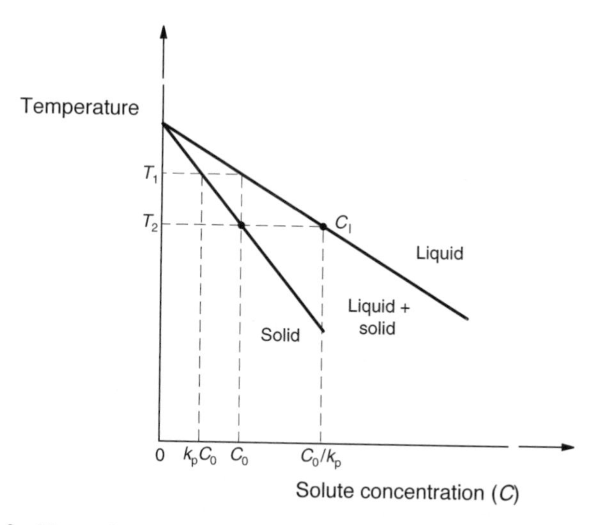

FIGURE 12B.2 Illustration of the partition coefficient using a partial binary phase diagram. (From Tiller, W. A., Jackson, K. A., Rutter, J. W., and Chalmers, B., *Acta Metallurgica*, Vol. 1, pp. 428–437. Copyright Elsevier, 1953.)

is reached. The distribution of solute in the liquid near the interface will then be constant. This is illustrated in Fig. 12B.3. Since the solidus and liquidus lines in most phase diagrams are slightly curved, k_p will generally vary. However, it may often be assumed to be constant. $k_p < 1$ when the solubility of solute in the solid is less than in the liquid, and it results in rejection of solute atoms by the solid as it forms.

To develop the governing equation for solute distribution in the liquid, let us go back to Fig. 12B.1 and consider one-dimensional solidification of molten material in a cylindrical container. Furthermore, consider a control volume of the liquid close to the solid–liquid interface, of length Δx, and unit cross-sectional area (Fig. 12B.1b). Now let the origin of the coordinate system be attached to the interface, and further, let J_1 be the flux of solute atoms into the control volume per unit area per unit time, J_2 the flux out of the control volume, C the solute concentration into the control volume (solute atoms per unit volume), $C_{x+\Delta x}$ the solute concentration out of the control volume, C_l and C_s the solute concentrations of the liquid and solid, respectively, at the interface, D_l the diffusion coefficient of solute in the liquid, and SR the solidification rate.

Considering the conservation of solute atoms into and out of the control volume per unit time, the change in concentration of the control volume per unit time is equal to the net flux of solute atoms into the control volume. Thus, since the interface is

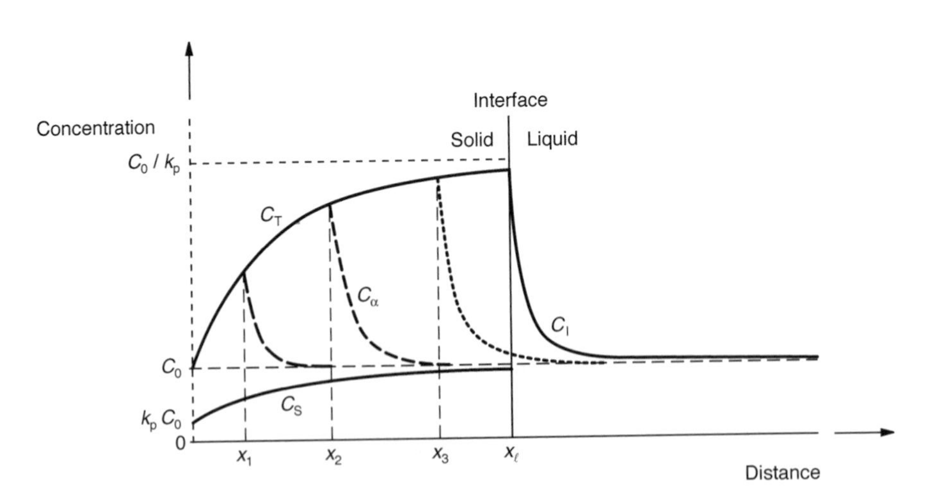

FIGURE 12B.3 Distribution of solute in liquid and solid. C_l is the current solute distribution in the liquid; C_s is the solute distribution in the solid; C_T is the solute concentration at the growing interface; and C_α is the solute distribution in the liquid in earlier stages of the solidification process. (From Tiller, W. A., Jackson, K. A., Rutter, J. W., and Chalmers, B., *Acta Metallurgica*, Vol. 1, pp. 428–437. Copyright Elsevier, 1953.)

moving at a speed equal to SR, we have

$$\frac{\Delta C}{\Delta t} \cdot \Delta x = (J_1 - J_2) - \mathrm{SR}(C - C_{x+\Delta x}) \tag{12B.1}$$

The first term on the right-hand side is due to flux by diffusion of atoms, and the second term results from flux by convection due to motion of the interface. Dividing through by Δx gives

$$\frac{\Delta C}{\Delta t} = \frac{(J_1 - J_2)}{\Delta x} - \mathrm{SR}\frac{(C - C_{x+\Delta x})}{\Delta x} \tag{12B.2}$$

Now considering only first-order terms in a Taylor series, we have

$$J_2 = J_1 + \frac{\partial J_1}{\partial x}\Delta x$$

and

$$C_{x+\Delta x} = C + \frac{\partial C}{\partial x}\Delta x$$

Thus,

$$\frac{\Delta C}{\Delta t} = \frac{J_1 - \left(J_1 + \frac{\partial J_1}{\partial x}\Delta x\right)}{\Delta x} - \mathrm{SR}\left[\frac{C - \left(C + \frac{\partial C}{\partial x}\Delta x\right)}{\Delta x}\right] \tag{12B.3}$$

and in the limit,

$$\frac{\partial C}{\partial t} = -\frac{\partial J_1}{\partial x} + \mathrm{SR}\frac{\partial C}{\partial x} \tag{12B.4}$$

Now the rate at which atoms diffuse in a material can be measured by J, which is defined as the number of atoms passing through a plane of unit area per unit time. Fick's first law explains the net flux of atoms and is given by

$$J = -D_l\frac{\partial C}{\partial x}$$

Thus,

$$\frac{\partial C}{\partial t} = \frac{\partial}{\partial x}\left(D_l\frac{\partial C}{\partial x}\right) + \mathrm{SR}\frac{\partial C}{\partial x} \tag{12B.5}$$

This is the general one-dimensional form of the solute diffusion in a medium with a concentration gradient, and moving in the negative x-direction, that is, from right to left, or the solidification front moves from left to right in the positive x-direction.

The analysis that follows is simplified by considering the steady-state condition. The control volume is considered to be at the solid–liquid interface. Further simplifications are made by starting with equation (12B.1) and eliminating the time-dependent term on the left-hand side of the equation since only the steady-state case is of interest, reducing it to the form

$$(J_1 - J_2) - \text{SR}\ (C_s - C_1) = 0 \tag{12B.6}$$

Also, the net flux of solute atoms into the control volume is simplified to

$$(J_1 - J_2) = D_1 \frac{dC_1}{dx} \tag{12B.7}$$

Thus,

$$D_1 \frac{dC_1}{dx} = \text{SR}\ (C_s - C_1) \tag{12B.8}$$

The solution of this first-order equation gives an exponential concentration profile of the solute in the liquid directly ahead of the interface, Fig. 12B.4a, and is given by

$$C_1 = C_0 \left[1 + \left(\frac{1}{k_p} - 1 \right) e^{-\frac{\text{SR}}{D_1} x} \right] \tag{12B.9}$$

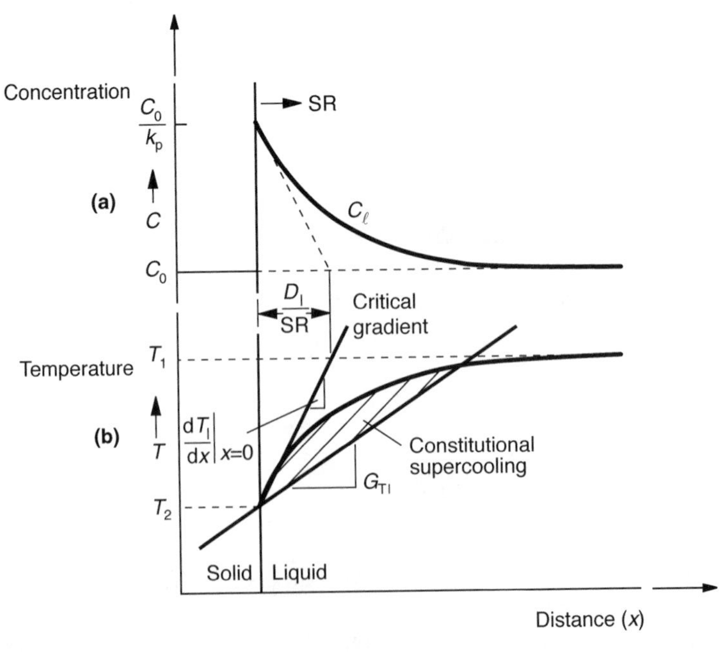

FIGURE 12B.4 (**a**) Solute distribution in the liquid during one-dimensional solidification. (**b**) Temperature profile near the solid/liquid interface. (From Easterling, K., *Introduction to the Physical Metallurgy of Welding*. Copyright Elsevier, 1983.)

Equation (12B.9) indicates that the decay constant is determined by the ratio of the growth rate to the diffusion constant. The corresponding temperature profile in the liquid is shown in Fig. 12B.4b. The concentration in the solid, C_s, on the contrary, rises from its initial value of $k_p C_0$ and approaches its equilibrium value, C_0 asymptotically, as solidification proceeds, Figs. 12B.1 and 12B.3, and is given by

$$C_s = C_0 \left[(1 - k_p) \left(1 - e^{-k_p \frac{\mathbf{SR}}{D_l} x'} \right) + k_p \right] \tag{12B.10}$$

where x' is the distance from the beginning of the crystal.

If we now consider flux of solute atoms directly at the solid/liquid interface, then equation (12B.8) can be further simplified to the form

$$D_l \left. \frac{dC_l}{dx} \right|_{x=0} = \mathrm{SR}(C_s - C_l) \tag{12B.11}$$

but

$$\left. \frac{dT_l}{dx} \right|_{x=0} = \left. \frac{dT_l}{dC_l} \frac{dC_l}{dx} \right|_{x=0} \tag{12B.12}$$

Substituting equation (12B.11) into equation (12B.12) gives the equilibrium temperature gradient in the liquid at the interface as

$$\left. \frac{dT_l}{dx} \right|_{x=0} = \frac{\mathrm{SR}}{D_l} (C_s - C_l) \frac{dT_l}{dC_l} \tag{12B.13}$$

If the liquid near the interface is constitutionally supercooled, then its actual temperature near the interface will be less than the equilibrium (liquidus) temperature. In other words, the actual temperature gradient, G_{Tl}, will be smaller than the equilibrium temperature gradient, $dT_l/dx|_{x=0}$, given by equation (12B.13). This is illustrated in Fig. 12B.4b.

A planar interface breaks down to form either a cellular or dendritic interface only when there is constitutional supercooling, that is, when

$$G_{Tl} \leq \left. \frac{dT_l}{dx} \right|_{x=0} = \mathrm{SR} \frac{(C_s - C_l)}{D_l} \frac{dT_l}{dC_l} \tag{12B.14}$$

or

$$\frac{G_{Tl}}{\mathrm{SR}} \leq \frac{(C_s - C_l)}{D_l} \frac{dT_l}{dC_l} \tag{12B.15}$$

which can also be expressed as

$$\frac{G_{Tl}}{\mathrm{SR}} \leq C_0 \frac{(1 - k_p)}{k_p D_l} \frac{dT_l}{dC_l} \tag{12B.16}$$

From the partial binary phase diagram shown in Fig. 12B.2, it is evident that $(C_s - C_l) dT_l/dC_l$ in equation (12B.15) can be approximated by $\Delta T = T_1 - T_2$ if the liquidus curve is assumed to be a line. T_1 is the liquidus temperature at concentration C_0, while T_2 is the corresponding solidus temperature. Thus, from equation (12B.15), we have the criterion for the formation of either cellular or dendritic structure as

$$\frac{G_{\mathrm{Tl}}}{\mathrm{SR}} \leq \frac{\Delta T}{D_{\mathrm{l}}} \tag{12B.17}$$

PROBLEMS

12.1. Show that the morphological stability criterion, equation (12.13a), reduces to the constitutional supercooling criterion, equation (12.14), in the limiting case when the interface velocity is very low.

12.2. Show that at high interface velocities, and for a stabilizing thermal field, the morphological stability criterion reduces to

$$m_{\mathrm{g}} G_{\mathrm{c}} \left(\frac{D_{\mathrm{l}}^2}{u^2 k_{\mathrm{p}}} \right) \omega^2 - \frac{T_{\mathrm{m}} \gamma_{\mathrm{e}} \omega^2}{L_{\mathrm{v}}} < 0$$

12.3. Given

$$\frac{G_{\mathrm{Tl}}}{\mathrm{SR}} \leq \frac{(C_{\mathrm{s}} - C_{\mathrm{l}})}{D_{\mathrm{l}}} \frac{\mathrm{d}T_{\mathrm{l}}}{\mathrm{d}C_{\mathrm{l}}}$$

Show that

$$\frac{G_{\mathrm{Tl}}}{\mathrm{SR}} \leq C_0 \frac{(1 - k_{\mathrm{p}})}{k_{\mathrm{p}} D_{\mathrm{l}}} \frac{\mathrm{d}T_{\mathrm{l}}}{\mathrm{d}C_{\mathrm{l}}}$$

12.4. Given that the concentration of solute atoms in the liquid in front of the solid–liquid interface during one-dimensional solidification in the x-direction is described by the following first-order differential equation

$$D_{\mathrm{l}} \frac{\mathrm{d}C_{\mathrm{l}}}{\mathrm{d}x} = \mathrm{SR}\ (C_{\mathrm{s}} - C_{\mathrm{l}})$$

Show that the concentration profile is given by

$$C_{\mathrm{l}} = C_0 \left[1 + \left(\frac{1}{k_{\mathrm{p}}} - 1 \right) \mathrm{e}^{-\frac{\mathrm{SR}}{D_{\mathrm{l}}} x} \right]$$

where C_0 is the original composition of the alloy in the liquid state, and the diffusion coefficient, D_{l}, is assumed constant.

13 Residual Stresses and Distortion

Residual stresses are stresses that continue to exist in a material even when there are no external forces acting on it. They are produced in a number of ways, for example, during welding, grinding, forming, and heat treatment (especially when martensite is formed due to volumetric change, as austenite transforms into martensite). Residual stresses are important because they affect the strength of the part, either positively or negatively, and may also result in distortion.

Our discussion in this chapter starts with the fundamental causes of residual stresses, focusing on thermal stresses and nonuniform plastic deformation. This is followed by a basic stress analysis to explain the underlying mechanisms of residual stresses. Discussion on this includes equilibrium conditions, strain–displacement relations, stress–strain relations, plane stress and plane strain, and the compatibility equation. The effects of residual stresses are then outlined, specifically strength effects and distortion. A discussion on techniques for measuring residual stresses (stress relaxation, X-ray diffraction, and neutron diffraction) then follows, and then finally, methods for relieving residual stresses and distortion.

13.1 CAUSES OF RESIDUAL STRESSES

Two principal causes of residual stresses in materials processing are

1. Thermal stresses.
2. Nonuniform plastic deformation.

Our discussion will focus on these two causes. Other causes of residual stresses are nuclear radiation and electromagnetic field exposure.

13.1.1 Thermal Stresses

To illustrate the development of residual stresses as a result of temperature changes, let us consider three rods of equal length that are rigidly connected together at the top and bottom (Fig. 13.1). Now, let the middle rod 2 be heated uniformly to a high

---

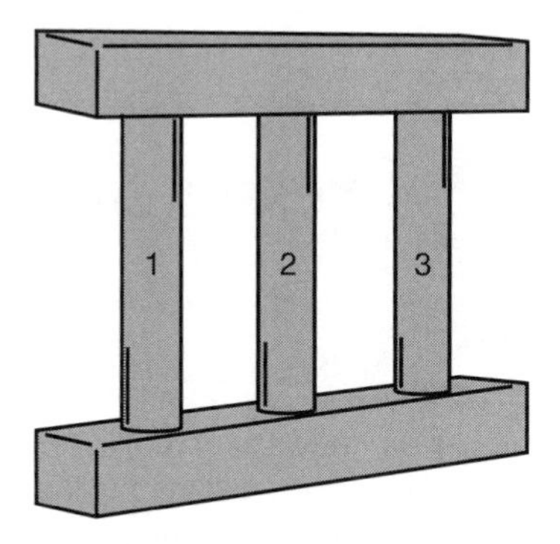

FIGURE 13.1 Three-rod frame with rods of equal length that are rigidly connected together at the top and bottom. (Copyright American Welding Society, *Welding Handbook*, Vol. 1, 9th edition, 2001. Used with permission.)

temperature and cooled. As it is heated, it tends to expand. If it were free, it would expand by an amount ΔL, given by

$$\Delta L = \beta_l L_o \Delta T \tag{13.1}$$

where L_o is the original length of the rod, β_l is the linear coefficient of thermal expansion of the rod material $(/K)$, and ΔT is the temperature rise in the rod.

However, it is prevented from doing so by the other two unheated rods. This results in compressive stress being induced in 2 while tensile stresses are induced in 1 and 3. Before the yield strength of the material is reached, the elastic strain (ε^e) induced in 2 is given by

$$\varepsilon^e = \frac{\Delta L}{L_o} = \beta_l \Delta T \tag{13.2a}$$

which results in a compressive elastic stress of

$$\sigma = E_m \beta_l \Delta T \tag{13.2b}$$

where E_m is Young's modulus.

The stress in 2 keeps increasing with temperature until the yield stress in compression is reached (point Q) (Fig. 13.2). At this time, plastic flow occurs. Beyond this point, the additional strain that is experienced, the plastic strain, ε^p, is given by

$$\varepsilon^p = \beta_l \Delta T - \frac{S_y}{E_m} \tag{13.3}$$

where S_y is the yield strength of the material.

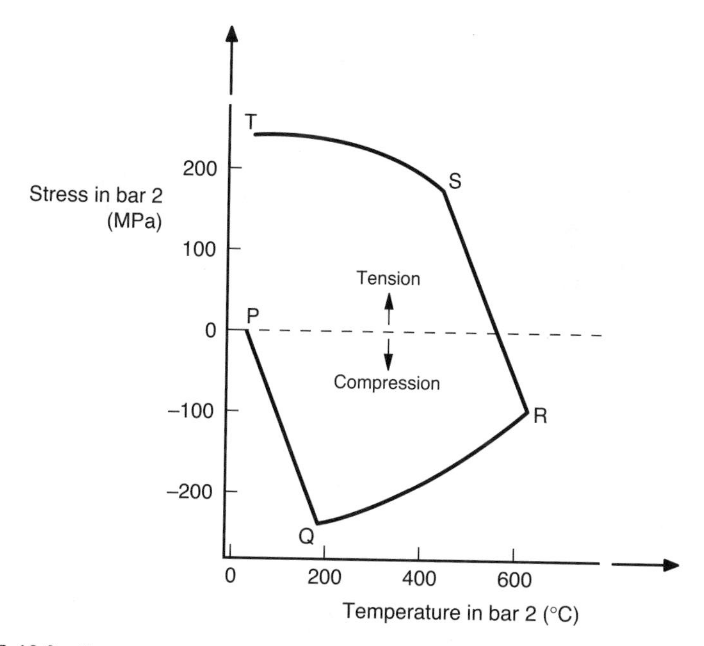

FIGURE 13.2 Stress–temperature curve for the middle rod of a three-rod frame. (Copyright American Welding Society, *Welding Handbook*, Vol. 1, 9th edition, 2001. Used with permission.)

Since the yield strength decreases with increasing temperature, the stress in 2 decreases with further increase in temperature (QR).

On cooling, 2 tends to contract and again is restrained by 1 and 3. Since it is cooling, the yield strength increases as the temperature reduces further, and the stress in the rod becomes elastic. It quickly reduces in compressive stress and then becomes tensile, the tensile stress increasing elastically with further cooling until yielding (point S). Thus, at room temperature, rod 2 will be subjected to tensile residual stress while 1 and 3 will be in compression. The residual stresses so produced are also called thermal stresses.

For a specific example in laser processing, we consider the development of residual stresses during welding. A schematic of a weld bead in progress along with temperature and stress distribution at various points along the bead are shown in Fig. 13.3 at points A, B, C, and D.

Far ahead of the heat source at A–A, the temperature and stress distributions are zero. At section B–B, which passes through the heat source, very high temperatures exist in the immediate vicinity of the weld, and this tapers off to low temperatures away from the weld. Since the molten metal cannot endure stress, stresses in the weld are zero. Slightly farther off, there are low compressive stresses since the heated material close to the weld is prevented from expanding. Also, the yield stress is low at the high temperatures and that makes the stresses low. Yet farther

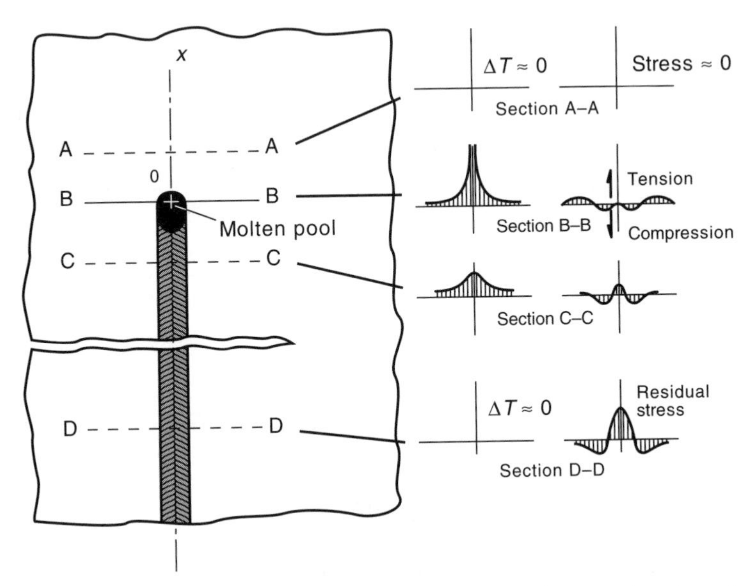

FIGURE 13.3 Stress and temperature changes during welding. (Copyright American Welding Society, *Welding Handbook*, Vol. 1, 9th edition, 2001. Used with permission.)

away, tensile stresses are developed that balance the compressive stresses close to the weld.

Behind the heat source, section C–C, lower temperatures exist in the weld bead, which then has a more uniform temperature distribution. The cooling weld bead and material close to it tend to contract and this is prevented by the rest of the material. The result is tensile stresses in the material within and close to the bead and balancing compressive stresses further away.

Even further behind the heat source at section D–D, the temperatures are much lower, and the stresses become more pronounced. The stress distributions that result at room temperature, both longitudinal and transverse, are shown in Fig. 13.4. These are the residual stresses that continue to exist in the material after welding is completed. Important aspects of the longitudinal stress distribution (Fig. 13.4b) are

1. Maximum tensile stress, σ_m
2. Width of the tensile region, b.

The nature of the transverse distribution (Fig. 13.4c) is due to the fact that the ends are relatively free to expand and contract. However, the central position is restrained and therefore experiences tensile stresses. These are balanced by compressive stresses on the ends.

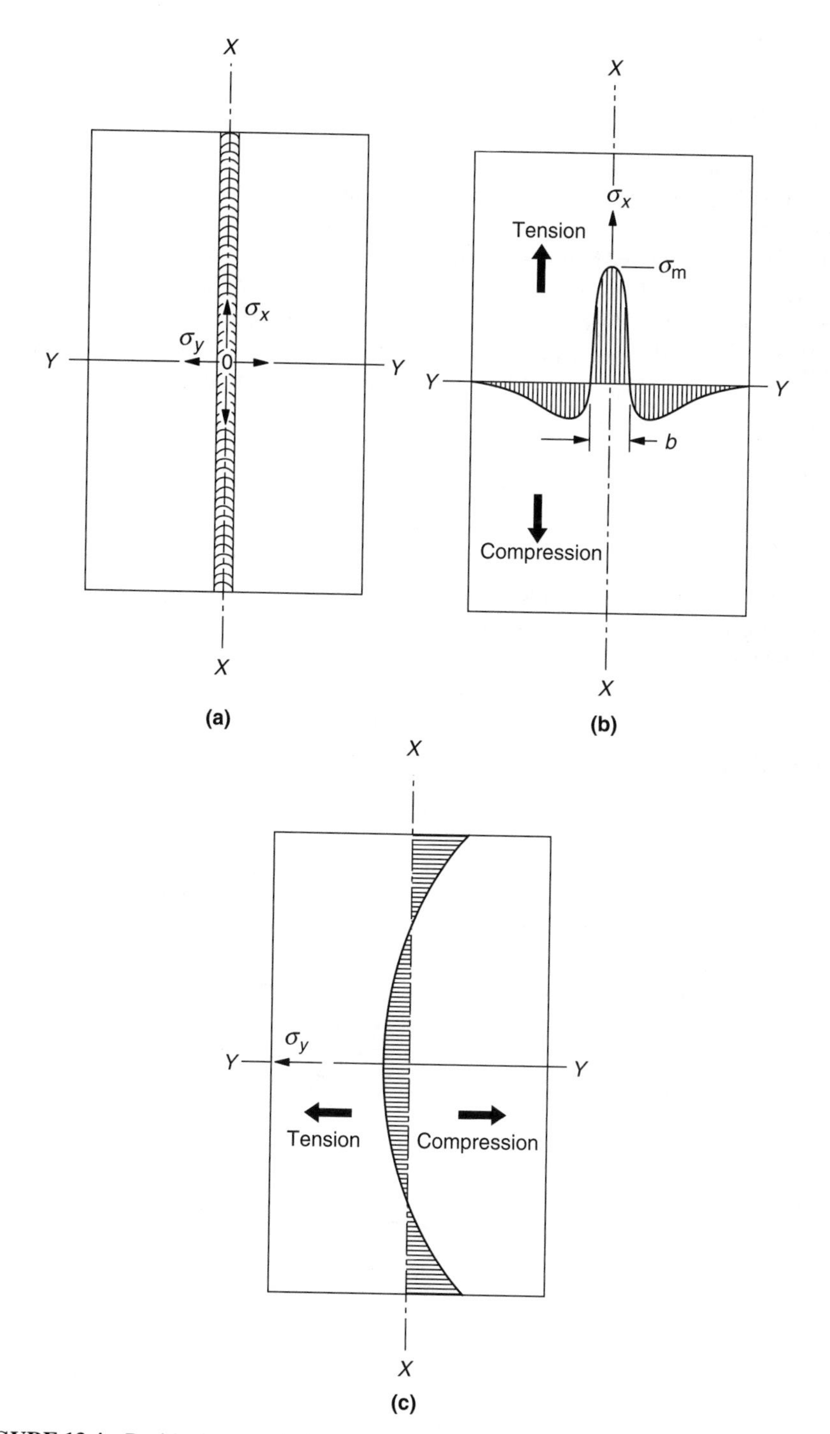

FIGURE 13.4 Residual stress distribution in a weld. (a) Weld configuration. (b) Distribution of longitudinal stresses. (c) Distribution of transverse stresses. (Copyright American Welding Society, *Welding Handbook*, Vol. 1, 9th edition, 2001. Used with permission.)

13.1.2 Nonuniform Plastic Deformation

To understand how residual stresses are introduced as a result of plastic deformation, let us first consider a rod that is subjected to simple tension where the loading is high enough to produce uniform plastic deformation throughout the rod. If the applied load is subsequently released before necking occurs, there will be uniform elastic recovery throughout the rod. However, there will be permanent deformation within it after the load is removed (Fig. 13.5) but no residual stresses will exist.

Now consider the beam shown in Fig. 13.6a, which is subjected to bending. The bending moment, M_{init}, that will result in plastic deformation of the outer fibers is given by

$$M_{\text{init}} = \frac{S_y bh^2}{6} \tag{13.4}$$

where b is the width of the beam and h is the thickness of the beam.

The stress distribution in the beam is then given by the dashed line (Fig. 13.6b). Assuming an elastic perfectly plastic material, the stress distribution that results after full yield is given by the solid line (Fig. 13.6b). The resulting force on either side of the beam (either the upper or lower half) is then given by $S_y bh/2$. Since the distance between the resulting forces is $h/2$, the bending moment M_y required to achieve full yield throughout the beam is

$$M_y = \frac{S_y bh}{2} \cdot \frac{h}{2} = \frac{S_y bh^2}{4} \tag{13.5}$$

If the external bending moment is now removed, elastic unloading occurs, and the stress distribution that results from elastic unloading is given by the dashed line in Fig. 13.7a. The resultant stress distribution after unloading is obtained by superimposing the plastic stress distribution and the unloading curves, giving the final stress distribution within the beam after the load is removed, the residual stress (given by the solid line in Fig. 13.7). The distribution in Fig. 13.7a is the ideal case. The actual distribution is more like the one shown in Fig. 13.7b.

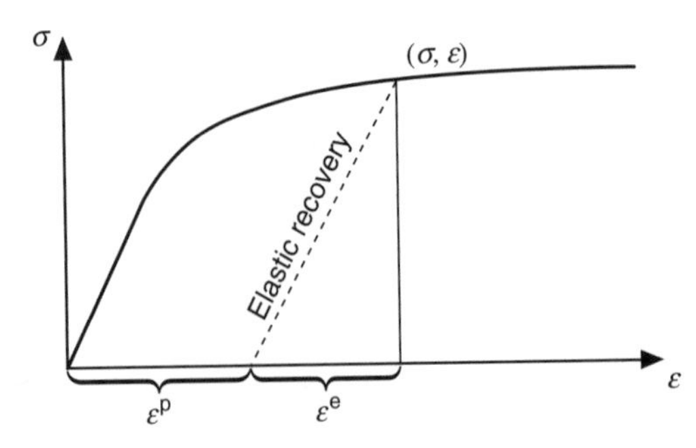

FIGURE 13.5 Stress–strain curve with elastic recovery and permanent set.

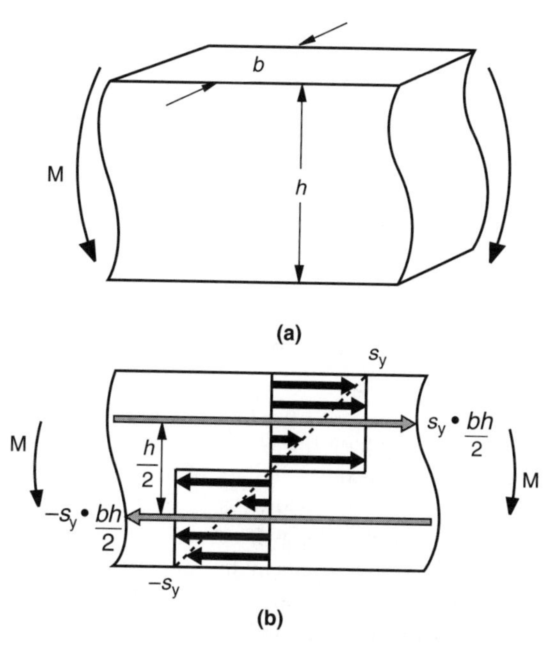

FIGURE 13.6 Beam bending and resulting stress distribution. (a) The beam. (b) The stress distribution.

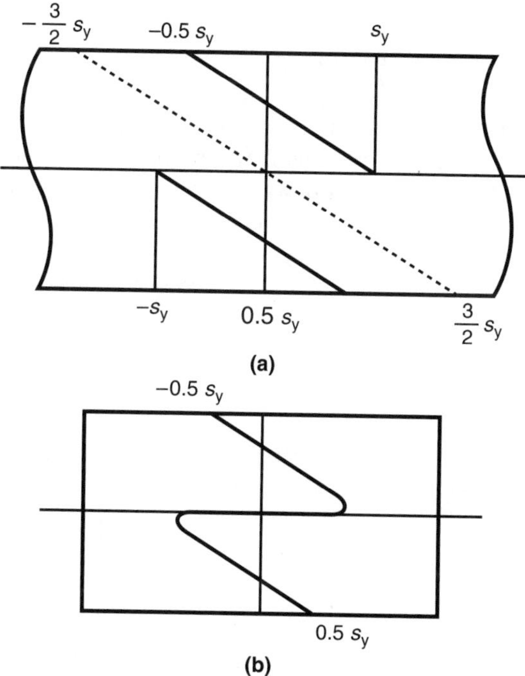

FIGURE 13.7 Stress distribution in a bent beam after unloading. (a) Ideal stress distribution. (b) Actual stress distribution.

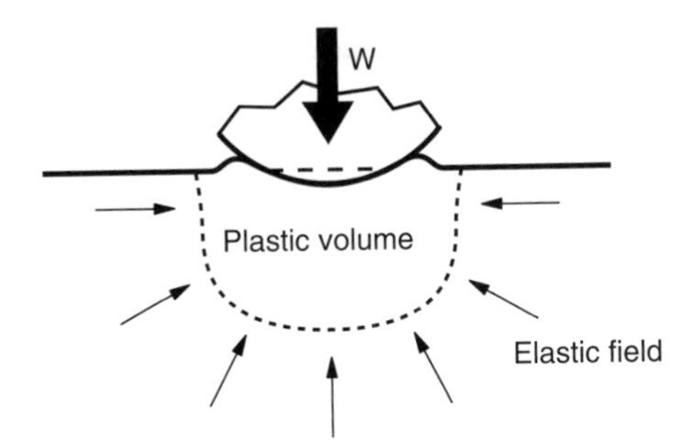

FIGURE 13.8 Indentation in a body.

Next, we consider a small region that is plastically deformed in a body by say, indentation (Fig. 13.8). The immediate vicinity of the indenter will be subjected to plastic flow, and this will be surrounded by a large compressive elastic field. On removing the indenter, elastic unloading will occur, but there will be a permanent set (plastic strain that remains), which results in the dent left behind. The end result will be elastic compressive stresses that are left after unloading. Residual stresses due to nonuniform plastic deformation may thus result from hammering, rolling, dull cutting tools, and eroding particles.

13.2 BASIC STRESS ANALYSIS

In this section, we discuss some of the basic concepts necessary for analyzing material behavior, and thus residual stresses that may develop in such materials. Consider a body of general shape that consists of a continuous, deformable material. Now let an equilibrium system of forces be applied to the surface of this body. The resulting stress state at any point in the body may be represented using Cartesian coordinates as shown in Fig. 13.9 for an element in the body. Normal or direct stresses σ_x, σ_y,

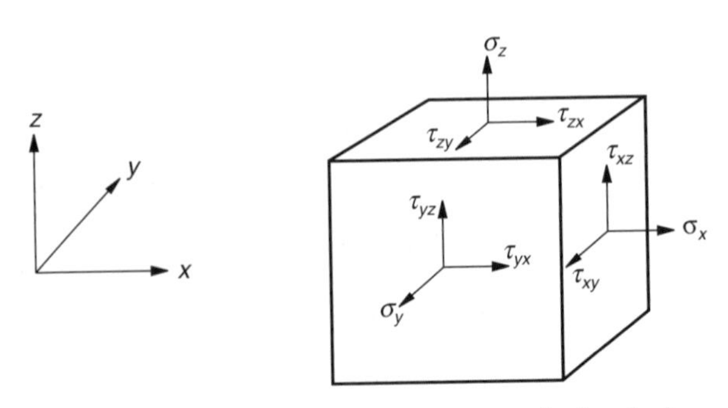

FIGURE 13.9 General stress state at a point in a body.

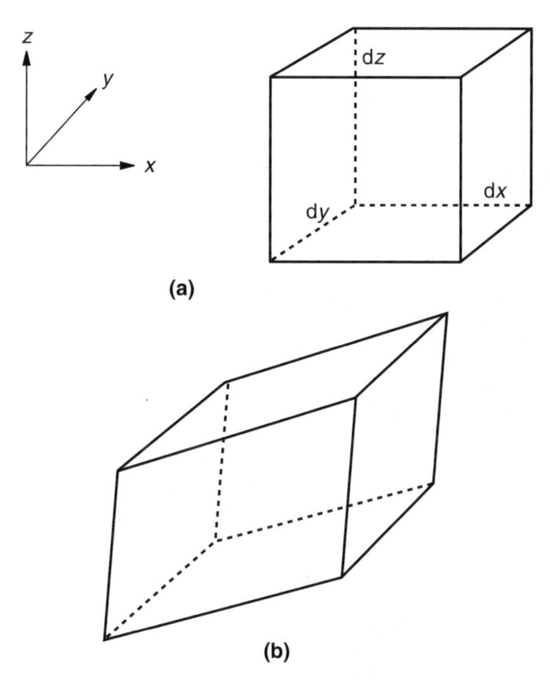

FIGURE 13.10 General strain state in a body. (a) Original body. (b) After deformation.

and σ_z and shear stresses τ_{xy}, τ_{yz}, and τ_{zx} act on the faces of the element. Since the stresses will normally vary throughout the body, the stresses have to satisfy certain equilibrium conditions. In addition, the stresses must satisfy equilibrium conditions with the surface traction (boundary conditions).

The general state of strain that the body undergoes as a result of deformation is illustrated in Fig. 13.10. The change in length of the sides are expressed by the normal strains ε_x, ε_y, ε_z. The shear strains γ_{xy}, γ_{yz}, γ_{zx} are given by the deviation of the angles from a right angle. Since changes in the strains cannot be arbitrary, they must satisfy strain–displacement relations.

Finally, stress–strain relations must be satisfied. These are related to the material properties, while the equilibrium conditions and strain–displacement relations are completely independent of the material properties, and must be satisfied whatever the properties. A problem of stress analysis may thus involve determining a set of stresses $\sigma_x, \sigma_y, \sigma_z, \tau_{xy}, \tau_{yz}, \tau_{zx}$ throughout the body which

1. satisfy the conditions of equilibrium in the interior;
2. satisfy the conditions of equilibrium with the surface tractions; and
3. lead through the stress–strain and strain–displacement relations to a set of continuous displacements u, v, and w.

Or the problem may involve determining a set of displacements which

1. are continuous throughout the body and satisfy any prescribed conditions of displacement at the surface;
2. lead through the stress–strain and strain–displacement relations to stresses that satisfy both equilibrium conditions in the interior of the body and boundary conditions on the surface.

The equilibrium conditions, strain–displacement relations, and stress–strain relations are briefly discussed in the following sections.

13.2.1 Equilibrium Conditions

The variation of stresses in the body are illustrated in Fig. 13.11. From this figure, the equilibrium conditions in the interior of the body can be obtained as

$$\frac{\partial \sigma_x}{\partial x} + \frac{\partial \tau_{xy}}{\partial y} + \frac{\partial \tau_{zx}}{\partial z} = 0 \tag{13.6a}$$

$$\frac{\partial \tau_{xy}}{\partial x} + \frac{\partial \sigma_y}{\partial y} + \frac{\partial \tau_{zy}}{\partial z} = 0 \tag{13.6b}$$

$$\frac{\partial \tau_{xz}}{\partial x} + \frac{\partial \tau_{yz}}{\partial y} + \frac{\partial \sigma_z}{\partial z} = 0 \tag{13.6c}$$

These assume that there are no body forces present. The equilibrium conditions to be satisfied at points on the surface of the body may be written as

$$\sigma_x \cos(n, x) + \tau_{yx} \cos(n, y) + \tau_{zx} \cos(n, z) = P_x \tag{13.7a}$$

$$\tau_{xy} \cos(n, x) + \sigma_y \cos(n, y) + \tau_{zy} \cos(n, z) = P_y \tag{13.7b}$$

$$\tau_{xz} \cos(n, x) + \tau_{yz} \cos(n, y) + \sigma_z \cos(n, z) = P_z \tag{13.7c}$$

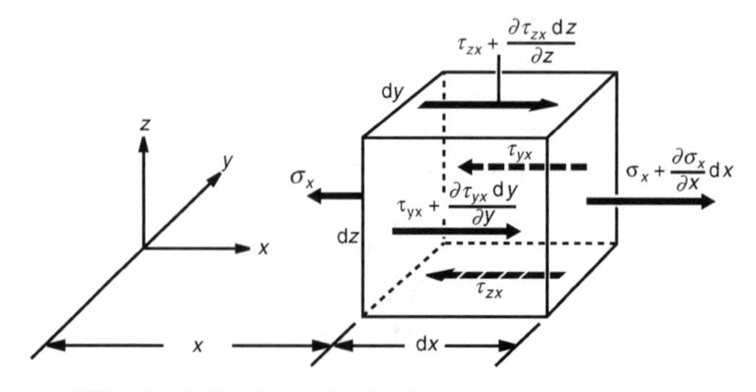

FIGURE 13.11 Stress distribution on an element in a body.

where P_x, P_y, P_z are the surface tractions *per unit area* (N/m^2), $\cos(n, x)$, $\cos(n, y)$, $\cos(n, z)$ are the direction cosines of the outward directed normal to the surface at the point of interest.

13.2.2 Strain–Displacement Relations

If the strains and displacements are small enough, then the relationships between them are given (Fig. 13.12) by

$$\varepsilon_x = \frac{\partial u}{\partial x} \tag{13.8a}$$

$$\varepsilon_y = \frac{\partial v}{\partial y} \tag{13.8b}$$

$$\varepsilon_z = \frac{\partial w}{\partial z} \tag{13.8c}$$

$$\gamma_{xy} = \frac{\partial v}{\partial x} + \frac{\partial u}{\partial y} \tag{13.8d}$$

$$\gamma_{yz} = \frac{\partial w}{\partial y} + \frac{\partial v}{\partial z} \tag{13.8e}$$

$$\gamma_{zx} = \frac{\partial u}{\partial z} + \frac{\partial w}{\partial x} \tag{13.8f}$$

where u, v, w are displacements in the x, y, z directions, respectively.

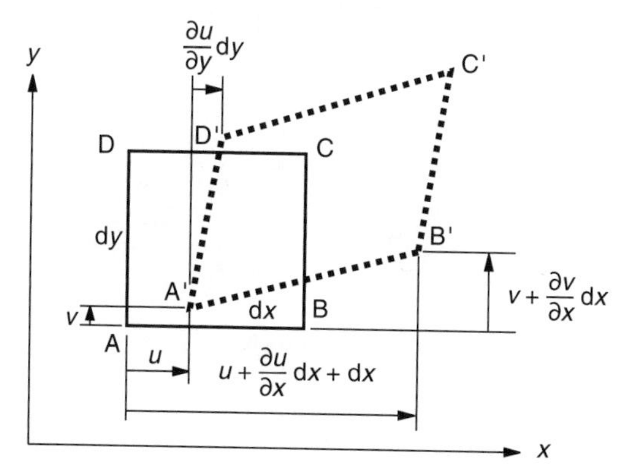

FIGURE 13.12 Displacements associated with an element in a body after deformation.

13.2.3 Stress–Strain Relations

Due to space constraints, we shall limit our discussion in this section to isotropic materials, and also to linear elastic behavior and plastic flow of metals.

13.2.3.1 Linear Elastic Behavior The stress–strain relations for linear elastic materials are given by Hooke's law:

$$\varepsilon_x = \frac{1}{E_\mathrm{m}}\left[\sigma_x - \nu_\mathrm{p}(\sigma_y + \sigma_z)\right] \tag{13.9a}$$

$$\varepsilon_y = \frac{1}{E_\mathrm{m}}\left[\sigma_y - \nu_\mathrm{p}(\sigma_z + \sigma_x)\right] \tag{13.9b}$$

$$\varepsilon_z = \frac{1}{E_\mathrm{m}}\left[\sigma_z - \nu_\mathrm{p}(\sigma_x + \sigma_y)\right] \tag{13.9c}$$

$$\gamma_{xy} = \frac{\tau_{xy}}{G} \tag{13.9d}$$

$$\gamma_{yz} = \frac{\tau_{yz}}{G} \tag{13.9e}$$

$$\gamma_{zx} = \frac{\tau_{zx}}{G} \tag{13.9f}$$

where E_m is Young's modulus (N/m^2), G is the shear modulus (N/m^2), and ν_p is Poisson's ratio (−).

Equations (13.9) may also be written as

$$\varepsilon_x = \frac{1}{3K}\left(\frac{\sigma_x + \sigma_y + \sigma_z}{3}\right) + \frac{1}{3G}\left[\sigma_x - \frac{1}{2}(\sigma_y + \sigma_z)\right] \tag{13.10a}$$

$$\varepsilon_y = \frac{1}{3K}\left(\frac{\sigma_x + \sigma_y + \sigma_z}{3}\right) + \frac{1}{3G}\left[\sigma_y - \frac{1}{2}(\sigma_z + \sigma_x)\right] \tag{13.10b}$$

$$\varepsilon_z = \frac{1}{3K}\left(\frac{\sigma_x + \sigma_y + \sigma_z}{3}\right) + \frac{1}{3G}\left[\sigma_z - \frac{1}{2}(\sigma_x + \sigma_y)\right] \tag{13.10c}$$

$$\gamma_{xy} = \frac{\tau_{xy}}{G} \tag{13.10d}$$

$$\gamma_{yz} = \frac{\tau_{yz}}{G} \tag{13.10e}$$

$$\gamma_{zx} = \frac{\tau_{zx}}{G} \tag{13.10f}$$

where K is the bulk modulus for the material (N/m^2).

E_m, G, K, and ν_p are related by

$$E_\mathrm{m} = 2G(1 + \nu_\mathrm{p})$$

$$E_\mathrm{m} = 3K(1 - 2\nu_\mathrm{p})$$

The first terms on the right-hand side of equations (13.10a)–(13.10c) are associated with the change in volume of the element. This becomes evident when the three equations are added to give the volume change:

$$\varepsilon_x + \varepsilon_y + \varepsilon_z = \frac{1}{K}\left(\frac{\sigma_x + \sigma_y + \sigma_z}{3}\right)$$

The mean of the three normal stresses $\left(\frac{\sigma_x+\sigma_y+\sigma_z}{3}\right)$ is referred to as the hydrostatic component of the state of stress. The second terms in the expressions are associated with the change in shape.

13.2.3.2 Plastic Flow of Metals In uniaxial loading such as occurs during a simple tensile test, plastic flow or permanent deformation is known to occur when the yield stress is exceeded. For a body that is subjected to a more general multiaxial stress state, a yield criterion is used to determine when plastic flow occurs. One that is commonly used is the von Mises yield criterion, which in simple terms states that

$$(\sigma_1 - \sigma_2)^2 + (\sigma_2 - \sigma_3)^2 + (\sigma_3 - \sigma_1)^2 = C_k \quad (13.11)$$

where σ_1, σ_2, and σ_3 are the principal stresses acting on any element and C_k is a constant (Pa^2).

In a uniaxial tensile test, σ_1 is positive and $\sigma_2 = \sigma_3 = 0$. Furthermore, yielding occurs when $\sigma_1 = S_y$, where S_y is the yield strength of the material. Thus, from equation (13.11), $C_k = 2{S_y}^2$, giving

$$(\sigma_1 - \sigma_2)^2 + (\sigma_2 - \sigma_3)^2 + (\sigma_3 - \sigma_1)^2 = 2{S_y}^2 \quad (13.12a)$$

or in more general terms,

$$(\sigma_x - \sigma_y)^2 + (\sigma_y - \sigma_z)^2 + (\sigma_z - \sigma_x)^2 + 6({\tau_{xy}}^2 + {\tau_{yz}}^2 + {\tau_{zx}}^2) = 2{S_y}^2 \quad (13.12b)$$

To facilitate analysis, an effective stress, $\bar{\sigma}$, is defined such that it has the same effect as the multiaxial stress state. Yielding or plastic flow occurs when its value reaches the yield strength, S_y. Thus, considering the von Mises yield criterion,

$$\bar{\sigma} = \frac{1}{\sqrt{2}}\left[(\sigma_1 - \sigma_2)^2 + (\sigma_2 - \sigma_3)^2 + (\sigma_3 - \sigma_1)^2\right]^{1/2} \quad (13.13)$$

or

$$\bar{\sigma} = \frac{1}{\sqrt{2}}\left[(\sigma_x - \sigma_y)^2 + (\sigma_y - \sigma_z)^2 + (\sigma_z - \sigma_x)^2 + 6({\tau_{xy}}^2 + {\tau_{yz}}^2 + {\tau_{zx}}^2)\right]^{1/2} \quad (13.14)$$

Yielding thus occurs when

$$\bar{\sigma} = S_y \quad (13.15)$$

The relationship between the plastic strains and the associated stresses that represent plastic flow of metals reasonably well is given by the Levy–Mises equations:

$$d\varepsilon_x{}^{\mathrm{p}} = \frac{d\bar{\varepsilon}^{\mathrm{p}}}{\bar{\sigma}}\left[\sigma_x - \frac{1}{2}(\sigma_y + \sigma_z)\right] \tag{13.16a}$$

$$d\varepsilon_y{}^{\mathrm{p}} = \frac{d\bar{\varepsilon}^{\mathrm{p}}}{\bar{\sigma}}\left[\sigma_y - \frac{1}{2}(\sigma_z + \sigma_x)\right] \tag{13.16b}$$

$$d\varepsilon_z{}^{\mathrm{p}} = \frac{d\bar{\varepsilon}^{\mathrm{p}}}{\bar{\sigma}}\left[\sigma_z - \frac{1}{2}(\sigma_x + \sigma_y)\right] \tag{13.16c}$$

$$d\gamma_{xy}{}^{\mathrm{p}} = \frac{d\bar{\varepsilon}^{\mathrm{p}}}{\bar{\sigma}} \cdot 3\tau_{xy} \tag{13.16d}$$

$$d\gamma_{yz}{}^{\mathrm{p}} = \frac{d\bar{\varepsilon}^{\mathrm{p}}}{\bar{\sigma}} \cdot 3\tau_{yz} \tag{13.16e}$$

$$d\gamma_{zx}{}^{\mathrm{p}} = \frac{d\bar{\varepsilon}^{\mathrm{p}}}{\bar{\sigma}} \cdot 3\tau_{zx} \tag{13.16f}$$

where $d\varepsilon_i{}^{\mathrm{p}}$, $d\gamma_{ij}{}^{\mathrm{p}}$ indicate the plastic strain increments.

$d\bar{\varepsilon}^{\mathrm{p}}$ refers to an increment of effective plastic strain that is defined by

$$d\bar{\varepsilon}^{\mathrm{p}} = \left[\frac{2}{9}\left\{(d\varepsilon_x{}^{\mathrm{p}} - d\varepsilon_y{}^{\mathrm{p}})^2 + (d\varepsilon_y{}^{\mathrm{p}} - d\varepsilon_z{}^{\mathrm{p}})^2 + (d\varepsilon_z{}^{\mathrm{p}} - d\varepsilon_x{}^{\mathrm{p}})^2\right\} + \frac{1}{3}\left\{(d\gamma_{xy}{}^{\mathrm{p}})^2 + (d\gamma_{yz}{}^{\mathrm{p}})^2 + (d\gamma_{zx}{}^{\mathrm{p}})^2\right\}\right]^{1/2} \tag{13.17}$$

In linear elasticity, the current state of stress completely defines the state of strain. However, the current state of stress cannot define the plastic strains measured from the virgin (e.g., annealed) condition of the material. Any plastic deformation done before the current stresses were applied would have to be known before the total plastic strains could be defined. The value of $\bar{\sigma}$ necessary to maintain plastic flow thus depends on the sum of the effective plastic strain increments accrued over the entire load path from the virgin condition of the material, that is $\bar{\sigma}$ is a function of $\int d\bar{\varepsilon_{\mathrm{p}}}$. When $\bar{\sigma}$ for a general state of stress is increased to $\bar{\sigma} + d\bar{\sigma}$, the corresponding increase in effective plastic strain $d\bar{\varepsilon_{\mathrm{p}}}$ is read directly from the stress–strain curve and used with $\bar{\sigma}$ and the stresses in equation (13.16) to compute the plastic strain increments.

13.2.3.3 Elastic–Plastic Conditions For a body that is subjected to plastic flow either wholly or in part, any change in the applied loading will result in a change in stress throughout the body. This will produce a change in both elastic and plastic strains. For changes in stress from σ_x to $\sigma_x + d\sigma_x$, and so on the corresponding increments of total strain $d\varepsilon_x$, and so on (elastic increment + plastic increment) are given by the Prandtl–Reuss equations:

$$d\varepsilon_x = \frac{1}{E_m}\left[d\sigma_x - \nu_p(d\sigma_y + d\sigma_z)\right] + \frac{d\bar{\varepsilon}^p}{\bar{\sigma}}\left[\sigma_x - \frac{1}{2}(\sigma_y + \sigma_z)\right] \quad (13.18a)$$

$$d\varepsilon_y = \frac{1}{E_m}\left[d\sigma_y - \nu_p(d\sigma_z + d\sigma_x)\right] + \frac{d\bar{\varepsilon}^p}{\bar{\sigma}}\left[\sigma_y - \frac{1}{2}(\sigma_z + \sigma_x)\right] \quad (13.18b)$$

$$d\varepsilon_z = \frac{1}{E_m}\left[d\sigma_z - \nu_p(d\sigma_x + d\sigma_y)\right] + \frac{d\bar{\varepsilon}^p}{\bar{\sigma}}\left[\sigma_z - \frac{1}{2}(\sigma_x + \sigma_y)\right] \quad (13.18c)$$

$$d\gamma_{xy} = \frac{d\tau_{xy}}{G} + \frac{d\bar{\varepsilon}^p}{\bar{\sigma}} \cdot 3\tau_{xy} \quad (13.18d)$$

$$d\gamma_{yz} = \frac{d\tau_{yz}}{G} + \frac{d\bar{\varepsilon}^p}{\bar{\sigma}} \cdot 3\tau_{yz} \quad (13.18e)$$

$$d\gamma_{zx} = \frac{d\tau_{zx}}{G} + \frac{d\bar{\varepsilon}^p}{\bar{\sigma}} \cdot 3\tau_{zx} \quad (13.18f)$$

The first terms on the right-hand side of equations (13.18) are the elastic strain components, and the second terms are the plastic strain components.

13.2.4 Plane Stress and Plane Strain

The discussion in the preceding sections is based on a general three-dimensional analysis. To simplify subsequent discussions with respect to residual stresses, we outline the concepts of plane stress and plane strain, which are valid for a number of engineering applications.

13.2.4.1 Plane Stress An example of a body under plane stress conditions is a relatively thin plate that is subjected to a two-dimensional stress field. Consider a thin flat plate of uniform thickness that is loaded such that the applied loads are parallel to the plane of the plate and are uniformly distributed through its thickness (Fig. 13.13). The state of stress on the top and bottom free surfaces are such that

$$\sigma_z = \tau_{zx} = \tau_{zy} = 0 \quad (13.19)$$

Since the plate is assumed to be thin, this condition is considered to hold throughout the plate. Thus, the only stresses that exist within the plate are σ_x, σ_y, and τ_{xy}, and are functions of x and y alone. These are the assumptions of generalized plane stress.

13.2.4.2 Plane Strain Let us now consider a prismatic cylinder that has the same cross-sectional area as the plate referred to in the preceding section, and let it be subjected to the same uniform loading per unit length in the z-direction (Fig. 13.14). However, its thickness in the z-direction is much greater. It is assumed that

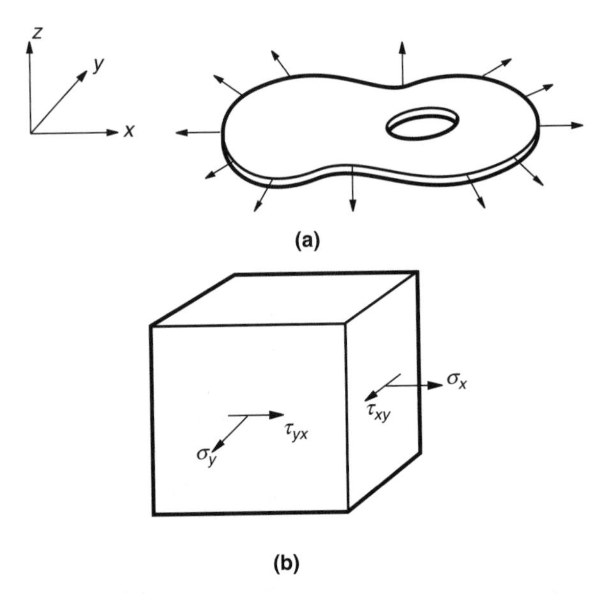

FIGURE 13.13 Plane stress state in a body. (a) Thin plate subjected to external loading parallel to the plane of the plate. (b) State of plane stress in the plate.

1. Normal cross sections that were plane before loading remain plane after loading, that is, do not distort in the z-direction. In other words, the displacement w in the z-direction is a linear function of z alone.
2. All normal cross sections well away from the ends undergo the same deformation in the x–y plane. In other words, the displacements u and v are functions of x and y alone.

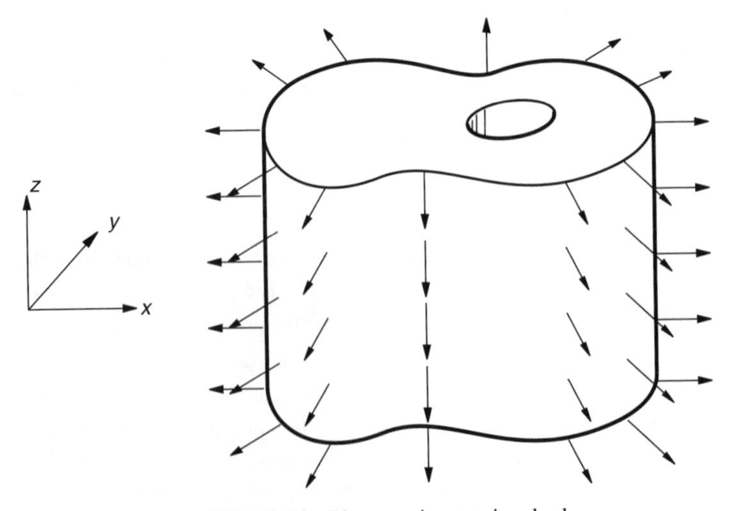

FIGURE 13.14 Plane strain state in a body.

These are the fundamental assumptions of plane strain, and lead to the following equations:

$$\varepsilon_z = \frac{\partial w(z)}{\partial z} = \text{constant at any } z \tag{13.20a}$$

$$\gamma_{yz} = \frac{\partial w(z)}{\partial y} + \frac{\partial v(x, y)}{\partial z} = 0 \tag{13.20b}$$

$$\gamma_{zx} = \frac{\partial u(x, y)}{\partial z} + \frac{\partial w(z)}{\partial x} = 0 \tag{13.20c}$$

Thus, it can be shown that all the stresses are functions of x and y alone, and that

$$\tau_{zx} = \tau_{zy} = 0 \tag{13.20d}$$

13.2.4.3 Plane Stress/Plane Strain Equations Using the simplifications in the preceding sections, the equilibrium equations for both plane stress and plane strain (obtained from equations (13.6a)–(13.6c)) reduce to

$$\frac{\partial \sigma_x}{\partial x} + \frac{\partial \tau_{xy}}{\partial y} = 0 \tag{13.21a}$$

$$\frac{\partial \tau_{xy}}{\partial x} + \frac{\partial \sigma_y}{\partial y} = 0 \tag{13.21b}$$

And the strain–displacement equations (obtained from equations (13.8a)–(13.8f)) become

$$\varepsilon_x = \frac{\partial u}{\partial x} \tag{13.22a}$$

$$\varepsilon_y = \frac{\partial v}{\partial y} \tag{13.22b}$$

$$\gamma_{xy} = \frac{\partial v}{\partial x} + \frac{\partial u}{\partial y} \tag{13.22c}$$

13.2.4.4 Compatibility Equation The three strain components in equations (13.22), ε_x, ε_y, and γ_{xy}, are functions of only two variables, u and v, and are therefore not independent. The compatibility equation that the total strains must satisfy is given by

$$\frac{\partial^2 \varepsilon_x}{\partial y^2} + \frac{\partial^2 \varepsilon_y}{\partial x^2} = \frac{\partial^2 \gamma_{xy}}{\partial x \partial y} \tag{13.23}$$

Residual stresses exist when the nonelastic (plastic, thermal, etc.) strain components do not satisfy equation (13.23). Thus, an uneven distribution of plastic strains in a material results in residual stresses.

13.2.4.5 Stress–Strain Relations for Plane Stress/Plane Strain Here, we refocus our attention on elastic deformation. The stress–strain relations for generalized plane stress and plane strain (obtained from equations (13.9a)–(13.9f)) are

Plane Stress:

$$2G\varepsilon_x = \sigma_x - \frac{\nu_p}{1+\nu_p}(\sigma_x + \sigma_y) \tag{13.24a}$$

$$2G\varepsilon_y = \sigma_y - \frac{\nu_p}{1+\nu_p}(\sigma_x + \sigma_y) \tag{13.24b}$$

$$G\gamma_{xy} = \tau_{xy} \tag{13.24c}$$

Plane Strain:

$$2G\varepsilon_x = \sigma_x - \nu_p(\sigma_x + \sigma_y) \tag{13.25a}$$

$$2G\varepsilon_y = \sigma_y - \nu_p(\sigma_x + \sigma_y) \tag{13.25b}$$

$$\sigma_z = E_m\varepsilon_z + \nu_p(\sigma_x + \sigma_y) \tag{13.25c}$$

$$G\gamma_{xy} = \tau_{xy} \tag{13.25d}$$

From equations (13.24) and (13.25), it is obvious that a solution for stresses σ_x and σ_y under plane stress conditions can be applied to plane strain if ν_p is replaced by $\nu_p/(1-\nu_p)$.

13.2.5 Solution Methods

Due to space constraints, we only consider the case of plane stress. For this case, the problem involves determining σ_x, σ_y, and τ_{xy}, which satisfy the necessary field equations and boundary conditions. We start by obtaining the stress compatibility equation:

$$\frac{\partial^2}{\partial y^2}(\sigma_x - \nu_p\sigma_y) + \frac{\partial^2}{\partial x^2}(\sigma_y - \nu_p\sigma_x) - 2(1+\nu_p)\frac{\partial^2\tau_{xy}}{\partial x\partial y} = 0 \tag{13.26}$$

This reduces the problem to that of three field equations: (13.21a), (13.21b), and (13.26), which must be satisfied by the three stresses: σ_x, σ_y, and τ_{xy}.

Now let us represent the stresses by a stress function ψ such that

$$\sigma_x = \frac{\partial^2\psi}{\partial y^2} \tag{13.27a}$$

$$\sigma_y = \frac{\partial^2\psi}{\partial x^2} \tag{13.27b}$$

$$\tau_{xy} = -\frac{\partial^2\psi}{\partial x\partial y} \tag{13.27c}$$

Substituting equations (13.27) into (13.21) shows that the equilibrium equations (13.21) are satisfied. Further substituting into the stress compatibility equation (13.26) gives

$$\frac{\partial^4 \psi}{\partial x^4} + 2\frac{\partial^4 \psi}{\partial x^2 \partial y^2} + \frac{\partial^4 \psi}{\partial y^4} = 0 \tag{13.28}$$

which can also be written as

$$\nabla^4 \psi = 0 \tag{13.29}$$

where

$$\nabla^2 = \frac{\partial^2}{\partial x^2} + \frac{\partial^2}{\partial y^2} \tag{13.30}$$

Equation (13.28) is also valid for plane strain conditions. ψ is known as Airy's stress function. If ψ is obtained that satisfies equation (13.28), then the equilibrium and compatibility equations are automatically satisfied.

Thus, for a two-dimensional elastic problem where body forces can be ignored, we only need to obtain a solution for equation (13.28) that satisfies the appropriate boundary conditions. Quite often, an inverse approach is used by selecting a stress function and determining the problem that it solves. Numerical methods are normally used when the boundary conditions are complex.

Example 13.1 Residual stresses obtained by measuring elastic strains recovered after a part is cut are calculated for a plane stress, using Hooke's law as follows:

$$\sigma_x = -\frac{E_{\mathrm{m}}}{1 - \nu_{\mathrm{p}}{}^2}[\varepsilon_x{}^{\mathrm{e}} + \varepsilon_y{}^{\mathrm{e}}]$$

What is the reason for the negative sign in the expression?

Solution:

Because a region that is under tensile residual stress will contract when cut since the tensile stresses arise because that region is prevented from contracting.

13.3 EFFECTS OF RESIDUAL STRESSES

The two primary effects of residual stresses are as follows:

1. Apparent change in the strength of a part.
2. Distortion of the part.

13.3.1 Apparent Change in Strength

Consider a relatively large diameter steel rod that is heated into the austenite range for a long enough time for the temperature to become uniform throughout. If it is then quenched, tensile residual stresses will be developed on the surface (Fig. 13.15a) as a result of volume change associated with the austenite to martensite (density of austenite is higher than that of martensite) and other phase transformations that occur during quenching. Since tensile residual stresses then already exist in the rod after heat treatment, if it is subsequently loaded in tension, it will yield at applied stresses that are lower than the yield strength of the material. This is because the total stress on the rod surface will be the sum of the existing residual stresses and the applied stress (Fig. 13.15b). In essence, in the elastic range, the residual stress can be added to the stress due to the applied external load. However, in the plastic region, the residual stress relaxes. This may result in a more uniform redistribution of the stresses (Fig. 13.16).

If a glass plate is uniformly heated and the surfaces subsequently air-cooled, compressive residual stresses will be developed close to the surfaces (Fig. 13.17) resulting in tempered glass. In this case, there is no phase transformation and thus no volume change during cooling. The mechanism is therefore similar to that discussed in Section 13.1.1. Such a plate can sustain higher bending moments since the tensile stresses on the outer surface that result from bending will be reduced by the existing compressive residual stresses.

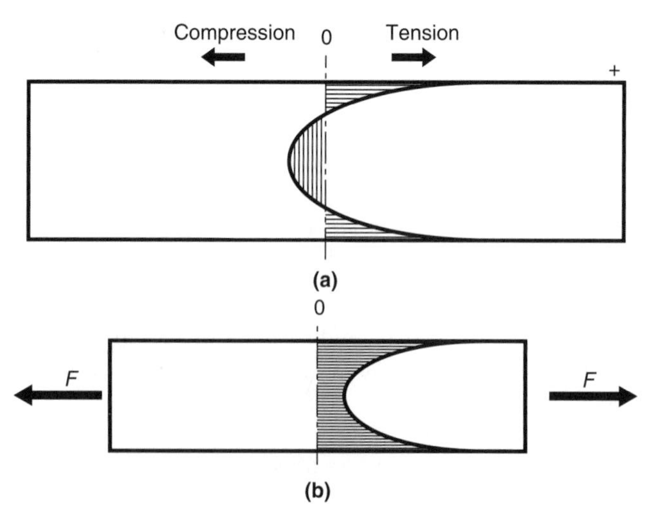

FIGURE 13.15 Residual stresses developed in a steel rod after quenching. (a) Possible residual stress distribution. (b) Stress state in the quenched bar after a tensile external load is applied.

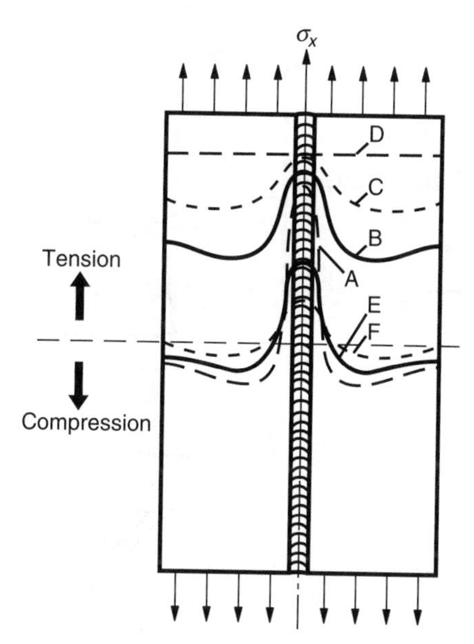

FIGURE 13.16 Redistribution of residual stresses in a weldment after loading into the plastic range. (Copyright American Welding Society, *Welding Handbook*, Vol. 1, 9th edition, 2001. Used with permission.)

13.3.2 Distortion

When a material is subjected to heating and cooling during processing, thermal strains occur in the vicinity of the heat source. The stresses that result from these strains combine and react to produce internal forces that cause distortion, which is often evident as bending, buckling, and rotation (Fig. 13.18). Principal changes that result in distortion of a plate, for example, include the following:

1. Transverse shrinkage perpendicular to the line along which the heat source moves (Fig. 13.18a).

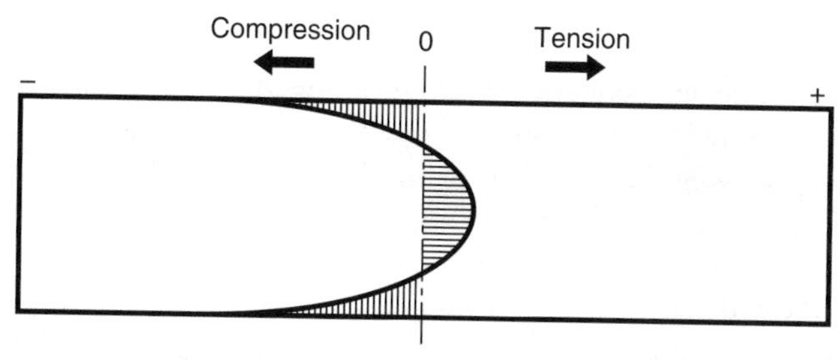

FIGURE 13.17 Residual stresses developed in a glass plate after uniform heating and subsequent air cooling of the surfaces.

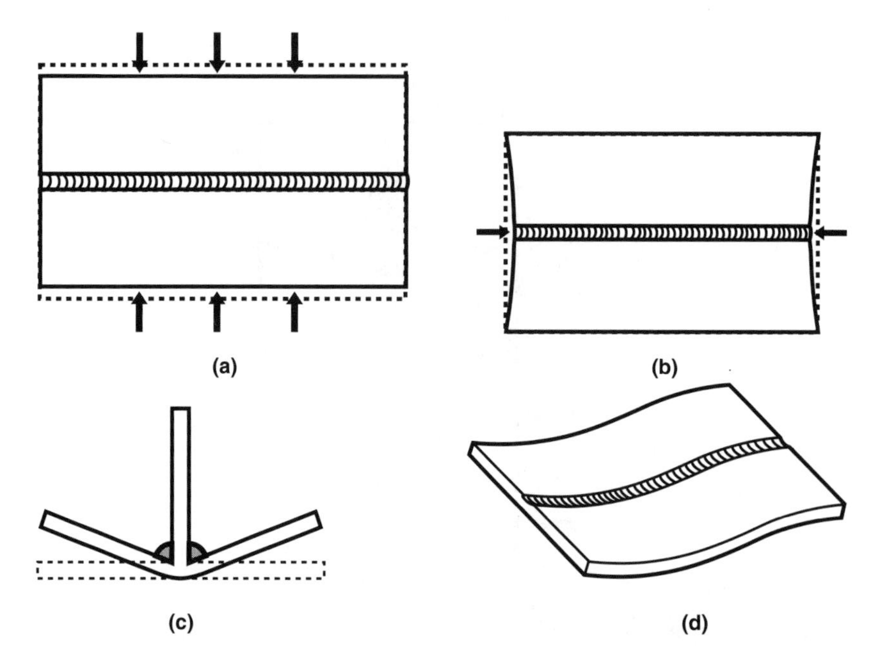

FIGURE 13.18 Various modes of distortion in weldments. (a) Transverse shrinkage. (b) Longitudinal shrinkage. (c) Angular change. (d) Buckling distortion. (From Masubuchi, K., 1980, *Analysis of Welded Structures*, Pergamon Press, Oxford, UK.)

2. Longitudinal shrinkage parallel to the heat source line of motion (Fig. 13.18b).
3. Angular rotation about the heat source line of motion, (Fig. 13.18c). This is due primarily to nonuniformity of transverse shrinkage in the z-direction.
4. Buckling distortion (Fig. 13.18d). Thin plates may be subjected to buckling instability when thermal compressive stresses are induced in them.

Actual distortion of structures tends to be more complex. Analysis of distortion is a coupled problem that involves three basic issues:

1. Analysis of the temperature field. This is essentially a three-dimensional conduction heat transfer problem involving a moving heat source.
2. Analysis of the plastic strains using the appropriate constitutive laws, incorporating temperature-dependent thermal expansion coefficient, elastic modulus, and flow stress. This would also involve a suitable yield criterion and restraint due to the geometry of the structure.
3. Analysis of the distortion that results from the plastic strains. This involves consideration of force and moment equilibrium, strain hardening, and flow stress.

In Section 13.2, we presented the basic concepts necessary for analyzing the plastic strains that form the basis for distortion. The final solution that is obtained is determined, to a large extent, by the degree of restraint, which is structure-specific, and can only be discussed on a case by case basis. We now discuss some of the common methods used in measuring residual stresses.

13.4 MEASUREMENT OF RESIDUAL STRESSES

Various methods are available for measuring residual stresses. The major ones include

1. Stress relaxation.
2. X-ray diffraction.
3. Neutron diffraction.

Others are the ultrasonic and magnetic techniques. We shall limit our discussion to the three techniques listed. All three are based on the measurement of residual elastic strain. Residual stresses are then computed from the measured strains. The discussion on the measurement of residual stresses will be followed by a brief discussion on residual stress equilibrium, which can be used to check the validity of any measurements that are made.

13.4.1 Stress Relaxation Techniques

The stress relaxation techniques involve measurement of elastic strain recovered when the part is either cut into pieces or only parts are removed. Once the elastic strains are measured, the stresses that were acting under those strains can be calculated using Hooke's Law (equations (13.9), (13.24), or (13.25)) and considering the recovery or unloading process through the stress–strain curve (Fig. 13.5). Using equation (13.24) for the plane stress condition, we can express the residual stress in terms of the measured strains as

$$\sigma_x = -\frac{E_\mathrm{m}}{1-{\nu_\mathrm{p}}^2}({\varepsilon_x}^\mathrm{e} + \nu_\mathrm{p}{\varepsilon_y}^\mathrm{e}) \tag{13.31a}$$

$$\sigma_y = -\frac{E_\mathrm{m}}{1-{\nu_\mathrm{p}}^2}({\varepsilon_y}^\mathrm{e} + \nu_\mathrm{p}{\varepsilon_x}^\mathrm{e}) \tag{13.31b}$$

$$\tau_{xy} = -G\gamma_{xy} \tag{13.31c}$$

The negative signs indicate that a part of the material that is subjected to tensile residual stresses will contract when relaxed, that is residual stresses are removed. The strains are measured using strain gauges. The disadvantage with these techniques, however, is that they are destructive, that is the part must be cut to some extent.

However, they are very common and reliable methods. In this section, we shall only discuss two primary types of the stress relaxation technique:

1. The sectioning technique.
2. The drilling technique.

The sectioning and drilling techniques are relatively rapid and less destructive, compared to other techniques such as the layer removal technique.

13.4.1.1 Sectioning Technique This is also sometimes referred to as the ring core method. With this method, electrical resistance strain gages are mounted on the surface of the structure. A circular disc piece that surrounds the strain gages is then removed by trepanning (Fig. 13.19). The internal diameter of the piece removed may range between 15 and 150 mm, and the depth between 25 and 150% of its internal diameter. The change in strains after removal is measured. For a small enough disc, the removal may be assumed to have resulted in complete elastic recovery. Thus, the measured strains are the elastic strains. Residual stresses can then be calculated using Hooke's laws.

The sectioning technique is more sensitive than the drilling technique (see the next section) since it almost relieves the surface strains entirely. In addition, it is insensitive to small errors in the diameter or location of the annular hole relative to the strain gages. However, since the diameter is relatively large, the measured stresses are not very localized. Furthermore, more damage is done to the workpiece.

13.4.1.2 Drilling Technique The drilling technique is also known as the Mathar–Soete technique. In this case, strain gages are placed in a star or rectangular arrangement, and a small hole then drilled in the center of the arrangement (Fig. 13.20). The hole diameter typically ranges between 1 and 4 mm, and is drilled to a depth that is approximately equal to the diameter. Strain changes as measured by the three gages are then used to calculate the stresses.

This method is less sensitive, since the strains that are relieved decay quite rapidly with distance from the edge of the hole. Thus, the measured strains are only about 25–40% of the original residual strains at the hole location. Furthermore, the calculated stresses are significantly affected by errors in hole diameter or location. However, the hole drilling technique is more commonly used since the measured stresses are more localized, and less damage is done to the workpiece. In the next section, we discuss how the measured strains can be converted to strains in cartesian coordinates.

13.4.1.3 Strain Analysis Consider the general case where strain gauges A, B, C are mounted on the surface of a part at angles θ_A, θ_B, and θ_C with respect to the x-axis (Fig. 13.21).

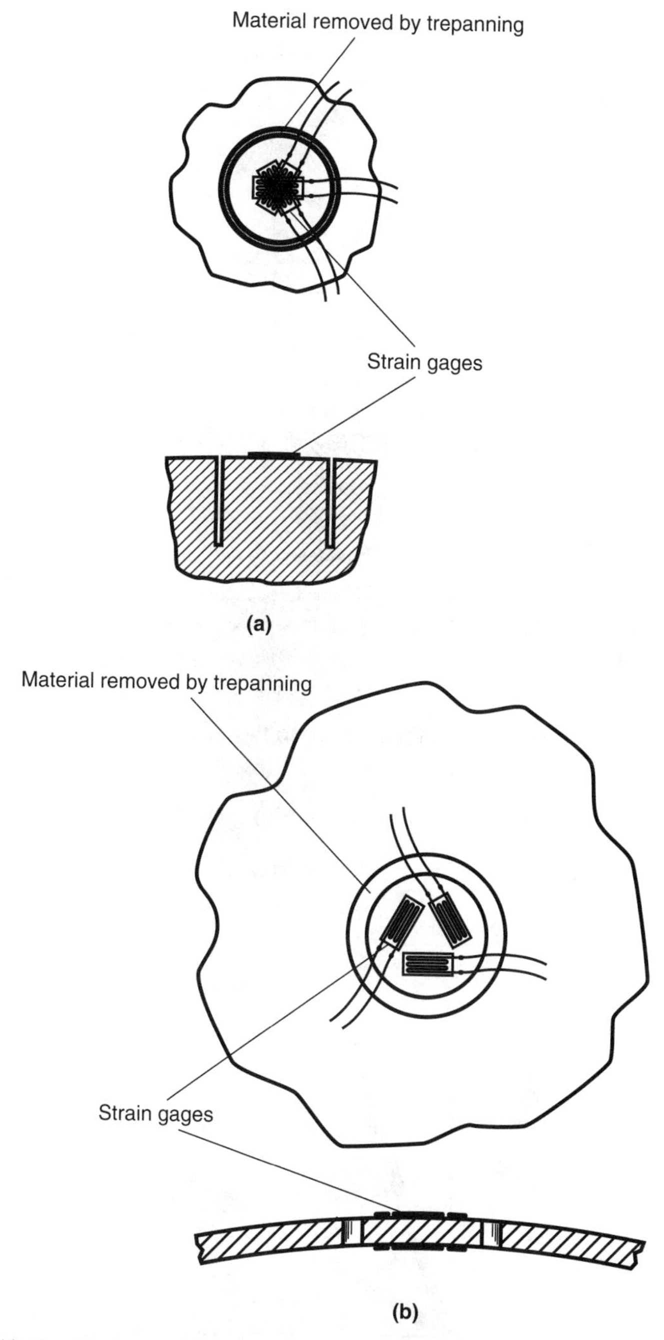

FIGURE 13.19 The sectioning method of residual stress measurement. (a) Thick plate. (b) Thin plate. (Reprinted from *Handbook of Measurement of Residual Stresses* by permission of Fairmont Press, Inc., Lilburn, GA (www.fairmontpress.com).)

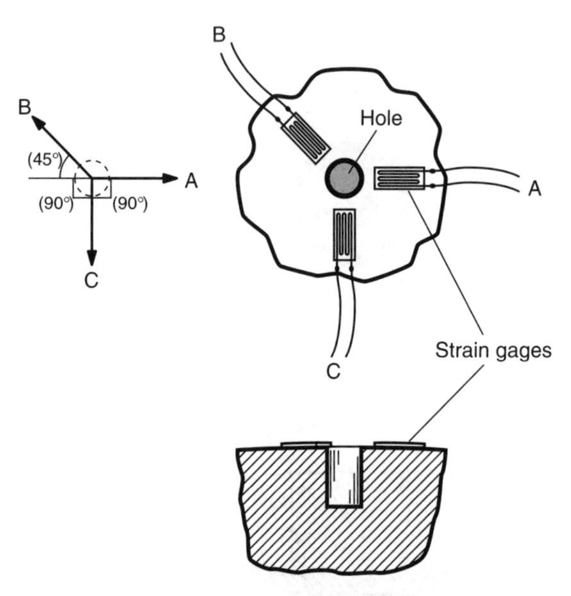

FIGURE 13.20 The drilling method of residual stress measurement. (Reprinted from *Handbook of Measurement of Residual Stresses* by permission of Fairmont Press, Inc., Lilburn, GA (www.fairmontpress.com).)

The strains ε_x, ε_y, and γ_{xy} are then related to those measured by the strain gages, ε_A, ε_B, and ε_C as

$$\varepsilon_A = \varepsilon_x \cos^2\theta_A + \varepsilon_y \sin^2\theta_A + \gamma_{xy} \sin\theta_A \cos\theta_A \tag{13.32a}$$

$$\varepsilon_B = \varepsilon_x \cos^2\theta_B + \varepsilon_y \sin^2\theta_B + \gamma_{xy} \sin\theta_B \cos\theta_B \tag{13.32b}$$

$$\varepsilon_C = \varepsilon_x \cos^2\theta_C + \varepsilon_y \sin^2\theta_C + \gamma_{xy} \sin\theta_C \cos\theta_C \tag{13.32c}$$

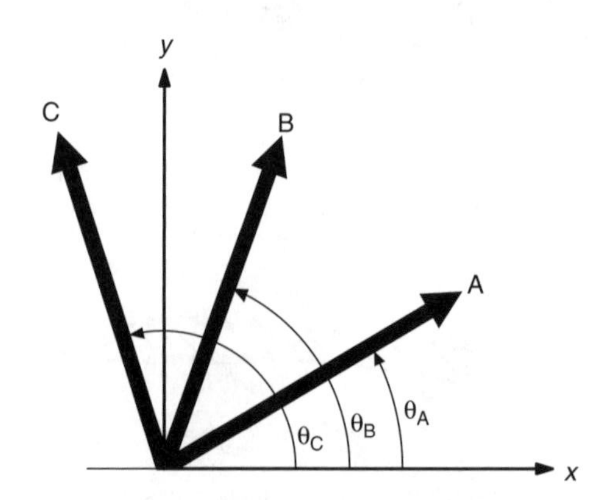

FIGURE 13.21 A general arrangement of strain gages.

ε_x, ε_y, and γ_{xy} can be calculated from equations (13.32), from which the principal strains, ε_1 and ε_2, are obtained as

$$\varepsilon_{1,2} = \frac{1}{2}(\varepsilon_x + \varepsilon_y) \pm \frac{1}{2}\sqrt{(\varepsilon_x - \varepsilon_y)^2 + {\gamma_{xy}}^2} \tag{13.33a}$$

$$\tan 2\phi = \frac{\gamma_{xy}}{\varepsilon_x - \varepsilon_y} \tag{13.33b}$$

where ϕ is the angle between the maximum principal strain ε_1 (or maximum principal stress σ_1 if the material is isotropic) and the x-axis.

Knowing the measured strains, the residual stresses can be estimated using equations (13.31).

Example 13.2 The residual stresses at a point in a thin titanium plate that has been formed using a laser are measured using a stress relaxation technique. The strain measurements obtained are shown in the figure below. Assuming Poisson's ratio to be 0.3, determine the maximum normal and shear stresses acting at that point.

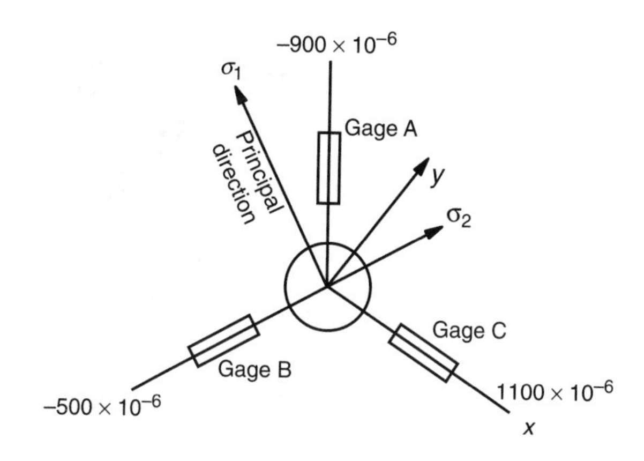

Solution:

From Appendix 13B, Young's modulus for titanium is 116 GPa. To simplify things, we set up a coordinate system such that the x-axis is along Gage C. Then $\theta_C = 0°$, $\theta_A = 120°$, $\theta_B = 240°$. Thus,

$$\varepsilon_C = \varepsilon_x = 1100 \times 10^{-6}$$

Now using equations (13.32), we have

$$\varepsilon_A = \frac{1}{4}\left(\varepsilon_x + 3\varepsilon_y - \sqrt{3}\gamma_{xy}\right)$$

$$\varepsilon_B = \frac{1}{4}\left(\varepsilon_x + 3\varepsilon_y + \sqrt{3}\gamma_{xy}\right)$$

Thus,

$$0.25 \times \left(1100 + 3\varepsilon_y - \sqrt{3}\gamma_{xy}\right) = -900$$

$$0.25 \times \left(1100 + 3\varepsilon_y + \sqrt{3}\gamma_{xy}\right) = -500$$

Giving

$$\varepsilon_y = -1300 \times 10^{-6}, \qquad \gamma_{xy} = 462 \times 10^{-6}$$

Now from equation (13.33a), we obtain the principal strains as

$$\varepsilon_{1,2} = \frac{1}{2}(1100 - 1300) \pm \frac{1}{2}\sqrt{(1100 + 1300)^2 + 462^2}$$
$$= \frac{1}{2}(-200 \pm 2444) = -1322 \text{ or } 1122 \times 10^{-6}$$

The principal stresses are then obtained using the equivalent of equations (13.31a) and (13.31b).

$$\sigma_1 = -\frac{E_m}{1 - \nu_p{}^2}(\varepsilon_1 + \nu_p\varepsilon_2)$$
$$= -\frac{116 \times 10^9}{1 - 0.3^2}(1122 - 0.3 \times 1322) \times 10^{-6}$$
$$= -92.5 \text{ MPa} \quad \text{(Minimum normal stress)}$$

$$\sigma_2 = -\frac{E_m}{1 - \nu_p{}^2}(\varepsilon_2 + \nu_p\varepsilon_1)$$
$$= -\frac{116 \times 10^9}{1 - 0.3^2}(-1322 + 0.3 \times 1122) \times 10^{-6}$$
$$= 125.6 \text{ MPa} \quad \text{(Maximum normal stress)}$$

And the maximum shear stress is

$$\tau_{\max} = \frac{\sigma_1 - \sigma_2}{2} = \frac{125.6 - (-92.5)}{2}$$

$$= 109.1 \text{ MPa} \quad \text{(Maximum shear stress)}.$$

13.4.2 X-Ray Diffraction Technique

The X-ray diffraction technique is primarily used to measure surface stresses, up to a depth of about 50 μm. It is mainly for materials with a crystalline structure. The unit cell of each crystalline structure has lattice constants (lengths of the sides of the unit cell) that are characteristic of that material. When the material is subjected to elastic strains, that results in changes in the lattice constants. Plastic strains do not affect them since they result from dislocation motion. Thus, by measuring the lattice constants of the stressed material, the corresponding elastic strains can be determined. The lattice constants of the unstressed material are either known or are easily determined.

Measurement of the lattice constants is commonly made by X-ray diffraction. Advantages of this method are as follows:

1. It is capable of measuring strains in areas as small as 3 μm in diameter.
2. It is nondestructive.
3. Even though earlier measurement systems were very slow, more recent systems can make measurements in a few seconds or less.

However, it has the following disadvantages:

1. It is limited to surface stresses, up to about 2 to 50 μm from the surface.
2. Its accuracy is relatively low, especially for heat-treated parts with a distorted structure.

13.4.2.1 Principle of the X-Ray Diffraction Technique This technique is based on Bragg's Law. A beam of monochromatic radiation (AB, DF) incident on parallel atomic planes will be reflected in the direction BC, FH as shown in Fig. 13.22. For the waves reflected from adjacent parallel atomic planes to reinforce each other, they must be out of phase by an integral (m) number of wavelengths. Thus, for diffraction, that is, reinforcement,

$$\text{EFG} = m\lambda \tag{13.34}$$

where λ = wavelength of incident beam (X-ray) (μm).

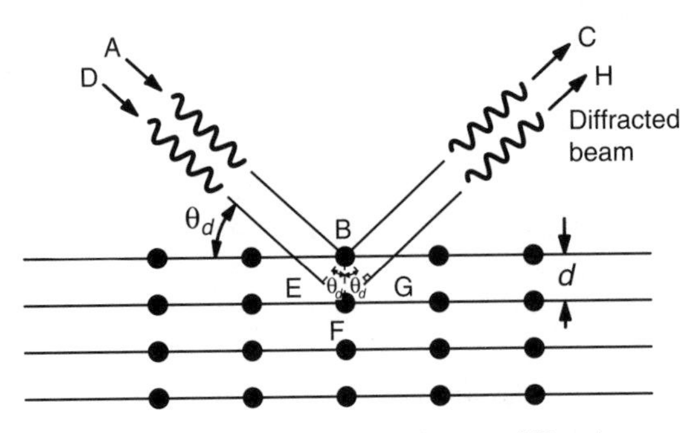

FIGURE 13.22 Principle of X-ray diffraction.

Or we can say that

$$\mathrm{EF} = m\frac{\lambda}{2} \tag{13.35}$$

However,

$$\mathrm{EF} = d \sin \theta_d \tag{13.36}$$

where d is the interplanar spacing and θ_d is the angle between incident beam and atomic plane.

Therefore,

$$m\frac{\lambda}{2} = d \sin \theta_d \tag{13.37}$$

or

$$m\lambda = 2d \sin \theta_d \tag{13.38}$$

Equation (13.38) is known as Bragg's Law. From it, the interplanar spacing d can be calculated if λ and θ_d are known. Determining d also determines the lattice constants, a_i.

For a metal with a cubic lattice such as a face-centered cubic or body-centered cubic structure, the lattice constants are the same along all crystal axes, that is, $a_x = a_y = a_z = a$. Then, we have

$$d = \frac{a}{\sqrt{h_\mathrm{m}{}^2 + k_\mathrm{m}{}^2 + l_\mathrm{m}{}^2}} \tag{13.39}$$

where $h_\mathrm{m}, k_\mathrm{m}, l_\mathrm{m}$ are the Miller indices of the reflecting plane.

$h_\mathrm{m}, k_\mathrm{m}, l_\mathrm{m}$ incorporate the integer m and give a measure of the intersection of the reflecting plane on the crystal axes. In other words, m in equation (13.38) can be taken as 1. In reality, the Miller indices are the reciprocals of the intercepts of the plane with the x-, y-, and z-axes.

Constructive interference (diffraction) is obtained for an X-ray beam incident on the surface of a crystalline material when the beam meets lattice planes that are oriented in such a way that Bragg's law is fulfilled. Generally, a polycrystalline material will have some grains oriented suitably enough to produce a diffracted beam.

Let the lattice spacing for a fine grained and stress-free polycrystalline specimen of a given material be d_o, and that for a stressed specimen of the same material be d. The elastic strain to which the specimen has been subjected can then be inferred from the change in crystal lattice as

$$\varepsilon = \frac{d - d_o}{d_o} = -(\theta_d - \theta_{do}) \cot \theta_{do} \tag{13.40}$$

where θ_d is the position of the diffraction peak corresponding to the extended lattice spacing d, θ_{do} is the position of the diffraction peak corresponding to the original lattice spacing d_o.

Two of the methods on which the X-ray diffraction technique is based are

1. The film technique.
2. The diffractometer technique.

13.4.2.2 The Film Technique With this method, the X-ray beam is passed through a collimator and then directed toward the specimen (Fig. 13.23). The reflected (or diffracted) beam is recorded using a film. When the film is developed, it shows near-

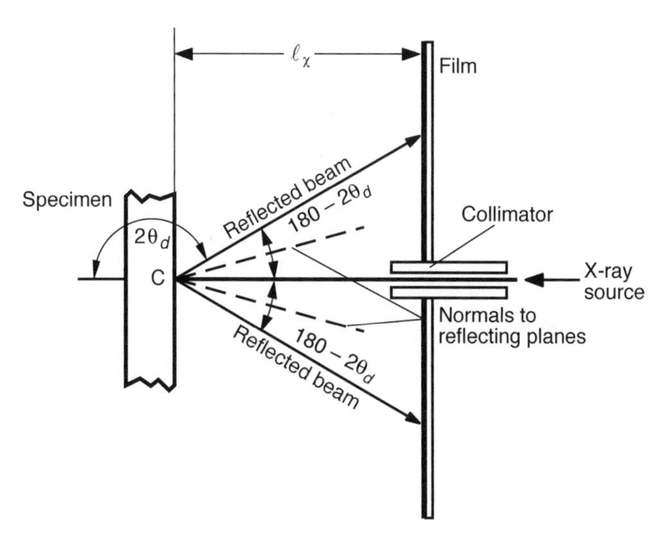

FIGURE 13.23 Schematic of the X-ray film technique. (From Masubuchi, K., 1980, *Analysis of Welded Structures*, Pergamon Press, Oxford, UK.)

circular rings, with each set of rings corresponding to an interplanar distance, d_{hkl}. For each ring, θ_d from Bragg's law can be obtained using the following relation:

$$\frac{H_0}{l_x} = 2\tan(180 - 2\theta_d) \tag{13.41}$$

where l_x is the distance from the film to the specimen, H_0 is the diameter of the diffraction ring.

The strain is then obtained using equation (13.40). The film technique, though accurate, is relatively slow, and thus not extensively used anymore.

13.4.2.3 The Diffractometer Technique In this case, the diffraction angle is measured by moving a counter and receiving slit along a goniometer circle to measure the reflected beam intensity (Fig. 13.24). The position of maximum beam intensity gives the diffraction angle. The major drawback of the diffractometer technique is that the specimen size is limited by equipment geometry. However, it is more accurate and quicker than the film technique. A measurement can be made in a few seconds or less. Thus, it is more widely used.

13.4.3 Neutron Diffraction Technique

The neutron diffraction technique is similar in principle to the X-ray diffraction technique. It is also based on the measurement of lattice strains. The main difference is that due to the ability of neutrons to penetrate more deeply into materials, neutron diffraction enables residual stresses to be measured much deeper into the material, that is, within the volume of the material. Even though the neutron beam may penetrate the entire volume, slits are used to define the local volume (Fig. 13.25).

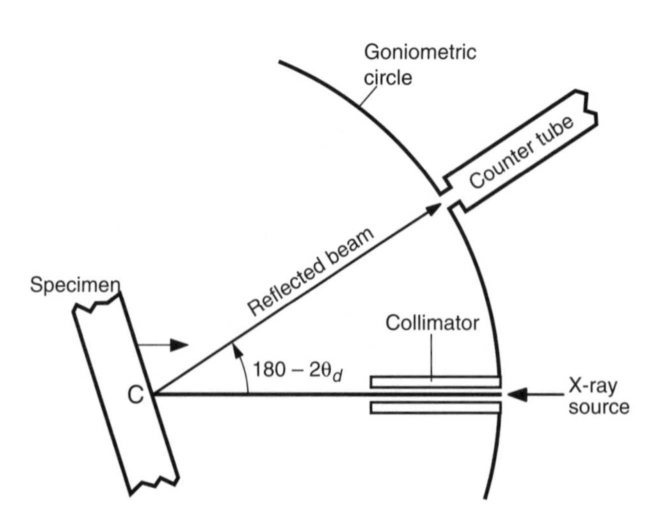

FIGURE 13.24 Schematic of the diffractometer technique. (From Masubuchi, K., 1980, *Analysis of Welded Structures*, Pergamon Press, Oxford, UK.)

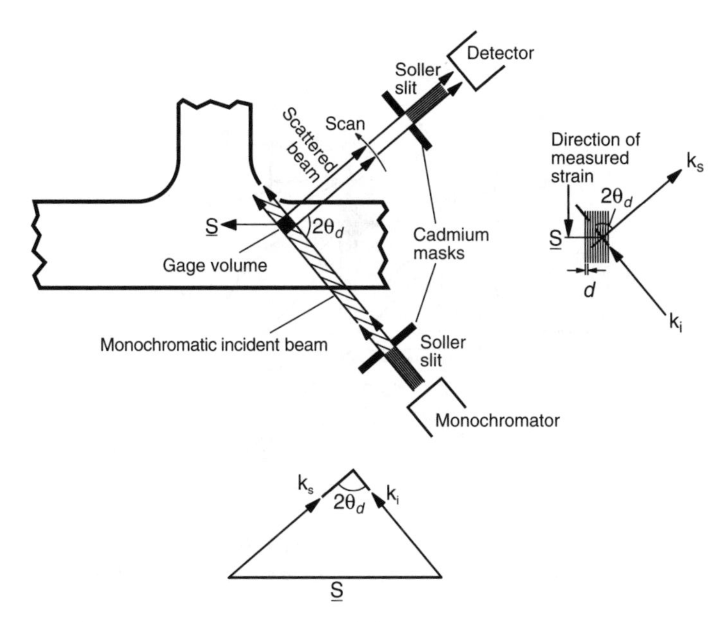

FIGURE 13.25 Principle of the neutron diffraction measurement technique. (From *Measurement of Residual and Applied Stress Using Neutron Diffraction*, 1992, Kluwer Academic Publishers, Dordrecht, Netherlands.)

With this technique, a neutron beam from a reactor is first converted into a monochromatic beam of wavelength λ, using, for example, a germanium crystal. A specific direction is imparted to the beam using a soller slit collimator. The cross-sectional area of the beam as it enters the sample and also as it enters the detector is determined by the aperture of neutron absorbing cadmium masks (Fig. 13.25) that are placed before and after the sample. The intersection of the incident and diffracted beams is referred to as the gage volume. Measurement is obtained of the average strain in grains that are properly oriented in a sample that is positioned completely within this volume. Larger samples may be moved through the gage volume to obtain the strain at different locations within it. The diffraction rings that result correspond to lattice planes that are oriented at an angle θ_d to the incident beam. These are also the planes whose normals lie along the scattering vector. The scattering vector, **s**, bisects the included angle between the incident and scattered beams. It is defined by the wavevectors of the incident and scattered beams:

$$\mathbf{s} = \mathbf{k_i} - \mathbf{k_s} \tag{13.42}$$

where $\mathbf{k_i}$ is the wavevector of the incident beam and $\mathbf{k_s}$ is the wavevector of the scattered beam.

When θ_d is obtained, the strain is then determined using equation (13.40). Measurements are usually made using either a crystal spectrometer or a time-of-flight diffractometer. A typical diffractometer is illustrated in Fig. 13.26.

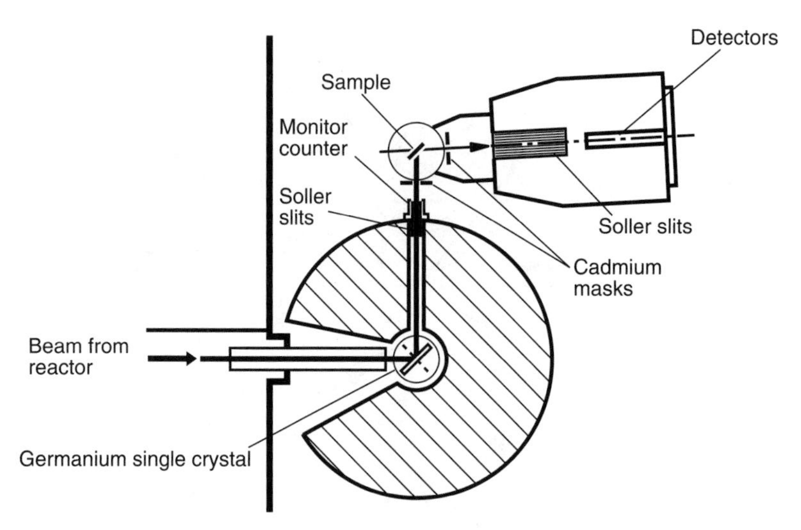

FIGURE 13.26 Schematic of a neutron diffractometer. (From *Measurement of Residual and Applied Stress Using Neutron Diffraction*, 1992, Kluwer Academic Publishers, Dordrecht, Netherlands.)

13.4.4 Residual Stress Equilibrium

Since residual stresses exist within a body even when there are no external loads acting on the body, the resultant force and moment of the residual stresses in a body must be zero. Thus, on any plane section in the body, the following equations must hold:

$$\int \sigma \cdot \mathrm{d}A = 0 \tag{13.43a}$$

$$\int \mathrm{d}M = 0 \tag{13.43b}$$

where A is the area, M is the moment at any location, and σ is the stress at any point.

These equations can be used to validate experimentally determined residual stress data.

Example 13.3 The X-ray diffraction method is used to measure the residual stresses at a point in a thin steel plate subjected to laser heat treatment. The material has a lattice constant of 3.008×10^{-7} mm. The wavelength of the X-ray is 1.54×10^{-7} mm, and is diffracted on the {101} planes. If uniaxial compressive stresses of 266 MPa are measured at a particular point, calculate the diffraction angle θ_d obtained during measurement in degrees.

Solution:

Young's Modulus for steel is approximated as being the same as that for iron that, from Appendix 13B, is given by $E_m = 208$ GPa.

$$\sigma = E_m \varepsilon$$

$$\Rightarrow -266 \times 10^6 = 208 \times 10^9 \times \varepsilon$$

or

$$\varepsilon = -1.279 \times 10^{-3}$$

Now

$$\varepsilon = \frac{a - a_0}{a_0} = \frac{a - 3.008}{3.008} = -1.279 \times 10^{-3}$$

Thus,

$$a = 3.00415 \times 10^{-7} \text{mm}$$

But from equation (13.39),

$$d = \frac{a}{\sqrt{h_m{}^2 + k_m{}^2 + l_m{}^2}} = \frac{3.00415 \times 10^{-7}}{\sqrt{1^2 + 0^2 + 1^2}}$$
$$= 2.124 \times 10^{-7} \text{ mm}$$

Also, from equation (13.38),

$$m\lambda = 2d \sin \theta_d$$

Therefore,

$$1 \times 1.54 \times 10^{-7} = 2 \times 2.124 \times 10^{-7} \sin \theta_d$$

Giving

$$\theta_d = 21.26^\circ$$

13.5 RELIEF OF RESIDUAL STRESSES AND DISTORTION

Methods for relieving residual stresses and distortion can be grouped into two general categories:

1. Thermal Treatments.
2. Mechanical Treatments.

13.5.1 Thermal Treatments

Thermal treatments that are normally used to relieve residual stresses in welding for example, include

1. Preheating.
2. Postheating.

13.5.1.1 Preheating Preheating involves heating the workpiece to a reasonably high temperature before it is processed. The preheating temperature depends on the workpiece material. For example, for ductile iron, this may be 650°C, while for hardenable steels, it is normally 300°C. A number of methods are available for preheating. These include

1. Heating in a furnace.
2. Induction coil heating.
3. Using a hot air blast.
4. Using another laser beam or splitting the original beam (see Section 16.7.1.1).

The preheated material results in less contraction stresses since it is softer, and thus, plastic flow occurs more easily under the stresses that are developed, rather than shrinkage. Preheating also reduces cooling rates that result from the process, thereby minimizing the formation of hard phases.

However, it has the disadvantage that if the preheating temperature is too high, it may

1. Cause corrosion in stainless steels.
2. Soften the material being processed.
3. Reduce the strength of a workpiece that has already been hardened by heat treatment.

13.5.1.2 Postheating The use of postheating to relieve residual stresses is essentially a recovery/recrystallization process. The workpiece is heated to a temperature that depends on the material, for a specified time. Any residual stresses that exist in the material are reduced to the yield strength of the material at the treatment temperature. Subsequent cooling must be uniform throughout the workpiece to prevent the development of additional thermal stresses. If the postheating temperature is too high, undesirable properties may result. For carbon and low-alloy steels, stress relief is commonly performed between 1100° and 1350°F. Postweld heat treatment is the most common method for relieving residual stresses in weldments.

13.5.2 Mechanical Treatments

Common mechanical treatments that are used in relieving residual stresses include

1. Peening.
2. Proof stressing.
3. Vibratory stress relief.

13.5.2.1 Peening Compressive residual stresses are induced in the surface layer of a material by applying impact loading. Peening has traditionally been done by directing a hammer or stream of small, high-velocity metal balls into the surface of the workpiece (shot peening). This is a cold working process. In recent years, lasers have been used to induce the same effect (laser shock peening, see Section 17.5). In addition to reducing residual stresses and distortion, peening may also improve the fatigue strength of the part, as a result of the compressive stresses that are induced. However, it has the disadvantage that it may conceal or introduce cracks, and may thereby reduce the fracture toughness of the material.

13.5.2.2 Proof Stressing Proof stressing involves applying uniform loading to the entire workpiece. This results in a reduction of the residual stresses and a more uniform distribution after unloading (Fig. 13.16). It serves as a means for stress relief and a test of structural integrity.

13.5.2.3 Vibratory Stress Relief This normally involves vibration of the part, usually at the natural frequency, for a period that ranges from 10 to 30 min depending on the weight of the workpiece. It is more economical than the thermal stress-relieving processes. The process induces yielding in parts of the structure and thereby relieves stresses and reduces distortion. However, the underlying mechanism is not fully understood.

13.6 SUMMARY

Our discussion in this chapter started with the basic causes of residual stresses, where thermal stresses and nonuniform plastic deformation were identified as two of the major causes, with others being nuclear radiation and exposure to electromagnetic fields. The residual stresses that result during laser welding, for example, are thermal stresses, and are tensile along the weld centerline, reducing to balancing compressive stresses as one moves further away from the centerline. This was followed by a basic discussion on stress analysis to lay the analytical framework for understanding residual stresses. This included discussions on the general stress state in a body; equilibrium conditions; strain–displacement relations; stress–strain relations; and conditions of plane stress and plane strain. Both elastic and plastic behavior were presented, culminating in the compatibility equation, which determines the conditions under which residual stresses would exist in a body. In linear elasticity, the current state of stress completely defines the state of strain. However, the current state of stress cannot define the plastic strains measured from the virgin (e.g., annealed) condition of the material. Any plastic deformation done before the current stresses were applied would have to be known before the total plastic strains could be defined.

The impact that residual stresses have on a part were then discussed, focusing on two primary effects: apparent change in strength and distortion. Depending on the nature of the residual stresses (compressive or tensile), and the external loading that is applied, the strength of the part may either be enhanced or reduced as the stresses due to external loading are superimposed on the existing residual stresses. Following this, some of the major methods for measuring residual stresses were discussed, focusing on the stress relaxation, X-ray diffraction, and neutron diffraction methods. Under stress relaxation, which is destructive, the sectioning and drilling techniques were discussed. Under X-ray diffraction, the basic principles leading up to the film and diffractometer techniques were discussed, where the latter was identified as being the most commonly used method due to its accuracy and speed of implementation. Since residual stresses exist within a body even when there are no external loads acting on it, the resultant force and moment of the residual stresses in the body must be zero. This may be used to check the efficacy of measurements that are made. Finally, different methods for relieving residual stresses in a part were outlined, with the primary methods being identified as either thermal or mechanical based. The thermal methods include pre- and postheating, while the mechanical methods include peening, proof stressing, and vibratory stress relief.

REFERENCES

Barrett, C. R., Nix, W. D., and Tetelman, A. S., 1973, *The Principles of Engineering Materials*, Prentice Hall, Englewood Cliffs, NJ.

Bardes, B. P., editor, *Metals Handbook*, 1978a, *Properties and Selection: Irons and Steels*, Vol. 1, 9th edition, American Society for Metals, OH.

Bardes, B. P., *Metals Handbook*, 1978b, *Properties and Selection: Nonferrous Alloys and Pure Metals*, Vol. 2, 9th edition, American Society for Metals, OH.

Dally, J. W., and Riley, W. F., 1991, Strain-Analysis Methods, *Experimental Stress Analysis*, 3rd edition, McGraw Hill, Inc., NY, pp. 311–340.

Felbeck, D. K., and Atkins, A. G., 1996, *Strength and Fracture of Engineering Solids*, 2nd edition, Prentice Hall, Englewood Cliffs, NJ.

Finnie, I., 1975, Unpublished Manual, Berkeley, CA.

Holden, T. M., and Roy, G., 1996, The application of neutron diffraction to the measurement of residual stress and strain, *Handbook of Measurement of Residual Stresses*, J. Lu, editor, The Fairmont Press, Inc, Lilburn, GA, pp. 133–178.

Hutchings, M. T., 1992, Neutron diffraction measurement of residual stress fields: overview and points for discussion, *Measurement of Residual and Applied Stress Using Neutron Diffraction*, M. T. Hutchings and A. D. Krawitz, editors, Kluwer Academic Publishers, Dordrecht, Netherlands, pp. 3–18.

Jenney, C. L., and O'Brien, A., editors, 2001, *Welding Handbook*, 9th edition, American Welding Society, Miami, FL.

Lu, J., editor, 1996, *Handbook of Measurement of Residual Stresses*, The Fairmont Press, Inc., Lilburn, GA.

Ludema, K., 1998, Unpublished Manual, Ann Arbor, MI.

Masubuchi, K., 1980, *Analysis of Welded Structures*, Pergamon Press, Oxford, UK.

Noyan, I. C., and Cohen, J. B., 1987, *Residual Stress: Measurement by Diffraction and Interpretation*, Springer-Verlag, NY.

Schajer, G. S., Roy, G., Flaman, M. T., and Lu, J., 1996, Hole-drilling and ring core methods, *Handbook of Measurement of Residual Stresses*, J. Lu, editor, The Fairmont Press, Inc, Lilburn, GA, pp. 5–34.

Soete, W., 1949, Measurement and relaxation of residual stresses, *Welding Journal*, Vol. 28, pp. 354s–364s.

Webster, G. A., 1992, Role of neutron diffraction in engineering stress analysis, *Measurement of Residual and Applied Stress Using Neutron Diffraction*, M. T. Hutchings and A. D. Krawitz, editors, Kluwer Academic Publishers, Dordrecht, Netherlands, pp. 21–35.

APPENDIX 13A

List of symbols used in the chapter.

Symbol	Parameter	Units
a, a_i	lattice constants	m
$\cos(n, i)$	direction cosine of outward directed normal	—
C_k	constant	Pa^2
d	interplanar spacing of a crystal structure	m
d_o	interplanar spacing of undeformed material	m
h_m, k_m, l_m	Miller indices of a reflecting plane	—
H_0	diameter of diffraction ring	m
$\mathbf{k_i}$	wavevector of incident beam	/m
$\mathbf{k_s}$	wavevector of scattered beam	/m
K	bulk modulus	Pa (N/m^2)
l_x	distance from film to specimen	m
L_o	original length of specimen	m
ΔL	extension or contraction	m
M_{init}	moment required to achieve yielding in outer fibers	N m
M_y	moment required to achieve full yield	N m
$\mathbf{s}$	scattering vector	/m
ΔT	temperature change	°C
$\bar{\varepsilon}$	effective strain	—
ν_p	Poisson's ratio	—
ϕ	angle between maximum principal stress and X-axis	°
ψ	stress function	N
$\bar{\sigma}$	effective stress	N/m^2
θ_d	diffraction peak corresponding to lattice spacing d	°
θ_{do}	diffraction peak corresponding to lattice spacing d_o	°

APPENDIX 13B

Mechanical properties of some pure annealed materials at 20°C (Bardes, 1978a, 1978b; Barrett, Nix, and Tetelman, 1973; Felbeck and Atkins, 1996; MatWeb)

Material	Young's Modulus GPa	Yield Strength MPa	Tensile Strength MPa	Percent Elongation %
Aluminum (Al)	62	15–20	40–50	50–70
Chromium (Cr)	248	–	83	0
Copper (Cu)	128	33.3	209	60
Gold (Au)	78	–	103	30
Iron (Fe)	208	276	414	18
Lead (Pb)	14	–	–	–
Magnesium (Mg)	44	21	90	2–6
Manganese (Mn)	191	241	496	40
Molybdenum (Mo)	324	≈ 400	≈ 500	–
Nickel (Ni)	207	59	317	30
Niobium (Nb)	103	207	275	30
Silicon (Si)	106–113	–	–	–
Silver (Ag)	76	–	125	–
Tin (Sn)	43	–	–	–
Titanium (Ti)	116	140	235	54
Tungsten (W)	345	–	–	–
Vanadium (V)	131	–	–	–
Zinc (Zn)	92	–	–	–
Zirconium (Zr)	94	≈ 230	–	20–35

PROBLEMS

13.1. Derive the equilibrium conditions that are given by

$$\frac{\partial \sigma_x}{\partial x} + \frac{\partial \tau_{xy}}{\partial y} + \frac{\partial \tau_{zx}}{\partial z} = 0 \tag{13.6a}$$

$$\frac{\partial \tau_{xy}}{\partial x} + \frac{\partial \sigma_y}{\partial y} + \frac{\partial \tau_{zy}}{\partial z} = 0 \tag{13.6b}$$

$$\frac{\partial \tau_{xz}}{\partial x} + \frac{\partial \tau_{yz}}{\partial y} + \frac{\partial \sigma_z}{\partial z} = 0 \tag{13.6c}$$

13.2. Derive the boundary conditions that are given by

$$\sigma_x \cos(n, x) + \tau_{yx} \cos(n, y) + \tau_{zx} \cos(n, z) = P_x \tag{13.7a}$$

$$\tau_{xy}\cos(n,x) + \sigma_y\cos(n,y) + \tau_{zy}\cos(n,z) = P_y \tag{13.7b}$$

$$\tau_{xz}\cos(n,x) + \tau_{yz}\cos(n,y) + \sigma_z\cos(n,z) = P_z \tag{13.7c}$$

13.3. For the two-dimensional element A B C D shown in the figure below, obtain expressions for the strains ε_x, ε_y, and γ_{xy} that the element experiences when deformed to the shape A′ B′ C′ D′, in terms of the displacements u, and v in the x and y directions, respectively. Start from first principles.

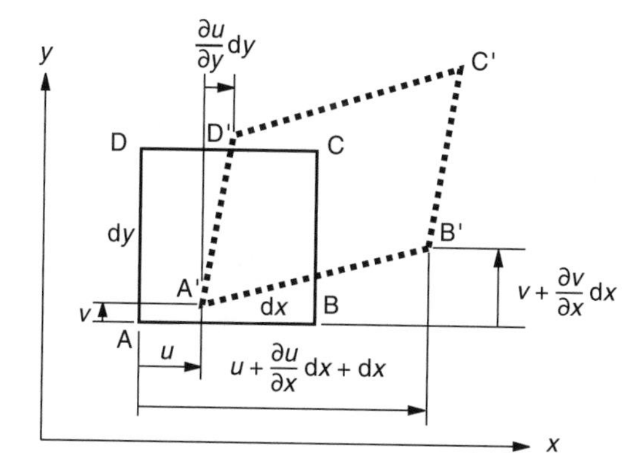

13.4. Given the strain–displacement equations

$$\varepsilon_x = \frac{\partial u}{\partial x}$$

$$\varepsilon_y = \frac{\partial v}{\partial y}$$

$$\gamma_{xy} = \frac{\partial v}{\partial x} + \frac{\partial u}{\partial y}$$

Derive the compatibility equation

$$\frac{\partial^2 \varepsilon_x}{\partial y^2} + \frac{\partial^2 \varepsilon_y}{\partial x^2} = \frac{\partial^2 \gamma_{xy}}{\partial x \partial y}$$

13.5. Show that a solution for stresses σ_x and σ_y under plane stress conditions can be applied to plane strain if ν_p is replaced by $\nu_p/(1 - \nu_p)$.

13.6. Starting with the strain compatibility equation, obtain the stress compatibility equation:

$$\frac{\partial^2(\sigma_x - \nu_p\sigma_y)}{\partial y^2} + \frac{\partial^2}{\partial x^2}(\sigma_y - \nu_p\sigma_x) = 2(1+\nu_p)\frac{\partial^2\tau_{xy}}{\partial x\partial y}$$

13.7. Obtain expressions for ε_x, ε_y, and γ_{xy} in terms of ε_A, ε_B, and ε_C for the delta rosette shown below.

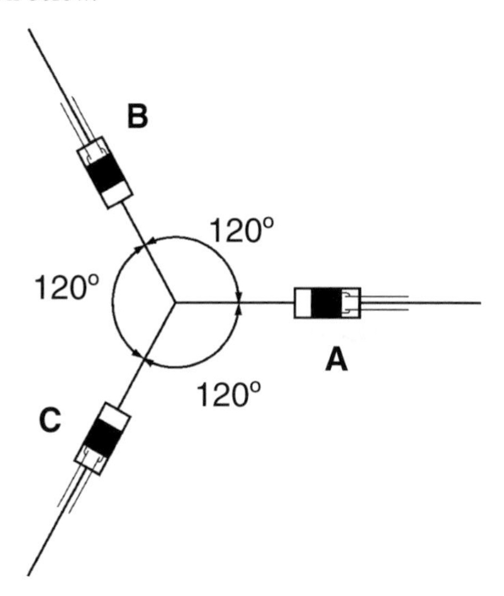

13.8. Obtain expressions for ε_x, ε_y, and γ_{xy} in terms of ε_A, ε_B, and ε_C for the rectangular rosette shown below.

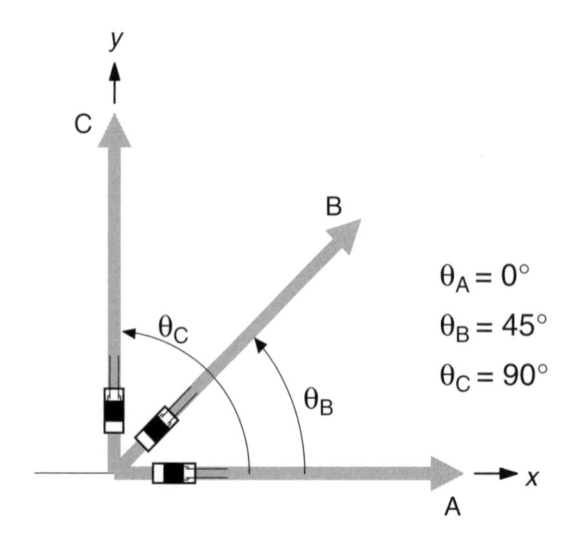

Obtain expressions for the principal stresses, σ_1 and σ_2, in terms of ε_A, ε_B, and ε_C.

13.9. The middle member (2) of the fixture shown below is heated to a temperature close to its melting temperature, about 100°C below the melting temperature. On a stress-temperature graph, plot the variation of stress in member 1 while member 2 is first being heated and then cooled back to room temperature. Label the heating and cooling cycles.

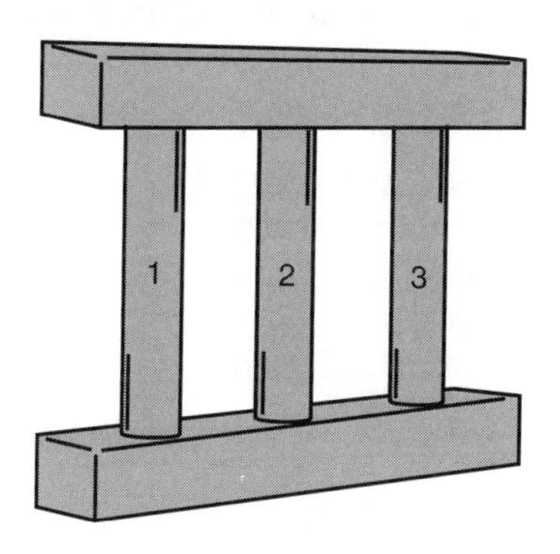

13.10. (a) A rectangular strain gauge rosette arranged as shown in the figure below is used to measure the residual stresses in a weldment. If the strain gauges measure strains in the directions **A, B, C,** express the strains ε_x, ε_y, and γ_{xy} in terms of the measured strains ε_A, ε_B, and ε_C.

(b) If ε_A, ε_B, and ε_C are respectively 1000×10^{-6}, -200×10^{-6}, and 750×10^{-6}, determine the maximum normal stress at that point if Young's modulus for the material is 200 GPa and the Poisson ratio is 0.3.

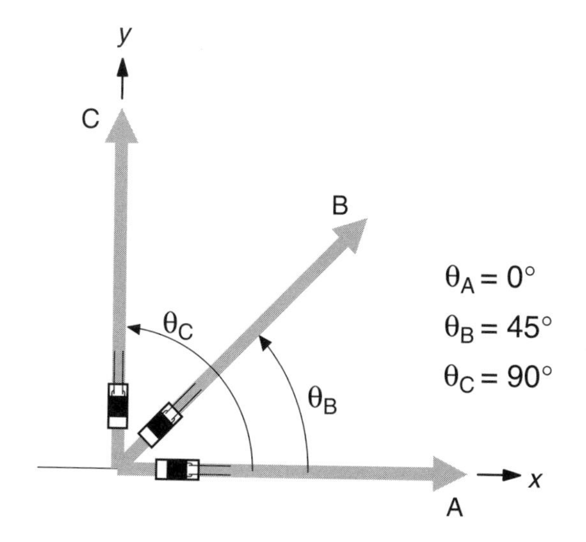

13.11. (a) After a part is processed using a laser, a small square piece is cut from a region that is subjected to compressive residual stresses. Will the piece expand or contract after it is cut? Why?

(b) Explain why preheating would reduce distortion in a laser-welded part.

13.12. The residual stresses at a point in an aluminum plate that has been subjected to some form of laser treatment are measured by the X-ray diffraction method. The X-rays, of wavelength 1.54×10^{-7} mm, are diffracted $2\theta_d = 38.4°$ on the {111} planes. If the lattice constant of undeformed aluminum is 4.049×10^{-7} mm, estimate the residual stresses that exist at that point. Consider the stresses to act in only one direction.

13.13. A consultant claims to have developed a new method for measuring residual stresses using a three-element rectangular rosette as shown in the figure below. If the strains measured by the gages A, B, and C are $-500,\ 300,\ 150$ μm/m respectively, calculate the maximum stress at the point being measured, if the material is steel with a Poisson's ratio of 0.3.

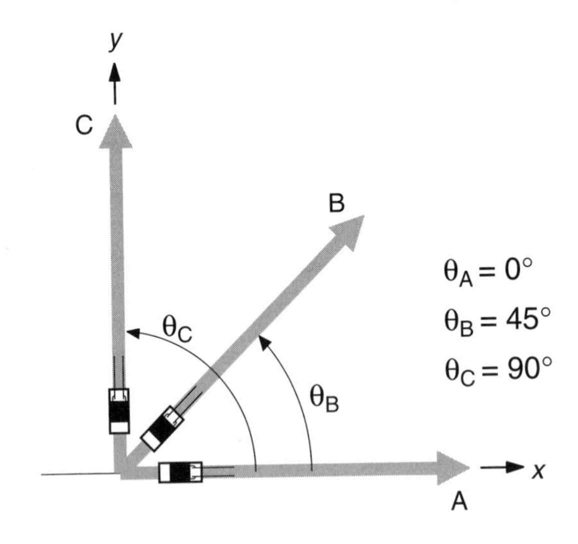

13.14. In measuring the residual stresses in a thin copper plate that has been subjected to laser forming using the X-ray diffraction method, the diffraction angle is obtained as 42° for the plane with the Miller indices {111}. If the lattice constant of copper is 3.615×10^{-7} mm, what is the wavelength of the X-rays used?

13.15. The drilling technique is used to measure the residual stresses at a point in a laser welded thin steel plate. The strain measurements obtained are shown in the figure below.

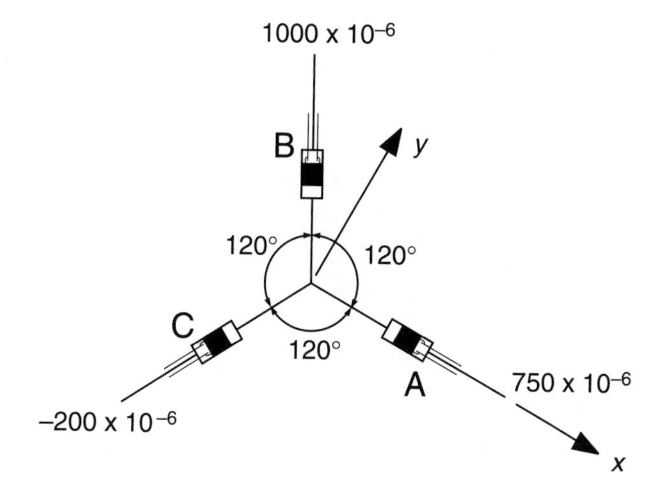

Assuming a Poisson's ratio of 0.3, determine the following stresses that act at that point:

1. The maximum normal stress.
2. The maximum shear stress.

PART III
Laser Materials Processing

14 Background on Laser Processing

The laser beam is the heat source in laser materials processing. Even though the laser is normally considered to be a light source, it is also a form of energy and as such can be a useful source of intense heat when concentrated by focusing. Lasers are able to produce high energy concentrations because of their monochromatic, coherent, and low divergence properties compared to an ordinary light source (Figs 7.1 and 7.2). As a result, they can be used to heat, melt, and vaporize most materials. The processes for which lasers are commonly used include welding, cutting, surface modification (including heat treatment), and forming. The power densities and exposure times necessary for various processes are shown in Fig. 14.1. To provide a frame of reference, a comparison of various heat sources is provided in Table 14.1. For the laser beam, the highest power densities are achieved with the beam in the Gaussian (TEM_{00}) mode (see Sections 2.2.1 and 14.1.3).

In this chapter, we shall recap our discussions in Part I on the characteristics of lasers and relate them to issues that are common to all the processes that will be discussed in subsequent chapters. We start with a general discussion on system-related parameters that have the most significant effect on process output. The major stages of energy loss, from the power drawn from the mains supply to the actual power used in the process, are then discussed in terms of process efficiency. This is followed by a discussion on disturbances that affect process quality, and finally, the advantages and disadvantages of laser processing in general are briefly presented.

14.1 SYSTEM-RELATED PARAMETERS

The principal system parameters that affect process quality include the beam power and power density at the focal point, wavelength and focusing system, beam mode (distribution), beam form (CW or pulsed), and beam quality. Others include beam absorption, beam delivery, beam alignment, and the motion unit used for positioning the beam or workpiece. Our discussion follows in that order. Some parameters such as beam polarization, assist and/or shielding gas, focal point location, traverse speed, and so on are more process specific, and are thus discussed in relation to individual processes in subsequent chapters.

Principles of Laser Materials Processing, by Elijah Kannatey-Asibu, Jr.

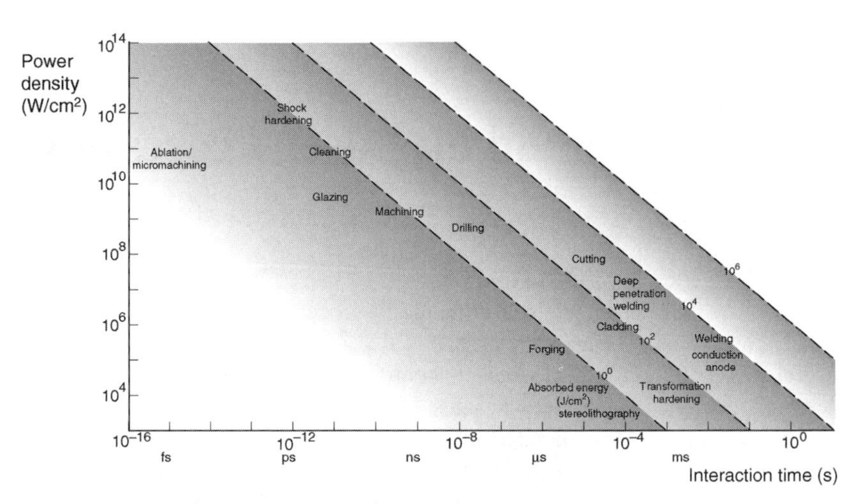

FIGURE 14.1 Power densities and interaction times for various laser processes.

14.1.1 Power and Power Density

The power density at the focal point is determined by the incident power and the size of the focused beam. For a Gaussian distributed laser beam, the beam size is defined as the radial distance at which the intensity of the beam reduces to $1/e^2$ or 13.5% the peak intensity. This is also the area that has approximately 87.5% of the total beam energy. The power density (intensity or irradiance), I, is given by

$$I = \frac{q}{\pi w^2} \tag{14.1}$$

where q is the power of the laser beam (W) and w is the beam radius (mm).

14.1.2 Wavelength and Focusing

The wavelength of the beam influences the process in several ways, including absorption of the beam and the extent of focusing. In this section, we shall discuss

TABLE 14.1 Comparison of Power Densities for Various Heat Sources

Heat Source	Power Density (W/cm^2)
Blackbody radiation (3000°C)	6.45×10^4
Oxyacetylene flame	10^5
Transferred arc plasma torch	10^7
Electron beam	10^{11}
Laser (continuous CO_2)	2.5×10^8
Laser (pulsed Nd:YAG)	$>10^{14}$
Laser (ultrashort pulse)	$>10^{21}$

From Mazumder, J., 1987, *SPIE High Power Lasers*.

its influence on focusing. Its effect on beam absorption is separately addressed in Section 14.1.6.

From equation (7.20a), we know that when a Gaussian beam of radius w_0 and wavelength λ is incident on a lens of focal length f_l, it is focused to a radius w_f given by

$$w_f = \frac{1}{\pi}\left(\frac{f_l \lambda}{w_0}\right) \tag{14.2}$$

where w_0 is the incident beam radius at the lens.

Equation (14.2) indicates that the power density at the focal point is inversely proportional to the square of the wavelength. Thus, the smaller the wavelength of the beam, the higher the output power density, all other conditions being equal. The shorter wavelength of the Nd:YAG laser (1.06 μm) thus gives it some advantages over the CO_2 laser (10.6 μm). For the same input power, beam mode, beam size, and optics, the power density ratio of an Nd:YAG laser to a CO_2 laser is 100. Equation (14.2) also indicates that the greater the original beam diameter, the smaller the focused beam radius, and thus the greater the power density of the focused beam. Therefore, an expander is often used to increase the diameter of the beam incident on the focusing lens.

It is also evident from equation (14.2) that the shorter the focal length of the lens, the smaller the size of the focused beam, and thus the greater the power density. Thus in laser welding, the use of a shorter focal length lens results in deeper penetration for the same beam power, especially at lower power levels. At much higher power levels of say, 20 kW, there is little difference, if any, on the impact of the focal length on penetration. The use of very short focal length lenses, however, has two disadvantages:

1. The depth of focus (see next section) is correspondingly small, (Fig. 14.2), requiring greater precision in maintaining the lens to workpiece distance.
2. In welding, the shorter focal length lens is more liable to be damaged by vapor and molten metal ejected from the weld.

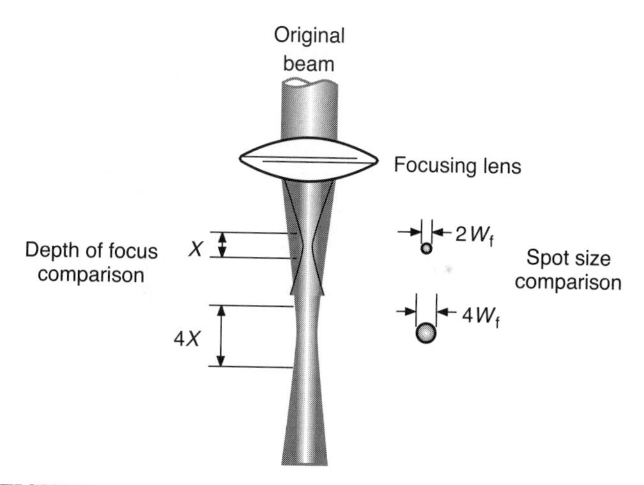

FIGURE 14.2 Lens focal length and corresponding depth of focus.

Focal lengths normally used range from 2.5 in. (65 mm) to 10 in. (254 mm), but 7.5 in. (190 mm) lenses are about the standard. Common lens materials include ZnSe, KCl, GaAs, and CdTe, and a typical beam radius at the focal point would be 100 μm. For very long focal lengths (say, 200 mm and over), a reflective parabolic copper mirror is often used for focusing.

14.1.2.1 Determining the Focal Position There are various ways in which the focal position can be determined. In one case, an acrylic sheet mounted at 80° to the horizontal is traversed horizontally to cross the path of the vertical beam. The imprint of the beam on the sheet identifies the location of the focal point (Fig. 14.3). Another technique involves simply laying the acrylic or cardboard sheet horizontally on a table that has a vertical height adjustment and making imprints on the sheet, while adjusting the height.

14.1.2.2 Depth of Focus The depth of focus z_f of a focusing system is the distance over which the workpiece can be moved from the original focal point without a significant change in the focused beam radius (Fig. 14.2). A long depth of focus is good for laser cutting, for example, where parallel-sided cuts or holes are needed.

Using equations (2.61) and (2.62), the depth of focus, z_f, over which 80% of the power density at the focal point is maintained can be shown to be given approximately (see Problem 14.2) by

$$z_f = \frac{\pi \times {d_f}^2}{4 \times \lambda} \tag{14.3}$$

where d_f is the focused beam diameter.

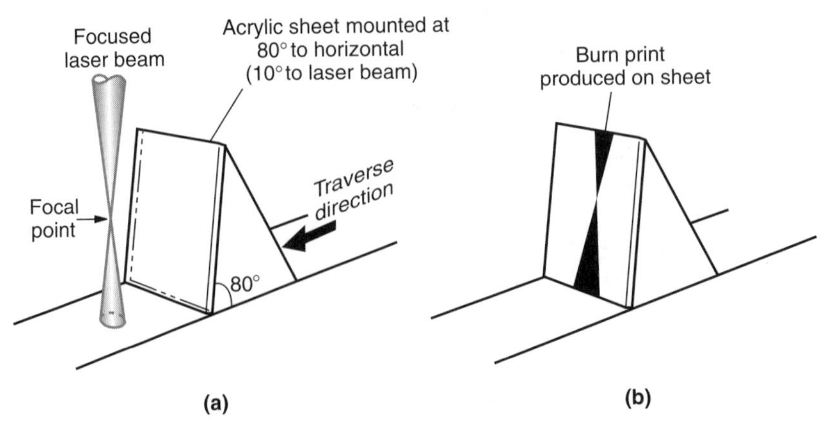

FIGURE 14.3 Determination of the focal point of a laser beam. (From Watson, M. N., Oakley, P. J., and Dawes, C. J., 1985, *Metal Construction*. Reproduced by permission, TWI Ltd.)

14.1.3 Beam Mode

The basics of the beam mode or profile are extensively discussed in Chapter 2. In simple terms, the profile is the intensity (irradiance or power per unit area) distribution in a plane perpendicular to the beam propagation axis. It is an indication of how the energy intensity is distributed over the beam cross section (see Section 2.2.1). The profile varies with propagation distance from the output mirror. The near-field profile may be distorted due to diffraction effects, but the profile tends to assume a constant, symmetrical shape in the far field. Far field is reliably detected after five Rayleigh ranges (see Sections 2.3 and 7.1). Depending on the type of beam, this distance can be as little as a few millimeters or as much as a number of meters.

Profiles of a Gaussian (TEM_{00}) (low order) mode and a multimode (TEM_{mn}) (high order) are shown schematically in Fig. 14.4. The Gaussian or low-order mode (Fig. 14.4a), can be focused to the laser's theoretical minimum radius. The higher order or multimode beams, however, are more spread out and are not as well focused, resulting in larger focal spot sizes and thus lower power density for the same output power. These are more common with higher power lasers.

A Gaussian mode beam is thus normally preferred for cutting, for example, since it results in the smallest possible focused beam radius and thus highest power density for a given incident power. However, for butt welding, a higher order multimode beam is preferable since the larger beam diameter means less stringent joint fit-up requirements.

Nd:YAG lasers normally generate a multimode beam (Fig. 14.4b), since that ensures the most efficient power extraction from the rod used in the laser source. CO_2 lasers, however, can generate either a low-order or multimode beam, with the low-order mode being more common at power levels less than 5 kW. However, the output of high-power CO_2 lasers is normally multimode since the high-power lasers are usually transverse flow machines that inherently produce multimode beams.

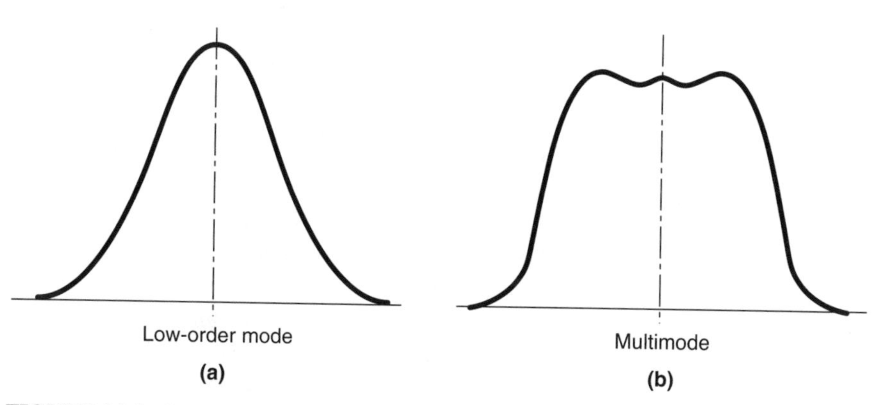

FIGURE 14.4 Beam distributions. (a) Low-order (Gaussian) mode. (b) High-order (multimode) mode.

14.1.4 Beam Form

The laser output is usually of two forms:

1. Continuous wave (CW)—where the output power is essentially constant.
2. Pulsed—where the output power varies periodically, in a manner similar to a clocked signal.

The CW output is used quite often. However, the pulsed output of a laser has advantages in a number of situations. For example, variation of the on/off ratio or duty cycle of the pulsed beam enables applications such as drilling and cutting of sharp-edged contours to be effectively controlled. The reduced total heat input also makes the pulsed output suitable for parts that are sensitive to distortion. However, the lower overall average power of the pulsed beam results in a reduced processing speed. Furthermore, in processes such as cutting and seam welding, where there is a need for continuity, the pulse frequency, when a pulsed beam is used, has to be high enough for the pulses to overlap.

Pulsed beams can be classified into three forms (Fig. 14.5):

- *Gated pulsing*—which is obtained by switching the discharge current on and off to produce a pulsed output (Fig. 14.5a). The peak output power is about the same as the nominal CW output power.
- *Superpulsing*—where the peak output power is about two to three times that of the CW output, Fig. 14.5b. However, the duty cycle is correspondingly reduced. Superpulsing enables low to medium power CW lasers to produce relatively high output powers over short periods of time and is similar to Q-switching except that it is achieved electronically.
- *Hyperpulsing*—where high-peak power pulses are superposed on the CW output (Fig. 14.5c).

Both superpulsing and hyperpulsing are useful for minimizing dross formation at the lower surface during cutting of some materials such as stainless steels. This is because much higher temperatures are achieved in the molten metal when either superpulsing or hyperpulsing is used. That results in a lower viscosity of the material, making it easier to eject with the same level of gas pressure. Both pulsing modes enable deeper weld penetration or greater cut thickness (up to about 50%) to be achieved for the same average power as a CW output. They also make it easier to process highly reflective and/or high thermal conductivity materials.

The shape of a pulsed beam affects the resulting temperature distribution in the workpiece. For a given total pulse energy, a shorter but higher power density pulse will result in higher peak surface temperature than a longer, lower power density pulse. However, the lower power density pulse produces a more uniform temperature distribution and a greater heat penetration at the end of the pulse.

As Fig. 14.6 illustrates, less energy is required to reach a given temperature at the surface if high power densities and short interaction times are employed. In the figure, the dashed curves represent the energy supplied per unit area, while the solid curves

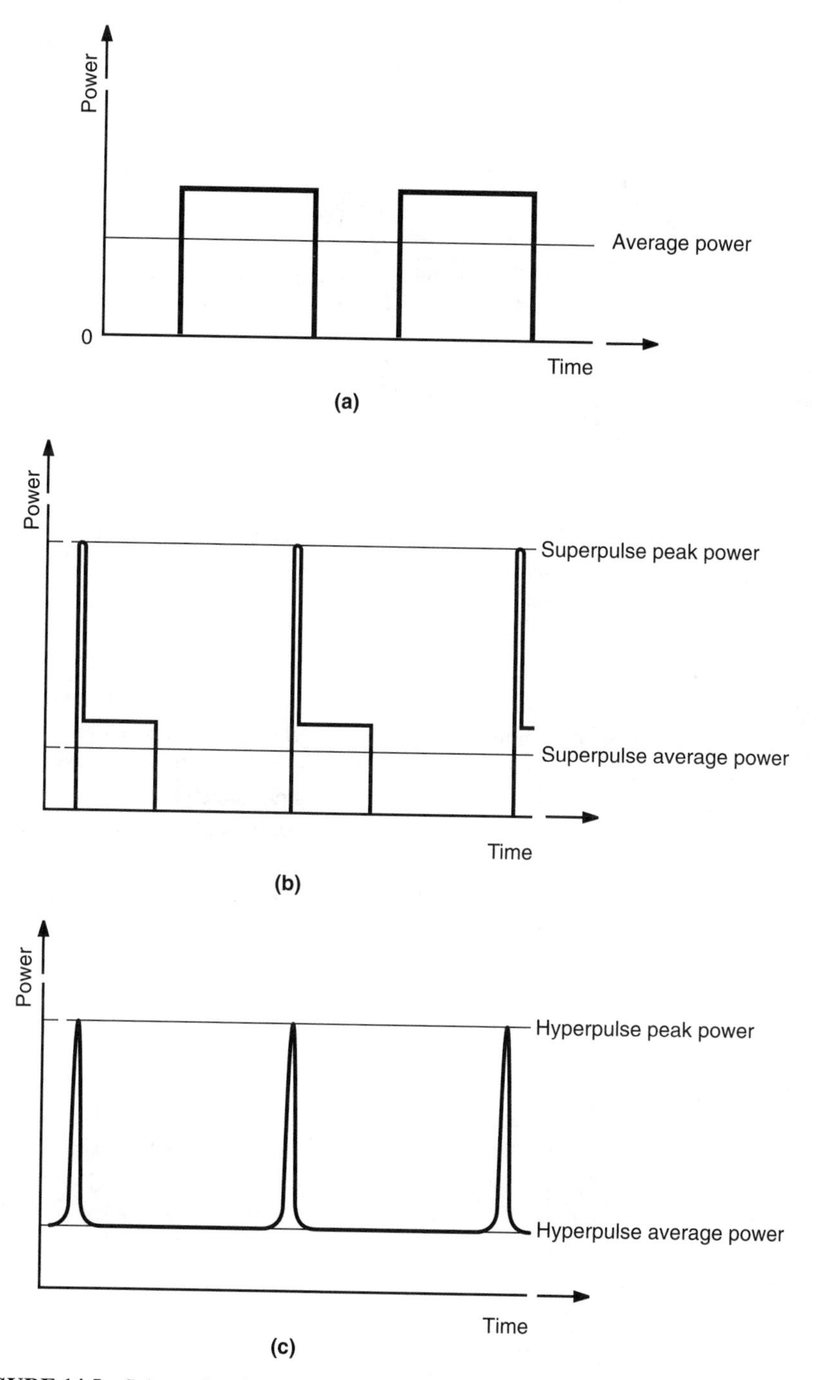

FIGURE 14.5 Schematic of pulsed outputs. (a) Gated pulsing. (b) Superpulsing. (c) Hyperpulsing.

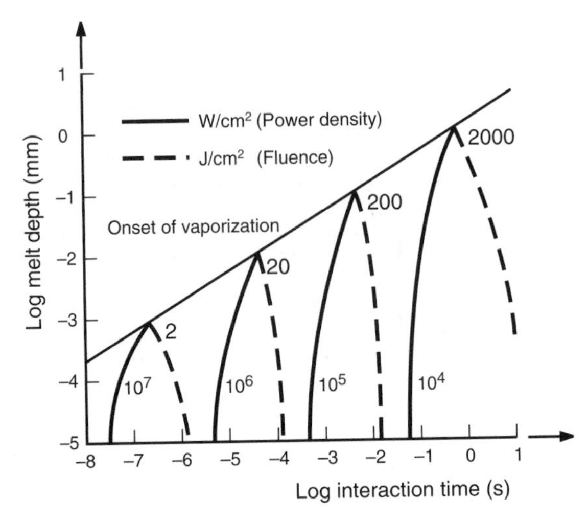

FIGURE 14.6 Illustration of transient surface melting characteristics. (From Gutierrez-Miravete, E., 1993, *Rapid Solidification Technology: An Engineering Guide,* T. S. Srivatsan and T. S. Sudarshan, editors, Technomic Publishing Co., Lancaster.)

represent the power per unit area. Above the straight line, vaporization occurs. As the figure shows, for a fixed total input energy of say 2 J/cm^2, the melt depth increases as the interaction time decreases since the resulting power density then increases. Also, for a given power density, say 10^7 W/cm^2, the melt depth increases as the interaction time increases. More energy is then made available to the material. Furthermore, at high power densities and short interaction times, vaporization occurs when only a small amount of material has melted. However, it takes much longer for vaporization to occur even for a high-energy input, if the interaction time is long, since the power density is then relatively low.

Due to the nonuniform nature of pulsed beams, they are commonly classified using the total energy content rather than the power. Energy units (J) are thus normally used for pulsed beams.

For applications such as microwelding, trimming, and drilling, the pulsed beam is found to be most useful, whereas for applications such as cutting and seam welding where a continuity of heat source is essential, the CW output is most suitable. Nd:YAG lasers normally produce pulsed outputs. However, in recent years, high-power Nd:YAG lasers with CW output have been developed. CO_2 lasers, however, have traditionally been used in the CW mode. However, pulsed outputs can also be obtained from CO_2 lasers.

14.1.5 Beam Quality

The quality of a laser beam is normally characterized in terms of the K and M^2 factors, which are dimensionless. They indicate the resemblance of the beam to a Gaussian beam. The two are related as follows:

$$M^2 = \frac{1}{K} \tag{14.4}$$

From equations (2.61) and (2.62), the spot size (beam radius), $w(x)$, of a propagating diffraction-limited Gaussian beam at any general location along the cavity axis is expressed as

$$w(x) = w_w \left[1 + \left(\frac{2x}{L}\right)^2\right]^{1/2} \tag{14.5}$$

where

$$w_w = \left(\frac{L\lambda}{2\pi}\right)^{1/2} \tag{14.6}$$

x is the distance measured from the beam waist along the axis of beam propagation, and L is the distance between the cavity mirrors.

The propagation of a non-Gaussian laser beam can be expressed in a similar manner with w_{w} modified as

$$w_{\mathrm{w}} = \left(\frac{L\lambda}{2\pi K}\right)^{1/2} \tag{14.7}$$

where K is the K-factor. Both the M^2 and K factors are unity for diffraction-limited Gaussian beams. However, for non-Gaussian beams, the K-factor is smaller (less than unity), while the M^2-factor is higher (greater than unity). A beam with an $M = 3$ has a waist almost three times larger than the Gaussian beam and diverges slightly more. Comparison of the two types of beam propagation is illustrated in Fig. 14.7a. The M-factor can be obtained by first determining the beam diameter at three positions of the beam (Fig. 14.7a and b) and using equation (14.7). For example, most high-power Nd:YAG lasers have beam quality of M^2 between 50 and 100.

14.1.6 Beam Absorption

In analyzing the thermal effect of a laser beam on a workpiece, the amount of power that actually enters or is absorbed by the workpiece needs to be considered. When the beam is incident on a solid surface, absorption occurs in a thin surface layer called the skin depth (of about 0.01 μm depth for a CO_2 laser on metals) and transferred to the inner part by conduction. The skin depth, δ', of a substance is the distance to which incident electromagnetic radiation penetrates that substance, during which the

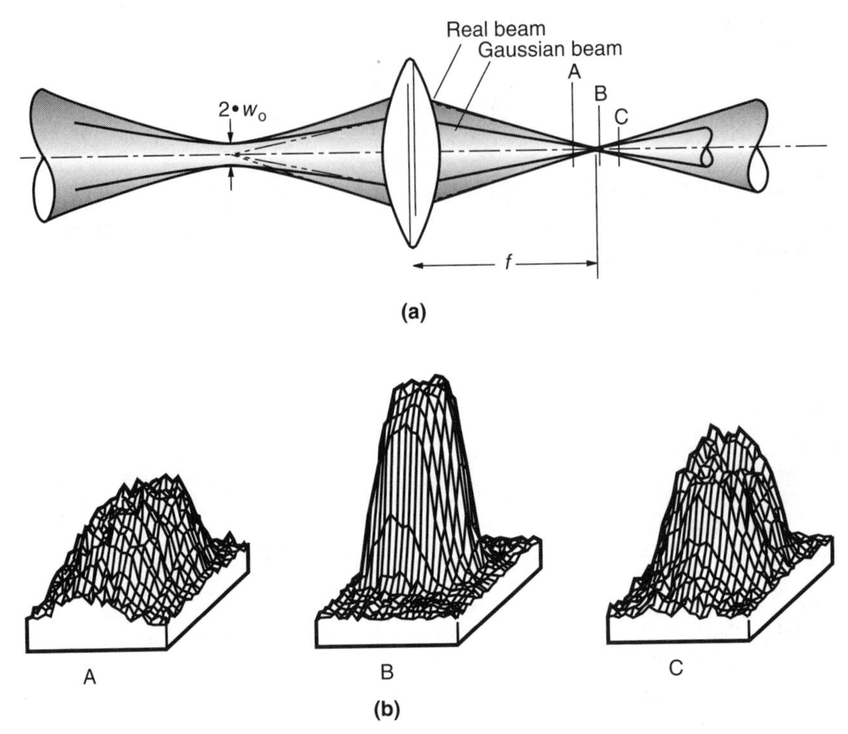

FIGURE 14.7 (a) Comparison of Gaussian and non-Gaussian beam propagation. (b) Beam profiles at three different locations. (From ICALEO 1990 Proceedings, Vol. 71. Copyright 1991, Laser Institute of America. All rights reserved.)

light intensity drops by a factor *e* from its initial value, and is expressed as

$$\delta' = \frac{1}{\sqrt{\pi \sigma_e \mu_m \nu}} \tag{14.8}$$

where σ_e is the electrical conductivity (S/m), μ_m is the magnetic permeability (H/m), and ν is the frequency (Hz).

For a beam of power q_i, incident on a surface of reflectance R_s, the power q_a absorbed by the workpiece is given by

$$q_a = q_i(1 - R_s) = a_\alpha q_i \tag{14.9}$$

The reflectance (R_s) or absorptivity ($a_\alpha = 1 - R_s$) depends on both the material on which the beam is incident and the beam wavelength (Fig. 14.8a). For metals, the absorptivity increases with decreasing wavelength, while for insulators, it increases with increasing wavelength (Fig. 14.8a). Metals thus generally tend to have better absorption at the shorter 1.06 μm wavelength characteristic of Nd:YAG lasers than at

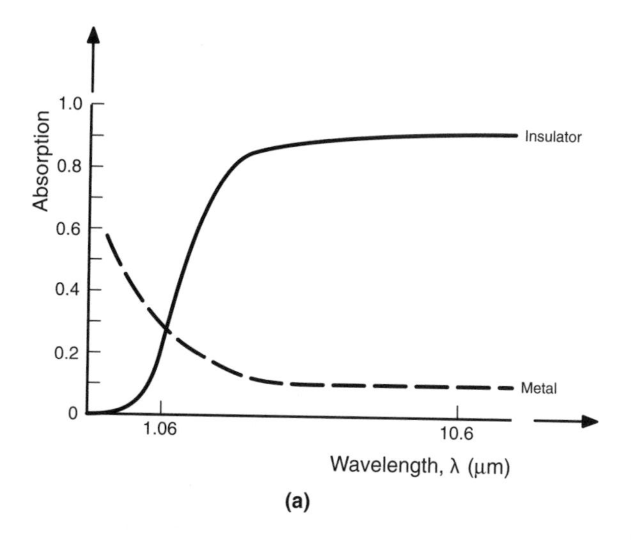

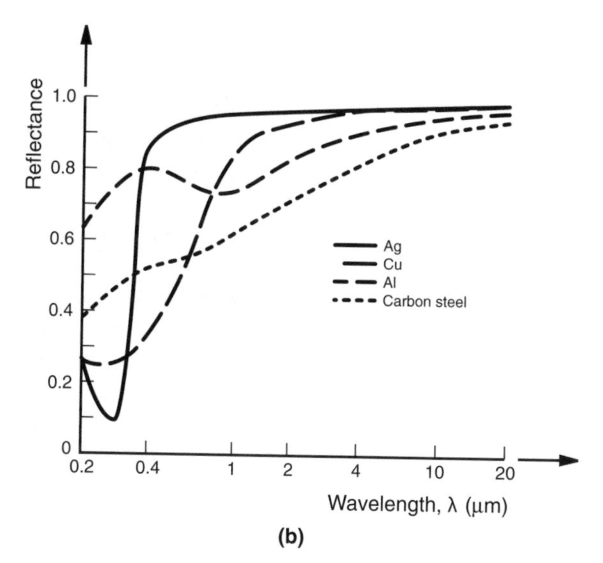

FIGURE 14.8 (a) Wavelength dependence of absorption for different materials (polished). (b) Wavelength dependence of reflectance for different metals. (From Wilson, J., and Hawkes, J. F. B., 1987, *Lasers, Principles and Applications*. Reprinted by permission of Pearson Education, Inc., Harlow, Essex, UK.)

the 10.6 μm wavelength of CO_2 lasers at temperatures below their melting points. The reverse is true of nonmetallic materials such as glass, quartz, plastics, and insulators in general. For processes that do not require melting, such as surface heat treatment, such differences in absorption can be very significant. Figure 14.8b shows the reflectance values for different metals. The absorptivity also increases with an increase in

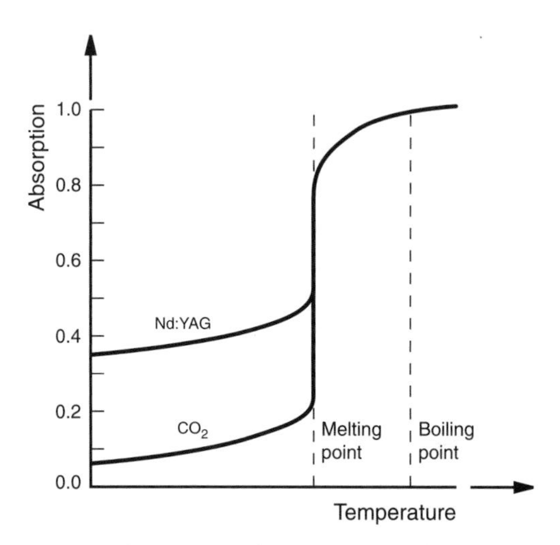

FIGURE 14.9 Temperature dependence of absorption for Nd:YAG and CO_2 laser beams on metals. (From Wilson, J., and Hawkes, J. F. B., 1987, *Lasers, Principles and Applications*. Reprinted by permission of Pearson Education, Inc., Harlow, Essex, UK.)

temperature for a given wavelength. Once the material melts, the absorptivity increases dramatically, and essentially becomes independent of the wavelength (Fig. 14.9).

Initial absorption of the laser beam by the workpiece is also affected by other factors such as the nature of the surface and the joint geometry. Absorption on a metallic surface can be enhanced by appropriate surface modification such as anodizing, sand blasting, and surface coating. However, as indicated in the preceding paragraph, the absorptivity of metals increases (i.e., reflectivity decreases) with increasing temperature and becomes even more significant in the molten state. Thus, for processes that involve melting and/or vaporization, such as welding and cutting, the high initial reflectivity need not be a major problem, except in the initial stages of the process, where a high initial power is required for highly reflective materials. However, for processes that do not involve melting, such as heat treatment and forming, high reflectivity will result in significant power loss.

In addition to its influence on beam absorption, reflectivity also determines the amount of energy that is reflected back to the beam delivery system (backscattering) and may result in optical feedback into the laser resonator, affecting both the beam generation mechanism and the optical elements.

14.1.6.1 Measurement of Absorptivity The absorptivity of a material can be measured using the simple calorimetric setup shown in Fig. 14.10. A specimen of the material is attached to one side of a calorimeter through which water flows. When the laser beam is incident on the specimen, the energy it absorbs is transferred to the water flowing beneath it. Measurement of the inlet and outlet temperatures of

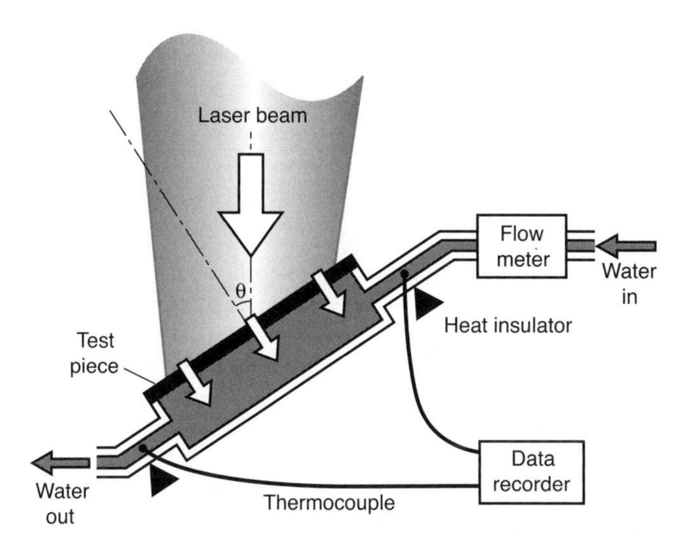

FIGURE 14.10 Setup for measuring absorptivity calorimetrically. (From *ICALEO 1991 Proceedings*, Vol. 74. Copyright 1992, Laser Institute of America. All rights reserved.)

the water, along with the flow rate, enables the power absorbed to be determined as follows:

$$\text{Power absorbed by the water} = \dot{m}c_p(T_{\text{out}} - T_{\text{in}}) \tag{14.10}$$

where $\dot{m}$ is the mass flow rate of the water (kg/s), c_p is the specific heat (J/kg K), T_{in} is the inlet temperature (K), and T_{out} is the outlet temperature (K).

The absorptivity of the material is then given by

$$\text{Absorptivity} = \frac{\text{power absorbed by the water}}{\text{incident laser power}} \tag{14.11}$$

14.1.6.2 Beam–Plasma Interaction As will be shown in subsequent Chapters, especially Chapter 16 on laser welding, when the beam power is high enough, a plasma cloud may be formed above the workpiece. The extent to which the plasma affects the process depends on its interaction with the laser beam. Propagation of the laser beam through the plasma is influenced by the plasma oscillation, since the beam is an electromagnetic wave. Plasma oscillation may be considered either in terms of electron oscillation or ion oscillation. The frequency of electron oscillation is always greater than that of ion oscillation due to the heavier mass of the ion. However, it suffices to characterize the beam–plasma interaction by only one of these oscillations. The frequency of plasma electron oscillation, ω_e, is given by

$$\omega_e{}^2 = \frac{n_e e_c{}^2}{\mu_m m_e} \tag{14.12}$$

where n_e = electron number density (/m^3), m_e = electron mass (kg), and e_c = electron charge (C).

When the beam frequency is greater than the plasma oscillation frequency, the incident beam propagates through the plasma, while the beam is reflected if its frequency is lower than the plasma frequency. Resonant absorption of the beam occurs when its frequency is the same as that of the plasma. However, the oscillation frequency of typical plasmas generated during material processing is significantly less than the frequencies of commonly used laser beams. Thus, one would expect the beam to propagate through the plasma.

Absorption of the incident beam energy by the plasma thus occurs primarily by the inverse bremsstrahlung process (photon absorption by electrons—see also Section 16.3.2.1). The absorption coefficient, α, under these circumstances is then given by

$$\alpha = k_\alpha \frac{e_c{}^6 n_e{}^2}{\omega^2 c\,(2\pi m_e k_B T)^{3/2} \sqrt{1 - \left(\frac{\omega_e}{\omega}\right)^2}} \tag{14.13}$$

where ω the angular frequency of the incident wave (rad/s) and k_α is a constant that depends on plasma characteristics.

As equation (14.13) indicates, absorption due to the inverse bremsstrahlung process increases as the temperature or beam frequency decreases, and it becomes infinitely large as the frequency approaches the plasma oscillation frequency.

Part of the incident beam may also be lost by Rayleigh scattering. Since the scattering loss is inversely proportional to the fourth power of the wavelength, the shorter wavelength beams are liable to suffer greater scattering losses. Such scattering can occur in a plasma when ultrafine particles (about tens of nanometers in size) are formed due to condensation of evaporated atoms.

The principal mechanisms for energy dissipation in a laser-induced plasma are thus the inverse bremsstrahlung process and Rayleigh scattering. The former is predominant at longer wavelengths, while the latter is predominant at shorter wavelengths.

14.1.7 Beam Alignment

To ensure that the laser beam is delivered from the generator to the appropriate spot on the workpiece, the beam is always aligned with the optical and beam delivery systems. Since the infrared radiation is not visible, the alignment process is normally facilitated by installing in the high-power laser, a low-power HeNe laser to generate a visible red beam that is coaxial with the invisible primary infrared beam. With the two beams being coaxial, alignment of the visible beam automatically ensures alignment of the invisible beam. For such systems, the output lens must be transmissive to both infrared and visible radiation. ZnSe is one such suitable material for CO_2 lasers.

The alignment process essentially involves adjustment of the beam bender (mirror) orientation using the mounting screws to redirect the beam. A crosshair assembly is mounted at the output of either the beam bender or focusing assembly. The mirror orientation is then adjusted until the beam center coincides with the crosshair center when observed on a screen held behind the crosshair assembly, at the location where the workpiece would normally be positioned (Fig. 14.11).

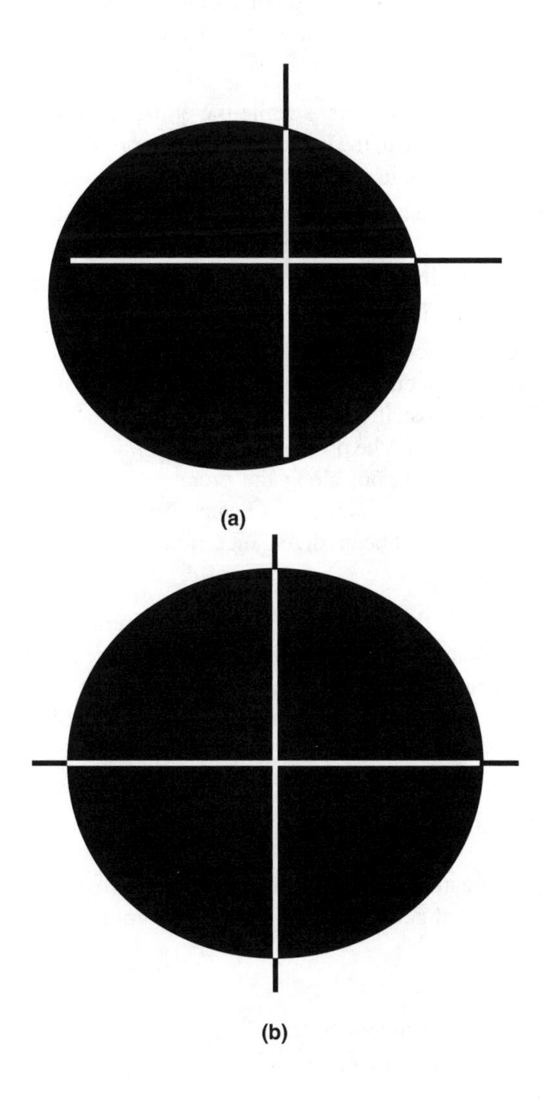

FIGURE 14.11 Beam alignment using a crosshair. (a) Misaligned beam. (b) Aligned beam.

For systems that do not have a visible beam installed, essentially the same procedure is followed except that instead of observing the visible radiation on the screen, the laser is given a single pulse at low power with a sheet of cardboard paper held in place of the screen. The beam will burn a pattern on the paper that indicates the crosshair position relative to the beam center.

14.1.8 Motion Unit

The motion unit is the means by which relative motion is achieved between the laser beam and the workpiece. There are four ways in which this can be done:

1. The workpiece is kept stationary while the entire laser and its beam delivery system is moved relative to the workpiece. This approach is more suited to large workpieces, and is often inconvenient, given the typical size of industrial lasers.
2. The laser and its entire beam delivery system is kept fixed while the workpiece is moved with respect to it. This method is suitable for processing pieces of limited size. Processing of large pieces requires significantly large floor space.
3. The use of a fiber optic system (see Section 9.9.2).
4. The use of flying optics. In this case, the laser generator is kept fixed while the beam delivery components (focusing assembly and mirrors) are moved relative to the workpiece. It usually involves the use of a robot that incorporates the beam delivery elements. The principal setback of this approach is the change in the size of the focused spot along the processed path. This is due to beam divergence that results in increase in beam size with distance from the laser output window. Minimal beam divergence is thus essential.

Either the fiber optic or flying optics arrangement minimizes floor space, and the size of workpiece that can be processed is unlimited. Another advantage is the ability to keep the beam orientation normal to the workpiece surface and to keep the nozzle to workpiece distance constant, even when processing parts with complex contours.

14.2 PROCESS EFFICIENCY

Laser processing applications involve transfer of energy from the generator to the workpiece. In the course of generation and transfer, part of the energy is lost. Sources of energy loss include

1. Absorption in the optical components.
2. Convection through the shielding gas.
3. Radiation from the workpiece surface to the surroundings.
4. Conduction into the workpiece.

Thus, only a fraction of the power generated is available for processing. The overall efficiency of a laser process consists of four efficiency components that determine the amount of power that is actually used in the process. These are

1. Laser generator efficiency.
2. Beam transmission efficiency.
3. Heat transfer efficiency.
4. Processing efficiency.

- *Laser Generator Efficiency,* η_g: This is discussed in detail in Section 3.1.3 and is given by the ratio of the power output of the laser generator to the power input to the generator. It is expressed as

$$\eta_g = \frac{\text{power output of the laser}}{\text{power input to the laser generator}} \tag{14.14}$$

It typically ranges approximately from 5 to 30% for CO_2 lasers, 1 to 3% for Nd:YAG lasers, 1 to 4% for excimer lasers, and 30 to 50% for diode lasers (Table 8.1).

- *Beam Transmission Efficiency,* η_t: The beam transmission efficiency depends on the transmission components being used. These are further discussed in Chapter 9. The efficiency is defined as the ratio of the energy coming out of the beam delivery system to the energy input to the beam delivery system from the laser and is expressed as

$$\eta_t = \frac{\text{power output from beam delivery system}}{\text{power input from laser to beam delivery system (power output of the laser)}} \tag{14.15}$$

- *Heat Transfer Efficiency,* η_{th}: The actual power available for the process is determined by the efficiency of heat transfer, η_{th}, which is the efficiency of energy transmission from the source to the workpiece (absorption), that is,

$$\eta_{th} = \frac{\text{power actually absorbed by the workpiece } (q_a)}{\text{total power available at the source (power output from beam delivery system) } (q_i)} \tag{14.16}$$

Thus,

$$q_a = \eta_{th} \times q_i \tag{14.17}$$

For laser welding, this typically ranges from 50 to 95%, once the welding process is initiated and a keyhole is formed. However, before any melting occurs, this may be as low as 1% for polished aluminum or copper and about 40% for stainless steel when using a CO_2 laser. Nd:YAG lasers result in higher absorption efficiencies in solid metals. Power losses at this stage result primarily from reflection from the workpiece surface and plasma absorption when the material melts.

- *Processing Efficiency,* η_p: The processing efficiency is the thermal efficiency of energy conversion in the workpiece, and is defined as the ratio of the power required for heating and/or melting to the power actually transferred to the workpiece.

  For processes that involve melting, the processing efficiency, η_p, is given by

$$\eta_p = \frac{\text{power required to melt a given volume of material}}{\text{Power actually absorbed by the workpiece}} = \frac{Q_1 A_w u_x}{q_a} \tag{14.18}$$

where $Q_1 = \rho_w[L_m + c_p(T_m - T_0)]$ (J/m^3) that is the heat required to melt a given volume of material from room temperature, A_w is the total cross-sectional area of fusion zone (m^2), ρ_w is the density of fused metal (kg/m^3), and u_x is the velocity of the moving heat source (m/s).

The processing efficiency depends on the processing conditions, material (workpiece), and joint configuration. Materials with higher conductivity have lower efficiency, since a significant amount of the input heat is lost by conduction. It increases with the intensity of the heat source. Higher efficiencies are achieved at higher power levels and speeds for lower thermal conductivity materials, since less incident power is then lost by conduction into the workpiece. Power is also lost by radiation and convection from the surface.

- *Overall Efficiency,* η: The overall efficiency of the process is then given by

$$\eta = \eta_g \eta_t \eta_{th} \eta_p = \frac{\text{power used in processing the material}}{\text{power input to the laser generator}} \tag{14.19}$$

Typical values of overall efficiency for arc, laser, and electron beam welding are listed in Table 16.3 for comparison.

14.3 DISTURBANCES THAT AFFECT PROCESS QUALITY

High quality in the production environment can be maintained by minimizing the impact of disturbances that affect the production system. Common sources of disturbance during laser processing include

1. Fluctuations in power line.
2. Temperature and humidity variations in the production environment.
3. Condensation that may result from cooling of optical components with tap water. Air cooling alleviates this problem.

4. Contamination of the output window by dust and vapor from the atmosphere. This requires prevention of contact with surrounding atmosphere.
5. Vibrations from the surroundings.
6. Backscattering, a phenomenon that results from reflection of the incident laser beam off the workpiece. If the nozzle orifice is large enough, a significant portion of this reflected beam may propagate backward along the transmission path. In extreme cases, this may change the state of polarization of the laser beam, cause uncontrolled amplification of the laser process, and possibly damage the optical elements and laser resonator. The effect of backscattering on the process may be minimized by tilting the workpiece at such an angle that any beam that is reflected from it is unable to propagate back into the beam delivery system.

14.4 GENERAL ADVANTAGES AND DISADVANTAGES OF LASER PROCESSING

Lasers have several advantages over conventional techniques that are normally used in materials processing. At the same time, there are disadvantages that may make lasers less suitable for certain applications. These are outlined in the following paragraphs and further elaborated on with respect to specific processes as they are discussed in subsequent chapters.

14.4.1 Advantages

1. The ability to focus the beam to a small size, thereby attaining a high intensity and highly localized heating source with minimal effect on surrounding areas. Spot sizes may vary approximately from 0.001 to 10 mm.
2. The ease with which the beam power can be controlled by regulating the current through the electric discharge.
3. Minimal contamination of the process.
4. Ability to manipulate the beam into ordinarily inaccessible areas using mirrors and fiberoptic cables.
5. Minimal heat-affected zone and distortion.
6. Noncontact nature of the process.

14.4.2 Disadvantages

1. Relatively high capital cost of equipment.
2. High reflectivity of laser beam on metals.
3. Low efficiency of lasers.
4. Energy waste by beam dumping when the laser is not used continuously.

14.5 SUMMARY

Power densities generated by pulsed lasers generally exceed 10^{14} W/cm^2 and can be as high as 10^{21} W/cm^2, much greater than the outputs of commonly available heat sources such as the electron beam, arc, and blackbody radiation. For materials processing, ablation requires the highest power densities and shortest interaction times, while transformation hardening requires the lowest power densities and longest interaction times. The principal system parameters that affect process quality include the beam power and power density at the focal point, wavelength and focusing system, beam mode (distribution), beam form (CW or pulsed), and beam quality. Others include beam absorption, beam delivery, beam alignment, and the motion unit used for positioning the beam or workpiece.

The power density of a beam at the focal point is inversely proportional to the square of the beam wavelength and lens focal length and directly proportional to the square of the original beam diameter. Common lens materials that are used for focusing include ZnSe, KCl, GaAs, and CdTe. The depth of focus of a focusing system is the distance over which the workpiece can be moved from the original focal point without a significant change in the focused beam radius.

The Gaussian mode is the lowest order mode a beam can have, and it can be focused to the laser's theoretical minimum radius. The size of a Gaussian beam is defined as the radial distance at which the beam intensity reduces to $1/e^2$ or 13.5% of the peak intensity. The higher order or multimode beams, however, are more spread out and are not as well focused. The form of the laser output can be either continuous or pulsed, with the pulsed output being one of the three types: gated, superpulsed, or hyperpulsed. For a given total pulse energy, a shorter but higher power density pulse will result in higher peak surface temperature than a longer, lower power density pulse. However, the lower power density pulse produces a more uniform temperature distribution and a greater depth of penetration of the heat at the end of the pulse. The quality of a laser beam is characterized by either the M^2 or K $(= 1/M^2)$ factor, with $M = K = 1$ for a diffraction-limited Gaussian beam.

When a beam is incident on a solid surface, absorption occurs in a thin surface layer, the skin depth, and transferred to the inner part by conduction. For metals, the absorptivity increases with decreasing wavelength, while for insulators, it increases with increasing wavelength. The absorptivity also increases with an increase in temperature for a given wavelength. Interaction of a laser beam with plasma depends on the frequencies of the beam and plasma. When the beam frequency is greater than the plasma oscillation frequency, the incident beam propagates through the plasma, while the beam is reflected if its frequency is lower than the plasma frequency. Resonant absorption of the beam occurs when its frequency is the same as that of the plasma.

As the laser beam propagates from the generator to the workpiece, the major sources of energy loss include absorption in the optical components, convection through the shielding gas, radiation from the workpiece surface to the surroundings, and conduction into the workpiece. The overall efficiency of a laser process consists of four efficiency components. These are laser generator efficiency, beam transmission efficiency, heat transfer efficiency, and processing efficiency.

REFERENCES

Gregersen, O., and Olsen, F. O., 1990, Beam analyzing system for CO_2 lasers, *Proceedings of the 7th International Conference on Applications of Lasers and Electro Optics, ICALEO'90*, pp. 28–35.

Gutierrez-Miravete, E., 1993, Mathematical modeling of rapid solidification, *Rapid Solidification Technology: An Engineering Guide,* T.S. Srivatsan and T.S. Sudarshan, editors, Technomic Publishing Co., Lancaster, pp. 3–70.

Harry, J., 1974, *Industrial Lasers and Their Applications*, McGraw Hill, London.

Lewis, G. K., and Dixon, R. D., 1985, Plasma monitoring of laser beam welds, *Welding Journal*, Vol. 64, pp. 49s–54s.

Matsunawa, A., 1990, Physical phenomena and their interpretation in laser materials processing, *Proceedings, of the 7th International Conference on Applications of Lasers and Electro Optics, ICALEO'90*, p. 313.

Mazumder, J., 1987, An overview of melt dynamics in laser processing, *SPIE, High Power Lasers*, Vol. 801, pp. 228–241.

Ready, J. F., 1965, Effects due to absorption of laser radiation, *Journal of Applied Physics,* Vol. 36, (2), pp. 462–468.

Rofin-Sinar, *Technical Note*: CO_2 *Laser Welding*, Rofin Sinar, Plymouth, MI.

Shibata, K., Sakamoto, H., and Matsuyama, H., 1991, Absorptivity of polarized beam during laser hardening, *Proceedings, of the 7th International Conference on Applications of Lasers and Electro Optics, ICALEO'91*, pp. 409–413.

Schuocker, D., 1986, Laser cutting, *SPIE, The Industrial Laser Annual Handbook*, Vol. 629, pp. 87–107.

Thompson, A., 1986, CO_2 laser cutting of highly reflective materials, *SPIE, The Industrial Laser Annual Handbook*, Vol. 629, pp. 149–153.

von Allmen, M., and Blatter, A., 1994, *Laser-Beam Interactions with Materials*, Springer-Verlag, Berlin.

Watson, M. N., Oakley, P. J., and Dawes, C. J., 1985, Laser welding: techniques and testing, *Metal Construction*, pp. 25–28.

Wilson, J., and Hawkes, J. F. B., 1987, *Lasers: Principles and Applications*, Prentice-Hall International Series in Optoelectronics, New York.

APPENDIX 14A

List of symbols used in the chapter.

Symbol	Parameter	Units
a_α	absorptivity	–
A_w	cross-sectional area of fusion zone	m^2
d_f	focused beam diameter	m
k_α	constant that depends on a number of plasma characteristics	–
q_a	absorbed power	W
q_i	incident power	W

Q_l	heat required to melt a given volume of material from room temperature	J/mm^3
R_s	reflection coefficient of surface	–
S_R	Rayleigh scattering loss	dB/km
T_{in}	inlet temperature	K
T_{out}	outlet temperature	K
z_f	depth of focus	m
δ'	skin depth	μm
ϵ_r	dielectric constant	–
η_g	generator efficiency	%
η_p	process efficiency	%
η_t	beam transmission efficiency	%
η_{th}	heat transfer efficiency	%
ω_e	angular frequency of the plasma electron oscillation	rad/s
ρ_w	density of fused metal	kg/m^3

PROBLEMS

14.1. How would you expect the processing efficiency to vary with the thermal conductivity of the base metal?

14.2. Using equations (2.61) and (2.62), show that the depth of focus, z_f, over which 80% of the power density at the focal point is maintained is given approximately by

$$z_f = \frac{\pi \times d_f{}^2}{4 \times \lambda}$$

where d_f = focused beam diameter.

14.3. A CO_2 laser beam with a Gaussian mode is focused using a 150 mm lens. If the beam radius as it enters the lens is 12.5 mm, determine:

(a) The focused spot radius

(b) The depth of focus for a spot radius increase of 7.5%.

14.4. (a) Two laser beams of the same power are incident on a workpiece. One is from a Nd:YAG laser and the other is from a CO_2 laser. Which one would result in the most intense heat at the point of focus? Explain.

(b) The maximum power of a laser available in a plant is found to be inadequate to weld a new material that requires joining. Then a self-styled consultant (qualifications unknown) suggests that the beam diameter should be increased before it gets to the focusing lens. Is this a knowledgeable or quack consultant? Explain.

14.5. A given material has the same absorption coefficient for both CO_2 and Nd:YAG lasers. Disregarding cost, which of the two lasers would you select for welding the material? Explain.

15 Laser Cutting and Drilling

In laser cutting and drilling, the focused laser beam is directed onto the surface of the workpiece to rapidly heat it up, resulting in melting and/or vaporization, depending on the beam intensity and workpiece material (Fig. 15.1). The molten metal and/or vapor is then blown away using an assist gas. The power density required is typically of the order of 10^6–10^7 W/cm^2 for metals. Lasers can be used to effectively cut metal plates of thicknesses up to about 10 cm. The cut surfaces are roughly parallel and straight edged. In reality, the diverging nature of the beam results in a slight taper being produced. The workpiece thickness that can be cut with parallel sides is determined by the depth of focus (see Section 14.1.2.2). Plates that are thicker than the depth of focus normally result in tapered surfaces. Furthermore, for effective removal of material from the cutting zone, the ratio of workpiece thickness to kerf width needs to be less than 20:1 for most metals, and less than 40:1 for ceramics. The kerf is the slot or opening created by the cutting process.

The total heat input required for laser cutting is relatively small. This results in a small heat-affected zone size, of the order of 0.1 mm. Also, the small size of the focused beam results in very narrow kerf sizes, typically about 0.05–1 mm. Laser cutting is used for both straight and contour cutting of sheet and plate stock in a wide variety of materials.

Laser cutting and laser drilling are discussed in two separate sections, because of the unique characteristics of the two processes. The chapter as a whole begins with a general discussion on the different ways in which laser cutting may take place. Components of a laser-cutting system are then outlined. The processing conditions that are normally used are then presented. This is followed by a discussion on the principles of the cutting process, which leads to common defects that may result during the process. The specific characteristics of individual materials are then discussed, followed by the advantages and disadvantages of laser cutting, and a comparison with conventional processes. The section on laser drilling follows a similar outline, starting with the different forms of laser drilling. The essential process parameters are then discussed, followed by an analysis of the material removal process. The advantages and disadvantages of the process are then presented, followed by some of the major applications of laser drilling. Finally, new developments such as micromachining and laser-assisted machining are discussed.

Principles of Laser Materials Processing, by Elijah Kannatey-Asibu, Jr.

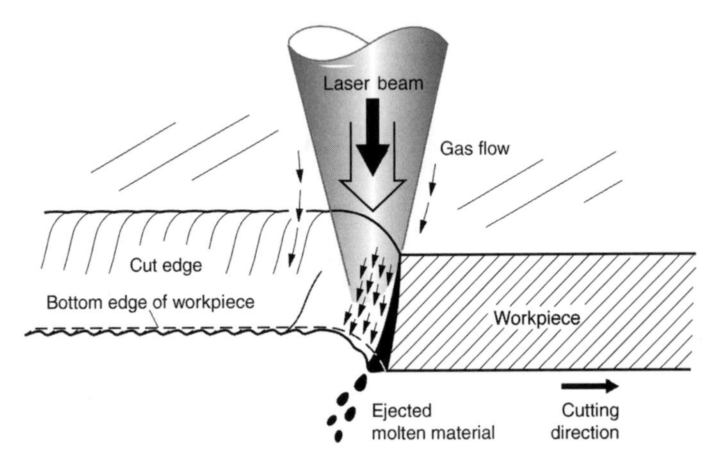

FIGURE 15.1 Schematic of the laser cutting process. (From *LIA Handbook of Laser Materials Processing*. Copyright 2001, Laser Institute of America. All rights reserved.)

15.1 LASER CUTTING

15.1.1 Forms of Laser Cutting

The process of laser cutting can occur in one of the three forms:

1. Fusion cutting.
2. Sublimation cutting.
3. Photochemical ablation.

15.1.1.1 Fusion Cutting Fusion cutting involves melting of the base material, which is then ejected using a high-pressure assist gas. The assist gas may be an inert gas, in which case the energy for melting is provided entirely by the laser beam. It may also be oxygen (or air), which reacts with the base metal, and the resulting exothermic reaction provides additional energy to enhance the process. The term fusion or clean cutting is sometimes used to indicate inert gas assisted cutting, while the process involving exothermic reaction is then referred to as gas cutting. A major problem of fusion cutting is the formation of striations (valleys and peaks that run along the thickness (see Section 15.1.5.1)) on the cut surface and dross (molten material that clings to and solidifies on the underside of the cut edge as burr (see Section 15.1.5.2)) at the lower cut edge. However, the fusion cutting process is more efficient, requiring less energy per unit volume of material removed as compared to the other methods.

15.1.1.2 Sublimation Cutting In sublimation cutting, the workpiece material is vaporized along the cutting seam. This is often achieved using a pulsed beam, and a jet of inert assist gas that is coaxial with the beam is used to blow away the vapor produced. It is limited to thin sections since more energy is required to remove a unit volume of material as compared to fusion cutting. However, it has the advantage of a narrower kerf width and higher quality surface. Pulsed beams with high peak power may be necessary when surface quality is critical.

15.1.1.3 Photochemical Ablation Organic materials, ceramics or difficult-to-cut materials in general are normally cut by this method or by sublimation cutting. Organic compounds tend to absorb ultraviolet radiation in an efficient manner. The photon energy levels of lasers based on ultraviolet radiation range between 3.5 and 6.5 eV. This corresponds with the energy levels required for molecular bonding. For example, the energy associated with the C–C bond is roughly 4.6 eV, while that of the C–H bond is about 4.2 eV. Compare with the photon energy levels associated with CO_2 and Nd:YAG lasers of about 0.12 and 1.2 eV, respectively.

As a result, when an organic material is irradiated with an ultraviolet beam, the material absorbs the beam's energy in a very thin layer near the surface, of the order of submicrons. This breaks molecular bonds, causing ablative decomposition of the irradiated area. The process occurs almost instantaneously (about 20 ns duration), and since the thermal conductivity of organic materials is relatively low, the resulting edges are well defined, with minimal thermal damage to the surrounding area. Thus the cut region is cleaner and smoother compared to that obtained using CO_2 and Nd:YAG lasers. The process is sometimes referred to as cold cutting since little heat is generated.

15.1.2 Components of a Laser Cutting System

The basic components of a cutting system are illustrated in Fig. 15.2 and include

1. The laser generator that produces the beam.
2. A beam delivery system for directing the beam to the workpiece.
3. A nozzle assembly, usually integral with the focusing assembly and coaxial with the beam, for directing the assist gas to the workpiece.
4. A motion unit for providing relative motion between the laser beam and the workpiece.
5. An exhaust for the waste material.

The beam normally emerges from the generator horizontally and is deflected vertically downward by a bending mirror. The vertical orientation is used to minimize trapping of the beam and molten material blown through the cut. The beam is then focused by the lens onto the workpiece. At the same time, a gas jet is directed through a nozzle attached to the tip of the focusing assembly, onto the workpiece. A typical nozzle diameter would be about 1–2 mm. The delivery pressures are normally maintained at about 3–4 bar (45–60 psi or 0.3–0.4 MPa) in the gas jet nozzle for cutting thin materials at high speeds. At high gas pressures, it is often necessary to use relatively thick lenses that can withstand the pressure. However, care has to be taken since thermal deformation of the lens increases with its thickness. Thus, at high

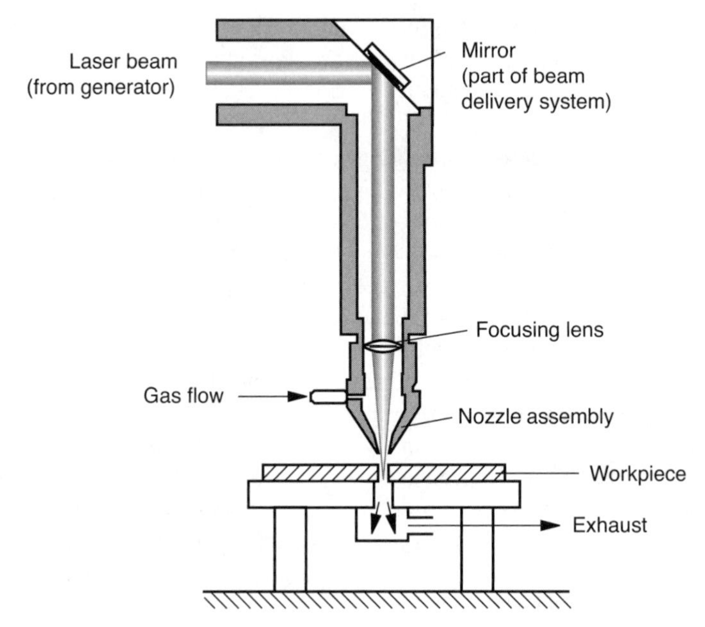

FIGURE 15.2 Components of a laser cutting system.

power levels, say 10 kW, it is preferable to use reflective optics (mirrors) rather than transmissive optics (lenses).

The beam is generally oriented in a direction almost normal to the workpiece surface, and there is a deterioration in product quality as the beam orientation deviates significantly from the normal direction. The distance from the nozzle tip to the workpiece surface is typically maintained constant at about 0.3 mm to minimize expansion of the gas flow. Capacitive or light sensors may be installed to measure this distance (see Section "Capacitive Transducers" in Chapter 21). The focusing unit may then be moved using a computer numerically controlled (CNC) third axis or a robot. A simpler but less flexible approach involves mounting the focusing unit on rollers that glide on the workpiece surface.

Directly opposite the nozzle, on the other side of the workpiece, an exhaust system is provided to absorb the transmitted beam, molten debris from the cut, and exhaust vapors. This normally consists of a vacuum pump that draws the exhaust to a disposal unit.

15.1.3 Processing Conditions

The principal parameters that affect the laser cutting process include the following:

1. Beam power.
2. Beam characteristics.

3. Traverse speed.
4. Assist gas type and flow.
5. Location of focal point relative to the workpiece surface.

Table 15.1 gives sample conditions for laser cutting of various materials. Each of these parameters are further discussed in the following sections.

15.1.3.1 Beam Power The power is the most significant of the parameters listed. An increase in power increases the maximum thickness that can be cut and/or speed at which it can be cut.

15.1.3.2 Beam Characteristics The most important beam characteristics in laser cutting operations are

1. Beam mode
2. Stability
3. Polarization
4. Beam form (pulsed or continuous wave).

Beam Mode As indicated in Section 14.1.3, the beam mode is an indication of how the energy intensity is distributed over the beam cross section. The mode relates to the beam's ability to be focused, and may be compared with the sharpness of a cutting tool. In laser cutting, it is desirable to have the beam distribution as close as possible

TABLE 15.1 Possible Cutting Conditions for CO_2 Laser Cutting of Different Materials (Stainless steel, nitrogen assist gas; aluminum, nitrogen assist gas; titanium, argon assist gas).

Material	Thickness (mm)	Power (W)	Cutting Speed (m/min)	Gas Pressure (kPa)
Carbon steel	0.5	250	3.5	
Carbon steel	1.5	400	4.0	
Carbon steel	3.0	600	3.0	
Carbon steel	6.0	1200	1.5	
Stainless steel	1.0	1000	3.5	600
Stainless steel	1.5	1500	3.5	700
Stainless steel	3.0	1800	2.0	800
Stainless steel	6.0	2000	1.0	1200
Aluminum	1.0	1200	3.0	600
Aluminum	1.5	1500	2.5	800
Aluminum	3.0	1800	1.0	1000
Titanium	1.0	800	3.5	600
Titanium	1.5	900	3.0	700

From Tables 2–5, Chapter 12, LIA Handbook of Laser Materials Processing, 2001.

to the fundamental or Gaussian distributed TEM_{00} mode. This is the mode that can be focused to the laser's theoretically smallest possible focal size and thus the highest density for a given power. This reduces the kerf width, and increases cutting speeds and thickness of materials that can be cut. Since higher order or multimode beams are more spread out, (Fig. 14.4), they result in larger focal spot sizes and thus lower power density for the same output power.

Beam Stability Stability of the beam is necessary to ensure that the beam power, mode, and direction (pointing stability) remain constant with time. An unstable beam affects the tolerances and surface finish achievable with laser cutting. A stable beam thus reduces variations in product output and enhances quality. The stability of the beam is a characteristic of the laser generator, and depends on its design (see Section 7.5).

Effect of Beam Polarization The basic concepts of polarization are discussed in Section 9.5. When the output of a laser beam is randomly polarized, it shows up in the quality of the cut workpiece as variations in kerf, edge smoothness, and perpendicularity. This is mainly due to the impact of polarization on beam absorption by the material. Absorption of the laser beam during cutting is determined using the Fresnel relationships (see equations (16.3) and (16.7)), which indicate that the absorption depends on both the angle of incidence and polarization. Experimental results indicate that in CO_2 laser cutting, beams which are plane-polarized in a plane parallel to the cutting direction result in cutting speeds which can be up to 50% greater than those which are plane-polarized in the plane normal to the cutting direction. Generally, the circularly polarized beam results in higher cutting speeds at higher power levels. For beams which are plane-polarized in the cutting plane, the beam is found to be more effectively absorbed at angles of incidence greater than 80° and less than 90°.

The cut quality varies with the orientation of the plane of polarization, and a circularly polarized beam generally results in better cut quality when the cutting direction changes. Thus, profile or contour cutting is preferably done using a circularly polarized beam. The impact of the orientation of a linearly polarized beam on the cut quality is illustrated schematically in Fig. 15.3. As the figure shows, when the beam is polarized in the cutting direction, the resulting cut may have a narrow kerf with sharp, straight edges. However, as the plane of polarization is oriented away from the cutting direction, the energy absorption decreases, and as a result, the cutting speed is reduced, the kerf becomes wider, and the edges rougher and not square to the material's surface.

Beam Form Both pulsed and continuous wave (CW) beams can be used for laser cutting, with CW beams being more common.

For pulsed beams, the quality of cut is affected by the pulse duty, with the surface roughness decreasing with increasing pulse duty. The pulse duty is the ratio of the length of time during which the laser beam is on in one cycle to the total length of the cycle time. In addition, the maximum cutting speeds that can be achieved are

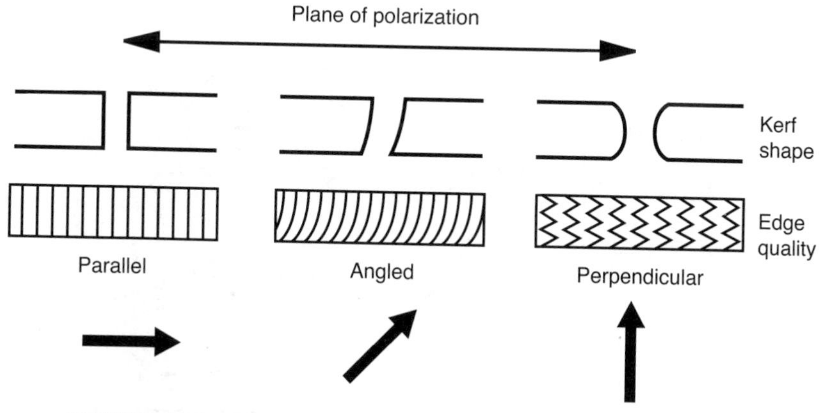

FIGURE 15.3 Effect of plane-polarized laser orientation on cut quality.

significantly reduced at low pulse rates (number of pulses per second). This is due to the fact that at low pulse rates, there is enough time between pulses for the material to substantially cool down. This helps extinguish the exothermic oxidation reaction, thereby reducing the overall process efficiency. Furthermore, as the material cools down between pulses at low pulse frequencies, there is a greater likelihood of forming dross (see Section 15.1.5.2). The resulting lower average temperature increases the surface tension or viscosity of the molten material, making it more difficult to flow out of the reaction zone.

As discussed in Section 14.1.4, superpulsing and hyperpulsing can also be used to obtain significant improvements in process performance such as

1. Minimizing dross formation at the lower surface during cutting of some materials.
2. Greater cut thickness (up to about 100%) to be achieved for the same average power as a CW output.
3. Increased processing speeds (20–33%) for the same average power as a CW output.
4. Making it easier to process highly reflective and/or high thermal conductivity materials.

15.1.3.3 Traverse Speed The maximum speed that can be achieved for a given laser power decreases with increasing thickness of the workpiece (Fig. 15.4). For a given power, a plot of the variation of cutting speed with thickness will generally have two limiting curves (Fig. 15.5). The upper curve indicates the maximum speed that can be achieved for a given thickness. Above this curve, cutting is incomplete. Below the lower curve, self-burning occurs. In other words, the material continues to burn without the aid of the laser. This often widens the kerf and produces a rough surface.

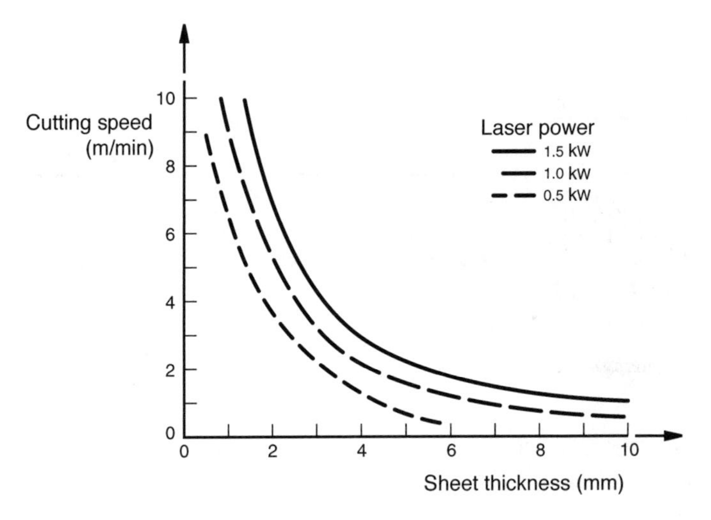

FIGURE 15.4 Variation of cutting speed (maximum cutting rates) with workpiece thickness in oxygen-assisted laser cutting of steel plate.

15.1.3.4 Assist Gas Functions The assist gas used in laser cutting serves one or more of the following functions:

1. Facilitates ejection of the molten metal through the backside of the workpiece.
2. Protects the lens from spatter.

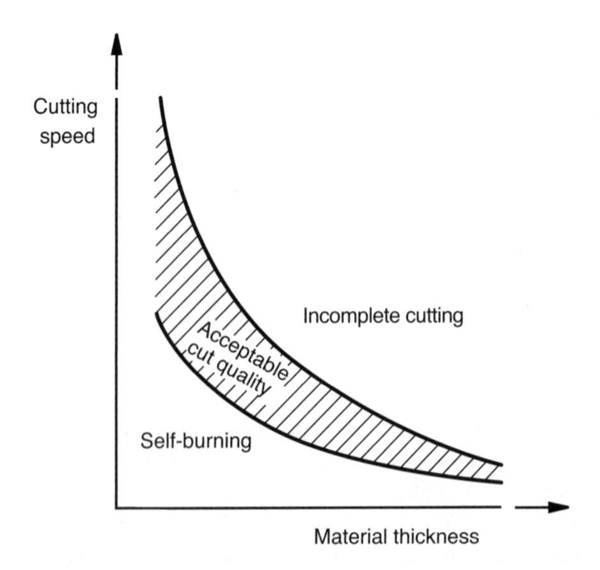

FIGURE 15.5 Limiting curves for cutting speed with workpiece thickness variation in laser cutting.

3. Acts as a heat source where it results in an exothermic reaction that aids in cutting, such as may occur in oxygen-assisted cutting of steel.

Effect of Different Types of Assist Gases Common gases used include oxygen, inert gases, nitrogen, and air. Oxygen or air is used for exothermic reaction while cutting and therefore they improve the cutting efficiency. Otherwise, an inert gas (usually argon) is used to assist in ejecting the molten metal without oxidation. In cases where the exothermic reaction is desired, the decision to use oxygen or air depends primarily on economics. Air is cheaper, but would require higher flow rates for the same amount of thermal energy produced. Furthermore, air may introduce other gases such as nitrogen into the cut surface, making it more brittle.

One setback of oxygen-assisted cutting is the deposition of an oxide layer on the cut surface, giving it a dark appearance. Depending on the subsequent use of the cut parts, it may be necessary to clean off this oxide layer (by grinding or wire brushing). The use of inert gases or nitrogen eliminates the formation of the oxide layer. However, this may significantly reduce the cutting speeds that can be achieved. Furthermore, higher pressures are then necessary to reduce dross formation. Dross is molten material that clings to and solidifies on the underside of the cut edge as burr (see Section 15.1.5.2).

Small levels of impurity can cause significant deviations in cutting performance (such as a reduction in maximum cutting speed or increase in dross adhesion) compared to that of the pure gas, be it oxygen or inert gas. This is illustrated in Fig. 15.6, which shows a variation of the cutting speed for various levels of oxygen gas purity. This sensitivity to contamination is due to the build up of a boundary layer of the

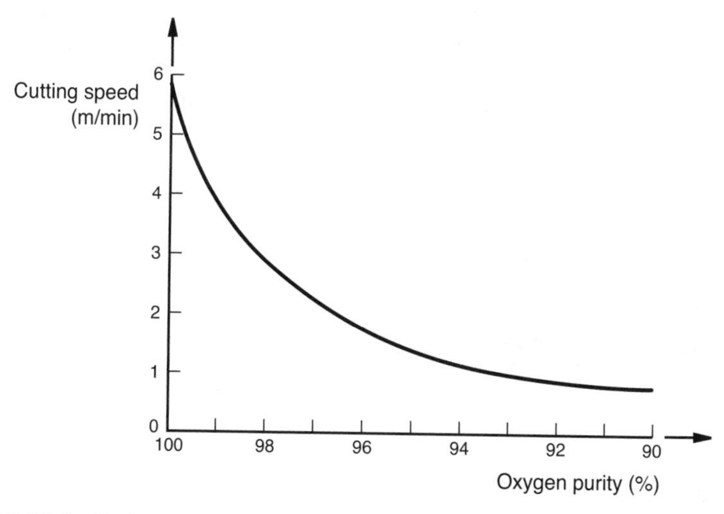

FIGURE 15.6 Cutting speed as a function of oxygen gas purity. Mild steel of thickness 2 mm, which is cut using 800 W at 2.5 bar pressure. (From ICALEO 1992 Proceedings, Vol. 75. Copyright 1993, Laser Institute of America. All rights reserved).

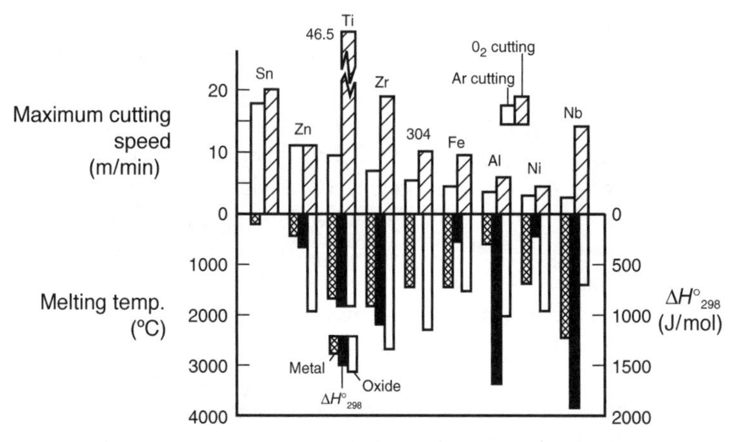

FIGURE 15.7 Comparison of maximum cutting speed with O_2 and Ar assist gases for different metals. ΔH° is the energy associated with exothermic reaction of the process. (From Miyamoto, I., and Maruo, H., 1991, *Welding in the World*, Vol. 29, No. 9/10, pp. 12–23.)

contaminant at the liquid–cut front interface. It decreases the oxidation rate of the material, thereby lowering the energy input to the cut zone.

Oxygen-Assisted Cutting To summarize the impact of oxygen assist in cutting of various materials, Fig. 15.7 illustrates the maximum cutting speeds achievable for different materials when cut using inert gas and also oxygen assist. In addition to the exothermic reaction that results, the use of oxygen assist in cutting also reduces the viscosity and surface tension for some metals, making it easier for the molten metal to flow. Table 15.2 compares the surface tension of some metals and oxides. The oxide film formed also tends to increase beam absorption.

The maximum cutting speeds achieved depend on the thermal properties of the metal. For inert gas-assisted cutting, higher speeds are obtained for low-melting and low-thermal conductivity metals. Maximum cutting speeds are higher for oxygen-assisted cutting, compared to inert gas-assisted cutting of titanium, zirconium, and niobium due to the relatively high exothermic energy associated with these metals.

TABLE 15.2 Surface Tension Values for Some Metals and Oxides

Material	Melting Temperature (°C)	Surface Tension (dyne/cm)
Cu	1085	1350
Fe	1538	1700–1800
Al_2O_3	2050–2400	360–570
Cr_2O_3	2350–2500	810
FeO		580

Table 1, Miyamoto and Maruo, *Welding in the World*, 1991.

However, the surface quality obtained is relatively poor. This is due to the fact that the oxidation region cannot be limited to the beam irradiating region as a result of the high exothermic energy. The relatively low cutting speeds achieved with aluminum and zinc is due to the high melting temperature of their oxides. The exothermic reaction associated with steel is not very high, when compared to that of other metals such as titanium, zirconium, and niobium. Thus, the cutting speeds are not very high, but the quality of cut is relatively good since the relatively low exothermic reaction results in a reaction zone that is limited to the beam irradiated region.

Care must be used in oxygen-assisted cutting since excess oxygen may result in overreaction or uncontrollable burning away from the main cutting direction, especially for thick materials. That may increase striation formation, and thus, roughness.

Gas Nozzles Some of the nozzle designs that are commonly used for coaxial application of a gas jet during laser cutting are shown in Fig. 15.8. The most commonly used ones are the conical, convergent, and convergent–divergent designs.

Low or subsonic flow rates from a coaxial nozzle are found to produce repeatable results, especially when the nozzle is positioned close to the workpiece, that is, with a standoff (nozzle to work) distance of about 0.1–1.5 mm. Short standoff distances are used despite the fact that the effective cutting pressure decreases rather slowly with distance from the nozzle tip (about 15% decrease over a nozzle gap of 10 mm). This is because the jet direction is highly sensitive to manufacturing imperfection or tip damage. Even though higher pressures or flow rates result in higher cutting speeds and quality, they do not necessarily ensure repeatability of the process due to turbulence in the gas flow.

A version of the ring nozzle design that has been shown to produce drossles, oxide-free edge when it is used to cut metals, especially stainless steel and aluminum of thicknesses up to about 3.4 mm, is shown in Fig. 15.9a. The resulting process has been

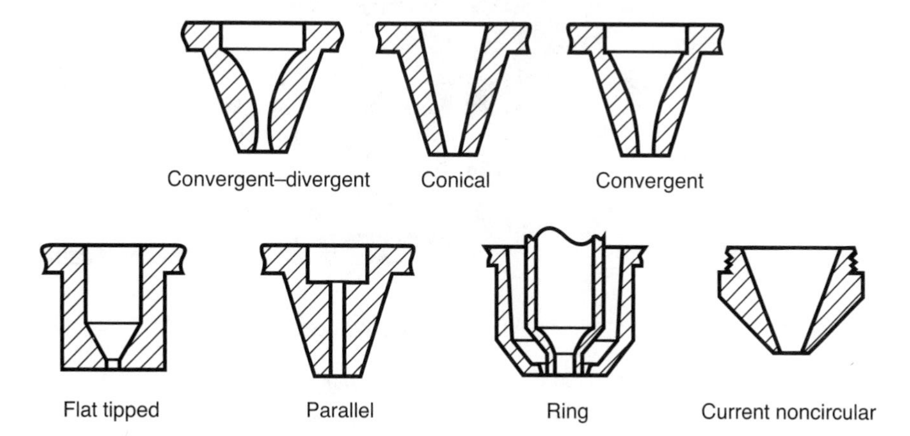

FIGURE 15.8 Nozzle designs for laser cutting. (From Fieret, J., Terry, M. J., and Ward, B. A., 1986, *SPIE, Laser Processing: Fundamentals, Applications, and Systems Engineering*, Vol. 668, pp. 53–62.)

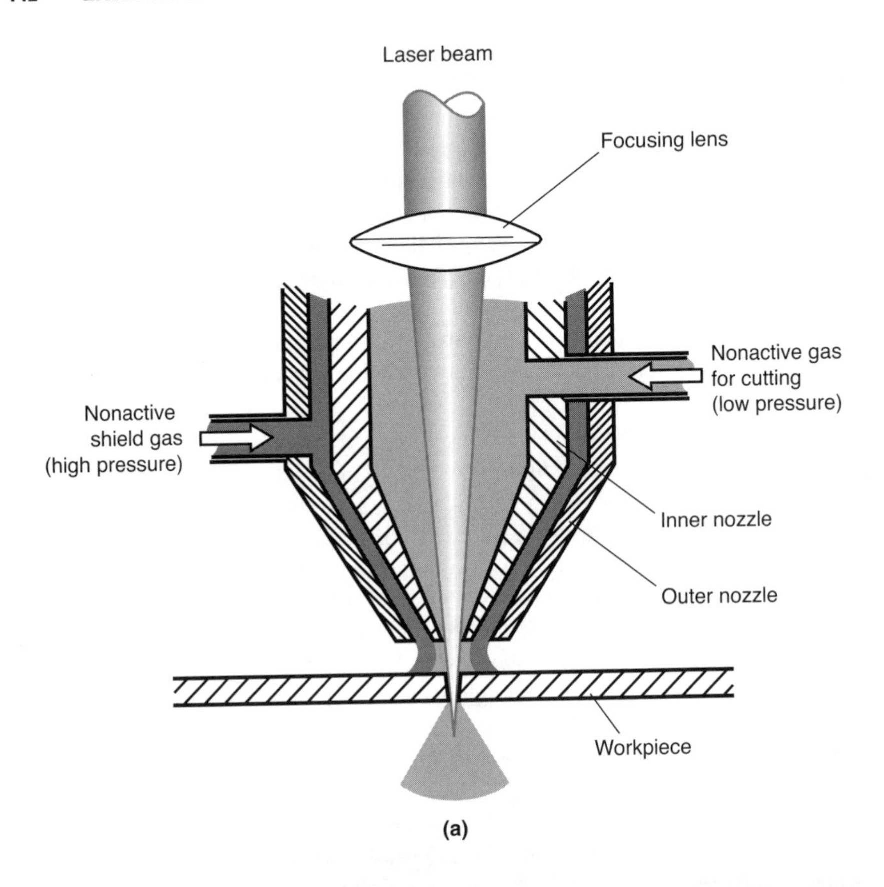

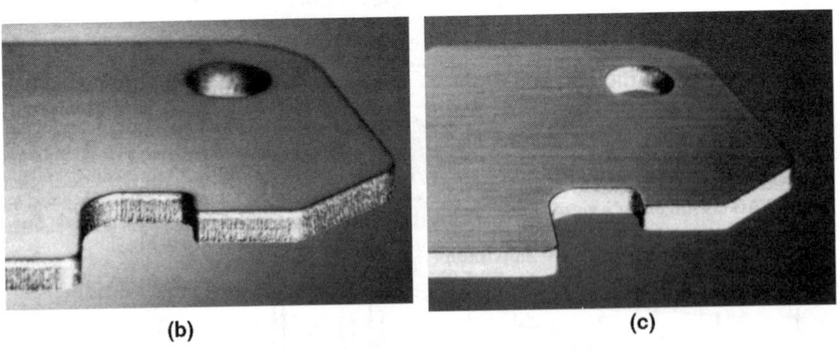

FIGURE 15.9 (a) Schematic of cutting using the ring design. (b) Stainless steel plate cut using a traditional oxygen gas assist. (c) Stainless steel plate cut using the "clean-cut" method. (From Kawasumi, H., 1990, *Industrial Laser Annual Handbook*, Penwell Books, pp. 141–143.)

referred to as the "clean-cut" technique. The low-pressure (about 1 atm) nonoxide gases flowing through the inner nozzle protects the lens from the vapor plume, while the high-pressure (about 5 atm) nonoxide gases flowing through the outer nozzle remove viscous material. Figure 15.9b and c compare the cut quality obtained for a 2 mm thick stainless steel plate cut using a traditional oxygen gas assist and the "clean-cut" method.

15.1.3.5 Effect of Focal Position Due to divergence of the beam, it is essential to have consistency in the location of the focal point relative to the workpiece surface. Best results (minimum kerf width) are obtained when the focal point is either on the workpiece surface or just below it (Fig. 15.10). For thick plates, it might be preferable to have the focus positioned one-thirds of the plate thickness below its surface.

15.1.4 Laser Cutting Principles

The laser fusion cutting process is illustrated schematically in Fig. 15.1. Both the laser beam and gas jet impinge on the workpiece surface. Sources of heat for the cutting process include

1. Absorbed laser radiation.
2. Energy due to exothermic reaction between the base material and assist oxygen gas.

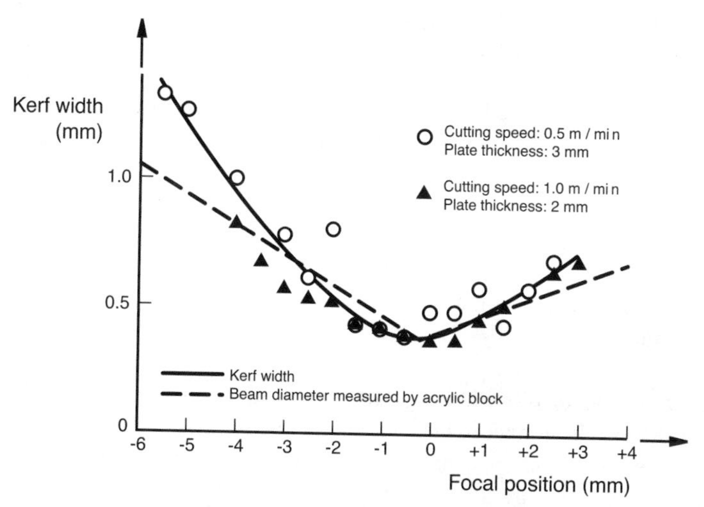

FIGURE 15.10 Effect of focal position on kerf width for a 7075-T6 aluminum alloy cut with a 1.8 kW CO_2 laser power and argon assist gas. (From Masumoto, I., Kutsuna, M., and Ichikawa, K., 1992, *Transactions of the Japan Welding Society*, Vol. 23, No. 2, pp. 7–14.)

The energy produced melts, and may partly evaporate material in front of the beam. The pressurized gas jet then ejects the molten material from the lower surface of the workpiece. Material removal thus occurs by

1. Evaporation from the surface of the molten layer.
2. Ejection from the lower surface of the workpiece due to friction between the gas flow and the surface of the molten layer.

Energy is lost from the process by

1. Heat conduction.
2. Evaporation from the erosion front.
3. Melting of solid metal.
4. Ejection of the molten metal.
5. Reflection, radiation, and convection cooling by the gas flow. For subsonic gas flow, the convection cooling effect is found to be negligible.

In the initial stages of the process, the entire beam and gas jet hit the workpiece surface. However, once the process is initiated, a kerf is formed, and only a portion of the gas and laser beam impinge on the top surface of the workpiece directly ahead of the kerf, and that portion may be reflected back. The remaining portion of the beam propagates downward into the kerf, and is partly absorbed at the front end of the kerf (erosion front), which is slightly inclined to the vertical. For a beam that is linearly polarized in the cutting direction, most of the beam power is absorbed at the erosion front. A thin layer of molten material forms at the interface between the kerf and the solid base material in front of it, and is ejected by the mechanisms listed earlier. The thickness of the melt increases with increasing cutting speed since more molten metal is then produced per unit time. However, an increase in assist gas velocity decreases the melt thickness since it results in a more rapid ejection of molten material (Fig. 15.11).

For a given laser power, the average temperatures attained in the molten layer decrease as the thickness of the workpiece increases, until above a certain thickness, the average temperature falls below the melting temperature. This is the maximum thickness that can be cut with a laser of that power level, all other laser parameters being the same.

15.1.4.1 Beam Absorption During Laser Cutting Absorption of the laser beam during cutting may be enhanced by a number of phenomena, including surface roughness, oxide formation, and plasma formation. Absorption may be by Fresnel absorption or by inverse bremsstrahlung (see Section 16.3.2.1). However, since the direction of beam polarization strongly influences the cutting efficiency, this suggests that Fresnel absorption is the principal mechanism. Since a significant portion of the energy generated is removed with the ejected molten metal, temperatures in the cutting region may not be high enough to generate the high vapor densities essential for plasma

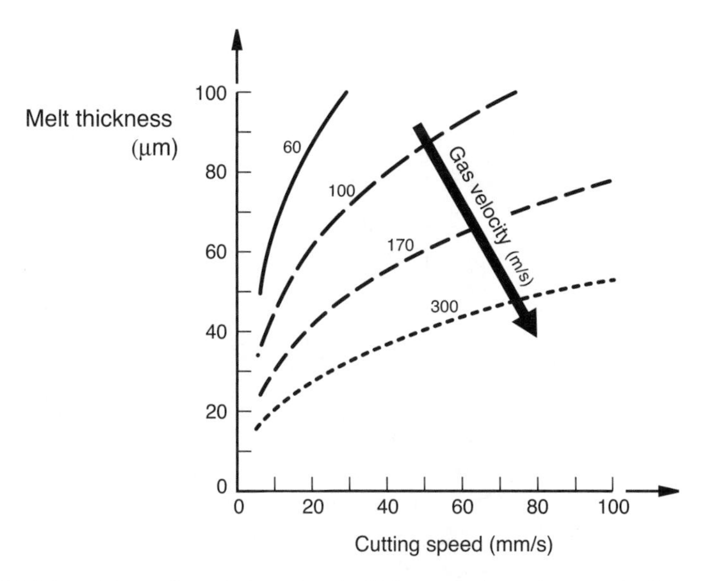

FIGURE 15.11 Schematic variation of melt thickness with cutting speed. (From Vicanek, M., Simon, G., Urbassek, H. M., and Decker, I., 1987, *Journal of Physics D*: Applied Physics, Vol. 20, IOP Publishing Ltd.)

formation. The role of the plasma in the cutting process (and therefore absorption by inverse bremsstrahlung) is thus minimal, as opposed to the case of welding and drilling.

As illustrated in Fig. 9.3c, for a p-polarized light (plane-polarized light parallel to the plane of incidence) which is incident on metallic surfaces, the maximum value of the absorption coefficient occurs at high angles of incidence, θ (angle between the laser beam and the surface normal), almost glancing incidence (i.e., when the laser beam is almost parallel to the surface on which it is incident) (Fig. 15.12). This is supported by the fact that the inclination of the cut surface on which the laser beam is incident is such that the angle of incidence is only a few degrees below 90°. It is also about an order of magnitude greater than that at normal incidence (when the laser beam is normal to the surface on which it is incident). Furthermore a p-polarized beam is absorbed much more strongly than an s-polarized light. Thus a p-polarized beam is more efficient for laser cutting in a straight line.

Absorption efficiency of the laser beam in cutting varies with the beam intensity, being low at low intensities, increasing with increasing beam intensity to a maximum, and then decreasing with further increases in the intensity (Fig. 15.13). This is due to the dependence of absorption on the angle of incidence. At low intensities, the cutting front is relatively flat, and that results in low absorption since the angle of incidence is then almost zero, that is, the laser beam is almost normal to the workpiece surface. With increasing intensity, the angle of incidence increases, further increasing the absorption as described in Section 9.4. At very high intensities, the cutting front is

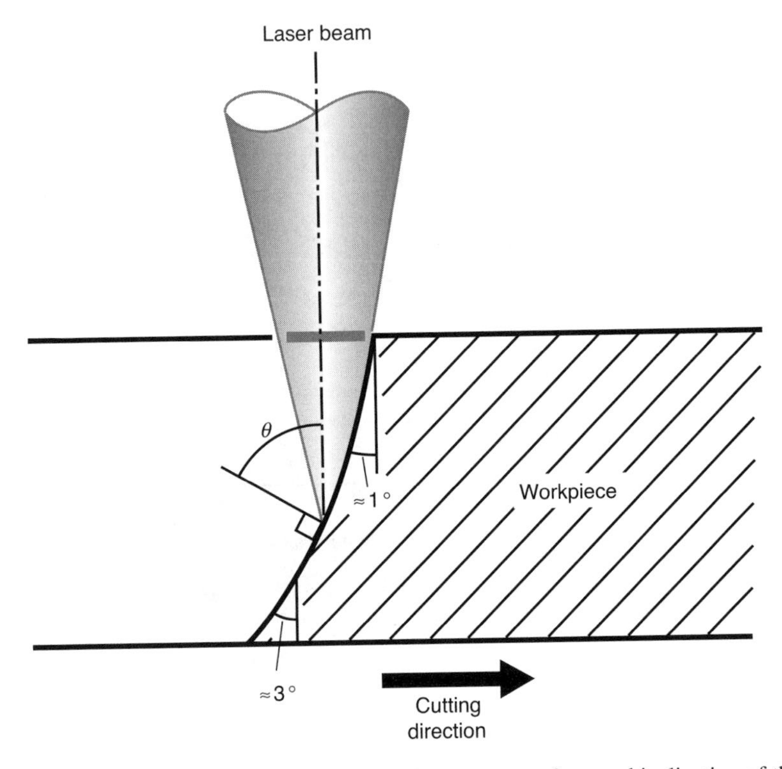

FIGURE 15.12 Angle of incidence of laser beam on cut surface, and inclination of the cut surface with the vertical.

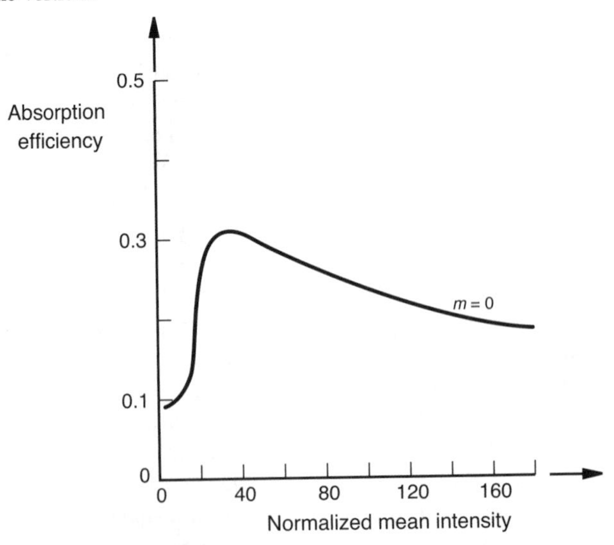

FIGURE 15.13 Variation of absorption efficiency of a p-polarized beam with beam intensity for a Gaussian beam ($m = 0$). (From Schulz, W., Simon, G., Urbassek, H. M., and Decker, I., 1987, *Journal of Physics D: Applied Physics*, Vol. 20, IOP Publishing Ltd.)

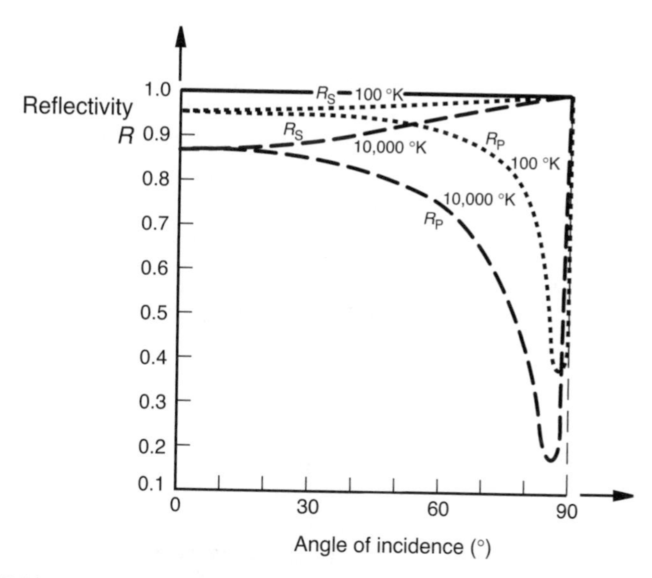

FIGURE 15.14 Variation of reflectivity of mild steel with beam incidence angle. R_p is the reflectivity of p-polarized light, and R_s is the reflectivity of s-polarized light. (From ICALEO 1991 Proceedings, Vol. 74. Copyright 1992, Laser Institute of America. All rights reserved.)

almost vertical, resulting in almost 90° incidence, at which the absorption is relatively low (Fig. 15.14). For typical cutting conditions, the inclination of the cutting front to the vertical is less than 1° close to the top of the workpiece and increases to an inclination of 3°–5° toward the bottom of the workpiece.

15.1.4.2 Process Modeling The laser cutting process is modeled in this section to relate process inputs such as gas velocity, cutting velocity, laser power, and material properties to the process outputs such as the melt thickness and quality of the cut surface. The gas jet in laser cutting significantly influences the process dynamics. Modeling of the process thus involves consideration of the following:

1. Gas flow.
2. Flow of the molten metal.
3. Heat flow in the solid material.

These flow regimes are now addressed in the following sections.

Gas Flow Model To provide a basic understanding of the process without extensive analysis, we consider a nonreactive inert assist gas, thereby neglecting exothermic reaction, and make the following simplifying assumptions:

1. Gas flow is subsonic and laminar.
2. The flow regime consists of a region of irrotational motion and a boundary layer region at the cutting front.

3. The cutting front is an inclined plane.
4. The flow only separates at the lower edge.
5. The gas stream is bounded by free streamlines. This is an oversimplification since the gas is surrounded to a large extent by air of almost the same density as the gas. Turbulent mixing of the gas stream with the surrounding air increases downstream from the upper edge. Thus for relatively thick workpieces, the assumption will only hold for the upper portion of the cutting front.

With this background, and considering the schematic in Fig. 15.15, we can obtain an expression for the pressure distribution, P_g, in the gas stream using Bernoulli's equation:

$$\frac{P_1}{\rho g} + \frac{{u_1}^2}{2g} + z_1 = \frac{P_2}{\rho g} + \frac{{u_2}^2}{2g} + z_2 + \text{friction} \tag{15.1a}$$

where g is the acceleration due to gravity (m/s^2), P is the pressure at a given location (Pa), u is the velocity at that location (m/s), z is the height of the location of interest and subscripts 1 and 2 refer to two different locations.

Neglecting friction and the small height difference in the laser cutting area and considering the reference gage pressure at the nozzle to be zero (atmospheric), we have

$$P_g = \frac{1}{2}\rho_g {u_{go}}^2 - \frac{1}{2}\rho_g {u_g}^2 \tag{15.1}$$

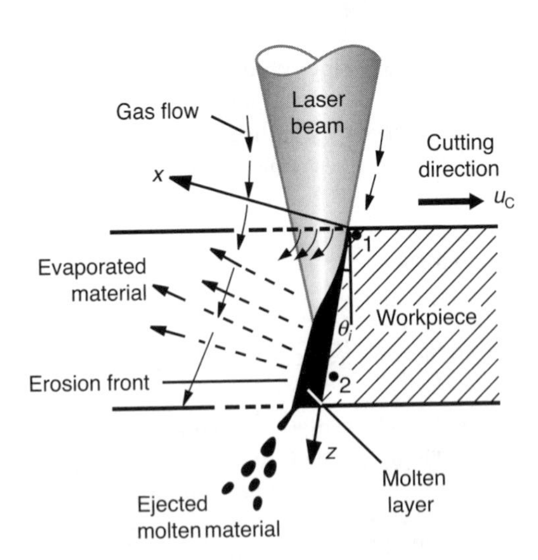

FIGURE 15.15 A two-dimensional view of the cutting process. (From Vicanek, M., and Simon, G., 1987, *Journal of Physics D: Applied Physics*, Vol. 20, IOP Publishing Ltd.)

where P_g is the pressure distribution in the gas stream, u_g is the gas velocity distribution, u_{go} is the gas velocity at the nozzle, and ρ_g is the room temperature gas density, which is assumed to be constant with pressure (i.e., incompressible).

The velocity distribution is needed to obtain the pressure distribution from equation (15.1).

For the shear stress distribution, we first consider the coordinate system shown in Fig. 15.15 and then make the following assumptions:

1. Flow is Newtonian.
2. The thickness of the molten layer can be neglected.

Now considering the fact that the shear stress is proportional to the shear strain rate since flow is Newtonian, the shear stress distribution at the cutting front can be expressed as

$$\tau = \mu_g \frac{\partial u_{gz}}{\partial x}\bigg|_{x=0} \tag{15.2}$$

The velocity field is obtained from the boundary layer equation:

$$\rho_g \left(u_{gz}\frac{\partial u_{gz}}{\partial z} + u_{gx}\frac{\partial u_{gz}}{\partial x} \right) = -\frac{\partial P_g}{\partial z} + \frac{\partial}{\partial x}\left(\mu_g \frac{\partial u_{gz}}{\partial x} \right) \tag{15.3}$$

where μ_g is the gas viscosity ($Pa\ s$), τ is the shear stress distribution at the cutting front (N/m^2), u_{gz} is the gas velocity in the z-direction, and u_{gx} is the gas velocity in the x-direction.

The pressure gradient in equation (15.3) is assumed to be known. The resulting pressure and shear stress distributions along the cutting front, obtained from equations (15.1)–(15.3) are of the form illustrated in Figs 15.16 and 15.17, respectively. The pressure distribution has a peak just below the upper edge of the cutting front, and then reduces monotonically to zero at the lower edge. The pressure also increases with increasing inclination (θ_i) of the cutting front from the vertical. The smaller the cutting front slope, the lower the resistance to flow, and thus the smaller the pressure. The shear stress, however, remains relatively constant over most of the cutting front, and is less sensitive to the inclination.

Modeling the Melt Flow Since most of the molten material is forcibly ejected by the gas jet, only a thin layer of melt remains on the cutting front, with a thickness of the order of 40 μm. Thus, flow may be assumed, with a reasonable degree of accuracy, to be two dimensional. In other words, this is a boundary layer problem. Using the coordinate system shown in Fig. 15.15 and assuming incompressible flow, the continuity equation can be expressed as

$$\frac{\partial u_{lz}}{\partial z} + \frac{\partial u_{lx}}{\partial x} = 0 \tag{15.4}$$

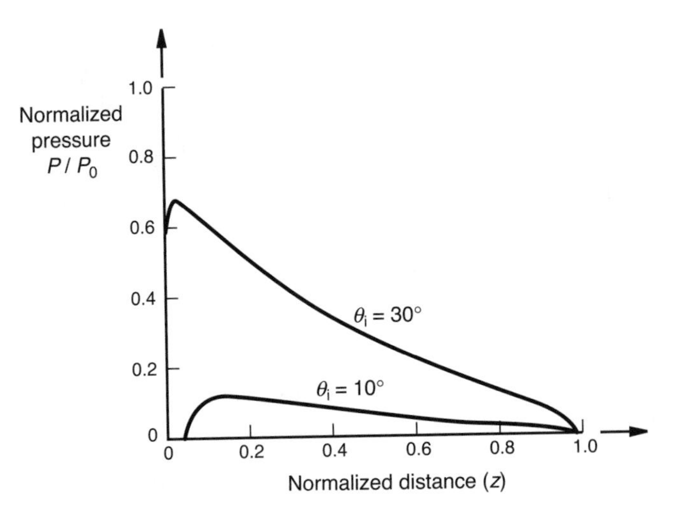

FIGURE 15.16 Schematic of pressure distribution along the cutting front. P_0 is the reference pressure at the stagnation point. (*Source:* From Vicanek, M., and Simon, G., 1987, *Journal of Physics D: Applied Physics*, Vol. 20, IOP Publishing Ltd).

Likewise, the Navier–Stokes equation is given by

$$\rho_l \left(\frac{\partial u_{lz}}{\partial t} + u_{lz} \frac{\partial u_{lz}}{\partial z} + u_{lx} \frac{\partial u_{lz}}{\partial x} \right) = - \frac{\partial P_l}{\partial z} + \mu_l \left(\frac{\partial^2 u_{lz}}{\partial z^2} + \frac{\partial^2 u_{lz}}{\partial x^2} \right) \tag{15.5}$$

where P_l is the pressure in the molten metal, and is assumed to be independent of x since this is a boundary layer problem, u_{lz}, u_{lx} are the molten material velocities in

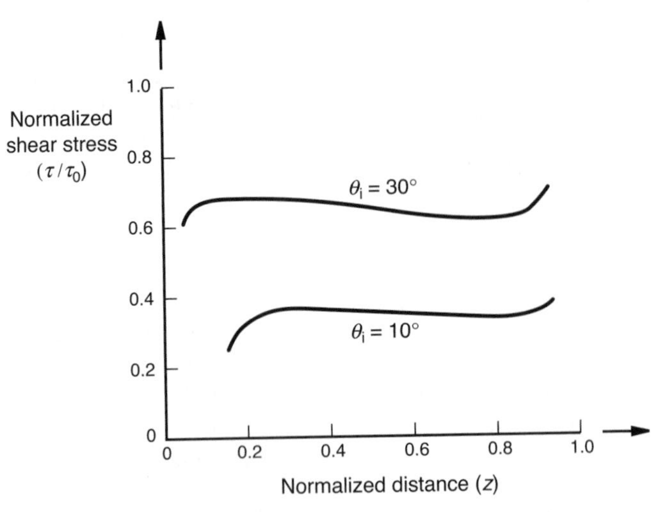

FIGURE 15.17 Schematic of shear stress distribution along the cutting front. τ_o is a reference shear stress. (From Vicanek, M., and Simon, G., 1987, *Journal of Physics D: Applied Physics*, Vol. 20, IOP Publishing Ltd.)

the z and x directions, respectively, μ_l is the dynamic viscosity of the molten metal, and ρ_l is the density of the molten metal.

Boundary conditions have to be specified at the solid/liquid interface, $x = 0$, and also at the surface of the liquid layer, $x = h_x(z, t)$, where h_x, the melt thickness, is a function of both position and time. At the solid/liquid interface, material is uniformly moved with the cutting velocity u_c, Fig. 15.15, and thus continuity there requires that

$$u_{lz} = u_c \, sin \, \theta_i \quad \text{at} \quad x = 0 \tag{15.6a}$$

$$u_{lx} = u_c \cos\theta_i \quad \text{at} \quad x = 0 \tag{15.6b}$$

where u_c is the cutting velocity and, θ_i is the inclination of the cut surface with the vertical.

There are normal and tangential stresses, P_g and τ, respectively, at the melt surface due to the gas jet. For equilibrium, we have

$$P_l = P_g + \frac{\gamma_l}{r_c} \quad \text{at} \quad x = h_x \tag{15.7a}$$

$$\mu_l \frac{\partial u_{lz}}{\partial x} = \tau_{gl} \quad \text{at} \quad x = h_x \tag{15.7b}$$

where h_x is the melt thickness, r_c is the local radius of surface curvature, γ_l is the surface tension coefficient of the molten material, and τ_{gl} is the shear stress due to friction at the interface between the gas and molten material.

The following kinematic condition ensures that there is no mass transfer through the surface:

$$\frac{\partial h_x}{\partial t} + u_{lz} \frac{\partial h_x}{\partial z} = u_{lx} \quad \text{at} \quad x = h_x \tag{15.8}$$

Evaporation effects are neglected in equation (15.8) for simplicity. Other conditions that can be applied are

$$u_{lz} = 0 \quad \text{when} \quad t = 0 \tag{15.9a}$$

$$h_x = 0 \quad \text{when} \quad t = 0 \tag{15.9b}$$

$$h_x = 0 \quad \text{at} \quad z = 0 \tag{15.9c}$$

Knowing P_g and τ from the solution to the gas flow equations, equations (15.4)–(15.9) can be solved numerically for the pressure, velocity, and height of the molten layer. Typical trends obtained from solution of these equations are illustrated in Fig. 15.11, which shows that increasing the cutting velocity u_c increases the melt thickness h_x, and that increasing the gas jet velocity decreases the melt thickness.

The Heat Flow Effect The analyses in the preceding sections neglect the effect of heat flow. A more complete solution of the pressure and stress fields requires consideration of the thermal effect. The governing equation for heat flow is given by equation (10.2) as

$$\rho c_p \frac{dT}{dt} = \frac{\partial}{\partial x}\left(k\frac{\partial T}{\partial x}\right) + \frac{\partial}{\partial y}\left(k\frac{\partial T}{\partial y}\right) + \frac{\partial}{\partial z}\left(k\frac{\partial T}{\partial z}\right) + q_s \tag{10.2a}$$

In Chapter 10, this equation was first simplified by neglecting the internal energy generation term, q_s. The resulting equations associated with the line heat source model, equations (10.16), (10.18), and (10.21), for the temperature distribution, peak temperatures, and cooling rates, are still applicable as initial approximations for conduction heat flow in the workpiece when an inert gas or nitrogen is used as the assist gas. When oxygen or air is used as the assist gas, however, q_s cannot be neglected, as the exothermic reaction is then significant enough to influence the resulting temperatures.

For a more accurate analysis, however, the continuity equation, Navier–Stokes equation, and the thermal equation will have to be solved numerically, with the appropriate boundary conditions for the gas flow, melt flow, and solid workpiece regimes.

Example 15.1 A 2.5 mm thick aluminum plate is cut using a 1.5 kW laser. If the resulting kerf width is 0.3 mm; the ambient temperature is 25°C; and 5% of the incident laser beam is absorbed by the plate, determine the cutting speed. You may assume that there is no vaporization, and that there are no energy losses by conduction, convection, or radiation.

Solution:

We apply the discussion in Chapter 10 to this problem. The material properties are obtained from Appendices 10D and 10E as

Average density, $\rho = 2700\ \mathrm{kg/m^3}$
Average specific Heat, $c = 900\ \mathrm{J/kgK}$
Latent heat of fusion, $L = 397\ \mathrm{kJ/kg}$
Melting temperature, $T_m = 660.4^\circ\ \mathrm{C}$

From equation (10.1), we have

$$Q_a = q \times t = m_a c_p \Delta T_m + m_a L_m + m_a c_p \Delta T_v + m_a L_v + Q_l$$

where q is the power supplied, and t is the cutting time. Since it is assumed that there is no vaporization, and that there are no energy losses by conduction, convection, or radiation, we have

$$q \times t = m_a c_p \Delta T_m + m_a L_m = \rho V (c_p \Delta T_m + L_m)$$

or

$$q \times t = \rho \times l \times w_{\mathrm{k}} \times h(c_p \Delta T_{\mathrm{m}} + L_{\mathrm{m}})$$

where l is the length of the cut that is made, w_k is the kerf width, and h is the plate thickness. Thus the average cutting speed, u_{c}, is

$$\begin{aligned} u_{\mathrm{c}} &= \frac{l}{t} = \frac{q}{\rho \times w_{\mathrm{k}} \times h(c_p \Delta T_{\mathrm{m}} + L_{\mathrm{m}})} \\ &= \frac{1500 \times 0.05}{2.7 \times 10^{-6} \times 0.3 \times 2.5(900 \times (660.4 - 25) + 397 \times 10^3)} \\ &= 38.23 \ \mathrm{mm/s} = 2.3 \ \mathrm{m/min} \end{aligned}$$

15.1.5 Quality of Cut Part

The major factors that determine the quality of the cut part are the surface striations, dross formation, and cracking. In the following sections, we discuss the mechanisms of striation and dross formation. Cracking is discussed in relation to specific materials in Section 15.1.6, and also more extensively in Section 11.2.2.

15.1.5.1 Striations of the Cut Surface The surfaces cut using a laser beam normally have a nearly periodic striation pattern, Fig. 15.18a, that results in a surface roughness that may also vary in the thickness direction (Fig. 15.18b), depending on the processing conditions.

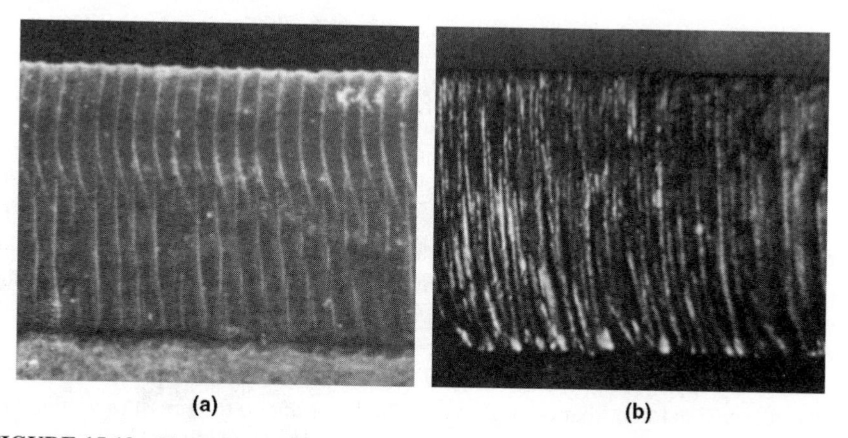

FIGURE 15.18 Illustration of striations formed on a surface after laser cutting. (a) Periodic nature of the striation pattern. (b) Variation of striation pattern in the thickness direction for relatively thick materials. (From (a)*LIA Handbook of Laser Materials Processing*, 2001, Laser Institute of America, Orlando, FL; (b) Powell, J., Frass, K., Menzies, I. A., and Fuhr, H., 1988, *SPIE, High Power* CO_2 *Laser Systems and Applications*, Vol. 1020, pp. 156–163.)

Primary causes of these striations are the following:

1. Vibrations in the motion unit.
2. Fluctuations in the laser power.
3. Fluctuations in the gas flow.
4. Hydrodynamics of the molten metal flow.

Vibrations in the motion unit is a classical dynamics or robotics problem and will not be further considered. Fluctuations in the absorbed laser power may be due to changes in absorption that result from plasma formation. They may also result from back reflections into the resonator, dynamics of the root blowers pumping the lasing gas in a CO_2 laser, dynamics of switching in a gas laser, and dirty optics. The effect of melt hydrodynamics may be analyzed by considering small perturbations in the stationary behavior of the system. We simplify the analysis by making the assumption that the properties of the molten material are independent of z. This is an appropriate assumption when the workpiece thickness h is much greater than the wavelength of the disturbance, that is, $h >> 1/k_z$, where $k_z = 2\pi/\lambda_z$ is the wavenumber in the z-direction. λ_z is the wavelength in the z-direction. Flow in the x-direction is neglected since $u_z >> u_x$. On this basis, then, the perturbations can be approximated as plane waves:

$$\delta u_z = U_z(x)e^{ik_z z}e^{\chi t} \tag{15.10}$$

$$\delta u_x = U_x(x)e^{ik_z z}e^{\chi t} \tag{15.11}$$

$$\delta h = He^{ik_z z}e^{\chi t} \tag{15.12}$$

where U_z, U_x, H are the amplitudes of oscillations in the z-velocity component, x-velocity component, and melt thickness, respectively, and χ is the complex growth rate and is a function of the wavenumber k_z, which is real valued.

The behavior of the melt under such small disturbances can be obtained by substituting equations (15.10)–(15.12) into equations (15.4)–(15.6) and analyzing. However, this is still the subject of research and thus beyond the scope of this book. Here, a qualitative discussion of the subject is provided. When the system is subjected to infinitesimal disturbances, then for wavenumbers that result in a growth rate χ with a positive real part, the flow is unstable, resulting in striations being formed. On the contrary, when the real part of χ is negative, the perturbations, δu_z, δu_x, and δh, will die out, resulting in a smoother surface.

Using the approximations outlined, it can be shown that the process is stable under infinitesimal disturbances only when the force due to the pressure gradient of the gas flow is much smaller than the shear force due to friction. However, since the two parameters are normally about the same order of magnitude, instability is almost always inevitable. However, an increase in surface tension tends to stabilize the process by smoothening the surface and thus minimizing perturbations. Experimental results further show that the system can be made more stable by increasing the assist gas velocity or decreasing the cutting velocity when the cutting velocity is high. On the contraly, at low cutting velocities, an increase in the assist gas velocity or decrease in cutting velocity makes the system less stable.

A unique aspect of the striations is the two distinct patterns evident on the surface (Fig. 15.15). One pattern is closer to the upper surface of the workpiece, with relatively fine striations. The other pattern is on the lower portion of the cut surface, with relatively coarse striations. The two patterns are separated by a distinct line that is almost parallel to the workpiece surface. The two striation patterns result from the temperature distribution in the molten layer in the vertical direction, being higher in the upper portion compared to the lower portion. The speed of the assist gas as it enters the kerf is generally subsonic. However, its temperature increases as it penetrates deeper into the kerf, resulting in a decrease in the gas density. In a sense, then, the cut acts as a converging nozzle, increasing the gas flow velocity. When the flow reaches sonic velocities, it becomes turbulent and thus changes the dynamics of the molten layer. The turbulence in the gas flow also cools the molten layer more effectively, and since the oscillation frequency of the molten layer when subjected to perturbations is higher for higher temperatures, the frequency of the striation pattern is lower for the lower portion of the cut surface. The characteristic frequencies of the oscillations depend on the workpiece thickness, and may be of the order of 10^3 cycles/s, increasing with cutting speed. As the thickness of the cut piece increases, the striations tend to become irregular, and the average roughness of the cut surface increases (Table 15.3).

The increase in roughness of the lower portions of the cut surface with workpiece thickness is also partly due to the inability to provide adequate energy on the lower portion of the workpiece to produce molten metal of very low viscosity.

15.1.5.2 Dross Formation Dross is essentially material that clings to the lower edge of the workpiece, and appears as solidified drops after laser cutting (Fig. 15.19). The formation of dross depends on the surface tension and viscosity of the molten material. The higher the surface tension or viscosity, the greater the tendency to form dross, since that prevents smooth flowing of the molten material out of the

TABLE 15.3 Illustration of Roughness Variation of Cut Surface with Workpiece Thickness for a Steel Plate

Thickness (mm)	Upper Roughness (μm)	Lower Roughness (μm)
1	3	3
3	9	11
5	24	30
7	40	55
9	70	60
10	85	110

Table 3, Schoucker, *Industrial Laser Annual Handbook*, 1986.

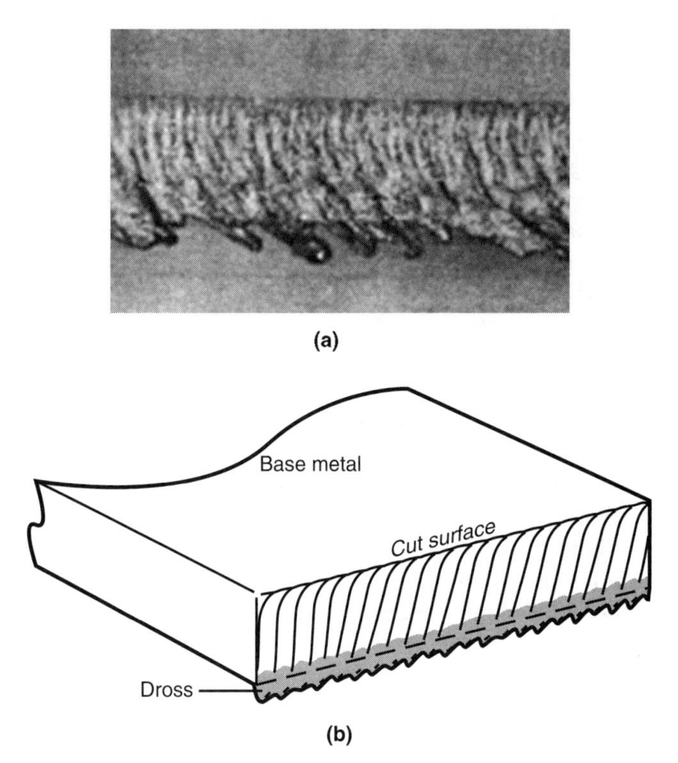

FIGURE 15.19 Dross formed on the lower edge of the workpiece after laser cutting. (a) Photograph of dross. (b) Schematic of dross. (From Powell, J., Frass, K., Menzies, I. A., and Fuhr, H., 1988a, *SPIE, High Power CO_2 Laser Systems and Applications*, Vol. 1020, pp. 156–163.)

reaction zone. Inert gas-assisted cutting has a greater tendency to form dross as compared to oxygen-assisted cutting of the same material since the surface tension of the pure metal is generally greater than that of its oxide. Thus, higher pressures (over 1 MPa or 10 bars) are required to achieve dross-free cutting with inert gas assist than with oxygen assist. Furthermore, the dross formed during inert gas assisted cutting is usually more difficult to remove than the more brittle oxide dross formed when oxygen assist is used.

15.1.6 Material Considerations

The effectiveness of a laser in processing a given material depends on the properties of the material, for example, absorptivity (1 - reflectance) at that wavelength, thermal conductivity, melting and boiling points, heat of reaction, and surface tension of the molten material. The reflectance, thermal, and physical properties of some common metals are summarized in Appendices 10D–10F. In the following sections, we shall consider the behavior of both metals and nonmetals when cut with a laser.

15.1.6.1 Metals

Plain Carbon Steels Oxygen-assisted cutting is normally used for carbon steels where the exothermic reaction between oxygen and iron aids in the cutting process, resulting in cutting speeds, which are much higher than those achieved when inert gas is used. The exothermic reaction is given by the following equation:

$$\mathrm{Fe} + \frac{1}{2}\mathrm{O_2} = \mathrm{FeO} + \text{Energy (258 kJ/mol)} \quad \text{at} \quad 2000\,\mathrm{K} \tag{15.13}$$

$$2\mathrm{Fe} + \frac{3}{2}\mathrm{O_2} = \mathrm{Fe_2O_3} + \text{Energy (827 kJ/mol)} \quad \text{at} \quad 2000\,\mathrm{K} \tag{15.14}$$

Oxygen-assisted cutting also helps reduce dross formation, since the viscosity of the molten metal decreases when oxidized, making it easier to be removed by the gas jet. The presence of phosphorus and sulfur in the carbon steel can cause burnout along the cut edge. An increase in carbon content of the steel tends to improve the edge quality, but with a greater tendency for cracking in the heat-affected zone (see Section 11.2.2).

Galvanized Steel Laser cutting of galvanized steel often results in rough edges with extensive dross. This is most likely due to the formation of zinc oxide which does not flow easily. To obtain reasonable quality, lower cutting speeds are normally used, compared to the speeds used for uncoated steels.

Stainless Steels The small heat-affected zone associated with laser cutting minimizes any impact that the cutting process may have on its corrosion resistance. In using oxygen assist, the heat of reaction that results is not as significant as that associated with plain carbon steels. Thus the cutting speeds achieved for stainless steels are relatively lower than those for carbon steels. As a result, oxygen assist is normally not recommended for cutting of stainless steels, especially since the burning caused by the assist gas may be detrimental to the cut quality. Ferritic and martensitic stainless steels tend to produce clean and smooth-edged cuts, while austenitic stainless steels tend to have dross sticking out from the lower surface of the workpiece. This is primarily due to the high viscosity of molten nickel.

Alloy and Tool Steels The edges produced in laser cutting of high alloy steels such as AISI 4340 steel are often clean and square. This is essentially due to the more precise control of the alloying element contents of such materials. Likewise, a number of tool steels are also good candidates for laser cutting. However, the tungsten-based tool steels, Groups T and H, retain a lot of heat in the molten state, resulting in often burned out cuts.

Aluminum and Its Alloys Aluminum and its alloys are highly reflective (up to 97%) of the high 10.6 μm CO_2 laser beam wavelength. This, coupled with their relatively high thermal conductivity ($\approx$ 247 W/m K), makes it difficult to initiate cutting in aluminum alloys. High beam intensities, achieved using either superpulsing or hyperpulsing (Sections 14.1.4 and "Beam Form"), are thus necessary to cut aluminum alloys. Absorption of the laser beam at the top surface can be enhanced by using any of the coating techniques discussed in Section 17.1.2.3, but anodizing aluminum has been found to be effective. This is done by coating the base material with a thin layer (approximately 20 μm) of Al_2O_3. Al_2O_3 is highly absorptive (≈100%) of the infrared CO_2 laser beam at thicknesses greater than 5 μm. Absorptivity can also be improved by graphite coating.

Aluminum undergoes a highly exothermic reaction with oxygen as

$$4Al + 3O_2 = 2Al_2O_3 + \text{Energy (1670 kJ/mol)} \quad \text{at} \quad 293\ \text{K} \tag{15.15}$$

which can increase cutting speeds to some extent. However, the exothermic reaction is not as effective in increasing the cutting speeds for the same amount of released energy as compared to the case of iron. This is because the resulting oxide layer, aluminum oxide (Al_2O_3), reduces the diffusion of atoms to the reaction front, thereby reducing the oxidation rate. The oxide layer is periodically disrupted during the cutting process by the pressurized oxygen jet, permitting some amount of oxidation to occur. The function of assist gas in cutting these materials is thus primarily to eject the molten metal. As a result, oxygen assist is normally not recommended for cutting aluminum.

Dross formation can be a problem in cutting aluminum. This is because the high thermal conductivity of the base material results in a low temperature melt. This results in the surface tension of the molten layer being relatively high, and thus more difficult to eject, leaving dross at the bottom edge of the cut. The dross formed is easily removed by mechanical means, though. The high thermal conductivity of the base material may sometimes lead to overheating of small components during cutting. This, combined with the low melting temperature, may even result in widescale melting. This can be prevented by spraying either water or soluble oil on the component during cutting. The water is kept away by the pressure of the gas jet, and cools the component by evaporation.

Laser cutting sometimes produces intergranular cracks (cracks that preferentially propagate along grain boundaries) on the surface of some aluminum alloys, making them unsuitable for aircraft structural components.

Copper and Its Alloys Both the conductivity (≈ 398 W/m K) and reflectivity (up to 98.4%) of copper are higher than those of aluminum, making it even more difficult to process with a CO_2 laser. Even though the exothermic reaction associated with the process can increase the cutting speeds, the energy released is relatively small:

$$2Cu + \frac{1}{2}O_2 = Cu_2O + \text{Energy (167 kJ/mol)} \quad \text{at} \quad 293\ \text{K} \tag{15.16}$$

$$Cu + \frac{1}{2}O_2 = CuO + \text{Energy (155 kJ/mol)} \quad \text{at} \quad 293\ \text{K} \tag{15.17}$$

Any increase in cutting speed is primarily due to increased absorption that results from the formation of the highly absorptive oxide layer in the vicinity of the cut.

A number of copper alloys such as brass and bronze are more effectively processed due to a reduced reflectivity and thermal conduction.

Titanium and Its Alloys Titanium is highly effective in absorbing the CO_2 laser beam. The exothermic reaction between oxygen and titanium when oxygen assist gas is used greatly enhances the cutting efficiency, thereby increasing the cutting speeds. The reaction with oxygen may be expressed as

$$\mathrm{Ti} + \mathrm{O_2} = \mathrm{TiO_2} + \text{Energy } (912\ \mathrm{kJ/mol}) \quad \text{at} \quad 293\,\mathrm{K} \tag{15.18}$$

A relatively thick and brittle oxide layer (of the order of 1–3 mm) is normally produced along the cut edge, since the reaction is very rapid at the high cutting temperatures. When a Ti6Al4V titanium alloy is cut with a laser using high-pressure inert gas, the resulting variation of oxygen content with depth from the cutting edge is illustrated in Fig. 15.20.

This process is mainly useful in applications where the oxide layer is not detrimental, such as low service life or electrical applications where titanium may be used for corrosion resistance, rigidity, and its lightweight characteristics. However, for aerospace applications where they are extensively used for their toughness, fatigue resistance, and high strength-to-weight ratio, these properties are highly degraded by the oxide layer formed. Cutting titanium with oxygen assist for such applications

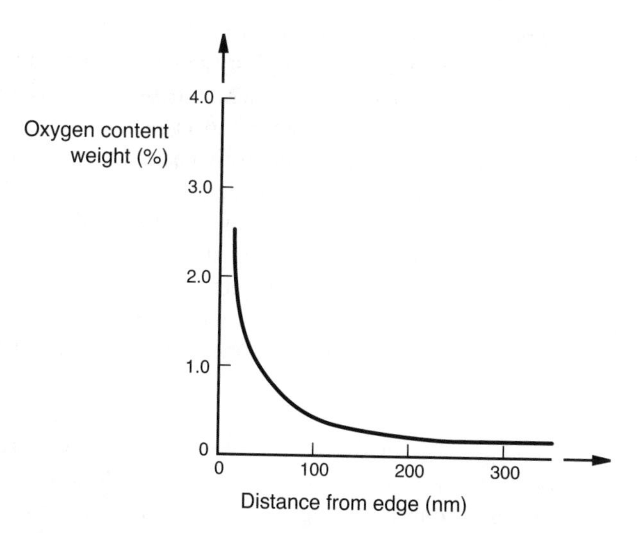

FIGURE 15.20 Illustration of oxygen content variation with distance from the cut edge in laser cutting of a titanium alloy, Ti6Al4V, using high-pressure argon assist gas. (From Powell, J., Frass, K., Menzies, I. A., and Fuhr, H., 1988, *SPIE, High Power CO_2 Laser Systems and Applications*.)

is therefore undesirable. Thus, it may be preferable to use an inert gas assist when cutting titanium alloys for aerospace applications, but at a reduced cutting rate which may be an order of magnitude lower than that attainable when using oxygen assist. However, there is a tendency for dross to form on the underside of the cut edge under such circumstances. As an example, high cutting speeds of about 3.6 m/min can be obtained in cutting a 6.4 mm thick plate with 500 W power using oxygen assist.

Nickel and its Alloys Pure nickel has a relatively low thermal conductivity ($\approx$ 59 $W/m\,K$) and reflectivity (up to 94%) compared to those of copper and aluminum, making it easier to cut using lasers. The use of oxygen assist does improve cutting speeds as a result of the exothermic reaction:

$$\mathrm{Ni} + \frac{1}{2}\mathrm{O_2} = \mathrm{NiO} + \mathrm{Energy}\ (244\,\mathrm{kJ/mol}) \quad \mathrm{at} \quad 293\,\mathrm{K} \tag{15.19}$$

This is more effective when cutting Ni–Fe alloys. Some nickel alloys, however, have reflectivities and conductivities that make them more difficult to process with lasers. Furthermore, the high viscosity of molten nickel results in dross formation.

Despite the relative ease with which a number of nickel alloys can be processed with a laser, the metallurgical effects of the process, even in the small heat affected zone, precludes its use in some aerospace applications such as jet engines where the working environment can be aggressive.

15.1.6.2 Nonmetals

Polymers In cutting polymers, the intense laser beam tends to break down the polymer chains. Edges produced in cutting thermoplastics appear polished as a result of resolidified melting. As the strength of the polymer increases, there is a tendency for charring to occur along the cut edge, since more energy is then required to break down the bonds. Polyester and polycarbonate are relatively easier to cut, while with PVC (polyvinyl chloride), phenolic, and polyimide, a significant amount of decomposed material may be found along the cut edge. Care must always be taken to contain hazardous and/or corrosive fumes that may be generated in the processing of polymers. It is generally preferable to use air rather than oxygen in cutting polymeric materials.

Both natural and synthetic rubber are effectively cut in thicknesses up to 19 mm using lasers, and they tend to exhibit slight stickiness along the edge when freshly cut.

Composites Since composites consist of different materials, the cutting conditions are determined by only one of the component materials, which may result in degradation of the other materials during cutting. One approach to this problem is to cut at higher speeds using higher power lasers. Thin sheets (say, less than 0.5 mm) of laminated composites with a polymeric material as the matrix can be trimmed using lasers before they are cured. Thicker sections or fully cured composites often result in charring, delamination, and thermal damage along the cut edge. When the properties of the constituent materials are similar, the composite material is then easier to process using a laser. In all cases, care must also be exercised in containing hazardous fumes and material.

Quartz and Glass The relatively low coefficient of expansion of quartz makes it easier to cut in comparison to glass. The high coefficient of thermal expansion of glass results in thermal shock that tends to form edge cracks.

One approach to cutting such brittle materials is by first laser scribing to remove a small amount of material from the surface of the glass, and then applying force by flexing, to fracture the material along the line of scribing. This requires only a small power density, resulting in a very small heat-affected zone size. Another approach involves controlled thermal fracture, which takes advantage of the localized thermal stresses that are set up due to the high coefficient of thermal expansion and high beam intensity to fracture the material. The process is quite controllable, with the fracture essentially following the path of the beam.

Ceramics The high transverse temperature gradients associated with laser processing can generate relatively high thermal stresses that cause cracking in ceramic materials, which are inherently brittle. Furthermore, the thermal conductivity decreases rapidly with increasing temperature in such materials, significantly reducing thermal diffusion into the body of the workpiece. Preheating the workpiece to a relatively high temperature before cutting, or preheating the region directly ahead of the laser beam reduces the transverse temperature gradient, and thus the tendency to crack.

Cutting of alumina often results in cracking. Cracks also develop during cutting of silicon carbide of thickness greater than 3 mm. Silicon nitride, on the other hand, can be cut to 4 mm thickness at about 10 mm/s. The surface roughness obtained when zirconia is cut is about 10 μm, while silicon nitride results in a surface roughness of about 100 μm.

Textiles and Fabrics Lasers are also used in the textile industry to cut fabric. To increase productivity, several layers of fabric may be stacked up and cut at the same time. However, this causes the edges of the individual layers to become welded together. This may be prevented by pressing the layers together to prevent the cutting gas from entering sideways between them.

Wood Wood can generally be cut using CO_2 lasers. The cutting rate tends to increase as the wood's density and moisture content decreases, but appears to be independent of the grain direction. One problem that is often encountered is charring of the cut edge during cutting. However, the extent of charring is usually very small, of the order of microns in depth, and reduces with increasing cutting speed.

Paper and Cardboard Laser cutting of paper and cardboard has an advantage over conventional cutting since it does not result in broken fibers which may remain attached to the paper and interfere with printing.

Example 15.2 A 1.0 kW CO_2 laser is used to cut a 4.5 mm thick titanium alloy plate using an inert gas assist. The resulting cutting speed and kerf width are 3.0 m/min and 400 μm, respectively.

(a) Calculate the surface temperature at a point 4.0 mm behind the laser beam and 3.5 mm to one side, if the heat transfer efficiency is 90%, and the ambient temperature is 25°C.
(b) Estimate the size of the heat-affected zone for this process if a cold rolled alloy recrystallizes at a temperature equal to half the absolute melting temperature of pure titanium.
(c) Obtain the surface temperature for the point in (a) if oxygen assist was used, and 50% of the exothermic reaction contributed to the cutting process.

Solution:

Again, we apply the discussion in Chapter 10 to this problem. The material properties (approximated to be the same as for pure titanium) are obtained from Appendices 10D and 10E as

Average density, $\rho = 4507\ \text{kg/m}^3 = 4.507 \times 10^{-6}\text{kg/mm}^3$
Average specific heat, $c_p = 99.3\ \text{J/kg K}$
Thermal conductivity, $k = 11.4\ \text{W/m K} = 0.0114\ \text{W/mm K}$.
Melting temperature, $T_m = 1668°\text{C} = 1941\ \text{K}$.

Other data

Atomic weight of titanium is 47.9 g/mol
Cutting speed is 3.0 m/min = 50 mm/s

The amount of material, M_t, removed per unit time is

$$M_t = \rho \times V = 4.507 \times 10^{-6} \times 50 \times 4.5 \times 0.4 = 405.6 \times 10^{-6}\text{kg/s} = 0.405\text{g/s}$$

Now the laser power, q_1, available at the workpiece is

$$q_1 = 1000 \times 0.9 = 900\ \text{W}$$

When oxygen assist is used, additional heat available from the exothermic reaction (see equation (15.18)) is

$$q_2 = 0.5 \times 912\ \text{kJ/mol} = 0.5 \times \frac{912 \times 10^3}{47.9} \times 0.405\ \text{J/s} = 3855.5\ \text{W}$$

Thus, the total power available for cutting is

$$q = q_1 + q_2 = 900 + 3855.5 = 4755.5 \text{ W}$$

Since in cutting, the heat source goes all the way through the thickness, it is considered to be a line source, and thus the heat flow is two dimensional. The distance, r, of the point of interest from the heat source is thus given by

$$r = \sqrt{\xi^2 + y^2} = \sqrt{(-4)^2 + 3.5^2} = 5.315 \text{ mm}$$

The thermal diffusivity, κ, of the material is

$$\kappa = \frac{k}{\rho c} = \frac{0.0114}{4.507 \times 10^{-6} \times 99.3} = 25.47 \text{ mm}^2/\text{s}$$

(a) For the temperature at the point (−4, 3.5), we use equation (10.22) for a thin plate:

$$T - T_0 = \frac{q}{2\pi kh} \exp\left(-\frac{u_x \xi}{2\kappa}\right) \times K_0\left(\frac{u_x r}{2\kappa}\right)$$

$$\Rightarrow T - 25 = \frac{900}{2\pi \times 0.0114 \times 4.5} \exp\left(-\frac{50 \times (-4)}{2 \times 25.47}\right) \times K_0\left(\frac{50 \times 5.315}{2 \times 25.47}\right)$$

$$\exp\left(-\frac{u_x \xi}{2\kappa}\right) = \exp\left(-\frac{50 \times (-4)}{2 \times 25.47}\right) = 50.71$$

And from Appendix 10G,

$$K_0\left(\frac{u_x r}{2\kappa}\right) = K_0\left(\frac{50 \times 5.315}{2 \times 25.47}\right) = K_0(5.217) = 0.002902$$

Thus,

$$T = 436°\text{C}.$$

(b) For the heat-affected zone size, we use equation (10.24) for a thin plate, and since the recrystallization temperature is half the absolute melting temperature, the peak temperature to be used in the equation is 697.5°C:

$$\frac{1}{T_p - T_0} = \frac{\sqrt{2\pi e}\rho c_p h u_x Y}{q} + \frac{1}{T_m - T_0} \Rightarrow \frac{1}{697.5 - 25}$$

$$= \frac{\sqrt{2\pi \times 2.71828} \times 4.507 \times 10^{-6} \times 99.3 \times 4.5 \times 50 \times Y}{900}$$

$$+ \frac{1}{1668 - 25}$$

And the heat-affected zone size is

$$Y = 1.9 \text{ mm}.$$

(c) The heat input is now 4755.5 kW. Thus,

$$\Rightarrow T - 25 = \frac{4755.5}{2\pi \times 0.0114 \times 4.5} \exp\left(-\frac{50 \times (-4)}{2 \times 25.47}\right) \times K_0\left(\frac{50 \times 5.315}{2 \times 25.47}\right)$$

or

$$T = 2196^{\circ}\text{C}.$$

This is higher than the melting point of titanium. That means the kerf width will be greater than 7 mm.

15.1.7 Advantages and Disadvantages of Laser Cutting

15.1.7.1 Advantages The general advantages of laser cutting in comparison with conventional cutting processes such as plasma arc cutting, electrical discharge machining, oxyacetylene flame cutting, and mechanical cutting are the following:

1. It results in a narrow kerf width, thereby reducing waste. The reduced kerf width also makes the production of arbitrary contours more feasible.
2. Relatively high cutting speeds.
3. Small heat-affected zone due to the relatively small total heat input. Thus, there is very little damage to the base material, making it suitable for heat sensitive and burnable materials. There is also very little residual stress and distortion.
4. It is a noncontact process, and thus there is no tool wear and no mechanical forces that could damage delicate workpieces. The lack of mechanical forces also implies less complex fixtures to hold the workpiece in place.
5. Good for both very soft (highly deformable) materials such as paper and very hard (difficult to cut) materials such as diamond.
6. High degree of flexibility (which may facilitate the cutting of complex geometries) and low level of noise.
7. It results in cut edges that are square, and not rounded, as occurs in many other thermal cutting methods.

15.1.7.2 Disadvantages The principal disadvantages of laser cutting are the following:

1. Highly reflective and conductive materials such as gold and silver are difficult to cut using lasers. Methods for improving the absorptivity of metals in general for the laser beam are discussed in Section 17.1.2.3, while those for facilitating

cutting in highly reflective materials are discussed in Sections 14.1.4 and "Bean Form".

2. The melting and rapid quenching associated with the process result in a hard edge of the cut piece for hardenable materials.
3. Laser cutting has traditionally been limited to cutting thin materials, that is, less than a few millimeters thick. Furthermore, it is generally limited to cutting through the materials. Thus blind slots, pockets, or holes are difficult to cut accurately using a laser.
4. The initial capital cost of a laser cutting system is relatively high, about two orders of magnitude higher than that of an oxy-fuel system.
5. Processing of certain materials (such as polymers) may result in the production of dangerous exhaust fumes.

15.1.8 Specific Comparison with Conventional Processes

In the following sections, laser cutting is compared with specific alternate processes. Figure 15.21 starts with a numerical comparison of the laser, oxyacetylene, and plasma arc cutting processes. This is followed in Fig. 15.22 with a graphical comparison of electrical discharge machining (EDM), laser cutting, and plasma arc cutting with respect to kerf size, accuracy, maximum cutting speed, and heat-affected zone size. Additional comparison is then made with abrasive waterjet machining and punching/nibbling.

15.1.8.1 Laser, Plasma Arc, and Oxyacetylene (Oxy-Fuel) Cutting Each of these three processes uses heat to melt and remove material. Two attributes that are compared in Fig. 15.21 for these processes are the kerf width and heat-affected zone size (HAZ). For both attributes, laser cutting is most advantageous, offering the least kerf width and heat affected zone size.

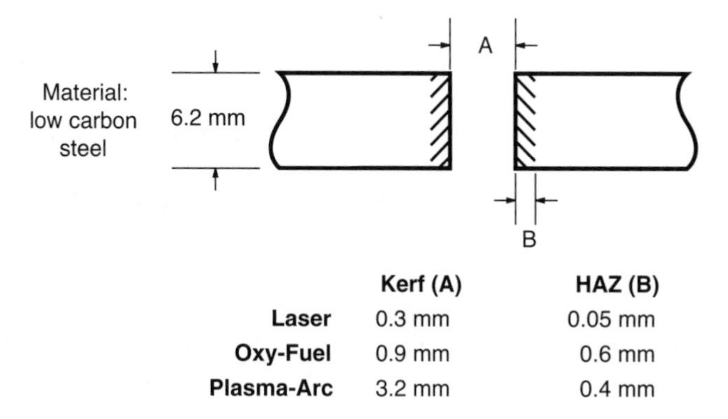

	Kerf (A)	HAZ (B)
Laser	0.3 mm	0.05 mm
Oxy-Fuel	0.9 mm	0.6 mm
Plasma-Arc	3.2 mm	0.4 mm

FIGURE 15.21 Comparison of thermal cutting processes.

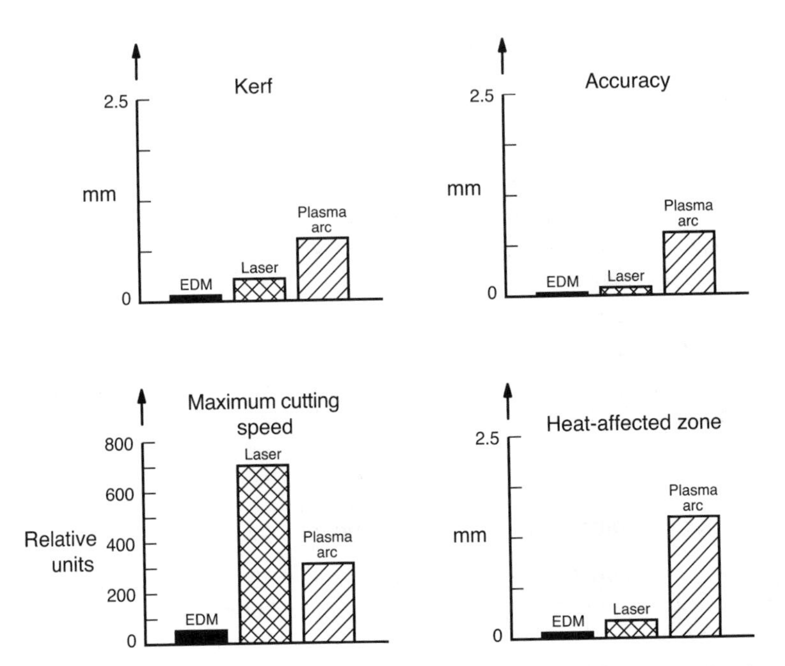

FIGURE 15.22 Comparison of electrical discharge machining (EDM), laser cutting, and plasma arc cutting processes.

15.1.8.2 Laser Cutting and Electrical Discharge Machining Unlike the previous case, each of these two processes has certain advantages over the other, Fig. 15.22. EDM results in better accuracy and smaller kerf width and heat-affected zone size than laser cutting. Furthermore, EDM is able to cut through relatively thick material. On the contrary, laser cutting is performed at much higher speeds. In addition, the accuracy, kerf width, and HAZ size advantages of EDM are only slight.

15.1.8.3 Laser Cutting and Abrasive Waterjet Machining Abrasive waterjet machining involves cutting through a material using high-pressure water jet. Thus, it does not produce a heat affected zone. It is also able to cut much thicker materials than does laser cutting. However, the kerf width produced in waterjet machining is less accurate than that associated with laser cutting. Furthermore, waterjet machining is a contact process, and as such, may cause workpiece deflection for weaker materials.

Waterjet machining is often recommended as the process of choice in cutting composites, mainly due to the thermal effect of laser cutting.

15.1.8.4 Laser Cutting and Punching/Nibbling The primary advantage of laser cutting over punching or nibbling is that it is more flexible, since it does not require tooling. Laser cutting has a definite advantage for small production runs, while at higher production volumes, it loses its advantage, since traditional punching operations are then much faster.

15.1.9 Special Techniques

One special technique that has been developed to enhance the capability of laser cutting is the laser-assisted oxygen cutting process (Lasox). Conventional laser cutting is limited to thicknesses of up to 25 mm, which can be achieved with a laser power of about 3 kW in steels. As discussed in Section 15.1.5.1, the roughness of the cut surface increases as the workpiece thickness increases, as a result of striations which are formed in the thicker material.

The Lasox process is designed to significantly increase the thicknesses that can be cut with a laser of given power without deterioration in quality. The principle is similar to traditional laser cutting and is illustrated schematically in Fig. 15.23.

In this process, the laser beam preheats the surface of the workpiece to a temperature close to its melting temperature ($>1000^{\circ}$C for steels), and this facilitates melting of the workpiece by an oxygen gas jet which flows coaxially with the beam. Once the reaction is initiated on the surface, it continues through the workpiece thickness. The cutting action is driven primarily by the gas jet. The footprint of the laser beam is made to cover an area which is just slightly larger than the footprint of the oxygen jet. This is achieved by using a short focal length lens to maximize beam divergence on

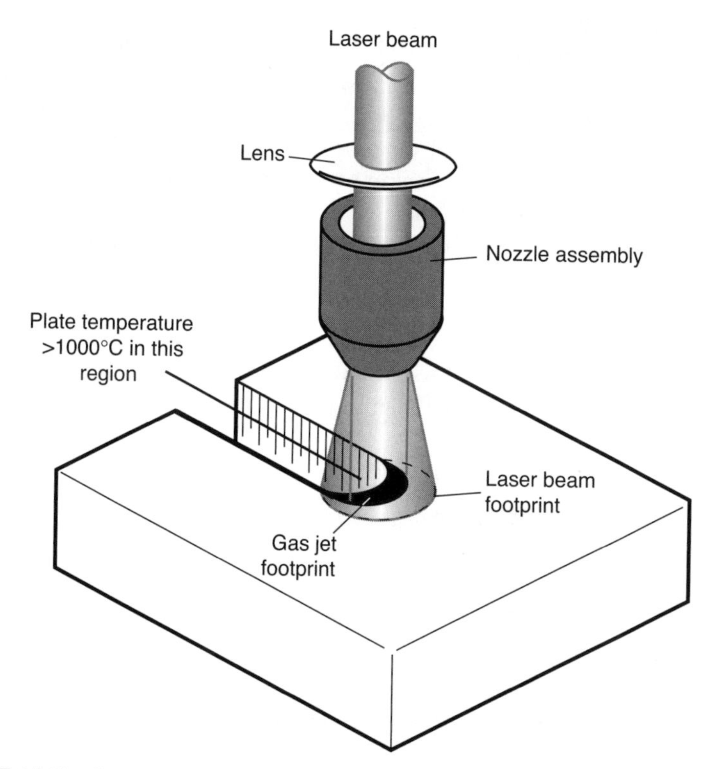

FIGURE 15.23 Schematic of the laser-assisted oxygen cutting process. (From O'Neill, W., and Gabzdyl, J. T., 2000, *Optics and Lasers in Engineering*, Vol. 34, pp. 355–367.)

leaving the nozzle, and yet maintain reasonable standoff distances. As an illustration, 20–50 mm steel plates can be cut with laser power levels of 700–1100 W at speeds of 0.15–0.5 m/min if the beam diameter on the workpiece surface is 4 mm, gas jet diameter is 3 mm, nozzle exit diameter is 2.5 mm, gas pressure is about 8 bar, and the beam absorptivity is 30%.

The surface quality obtained under these conditions is much better than that obtained by either oxyacetylene cutting alone or by traditional laser cutting, which would require much higher power. The taper that results is also less than 1°, and the top edge is square.

15.2 LASER DRILLING

Laser drilling is used in such applications as jet engine turbine airfoils, injector nozzles, watches, and so on. The process involves applying a laser beam to heat up the material to its melting point or vaporization temperature. When vaporization occurs, it generates a keyhole that results in increased absorptivity, further increasing the hole depth. The molten material or vapor formed is blown away using an assist gas. When the vapor or molten material is ejected out into the surrounding atmosphere, some of it may condense on the workpiece surface as spatter. Very small diameter holes (of the order of microns) can be drilled using a laser, resulting in high aspect ratios.

Our discussion on laser drilling starts with an outline of the different ways in which a laser may be used to drill a hole, followed by a discussion of the basic process parameters. An analysis of the process is then presented, first mentioning the governing equations that will need to be used for a rigorous analysis, followed by an approximate analysis that enables the drilling velocity to be estimated. The advantages and disadvantages of laser drilling are then presented, and finally, potential applications.

15.2.1 Forms of Laser Drilling

There are three ways in which a hole can be produced by laser drilling (Fig. 15.24):

1. Single-pulse (or on-center) drilling.
2. Multipulse percussion drilling.
3. Trepanning.

15.2.1.1 Single-Pulse Drilling This form of drilling involves using a single pulse of laser beam, with a pulse duration of about 1 ms, and energy of several Joules, resulting in a peak power of the order of 10–100 kW. The material in contact with the laser beam is partly melted and partly evaporated, and blown out backward using a gas jet to form the hole. The exceedingly high energy required for drilling very deep holes may result in poor tolerances; reduce the pulse frequency obtainable with the laser, thereby reducing productivity; and increase plasma formation. Plasma formation may

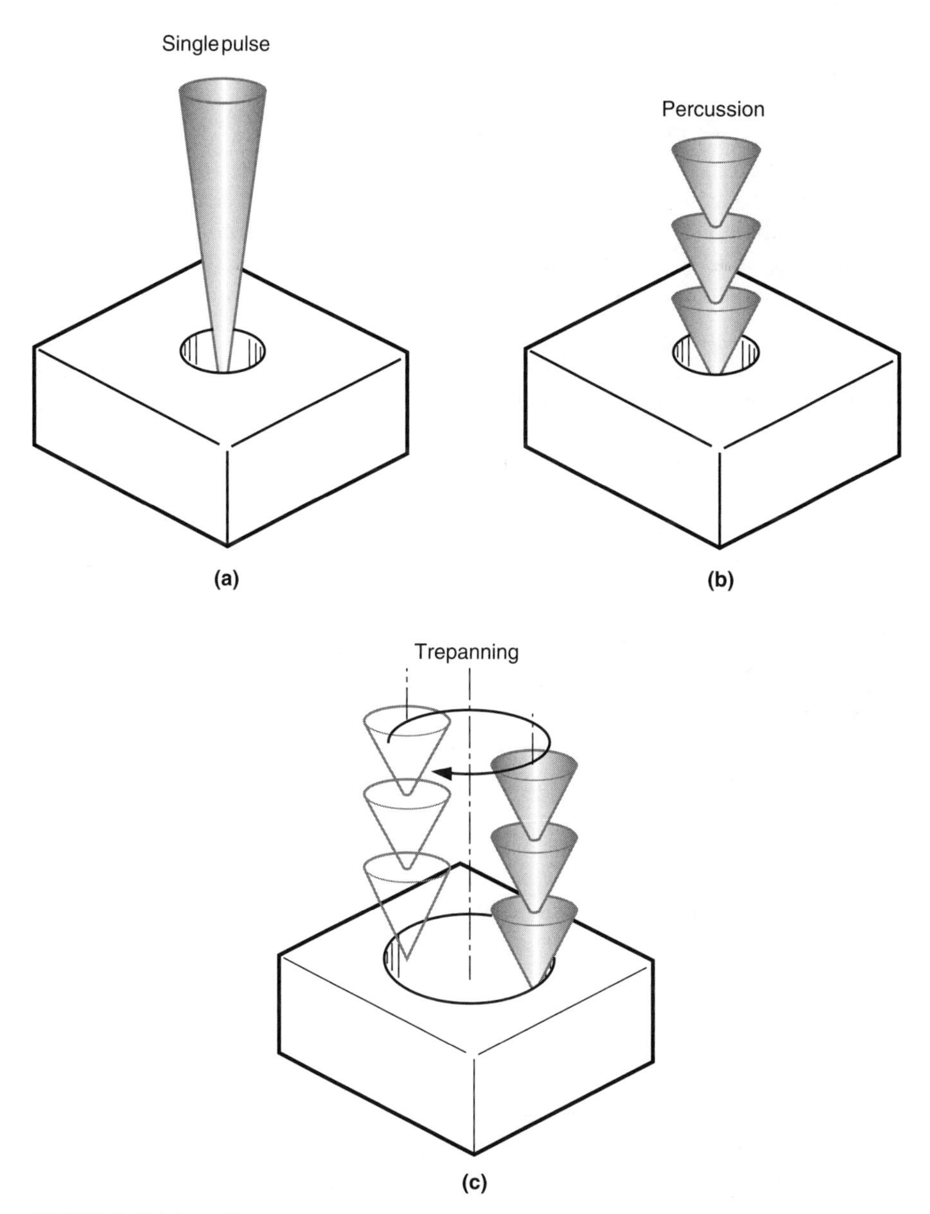

FIGURE 15.24 Different ways in which a hole may be produced by laser drilling. (a) Single-pulse drilling. (b) Multipulse percussion drilling. (c) Trepanning.

affect the hole quality as the laser beam is scattered and/or absorbed and reradiated to the walls.

15.2.1.2 Multipulse Percussion Drilling Multipulse percussion drilling involves drilling a hole using a series of pulses. The hole quality is improved (e.g., reduced tapering along the hole length) since the peak power required is reduced, and with it, the amount of molten material and plasma produced during any single pulse period. However, the series of pulses used reduces the production rate. Typical pulse duration ranges from about 100 fs to 2 ms.

In both percussion modes, the molten and/or evaporated material is ejected out of the hole by the excess pressure of the gas or plasma in the hole over ambient, and this increases with the beam intensity. With increasing hole depth, the pressure required to blow out the molten material increases accordingly. One disadvantage of percussion drilling is that the material that is removed is blown backward toward the lens.

15.2.1.3 Trepanning Trepanning involves cutting out the hole by performing a relative motion (circular or noncircular) between the laser beam and the workpiece. The process is essentially one of laser cutting. Thus large diameter holes can be produced, with improved hole quality and repeatability. Another advantage of the process is that the material that is removed is blown away from the lens. However, the process is time consuming, typically requiring about 1 s for producing a hole. Holes drilled by this method are often limited to depths less than 12 mm.

15.2.2 Process Parameters

The type of laser selected for a given application is determined largely by the drilling technique used, the material, and geometry of the hole to be drilled. Nd:YAG lasers are more extensively used in drilling metals than CO_2 lasers when thicknesses exceed 0.5 mm, since absorption of the shorter wavelength 1.06 μm energy by the plasma formed is relatively small, compared to absorption at the 10.6 μm wavelength of the CO_2 laser. Furthermore, higher peak pulse powers are attainable with the Nd:YAG laser. However, both the capital and running costs of CO_2 lasers are lower than those of the Nd:YAG lasers of equivalent power.

The principal parameters associated with percussion laser drilling may be broadly categorized as the beam characteristics (inputs), process characteristics (outputs), and process defects. These are further discussed in the following paragraphs.

15.2.2.1 Beam Characteristics The principal beam characteristics that influence the drilling process include pulse energy, pulse duration, number of pulses, and beam quality. A typical value of the power density used would be 5×10^7 W/cm^2 at a peak power of 8 kW, using a lens of focal length 100 mm.

Pulse Energy: The hole depth generally increases with increasing pulse energy. At much higher energy levels, however, there tends to be greater deformation at the top surface.

Pulse Duration: Laser drilling pulses typically fall in the range of 100 fs to 2 ms. Higher quality holes, (e.g., reduced taper and distortion), are obtained with shorter beam pulses. However, the number of pulses required to drill a given hole then increases.

Number of Pulses: Generally, hole taper can be reduced in holes with a high depth-to-diameter ratio by using a relatively large number of pulses, while for low depth-to-diameter holes, fewer pulses result in a better quality hole.

Beam Quality: Beams of lower M^2 values or higher quality can be focused to a smaller diameter and also have a longer depth of focus. This increases penetration depth for drilling and cutting, while improving hole quality (high depth-to-diameter ratio with reduced taper), especially for small diameter holes.

15.2.2.2 Drilling Characteristics The principal characteristics of the laser drilling process include the hole diameter, depth, and drilling angle.

Hole Diameter: Typical diameters of laser drilled holes range in size from 1 μm to 1.5 mm. Hole sizes smaller than 1 μm have been produced using femtosecond lasers. Maintaining a reasonable depth of focus for holes smaller than 12.7 μm is difficult with traditional lasers. For holes of diameter greater than 1.5 mm, it might be more appropriate to use trepanning, which is a relatively slower process.

Hole Depth: In using Nd:YAG lasers, depth to diameter ratios as high as 30:1 can be achieved.

Drilling Angle: This is the angle that the hole or beam axis makes with the workpiece surface. The power density incident on the workpiece surface is maximum when the workpiece is normal to the beam axis. This configuration also permits the widest range of focal lengths and focusing angles. The drawback, however, is that it results in any reflected beam being redirected back into the optical system. Furthermore, material expelled during the process may be deposited on the lens. This problem may be mitigated by the use of a coaxial gas flow or protective transparent cover plate for the lens.

Increasing the angle between the hole axis and the normal to the workpiece surface reduces the amount of reflected beam and debris that is directed at the lens. However, short focal length lenses are then more difficult to use, and so are gas nozzles. Furthermore, since the beam is then spread over a wider area, the power density is reduced.

15.2.2.3 Process Defects The product quality is affected by defects such as taper of the hole walls, recast, and microcracking.

Taper: Erosion of molten and vaporized material from the hole, and the conical shape of the focused beam, together result in tapering of the drilled hole. The extent of taper is influenced by the pulse energy, number of pulses, and optical system design. It can be minimized by using long focal length lens, which tend to have longer depth of focus. Tapering is usually not a problem for very small hole depths, say, less than 0.25 mm.

Recast: Recast is the result of resolidified excess molten or vaporized material that is not completely removed. It essentially remains on the newly created surface.

Microcracking: Microcracking normally arises when drilling brittle or high hardenability materials, and often results from high cooling rates or temperature gradients.

15.2.3 Analysis of Material Removal During Drilling

As indicated in Section 15.1, when a pulsed beam of adequate intensity is incident on a substrate, the absorbed energy may melt and/or vaporize the substrate material. These are the mechanisms by which drilling takes place. Depending on the energy source and the substrate material, this may occur by photochemical ablation or photothermal ablation. Photochemical ablation is briefly discussed in Section 15.1.1.3. In this section, we focus on photothermal ablation, which may occur by normal vaporization, normal boiling, or phase explosion.

Normal Vaporization: refers to the transformation from the condensed phase (solid or liquid) to the vapor phase as atoms or molecules are emitted from the extreme outer surface. It can occur at any fluence and pulse duration, and there is no temperature threshold. The process does not involve nucleation, and the surface temperature is not constant since the vapor pressure is not zero. The contribution of normal vaporization to ablation is insignificant at time scales shorter than 1 ns, and also for very low temperatures. The rate at which the drilled surface recedes during normal vaporization, that is, the drilling velocity, u_d, can be obtained from the Hertz–Knudsen equation as

$$u_d = \left.\frac{\partial z}{\partial t}\right|_{z=0} = \frac{k_c P_s \lambda_a{}^3}{\sqrt{(2\pi m_m k_B T)}} \approx \frac{k_c P_a \lambda_a{}^3}{\sqrt{(2\pi m_m k_B T)}} \exp\left[\frac{L_v m_m}{k_B}\left(\frac{1}{T_v} - \frac{1}{T}\right)\right] \tag{15.20}$$

where k_c = condensation (or vaporization) coefficient, L_v = latent heat of vaporization (J/kg), m_m = mass of the atom or molecule (particle)(kg), P_a = ambient gas pressure (boiling pressure) (Pa), P_s = saturated (equilibrium) vapor pressure (Pa), T_v = vaporization (boiling) temperature corresponding to P_a, with the assumption that there is no vapor present in the ambient, and no recondensation (K), and λ_a = mean atomic spacing of the target (μm).

Normal Boiling: requires a relatively long pulse period, and involves the nucleation of heterogeneous vapor bubbles which may form at the outer surface of the liquid; in the bulk of the liquid; or at the interface between the liquid and surrounding solid. The normal boiling process occurs within the absorption depth ($1/\alpha$, where α is the absorption coefficient). The surface temperature is constant and the same as the vaporization temperature corresponding to the pressure at the surface. The temperature gradient at the surface and also directly beneath it is zero (i.e., $\partial T/\partial z = 0$).

For *phase explosion*: (also known as explosive boiling) to occur, the laser fluence has to be sufficiently high and the pulse duration sufficiently short such that the temperature of the surface and the region immediately beneath it reaches about 90% of the thermodynamic critical temperature (i.e. $0.90T_{ct}$). That results in homogeneous bubble nucleation, and the material undergoes a rapid transition from a superheated liquid to a vapor/liquid droplet mixture. Homogeneous nucleation is feasible since the rate at which it occurs increases significantly near the critical temperature. Here too, the temperature gradient at the surface and also directly beneath it is zero.

Now we focus again on normal vaporization where the substrate material is first heated through the melting point to the vaporization temperature. In this case, part of the beam energy is used to melt the substrate, and the molten material (liquid layer) formed is in direct contact with the substrate. Part of this molten material is further vaporized by the beam. The vapor produced creates a vapor or recoil pressure which pushes the vapor away from the target and also exerts a force on the molten material and expels it sideways.

Material removal may thus occur in both liquid and vapor forms. In the liquid form, the thickness of the liquid layer is estimated to be of the order of 10 μm, and decreases with increasing power. In the vapor form, the vaporization rate increases with an increase in power. However, the liquid expulsion rate increases to a maximum, and then decreases. At lower power, the vapor temperature is low. Consequently, the expulsion rate is low because of low recoil pressure. At high power, the vapor temperature is high. However, the thickness of the liquid layer is then so small that the liquid expulsion rate is also reduced. Generally, at low beam power, liquid expulsion is the predominant form of material removal, while vaporization is the predominant form at high power. The process of liquid expulsion is illustrated in Fig. 15.25. For beams of much greater intensity and shorter pulse duration, the resulting recoil pressure may be so high as to induce shock waves which are transmitted through the material. Shock waves may also be generated as a result of the interaction of the laser with the plasma. This phenomenon resembles that of combustion waves.

In the next section, a fundamental approach for analyzing the laser drilling process is outlined. Since the governing equations on which this approach is based are nonlinear, they can only be fully solved using numerical methods. Therefore, in the subsequent section, we present an approximate method for obtaining a quick estimate of the rate at which laser drilling takes place.

15.2.3.1 Basic Analysis The governing equations necessary for analyzing the laser drilling process are the basic continuity, momentum, and energy equations, with the appropriate boundary conditions. There are two interfaces associated with the material removal process. One is the solid–liquid interface that can be treated as a conventional Stefan problem, (Section 12.1.1). The other is the liquid–vapor interface, adjacent to which may exist a Knudsen layer with a thickness of a few molecular mean free paths.

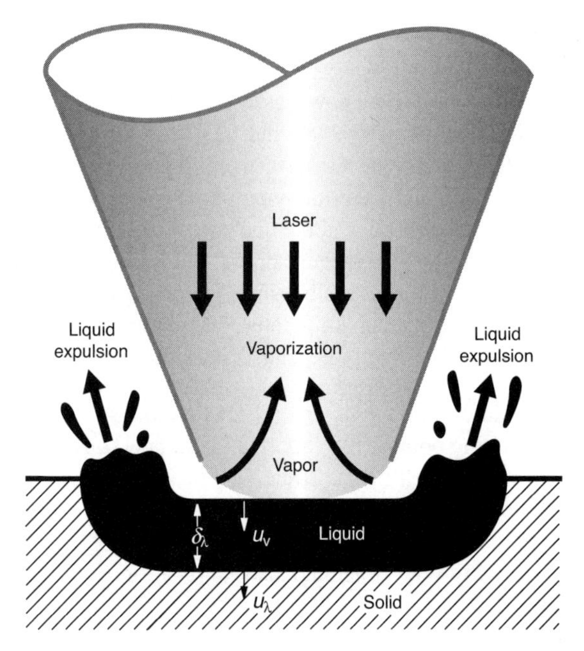

FIGURE 15.25 Schematic of material removal by a pulsed beam. (From Chan, C., and Mazumder, J., 1987, *Journal of Applied Physics*, Vol. 62, pp. 4579–4586.)

Across this layer, continuum theories are no longer valid since discontinuities exist in temperature, density, and pressure. The discontinuity across the Knudsen layer is the basis for the shock wave. To understand the system's behavior under these conditions, we focus on the process of surface evaporation, which is illustrated schematically in Figs 15.26 and 15.27.

To simplify the analysis, we make the following assumptions:

1. The process occurs under quasi-equilibrium conditions involving evaporation from a plane liquid–vapor interface into air at ambient conditions.
2. The vapor is an ideal gas.
3. Thermodynamic equilibrium exists in the liquid.
4. Diffusive evaporation is negligible compared to convective evaporation.
5. Immediately in contact with the liquid surface is a thin layer of vapor whose thickness may be several mean free paths in size, known as the Knudsen layer, which is a discontinuity region where the change in temperature, pressure, and density across the evaporation front are discontinuous.
6. Flow just outside the Knudsen layer is subsonic.

Using the conservation of mass, momentum, and translational kinetic energy, the temperature, pressure, and density in the vapor at the edge of the Knudsen layer (subscript v) can be related to the corresponding properties at the liquid surface

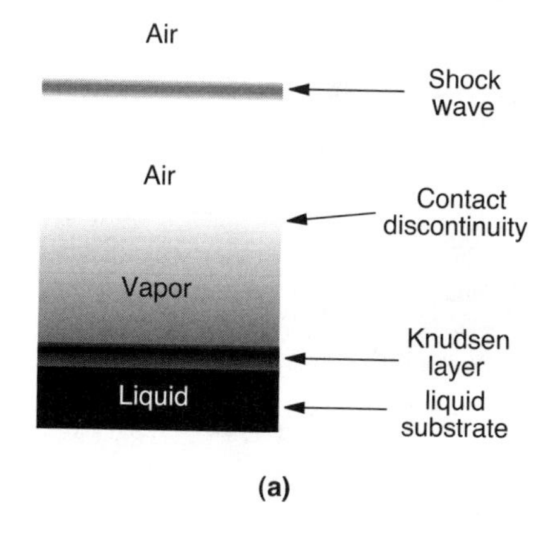

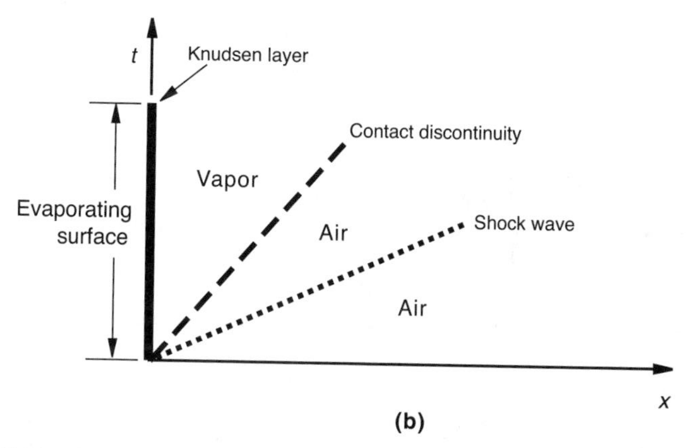

FIGURE 15.26 Schematic of the evaporation process under subsonic conditions. (a) The different zones. (b) Variation of the different zones with time. (From Knight, C. J., 1979, *AIAA Journal*, Vol. 17, pp. 519–523. Reprinted by permission of the American Institute of Aeronautics and Astronautics, Inc.)

(subscript l) and saturated vapor conditions (subscript s), taking into account the discontinuity across the Knudsen layer. The relationships are given by

$$\frac{T_{vk}}{T_l} = \left[\sqrt{1 + \pi\left(\frac{m_u(\gamma_v - 1)}{2(\gamma_v + 1)}\right)^2} - \sqrt{\pi}\frac{m_u(\gamma_v - 1)}{2(\gamma_v + 1)}\right]^2 \tag{15.21}$$

$$\frac{\rho_v}{\rho_s} = \sqrt{\frac{T_l}{T_{vk}}}\left[(m_u{}^2 + 1/2)e^{m_u{}^2}\mathrm{erfc}(m_u) - \frac{m_u}{\sqrt{\pi}}\right] + \frac{T_l}{2T_{vk}}\left[1 - \sqrt{\pi}m_u e^{m_u{}^2}\mathrm{erfc}(m_u)\right] \tag{15.22}$$

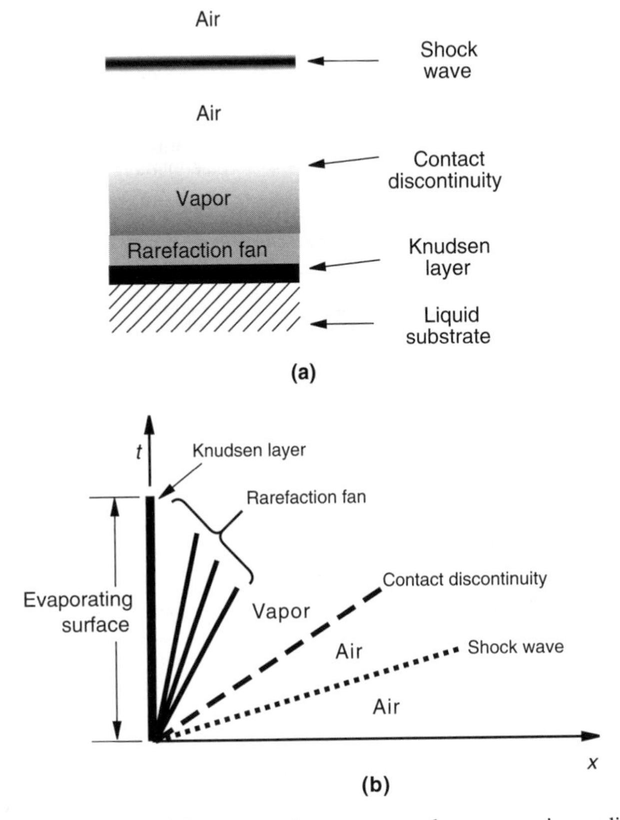

FIGURE 15.27 Schematic of the evaporation process under supersonic conditions. (a) The different zones. (b) Variation of the different zones with time. (From Knight, C. J., 1979, *AIAA Journal*, Vol. 17, pp. 519–523. Reprinted by permission of the American Institute of Aeronautics and Astronautics, Inc.)

There is a pressure rise at the liquid–vapor interface, and this propagates as a pressure wave. The pressure change across the wave front, which may be considered as a pressure discontinuity, is given by

$$\frac{P_\mathrm{v}}{P_\mathrm{a}} = 1 + \gamma_\mathrm{a} M_\mathrm{h} \frac{u_\mathrm{v}}{u_\mathrm{a}} \left[\frac{\gamma_\mathrm{a}+1}{4} M_\mathrm{h} \frac{u_\mathrm{v}}{u_\mathrm{a}} + \sqrt{1 + \left(\frac{\gamma_\mathrm{a}+1}{4} M_\mathrm{h} \frac{u_\mathrm{v}}{u_\mathrm{a}} \right)^2} \right] \tag{15.23a}$$

$$M_\mathrm{h} = m_\mathrm{u} \sqrt{\frac{2}{\gamma_\mathrm{v}}} \tag{15.23b}$$

where c_p and c_v are the specific heats at constant pressure and volume, respectively (J/kg K); $\mathrm{erfc}(m_\mathrm{u}) = \frac{2}{\sqrt{\pi}} \int_{m_u}^{\infty} \mathrm{e}^{-x^2}\,\mathrm{d}x$ = complementary error function; m_u is the

$\frac{u_k}{\sqrt{2R_{gv}T_{vk}}}$ (m kg$^{1/2}$/sJ$^{1/2}$); M_h is the flow Mach number of the vapor leaving the Knudsen layer; m_g is the molecular weight of the ambient gas (kg/mol); m_v is the molecular weight of the vapor (kg/mol); P_a is the ambient pressure (Pa); P_s is the saturated vapor pressure (Pa); P_v is the vapor pressure (Pa); P_{vk} is the vapor pressure at the edge of the Knudsen layer (Pa); $R_g = 8.314$ is the universal gas constant (J/mol K); $R_{ga} = \frac{R_g}{m_g}$ is the gas constant for the ambient gas (J/kg K); $R_{gv} = \frac{R_g}{m_v}$ is the gas constant for the vapor (J/kg K); T_a is the ambient temperature (K); T_l is the liquid temperature at the vaporizing surface (K); T_{vk} is the vapor temperature at the edge of the Knudsen layer (K); u_k is the mean vapor velocity at the edge of the Knudsen layer (m/s); u_a is the speed of sound in the ambient gas (m/s); u_v is the speed of sound in the vapor (m/s); ΔH_{vm} is the molar enthalpy of vaporization (J/mol); γ_a is the ratio of specific heats for the ambient gas; $\gamma_v = c_p/c_v$ is the ratio of specific heats for the vapor, considering the vapor to be a monatomic gas; ρ_s refers to the saturated vapor density at the liquid temperature, T_l (kg/m^3); ρ_v refers to the vapor density at the edge of the Knudsen layer (kg/m^3).

In analyzing the vapor pressure, two cases have to be distinguished:

1. Subsonic flow just outside the Knudsen layer, Fig. 15.26a and b
2. Supersonic flow just outside the Knudsen layer, Fig. 15.27a and b.

Subsonic Flow: The Mach number for flow outside the Knudsen layer is less than one, that is $M_h < 1$. The following conditions then exist:

$$\frac{P_s}{P_a} = \frac{P_v}{P_a} \Big/ \frac{P_v}{P_s} \tag{15.24a}$$

$$\frac{P_s}{P_v} = \frac{\rho_s}{\rho_v}\frac{T_l}{T_{vk}} \tag{15.24b}$$

$$u_v = \sqrt{\gamma_v R_{gv} T_{vk}} \qquad u_a = \sqrt{\gamma_a R_{ga} T_a} \tag{15.24c}$$

Supersonic Flow: The Mach number for flow outside the Knudsen layer is greater than one, that is $M_h > 1$. However, flow within, and at the outside edge of the Knudsen layer is one, that is, $M_h = 1$. The following conditions then exist:

$$\frac{P_s}{P_a} = \frac{P_{vk}}{P_v}\frac{P_v}{P_a} \Big/ \frac{P_{vk}}{P_s} \tag{15.25a}$$

$$\frac{P_{vk}}{P_v} = \left[\frac{2}{\gamma_v + 1} + \frac{\gamma_v - 1}{\gamma_v + 1} M_h\right]^{\frac{2\gamma_v}{\gamma_v - 1}} \tag{15.25b}$$

$$\frac{u_v}{u_a} = \sqrt{\left[\frac{\gamma_v R_{gv} T_l T_{vk}}{\gamma_a R_{ga} T_a T_l}\right]} \left[\frac{2}{\gamma_v + 1} + \frac{\gamma_v - 1}{\gamma_v + 1} M_h\right]^{-1} \tag{15.25c}$$

$$\frac{P_{vk}}{P_s} = 0.206 \qquad \frac{T_{vk}}{T_l} = 0.669 \tag{15.25d}$$

Considering the subsonic case, equations (15.21)–(15.23) contain eight unknowns, namely, T_{vk}, T_l, ρ_v, ρ_s, P_v, P_s, m_u, and M_h. The equation of state

$$P = \rho R_{gv} T \tag{15.26}$$

can be used for saturated and expanding vapor states, along with a Clausius-Clapeyron equation:

$$P_s = k_c e^{-\frac{\Delta H_{vm}}{R_g T_l}} \tag{15.27}$$

to express all the variables in terms of T_l. In equation (15.27), k_c is a constant. Energy balance at the evaporation front provides the missing equation to determine the thermodynamic state of the saturated vapor.

15.2.3.2 Approximate Analysis In this section, we develop a simple extension of the Hertz-Knudsen equation for estimating the rate (velocity) of the drilling process, which takes into account the contribution of the ejected molten material. The following assumptions are made:

1. Heat conduction and vapor absorption losses are negligible.
2. The thermal properties of the material are constant, and thermal expansion is negligible.
3. The skin depth ($\approx 10^{-5}$ mm) (see Section 14.1.6) is negligible.

If we consider the energy flux carried away with the expulsed material to be equal to the irradiance or power density (power per unit area) absorbed by the material, I_a, then we can write (cf. equation (10.1)):

$$I_a = \phi_v \Delta H_v + \phi_l \Delta H_l \tag{15.28}$$

where ΔH = specific energy absorbed by the expulsed material (J/kg), ϕ = expulsion rate (kg/m^2 s) and the subscripts v and l refer to the vapor and liquid, respectively.

The drilling velocity, u_d, can be expressed in terms of ϕ_v and ϕ_l as

$$u_d = \frac{1}{\rho}(\phi_v + \phi_l) \tag{15.29}$$

In developing an expression for ϕ_l, one may consider a simple one-dimensional conduction heat flow for a system with a constant moving heat source. The quasistatic temperature distribution, T, in the medium for this case, neglecting any phase changes and lateral heat conduction losses, may be approximated as

$$T = \frac{I_a'}{\rho c_p u_d} \exp\left(-\frac{\xi u_d}{\kappa}\right) \tag{15.30}$$

where $I_a' = I_a - \phi_v L_v$ is the difference between the absorbed power density and that carried away from the absorbing layer by the evaporated material (W/m^2); L_v is the latent heat of vaporization (J/kg); u_d is the drilling velocity (m/s); κ is the thermal diffusivity of the material (m^2/s); ξ is the moving coordinate pointing in the direction of u_d.

The rate of evaporation, ϕ_v, of a hot surface at a temperature T_h is obtained from the Hertz–Knudsen equation as

$$\phi_v = (1 - R)P_s \left(\frac{m_m}{2\pi k_B T_h}\right)^{1/2} \tag{15.31}$$

where k_B is Boltzmann's constant (J/K), m_m is the mass of the atom or molecule (particle) (kg), and $R \approx 0.2$ is the mean particle reflection coefficient of the metal surface.

The saturated vapor pressure, P_s, may be obtained from equation (15.24), or may be approximated by

$$P_s = P_a \exp\left[\frac{L_p}{k_B T_v}\left(1 - \frac{T_v}{T_h}\right)\right] \tag{15.32}$$

where L_p is the heat of evaporation per particle (J/particle), P_a is the ambient pressure (Pa), T_h is the hot surface temperature (K), and T_v is the evaporation temperature (K).

The hot surface temperature, T_h, may be estimated as the temperature at $\xi = 0$ in equation (15.30). Then again considering heat balance, but only for the molten material, we have

$$\int_{T_0}^{T_h} c_p \mathrm{d}T + L_m = \frac{I_a'}{\rho u_d} \tag{15.33}$$

where L_m is the latent heat of melting (J/kg) and $T_0 = T_a$ is the ambient temperature (K).

The thickness, δ_l, of the liquid layer formed by heat conduction may be estimated using equation (15.30) as

$$\delta_l = \left(\frac{\kappa}{u_d}\right) \ln\left(\frac{T_h}{T_m}\right) \tag{15.34}$$

where T_m is the melting temperature.

The pressure, P, of the evaporating surface acting on the liquid layer pushes the liquid radially outward with a velocity, u_l, determined using Bernoulli's equation (15.1a) as

$$u_l = \left(\frac{2P}{\rho}\right)^{1/2} \tag{15.35}$$

The expelled liquid escapes through an opening (Fig. 15.25) that is determined by the beam radius, w, of an area A_e given by

$$A_e = 2\pi w \delta_l \tag{15.36}$$

Since the actual pressure P is about half the saturation pressure P_s (equation (15.24) or (15.32)), the liquid expulsion rate, ϕ_l, with respect to the irradiated surface is obtained from equations (15.35)–(15.37) as

$$\phi_l = \left[\left(\frac{2\kappa}{w}\right) \ln\left(\frac{T_h}{T_m}\right)\right]^{1/2} P_s{}^{1/4} \rho^{3/4} \tag{15.37}$$

Equations (15.31), (15.32), and (15.37) give the expulsion rates of the vapor and liquid in terms of the surface temperature, T_h. However, T_h, in turn, depends on ϕ_l and ϕ_v, (equation (15.33)). Thus, actual calculation of the expulsion rates will need to be done iteratively. However, for a first approximation, the surface temperature, T_h, may be estimated from equation (15.30) and assumed constant.

It must be noted that the drilling speed calculated by this approach refers to the average rate at which material is removed during the laser pulse. It does not give the overall production rate. The latter can be obtained simply by determining the depth of material removed for each pulse and multiplying that by the pulse repetition rate.

Example 15.3 An Nd:YAG laser generates pulses of intensity 60 MW/cm^2 for drilling 200 μm diameter holes in a tungsten plate. Estimate the drilling speed that can be achieved for this operation. Assume ambient temperature and pressure of 25°C and 0.1 MPa, respectively, and that the diameter of the hole is the same as the beam diameter.

Solution:

Some of the material properties are obtained from Appendices 10D and 10E as

Average density, $\rho = 19254 \text{ kg/m}^3 = 19.254 \times 10^{-6} \text{ kg/mm}^3$
Average specific heat, $c_p = 134 \text{ J/kgK}$
Latent heat of fusion, $L_m = 220 \text{ kJ/kg} = 220,000 \text{ J/kg}$
Melting temperature, $T_m = 3410°\text{C} = 3683 \text{ K}$
Latent heat of vaporization, $L_v = 4815 \text{ kJ/kg} = 4815 \times 10^3 \text{ J/kg}$

Evaporation temperature, $T_v = 5700°C = 5973$ K
Thermal conductivity, $k = 151.2$ W/mK $= 0.1512$ W/mm K.

Other needed properties are

Boltzmann's constant, $k_B = 1.38 \times 10^{-23}$ J/K
Avogadro's number, $N_0 = 6.023 \times 10^{23}$ molecules
Atomic weight, $W_a = 183.8$ g/mol $= 0.1838$ kg/mol

Calculated properties

Particle mass, $m_p = 0.1838$ kg/$6.023 \times 10^{23} = 3.05 \times 10^{-25}$ kg/particle
Heat of vaporization per particle, $L_p = 4815 \times 10^3$ J/kg $\times 3.05 \times 10^{-25} = 14.69 \times 10^{-19}$ J/particle
Incident power density is given by $I_a = 60$ MW/cm$^2 = 60 \times 10^4$ W/mm^2
Thermal diffusivity, $\kappa = \frac{k}{\rho c_p} = \frac{0.1512}{19.254 \times 10^{-6} \times 134} = 58.6$ mm^2/s $= 58.6 \times 10^{-6}$ m^2/s.

We assume an initial value for the drilling speed of 2.5×10^4 mm/s $= 25$ m/s, which is used to estimate the hot surface temperature using equation (15.30):

$$T - T_0 = \frac{I_a'}{\rho c_p u_d} \exp\left(-\frac{\xi u_d}{\kappa}\right)$$

but

$$I_a' = I_a - \phi_v L_v$$

Since ϕ_v is not known, we initially use $I_a' = I_a$ to estimate the hot surface temperature.

From equation (15.33), assuming a constant specific heat, the hot surface temperature can be expressed as

$$T_h = T_0 + \frac{I_a}{\rho c_p u_d} - \frac{L_m}{c_p}$$

or

$$T_h = 25 + \frac{60 \times 10^4}{19.254 \times 10^{-6} \times 134 \times 2.5 \times 10^4} - \frac{22 \times 10^4}{134}$$

$$\Rightarrow T_h = 7686°C = 7959\,\text{K}$$

Now from equation (15.32), the saturation pressure is given by

$$P_s = P_a \exp\left[\frac{L_p}{k_B T_v}\left(1 - \frac{T_v}{T_h}\right)\right]$$

Now $P_a = 10^5\ \text{N/m}^2 = 0.1\ \text{N/mm}^2$. Thus,

$$P_s = 0.1 \times \exp\left[\frac{14.69 \times 10^{-19}}{1.38 \times 10^{-23} \times 5973}\left(1 - \frac{5973}{7959}\right)\right]$$
$$= 8.537\ \text{MPa}\ (\text{N/mm}^2) = 8.537 \times 10^6\ \text{Pa}\ (\text{N/m}^2)$$

Therefore, from equation (15.31),

$$\phi_v = (1 - R)P_s\left(\frac{m_p}{2\pi k_B T_h}\right)^{1/2}$$
$$= (1 - 0.2) \times 8.537 \times 10^6\left(\frac{3.05 \times 10^{-25}}{2\pi \times 1.38 \times 10^{-23} \times 7959}\right)^{1/2}$$
$$= 4540\ \text{kg/m}^2 s$$

And from equation (15.37)

$$\phi_l = \left[\left(\frac{2\kappa}{w}\right)\ln\left(\frac{T_h}{T_m}\right)\right]^{1/2} {P_s}^{1/4}\rho^{3/4}$$
$$= \left[\left(\frac{2 \times 58.6 \times 10^{-6}}{0.100 \times 10^{-3}}\right)\ln\left(\frac{7959}{3683}\right)\right]^{1/2} \times [8.537 \times 10^6]^{1/4} \times [19254]^{3/4}$$
$$= 8.4 \times 10^4\ \text{kg/m}^2\text{s}$$

This indicates that under these conditions, melting is the predominant mode of material removal. Now from equation (15.29), the drilling velocity is given by

$$u_d = \frac{1}{\rho}(\phi_v + \phi_l)$$
$$u_d = \frac{1}{19254}(84000 + 4540)$$
$$= 4.6\ \text{m/s} = 4.6 \times 10^3\ \text{mm/s}$$

This is about an order of magnitude lower than the original estimate. Thus an iteration is necessary. The calculated value for ϕ_v can be used to update the value of $I_a{}'$ during the iteration process.

15.2.4 Advantages and Disadvantages of Laser Drilling

15.2.4.1 Advantages The advantages that laser drilling has over conventional drilling are the following:

1. It is a noncontact process. There are no machining forces, and thus no deflection of a tool and/or workpiece that would result in dimensional errors in the part produced, or wear of the tool.
2. Accurate location of holes. In conventional drilling, the drill often deflects on coming in contact with the workpiece, resulting in drill "wandering". This affects accuracy of hole location and orientation.
3. There is no chip formation, and thus no need for elaborate chip disposal systems. However, an appropriate exhaust system is often necessary for the vaporized material.
4. Laser drilling results in high depth to diameter aspect ratios, but primarily for holes of relatively small diameter, and depths of a few millimeters.
5. Ability to drill difficult-to-machine materials such as diamond, ceramics, and highly refractory metals, on which conventional drilling performs poorly.
6. Ability to drill holes with difficult entrance angles. Drilling of holes with low entrance angles is quite difficult with conventional drills. However, with laser drilling, entrance angles as low as 10° are possible.
7. Ability to drill very small holes. Small diameter conventional drills are liable to break easily. On the contrary, the lower end of hole sizes that can be drilled using a laser is limited by the beam size at the focal point, with hole sizes of the order of microns being achievable.
8. High degree of flexibility, enabling a number of operations to be combined in a single setup.
9. High duty cycles and high production rates. Since lasers can be used continuously without interruption, high duty cycles are achieved, resulting in relatively high production rates.

15.2.4.2 Disadvantages The primary limitations of the laser drilling process include the following:

1. Inability to produce precision blind holes. Lasers are more suited to producing through holes.
2. Inability to drill deep holes. The hole depths that are achieved by laser drilling are relatively small, being typically a few millimeters deep. As the hole depth increases, the diameter begins to enlarge due to beam expansion, resulting in a tapered hole.
3. Holes that are close to an edge, and those that intersect other holes are often not effectively produced by laser drilling.

15.2.5 Applications

Typical applications for holes drilled using Nd:YAG lasers include holes in jet engine turbine airfoils and combustors (hole sizes about 180 μm −1 mm); injector nozzles; oil passages in automotive engine or gear components. Since CO_2 lasers are more effectively used on non-metals, typical drilling applications involving CO_2 lasers include sound suppression holes in aircraft engine liners made of polymers or polymeric composites; drilling holes in ceramic substrates in the microelectronic industry; and cigarette paper perforating.

15.3 NEW DEVELOPMENTS

15.3.1 Micromachining

Micromachining involves machining applications that result in feature sizes on the order of tens of micrometers or less. As indicated in the section on laser drilling, hole diameters of the order of a few microns can be produced using traditional lasers, with pulse durations ranging from continuous wave to hundreds of picoseconds. Common lasers used for this purpose include the CO_2, Nd:YAG, copper vapor, and excimer lasers. However, the mechanism of material removal with such long pulse lasers involves both melting and evaporation (Fig. 15.25). This results in significant dross formation (Section 15.1.5.2), which affects hole quality. Excimer lasers, being the exception, result in direct photochemical ablation (Section 15.1.1.3) due to the ultraviolet wavelength of the beam, and that improves the hole quality.

When a laser beam is incident on a material, the photon energy is first absorbed by electrons which then get heated to high temperatures rapidly. The electron energy gets transferred to the lattice through electron–phonon collisions (Section 10.2.6), resulting in heating of the lattice. The extent of lattice heating depends on the pulse duration and the electron–phonon coupling. For long pulse durations, the electrons and lattice can reach thermal equilibrium, and a relatively large portion of the lattice gets heated by diffusion, with a resulting peak temperature that is relatively low. The material is then first heated through the melting temperature to the vaporization temperature (if the power is high enough). A relatively large volume around the laser focus melts, but only a relatively small layer of material reaches the vaporization temperature.

As the pulse duration decreases, the laser field intensity increases, and the interaction time gets shorter. Thus, for much shorter pulse durations, the intensity is much higher, resulting in relatively high electron temperatures (of the order of tens of electron volts) during the pulse. Most of the electron–phonon coupling takes place after the pulse, and often results in lattice temperatures that are higher than are achieved with long pulses for the same total energy (Section 14.1.4). Most of the heating is confined to the interaction zone and passes rapidly through the melting temperature to the vaporization temperature. Thus, a large fraction of the material in the interaction volume is vaporized, going through a melt phase very quickly. With ultrashort pulses,

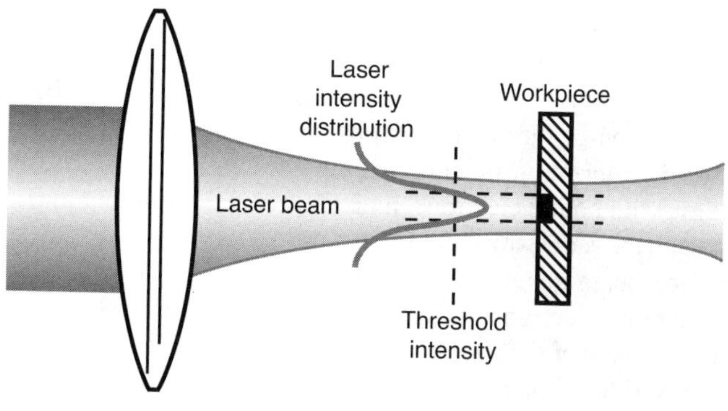

FIGURE 15.28 Setting the breakdown threshold to a specific level of beam mode. (From Liu, X., Du, D., and Mourou, G., 1997, *IEEE Journal of Quantum Electronics*, Vol. 33, No. 10, pp. 1706–1716.)

when the laser fluence just exceeds the threshold value, the region that is vaporized can be smaller than the focal area (Fig. 15.28).

Ultrashort pulse lasers ($\leq$ 1 ps) with wavelengths in the visible and near infrared range therefore result in high precision micromachining, with clean and uniform holes or cuts. The breakdown threshold of such lasers can be determined with high accuracy since it is deterministic. Since the beam intensity profile can be controlled such that only a specified portion of the focused beam spot is above the threshold (Fig. 15.28), a region of diameter smaller (1 μm or less) than the focused beam size (5 μm or less) can be ablated, with barely any heat being conducted out of the processing zone. The mechanics of laser–material interaction that results in ablation under such conditions are discussed in the following paragraphs.

Absorption of a laser beam that is incident on a material occurs by linear or nonlinear processes. The absorption results in heating of material in the focal region first to the melting temperature and then to the vaporization temperature, depending on the beam intensity and pulse duration. For radiation outside the ultraviolet range, the absorption mechanisms are different for materials such as metals and semiconductors which are absorbing, and glasses and plastics which are transparent dielectric materials. Absorption also depends on laser intensity (pulse duration and fluence). For opaque materials, absorption is linear when the beam pulse duration is long and the intensity low, while nonlinear absorption becomes dominant when the pulse duration is ultrashort and the intensity high.

15.3.1.1 Transparent Dielectric Materials Absorption for transparent materials results mainly from nonlinear processes. Avalanche ionization and multiphoton ionization are the nonlinear mechanisms that cause breakdown.

Avalanche Ionization The bound valence electrons of transparent dielectric materials have an ionization potential or bandgap which is greater than the laser photon energy (Section 8.4.1). Therefore at low beam intensities (typical of long pulse lasers), the laser energy is not absorbed by the bound electrons. However, a few free electrons (seed electrons) that exist in the material as a result of impurities do absorb laser energy when they collide with bound electrons and lattice (inverse Bremsstrahlung process). Acceleration of the seed electrons provide them with kinetic energy that is in excess of the ionization potential of the bound electron. Thus, as these seed electrons collide with bound electrons, ionization occurs, resulting in additional free electrons, in much the same way that a photon interacting with an excited atom results in another photon being generated. Repetition of this process leads to an avalanche that increases the free-electron density exponentially. At a critical density ($10^{18}/\text{cm}^3$ for pulse durations greater than a nanosecond), the transparent material breaks down and becomes absorbing. The absorption at the critical density is significant enough that irreversible damage takes place. At the critical density, the plasma oscillation frequency equals the laser frequency, and the transparent material becomes totally opaque. For ultrashort pulses, the critical density, ρ_e, is given by

$$\rho_\text{e} = \frac{m_\text{e}\omega^2}{4\pi e^2} \tag{15.38}$$

where m_e = electron mass and ω = laser frequency.

The critical electron density is then about $10^{21}/\text{cm}^3$ for lasers in the visible and near-infrared wavelength range. The breakdown process typically results in the generation of acoustic waves and radiation of optical plasma or spark at the focus.

Multiphoton Ionization At very high beam intensities, bound electrons of transparent materials are directly ionized (i.e., lifted from the valence band to the conduction band or free energy level) by multiphoton absorption. This is normally the case with ultrashort pulse beams, and involves simultaneous absorption of m photons during the pulse such that

$$mh_\text{p}\nu \geq E_\text{g} \tag{15.39}$$

where E_g is the bandgap energy or ionization potential.

Generally for the critical density to be achieved that results in irreversible breakdown and therefore ablation, the laser fluence must exceed a threshold value, which depends on the pulse duration. For pulse durations longer than a few tens of picoseconds, the threshold fluence is proportional to the square root of the pulse duration. Furthermore, there is significant scatter associated with the threshold fluence, making it more stochastic in nature. Avalanche ionization is the principal mechanism for breakdown in this pulse duration range. For ultrashort pulses, the seed electrons are created by multiphoton ionization, but even though the actual breakdown process is primarily accomplished through avalanche ionization, the threshold is more deterministic or precise.

15.3.1.2 Metals and Semiconductors Materials that are absorbing (such as metals and semiconductors) generally have a significant number of free and valence electrons whose ionization potential (bandgap) is smaller than the laser photon energy. Linear absorption is therefore the predominant mode when the laser pulse duration is long, resulting in Joule heating. Here too breakdown, which results in melting and vaporization, is accompanied by the generation of acoustic waves and radiation of optical plasma or spark at the focus.

There are two characteristic lengths associated with absorption of the laser beam. One is the skin depth, δ' (Section 14.1.6), which is the distance within which the photon energy is absorbed, and may be expressed as

$$\delta' = \frac{1}{\alpha} \tag{15.40}$$

The other is the heat diffusion length, l_{h}, which is the distance over which the thermal energy is conducted during the pulse, and can be expressed as

$$l_{\mathrm{h}} = \sqrt{\kappa \tau_{\mathrm{p}}} \tag{15.41}$$

where τ_{p} is the pulse duration.

For long pulses, the fluence threshold varies as the square root of the pulse duration since $l_{\mathrm{h}} > \delta'$, and therefore the volume of material heated by the beam pulse depends on the heat diffusion length. There is then significant scatter associated with the threshold. Since the volume of material that is heated is relatively large, significant amount of melting occurs, and during ablation, material is expelled in both liquid and vapor form by the recoil (vapor) pressure associated with the ablation process (Section 15.2.3). The unexpelled molten material resolidifies around the ablated region, resulting in dross formation, (Fig. 15.29a). The thickness of melt layer that results for long pulse lasers can be much as 10 times that of the evaporation layer. That means the feature sizes that can be obtained are relatively large.

When the pulse duration becomes short enough that $l_h < \delta'$, the threshold becomes independent of the pulse duration, and is therefore more deterministic. Furthermore, since the energy absorbed is highly localized, the material in this region is quickly heated through the melting temperature to the vaporization temperature. Most of the ablation process then occurs by direct vaporization, resulting in more precise and cleaner cuts (Fig. 15.29b). A significant portion of the material is removed after the laser pulse has ended. Even though heat diffusion results in heating of the surrounding material, the amount of material that melts as a result is relatively small. This is due in part to the rapid cooling that follows the heating process, and also to the significant amount of energy that is lost by vaporization. With ultrashort pulse lasers, the melt thickness may only be about twice that of the evaporation layer. Thus, submicron feature sizes are achievable.

The dependence of the fluence threshold on the square root of the pulse duration when $l_{\mathrm{h}} > \delta'$ also applies to transparent materials.

Advantages of ultrashort pulse lasers in relation to material ablation can therefore be summarized as the following:

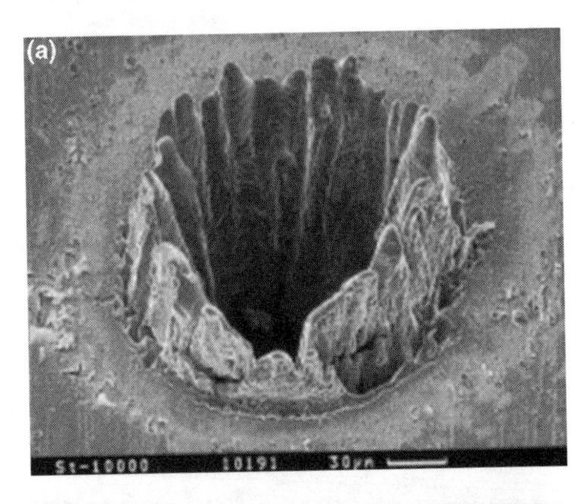

FIGURE 15.29 SEM photographs of holes drilled through a 100 μm steel foil using Ti:sapphire laser pulses at 780 nm. (a) Pulse duration = 3.3 ns and fluence = 4.2 J/cm^2 and (b) pulse duration = 200 fs and fluence = 0.5 J/cm^2. (From Momma, C., Nolte, S., Chichkov, B. N., Alvensleven, F. V., and Tunnermann, A., 1997, *Applied Surface Science*, Vol. 109/110, pp. 15–19.)

1. The precision of the ablation (breakdown) threshold (which depends on the pulse duration, being higher for the longer pulse duration beam).
2. Reduction in the laser fluence needed to initiate ablation.
3. Significant reduction in heat affected zone size.

In recent years, ultrashort pulse lasers have found increasing use in machining microchannels, such as those used in microfluidics. They have also been used to drill holes of diameter 35 μm in 1 mm thick stainless steel sheet, resulting in good quality holes with straight sides. Hole sizes as small as 0.3 μm have been produced in silver films using 200 fs pulses produced by a Ti:sapphire laser, which was focused to a spot size of 3 μ m. As with traditional lasers, femtosecond lasers can be used in machining materials ranging from diamond to biodegradable polymers.

15.3.2 Laser-Assisted Machining

Laser-assisted machining is a means for increasing productivity in metal removal for traditionally difficult-to-machine materials such as titanium alloys, nickel-based superalloys, and hardened steels. The metal removal rate and productivity achieved when machining such materials by conventional methods are limited by the significant reduction in tool life at high speeds.

The process involves using a laser beam to heat up a portion of a material directly in front of a cutting tool to a temperature lower than its melting point, but high enough to soften the material ahead of the cutting tool (Fig. 15.30). Increasing the temperature of a material normally decreases its flow stress and strain hardening rate. Ordinarily, it should be possible to reduce the flow stress simply by machining at high speeds. This is feasible with materials such as aluminum alloys whose flow stress decrease rapidly with increasing temperature. However, many of the difficult-to-machine materials tend to retain their strength at elevated temperatures. Thus high-speed machining of such materials increases tool failure rate. In conventional machining, the chip is hottest at the underside which is in contact with the cutting tool, while the opposite is true of laser assisted machining, even though in the latter case, the temperature of the underside of the chip may be the same as that obtained in conventional machining.

Laser-assisted machining is essentially an improvement of the process of using a gas torch or induction heating to facilitate the machining process. Gas torch and induction heating are not commercially acceptable because of the resulting large heat-affected zone and distortion of the machine and workpiece. The high heat input also affects the cutting tool. Some of the benefits of laser-assisted machining include

1. Reduction in cutting forces
2. Improvement in tool life
3. Improvement in surface finish.

Reduction in cutting forces increases the accuracy of cutting by decreasing the amount of workpiece and machine distortion. It also minimizes the tendency for tool fracture in interrupted cuts, especially when ceramic cutting tools are used. Looking at it from another viewpoint, by keeping the cutting forces constant, higher feed rates can be used, enabling higher material removal rates to be achieved. Improvement in surface finish of the workpiece improves its fatigue life as well as wear and corrosion resistance.

To minimize the effect of the applied heat on the cutting tool, the depth of heat penetration into the shear zone has to be limited to a region above the cutting tool. By doing so, the workpiece is softened without overheating the cutting tool.

In laser-assisted turning of 316 and 420 stainless steels and the nickel superalloy, udimet 700, using a pulsed laser beam of power 460 W, with a maximum average power density of 2.3 $\mathrm{MW/cm^2}$, it has been found that the use of laser assist produces a smoother workpiece surface. Even though the 420 stainless steel does not show a reduction in cutting forces, the improvement in surface finish obtained is significant. For the 316 stainless steel and 1018, 1040 plain carbon steels, force reductions in

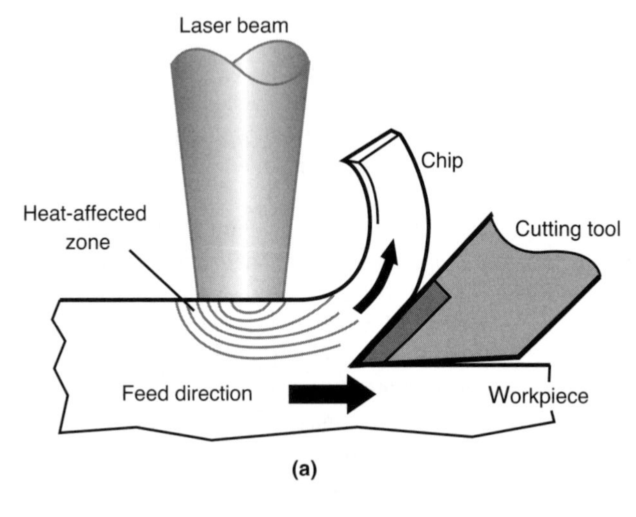

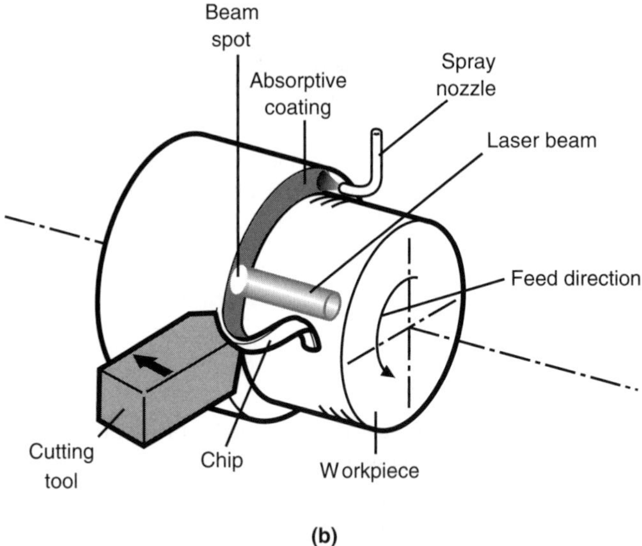

FIGURE 15.30 Schematic of the laser-assisted machining process. (a) Two-dimensional view and (b) three-dimensional view. (From Weck, M., and Kasperowski, S., 1997, *Production Engineering*, Vol. IV, No. 1, pp. 35–38. Permission of MIT Publishing Ltd. Production Engineering Solutions.)

the range of 25–50% can be expected. For 1090 steel, much higher forces (about 2–3 times) are obtained with the laser on than when it is off. This is most likely due to hardening of the workpiece by the laser beam through phase transformation. An improvement in surface quality, though, is obtained. It has been found that laser-assisted machining in general is not effective at very high machining speeds since the interaction time at an irradiated spot is too short for the metal to be heated at the high speeds. At very low speeds, however, the material cools between the time that it is irradiated and the time it arrives at the tool. Hardenable materials thus experience higher forces at low speeds. For a previously hardened material, the cutting forces are always reduced. For either type of material, the forces are reduced at intermediate speeds that result in the maximum amount of tempering of the material arriving at the tool.

Conventional machining of titanium alloys is normally done at relatively low speeds and exclusively with carbide cutting tools, due to the relative weakness of ceramics and the high reactivity of titanium. Also, the low thermal conductivity of titanium alloys tend to concentrate any heat generated in the immediate vicinity of the tool, accelerating tool failure. In laser-assisted machining of titanium alloys, it has been found that effective results are obtained with the beam spot oriented such that its minor axis is along the direction of rotation (Fig. 15.30b). The amount of time that any point on the workpiece is impinged by the beam is then minimal, thereby reducing heat penetration below the shear plane and into the cutting tool. This is especially true when carbide tools are used, since they are less resistant to high temperatures than ceramic tools. Due to the low speeds, absorptive coatings may not be necessary. At a laser power level of 12 kW, cutting speed of 0.7 m/s, feed rate 0.38–2.00 mm/rev, and depth of cut 2.5 mm, productivity is doubled by increasing the feed rate if the cutting forces and tool wear are kept constant. The low conductivity of titanium alloys results in a reduced heat penetration, thereby reducing the impact of the laser beam on tool wear. Since titanium alloys are normally machined at low speeds, much greater improvements can be expected with the application of laser-assisted machining.

In laser-assisted machining of inconel 718 using ceramic tools with a laser power of 12 kW, cutting speed of 2.0 m/s, feed rate 0.15 mm/rev, and depth of cut 1.3 mm, it has been found that impinging the beam at the top edge of the shoulder substantially reduces notch wear at the depth of cut line, but has little influence on nose wear. Nose wear could be reduced by focusing the beam near the root of the shoulder, but then that results in no improvement in the notch wear. In either case, however, flank wear is reduced. The forces have been found to decrease almost linearly with increasing laser power, with the rate being a little over 1% per kilowatt when no absorptive coatings are used on the workpiece surface. With absorptive coatings, the reductions average nearly 4% per kilowatt at power levels up to 4 kW. However, at power levels greater than 4 kW, the force curves level out, indicating a progressive decrease in absorptivity. At a speed of 0.5 m/s and power 14 kW, a 33% reduction in force can be obtained without a coating, but the forces may be reduced by an additional 33% when a potassium silicate coating is used. Overall, it can be deduced that metal removal rates could be increased by one third in laser-assisted machining, while at the same time reducing both tool wear and cutting forces.

Coating techniques are discussed in greater detail in Section 17.1.2.3. For laser-assisted machining, the necessary criteria to be satisfied by successful coatings include

1. Good dispersion with fine particle size to avoid intermittent disruption of spray.
2. Localized application.

3. Rapid bonding to workpiece surface.
4. Good thermal stability under the beam, with minimal volatilization or steaming.
5. High absorptivity.
6. Lack of abrasiveness.
7. Inertness to workpiece and ease of removal.

Process Parameters: The principal parameters that affect the process quality are the following:

1. Beam power and size (power density).
2. Distance between the beam and cutting tool. If the beam impinges on the workpiece very far from the tool, the thermal front can penetrate through to the tool and aggravate, rather than alleviate tool wear.
3. Cutting speed.
4. Depth of cut.

In turning operations, the beam may be positioned as shown in Fig. 15.30a. To minimize the penetration of heat below the shear plane, the distance between the beam and cutting tool has to be as small as possible. However, the actual distance depends on processing parameters such as the beam intensity, cutting speed, and depth of cut.

15.4 SUMMARY

Laser cutting and drilling both involve using a focused laser beam to rapidly heat up a material to the melting and/or vaporization temperature, and blowing away the molten material/vapor using an assist gas. The main difference is that in laser cutting, the ejected material is blown away from the laser optics, whereas in laser drilling, it is partly blown toward the optics.

The three principal forms of laser cutting are fusion cutting, sublimation cutting, and photochemical ablation. A laser cutting system in its simplest form consists of the laser generator and beam delivery system, a nozzle assembly for the assist gas, a motion unit, and an exhaust for the waste material. The key parameters that influence the process are the beam power and characteristics, traverse speed, assist gas, and relative location of the focal point. As an example, a 3 mm thick stainless steel plate can be cut with a CO_2 laser using an output power of 1.8 kW at a speed of 2 m/min and an 800 kPa nitrogen assist gas pressure. As the material thickness increases, the cutting speed for a given power decreases. Common assist gases used are oxygen, air, inert gases, and nitrogen. Oxygen and air result in an exothermic reaction that may increase the cutting speed, depending on the material.

A Gaussian beam is generally preferred for cutting since it results in the highest power density at the focal point. Even though beams which are plane polarized in a plane parallel to the cutting direction result in higher cutting speeds than beams which are plane polarized normal to the cutting direction, circularly polarized beams are generally used since they result in better cut quality when the cutting direction changes. It is also generally preferred to have the focal point either on the workpiece surface or just below it.

An analysis of the process shows that as the cutting velocity increases, the thickness of molten material at the cutting front increases, but on the contrary, increasing the gas jet velocity decreases the melt thickness. The quality of cut that is obtained is often determined by striations and dross that form on the cut surface. Thicker workpieces result in more irregular striations, culminating in a rougher cut surface. Dross formation is enhanced by higher surface tension or viscosity of the molten material. Thus, using oxygen assist tends to reduce dross formation, as opposed to when inert gas assist is used.

Lasers can be used to produce holes in parts in three ways: percussion drilling, which involves using a single-laser pulse; multipulse percussion drilling, where a series of pulses are used; and trepanning, which actually makes a circular cut to create the hole. The drilling process is influenced primarily by the pulse energy, pulse duration, number of pulses, and beam quality. Holes as small as 1 μm can be made, and depth to diameter ratios as high as 30:1 can be achieved. Common defects that affect product quality are tapering of the hole walls, recast, and microcracking. Typical applications of laser drilling include holes in jet engine turbine airfoils, and injector nozzles.

As the pulse duration of the laser beam decreases, so does the extent to which heat is diffused into the surrounding material, for the same input energy. Furthermore, the material is heated more rapidly through the melting range. Thus, cleaner cuts are obtained with ultrashort pulse lasers, compared to traditional long pulse lasers. The ablation threshold of ultrashort pulse lasers is also deterministic, enabling better control of the process.

Combining a laser with traditional machining processes can reduce cutting forces, while also improving surface finish and tool wear.

REFERENCES

Atanasov, P. A., and Gendjov, S. I., 1987, Laser cutting of glass tubing—a theoretical model, *Journal of Physics D: Applled. Physics*, Vol. 20, pp. 597–601.

Bass, M., Beck, D., and Copley, S. M., 1978, Laser assisted machining, *SPIE Fourth European Electro-Optics Conference*, Vol. 164, pp. 233–240.

Batanov, V. A., Bunkin, F. V., Prokhorov, A. M., and Fedorov, V. B., 1973, Evaporation of metallic targets caused by intense optical radiation, *Soviet Physics JETP*, Vol. 36, No. 2, pp. 311–322.

Beyer, E., and Petring, D., 1990, State of the art laser cutting with CO_2 lasers, *ICALEO Proceedings*, pp. 199–212.

Bronstad, B. M., 1993, Capability testing of laser cutting machines, *Welding in the World*, Vol. 31, pp. 27–30.

Bunting, K. A., and Cornfield, G., 1975, Toward a general theory of cutting: a relationship between the incident power density and the cut speed, *ASME Journal of Heat Transfer*, Vol. 97, pp. 116–121.

Caristan, C., 2004, *Laser Cutting Guide for Manufacturing*, Society of Manufacturing Engineers, Dearborn, MI.

Chan, C., and Mazumder, J., 1987, One-dimensional steady-state model for damage by vaporization and liquid expulsion due to laser-material interaction, *Journal of Applied Physics*, Vol. 62, pp. 4579–4586.

Chen, M. M., 1992, Generics structure of flow and temperature fields in welding and high energy beam processing, *Heat and Mass Transfer in Materials Processing*, pp. 365–382.

Chen, S. L., and Steen, W. M., 1991, The theoretical investigation of gas assisted laser cutting, *ICALEO Proceedings* , pp. 221–230.

Chryssolouris, G., 1991, *Laser Machining, Theory and Practice*, Springer-Verlag, Berlin.

Cronin, M. J., 1992, Metallurgical response of nickel base superalloys to laser machining, *ICALEO Proceedings*, pp. 469–479.

Edler, R., and Berger, P., 1991, New nozzle concept for cutting with high power lasers, *ICALEO Proceeding*, pp. 253–262.

Fieret, J., Terry, M.J., and Ward, B. A., 1986, Aerodynamic interactions during laser cutting, *SPIE, Laser Processing: Fundamentals, Applications, and Systems Engineering*, Vol. 668, pp. 53–62.

Gabzdyl, J. T., and Morgan, D. A., 1992, Assist gases for laser cutting of steels, *ICALEO Proceedings*, pp. 443–448.

Geiger, M., Bergmann, H. W., and Nuss, R., 1988, Laser cutting of sheet steels, *SPIE, Laser Assisted Processing*, Vol. 1022, pp. 20–33.

Geiger, M., and Gropp, A., 1993, Laser fusion cutting of stainless steel with a pulsed Nd:YAG laser, *Welding in the World*, Vol. 31, pp. 22–26.

Heglin, L. M., 1986, Introduction to laser drilling, in: Belforte, D., and Levitt, M., editors, *Industrial Laser Handbook, Annual Review of Laser Processing*, Pennwell Books, Tulsa OK, pp. 116–120.

Ignatiev, M., Okorokov, L., Smurov, I., Martino, V., Bertolon, G., and Flamant, G., 1994, Laser assisted machining: process control based on real-time surface temperature measurements, *Journal de Physique IV*, Vol. 4, pp. C4-65–C4-68.

Kawasumi, H., 1990, Laser Processing in Japan, *Industrial Laser Annual Handbook*, Pennwell Books, Tulsa, OK, pp. 141–143.

Kelly, R., and Miotello, A., 1996, Comments on explosive mechanisms of laser sputtering, *Applied Surface Science*, Vol. 96–98, pp. 205–215.

Ki, H., 2001, Modeling and measurement of processes with liquid-vapor interface created by high power density lasers, Ph.D. Thesis, University of Michigan, Ann Arbor.

Kim, C. -J., Kauh, S., Ro, S. T., and Lee, J. S., 1994, Parametric study of the two-dimensional keyhole model for high power density welding processes, *Journal of Heat Transfer*, Vol. 116, pp. 209–214.

Kitagawa, A., and Matsunawa, A., 1990, Three dimensional shaping of ceramics by using CO_2 laser and its optimum processing condition, *ICALEO Proceedings*, pp. 294–301.

Knight, C. J., 1979, Theoretical modeling of rapid surface vaporization with back pressure, *AIAA Journal*, Vol. 17, pp. 519–523.

Krokhin, O.N., 1972, Generation of high-temperature vapors and plasmas by laser radiation, in: *Laser Handbook*, Arecchi, F. T. and Schulz-Du Bois, E. O., editors, North-Holland Publishing Co., Amsterdam, pp. 1371–1407.

Landau, L. D., and Lifshitz, E.M., 1969, Phase Equilibrium, *Statistical Physics*, Pergamon Press, Oxford, p. 261.

Lei, S., Shin, Y. C., and Incropera, F. C., 2001, Experimental investigation of thermo-mechanical characteristics in laser-assisted machining of silicon nitride ceramics, *ASME Journal of Manufacturing Science and Engineering*, Vol. 123, pp. 639–646.

Liu, X., Du, D., and Mourou, G., 1997, Laser ablation and micromachining with ultrashort laser pulses, *IEEE Journal of Quantum Electronics*, Vol. 33, No. 10, pp. 1706–1716.

Marot, G., Fan, L. J., Tarrats, A., Cohen, P., and Longuemard, J. P., 1991, The workpiece material interaction and the laser assisted machining, *Annals of the CIRP*, Vol. 40, pp. 91–94.

Marot, G., Cohen, P., Salem, W. B., Ahdad, F., and Longuemard, J. P., 1992, Reduction in cutting force by laser assisted machining contribution of optimum condition research experiments, *Bulletin du Cercle d'betudes des mbetaux*, Vol. 16, No. 4, pp. 1–7.

Masumoto, I., Kutsuna, M., and Ichikawa, K., 1992, Relation between process parameters and cut quality in laser cutting of aluminum alloys, *Transactions of the Japan Welding Society*, Vol. 23, No. 2, pp. 7–14.

Miotello, A., and Kelly, R., 1999, Laser-induced phase explosion: new physical problems when a condensed phase approaches the thermodynamic critical temperature, *Applied Physics A—Materials Science and Processing*, Vol. 69, pp. S67–S73.

Miyamoto, I., and Maruo, H., 1991, The mechanism of laser cutting, *Welding in the World*, Vol. 29, No. 9/10, pp. 12–23.

Molian, P. A., 1993, Dual-beam CO_2 Laser cutting of thick metallic materials, *Journal of Materials Science*, Vol. 20, pp. 1738–1748.

Momma, C., Nolte, S., Chichkov, B. N., Alvensleven, F. V., and Tunnermann, A., 1997, Precise laser ablation with ultrashort pulses, *Applied Surface Science*, Vol. 109/110, pp. 15–19.

O'Neill, W., Gabzdyl, J. T., and Steen, W. M., 1992, The dynamical behavior of gas jets in laser cutting, *ICALEO Proceedings*, pp. 449–458.

O'Neill, W., and Steen, W. M., 1995, A three-dimensional analysis of gas entrainment operating during the laser-cutting process, *Journal of Physics D: Applied Physics*, Vol. 28, pp. 12–18.

O'Neill, W., and Gabzdyl, J.T., 2000, New developments in laser-assisted oxygen cutting, *Optics and Lasers in Engineering*, Vol. 34, pp. 355–367.

Olsen, F. O., and Heckermann, S., 1990, High speed laser drilling, *ICALEO Proceedings*, pp. 141–150.

Ortiz, A. L., 1990, On-the-fly drilling with a fiber delivered face pumped laser beam, *ICALEO Proceedings*, pp. 151–166.

Petring, D., Abels, P., and Beyer, E., 1988, Absorption distribution on idealized cutting front geometries and its significance for laser beam cutting, *SPIE, High Power CO_2 Laser Systems and Applications*, Vol. 1020, pp. 123–131.

Pfefferkorn, F. E., Shin, Y. C., and Tian, Y., 2004, Laser-assisted machining of magnesia-partially-stabilized zirconia, *ASME Journal of Manufacturing Science and Engineering*, Vol. 126, pp. 42–51.

Pizzi, P., 1987, Laser metalworking, in: Soares, O. D. D., and Perez-Amor, M., editors, *Applied Laser Tooling*, Martinus Nijhoff Publishers, Netherlands, pp. 213–233.

Powell, J., Frass, K., Menzies, I. A., and Fuhr, H., 1988a, CO_2 laser cutting of non-ferrous metals, *SPIE, High Power* CO_2 *Laser Systems and Applications*, Vol. 1020, pp. 156–163.

Powell, J., Jezioro, M., Menzies, I. A., and Scheyvearts, P. F., 1988b, CO_2 laser cutting of titanium alloys, *SPIE, Laser Technologies in Industry*, Vol. 952, pp. 609–617.

Powell, J., Ivarson, A., Kamalu, J. Broden, G., and Magnusson, C., 1992, The role of oxygen purity in laser cutting of mild steel, *ICALEO Proceedings*, pp. 433–442.

Precision Micro Machining Technology & Applications Conference, June 11–12, 2003, Minneapolis, MN.

Rajagopal, S., Plankenhorn, D. J., and Hill, V. L., 1982, Machining aerospace alloys with the aid of a 15 kW Laser, *Journal of Applied Metalworking*, Vol. 2, No. 3, pp. 170–184.

Ready, J. F., 1965, Effects due to absorption of laser radiation, *Journal of Applied Physics,* Vol. 36, No. 2, pp. 462–468.

Ready, J. F., chief editor, 2001, *LIA Handbook of Laser Materials Processing*, Laser Institute of America, Magnolia Publishing, Inc., Orlando, FL.

Ream, S. L., Bary, P., and Perozek, P. M., 1990, Heat input during high pressure laser cutting of inconel and titanium, *ICALEO Proceedings*, pp.213–224.

Rizvi, N. H., 2003, Femtosecond laser machining: current status and applications, *RIKEN Review*, No. 50, pp. 107–112.

Schock, W., Giesen, A., Hall, Th., Wittwer, W., Hugel, H., 1986, Transverse radio frequency discharge: a promising excitation technique for high power CO_2-lasers, *SPIE, Laser Processing: Fundamentals, Applications, and System Engineering*, Vol. 668, pp. 246–251.

Schreiner-Mohr, U., Dausinger, F., and Hugel, H., 1991, New aspects of cutting with CO_2 lasers, *ICALEO Proceedings*, pp. 263–272.

Schulz, W., Simon, G., Urbassek, H.M., and Decker, I., 1987, On laser fusion cutting of metals, *Journal of Physics. D: Applied Physics.*, Vol. 20, pp. 481–488.

Schulz, W., and Becker, D., 1989, On laser fusion cutting: The self-adjusting cutting kerf width, *SPIE, High Power Lasers and Laser Machining Technologies*, Vol. 1132, pp. 211–221.

Schulz, W., Becker, D., Franke, J., Kemmerling, R., and Herziger, G., 1993, Heat conduction losses in laser cutting of metals, *Journal of Physics D: Applied Physics*, Vol. 26, pp. 1357–1363.

Schuocker, D., 1986a, Dynamic phenomena in laser cutting and cut quality, *Applied Physics B, Photo-Physics and Laser Chemistry*, Vol. 40, pp. 9–14.

Schuocker, D., 1986b, Laser cutting, *SPIE, The Industrial Laser Annual Handbook*, Vol. 629, pp. 87–107.

Schoucker, D., and Muller, P., 1987, Dynamic effects in laser cutting and formation of periodic striations, *SPIE, High Power Lasers*, Vol. 801, pp. 258–264.

Schuocker, D., 1988, Heat conduction and mass transfer in laser cutting, *SPIE, Laser Technologies in Industry*, Vol. 952, pp. 592–599.

Schmidt, A. O., 1949, Hot milling, *Iron Age*, Vol. 163, pp. 66–70.

Sider, T., and Steffen, J., 1990, YAG-SLAB laser cutting performance, *SPIE, High-Powered Solid State Lasers and Applications*, Vol. 1277, pp. 232–243.

Simon, G., and Gratzke, U., 1989, Theoretical investigations of instabilities in laser gas cutting, *SPIE, High Power Lasers and Laser Machining Technologies*, Vol. 1132, pp. 204–210.

Takeno, S., Moriyasu, M., and Hiramoto, S., 1992, Laser drilling by high peak pulsed CO_2 laser, *ICALEO Proceedings*, pp. 459-468.

Tian, Y., and Shin, Y. C., 2006, Thermal modeling for laser-assisted machining of silicon nitride ceramics with complex features, *ASME Journal of Manufacturing Science and Engineering*, Vol. 128, pp. 425–434.

Tiffany, W. B., 1985, Drilling, marking, and application for industrial Nd:YAG lasers, *SPIE, Applications of High Powered Lasers*, Vol. 527, pp. 28–36.

Tour, S., and Fletcher, L. S., 1949, Hot spot machining, *Iron Age*, July.

Ursu, I., Nistor, L. C., Teodorescu, V. S., Milhailescu, I. N., and Nanu, L., 1984, Early oxidation stage of copper during CW CO_2 laser irradiation, *Applied Physics Letters*, Vol. 44, pp. 188–189.

van Krieken, A. H., Schaarsberg, Groote, J., and Raterink, H. J., 1988, Laser micro machining of material surfaces, *SPIE, Laser Assisted Processing*, Vol. 1022, pp. 34–37.

Vicanek, M., and Simon, G., 1987, Momentum and heat transfer of an inert gas jet to the melt in laser cutting, *Journal of Physics. D: Applied. Physics.*, Vol. 20, pp. 1191–1196.

Vicanek, M., Simon, G., Urbassek, H. M., and Decker, I., 1987, Hydrodynamical instability of melt flow in laser cutting, *Journal of Physics D: Applied Physics*, Vol. 20, pp. 140–145.

von Allmen, M., 1976, Laser drilling velocity in metals, *Journal of Applied Physics*, Vol. 47, pp. 5460–5463.

Webber, T., 1991, Laser cutting of galvanized steel: CO_2 Vs. CW Nd:YAG, *ICALEO Proceedings*, pp. 273–281.

Weck, M., and Kasperowski, S., 1997, Integration of lasers in machine tools for hot machining, *Production Engineering*, Vol. IV, No. 1, pp. 35–38.

Yaroslava, G., Yingling, G., and Garrison, B. J., 2003, Photochemical ablation of organic solids, *Nuclear Instruments and Methods in Physics Research B*, Vol. 202, pp. 188–194.

Yilbas, B. S., Davies, R., and Yilbas, Z., 1992, Study into penetration speed during CO_2 laser cutting of stainless steel, *Optics and Lasers in Engineering*, Vol. 17, pp. 69–82.

Znotins, T., Poulin, D., and Eisele, P., 1990, Excimer laser: continued growth in materials-processing applications, Belforte, D., and Levitt, M., editors, *The Industrial Laser Annual Handbook*, Pennwell Books, Tulsa, OK, pp. 102–105.

Zweig, A. D., 1991, A thermo-mechanical model for laser ablation, *Journal of Applied Physics*, Vol. 70, pp. 1684–1691.

APPENDIX 15A

List of Symbols Used in the Chapter.

Symbol	Parameter	Units
A_e	area of opening through which expelled liquid escapes	mm^2
h_x	melt thickness	µm
H	amplitude of oscillation of melt thickness	µm
H_l	specific energy absorbed by expelled liquid	kJ/kg
H_v	specific energy absorbed by expelled vapor	kJ/kg
I_a	irradiance absorbed by material	kW/m^2
k_c	constant	–
$k_z = \frac{2\pi}{\lambda_z}$	wavenumber in the z-direction	1/µm
l_h	heat diffusion length	m
L_p	heat of evaporation per particle	J/particle
m_m	mass of atom or molecule (particle)	g
$m_u = \frac{u}{\sqrt{2R_{gv}T_v}}$		$m - kg^{1/2}/sJ^{1/2}$
M_h	flow Mach number of the vapor leaving the Knudsen layer	–
m_g	molecular weight of the ambient gas	kg/mol
m_v	molecular weight of the vapor	kg/mol
P_a	ambient pressure	Pa
P_{vk}	pressure at edge of Knudsen layer	Pa
P_s	saturated vapor pressure	Pa
q_s	rate of local internal energy generated per unit volume	W/mm^3
r_c	local radius of curvature	mm
R	≈ 0.2 = mean particle reflection coefficient	–
$R_{ga} = \frac{R_g}{m_{ag}}$	gas constant for the ambient gas	J/mol K kg
$R_{gv} = \frac{R_g}{m_v}$	gas constantfor the vapor	J/mol K kg
T_{ct}	thermodynamic critical temperature	K
T_v	evaporation temperature	K
T_h	temperature of a hot surface	K
T_{vk}	temperature at edge of Knudsen layer	K
T_l	liquid temperature	K
T_{vk}	vapor temperature at edge of Knudsen layer	K
u	velocity	m/s
u_{go}	gas velocity at nozzle (reference)	m/s
u_a	sound velocity in ambient gas	m/s
u_c	cutting velocity	m/s
u_d	drilling velocity	m/s
u_i	velocity components (u_x, u_y, u_z)	m/s

u_k	mean vapor velocity at edge of Knudsen layer	m/s
u_v	sound velocity in vapor	m/s
U_i	amplitude of oscillation in the x, y, z directions	μm
z_i	height of location	mm
δ'	skin depth	m
δ_l	thickness of liquid layer formed by heat conduction	μm
ΔH	specific energy absorbed by the expulsed material	J/kg
ΔH_{vm}	molar enthalpy of vaporization	J/mol
γ_l	surface tension of the molten material	N/m
γ_a	ratio of specific heats for ambient gas	–
$\gamma_v = \frac{c_p}{c_v}$	ratio of specific heats for vapor	–
ϕ	expulsion rate	$kg/m^2 - s$
τ	shear stress distribution at the cutting front	Pa
τ_{gl}	shear stress at interface between gas and molten metal	Pa
τ_p	pulse duration	s
θ_i	inclination of the cut surface with the vertical	°
χ	complex growth rate	1/s

PROBLEMS

15.1. Given the quasistatic temperature distribution, T, as

$$T = \frac{I_a{}'}{\rho c_p u} \exp\left(-\frac{\xi u}{\kappa}\right)$$

Show that the thickness, δ_l, of the liquid layer formed by heat conduction may be expressed as

$$\delta_l = \left(\frac{\kappa}{u}\right) \ln\left(\frac{T_h}{T_m}\right)$$

15.2. Given that the thickness, δ_l, of the liquid layer formed during laser drilling is

$$\delta_l = \left(\frac{\kappa}{u}\right) \ln\left(\frac{T_0}{T_m}\right)$$

and that the radially outward velocity of the layer is

$$u_l = \left(\frac{2P}{\rho}\right)^{1/2}$$

Show that the liquid expulsion rate with respect to the irradiated surface is given by

$$\phi_l = \left[\left(\frac{2\kappa}{r}\right)\ln\left(\frac{T_0}{T_m}\right)\right]^{1/2} P_s^{1/4}\rho^{3/4}$$

15.3. Starting with Bernoulli's equation, show that the liquid layer is pushed radially outward with a velocity, u_l, given by

$$u_l = \left(\frac{2P}{\rho}\right)^{1/2}$$

15.4. A 5-mm thick plain carbon steel plate is cut using a 2.5 kW laser at a speed of 4.5 m/min.

(a) Calculate the surface temperature at a point 4 mm behind the laser beam and 1.5 mm to one side if the heat transfer efficiency is 80% and the ambient temperature is 30° C.

(b) Estimate the size of the heat-affected zone for this process.

15.5. A 15 × 10 mm rectangular piece is to be cut out of a polymer sheet of thickness 8 mm, using a laser. The output laser power available at the workpiece is 1.0 kW, and the part is to be cut at a speed of 2.0 m/min, resulting in a kerf width of 0.3 mm. If the reflectivity, R, of the polymer sheet is 25% and another 4.0% of the beam is lost and not used for cutting, determine the specific energy of the material.

15.6. (a) Which form of laser cutting involves melting of the base material, which is then ejected using assist gas?

(b) Which form of laser cutting uses the least energy?

(c) How does orientation of the plane of polarization of the laser beam relative to the cutting direction affect the quality of the cut surface?

(d) A plot of cutting speed versus workpiece thickness usually has two limiting curves. What happens when one selects speeds that are below the lower limiting curve?

15.7. A 5 mm thick nickel plate is cut using a 4 kW CO_2 laser at a speed of 1.5 m/min. If the resulting kerf width is 0.45 mm, the ambient temperature is 25° C, and 15% of the incident laser beam is absorbed by the plate, determine

(a) The amount of energy that is supplied by the laser while cutting a circular blank of diameter 0.5 m.

(b) The amount of material (in kg) that is removed or lost as a result of the process.

(c) The energy required to heat the material that is removed, to the melting point.

(d) The energy required to melt the material after it is heated to the melting point.

(e) The energy that is lost by conduction, convection, and radiation.

15.8. Estimate the drilling speed that can be expected while drilling 350 μm diameter holes in a low carbon steel plate using a 1.5 kW CO_2 laser. Assume ambient temperature and pressure of 25° C and 0.1 MPa, respectively.

15.9. A 2.5 mm thick aluminum plate is cut using a 1.5 kW laser. If the resulting kerf width is 0.3 mm, the ambient temperature is 25°, and 5% of the incident laser beam is absorbed by the plate, determine the cutting speed.

16 Laser Welding

Welding is one means for joining parts. Others include mechanical fastening that uses physical devices such as bolts and rivets to hold parts together; and adhesive bonding that uses nonmetallic materials to hold parts together, for example, gluing. Welding, on the contrary, uses heat and/or pressure to produce localized coalescence of material. It may also require filler metal, shielding, or fluxing. Welding processes can be categorized into two broad areas, namely, fusion welding processes where a portion of the pieces to be joined actually melts, and nonfusion welding processes. Examples of the former include arc welding, resistance welding, gas, laser, and electron beam welding processes, and for the latter, we have brazing, soldering, and solid-state welding processes. Our discussion in this chapter focuses on laser welding.

The most common form of lasers used for welding are CO_2 and Nd:YAG lasers. However, continuous welds have traditionally been done with CO_2 lasers since higher power continuous wave (CW) beams are obtained with CO_2 lasers. Recent developments in high-power CW Nd:YAG lasers have increased their application in this area.

The discussion in this chapter starts with a general outline of laser welding parameters. This is followed by a discussion on welding efficiency, and then the mechanism of the laser welding process. After that, the behavior of different materials, including some common metals and their alloys, ceramic materials, as well as dissimilar materials, when welded, are presented. Common discontinuities encountered during laser welding are then outlined, followed by advantages and disadvantages of the process. Special laser welding techniques, for example, multiple-beam welding and arc-augmented laser welding are then presented, and finally, specific applications such as microwelding and laser welded tailored blanks are discussed.

16.1 LASER WELDING PARAMETERS

The setup for laser welding is similar to the one used for laser cutting (Fig. 15.1). The primary difference is that in laser welding, the exhaust system is positioned above the beam delivery unit rather than the other side of the workpiece. The principal

Principles of Laser Materials Processing, by Elijah Kannatey-Asibu, Jr.

TABLE 16.1 Sample Conditions for Deep Penetration Laser Welding of Different Materials Using a CW CO_2 Laser

Material	Weld Type	Thickness (mm)	Welding Speed (m/min)	Laser Power (kW)
Carbon steel	Butt	12	0.8	9
Carbon steel	Butt	6	10	25
Stainless steel (302)	Butt	6	1.27	3.5
Stainless steel (304)	Butt	20	1.27	20
Aluminum (A5083)	–	2.7	3	3.7
Titanium	Butt	4.5	5.08	16

Source: Table 5.9, Harry, 1974; Section 11.4, Table 5, *Handbook of Laser Processing.*

parameters that affect the laser welding process include the following:

1. Beam power and traverse speed.
2. Beam characteristics (mode, stability, polarization, and form—pulsed or CW).
3. Shielding gas.
4. Location of focal point relative to the workpiece surface.
5. Joint configuration.

As an illustration of conditions that might be used in industry, Table 16.1 gives sample conditions for CO_2 laser welding of a variety of materials, while Table 16.2 provides additional details for a specific material. We now discuss each of the parameters listed above in detail in the following sections.

16.1.1 Beam Power and Traverse Speed

Figure 16.1 shows the effect of welding speed and laser power on weld penetration. The depth of penetration decreases almost exponentially with increasing traverse

TABLE 16.2 Typical Processing Parameters for Laser Welding of a Low Carbon Steel Sheet Using a CO_2 Laser. The Speed Varied with the Sheet thickness, but for a 1.5 mm Sheet, a Speed of 2.5 m/min was Used

Laser power	1700 W
Assist gas	Ar @ 50 l/min
Nozzle diameter	4 mm
Nozzle–workpiece distance	11 mm
Depth of focus	Upper 1/3rd in material thickness
Focusing optic	Reflective copper mirror with $f = 150$ mm
Beam mode	TEM_{01^*}

Source: Table 1, Betz, Retzbach, Alber, Prange, Uddin, Mombo-Caristan.

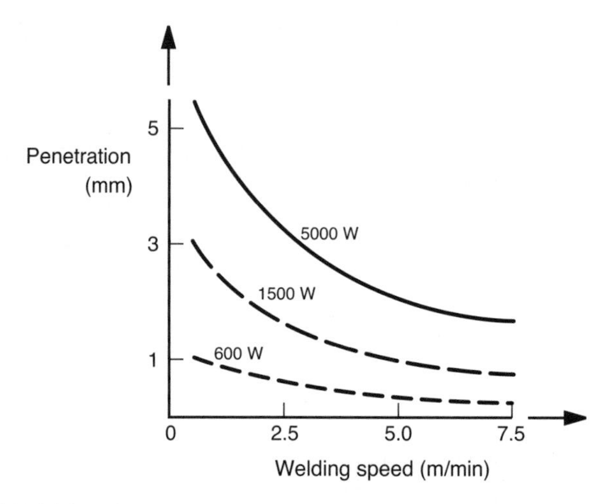

FIGURE 16.1 Effect of welding speed and laser power on weld penetration.

speed for a given power. Conversely, the depth-to-width ratio increases with increasing traverse speed, though not as dramatically (Fig. 16.2), and then levels off. For a given speed, the penetration depth increases with increasing power input. Generally, results obtained from low-power laser welding, say less than 5 kW cannot be extrapolated to high power welding, say 20 kW. This is due to differences caused by plasma shielding effects resulting from either high power densities or high evaporation rates that cause an enhanced plasma formation.

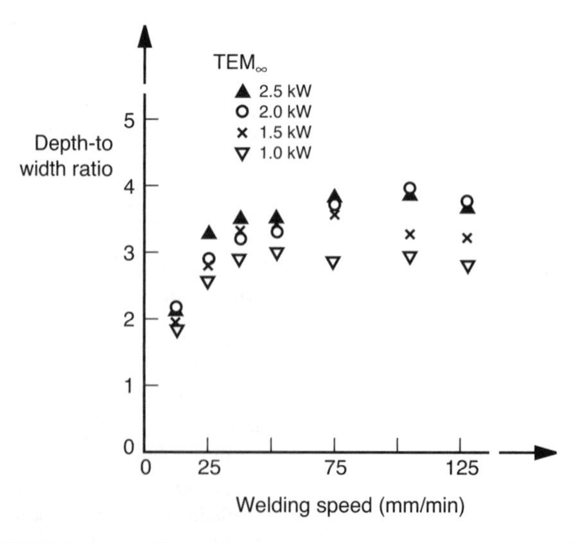

FIGURE 16.2 Effect of welding speed on depth-to-width ratio.

16.1.2 Effect of Beam Characteristics

The principal beam characteristics that affect the laser welding process include the beam mode, stability, polarization, and form (pulsed or continuous wave). The effect that these parameters have on the welding process are discussed in that order in the following subsections.

16.1.2.1 Beam Mode The beam mode (distribution of energy intensity, see Section 14.1.3) has a significant impact on the weld bead shape. A TEM_{00} mode results in a greater depth-to-width ratio than higher order modes. This is due primarily to the more concentrated shape of the TEM_{00} mode, which results in a smaller focused beam radius. The larger spot size of multimode beams, however, makes them more suited for butt welding applications since it reduces the tight fit-up requirements.

16.1.2.2 Beam Stability Repeatability or consistency of the weld quality is influenced by beam stability. Temporal variations of the beam characteristics such as the output power and mode structure, as well as variations in the absorption coefficient with temperature, do cause variations in the welding process, and thus the resulting weld quality. Beam stability is further discussed in Section 7.5.

16.1.2.3 Beam Polarization The basic concepts of beam polarization are discussed in Section 9.5. The influence of beam polarization on the welding process depends on both the material and welding conditions. In welding some steels, the depth of penetration is found to vary depending on the direction of beam polarization, being enhanced when the direction of polarization is aligned with the welding direction. This effect is only evident above a critical welding speed, below which the polarization direction has no impact on the penetration (Fig. 16.3).

This is different from the case of laser cutting where the process efficiency is found to be strongly influenced by the beam polarization. This difference in behavior is most likely the result of a loss in correlation between any polarization of the beam and the electromagnetic field in the keyhole due to the processes of scattering, reradiation, and reflection from the walls during keyhole welding (Section 16.3.2).

16.1.2.4 Pulsed Beams Both pulsed and CW laser beams (see Section 14.1.4) can be used for welding, with pulsed beams being more suited for spot welding while CW beams are more suited for continuous welds. Pulsed beams can also be used for continuous welds if the pulses are made to overlap. Additional process parameters that need to be considered when welding with pulsed beams are the pulse duration, shape, and pulse frequency (repetition rate).

In using a pulsed beam for keyhole welding, the pulses ought to have a high-enough energy to exert a significant thrust on the molten metal to maintain a keyhole once it is formed, and will have to be spaced closely enough to prevent freezing of the metal while the beam is off.

When large amounts of energy are supplied in short periods to a workpiece, they tend to concentrate near the surface as heat, since it takes time for the heat to be

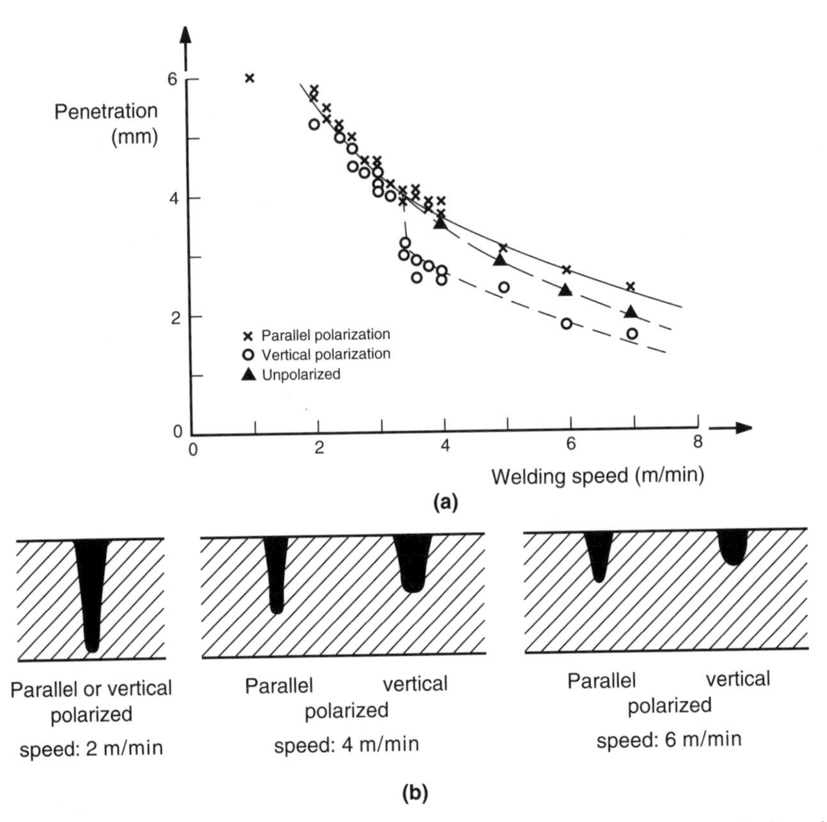

FIGURE 16.3 Influence of beam polarization on penetration. (a) Variation of penetration with speed for different modes of polarization. (b) Weld cross sections for parallel and vertically polarized beams for different speeds. Material—6 *mm* thick steel; laser power—5 kW; shielding gas—helium; lens f number—$f6$; focal diameter—280 μm. (From Beyer, E., Behler, K., and Herziger, G., 1988, *SPIE, High Power Lasers and Laser Machining Technology*, Vol. 1020, pp. 84–95.)

conducted into the bulk of the material, (see Section 14.1.4). This results in high-temperature gradients in the immediate vicinity of the heat source, producing high-cooling rates once the beam is switched off. Thus, Q-switching (see Section 6.2) is not normally used for deep penetration laser welding because the shorter pulse duration of the Q-switched laser results in lower penetration. It is more appropriate for microwelding.

16.1.3 Plasma Formation, Gas Shielding, and Effect of Ambient Pressure

16.1.3.1 Plasma Formation With high-power beams, a portion of the laser energy is expended in ionizing part of the shielding gas and/or vapor cloud that forms above

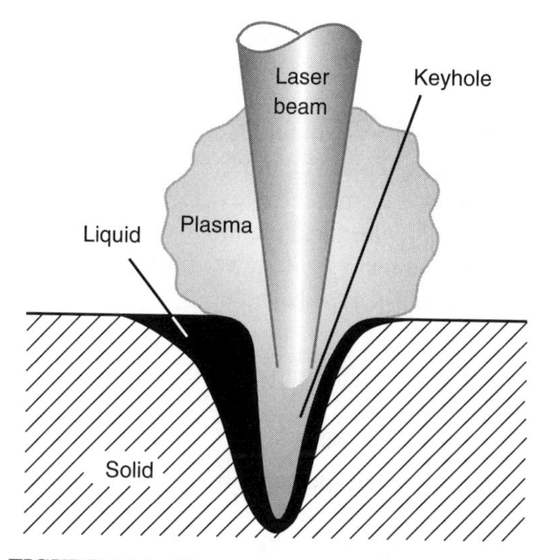

FIGURE 16.4 Plasma formation in keyhole welding.

the workpiece, to form a plasma (Fig. 16.4). This plasma may absorb part of the laser beam energy and reradiate it in all directions or scatter it, thereby reducing the amount of energy reaching the molten pool. The plasma cloud may also induce an optical lensing effect that changes the effective focusing of the laser beam. This laser-induced plasma normally develops in the intensity range of 10^6 W/cm^2 when welding steel with CO_2 lasers.

Even though initial absorption of a CO_2 laser beam on a metallic surface is normally less than 10%, the absorption can be as high as 100% once a plasma is formed. Such absorption can occur either above the workpiece or inside the keyhole. When it occurs above the workpiece as described above, it reduces the efficiency of the welding process as a significant portion of the beam power is then lost. On the contrary, absorption inside the keyhole may enhance the process efficiency. Plasma formation is of less significance when welding with an Nd:YAG laser. This is because the Nd:YAG laser beam is more efficiently absorbed at metallic surfaces while less is absorbed by the plasma.

The formation of the plasma has advantages and disadvantages. In the initial stages of the process, the formation of a shielding gas plasma reduces the amount of the laser energy that is reflected away as it absorbs a significant portion of the energy. Since the plasma formed is then relatively small and in close contact with the workpiece, it can enhance absorption of the beam energy by the workpiece. This it does by first absorbing the beam energy and then transferring it to the workpiece. This is referred to as plasma enhanced energy coupling and is most effective at the threshold or initial stages of plasma formation. However, as the beam intensity increases and the plasma builds up, if the plasma stays on top of the workpiece, then the energy it absorbs is not fully coupled to the workpiece. This reduces the amount of energy that reaches or is absorbed by the work, and thus, the process efficiency.

Variations in the plasma density can thus cause variations in beam absorption by the workpiece and, as a consequence, variations in weld quality. Plasma fluctuations that occur during laser welding are more severe at low beam intensity and high speed, especially in the initial stages of plasma formation. At higher beam intensity and lower speed, the plasma is more continuous with fewer fluctuations. This is especially the case when absorption occurs inside the keyhole.

16.1.3.2 Gas Shielding In laser welding, a shielding gas may be used for one or more purposes:

1. To blow away the vapor and plasma formed, enabling the beam to reach its target.
2. To provide a protective environment for the weld pool.
3. To protect the focusing lens.

The most common gases used for laser welding are argon and helium. Argon is more commonly used for low- to medium-power laser welding since it is cheaper than helium, and its higher density results in better shielding. The argon-shielded weld surface is also smoother than that obtained with helium shielding. Since the density of helium is lower, higher flow rates are required for effective shielding. However, for high-power or low-speed applications, helium is preferable since it has a higher ionization potential. Thus the threshold energy necessary to form a plasma when helium is used for shielding is relatively high. This reduces the amount of plasma formed and keeps the gas transparent to the laser beam, allowing the energy to reach the workpiece.

On the contrary, without the plasma, there will be no initial plasma energy coupling. This might make keyhole formation difficult, thus preventing deep penetration welding. This effect occurs, for example, in aluminum welding with a 2.5 kW laser at a speed of about 3 m/min. Under these conditions, a deep penetration weld can be obtained using nitrogen as a shielding gas, while very little molten metal is formed when helium shielding gas is used.

When argon is used in the higher power applications, its ionization results in a significant portion of the beam being absorbed. Thus less power reaches the weld, thereby reducing penetration. Generally, for a given set of conditions, penetration is found to be greater for helium shielding, followed by nitrogen, and then argon. Nitrogen is not as extensively used as argon and helium since it could form nitrides that might embrittle the weldment.

The weld quality thus depends on the type of shielding gas used. The effect is more significant at low welding speeds and at higher power densities since the local vapor density above the workpiece is then much higher. As is evident from Fig. 16.5,

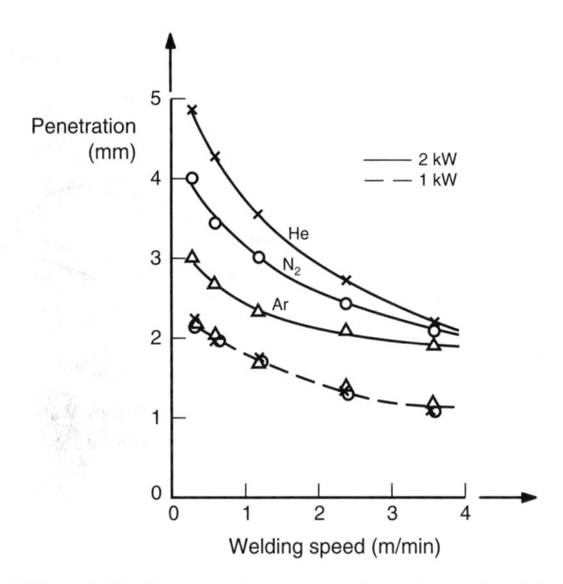

FIGURE 16.5 Effect of shielding gas, laser power, intensity, and welding speed on depth of penetration. Material—steel; shielding gas flow rate—20 L/min; Focused beam radius—100 μm. (From Beyer, E., Behler, K., and Herziger, G., 1988, *SPIE, High Power Lasers and Laser Machining Technology*, Vol. 1020, pp. 84–95.)

at low power levels, there is no difference between the impact of argon, helium, or nitrogen. However, at higher power levels, a distinct difference exists between the depths of penetration obtained using the various gases. As mentioned earlier, because of the high ionization potential of helium, it does not form a plasma easily. It is thus transparent (does not absorb the beam energy) to the incident laser beam over a wide range of conditions, especially at the lower welding speeds or high power densities. By comparison, other gases tend to form a plasma that absorbs the incident beam energy, under these conditions.

Another use of the shielding gas is protection of the focusing lens. For low power applications, say, up to 1500 W, the normal flow of shielding gas (about 10 L/min) for protecting the weld is also adequate to protect the lens, as long as the gas stream is coaxial with the laser beam. A typical nozzle diameter would be 3 mm. However, at higher powers, separate shielding or helium cross-flow that is provided by a plasma suppression device (Fig. 16.6) is often applied. The plasma suppression unit directs a high-velocity jet of inert gas, usually helium, at the top of the weld to suppress or blow away the plasma. Plasma suppression serves to protect the lens and also to enable the heat supplied to reach the weld.

The final form of the weld bead is affected by the mode of application of the shielding gas. For example, when welding steel with a 10 kW laser while using an axial gas nozzle of diameter 2 mm, no keyhole is produced at low helium flow rates of say, less than 5 L/min, since most of the beam energy is then absorbed by the plasma.

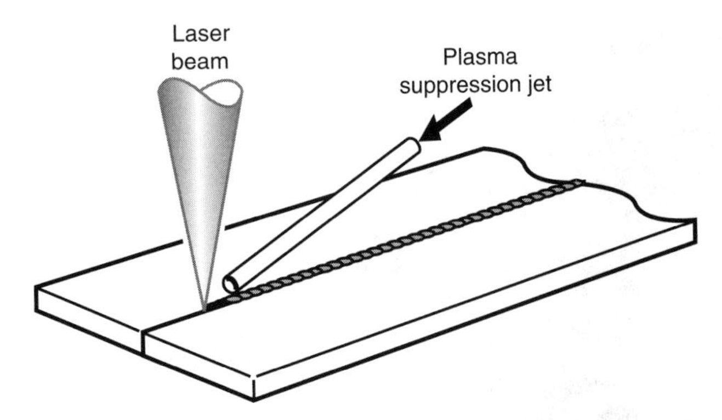

FIGURE 16.6 Schematic of a plasma suppression setup. (From Watson, M. N., Oakley, P. J., and Dawes, C. J., 1985, *Metal Construction*, pp. 25–28. Reproduced by permission, TWI Ltd.)

A small molten zone may result from conduction mode welding. With increasing gas flow rates, a point is reached when a transition occurs to keyhole formation, with a resulting deep penetration. Further increases in gas flow rate then produce a linear but slight increase in penetration. Beyond about 40 L/min, the gas flow may be vigorous enough to eject the molten metal out of the weld pool, resulting in an irregular and perhaps shallow bead shape. The general effect of suppressant gas flow rate on penetration is shown in Fig. 16.7.

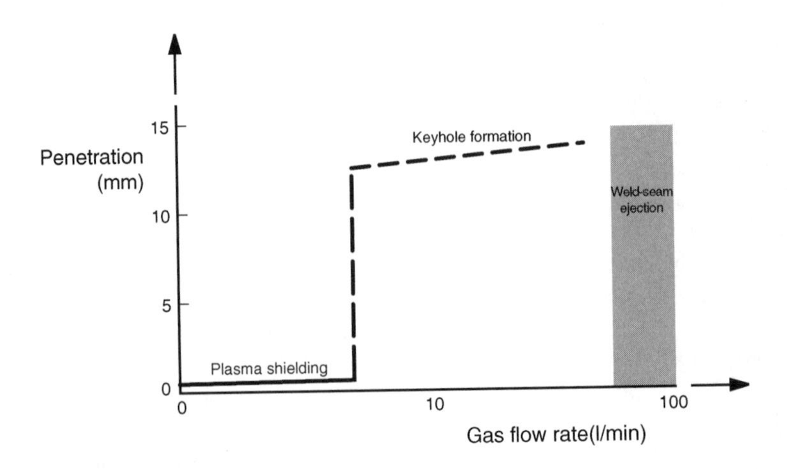

FIGURE 16.7 Effect of plasma suppression helium gas flow rate on the depth of penetration in plain carbon steel. Laser power—10 kW; shielding gas—helium; welding speed—1.2 m/min; lens focal length—476 mm. (From Behler, K., Beyer, E., and Schafer, R., 1988, *SPIE, High Power CO_2 Laser Systems and Applications*, Vol. 1020, p. 165.)

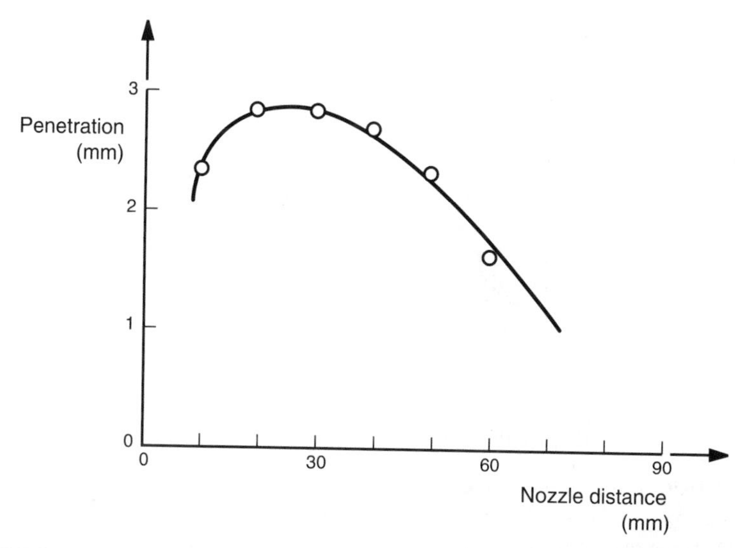

FIGURE 16.8 Effect of plasma suppression gas nozzle distance on the depth of penetration. Material—aluminum; shielding gas—nitrogen at a flow rate of 60 L/min; power—3.8 kW; welding speed—5.3 m/min; lens focal length—200 mm; nozzle inclination—40°. (From Behler, K., Beyer, E., and Schafer, R., 1988, *SPIE, High Power CO_2 Laser Systems and Applications*, Vol. 1020, p. 164.)

The angle of inclination of the plasma suppressant gas nozzle with the horizontal does not appear to have significant effect on the process in the range of 30–60°. However, increasing the distance between the gas nozzle and the laser-induced plasma tends to increase the penetration up to a maximum and then decrease with further increases in the distance (Fig. 16.8).

16.1.3.3 Effect of Ambient Pressure The amount or intensity of plasma generated decreases as the ambient welding pressure is reduced. Below a pressure of about 5 Torr (depending on the welding conditions), the plasma is so small in size that no significant change is observed with further reductions in the pressure. The pool width decreases, while the penetration increases with decreasing pressure. Laser welding in vacuum may increase the penetration about 30–50% more than that obtained when welding in the atmosphere. However, laser vacuum welding is not a common practice, due to the cost and other difficulties associated with setting up the vacuum welding system, especially for large workpieces.

16.1.4 Beam Size and Focal Point Location

For a given optical system (focusing lens), the location of the focal point relative to the workpiece surface determines the beam size on the workpiece surface,

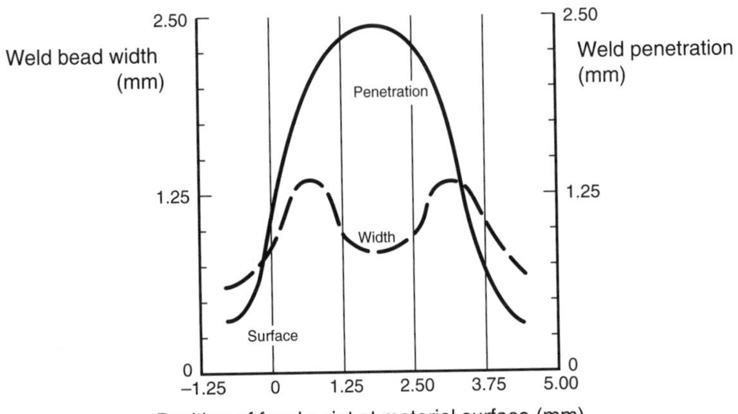

FIGURE 16.9 Effect of focal point positioning on penetration and bead width. Material—1018 steel; power—1.45 kW; welding speed—1.25 m/min. Positive position is downward into the workpiece. Negative is above the workpiece.

and this affects the bead width and depth of penetration (Fig. 16.9). Optimum location of the focus results in maximum depth-to-width ratio, that is, the maximum penetration depth and minimum bead width occur for almost the same focal position.

16.1.5 Joint Configuration

There are various types of joint configuration that can be used in welding. However, due to the small beam size, lasers are primarily used for lap or square butt joints. The rule of thumb is that the butt gap has to be less than 10% of the thinner plate's thickness (Fig. 16.10). Otherwise there is the likelihood that most or the entire beam will pass directly through the gap, resulting in inadequate weld or no weld at all.

To minimize the problem associated with the small beam size, the beam may be defocused at the joint to produce a relatively wide bead. However, the defocused beam is more sensitive to absorption by the material surface, since the power density is then low, and any small amount of power that is reflected is relatively significant. This may result in fluctuations in the energy absorbed by the workpiece. The use of Nd:YAG lasers for metals minimizes this fluctuation since the 1.06 μm wavelength is more efficiently absorbed by metals. Fluctuation in the energy absorbed may also be minimized by superimposing a pulsed beam over the CW welding beam. This is referred to as the rippled mode laser output or hyperpulse (see Section 14.1.4). The pulse rapidly melts the material surface, thereby increasing its absorption coefficient to a high enough value that the CW beam is efficiently and uniformly absorbed.

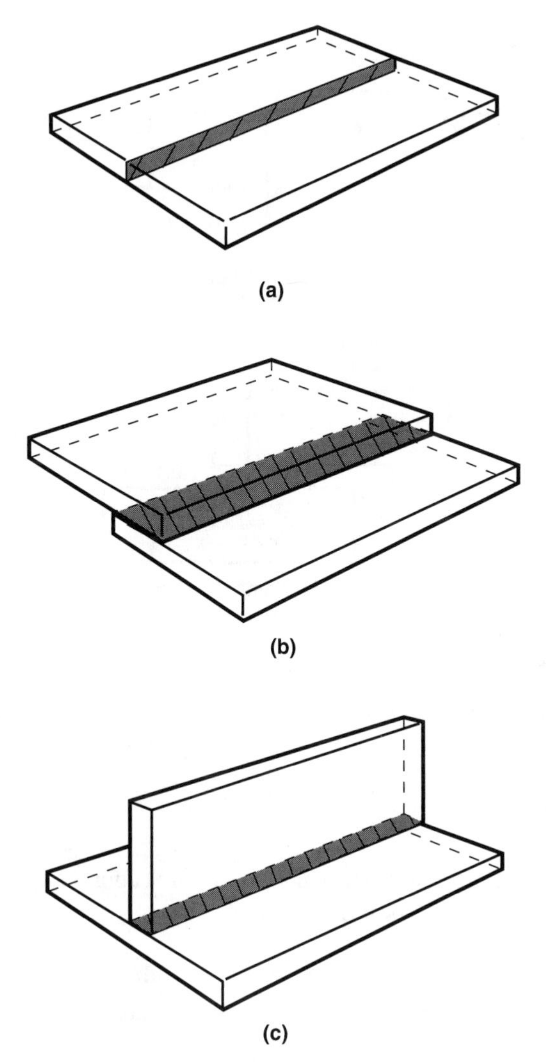

FIGURE 16.10 Some common joint designs. (a) Butt joint. (b) Lap joint. (c) T-joint.

The joint fitting requirements for lap joints are not as critical as for butt joints since in the lap joint, the laser beam is made to penetrate through one piece into the underlying component (Fig. 16.11). Lap joints are more appropriately used when the upper component is relatively thin (typically less than 1 mm in thickness).

Thus far, a significant number of laser welds that are made are autogenous, involving no filler metal. However, filler metal may be used where necessary, especially for nonsquare butt joints or when the butt gap is relatively large.

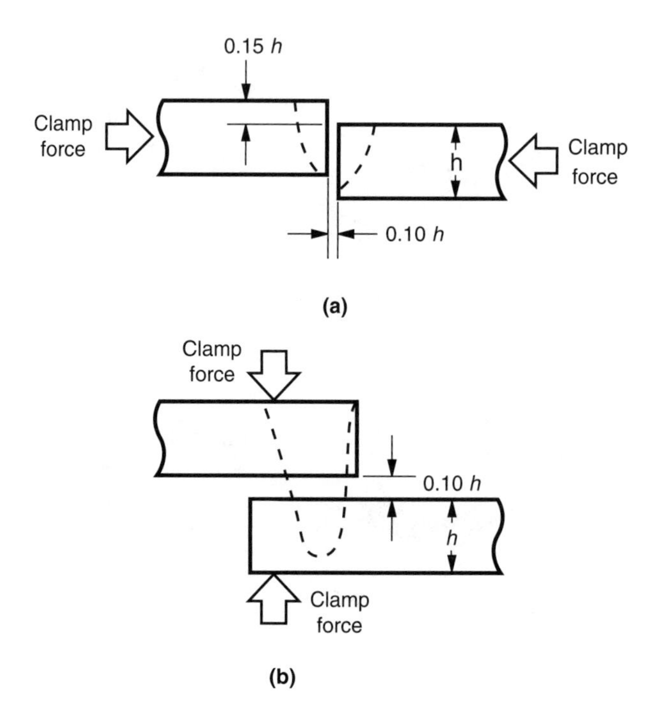

FIGURE 16.11 (a) Butt joint and (b) lap joint specifications in laser welding.

16.2 WELDING EFFICIENCY

Processing efficiency is discussed in general terms in Section 14.2. For laser welding, the overall efficiency, η, may be expressed as

$$\eta = \frac{\text{power used in melting solid}}{\text{power input to the laser generator}} \tag{16.1}$$

Typical values of overall efficiency for arc, laser, and electron beam welding are compared in Table 16.3.

TABLE 16.3 Comparison of Efficiency Values for Various Welding Processes

Weld Type	Generator η_g%	Transmission/ Heat Transfer η_{th}%	Processing η_p%	Overall η%
Laser	7	70	48	2.4
Arc (argon shield)	80	50	20	8
Electron beam	80	80	48	31

Source: From Abilsiitov, G. A., and Velikhov, E. P., 1984, *Optics and Laser Technology*, pp. 30–36.

16.3 MECHANISM OF LASER WELDING

When a laser beam is incident on a metallic surface, a sequence of events take place. A significant portion of the beam may initially be reflected away. The small percentage that is absorbed heats up the surface, raising its temperature. With increasing temperature, the surface absorptivity increases (see Section 14.1.6), which further increases the temperature. This may eventually result in localized melting and possible evaporation of the metal. Such vaporization, if it occurs, creates a vapor cavity in the metal. Laser welding may thus be one of the two forms:

1. Conduction mode (also known as conduction limited).
2. Keyhole (or deep penetration) welding.

Conduction mode welding is normally used for welding of foils and thin sheets, while keyhole mode welding is used for relatively thick sections.

16.3.1 Conduction Mode Welding

Conduction mode welding usually occurs at power densities lower than 10^6 W/cm^2, with vaporization of the workpiece being minimal. The laser power is first deposited on the surface and then transferred by conduction to the surroundings. The penetration is thus essentially controlled by conduction from the initial point of contact, that is, the surface of the workpiece, causing a small surface area to be heated above the melting point. Convection also plays a role once a weld pool is formed. The resulting weld is shallower with a wider heat-affected zone compared to that produced by keyhole welding. It is more common with low-power lasers, say below 1 kW. A typical configuration is shown in Fig. 16.12. The shape of the weld pool in conduction mode welding is influenced by flow in the weld pool and by the presence of surface active elements (see Section 10.3.3).

Approximate analysis of heat flow in conduction mode welding can be done using the point heat source analysis discussed in Section 10.2, while material in Section 10.3 will enable flow in the molten pool to be analyzed.

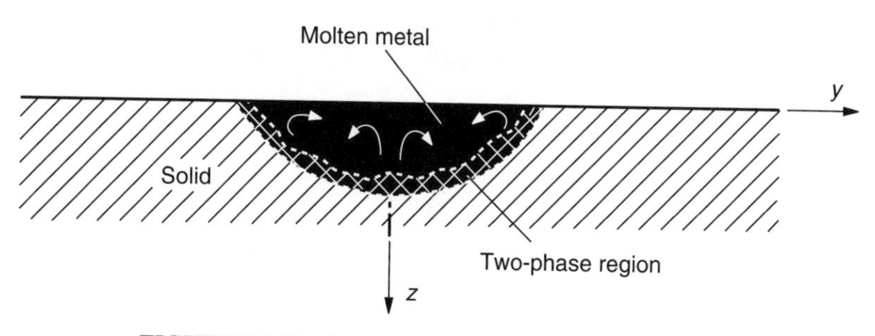

FIGURE 16.12 Schematic of conduction mode welding.

16.3.2 Keyhole Welding

At very high power densities, say, above 10^6 W/cm^2, part of the work material is vaporized to form a cavity, the *keyhole*, which is surrounded by molten metal, which in turn is surrounded by the solid material (Fig. 16.13). The molten material around the keyhole fills the cavity as the beam is traversed along the joint. The cavity contains vapor, plasma, or both. Forces at play that tend to collapse the keyhole cavity are as follows:

1. Surface tension at the interface between the molten metal and vapor or plasma.
2. Hydrostatic pressure of the molten metal.
3. Hydrodynamic pressure of the molten metal.

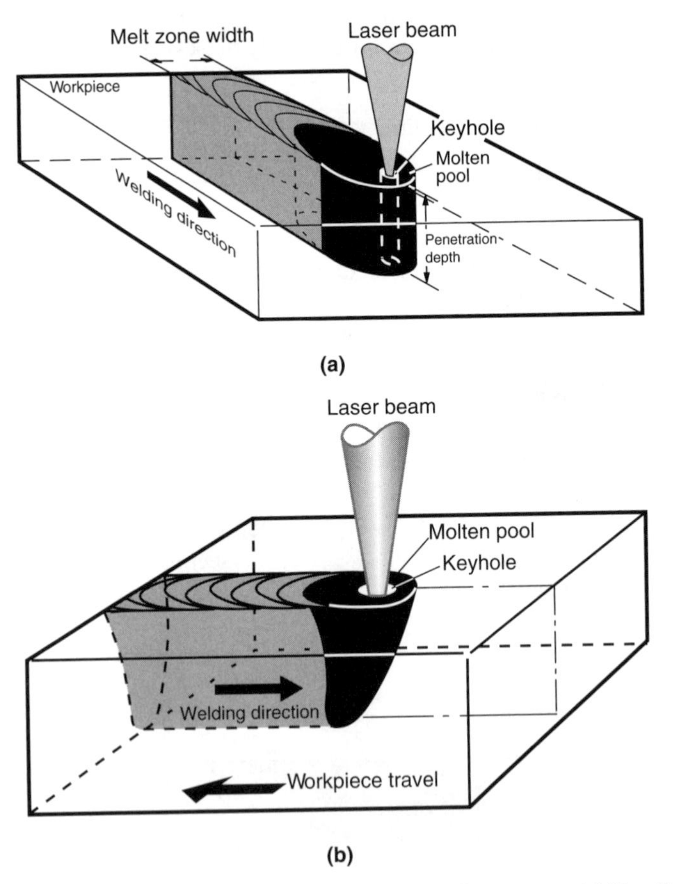

FIGURE 16.13 Schematic of the keyhole mode laser welding process. (a) Simplified sketch. (b) Illustration of bending of the keyhole in the direction of workpiece travel. (From Wilson, J., and Hawkes, J. F. B., 1987, *Lasers, Principles and Applications*. Reprinted by permission of Pearson Education, Inc., Essex, UK.)

These forces, however, are balanced by

1. Vapor pressure that exists within the cavity, which is greater than atmospheric pressure by about 10% and that pushes the molten metal toward the periphery.
2. Ablation pressure of material as it vaporizes from the inner surface of the keyhole, and which reinforces the vapor pressure.

Ablation is the sudden removal of material by melting and/or vaporization. The balancing pressure is maintained during the welding process, thus sustaining the keyhole.

Laser energy that enters the keyhole is trapped and carried deeper into the material, resulting in very high absorption, greater than 90% once it is established. The keyhole mode thus enables very narrow and deep penetration welds to be made. This feature makes it attractive for machined components and aerospace engine components, since distortion is then minimal.

If the laser beam is stationary, the continuous vaporization of material would cause the cavity to grow. Under such circumstances, a steady state cannot be achieved, and the keyhole would eventually collapse. On the contrary, the steady-state conditions can be achieved with a moving beam. The cavity moves along with the beam, at a rate determined by the beam. Since an adequate amount of vapor pressure needs to be maintained to prevent collapse of the cavity, a minimum traverse speed is necessary for steady-state conditions. As the cavity moves, material continuously moves from the front to the back of the cavity. This may occur either by flow of molten material around the cavity, or as vapor through the cavity or keyhole, with the bulk of the transport being through the molten material. The driving force for flow through the molten material is the ablation or recoil pressure of the evaporating material, or the surface tension due to temperature variation along the keyhole walls. The effect of surface tension as a driving force of flow is relatively small compared to the flow due to recoil pressure, since the interface temperature is close to the equilibrium evaporation temperature. The pressure in the melt at the vapor–liquid interface is a maximum directly in front of the laser beam, in the direction of welding, and reduces to a minimum on the side of the keyhole.

Under steady-state conditions, motion of the laser beam results in motion of the cavity and surrounding molten metal, along with all isotherms, at the same rate. The width of the molten region determines the width of the weld seam.

The simplest approximation to the keyhole configuration is that of a vertical cylinder (Fig. 16.13a). However, since the vapor pressure generally varies with depth, the radius is also a function of depth. The variation of pressure with depth results in outward flow of vapor, causing vapor to emerge from the end of the keyhole as an axial jet. Generally, the axis of the keyhole is bent in the direction of workpiece travel (Fig. 16.13b).

One typical shape of the keyhole is shown in Fig. 16.14a, and has a constriction near the cavity outlet. Other shapes can also develop, depending on the welding conditions. Some are relatively short with walls that are convex outward and rounded at the bottom, while others are long and thin, concave outward and have sharp points

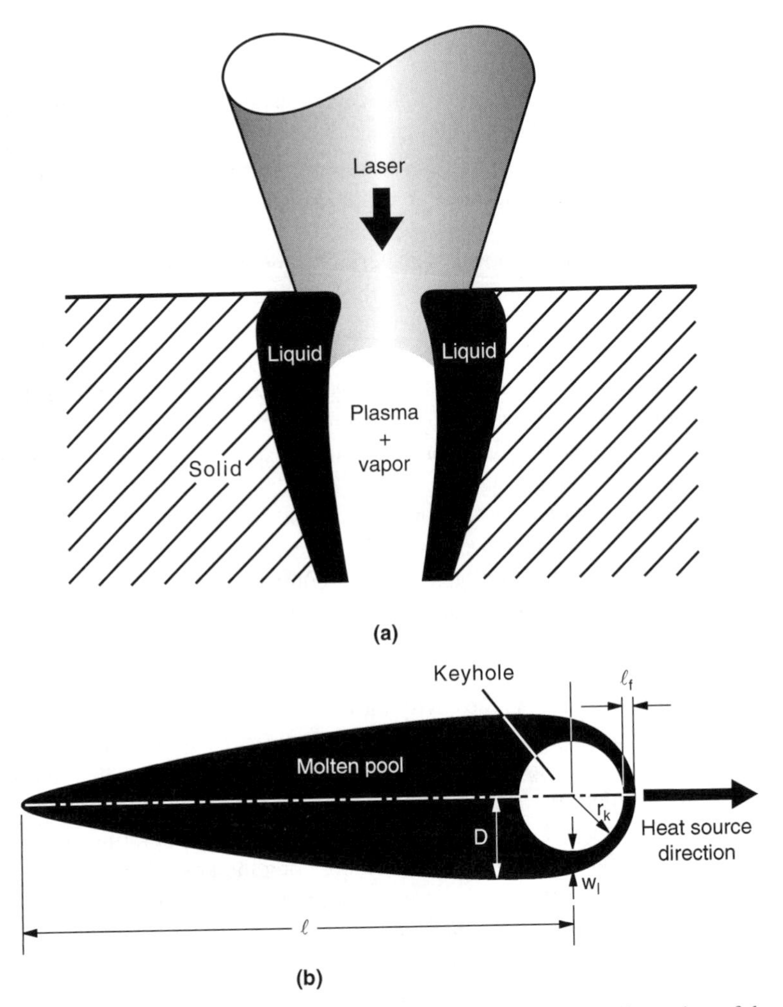

FIGURE 16.14 (a) Schematic of the keyhole cavity shape. (b) Typical top view of the weld pool during keyhole welding. (From Klemens, P. G., 1976, *Journal of Applied Physics*, Vol. 47, pp. 2165–2174. Reprinted by permission of American Institute of Physics, Melville, NY.)

at the bottom when blind. The shape shown in Fig. 16.14a tends to occur at relatively high welding speeds and larger penetration depths (or beam powers).

Due to the motion of the heat source, the top surface of the molten metal develops a teardrop shape (Fig. 16.14b), with a typical length-to-width ratio of about ten, and the width tails off toward the back. At low welding speeds, the molten pool is almost circular in cross section, but the teardrop shape becomes more pronounced as the speed increases (see also Fig. 10.6). The width of the molten pool decreases at higher welding speeds, forcing the molten metal to move at a higher velocity through the narrow channel between the vapor and the solid boundary.

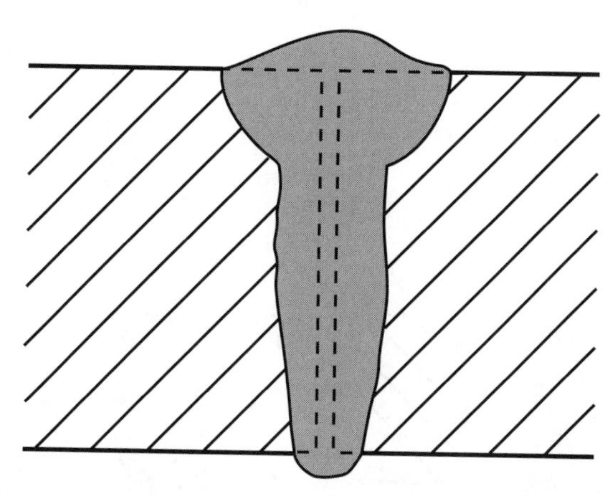

FIGURE 16.15 Schematic of the transverse cross-sectional shape of keyhole mode laser weld bead. (From Steen, W. M., Dowden, J., Davis, M., and Kapadia, P., 1988, *Journal of Physics D: Applied Physics*, Vol. 21, IOP Publishing Ltd.)

When the power level is low enough, the vapor pressure produced in the keyhole is much smaller than the fluid dynamic forces of the molten metal surrounding the keyhole. The keyhole thus collapses, and a sudden drop in penetration occurs, with the weld pool becoming roughly hemispherical in shape, resulting in conduction mode welding. Conduction mode welding is less efficient than keyhole welding, since the absence of a keyhole results in a much lower energy transfer efficiency of about 20%, as a significant portion of the incident beam is reflected away. Conduction mode welding typically results in a depth-to-width ratio of about 3:1 compared to about 10:1 or more for keyhole welding.

The overall transverse cross-sectional shape of the weld bead produced in keyhole welding often consists of two main sections. The lower part is nearly parallel sided, with a gradual narrowing from top to bottom. On top of that is a region of roughly semicircular shape, whose width is greater than that of the parallel-sided section (Fig. 16.15).

16.3.2.1 Power Absorption in the Keyhole Absorption of beam energy in the keyhole occurs by two principal mechanisms:

1. Inverse bremsstrahlung of the plasma.
2. Fresnel absorption (direct deposition) at the cavity walls.

Both forms of absorption normally occur simultaneously within the keyhole, and are illustrated in Fig. 16.16. Their relative magnitudes are still the subject of research. The absorbed beam becomes entrapped in the keyhole. This results in significantly higher absorption in deep penetration welding compared to conduction mode welding.

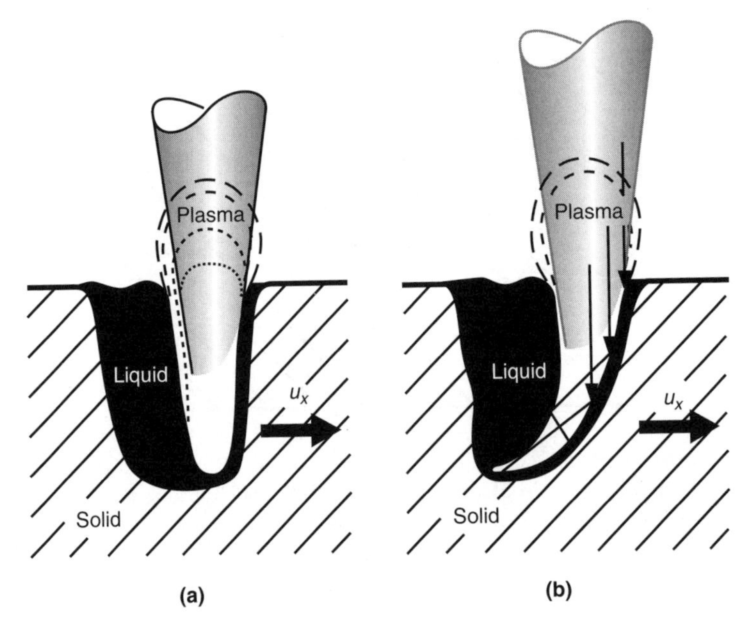

FIGURE 16.16 Absorption mechanisms inside the keyhole. (a) Inverse bresmsstrahlung absorption. (b) Fresnel absorption. (From Beyer, E., Behler, K., and Herziger, G., 1988, *SPIE, High Power Lasers and Laser Machining Technology*, Vol. 1020, pp. 84–95.)

Inverse Bremsstrahlung Absorption With inverse bremsstrahlung absorption, a portion of the vaporized metal or shielding gas is ionized by the high incident energy to form a plasma that may absorb part of the incident beam by the inverse bremsstrahlung effect inside the keyhole. Bremsstrahlung is the electromagnetic radiation produced by the sudden retardation of a charged particle in an intense electric field. Thus in simple terms, inverse bremsstrahlung is the absorption of electromagnetic radiation as a result of the sudden acceleration of a charged particle, or absorption of a photon by an electron in the field of a nucleus. The absorption thus depends on the presence of free electrons and occurs within the plasma, which consists of ionized particles (Fig. 16.16a). The absorption depends on temperature, partly because the electron density varies with temperature. As the vapor is heated in the presence of an intense beam, the absorption coefficient increases. Since the center of the keyhole is much hotter than the walls (which are at the vaporization temperature), most of the beam energy is absorbed within the cavity, and then coupled to the keyhole wall through conduction and radiation.

For inverse bremsstrahlung absorption, the corresponding absorption coefficient, α_{p}, is given by equation (14.13).

Fresnel Absorption The mechanism of Fresnel absorption results in the beam being spread along the entire keyhole length. It results from multiple reflections at the keyhole walls (Fig. 16.16b), where the beam energy that enters the vapor cavity is reflected repetitively at the walls of the keyhole until all the energy is eventually absorbed. For most applications, specular reflection (mirror-like) is assumed to be predominant, with diffuse reflection (similar to reflection from a rough surface) being neglected. Such multiple reflections, along with scattering processes, tend to randomize phase relations that may have existed in the original beam.

The absorption coefficient for Fresnel absorption can be estimated by first determining the coefficient for Fresnel reflection, R_F, which for a circularly polarized light, is given by

$$R_F(\theta) = \frac{R_p(\theta) + R_s(\theta)}{2} \tag{16.2}$$

$$R_p(\theta) = \frac{\epsilon^2 - 2\epsilon \cos\theta + 2\cos^2\theta}{\epsilon^2 + 2\epsilon \cos\theta + 2\cos^2\theta} \tag{16.3}$$

$$R_s(\theta) = \frac{1 + (1 - \epsilon \cos\theta)^2}{1 + (1 + \epsilon \cos\theta)^2} \tag{16.4}$$

$$\epsilon = \left[\frac{2\epsilon_2}{\epsilon_1 + \sqrt{\epsilon_1{}^2 + \Sigma^2}}\right]^{1/2} \tag{16.5}$$

$$\Sigma = \frac{\sigma_{st}}{\omega\epsilon_0} \tag{16.6}$$

where $R_p(\theta)$ is the reflection coefficient for p-polarized light (see sections 9.4 and 9.5) (−), $R_s(\theta)$ is the reflection coefficient for s-polarized light (−), ϵ_0 is the permittivity of free space (F/m), $\epsilon_{1,2}$ are the real parts of the dielectric constants of the material and plasma, respectively (−), ω is the frequency of the incident laser beam (rad/s), σ_{st} is the electrical conductance per unit depth (S/m), θ is the angle of incidence of the laser beam on the keyhole wall at any location, measured from the surface normal (°).

The absorption coefficient for Fresnel absorption is then given by

$$\alpha_F(\theta) = 1 - R_F(\theta) \tag{16.7}$$

16.3.2.2 Keyhole Characteristics Analysis of the keyhole is quite complex and is still the subject of intense research. The approach we shall adopt is to first present the governing equations that are needed for an in-depth analysis. We shall then follow it with a simplified analysis that will provide an approximate insight into the keyhole characteristics.

Governing Equations In this section, we shall outline the governing equations necessary for analyzing the keyhole welding process, focusing on flow in the molten material. The rarefied medium in the keyhole is considered in so far as it affects the boundary conditions.

The velocity and temperature fields in the molten material, assuming incompressible flow, are given by the following governing equations:

Mass Continuity

$$\nabla \cdot \mathbf{u_l} = 0 \tag{16.8}$$

Momentum Equation

$$\rho_l \left(\frac{\partial \mathbf{u_l}}{\partial t} + (\mathbf{u_l} \cdot \nabla)\mathbf{u_l} \right) = -\nabla P_l + \mu_l \nabla^2 \mathbf{u_l} + \mathbf{F_b} \tag{16.9}$$

Energy Equation

$$\rho_l c_{pl} \left(\frac{\partial T_l}{\partial t} + (\mathbf{u_l} \cdot \nabla) T_l \right) = k_l \nabla^2 T_l \tag{16.10}$$

where $\mathbf{F_b} = -\rho_l \mathbf{g} \beta_T (T - T_0)$ is the body force vector, incorporating the buoyancy forces, with components F_x, F_y, F_z. The negative sign results from the fact that the buoyancy force is opposite in direction to $\mathbf{g}$ (N/m^3), $\mathbf{g}$ is the gravitational acceleration vector (m/s^2), $\mathbf{u_l}$ is the velocity vector for the liquid phase (m/s), β_T is the volumetric thermal coefficient of expansion (/K), s, l, v are the subscripts for the solid, liquid, and vapor phases, respectively.

For the solid substrate, only the energy equation is required.

Boundary Conditions

At the solid–liquid interface, no-slip conditions are used, that is,

$$\mathbf{u} = 0$$

At the liquid–vapor interface, the boundary conditions for the momentum equation (16.9) are obtained by conserving momentum in the normal and tangential directions as follows:

Normal Component

$$\begin{aligned} m_l \mathbf{n} \cdot (\mathbf{u_l} - \mathbf{u_{iv}}) + m_v \mathbf{n} \cdot (\mathbf{u_v} - \mathbf{u_{iv}}) - \mathbf{n} \cdot [\mathbf{n} \cdot (\tau_l - \tau_v)] \\ + \mathbf{n} \cdot \nabla_{lv} \gamma_{lv} - (\nabla_{lv} \cdot \mathbf{n}) \gamma_{lv} \mathbf{n} \cdot \mathbf{n} = 0 \end{aligned} \tag{16.11}$$

Tangential Component

$$\mathbf{t} \cdot \nabla_{lv} \gamma_{lv} - \mathbf{t} \cdot [\mathbf{n} \cdot (\tau_l - \tau_v)] = 0 \tag{16.12}$$

Energy balance at the solid–liquid interface gives the following boundary condition for the energy equation (16.10):

$$k_s \mathbf{n} \cdot \nabla T_s - k_l \mathbf{n} \cdot \nabla T_l = \rho_s L_m \mathbf{n} \cdot \mathbf{u_{il}} \tag{16.13}$$

while energy balance at the liquid–vapor interface gives the following boundary condition:

$$k_l \mathbf{n} \cdot \nabla T_l - k_v \mathbf{n} \cdot \nabla T_v + (1 - R)\mathbf{I_0} \cdot \mathbf{n} = \rho_l L_v \mathbf{n} \cdot (\mathbf{u_l} - \mathbf{u_{iv}}) \tag{16.14}$$

where $\mathbf{I_0}$ is the intensity of the incident laser beam (W/cm^2), m_l and m_v are the mass flux of liquid and vapor, respectively (kg), $\mathbf{n}$ and $\mathbf{t}$ are unit normal and tangential vectors, respectively, R is the reflectivity (−), $\mathbf{u_v}$ and $\mathbf{u_{iv}}$ are the velocities of the vapor at the liquid–vapor interface and the free surface (vaporization interface), respectively (m/s), $\mathbf{u_{il}}$ is the velocity of the solid–liquid interface (melting interface) (m/s), γ_{lv} is the surface tension at the liquid–vapor interface (J/m^2 or N/m), τ is the stress tensor at the liquid–vapor interface (P$_a$), and ∇_{lv} is the surface gradient at the liquid–vapor interface.

The resulting equations (16.8)–(16.14) are highly nonlinear, time dependent and coupled, and can only be solved numerically. Further simplifications may still be necessary. Thus, in the following section, we present an approximate analysis that provides a simplified insight into the keyhole characteristics.

Approximate Analysis An estimate of heat flow in keyhole welding can be done using the line heat source analysis discussed in Section 10.2 to obtain the temperature distribution, cooling rates, and peak temperatures, without taking into account the fluid flow (liquid, vapor, and plasma). The discussions in Section 10.3 and the preceding section provide a basis for analyzing the molten pool that surrounds the keyhole, but do not address the gaseous plume (vapor/plasma) within the keyhole.

We add to these analyses by first looking at flow conditions in a plane parallel to the workpiece surface (Fig. 16.14b) and make the following assumptions:

1. All variations in the z-direction (normal to the workpiece surface) over a range comparable to the keyhole radius, r_k, are small. This assumption may not be particularly valid close to the surface.
2. Welding speeds are high enough that adiabatic heating conditions exist ahead of the weld pool. In other words, heat conducted outward in the forward direction is not lost, but available to subsequently melt the material, unless it is conducted sideways, out of the path of the molten metal, faster than the advance speed. Using the nomenclature in Fig. 16.14b, the average speed at which heat is conducted sideways, u_{cs}, may be approximated as

$$u_{cs} = \frac{\kappa_s}{r_k + w_l} \tag{16.15}$$

where r_k = keyhole radius (mm), w_l = average sideways width of the molten pool (mm), and κ_s = thermal diffusivity of the solid (mm^2/s).

Now if the welding speed, u_x, is greater than $\frac{\kappa_s}{r_k}$, only a small fraction of the heat conducted out ahead of the molten pool will be lost. Thus, $u_x > \frac{\kappa_s}{r_k}$ defines the "adiabatic" heating condition in laser welding.

Weld Pool Length Ahead of the Keyhole, l_f

The length of the molten material between the keyhole and unmelted solid in front of it can be obtained by a simple heat balance that gives

$$(c_{vol}T_m + L_{mv})u_x = k_l \frac{T_v - T_m}{l_f}\left(1 - \frac{u_x l_f}{\kappa_l}\right) \tag{16.16}$$

where $c_{vol} = \rho c_p$ is the specific heat per unit volume, assumed equal for both the solid and liquid phases (J/kg K m^3), l_f is the weld pool length ahead of the keyhole (m), L_{mv} is the latent heat of melting per unit volume (J/m^3), T_m and T_v are the melting and vaporization temperatures, respectively (K), and κ_l is the thermal diffusivity of the liquid (molten pool) (m^2/s).

The factor $\left(1 - \frac{u_x l_f}{\kappa_l}\right)$ corrects for the motion of the boundaries relative to the medium. Equation (16.16) is based on the concept that the heat conducted through the molten metal provides enough thermal energy to advance the melt front at a speed u_x. Since $k_l = c_{vol}\kappa_l$, equation (16.16) reduces to the form

$$l_f = \frac{\kappa_l}{u_x}\left(\frac{T_v - T_m}{T_v + L_{mv}/c_{vol}}\right) \tag{16.17}$$

from which we find that the length of the molten material between the keyhole and unmelted solid is inversely proportional to the welding speed.

Average Side Width of the Molten Pool, w_l

If we divide the material around the cavity into four quadrants (Fig. 16.14b) and consider the first two quadrants, then in each of those two quadrants, sufficient heat must be conducted into the molten region to melt cold material at such a rate that the melt boundary in the quadrant moves with a speed u_x. The average width of this melt is w_l, and the average temperature, $1/2(T_m + T_v)$. The heat conducted across the liquid–vapor interface, of projected area r_k (assuming unit depth), under the influence of a temperature gradient $T_v - T_m/w_l$, provides the energy required to heat and melt the new material. Thus a heat balance gives

$$k_l r_k\left(\frac{T_v - T_m}{w_l}\right) = u_x w_l[L_{mv} + \frac{1}{2}c_{vol}(T_m + T_v)] \tag{16.18}$$

$$\Rightarrow w_l^2 = \frac{\kappa_l}{u_x} r_k\left(\frac{T_v - T_m}{(T_v + T_m)/2 + L_{mv}/c_{vol}}\right)$$

$$\Rightarrow w_l \approx (l_f r_k)^{1/2} \tag{16.19}$$

Flow of Material Around the Cavity

As indicated earlier in Section 16.3.2, material is transported from the front to the back side of the cavity by flow of liquid around the cavity and/or flow of vapor across the cavity. Assuming that flow occurs in the horizontal plane, then when we consider continuity of mass, we get

$$\rho_v u_{vt} = \beta_v \rho_l u_x \tag{16.20}$$

where u_{vt} is the average velocity of vapor flow transversely across the cavity (m/s), β_v is the fraction of material transported across the cavity as vapor (−), ρ_l is the density of both liquid and solid, assumed equal (kg/m^3), and ρ_v is the vapor density at temperature T_v (kg/m^3).

The ratio of the solid or liquid density to the vapor density is typically of the order of 10^4. The excess pressure that results from flow of vaporized material from the front of the cavity is due to momentum transport per unit area and unit time, which can be expressed as

$$P = (\beta_v \rho_l u_x) u_{vt} \tag{16.21}$$

where the expression in brackets is the mass of the material removed per unit area per unit time. It is this pressure that pushes the molten metal around the keyhole cavity. The resulting velocity u_l of the molten metal can be obtained using Bernoulli's equation (15.1a), and is given by

$$P = \frac{1}{2}\rho_l u_l^2 \tag{16.22}$$

From equations (16.20)–(16.22), we get

$$u_l = \beta_v u_x \left(\frac{2\rho_l}{\rho_v}\right)^{1/2} \tag{16.23}$$

Now the volume of liquid (per unit time and unit cavity depth) that flows around the cavity is 2 $u_l l_f$, while the volume of material swept out by the cavity is $2r_k u_x$. Thus,

$$2u_l l_f = (1 - \beta_v) \cdot 2r_k u_x \tag{16.24}$$

which gives

$$\beta_v = \frac{1}{\left[1 + \frac{l_f}{r_k}\left(\frac{2\rho_l}{\rho_v}\right)^{1/2}\right]} \tag{16.25}$$

which is the fraction of material transported across the cavity as vapor. β_v typically ranges in value between 0.01 and 0.1. From the preceding analysis, it can be shown that $\beta_v \propto u_x$, $u_{vt} \propto u_x^2$, and $P \propto u_x^4$. The analysis breaks down at high enough welding speeds u_x, as u_{vt} then approaches the speed of sound. At that point, the pressure P becomes comparable to the ambient pressure, and β_v approaches 0.5. The assumption of a pressure that is uniform over the horizontal plane is then no longer valid. Also, horizontal and vertical flow can then no longer be treated as independent.

Molten Material Behind the Cavity

In estimating the dimensions of the molten pool behind the cavity, we consider the balance of energy in this region, which consists of the following components:

1. Latent heat of melting required to expand the molten material from a lateral distance of $r_k + w_l$ to D, the maximum width of the molten pool.
2. Heat liberated by condensation of the vapor transported across the cavity.
3. Heat released as the molten metal cools from a temperature of T_v behind the cavity to T_m at the edge of the molten pool.

The resulting heat balance, neglecting heat conduction from the cavity into the molten region, as well as that conducted from the liquid into the solid, which are assumed to approximately cancel out, gives

$$r_k[\beta_v L_{vol} + c_{vol}(T_v - T_m)] = L_{mv}(D - r_k - w_l) \tag{16.26}$$

where D is the maximum width of the molten zone (mm), L_{vol} is the latent heat of vaporization (or condensation), per unit volume of liquid (J/m^3).

The length of the molten pool can be approximated by considering the distance l that the beam advances during the time t_D that it takes for the melt to attain the width D, that is, the time it takes for heat to be conducted from the rear of the cavity over a distance D. l (see Fig. 16.14b) is then given by

$$l = u_x t_D = u_x \frac{D^2}{\kappa_l} \tag{16.27}$$

Thus, from equations (16.26) and (16.27), we have

$$l = u_x \frac{r_k{}^2}{\kappa_l}\left(1 + \frac{\beta_v L_{vol} + c_{vol}(T_v - T_m)}{L_{mv}} + \frac{w_l}{r_k}\right)^2 \tag{16.28}$$

Pressure Variation in the Keyhole

For the purposes of this discussion, pressure within the keyhole is considered to vary only along the length of the keyhole, that is, in the vertical direction and is uniform in a horizontal section. As indicated earlier, the cavity is kept open by an excess pressure of vapor in the keyhole, P_v, (the plasma plume) and ablation pressure, P_{ab}. The forces that tend to close it are due to surface tension, P_γ, hydrostatic pressure, P_{hs}, of the surrounding molten metal, and hydrodynamic pressure, P_{hd}. Force balance thus gives

$$P_v + P_{ab} = P_\gamma + P_{hs} + P_{hd} \tag{16.29a}$$

Assuming that the principal curvature is in the horizontal plane, that is, $\frac{d^2 r_k}{dz^2}$ is negligible compared to $1/r_k$, and neglecting hydrodynamic pressure, the gage pressure (right-hand side of equation (16.29a)), $P(z)$, at depth z below the liquid surface is given by

$$P(z) = \frac{\gamma_{lv}}{r_k(z)} + \rho_l g z \tag{16.29b}$$

where $r_k(z)$ is the keyhole radius at level z (m), γ_{lv} is the surface tension at the liquid–vapor interface (N/m), and ρ_l is the density of the molten metal (kg/m^3). This pressure in the liquid is balanced by the plume and ablation pressures.

Example 16.1 A 4 kW Nd:YAG laser is used to weld an aluminum plate at a welding speed of 6 m/min. Assume that the density of molten aluminum is $\rho_l = 2400\,\text{kg/m}^3$, and that adiabatic heating conditions exist ahead of the weld pool. Further assume keyhole mode welding with the keyhole radius, r_k, being equal to twice the focused beam radius, $w_f = 125\,\mu\text{m}$.

Determine:

(1) the weld pool length ahead of the keyhole, l_f,
(2) the average side width of the molten pool, w_l, and
(3) the length of the molten pool behind the cavity, l.

Solution:

(1) The weld pool length ahead of the keyhole, l_f, is given by equation (16.17):

$$l_f = \frac{\kappa_l}{u_x}\left(\frac{T_v - T_m}{T_v + L_{mv}/c_{vol}}\right)$$

Then from Appendixes 10D and 10E, we have

$$\kappa_l = \frac{k_l}{\rho_l c_p} = \frac{247}{2400 \times 900} = 114 \times 10^{-6}\,\text{m}^2/\text{s} = 114\,\text{mm}^2/\text{s}$$

$$L_{mv} = 397 \times 2700 = 1.072 \times 10^6\,\text{kJ/m}^3$$

$$c_{vol} = \rho c_p = 2700 \times 900 = 2.43 \times 10^6\,\text{J/m}^3 K$$

$$L_{vol} = 10.78 \times 10^3 \times 2400 = 25.9 \times 10^6\,\text{kJ/m}^3$$

therefore,

$$l_f = \frac{114 \times 10^{-6}}{6/60} \left(\frac{2494 - 660.4}{(2494 + 273) + \frac{1.072 \times 10^9}{2.43 \times 10^6}} \right)$$

$$\Rightarrow l_f = 0.65 \text{ mm}$$

(2) The average side width of the molten pool, w_l, is given by equation (16.19):

$$w_l \approx (l_f r_k)^{1/2} = (0.65 \times 0.250)^{1/2} = 0.403 \text{ mm} = 403 \text{ µm}$$

(3) The length of the molten pool behind the cavity, l, is obtained from equation (16.28):

$$l = u_x \frac{r_k{}^2}{\kappa_l} \left(1 + \frac{\beta_v L_{vol} + c_{vol}(T_v - T_m)}{L_{mv}} + \frac{w_l}{r_k} \right)^2$$

but

$$\beta_v = \frac{1}{\left[1 + \frac{l_f}{r_k} \left(\frac{2\rho_l}{\rho_v} \right)^{1/2} \right]} = \frac{1}{\left[1 + \frac{0.65}{0.250} \left(2 \times 10^4 \right)^{1/2} \right]} = 0.0027$$

therefore,

$$l = \frac{6000}{60} \times \frac{0.250^2}{114}$$

$$\times \left(1 + \frac{0.0027 \times 25.9 \times 10^9 + 2.43 \times 10^6 \times (2494 - 660.4)}{1.072 \times 10^9} + \frac{0.403}{0.250} \right)^2$$

$$= 2.6 \text{ mm}$$

16.4 MATERIAL CONSIDERATIONS

As indicated in Chapter 15, any form of laser processing is influenced by the properties of the material involved. The suitability of CO_2 laser welding to various materials is summarized in Table 16.4.

In the following sections, we shall consider the behavior of three groups of materials when laser welded:

(1) Steels.
(2) Nonferrous metals.
(3) Ceramic materials.

In addition, we shall also discuss welding of dissimilar materials.

TABLE 16.4 Suitability of Materials for CO_2 Laser Welding

Good	Fair	Poor
Low carbon steel	Aluminum	Galvanized steel
Stainless steel	Copper	Rimmed steel
Inconel 625	Kovar	Brass
Silicon bronze	Tool steel	Zinc
Titanium	Inconel 718	Silver
Tantalum	Medium-to-high	Gold
Zirconium	Carbon steel	

16.4.1 Steels

Plain carbon steels with low carbon content are generally easily welded using a laser beam. However, steels with carbon content greater than 0.25% may require preheating to prevent the formation of brittle microstructures due to the high cooling rates that normally occur during laser welding (see Section 11.1.3.3). On the contrary, the high cooling rates can be beneficial in steels that depend on the addition of alloying elements to produce fine-particle strengthening, such as the high-strength low alloy steels (HSLA). This is in contrast to conventional arc welding of HSLA steels that generally result in strength reduction due to the greater total heat input and consequent slower cooling rates. The presence of sulfur and phosphorus as alloying elements tend to promote hot cracking as usual (see Section 11.2.2.1). Some stainless steels (Section "Stainless Steels") normally result in good welds, especially most austenitic (300 series) types, except for those containing sulfur. Because of their relatively low thermal conductivities, higher depth-to-width ratios are obtained in stainless steel welds compared to carbon steels.

16.4.2 Nonferrous Alloys

Aluminum and copper require relatively high-power densities to sustain a molten puddle due to their high thermal conductivity and reflectivity. In the case of aluminum, these properties, coupled with the low melting temperature and low ionization energy, narrow the range of process parameters that can be used for welding, compared to steel. Filler material is often necessary for welding aluminum alloys. Series 1000 and 3000 are among the easiest to weld.

Of the 2000 series, 2219 exhibits the greatest resistance to cracking because of high Cu content (6.3%). 2014 and 2024 have lower Cu (4.5%) and contain a small amount of magnesium, which increases crack sensitivity by extending the solidification range.

The 5000 series aluminum alloys with magnesium as the principal alloying element (for strengthening) tend to crack in the weld bead when welded. This is primarily due to the long solidification range that induces hot cracking (see Section 11.2.2.1). The cracking problem is especially serious when the magnesium content in the weld metal is reduced to less than 3%. Silicon reduces the solidification range, thereby reducing hot-cracking. Silicon is more effective when its concentration is greater than 4%. In welding the 5000 series alloys, silicon can be introduced by using an Al–Si filler material. There is also a significant loss of magnesium from the weld bead during welding of the 5000 series. This tends to reduce the mechanical properties and occurs by vaporization, since the boiling temperature of magnesium is less than half that of aluminum. Another potential problem with the 5000 series is the formation of porosity. The pores are generally spherical in nature, indicating that they are formed by gas entrapment rather than shrinkage. Since hydrogen is the primary source of such porosity, ensuring that the shielding gas is of high purity, and further ensuring proper cleaning of the base metal to reduce surface contaminants and the oxide layer, will tend to minimize porosity formation. The 5083 and 5456 alloys are the strongest and easiest to weld in this group.

The 4000 series with silicon as the principal alloying element is less sensitive to cracks than the 5000 series. This is mainly because the range of composition over which the alloy is crack sensitive is narrower for the Al–Si alloys than for the Al–Mg alloys (see Fig. 16.17). Al–Mg–Si alloys that contain about 1% Mg_2Si are highly crack sensitive, since the Mg_2Si content is close to the peak of the hot cracking curve (Fig. 16.17). Examples are 6061 and 6063. The crack sensitivity in these alloys can be reduced by the addition of excess magnesium or silicon using Al–Mg or Al–Si filler materials, respectively. Of the 7000 series aluminum alloys, there are two main groups: (a) the Al–Zn–Mg types (e.g., X7005, X7106, and 7039), which are more resistant to

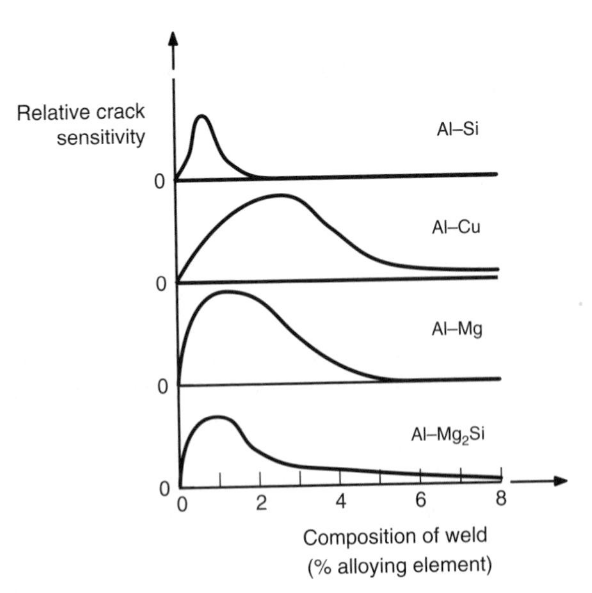

FIGURE 16.17 Variation of crack sensitivity with composition for various aluminum alloys. (From Dudas, J. H., and Collins, F. R., 1966, *Welding Journal*, Vol. 45, pp. 241s–249s. By permission of American Welding Society.)

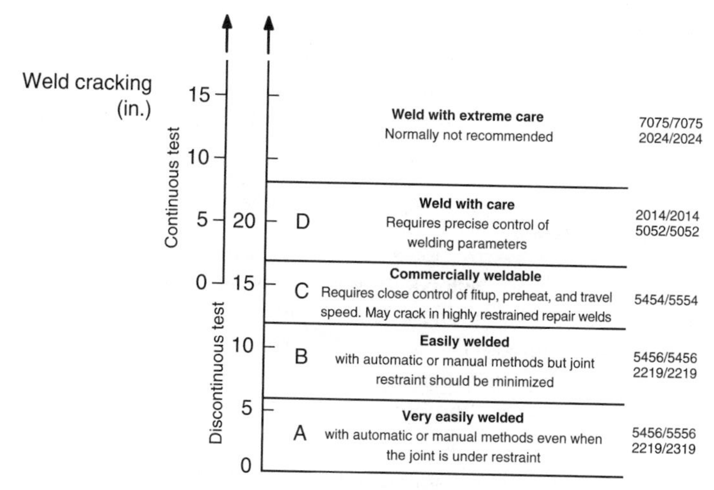

FIGURE 16.18 Weldability of different base metal/filler metal combinations of aluminum alloys. (From Dudas, J. H., and Collins, F. R., 1966, *Welding Journal*, Vol. 45, pp. 241s–249s. By permission of American Welding Society.)

weld cracking and (b) the Al–Zn–Mg–Cu types (e.g., 7075 and 7178), which are of higher strength, but less suitable for welding. This is because the addition of copper extends the coherence range. Crack sensitivity increases with copper content up to about 0.75% Cu, and then decreases. With the Al–Zn–Mg types, the crack sensitivity increases with magnesium content up to about 1% Mg, and then decreases with further increases in magnesium content.

The sensitivity of different aluminum alloys to cracking when welded, for a given base material and filler metal combination, is summarized in Fig. 16.18.

For copper, it is often necessary to coat the surface to enhance absorption (see Section 17.1.2.3). The volatile nature of the zinc in brasses makes them difficult to laser weld. Silver and gold are even more difficult to laser weld without prior treatment, because of their higher reflectivity to the laser beam, especially the CO_2 laser. Alloys of nickel generally tend to absorb the CO_2 laser beam effectively, resulting in good welds, even though alloys such as Hastelloy X and Kovar are subject to hot cracking. The reactive nature of tantalum, titanium, and zirconium alloys requires that they be welded in an inert atmosphere, since they are highly sensitive to oxidation.

16.4.3 Ceramic Materials

Laser welding of ceramic materials often presents problems because of the low thermal conductivity of ceramic materials, and also their inherent brittleness. The low thermal conductivity, coupled with the high intensity of the laser beam, results in high temperature gradients and thus high thermal stresses that cause the ceramic materials to crack. Welding of mullite ($3Al_2O_3.2SiO_2$) appears to work well, provided it is preheated to a temperature of 700°C to prevent cracking. Welding of alumina (Al_2O_3) does not present problems. On the contrary, problems may arise when silicon nitride (Si_3N_4) is welded. Other potential problems associated with welding of ceramic materials include porosity that results from vaporization associated with heating.

16.4.4 Dissimilar Metals

When different materials are welded, differences in composition between them can result in the formation of undesirable secondary phases if the solubility of the elements of the two materials is limited. This can produce an embrittled weld joint. Other phases such as eutectics and compounds that have low melting temperatures and/or are brittle can also form. Other sources of problems include the following:

(1) Differential thermal expansion resulting from different coefficients of expansion of the materials. This can create severe stresses during welding, post weld heat treatment, or in service.
(2) Differences in thermal conductivity. This can result in improper amounts of fused metal on the two pieces. The material with the higher conductivity will have a smaller nugget size on its side of the fused metal. This can be prevented by providing more heat to the higher thermal conductivity material or preheating it.
(3) Different melting points. This may also result in the weld bead solidifying while the lower melting base material is hot and not strong enough. This can produce hot tensile failure if the weldment is highly restrained.
(4) The formation of an electrolytic cell from the different materials. This can accelerate corrosion in service.

16.5 WELDMENT DISCONTINUITIES

A more detailed discussion on discontinuities is presented in Section 11.2. In this section, we discuss discontinuities that are commonly associated with laser welding, and these include porosity, cracking, humping, and spiking.

16.5.1 Porosity

Porosity formation in high-power CW laser welding of steels depends on the welding speed (Fig. 16.19). An increase in porosity with increasing speed is observed in deep penetration welding, up to a peak, and then begins to decrease at higher welding speeds. This is because in the initial stages, the increasing welding speed increases the depth-to-width ratio, which hinders the ability of the evolved gas to escape, thus increasing porosity. At much higher speeds, however, the reduced penetration enables the gas generated to escape more easily.

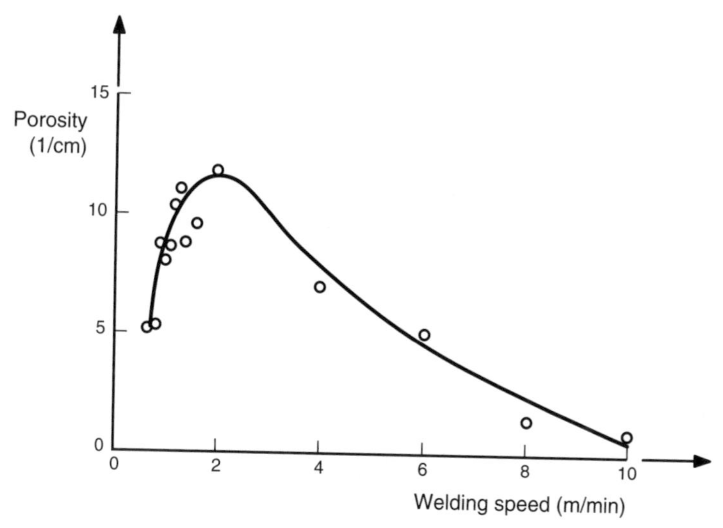

FIGURE 16.19 Variation of porosity formation with welding speed for a 20 mm thick low carbon steel welded with a 20 kW laser power. (From Funk, M., Kohler, U., Behler, K., and Beyer, E., 1989, *SPIE, High Power Lasers and Laser Machining Technology*, Vol. 1132, pp. 174–180.)

16.5.2 Humping

At very high welding speeds, say about 15 m/min depending on other process parameters, weld pool instability often develops, resulting in humping or slubbing (Fig. 16.23) and undercutting (see Section "Humping and Undercutting"), causing irregular weld bead formation. Even though this phenomenon is not fully understood, one approach to mitigating it in electron beam welding has involved the use of a second beam to control flow within the weld pool. Although there has not been an analytical basis to support this methodology, extensive experimental data indicate that it is an effective means of preventing humping. The use of a second beam in laser welding has also been shown to prevent humping or increase the speed at which humping occurs. Another approach involves reducing the power input. However, the latter approach does not appear to have significant enough supporting data to warrant its widespread use. Moreover, a reduction in power input, and thus welding speed, implies a reduction in productivity.

16.5.3 Spiking

There are other defects that are peculiar to high energy density electron or laser beam welding processes. One of these is the phenomenon of spiking, where the depth of penetration may be uniform for part of the weld length, but increases substantially at regular intervals (Fig. 16.20). Spiking is most likely due to instabilities in the process of keyhole formation. It results in variations in the weld bead cross-sectional area,

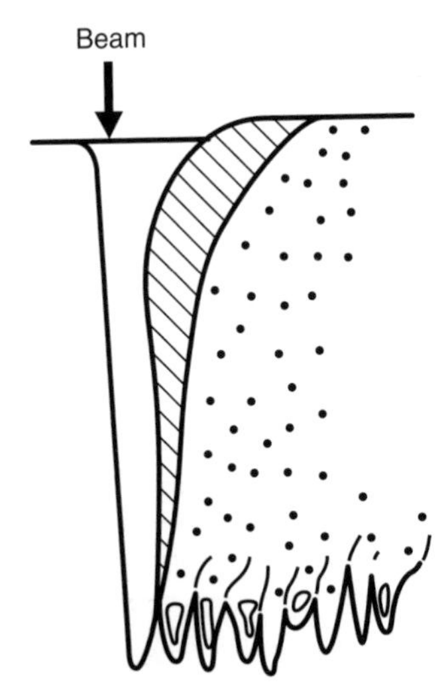

FIGURE 16.20 Schematic of the spiking phenomenon. (From Arata, Y., Nabegata, E., and Iwamoto, N., 1978, *Transactions of the Japan Welding Research Institute*, Vol. 7, No. 2, pp. 85–95.)

and thus the load carrying capacity of the weld. It may also be prevented by the use of a second beam to control flow within the weld pool (see Section "Porosity, Spiking, and Incomplete Fusion").

16.6 ADVANTAGES AND DISADVANTAGES OF LASER WELDING

16.6.1 Advantages

The primary advantages of laser welding relative to conventional processes such as arc welding include the following:

1. High depth-to-width ratio, ranging from 3:1 to 10:1.
2. Low and highly localized total heat input that results in minimal distortion and small heat-affected zone size.
3. Noncontact processing.
4. The beam is neither affected by a magnetic field nor by passage through air, and thus does not require a vacuum for welding as required for electron beam welding. It also does not generate radiation.
5. Easier access to the weld region since the beam can be easily directed to ordinarily inaccessible regions. In this regard, Nd:YAG lasers are more flexible than CO_2 lasers since Nd:YAG lasers are more easily delivered to the workpiece using fiber optic cables.

6. Ability to weld dissimilar materials since the heat input is limited to a very small region.
7. Ability to control the power density by focusing.
8. Easily automated process.
9. Relatively faster than most conventional welding processes.
10. Relatively high weld strength compared to most conventional welding processes.

16.6.2 Disadvantages

1. High cooling rates. The low energy input results in high cooling rates that may produce hardening in the weld bead and also in portions of the heat-affected zone. This is especially true of high-hardenability materials such as high carbon and some alloy steels. The formation of such undesirable microstructures is normally mitigated by either preheating before welding or postweld heat treatment.
2. High transverse temperature gradients that may result in microcracking.
3. Need for precise fitting of joints. The small size of the heat source and resulting molten pool require closely and accurately fitted joints, especially in the case of butt welding since any significant opening will result in a void, undercut, or power loss through the opening.
4. High initial capital investment cost of the laser.

16.7 SPECIAL TECHNIQUES

Some of the special techniques that have been developed to enhance the quality of laser welds include multiple-beam welding and arc-augmented welding.

16.7.1 Multiple-Beam Welding

Multiple-beam welding can be used either to control the microstructure in the solid material (by preheating and/or postweld heat treating) or to control flow in the weld pool.

16.7.1.1 Multiple-Beam Preheating and Postweld Heat Treatment As indicated in Section 16.6.2, one major disadvantage with laser welding of certain materials (e.g., high carbon and some alloy steels) is that the highly intense beam results in a relatively small total heat input. As a result, very high cooling rates are achieved that can form brittle phases in such materials, and consequently, cracking. To reduce the high cooling rates, either preheating or postweld heat treatment is used. However, conventional preheating and postweld heat treatment have several drawbacks. In the first place, they often involve separate heating of the workpiece, usually in a furnace.

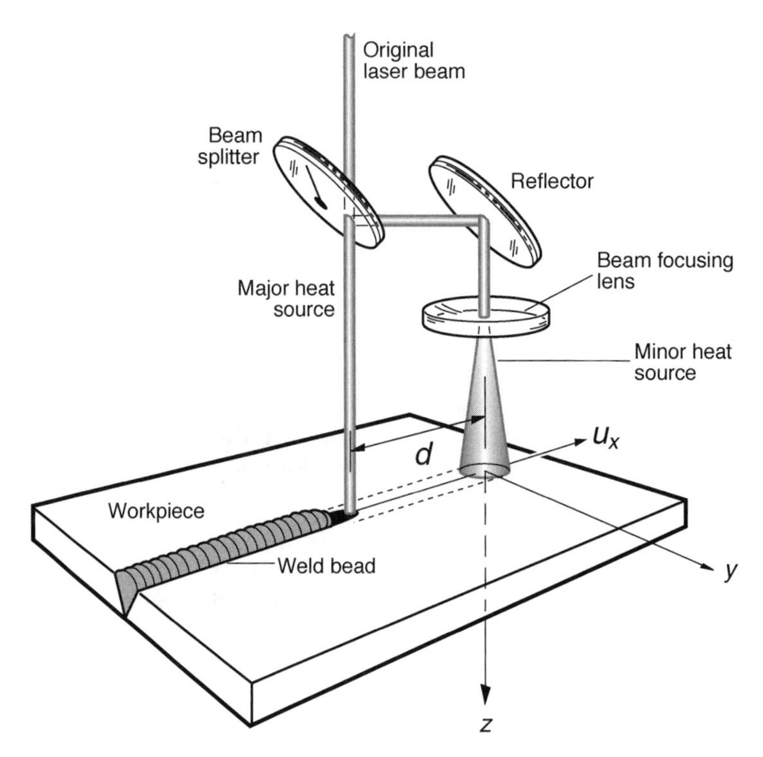

FIGURE 16.21 Schematic of the multiple-beam laser welding concept. (From Kannatey-Asibu Jr., E., 1991, *ASME Journal of Engineering Materials and Technology*, Vol. 113, pp. 215–221.)

This slows down the production rate, since it introduces an additional step into the process. Second, since the entire workpiece is often heated, energy is wasted, and also the heat applied may affect other portions of the workpiece. Finally, conventional preheating and postweld heat treatment are often either too expensive or impractical to use since the part may be complex in shape or bulky.

The drawbacks of traditional preheating and postweld heat treatment can be eliminated by using the multiple-beam configuration, which, in its simplest form, is illustrated in Fig. 16.21. In this configuration, there are two beams. One of the beams (the minor beam) is defocused, leads, and preheats the joint, while the major beam follows and welds the material. In another configuration, the defocused beam would follow the major beam, resulting in postweld heat treatment. Either arrangement reduces the cooling rates, thereby reducing or preventing the formation of hard microstructures. This is especially useful in welding high hardenability materials such as high carbon and/or alloy steels and some titanium alloys.

The analysis for the dual heat source system parallels that for the single source system discussed in Section 10.2. Since equation (10.3) is linear, we can use superposition for the two heat sources with the requirement that a common coordinate system be used. The solution for the single point source is given by equation (10.17) as

$$T_1 - T_{10} = \frac{q_1}{2\pi k r} e^{\left[-\frac{u_x(r+\xi)}{2\kappa}\right]} \tag{16.30}$$

This solution was obtained with the heat input at the origin of the moving coordinate system. For the minor heat source, which is displaced from the origin, we obtain the corresponding solution by performing a linear transformation along the x-axis, giving

$$T_2 - T_{20} = \frac{q_2}{2\pi k\sqrt{(\xi - d)^2 + y^2 + z^2}} e^{\left[-\frac{u_x[(\xi-d)+\sqrt{(\xi-d)^2+y^2+z^2}]}{2\kappa}\right]} \tag{16.31}$$

or

$$T_2 - T_{20} = \frac{q_2}{2\pi k r'} e^{\left[-\frac{u_x(\xi'+r')}{2\kappa}\right]} \tag{16.32}$$

where q_2 is the power input for the minor heat source, $\xi' = \xi - d$, and $r'^2 = (\xi - d)^2 + y^2 + z^2 = \xi'^2 + y^2 + z^2$.

Superposing the solutions for the two heat sources from equations (16.31) and (16.32), we get

$$\Theta = \frac{1}{2\pi k}\left[\frac{q_1}{r} e^{-\frac{u_x(\xi+r)}{2\kappa}} + \frac{q_2}{r'} e^{-\frac{u_x(\xi'+r')}{2\kappa}}\right] \tag{16.33}$$

where $\Theta = T_1 + T_2 - T_{10} - T_{20}$

In analyzing the cooling rate, we again consider only the worst case scenario, which occurs along the bead centerline, that is, $y = z = 0$. Thus from equation (16.33), we have

$$\Theta = \frac{1}{2\pi k}\left[\frac{q_1}{|\xi|} e^{-\frac{u_x(\xi+|\xi|)}{2\kappa}} + \frac{q_2}{|\xi'|} e^{-\frac{u_x(\xi'+|\xi'|)}{2\kappa}}\right] \tag{16.34}$$

With the moving heat source, cooling does not start until after the source passes the point of interest. In this dual heat source case, one would expect the cooling to start after the leading source passes the point. However, since the lead or minor source is only being used for preheating, and the major source will reheat points between the two sources, most likely to a temperature higher than that experienced as a result of the minor source, we are only interested in cooling rates along points behind the major source, that is, for $\xi < 0$. This reduces equation (16.34) to

$$\Theta = \frac{1}{2\pi k}\left(\frac{q_1}{\xi} + \frac{q_2}{\xi'}\right) \tag{16.35}$$

Differentiating with respect to time, we have

$$\frac{d\Theta}{dt} = \frac{1}{2\pi k}\left[q_1 \frac{d}{d\xi}\left(\frac{1}{\xi}\right)\frac{d\xi}{dt} + q_2 \frac{d}{d\xi'}\left(\frac{1}{\xi'}\right)\frac{d\xi'}{dt}\right] \tag{16.36}$$

but $d\xi/dt = d\xi\, dx/dx dt = u_x$. Likewise, $d\xi'/dt = u_x$. Thus the cooling rate becomes

$$\frac{d\Theta}{dt} = -\frac{u_x}{2\pi k}\left[\frac{q_1}{\xi^2} + \frac{q_2}{\xi'^2}\right] \tag{16.37}$$

Without loss of generality in the basic deductions to be made, we further consider the situation where the two heat sources are of the same intensity, which reduces equation (16.37) to the form

$$\frac{d\Theta}{dt} = -\frac{q_1 u_x}{2\pi k}\left[\frac{1}{\xi^2} + \frac{1}{\xi'^2}\right] \tag{16.38}$$

But $(2\pi k)^2\Theta^2 = {q_1}^2\left(\frac{1}{\xi^2} + \frac{1}{\xi'^2} + \frac{2}{\xi\xi'}\right)$, from equation (16.35).

Therefore,

$$\frac{d\Theta}{dt} = -\frac{q_1 u_x}{2\pi k}\left[\frac{(2\pi k)^2\Theta^2}{{q_1}^2} - \frac{2}{\xi\xi'}\right] = -\left(\frac{2\pi k u_x \Theta^2}{q_1} - \frac{q_1 u_x}{\pi k \xi\xi'}\right) \tag{16.39}$$

The first term on the right-hand side of equation (16.39) is equivalent to the cooling rate at a temperature Θ obtained for a single heat source. As is evident from the equation, introduction of the second heat source results in a reduction term in the cooling rate equation. In other words, at a given temperature, the dual heat source system reduces the cooling rate by an amount equal to the second term on the right-hand side. Thus, at points in a dual heat source system and a single source system that have the same temperature (not necessarily the same location), the cooling rate will be lower for the dual source system, even if the heat input for the single source system is the same as each of the dual system inputs.

The defocused preheating beam is useful when welding thin sheets. For relatively thick materials, it might be necessary to have both beams focused and being used for penetration welding. The leading beam then welds and at the same time preheats the material for the follow-up beam.

The use of the multiple-beam system also results in a reduction in the steep transverse temperature gradients ($\frac{dT}{dy}$) that arise during the welding process. This is especially useful in welding or cutting inherently brittle materials such as ceramics, where the steep temperature gradients can induce high thermal stresses and thus cause cracking of the workpiece.

The effect of the multiple-beam system on the weldment depends on the power ratio of the two beams, the interbeam spacing, and the distribution of the two beams.

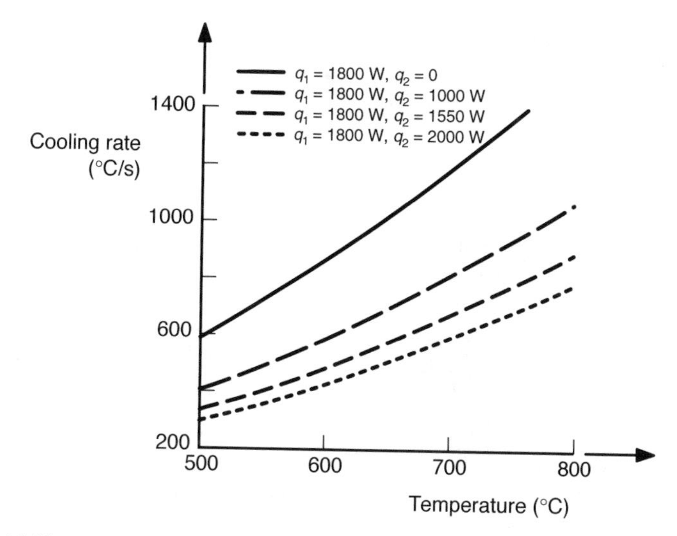

FIGURE 16.22 Effect of preheating with a dual beam on cooling rates. (From Liu, Y.-N., and Kannatey-Asibu Jr., E., 1993, *ASME Journal of Heat Transfer*, Vol. 115, pp. 34–41.

As Fig. 16.22 illustrates, for a given power of the major beam, increasing the power of the preheating beam reduces the cooling rate for any temperature. The transverse temperature gradient is also reduced. It can be shown that the lowest cooling rate is obtained when the preheating beam is behind the major beam, that is, the interbeam spacing is negative. That is essentially equivalent to postweld heat treatment, and implies that in using the multiple-beam system, postweld heat treatment would result in a lower cooling rate than can be obtained from preheating. On the contrary, the highest transverse temperature gradients are obtained when the interbeam spacing is negative. In other words, postweld heat treatment is not as effective for reducing the transverse temperature gradients as preheating.

Thus the configuration that is selected for a given application should be determined by the material to be welded. For example, for alloy steels where the martensitic transformation can be a problem, postweld heat treatment that effectively reduces cooling rates should be used, whereas for ceramics where the thermal stresses caused by transverse temperature gradients can present problems, preheating would be more appropriate.

The distribution of the leading beam intensity affects neither the cooling rate nor transverse temperature gradient as significantly as does the power ratio. However, the beam distribution tends to affect the transverse temperature gradient more than the cooling rate.

Example 16.2 A CO_2 laser output is split into two 3 kW beams and used for butt welding of two semiinfinite plates of AISI 1080 steel, each 6 mm thick. A dual-beam system is used to prevent the formation of a brittle microstructure in the weld bead. Assume that 20% of each beam is lost by reflection.

If the critical cooling rate to avoid the formation of martensite in AISI 1080 is about 150°C/s at a temperature of 550°C, and the minor heat source has been set at a distance $d = 4$ mm from the major heat source, what is the maximum welding speed, u_x, that can be used to avoid martensite formation?

Solution:

The cooling rate for a dual-beam system in a semi-infinite plate is given by equation (16.38)

$$\frac{d\Theta}{dt} = -\frac{q_1 u_x}{2\pi k}\left[\frac{1}{\xi^2} + \frac{1}{\xi'^2}\right]$$

Thus to determine the welding velocity, we first have to determine the positions at which the temperature of 550°C is reached. From equation (16.35), we have

$$\Theta = \frac{1}{2\pi k}\left(\frac{q_1}{\xi} + \frac{q_2}{\xi'}\right)$$

$$\Rightarrow 550 = \frac{1}{2\pi \times 73}\left(\frac{0.8 \times 3000}{|\xi|} + \frac{0.8 \times 3000}{|\xi - 0.004|}\right)$$

which gives

$$\xi = 21.2\,\text{mm},\ -17.2\,\text{mm}.$$

However, since we are interested in phenomena that occur during cooling of the weld, our focus is on temperatures behind the major heat source, that is, $\xi = -17.2$ mm. The cooling rate at this location should not be greater than 150°C/s. Thus we have

$$-150 = -\frac{0.8 \times 3000 \times u_x}{2\pi \times 73}\left[\frac{1}{(-0.0172)^2} + \frac{1}{(-0.0172 - 0.004)^2}\right]$$

$$\Rightarrow u_x = 0.025\,\text{m/s} = 25\,\text{mm/s}$$

16.7.1.2 Multiple-Beam Flow Control A number of discontinuities such as humping, undercutting, porosity, spiking, incomplete fusion, and spatter, that arise during high intensity welding processes, are associated with the nature of molten metal flow within the weld pool. Extensive research has been done using multiple beams to minimize these defects in electron beam welding. The following discussions are based on the electron beam research results. It is expected that in general, they will also apply to laser welding, given similarities in the beam intensities.

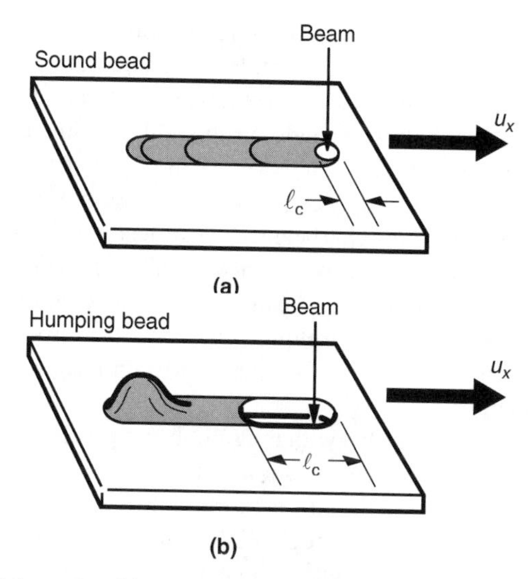

FIGURE 16.23 Schematic of humping during single-beam welding. (a) Sound bead. (b) Bead with a hump. (From Arata, Y., and Nabegata, E., 1978, *Transactions of Japan Welding Research Institute*, Vol. 7, No. 1, pp. 101–109.)

Humping and Undercutting At relatively high speeds, say greater than 15 m/min. in the keyhole mode, humping can occur (Fig. 16.23). The pool width decreases and the molten metal is forced to move at higher speeds through the narrow channel between the lateral walls and the keyhole (Fig. 16.24).

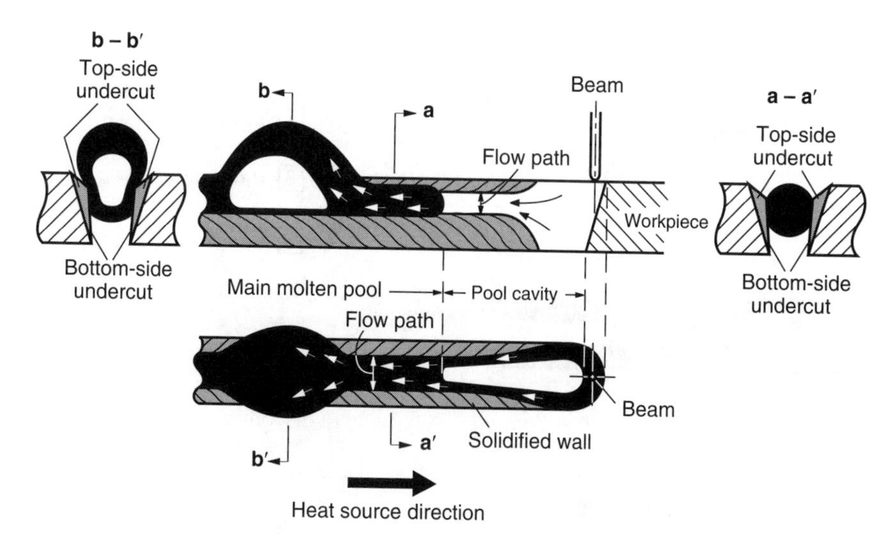

FIGURE 16.24 Mechanism of humping during single-beam welding. (From Arata, Y., and Nabegata, E., 1978, *Transactions of Japan Welding Research Institute*, Vol. 7, No. 1, pp. 101–109.)

In regions where the joint is not exposed, humping might appear to be aesthetically acceptable or could simply be machined off. However, beneath the hump often exists an opening or lack of penetration that might result in structural weakness. It is difficult to suppress these discontinuities in single-beam welding without compromising on the welding speed and penetration. However, such defects have been shown to be effectively curtailed or suppressed by using two beams in series along the welding direction in electron beam welding. One beam, B-1 (the leading or major beam, beam 1), is used as a heat source to melt the specimen with full penetration, while the second beam, B-2 (the trailing or minor beam, beam 2), is used to control flow of the molten metal (Fig. 16.25). Using the same welding conditions that produce a humped bead at 15 m/min with a single beam, speeds of up to 30 m/min can be achieved with the multiple-beam system.

Important parameters associated with multiple-beam processing include the interbeam spacing and power ratio. The location of B-2 with respect to the molten pool produced by B-1 affects the quality of the bead produced, and can be classified into three regions, region A, region B, and region C (Fig. 16.25). In region A, the interbeam spacing, l_b, is less than the cavity length, l_c, that is, $l_b < l_c(s)$, which means the second beam is located in the cavity formed by the leading beam. Humping and undercutting are enhanced under such circumstances. This is because the recoil pressure due to the vapor resulting from the second beam accelerates the flow rate of the molten metal in the pool cavity, thereby further elongating the cavity. In region B, l_b, is less than

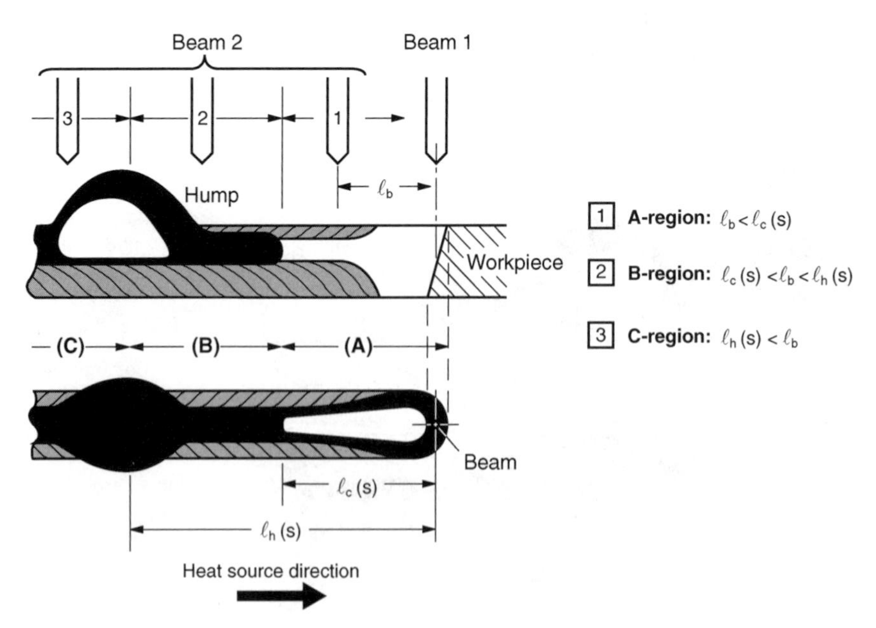

FIGURE 16.25 Prevention of humping using dual beam welding. (From Arata, Y., and Nabegata, E., 1978, *Transactions of Japan Welding Research Institute*, Vol. 7, No. 1, pp. 101–109.)

the distance between a single beam and its hump, $l_h(s)$, that is, $l_c(s) < l_b < l_h(s)$. In other words, the second beam impinges on the surface of the molten pool between the pool cavity and the hump that would otherwise be produced by the single beam. This is an effective way to prevent the hump formation. Under these circumstances, the flow direction of the molten metal is changed toward the lateral walls, resulting in melting of the solidified walls at the top and bottom of the lateral walls. This in turn broadens the flow in the molten pool and, consequently, reduces the inner pressure of the pool and prevents hump formation. With this configuration, the pool cavity length also becomes shorter. In region C, the second beam is located much further down beyond where the hump would start to form for the single beam and the bead formation again becomes irregular.

In addition to the location, proper selection of the beam power ratio is also important in achieving good weld quality. If the power ratio (B-2:B-1) is too low, broad undercutting results, and this is because the flow generated by the second beam is not adequate to fill up the top and bottom parts of the lateral walls with molten metal. On the contrary, if the power ratio is too large, an irregular weld bead with undercutting results, as the second beam then widens the pool width. The optimum power ratio of B-2:B-1 in terms of pool uniformity (size from the top to the root) and stability has been found to be about 10%.

Porosity, Spiking, and Incomplete Fusion Porosity, spiking, and incomplete fusion normally occur at the weld root when a high power density is used in deep penetration welding, resulting in a needle-like weld root with a small radius of curvature. The molten metal is then hardly deposited at the weld root. Thus these defects are not usually formed at low power densities that produce a wide and shallow bead. Spiking is associated with partial penetration welding, and tends to result in uneven penetration. In using a multiple-beam system to minimize the formation of such defects, a low power density beam B-2 of relatively large diameter is superimposed on a high power density beam B-1. With such an arrangement, B-2 penetrates over the depth created by B-1, resulting in a deep weld of relatively large root radius, where internal discontinuities are minimal. An important parameter in this situation is the power density of B-2. When the power density of B-2 is so low that the beam does not penetrate to the depth of B-1, then the discontinuities at the root will remain (Fig. 16.26a). As the density of B-2 increases, the number of discontinuities decreases as the wide beam gets deeper and deeper. With the large root radius resulting from this condition, the molten metal is able to penetrate to the weld root where the cooling rate is also relatively low, and eliminates the defects created by the first beam by reheating and remelting, as well as elongating and stabilizing the beam hole (Fig. 16.26b). The second beam is able to penetrate deeper than it ordinarily would, since it is incident on material already melted and preheated by the first beam. However, at a high-enough density of the second beam, the situation becomes tantamount to the single beam case, with the discontinuities again becoming significant (Fig. 16.26c).

The use of multiple-beam welding may also eliminate active zone porosity (which occurs in the central portion of the weld bead), while reducing root porosity (which is more common at the bead root), depending on the power ratio and interbeam spacing

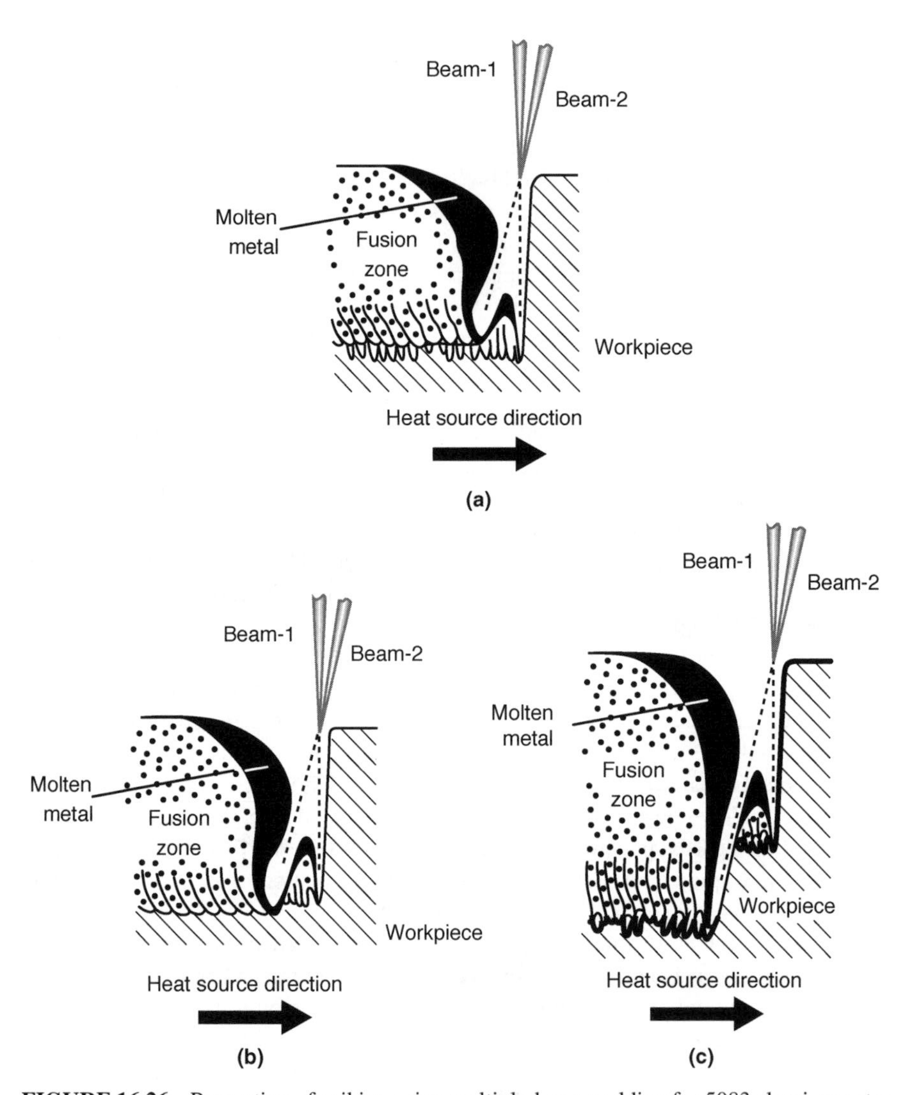

FIGURE 16.26 Prevention of spiking using multiple-beam welding for 5083 aluminum at a speed of 1.2 m/min. (a) Relatively low power density. (b) Intermediate power density. (c) High power density for beam 2. (From Arata, Y., Nabegata, E., and Iwamoto, N., 1978, *Transactions of Japan Welding Research Institute*, Vol. 7, No. 2, pp. 85–95.)

used. In deep penetration welding with a single beam, a neck is often observed at a depth of about one-third the penetration depth, and active zone porosities may be formed directly below it. Using multiple-beam welding eliminates the active zone porosity because the pool cross section is uniformly enlarged throughout the depth. The porosities formed by the leading beam are absorbed by the keyhole of the trailing beam. The flow pattern associated with single-beam welding is known to consist of several local flows. A circular flow that occurs near the middle of the molten pool tends to induce the active zone porosity. Such localized flow is not observed when a second beam is used.

16.7.2 Arc-Augmented Laser Welding

Laser welding can be significantly improved if combined with arc welding. Generally, for a given total energy input to the process, the augmented process is found to produce a greater penetration than either arc welding or laser welding by itself. For example, if a 5 kW laser beam is combined with 300 A gas tungsten arc welding (GTAW), the resulting depth of penetration is about 1.3–2 times than that obtained using the 5 kW laser alone. Similarly, higher penetration is obtained with the laser augmented gas metal arc welding (LAGMAW) process than with gas metal arc welding (GMAW) alone. The increase in penetration associated with LAGMAW is more significant at low arc welding currents (Fig. 16.27). A similar behavior is observed at lower speeds. The reduction in penetration at higher currents could be due to absorption of the laser beam by the increased plasma formation.

Possible arrangements of the setup are illustrated in Fig. 16.28. It is desirable to have the distance between the laser beam and the electrode as close as possible. In the case of GTAW, no significant increases in penetration is observed when the separation is less than 3 mm and the laser power is 2 kW with a current of 300 A, and a welding speed of 1 m/min. At greater distances, however, the penetration reduces with increasing separation distance. Even though the electric arc

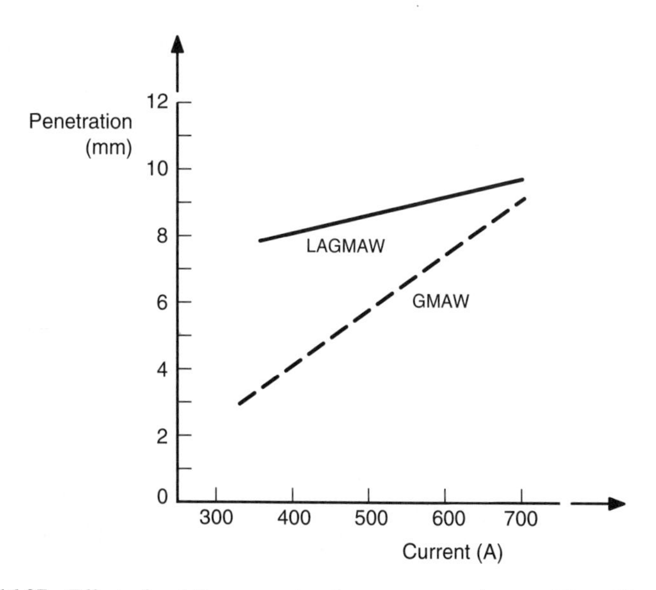

FIGURE 16.27 Effect of welding current on laser-augmented arc welding. (From *ICALEO 1990 Proceedings*, Vol. 71. Copyright 1991, Laser Institute of America.)

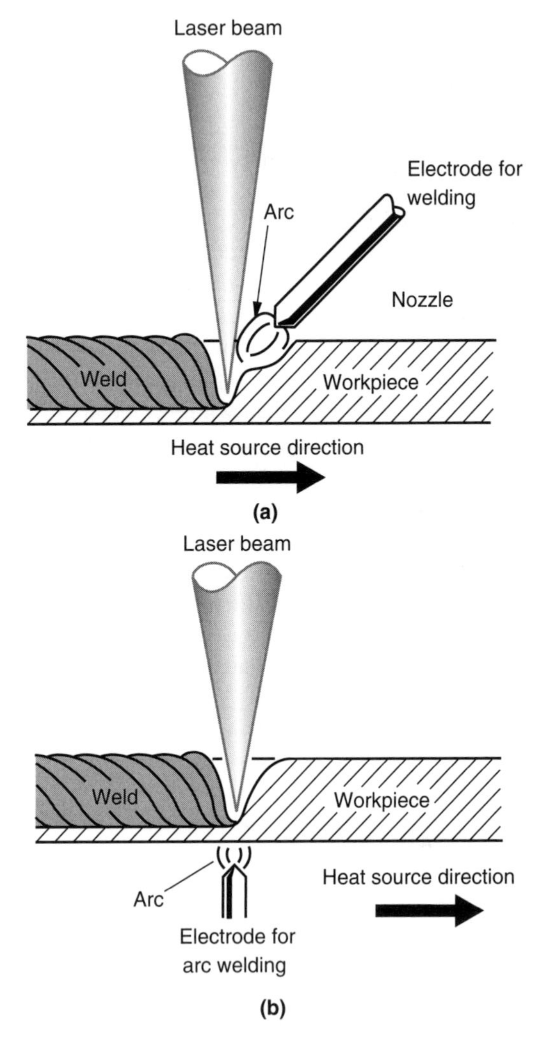

FIGURE 16.28 (a) Laser-arc welding system with both laser beam and arc on the same side of the workpiece. (b) Laser-arc welding system with the laser beam and arc on opposite sides of the workpiece. (From Duley, W. W., 1983, *Laser Processing and Analysis of Materials*. By permission of Springer Science and Business Media.)

can be located on either side of the workpiece, it has been found to be more effective on the side of the workpiece opposite the laser (Fig. 16.28b). In addition to increased penetration, another advantage of the augmented welding process is the elimination of undercut and humping that can occur at high speeds during GTAW alone. One disadvantage of arc-augmented welds is that they tend to have a larger heat-affected zone size than a weld made using a laser of the same total power.

When both the laser and arc are on the same side of the workpiece, the bead appearance is found to be better when the laser beam leads the arc in LAGMAW. This is because the suppression gas flow does not affect the molten pool created by the arc. However, a more stable arc is observed when the laser beam follows the arc.

16.8 SPECIFIC APPLICATIONS

Two of the more common and also rapidly growing applications of laser welding include microwelding and welding of tailored blanks.

16.8.1 Microwelding

The small size to which the heat source can be focused makes laser welding suitable for welding and soldering small components or joints as found in electronic components or in attaching a thermocouple to a part for temperature measurement. The ability to weld dissimilar metals using a laser beam and also the minimal thermal damage to surrounding material make it particularly attractive in this regard. For such applications as microwelding in electronic circuitry, the short wavelength 1.06 μm of the Nd:YAG laser makes it preferable to the CO_2 laser (10.6 μm wavelength). The shorter wavelength beam is least absorbed by insulating materials, and thus can be scanned from one solder joint to another without shutting off the beam. In contrast, the longer wavelength beam would need to be cycled on and off as the beam traverses from one joint to another as it can damage the intermediate insulating material.

Specific examples of microwelding applications include microspot welding of the Gillette Sensor razor head blades together and hermetic instruments such as heart pace makers and transducers.

16.8.2 Laser-Welded Tailored Blanks

Tailored blank welding essentially involves butt welding of sheet metal blanks of either different thicknesses (gages), coatings (bare or galvanized), or materials (grades). There are two primary ways in which tailored blanks are made:

1. Resistance mash-seam welding.
2. Laser welding.

This section focuses on laser-welded tailored blanks. Tailored blanks have replaced a number of units that are traditionally produced by first forming single components, which are then subsequently joined by welding, especially for the automotive body-in-white. One typical application of tailored blanks in this area is the use of a thicker or higher strength material to replace the reinforcement that would normally be required for strength or support. An example is the center pillar inner of a car body (Fig. 16.29) where a thicker material is used in the upper portion where a separate reinforcement would normally be needed and joined to the thinner material that is subjected to lower stresses, before forming.

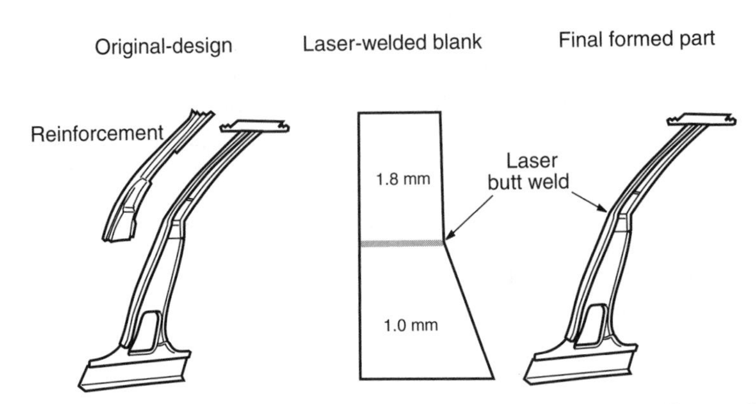

FIGURE 16.29 Cadillac center pillar inner. (From Uddin, N., *Photonics Spectra*, 1993.)

16.8.2.1 Advantages of Tailored Blank Welding The advantages of tailored blanks over conventionally produced body-in-white components include the following:

1. Weight reduction, achieved by welding a thinner gage to a thicker gage material that provides reinforcement only where needed and also by elimination of reinforcement panels. The elimination of weld flanges that are normally required for resistance spot welding of separately formed parts also reduces weight.
2. Improved part functional performance. This results from improved structural rigidity and body accuracy or dimensional control due to a reduction in assembly variation as a result of a decrease in the number of stampings being assembled. This eliminates tolerance stackup, since tailored blanks join sheets before they are stamped, rather than welding them after being stamped.
3. Cost reduction through reduced stamping and assembly costs and offal recovery. Offal is the scrap metal blank that falls off during sheet metal blanking from coil. Small pieces from the offal can be joined to form a functional single blank.
4. Reduced sealing needs.
5. Improved crash worthiness, since laser welds are stiffer than corresponding spot welds.

The advantages of tailored blank welding are further illustrated in Table 16.5, which compares two traditional methods of producing the body side frame with tailored blank welding. In Table 16.5, the separation (divided) method uses individual components (Fig. 16.30), which are first formed and then joined, usually by spot welding. In the integration (one-sheet) method, cuts are made in a large sheet panel to leave one large panel of the desired shape, which is then formed. In a sense, tailored blank welding combines the two approaches by first welding appropriate individual components to form the large panel of the desired shape (Fig. 16.31).

TABLE 16.5 Comparison of Two Traditional Methods of Producing Automotive Body Components with Components Produced by Tailored Blank Welding

	Divided Method	One-Sheet Method	Tailored Blanks
Appearance	Poor	Good	Good
Accuracy	Low	High	High
Yield	High (65%)	Low (40%)	High
Material flexibility	Selectable	Fixed	Selectable
Number of dies required	High	Low	Low

Source: Table 1, Nakagawa et al., SAE Publication #930522, 1993.

16.8.2.2 Disadvantages of Tailored Blank Welding The disadvantages of tailored blank welding include the following:

1. Formability performance. Due to the difference in thickness and/or strength of the two sheets, forming of tailored blanks without producing defects is more difficult than traditional forming of single sheets.
2. Required special edge preparation of the blanks. The gap between the blanks should not be more than 10% of the thickness of the thinner blank. This is often achieved by precision shearing, especially for long continuous welds of length greater than one meter. Precision shearing may not be necessary for shorter welds since they can often be processed with a die cut edge. The need for precision shearing, even for long welds, can also be eliminated by using filler wire or beam weaving. However, these techniques tend to slow the process and add to cost.
3. Potential for hot cracking in steels containing significant amounts of sulfur and phosphorus.
4. Potential for enhancing die wear. Since the weld bead is usually harder than the base metal, the welded sheet has a greater tendency to wear off die material than a single sheet that consists of the base metal.
5. Removal of protective coating in the heat-affected zone of the weld. Since the heat-affected zone size in laser welding is relatively small, the loss of coating may not be a significant factor.

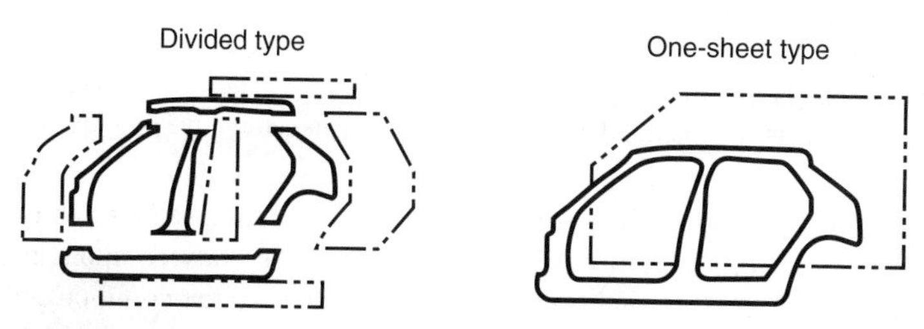

FIGURE 16.30 Illustration of the separation (divided) and integration (one-sheet) methods. (From Natsumi, F., Ikemoto, K., Sugiura, H., Yangfisawa, T., and Azuma, K., 1991, *JSAE Review*, Vol. 12, No. 3, pp. 58–63.)

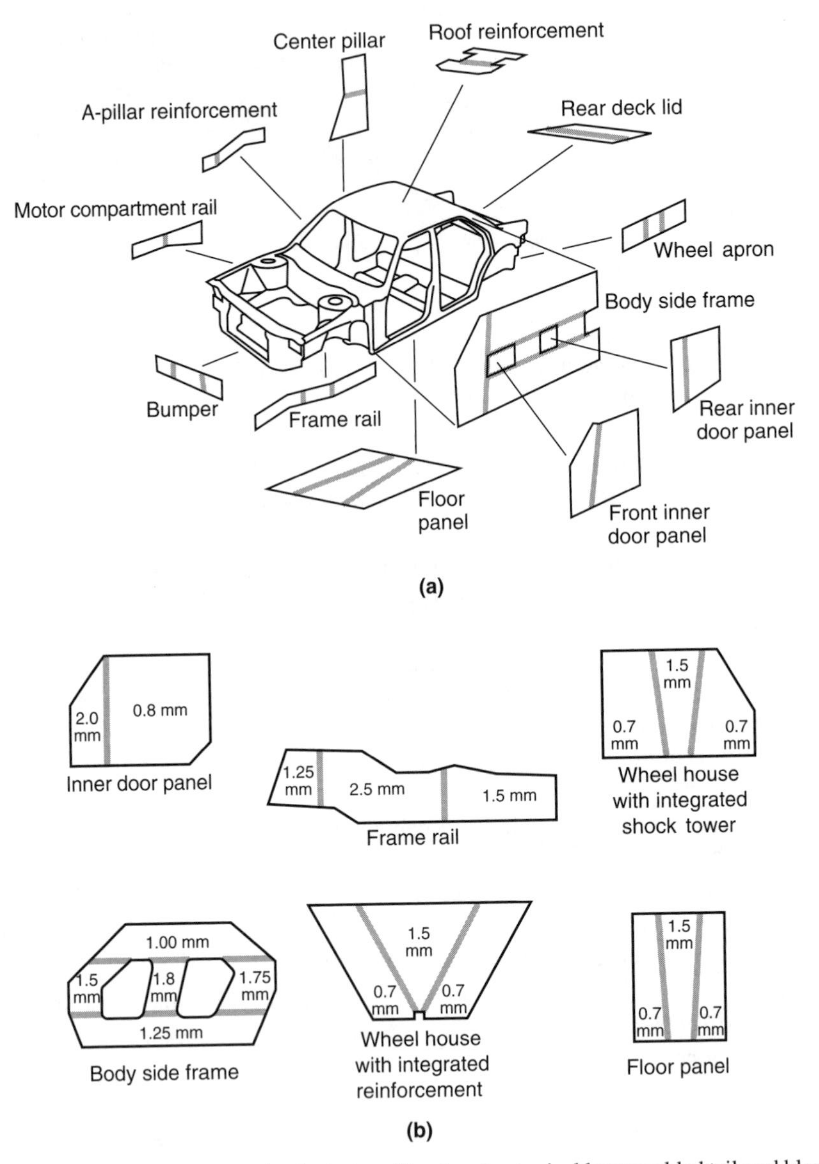

FIGURE 16.31 (a) Automotive body-in-white showing typical laser-welded tailored blanks. (b) Candidate laser-welded tailored blank components in an automotive body-in-white.

16.8.2.3 Applications of Laser-Welded Tailored Blanks Even though there are a number of applications for tailored blanks in automotive body components, not all components can be candidates for tailored blanks. Some of the common production applications in the automotive industry are shown in Fig. 16.31 and include center pillars, body side frames, upper and lower frame rails, roof reinforcement, rear deck lid, rear and front inner door panels, and floor panel. A typical vehicle may have about twenty components that are candidates for tailored blank welding. Multigage inner panels are considered to be among the tailor-welded blank applications that result in the highest benefits.

Some of the issues of concern that are often considered in setting up tailored blank applications include the welding costs; weld seam characteristics such as quality; reliability of the welding system; effect of welding on coatings; and weldment formability. Some of the important cost issues include cost of the welding process, work material, blanking, pallets, transportation, blank protection, inventory, and quality control. Critical factors that need consideration in application include the beam power and characteristics, gap width, edge preparation, and clamping. Processing speeds typically fall in the range of 4–10 m/min, depending on the laser power and sheet thickness.

The production system for laser-welded tailored blanks can be a fully automatic line with a number of workstations, or a manual single station operation. The latter would normally have a single welding system (which would include facilities for blank shearing, laser beam, and gantry), and two or more operators that perform the tasks of loading and unloading. The workstation may be either a single or two axis system. Single axis systems are more suitable for in-line or long continuous welds that would normally be over a meter in length. Two axis systems are appropriate for multiple, short welds in any planar orientation. In either case, the overall blank size that can be processed is determined by the gantry and fixturing used.

Components of a full-production system may include pallets; destacker; precision shearing for edge preparation; material handling; laser generator; inspection station; dimpling station for multi-gage blanks; and flipper when both left and right parts are made, for proper stacking of the parts on the pallet. Optional items may include blank washing; weld cleaning; oiling, and trimming. Some important factors to be considered in designing a production system include the following:

1. Sheet gage—single or multigage. This determines the type of clamping device to be used. Also, dimplers are required for the thinner sheet of multigage blanks, to ensure even stacking on the pallet.
2. Desired length of weld. This determines the gantry specifications.
3. Number of welds required per part.

A typical work cell for laser-welded tailored blank manufacturing is illustrated in Fig. 16.32.

16.8.2.4 Formability of Tailor-Welded Blanks Formability is a measure of the ease of forming a given material or body without failure. The formability of welded blanks is influenced by weld parameters such as the following:

1. Butt joint gap, which is the space between the two sheets (Fig. 16.33a).
2. Beam misalignment, which is the offset of the centerline of the laser beam from the centerline of the gap (Fig. 16.33b).
3. Shielding gas type and flow rate.

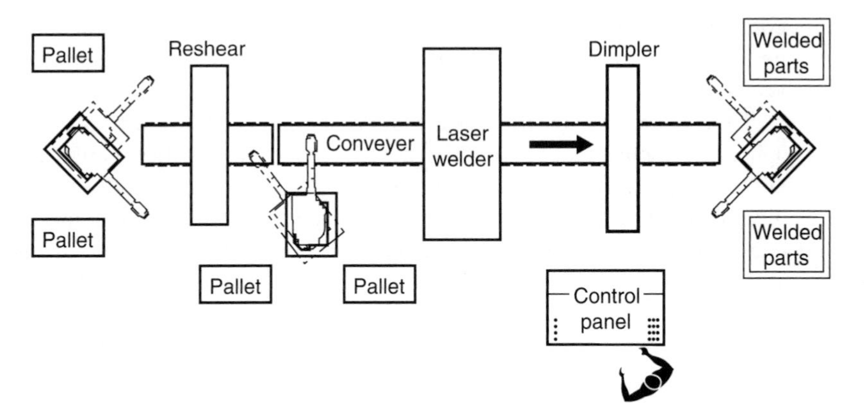

FIGURE 16.32 Example work cell for manufacturing medium-size laser-welded tailored blanks.

4. Focal point location.
5. Welding speed.
6. Laser power.

Some observations regarding the influence of process parameters on formability may be summarized as follows:

1. Formability increases with increasing welding speed, since the increased welding speed results in a reduced weld section, even though the hardness of the weld increases.
2. An increase in misalignment or joint gap reduces formability.

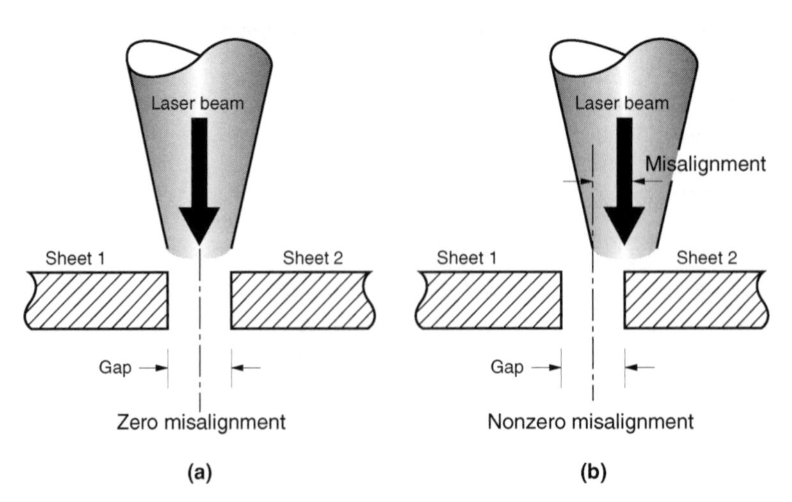

FIGURE 16.33 Illustration of (a) joint gap and (b) beam misalignment.

3. Misalignment or joint gap tends to promote high-weld concavity, and these have relatively high failure rates. Generally, misalignment and joint gap have to be less than 0.05 mm to enhance quality and formability.

Uniaxial tension tests provide the simplest approach to assess the formability of tailor-welded blanks. It is more appropriate to use longitudinal specimen, that is, those cut along the weld with the welding line in the center of the specimen (Fig. 16.34a) since these provide information on the mechanical properties of the welded blank. Transverse specimen (Fig. 16.34b) does not provide adequate formability information since deformation then occurs primarily in the base metal outside the weld. The higher yield strength of the weld then prevents plastic deformation in the weld. Transverse specimen thus primarily provide information on the weldment quality. It should be noted, however, that even for the longitudinal specimen, the tensile properties that are measured would depend on the ratio of the weld width to the entire tension specimen width. Thus the specimen widths that are used, typically about 6 mm, are preferably smaller than the standard widths used for tensile tests, that is, they are subsize specimen.

When the direction of principal strain is in the direction of the weld, failure normally occurs in the weld, with the crack running transversely (Fig. 16.35a). This form of failure may occur in welds made from materials with similar thicknesses/strengths or dissimilar thicknesses/strengths. The other form of failure involves a crack that runs parallel to and is adjacent to the weld and occurs primarily in welds made from dissimilar thicknesses/strengths (Fig. 16.35b). This normally occurs when the direction of principal strain is normal to the weld line, and fracture tends to occur in the lower strength or thinner material. Thus, optimum formability performance may be achieved by placing the weld line away from the major strain direction and/or away from the areas with high component strain level in the direction along the weld line, preferably perpendicular to the major strain direction.

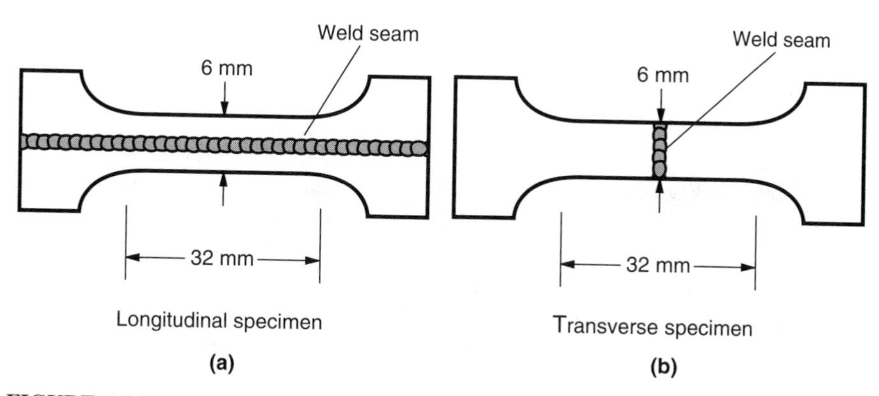

FIGURE 16.34 Tensile specimens of tailor-welded blanks. (a) Longitudinal specimen. (b) Transverse specimen.

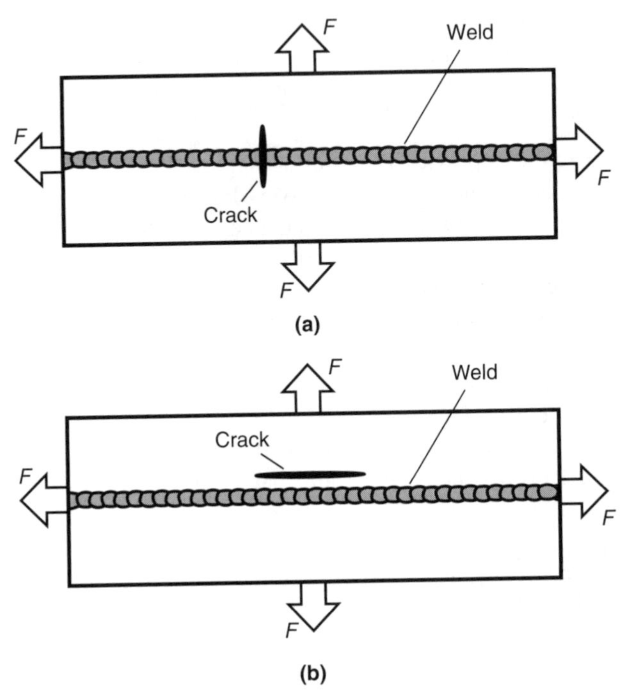

FIGURE 16.35 Cracking during forming of tailor-welded blanks. (a) Crack running transverse to weld. (b) Crack running parallel to weld.

A more elaborate approach to assessing formability involves measurement of limit strains or forming limit, that is, strains at the onset of localized necking. This can be done using one of the two methods:

1. The limiting dome height test (LDH), which involves out-of-plane stretching.
2. The Marciniak cup test, which involves in-plane stretching.

Both tests provide a relative measure of formability and are further discussed in the following sections.

Limiting Dome Height Test The limiting dome height test is a formability test that measures the ability of a sheet metal to uniformly distribute the strain induced during forming. This depends on both the base metal formability and interface friction between the sheet metal and tooling. The test measures the dome height at the onset of necking in the material. It tests the material in or near plane strain, and the outcome is affected by work hardening and ductility of the workpiece, as well as the frictional conditions. It does not distinguish between the effects of the individual variables. Furthermore, it cannot be used to specify desired material characteristics.

The rectangular specimen has two sides solidly clamped to a binder, while the unconstrained edges allow control of the strain state along the centerline (Fig. 16.36a).

Specimens of different widths are used, to determine the minimum dome height at fracture, referred to as the limiting dome height, (LDH_0) (Fig. 16.36b and c). Fracture occurs in or near plane strain.

The LDH test has been found to correlate well with the Rockwell hardness.

When the formability of a laser-welded blank and a single blank are compared using, say, the dome height test, it is normally found that the maximum dome height

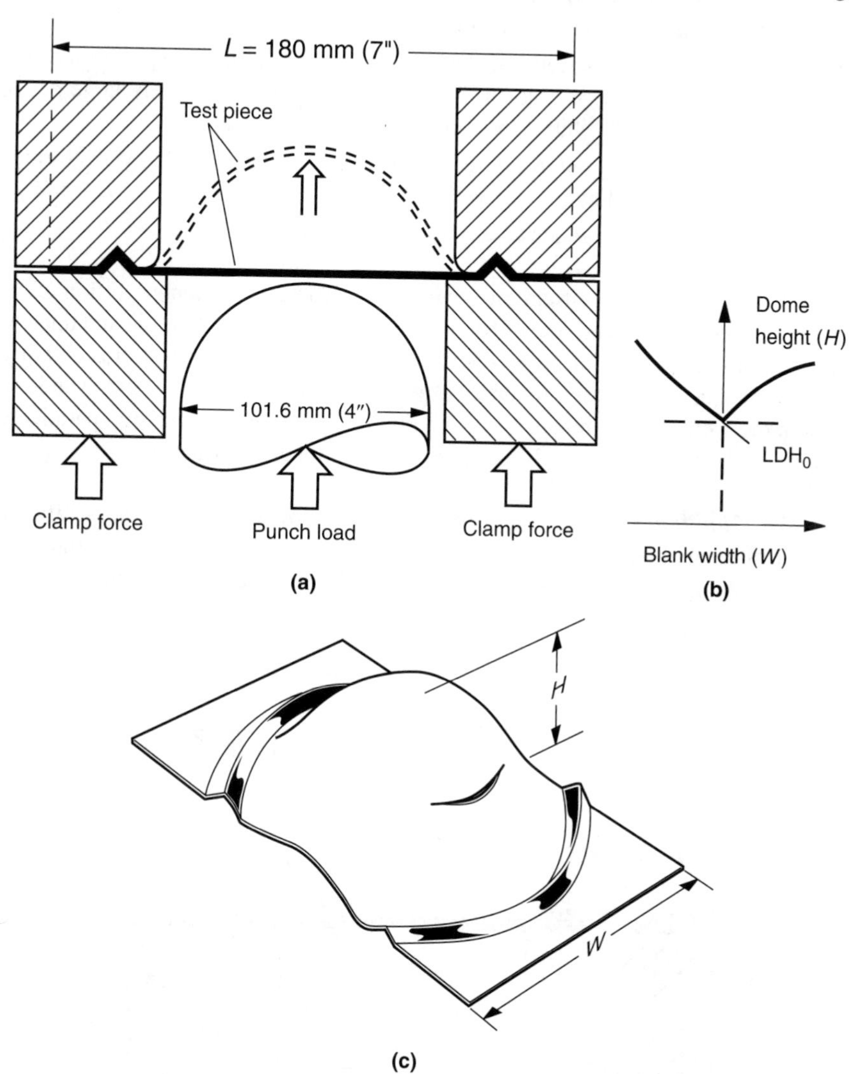

FIGURE 16.36 Illustration of the limiting dome height measurement. (a) Schematic of the limiting dome height test. (b) Plot of dome height at failure versus blank width. (c) Schematic of specimen after testing. (From Bayshore, J. K., Williamson, M. S., Adonyi, Y., and Milian, J. L., 1995, *Welding Journal*, Vol. 74, pp. 345s–352s. By permission of American Welding Society).

achieved for the laser-welded blank of materials of either the same strength or thickness is less than that of the single blank, and the height further decreases with an increase in either the strength or thickness ratio.

Test Limitations

1. Due to the fixed geometry and boundary conditions, the test does not cover all the strains found in real parts.
2. Low deformation speed and absence of bending strains. The performance of small, deep corners is not well simulated due to the relatively large punch radius and small relative motion between the punch and tool.
3. The single strain path is not representative of the one found in walls formed by large metal flow over corners.
4. Comparison between tests is only relative and not absolute.

Marciniak Cup Test In the Marciniak cup test (Fig. 16.37) the center portion of the cup bottom remains flat and free of frictional forces during deformation, and thus the strains in these areas are uniformly distributed. As a result, the limit strains obtained are only associated with the properties of the base material and the weld. The limit strain in the weld is determined by the longitudinal strain in the direction parallel to the welding line, since the weld undergoes unidirectional deformation.

16.8.2.5 Limiting Thickness or Strength Ratio When the thickness ratio or strength ratio of a tailor-welded blank exceeds a critical value, the material with the higher strength or thickness may be subjected to very little or no plastic defor-

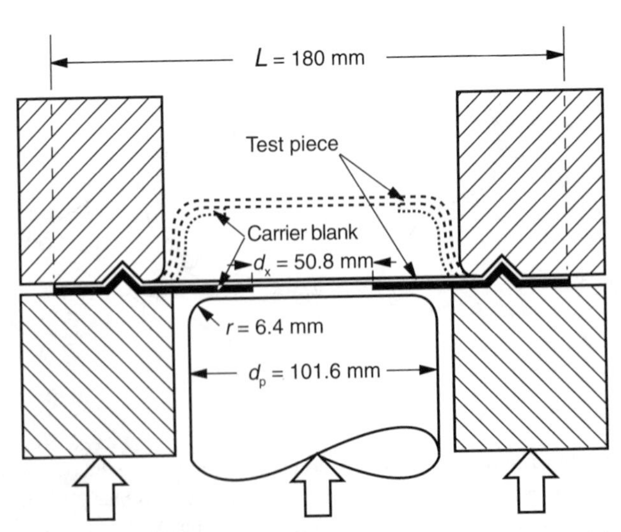

FIGURE 16.37 Schematic of the Marciniak cup test. (From Shi, M. F., Pickett, K. M., and Bhatt, K. K., 1993, *SAE Technical Paper #930278*, pp. 27–35.)

mation in the direction transverse to the weld direction, while the material of lower strength or thickness exceeds its forming limit near the weld. The limiting thickness ratio (LTR) (for same materials with different thicknesses) or limiting strength ratio (LSR) (for different materials with a prespecified thickness ratio) is the thickness or strength ratio at which one material just reaches initial yield strength when the other material reaches its forming limit. When this value is exceeded, failure may occur parallel to the weld, with cracking occurring on the side of the material of lower thickness or strength. When the thickness and strength ratios are less than the LTR and LSR, respectively, plastic deformation would occur in both materials.

To develop an expression for the LTR or LSR, let us consider Fig. 16.38, which shows two materials (1 and 2) that are joined by a laser weld. Neglecting friction effects, equilibrium in the transverse direction gives

$$F_1 = F_2 \tag{16.40}$$

where F is the force per unit width and the subscripts 1 and 2 represent the materials 1 and 2, respectively.

Equation (16.40) can be written as

$$\sigma_1 h_1 = \sigma_2 h_2 \tag{16.41}$$

where σ is the true stress and h is the instantaneous material thickness.

The true strain, ε_z, in the thickness direction is given by

$$\varepsilon_z = \ln\left(\frac{h}{h_0}\right) \tag{16.42}$$

where h_0 is the original thickness.

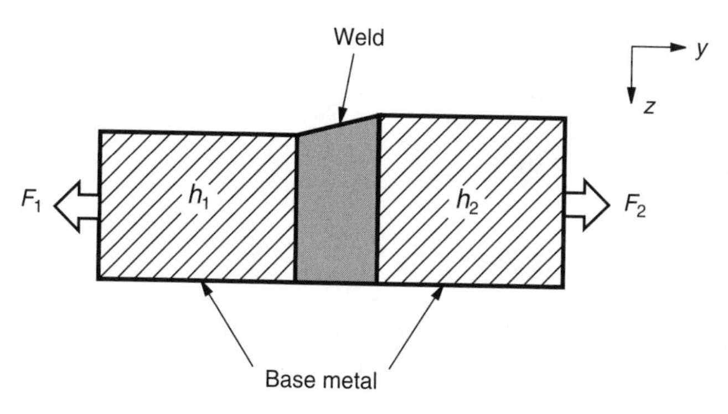

FIGURE 16.38 Force equilibrium in the transverse direction of a tailored weld.

Thus,

$$h = h_0 e^{\varepsilon_z} \tag{16.43}$$

Considering volume constancy during plastic deformation, we have

$$\varepsilon_x + \varepsilon_y + \varepsilon_z = 0 \tag{16.44}$$

where ε_x is the true strain in the welding direction (parallel to the weld in the sheet plane) and ε_y is the true strain in the transverse direction (normal to the weld in the sheet plane).

From equations (16.41), (16.43), and (16.44), we get

$$\sigma_1 h_{01} \mathrm{e}^{-(\varepsilon_x)_1 - (\varepsilon_y)_1} = \sigma_2 h_{02} \mathrm{e}^{-(\varepsilon_x)_2 - (\varepsilon_y)_2} \tag{16.45}$$

The strain component parallel to the weld (longitudinal strain) must be the same in both materials (1 and 2), that is,

$$(\varepsilon_x)_1 = (\varepsilon_x)_2 \tag{16.46}$$

Thus

$$\sigma_1 h_{01} \mathrm{e}^{-(\varepsilon_y)_1} = \sigma_2 h_{02} \mathrm{e}^{-(\varepsilon_y)_2} \tag{16.47}$$

or

$$S_1 h_{01} = S_2 h_{02} \tag{16.48}$$

where S is the engineering stress.

If two sheets of the same material but different thicknesses are being welded, then from equation (16.48), the limiting thickness ratio can be obtained from

$$S_T h_{01} = S_y h_{02} \tag{16.49}$$

where S_T is the tensile strength of the material and S_y is the yield strength of the material.

The LTR is then given by

$$\mathrm{LTR} = \text{limit of} \left(\frac{h_{02}}{h_{01}} \right) = \frac{S_\mathrm{T}}{S_\mathrm{y}} \tag{16.50}$$

If, on the contrary, the two sheets are made from different materials, but have a pre-specified thickness ratio, then equation (16.48) becomes

$$S_{\mathrm{T}1} h_{01} = S_{\mathrm{y}2} h_{02} \tag{16.51}$$

and the LSR is then given by

$$\text{LSR} = \text{limit of}\left(\frac{S_{T1}}{S_{y2}}\right) = \frac{h_{02}}{h_{01}} \tag{16.52}$$

The thickness ratio h_{02}/h_{01} normally has to be 1.25 or less since the formability of the part reduces for thickness ratios above 1.25.

16.9 SUMMARY

Laser welding can be done using both CO_2 and Nd:YAG lasers. The key parameters that influence the process are the beam power and characteristics, traverse speed, shielding gas, relative location of the focal point, and joint configuration. As an example, a 12 mm thick carbon steel plate can be welded with a CO_2 laser using an output power of 9 kW at a speed of 0.8 m/min, or a 6 mm thick carbon steel plate can be welded with a CO_2 laser using an output power of 25 kW at a speed of 10 m/min. As the cutting speed increases, the depth of penetration for a given power decreases.

A Gaussian beam mode results in a greater depth-to-width ratio than higher order modes. However, the larger spot size of multimode beams makes them more suited for butt welding applications since it reduces the tight fit-up requirements. The effect of polarization on the process depends on the material and welding conditions. For some steels, the depth of penetration is increased when the direction of polarization is aligned with the welding direction and when the welding speed is above a critical value.

Both pulsed and CW laser beams can be used for welding, with pulsed beams being more suited for spot welding while CW beams are more suited for continuous welds. Pulsed beams can also be used for continuous welds if the pulses are made to overlap. With high power beams, a portion of the laser energy is expended in ionizing part of the shielding gas and/or vapor to form a plasma. This plasma may absorb part of the laser beam, and may also induce an optical lensing effect that changes the effective focusing of the laser beam. Shielding gases are used to blow away the vapor and plasma formed, provide a protective environment for the weld pool, and protect the focusing lens. Common shielding gases used are argon and helium. Argon is more commonly used for low to medium power laser welding. It is also generally preferred to have the focal point either on the workpiece surface or just below it.

Laser welding may occur in the conduction mode (normally used for foils and thin sheets), or keyhole mode (for relatively thick sections). In conduction mode welding, the laser power is first deposited on the surface, and then transferred by conduction to the surroundings. Absorption in the keyhole occurs by inverse bremsstrahlung of the plasma and by Fresnel absorption (direct deposition) at the cavity walls.

Low carbon steels and some stainless steels are easily welded with a laser, while brass, silver, and galvanized steel are more difficult to weld. Common defects that affect product quality include porosity, cracking, humping, and spiking. A dual-beam system can be used either to control the cooling rates, and therefore the microstructure in the solid material (by preheating and/or postweld heat treating), or to control flow in the weld pool. Combining a laser beam with an arc welder increases the depth of penetration obtained compared to that for either process alone. Lasers are effectively used in microwelding and also in tailored blank welding, which involves butt welding of sheet metal blanks of either different thicknesses (gages), coatings (bare or galvanized), or materials (grades).

REFERENCES

Abilsiitov, G. A., and Velikhov, E. P., 1984, Application of CO_2 lasers in mechanical engineering technology in the USSR, *Optics and Laser Technology*, Vol. 16, No. 1, pp. 30–36.

Akhter, R., Davis, M., Dowden, J., Kapadia, P., Ley, M., and Steen, W.M., 1989, A method for calculating the fused zone profile of laser keyhole welds, *Journal of Physics D: Applied Physics*, Vol. 21, pp. 23–28.

Albright, C. E., and Chiang, S., 1988, High speed laser welding discontinuities, *Proceedings of 7th International Conference on Applications of Lasers and Electro Optics, ICALEO'88*, Santa Clara CA, pp. 207–213.

Arata, Y., and Nabegata, E., 1978, Tandem electron beam welding (Report-I), *Transactions of Japan Welding Research Institute*, Vol. 7, No. 1, pp. 101–109.

Arata, Y., Nabegata, E., and Iwamoto, N., 1978, Tandem electron beam welding (Report-II), *Transactions of Japan Welding Research Institute*, Vol. 7, No. 2, pp. 85–95.

Banas, C. M., 1992, Twin-spot lasers weld stainless tubing, *Man*, pp. 14–16.

Bayshore, J. K., Williamson, M. S., Adonyi, Y., and Milian, J. L., 1995, Laser beam welding and formability of tailored blanks, *Welding Journal*, Vol. 74, pp. 345s–352s.

Beyer, E., Behler, K., and Herziger, G., 1988, Plasma absorption effects in welding with CO_2 lasers, *SPIE, High Power Lasers and Laser Machining Technology*, Vol. 1020, pp. 84–95.

Chan, C., Mazumder, J., and Chen, M. M., 1984, A two-dimensional transient model for convection in laser melted pool, *Metallurgical Transactions A.*, Vol. 15A, pp. 2175–2184.

Chan, C., and Mazumder, J., 1987, One-dimensional steady-state model for damage by vaporization and liquid expulsion due to laser-material interaction, *Journal of Applied Physics*, Vol. 62, pp. 4579–4586.

Chande, T., and Mazumder, J., 1984, Estimating effects of processing conditions and variable properties upon pool shape, cooling rates, and absorption coefficient in laser welding, *Journal of Applied Physics*, Vol. 56, pp. 1981–1986.

Chen, T.-C., and Kannatey-Asibu Jr., E., 1995, Dual beam laser systems and their impact on conduction mode weld pool convection and surface deformation, *Transactions of NAMRI/SME*, Vol. 23, pp. 151–156.

Chen, T.-C., and Kannatey-Asibu Jr., E., 1996, Convection pattern and weld pool shape during conduction-mode dual beam laser welding, *Transactions of NAMRI/SME*, Vol. 24, pp. 259–265.

Cline, H. E., and Anthony, T. R., 1977, Heat treating and melting material with a scanning laser or electron beam, *Journal of Applied Physics*, Vol. 48, No. 9, pp. 3895–3900.

Conti, R. J., 1969, Carbon dioxide laser welding, *Welding Journal*, Vol. 48, pp. 800–806.

Dawes, C., 1992, *Laser Welding: A Practical Guide*, Abington Publishing, Cambridge, UK.

DebRoy, T., Basu, S., and Mundra, K., 1991, Probing laser induced metal vaporization by gas dynamics and liquid pool transport phenomena, *Journal of Applied Physics*, Vol. 70, pp. 1313–1319.

Douglas, D. M., Mazumder, J., and Nagarathnam, K., 1995, Laser welding of Al 6061-T6, *Proceedings of the 4th International Conference on Trends in Welding Research*, Gatlinburg, TN, June 5–8, 1995.

Douglas, D. M., and Mazumder, J., 1996, Mechanical properties of laser welded aluminum alloys, *ICALEO96*, pp. 31–38.

Dowden, J., Postacioglu, N., Davis, M., and Kapadia, P., 1987, A keyhole model in penetration welding with a laser, *Journal of Physics D: Applied Physics*, Vol. 20, pp. 36–44.

Dowden, J., Kapadia, P., and Postacioglu, N., 1989, An analysis of the laser–plasma interaction in laser keyhole welding, *Journal of Physics D: Applied Physics*, Vol. 22, pp. 741–749.

Dowden, J., Chang, W. S., Kapadia, P., and Strange, C., 1991, Dynamics of the vapour flow in the keyhole in penetration welding with a laser at medium welding speeds, *Journal of Physics D: Applied Physics*, Vol. 24, pp. 519–532.

Dudas, J. H., and Collins, F. R., 1966, Preventing weld cracks in high-strength aluminum alloys, *Welding Journal*, Vol. 45, pp. 241s–249s.

Duley, W. W., 1982, *Laser Processing and Analysis of Materials*, Plenum Press, New York.

Eisenmenger, M., Bhatt, K. K., and Shi, M. F., 1995, Influence of laser welding parameters on formability and robustness of blank manufacturing: an application to a body side frame, *SAE Technical Paper #950922*, pp. 171–182.

Elmer, J. W., Giedt, W. H., and Eagar, T. W., 1990, The transition from shallow to deep penetration during electron beam welding, *Welding Journal*, Vol. 69, pp. 167s–176s.

Fairbanks, R. H., and Adams, C. M., 1964, Laser beam fusion welding, *Welding Journal*, Vol. 43, pp. 97s–102s.

Funk, M., Kohler, U., Behler, K., and Beyer, E., 1989, Welding of steel with a CO_2 laser of 20 kW, *SPIE, High Power Lasers and Laser Machining Technology*, Vol. 1132, pp. 174–180.

Gatzweiler, W., Maischner, D., Faber, F. J., Derichs, C., and Beyer, E., 1989, The expansion of the plasmais examined by using different positions of the slit of the streak, *SPIE, High Power Lasers and Laser Machining Technology*, Vol. 1132, p. 157.

Grezev, A. N., Grigor'yants, A. G., Fedorov, V. G., and Ivanov, V. V., 1984, The structure and mechanical properties of laser welded joints between dissimilar metals, *Automatic Welding*, Vol. 63, pp. 29–31.

Harry, J. E., 1974, *Industrial Lasers and Their Applications*, McGraw Hill, London.

Irving, B., 1991, Blank welding forces automakers to sit up and take notice, *Welding Journal*, Vol. 70, pp. 39–45.

Irving, B., 1994, Automotive engineers plunge into tomorrow's joining problems, *Welding Journal*, Vol. 73, pp. 47–50.

Irving, B., and Baron, J., 1994, Mash-seam resistance welding fights it out with the laser beam, *Welding Journal*, Vol. 73, pp. 33–39.

Irving, B., 1995, Welding tailored blanks is hot issue for automakers, *Welding Journal*, Vol. 74, pp. 49–52.

Iwase, T., Shibata, K., Sakamoto, H., Dausinger, F., Hohenberger, B., Muller, M., Matsunawa, A., and Seto, N., 2000, Real time X-ray observation of dual focus beam welding of aluminum alloys, *ICALEO*, pp. C26–C34.

Kannatey-Asibu Jr., E., 1991, Thermal aspects of the split-beam laser welding concept, *ASME Journal of Engineering of Materials and Technology*, Vol. 113, pp. 215–221.

Kar, A., and Mazumder, J., 1995, Mathematical modeling of key-hole laser welding, *Journal of Applied Physics*, Vol. 78, No. 11, pp. 6353–6359.

Kim, C.-J., Kauh, S., Ro, S. T., and Lee, J. S., 1994, Parametric study of the two-dimensional keyhole model for high power density welding processes, *ASME Journal of Heat Transfer*, Vol. 116, pp. 209–214.

Klemens, P. G., 1976, Heat balance and flow conditions for electron beam and laser welding, *Journal of Applied Physics*, Vol. 47, No. 5, pp. 2165–2174.

Liu, Y.-N., and Kannatey-Asibu Jr., E., 1993, Laser beam welding with simultaneous gaussian laser preheating, *ASME Journal of Heat Transfer*, Vol. 115, pp. 34–41.

Liu, Y.-N., and Kannatey-Asibu Jr., E., 1998, Finite element analysis of heat flow in dual-beam laser welded tailored blanks, *ASME Journal of Manufacturing Science and Engineering*, Vol. 120, pp. 272–278.

Locke, E. V., Hoag, E. D., and Hella, R. A., 1972, Deep penetration welding with high-power CO_2 lasers, *IEEE Journal of Quantum Electronics*, Vol. QE-8, No. 2, pp. 132–135.

Lowry, J. F., Fink, J. H., and Schumacher, B. W., 1976, A major advance in high-power electron-beam welding in air, *Journal of Applied Physics*, Vol. 47, pp. 95–105.

Luxon, J. T., and Parker, D. E., 1985, *Industrial Lasers and Their Applications*, Prentice Hall, Englewood Cliffs, NJ.

Magee, K. H., Merchant, V. E., and Hyatt, C. V., 1990, Laser assisted gas metal arc weld characteristics, *ICALEO*, pp. 382–399.

Matsuda, J., Utsumi, A., Katsumura, M., Hamasaki, M., and Nagata, S., 1988, TIG or MIG arc augmented laser welding of thick mild steel plate, *Joining & Materials*, pp. 38–41.

Matsunawa, A., and Semak, V., 1997, The simulation of front keyhole wall dynamics during laser welding, *Journal of Physics D: Applied Physics*, Vol. 30, pp. 798–809.

Mazumder, J., and Steen, W. M., 1980, Heat transfer model for CW material processing, *Journal of Applied Physics*, Vol. 51, No. 2, pp. 941–947.

Mazumder, J., 1987, An overview of melt dynamics in laser processing, *High Power Lasers*, Vol. 801, pp. 228–241.

Metzbower, E. A., 1990, Laser beam welding: thermal profiles and HAZ hardness, *Welding Journal*, Vol. 69, pp. 272s–278s.

Mombo-Caristan, J.-C., Lobring, V., Prange, W., and Frings, A., 1993, Tailored welded blanks: a new alternative in automobile body design, *The Industrial Laser Handbook*, 1992–1993 edition, pp. 89–102.

Nakagawa, N., Ikura, S., Natsumi, F., and Iwata, N., 1993, Finite element simulation of stamping a laser-welded blank, *SAE Technical Paper #930522*, pp. 189–197.

Natsumi, F., Ikemoto, K., Sugiura, H., Yangfisawa, T., and Azuma, K., 1991, Laser welding technology for joining different sheet metals for one-piece stamping, *JSAE Review*, Vol. 12, No. 3, pp. 58–63.

Rayes, M. E., Walz, C., and Sepold, G., 2004, The influence of various hybrid welding parameters on bead geometry, *Welding Journal*, Vol. 83, pp. 147s–153s.

Rofin-Sinar, *Technical Note—CO_2 Laser Welding*, Rofin Sinar, Plymouth, MI.

Shi, M. F., Pickett, K. M., and Bhatt, K. K., 1993, Formability issues in the application of tailor welded blank sheets, *SAE Technical Paper #930278*, pp. 27–35.

Steen, W. M., and Eboo, M., 1979, Arc augmented laser-welding, *Metal Construction*, Vol. 11, No. 7, pp. 332–335.

Steen, W. M., 1980, Arc augmented laser processing of materials, *Journal of Applied Physics*, Vol. 51, No. 11, pp. 5636–5641.

Steen, W. M., 2003, *Laser Material Processing*, Springer-Verlag, Berlin.

Swift-Hook, D. T., and Gick, A. E. F., 1973, Penetration welding with lasers, *Welding Journal*, Vol. 52, pp. 492s–499s.

Thompson, R., 1993, The LDH test to evaluate sheet metal formability: final report of the LDH Committee of the North American Deep Drawing Research Group, *Sheet Metal and Stamping Symposium, SAE Technical Paper #930815*, pp. 291–301.

Utilase, Inc., 1994, Eliminate costs with laser welded blanks, Private Publication Leaflet.

Voldrich, C. B., 1947, Cold cracking in the heat-affected zone, *Welding Journal*, Vol. 26, pp. 152s–169s.

Wilson, J., and Hawkes, J. F. B., 1987, *Lasers: Principles and Applications*, Prentice Hall International Series in Optoelectronics, New York.

Yessik, M. J., 1978, Laser material processing, *Optical Engineering*, Vol. 17, No. 3, pp. 202–209.

Zacharia, T., David, S. A., Vitek, J. M., and Debroy, T., 1989, Weld pool development during GTA and laser beam welding of type 304 stainless steel, Parts 1 and 2, *Welding Journal*, Vol. 68, pp. 499s–509s, 510s–519s.

APPENDIX 16A

List of symbols used in the chapter.

Symbol	Parameter	Units
d	interbeam spacing	mm
D	maximum width of molten zone	μm
$e_c = 6.704 \times 10^{-19}$	electron charge	J
F	force per unit width	N/m
$\mathbf{F_b} = -\rho_l \mathbf{g} \beta_\mathbf{T}(T - T_0)$	body force vector	N/m^3
$\mathbf{F_i}$	body force components (F_x, F_y, F_z)	N/m^3
h_0	original thickness	mm
$\mathbf{I}_0$	intensity of incident laser beam	W/m^2
l	molten pool length	μm
l_f	molten pool length ahead of keyhole	μm
LDH	limiting dome height	mm
LTR	limiting thickness ratio	–
LSR	limiting strength ratio	–

P_{ab}	ablation pressure	Pa
P_{hd}	hydrodynamic pressure	Pa
P_{hs}	hydrostatic pressure	Pa
P_v	excess pressure of vapor in keyhole	Pa
P_γ	pressure due to surface tension	Pa
q_1	heat flux (power) input to the workpiece of major heat source	W
q_2	heat flux (power) input to the workpiece of minor heat source	W
r_k	keyhole radius	μm
R_F	coefficient for Fresnel reflection	–
R_p	reflection coefficient for p-polarized light	–
R_s	reflection coefficient for s-polarized light	–
S	engineering stress	Pa (N/m^2)
S_T	tensile strength	Pa (N/m^2)
S_y	yield strength	Pa (N/m^2)
t_D	time it takes for melt to attain width D	s
$\mathbf{t}$	unit tangential vector	–
T_1	absolute temperature of major heat source	K
T_2	absolute temperature of minor heat source	K
T_i	absolute temperature of ionized electrons	K
u_{cs}	average speed at which heat is conducted sideways	m/s
$\mathbf{u_{il}}$	velocity vector at solid–liquid interface	m/s
$\mathbf{u_l}$	velocity vector for the liquid phase	m/s
$\mathbf{u_v}$	velocity vector at liquid–vapor interface	m/s
$\mathbf{u}_{iv}$	velocity vector at free surface (vaporization interface)	m/s
u_{vt}	average velocity of vapor flow transversely across the cavity	m/s
$\xi' = \xi - d$	coordinate	–
w_l	average sideways width of molten pool	μm
z_i	ionized electrons per ion	–
α_F	coefficient for Fresnel absorption	/m
α_p	absorption coefficient of plasma	/m
β_v	fraction of material transported across cavity as vapor	–
∇_{lv}	surface gradient at the liquid–vapor interface	–
η_g	generator efficiency	%
η_p	process efficiency	%
η_{th}	transmission/heat transfer efficiency	%
$\epsilon = \left[\frac{2\epsilon_2}{\epsilon_1+\sqrt{\epsilon_1{}^2+\Sigma^2}}\right]^{1/2}$	dielectric constant	–

$\epsilon_{po} = 8.9 \times 10^{-12}$	permittivity of free space	F/M
ϵ_1	real part of dielectric constant of material	–
ϵ_2	real part of dielectric constant of plasma	–
$\Sigma = \frac{\sigma_{st}}{\omega \epsilon_0}$	–	–
σ_{st}	electrical conductance per unit depth	mho/m (Siemens/m)
γ_{lv}	surface tension at liquid-vapor interface	N/m
ρ_e	electron density	/mm^3
ρ_i	density of ions	/mm^3
ρ_v	density of the vapor at temperature T_v	kg/m^3
τ	stress tensor at liquid–vapor interface	Pa
θ	angle of incidence of the laser beam on the keyhole wall at any location, measured from the surface normal	°
Θ	temperature	K

PROBLEMS

16.1. Determine the welding speed above which heating in front of the molten pool is adiabatic, if the material being welded is a mild steel, and the keyhole diameter is 1 mm.

$$u_x > \frac{\kappa_s}{r_k}$$

16.2. An Nd:YAG laser with two output beams, each of power 3.5 kW is used to weld a pearlitic steel, with one beam being used for preheating to reduce the cooling rates. Determine the maximum welding speed, u_x, that can be used to avoid martensite formation, if the spacing between the two beams is $d = 5$ mm, and 90% of each beam is absorbed by the workpiece. Further assume that the density of molten steel is $\rho_l = 7407\,\text{kg/m}^3$, and that the critical cooling rate to avoid martensitic transformation in pearlitic steels is about 150°C/s at a temperature of 550°C.

16.3. For the conditions in Problem (16.2), what would be the maximum welding speed if a single beam of power 3.5 kW was used? Assume an ambient temperature of 0°C.

16.4. If the beam quality of the Nd:YAG laser used in Problem (16.3) is $M^2 = 20$, and the distance between the cavity mirrors is 400 mm, determine the radius, w_f, of the focused beam. Assume the focusing lens is positioned at a distance 2.5 m from the beam waist and that the lens focal length is 200 mm.

16.5. Using the welding speed, u_x, obtained from Problem (16.3), and assuming keyhole mode welding with the keyhole radius, r_k, being equal to the radius, w_f, of the focused beam, and further that the ratio of solid density to vapor density is approximately 10^4, estimate

(1) the weld pool length ahead of the keyhole, l_f,

(2) the average side width of the molten pool, w_l, and

(3) the length of the molten pool behind the cavity, l.

Comment on the results obtained.

16.6. A thin aluminum plate is welded using a laser beam with the following conditions:

Power input = 8 kW
Plate thickness = 3 mm
Welding speed = 75 mm/s
Initial plate temperature = 25°C
Heat transfer efficiency = 0.6.

Calculate the temperature on the top surface of the plate, 4.0 mm behind the laser beam and 1.5 mm to one side of it.

16.7. For the material and conditions in Problem 16.6, determine the cooling rate experienced at the projection of that point directly behind the heat source.

16.8. Given that

$$\beta_v = \frac{1}{\left[1 + \frac{l_f}{r_k}\left(\frac{2\rho_l}{\rho_v}\right)^{1/2}\right]}$$

Show that

(a) $\beta_v \propto u_{xo}$,

(b) $u_{vt} \propto u_{xo}^2$, and

(c) $p \propto u_{xo}^4$.

16.9. Show that equation (16.16) given by

$$(c_{vol} T_m + L_m) u_{xo} = k_l \frac{T_v - T_m}{l_f} \left(1 - \frac{u_{xo} l_f}{\kappa_l}\right)$$

reduces to the form:

$$l_f = \frac{\kappa_l}{u_{xo}} \frac{T_v - T_m}{T_v + L_m / c_{vol}}$$

Hint: Note that $k_l = c_{vol} \kappa_l$.

16.10. Show that the following equation

$$\Rightarrow w_l{}^2 = \frac{\kappa_l}{u_{xo}} r_k \left(\frac{T_v - T_m}{(T_v + T_m)/2 + L_m/c_{vol}} \right)$$

reduces to the form

$$\Rightarrow w_l \approx (l_f r_k)^{1/2}$$

Hint: Compare with equation (16.17) and ignore the small difference in the denominator.

16.11. Why is it more feasible to weld dissimilar materials using laser welding than with conventional arc welding?

16.12. Discuss the relevance of the derivation leading up to equation (16.48).

16.13. Show that the maximum value of the absorption coefficient occurs at high angles of incidence, almost glancing incidence.

17 Laser Surface Modification

Laser surface modification is used to change either the surface composition or the microstructure of a material to give it certain desired properties. This may involve hardening the surface to increase its resistance to wear, or inducing compressive residual stresses in the surface layers to enhance fatigue life.

There are five major forms of laser surface modification. These are the following:

1. Laser surface heat treatment.
2. Laser surface melting (skin melting or glazing).
3. Laser direct metal deposition (cladding, alloying, and hard coating).
4. Laser physical vapor deposition.
5. Laser shock peening.

In laser surface heat treatment, no melting takes place, while in laser surface melting, a thin surface layer of the workpiece is melted, and as a result of the rapid quenching, may form new structures that are harder. In laser deposition, a second material is applied to the surface and melted by the laser beam either to alloy with a thin surface layer of the base material, or to bond to the surface. However, in laser physical vapor deposition, the second material is applied in vapor form, and no melting takes place. Laser shock peening, however, generates shock waves, which induce compressive residual stresses on the surface of the material. These forms of surface modification are discussed in greater detail in the following sections according to the listed order.

17.1 LASER SURFACE HEAT TREATMENT

Surface heat treatment normally involves exposing the surface of a material to a thermal cycle of rapid heating and cooling such that the surface layers in the case of steels, for example, are first austenitized, and then quenched, to induce martensitic transformation. The process does not involve melting, and the transformations occur in the solid state.

The hardness of the martensite formed when steel is heat treated enhances the wear resistance of the surface. Examples of components that are hardened by laser

Principles of Laser Materials Processing, by Elijah Kannatey-Asibu, Jr.

heat treatment include laser-hardened wear tracks for power steering housing, cam shafts, automobile valve guides and seats, gear teeth, diesel cylinder liner bores, surface hardening of cylinder head units in aluminum car engines, hard-facing of car distribution shafts, and surface hardening of mill rollers.

The setup for laser surface heat treatment is illustrated schematically in Fig. 17.1. Rapid cooling results from mass or self-quenching into the bulk surrounding material, and may result in martensitic transformation even in some low-carbon steels. The cooling rates for laser heat treatment can be as high as 10^{7}°C/s (compare with cooling rates of about 300°C/s for arc welding) and produce relatively short austenitizing cycles (in the order of 0.01–1.0 s).

The essential steps involved in laser hardening normally comprise the following:

1. *Cleaning*—to remove surface contaminants such as oil.
2. *Masking*—to limit hardening to desired areas only. This step is not always necessary.
3. *Coating*—to increase beam absorption.
4. *Hardening*—the actual laser treatment.
5. *Tempering*—to reduce crack susceptibility of the fully martensitic case.
6. *Cleaning*—to remove any remaining coating.
7. *Inspection*—to ensure no defective parts are produced.

In this section, we first discuss criteria that are necessary for successful laser surface treatment. This is followed by a discussion on the key process parameters. A brief overview of the temperature field associated with laser surface treatment is then presented, followed by a discussion on microstructural changes in steels. Materials other than steels are then considered. Next, the hardness and residual stresses that

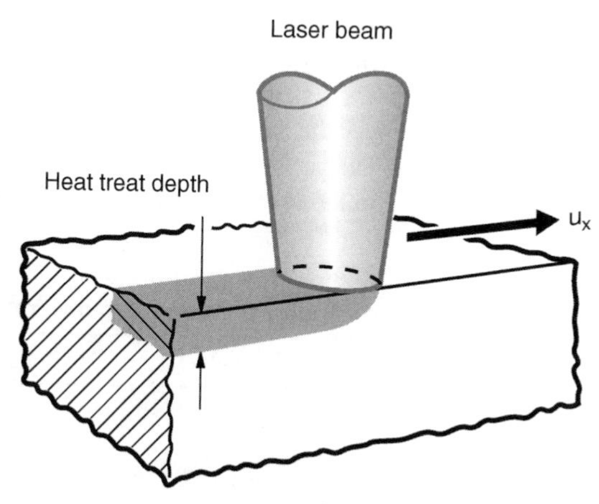

FIGURE 17.1 Setup for laser surface heat treatment. (From Mazumder, J., 1983, *Journal of Metals,* Vol. 35, pp. 18–26. By permission of TMS.)

result during laser surface treatment are discussed. Finally, the steps involved, and advantages and disadvantages of the process are presented.

17.1.1 Important Criteria

There are two criteria that are essential for a successful laser surface treatment operation:

1. For steels, the region to be hardened needs to be heated well into the austenite temperature range, and be maintained in that temperature range long enough for carbon diffusion to occur.
2. There must be adequate mass in contact with the region to be hardened to permit self-quenching by conduction into the bulk material.

17.1.2 Key Process Parameters

The principal processing parameters that affect laser surface heat treatment are the following:

1. Beam power, beam size, scanning speed, and shielding gas.
2. Beam mode.
3. Beam absorption by the workpiece.
4. Workpiece material properties and initial microstructure.

The process outputs include the hardness, depth of the hardened region, microstructure, and residual stresses induced.

These key process parameters are discussed in greater detail in the following sections.

17.1.2.1 Beam Power, Size, Speed, and Shielding Gas Typical process parameters associated with laser heat treatment are listed in Table 17.1.

A reduction in the scan rate reduces the surface hardness, but increases the depth of the hardened zone (Fig. 17.2). This is because reducing the scanning velocity increases the surface temperature and the reaction time for austenitization, resulting

TABLE 17.1 Typical Laser Heat Treatment Process Parameters

Parameters	Values	Units
Power	0.5–9	kW
Defocused beam diameter	2–20	mm
Power densities	1–100	kW/cm^2
Scanning velocities	5–400	mm/s
Hardened depths	0.2–2.5	mm

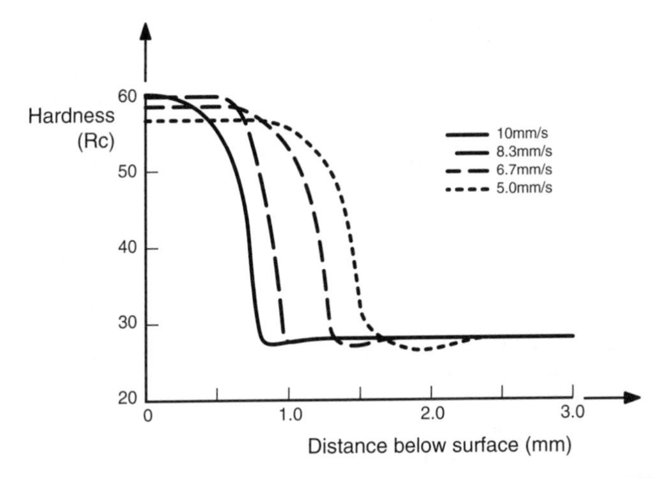

FIGURE 17.2 Hardness variation for different scanning velocities for AISI 4340 steel tempered at 649°C for 2 h and heat treated with a 1.8-kW CO_2 laser. (From Shiue, R. K., and Chen, C., 1992, *Metallurgical Transactions A,* Vol. 23A, pp. 163–169. By permission of Springer Science and Business Media.)

in a larger average austenite grain size. Shielding gas (usually argon or helium) may be used to protect the surface from oxidation with flow rates similar to those used during laser welding, about 10 L/min. Due to its higher thermal conductivity, helium results in higher hardness, all other parameters being constant.

Due to the relatively small size of the laser beam, it is often necessary to scan the surface a number of times, with the beam being shifted a specific amount for each scan. To ensure that the entire area of interest is treated, there has to be some amount of overlap between the scans. However, the overlap may anneal portions of the previously hardened structure. While such annealing may not significantly affect the wear properties of a material, it could have significant impact on the fatigue properties, with the softened regions providing preferential sites of stress relaxation at the surface.

17.1.2.2 ***Beam Mode*** In contrast to the case of laser cutting, low-order beam modes (e.g., Gaussian mode) are not particularly suitable for laser surface treatment because of the lack of uniformity in intensity distribution. A relatively wide and uniformly distributed beam is required to obtain uniform surface heating over a wide area and thus avoid localized surface melting. This is done using beam-shaping techniques.

There are four methods commonly available for shaping a laser beam into a more uniform distribution. These are the following:

1. Beam defocusing using a lens.
2. Optical integration.
3. Beam rastering or scanning.
4. Kaleidoscope or light pipe.

Beam Defocusing This approach is based on normal focusing using a lens. The beam shape is similar to that of the original beam, and the size is changed by varying the distance between the focusing lens and the workpiece or by using an appropriate lens. Thus, the focal point is positioned above the workpiece surface. A focusing system with a very high F number or long focal length may be more effective for this method. The F number is the ratio of the focal length of a lens to its diameter. Since beam defocusing does not necessarily produce a uniform beam distribution, the resulting penetration depth distribution across the heat-treated track is also nonuniform. Only about 20–30% of the track width may be of uniform penetration depth, depending on the original beam mode. Since the focusing lens is normally part of the beam delivery system, this method is the easiest to implement. It requires no additional effort or expense.

Optical Integration Optical integration involves segmenting the beam into a large number of portions and superposing the individual segments on the same focal plane using a beam integrator. A beam integrator has a number of small mirror segments mounted on a base plate, that is, a multifaceted mirror (Fig. 17.3a). Contiguous mirror segments are positioned to be in close contact with each other to minimize beam loss between the segments. The outcome of optical integration is illustrated in Fig. 17.3b where a Gaussian beam is segmented into four portions of equal size, and then superposed at the focal plane. Two of the segmented beams are of one orientation, while the other two are of opposite orientation, resulting in near-perfect intensity averaging. In normal cases where the beam is not necessarily symmetrical, as many mirror segments as possible are used.

One drawback of beam integrators is the interference effects that are induced in the integrated beam as a result of the mirror edges. Each edge results in diffraction effects and their superposition produces interference fringes. The higher the Fresnel number, the lower the diffraction losses, and thus the more uniform the integrated beam. The Fresnel number typically ranges between 6 and 12 (low) for infrared wavelength beams. A high Fresnel number would be about 100.

For a resonator consisting of mirrors of side $2a \times 2a$, with a spacing L between the mirrors (Fig. 17.4) the Fresnel number, N, is given by

$$N = \frac{a^2}{L\lambda} \tag{17.1}$$

The angular spread that results from diffraction of a plane electromagnetic wave of transverse dimensions $2a \times 2a$ is given by the semiangle:

$$\theta_{\mathrm{d}} \approx \frac{\lambda}{2a} \tag{17.1a}$$

However, the semiangle subtended by each of the mirrors at the center of the other is

$$\theta_{\mathrm{c}} \approx \frac{a}{L} \tag{17.1b}$$

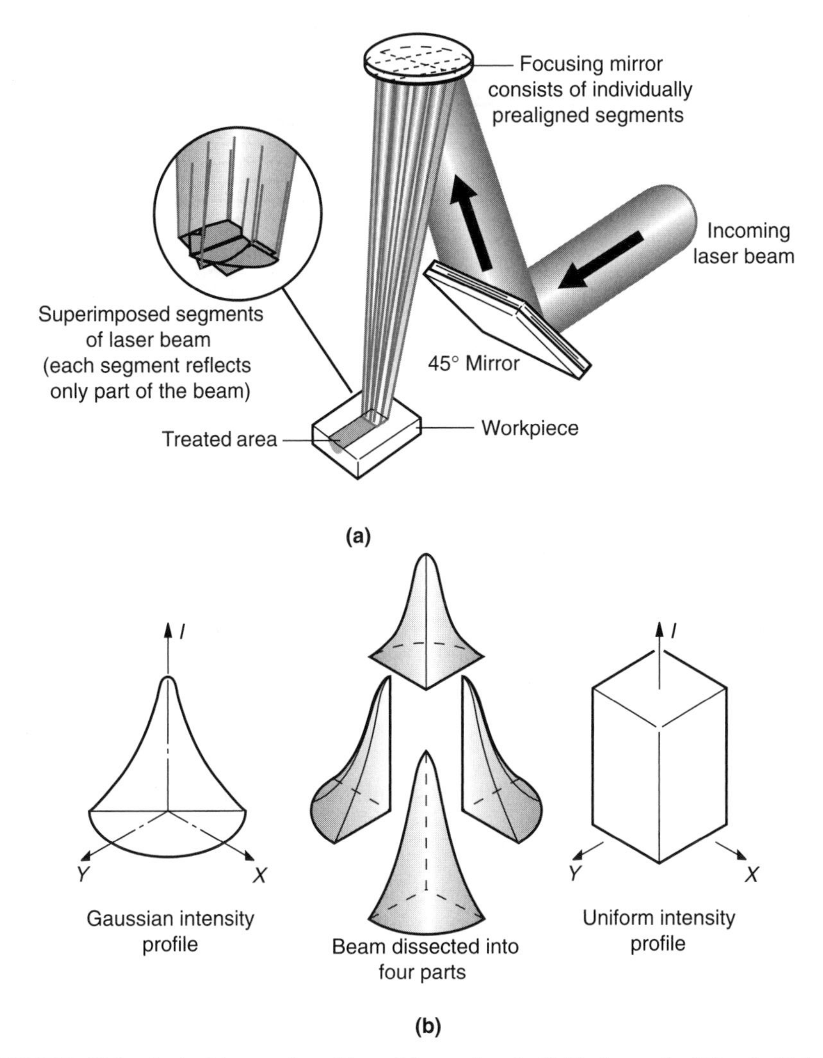

FIGURE 17.3 Optical integration. (From Mazumder, J., 1983, *Journal of Metals*, Vol. 35, pp. 18–26. By permission of TMS.)

Thus, the Fresnel number may be expressed as

$$N = \frac{\theta_c}{2\theta_d} \tag{17.2}$$

What this means is that, for the diffraction losses to be high, the Fresnel number has to be low.

Optical integration generally has the most uniform beam distribution, resulting in about 60–70% of the track width being of uniform penetration depth.

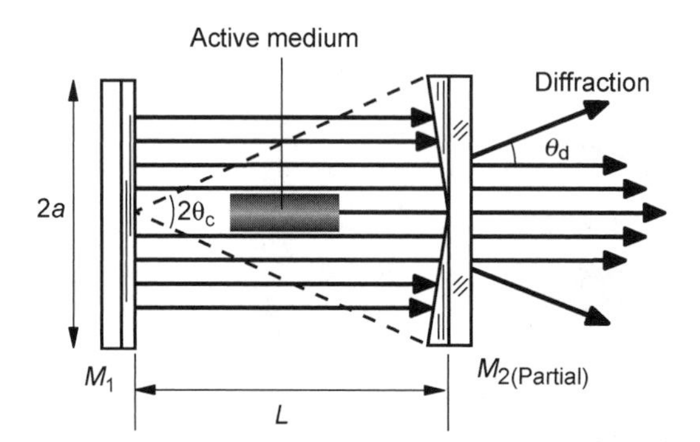

FIGURE 17.4 Schematic of the resonator geometry and diffraction of the output beam.

The optics for integrating an output laser beam generally consists of three main components:

1. A flat mirror for beam bending.
2. A spherical convex mirror for expanding the laser beam to fully illuminate the face of the beam integrator for efficient beam integration.
3. A spherical concave beam integrating mirror, which segments the beam into several small beams.

Installation of the optical integration system is easy, but requires more effort than beam defocusing.

Beam Rastering or Scanning A uniformly distributed beam can also be achieved by rastering a finely focused beam to cover a wide area (Fig. 17.5). The process involves vibrating two mirrors to get the beam to move back and forth at a high frequency to create the required pattern. Like the optical integration system, installation of the beam rastering system is also easy, but again requires more effort than beam defocusing.

Kaleidoscope or Light Pipe The kaleidoscope consists of four rectangular pieces of a highly reflective material, say brass, which are mounted to form a rectangular opening or pipe shape (Fig. 17.6). Typical dimensions of the pipe may be 30 mm × 55 mm × 270 mm. The focused beam is introduced at the top opening of the pipe, and through reflections off the walls, exits as an integrated beam. The uniformity achieved with the kaleidoscope may be about 45–50% of the track width. The kaleidoscope is also relatively easy to install.

Summary of Beam-Shaping Methods The characteristics of the four types of beam shaping devices are summarized in Table 17.2.

Figure 17.7a–c illustrate three common forms of beam shape that are used. The circular or elliptical pattern Gaussian beam can be broadened by defocusing; the

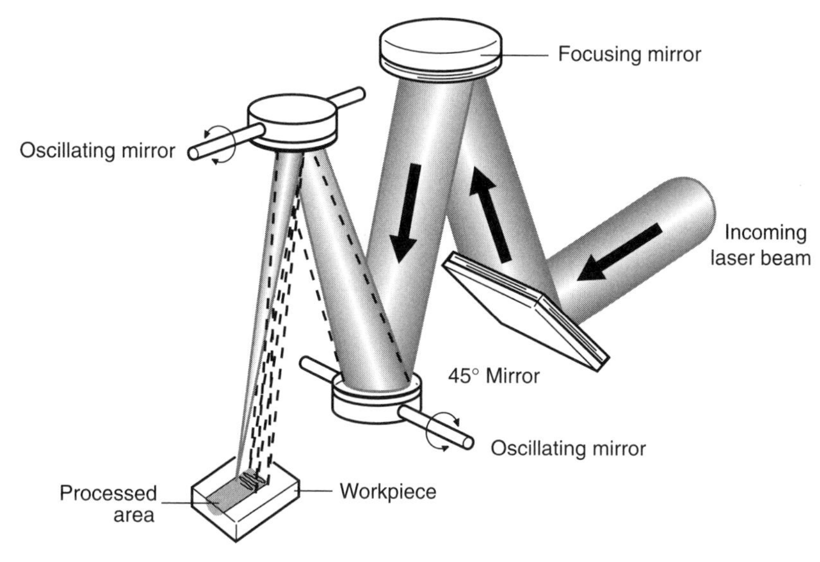

FIGURE 17.5 Beam rastering. (From Mazumder, J., 1983, *Journal of Metals,* Vol. 35, pp. 18–26. By permission of TMS.)

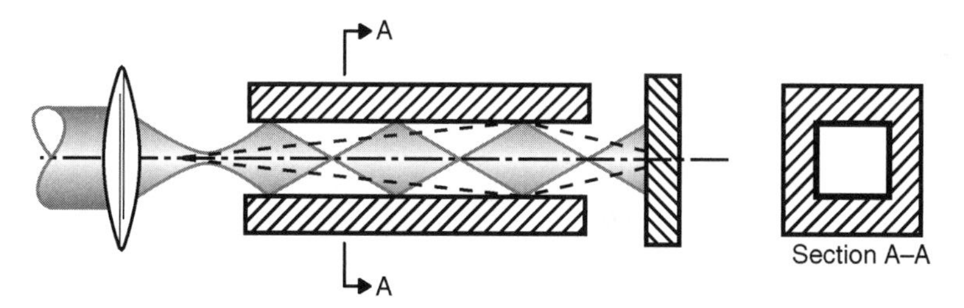

FIGURE 17.6 The kaleidoscope.

TABLE 17.2 Comparison of the Four Beam-Shaping Devices

	Focusing Lens	Integration	Kaleidoscope	Beam Rastering
Power Loss	Lowest (1–2%)	Moderate (7–8%)	High (18–20%)	Low (4–6%)
Energy Uniformity	Good	Better	Better	Variable
Cost	Moderate	High	Low	High
Penetration	High	Moderate	Moderate	Variable
Penetration Uniformity	Uneven (20–30%)	Even (60–70%)	Even (45–50%)	Even
Limitations	Fragility of lens	Device length, Fragility	Needs to be close to workpiece	Scan rate might limit travel speed
Installation	Very easy	Easy	Easy	Easy
Applications	All applications	Heat treating, cladding, alloying	Heat treating only	Heat treating cladding, alloying

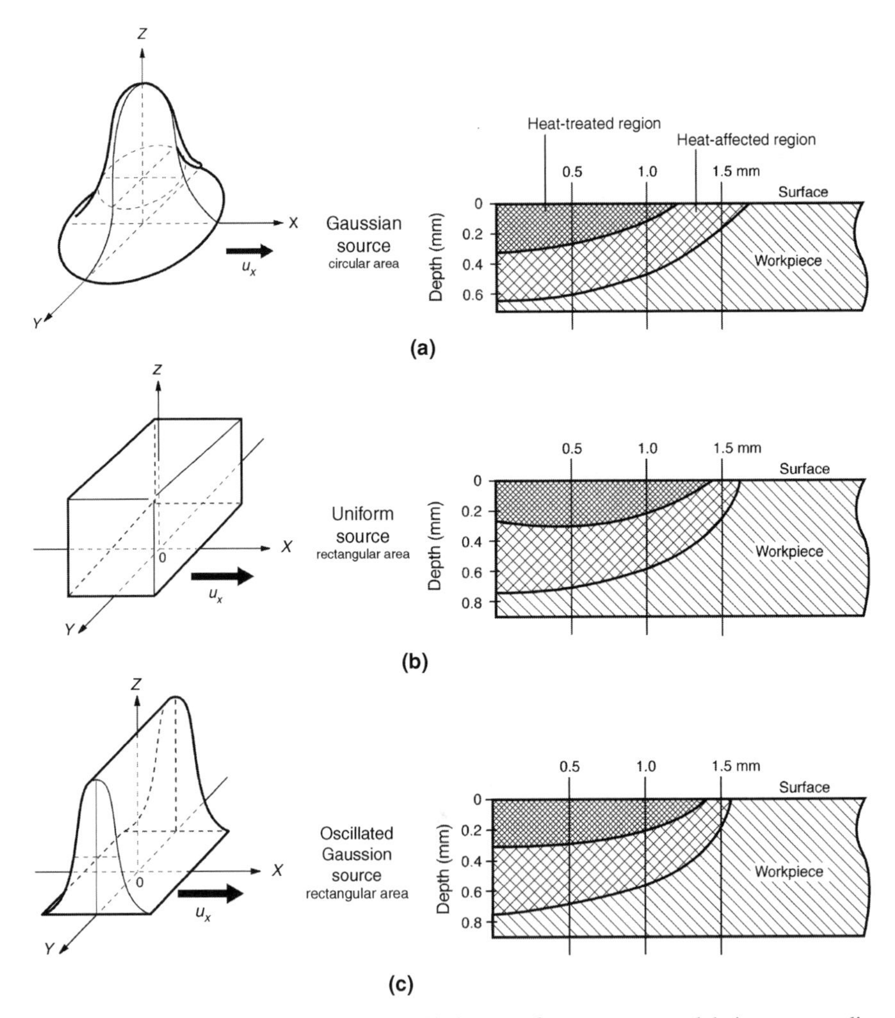

FIGURE 17.7 Common beam shapes used in laser surface treatment and their corresponding hardened layer geometry. (a) Circular or elliptical pattern Gaussian beam. (b) Rectangular beam. (c) Rectangular pattern Gaussian beam. (From Seaman, F. D., 1986, *Industrial Laser Handbook,* pp. 147–157.)

uniform pattern rectangular beam can be obtained by optical integration; and the rectangular pattern Gaussian beam is achieved by rastering or oscillating a Gaussian beam at a frequency of several hundred Hertz. The corresponding geometry of the hardened and heat-affected zones (HAZ) are also shown in the figures. It is obvious from the figures that the softened heat affected zone at the part surface is significantly wider for the circular Gaussian beam. The rectangular patterns reduce the width of this zone. However, the depth of the softened region is smaller for the circular Gaussian beam.

17.1.2.3 ***Beam Absorption*** The heat treatment process is strongly influenced by

1. Absorption of the beam, which depends on the beam wavelength.
2. Angle of incidence.
3. Material properties or core microstructure.
4. Workpiece surface conditions (finish, coating, and so on).

The high reflectivity of the laser beam when used on metals is a major disadvantage of laser surface treatment. This is especially true of the CO_2 laser beam when incident on metals such as copper, gold, and silver. Some of the common techniques that are used to improve absorptivity of reflective materials for effective processing include the following

1. Applying a thin layer of a highly absorbing coating to the material surface.
2. Using a linearly polarized laser beam on an uncoated surface.
3. Preheating the workpiece, since absorptivity increases with temperature.
4. Changing the composition of the material surface by oxidation before processing. However, this approach tends to be more expensive and may not be acceptable for aesthetic reasons. Thus, it will not be further discussed.

In the following sections, the first three methods of enhancing absorption are further discussed.

Surface Coating When a surface coating is used, it must have the following characteristics:

1. High-thermal stability.
2. Good adhesion to work surface.
3. Chemically passive to workpiece.
4. Easy to remove after treatment.

Examples of coatings normally used are the following:

1. A dispersion of carbon black in alcohol with a binder (colloidal graphite).
2. Chemically deposited copper selenide layer (about 2 μm thick).
3. Manganese, zinc, or iron phosphate (about 2–100 μm thick).
4. Black paint (about 10–20 μm thick).

For these, the fraction of the incident energy that is absorbed by the surface when coated is estimated to be in the range of 60–80% for a CO_2 laser, and this can vary during interaction of the laser beam with the absorptive coating. Graphite is easily applied and is inert, but is relatively more expensive. An excessive coating tends to

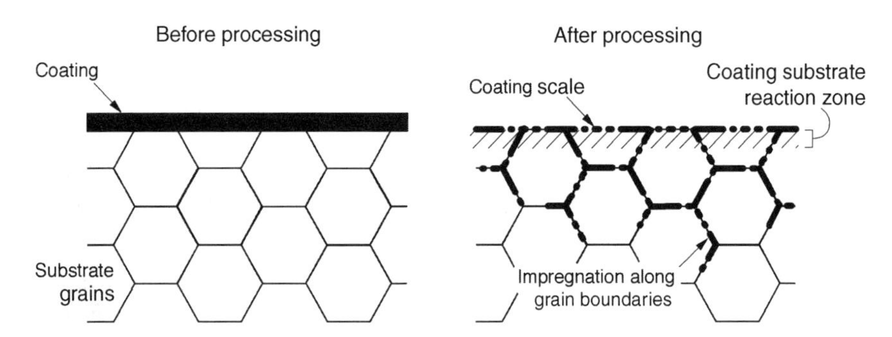

FIGURE 17.8 Grain boundary impregnation by coatings. (From Gnanamuthu, D. S., and Shankar, V. S., 1985, *SPIE, Applications of High Powered Lasers*, Vol. 572, pp. 56–72.)

reduce penetration, as some of the energy absorbed may not be easily transmitted to the base material.

In addition to increasing beam absorption, a carefully applied coating ensures reproducibility of the process. Essentially, all the power that enters the material is absorbed at the surface, and the depth and width of the transformed region depends on the resulting temperature distribution.

Certain coatings, especially the phosphates, have the disadvantage that during processing, they may impregnate along grain boundaries of the base material (Fig. 17.8), to form low-melting compounds where failure can be initiated. Surface coatings also have the disadvantage that they add to the cost of the process, and may prove to be an environmental hazard (especially paints) when they are vaporized during heating. Also, the absorptivity is influenced by variations in thickness or adhesion properties of a surface coating.

Linearly Polarized Beam As discussed in chapter 9, when an unpolarized beam is incident on a surface at the polarizing angle, the reflected beam is linearly or plane polarized. This is because the component of the electric vector parallel to the plane of incidence vanishes, while the perpendicular component is reflected (Fig. 9.3).

When the plane polarized beam is incident on a surface, depending on its orientation, it can have components parallel (p-polarized) and perpendicular (s-polarized) to the plane of incidence. The absorption coefficients for these two components are shown in Fig. 9.3c, which shows that the parallel component is efficiently absorbed at high angles of incidence while the perpendicular component is highly reflected. Thus, by using an angle of incidence greater than, say 60°, for a linearly polarized beam, the parallel component (p-polarized beam) will be efficiently absorbed by the surface. The depth of hardening will therefore be greater if a p-polarized beam is used at an incidence angle that corresponds with the angle of maximum absorptivity (Fig. 17.9). The drawback of this approach, however, is the significant power loss incurred in the conversion of an unpolarized laser beam to a linearly polarized beam.

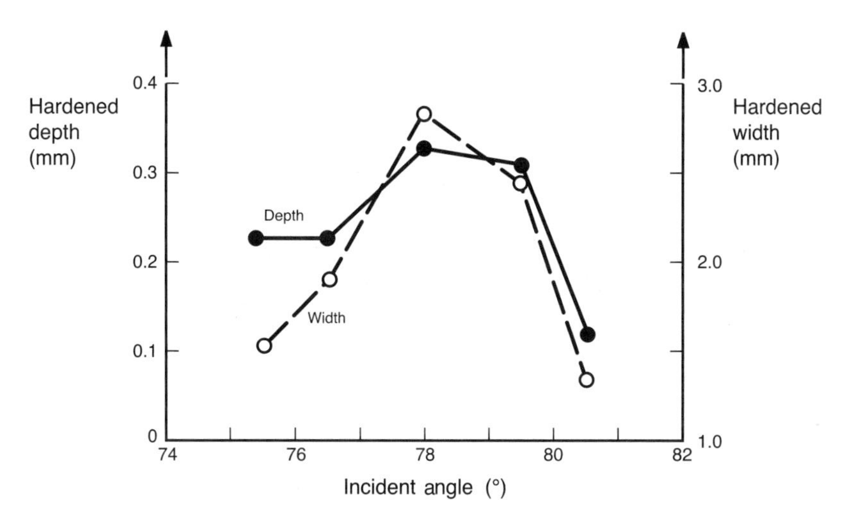

FIGURE 17.9 Variation of depth of hardening with angle of incidence of a p-polarized beam. (From *ICALEO 1991 Proceedings*, Vol. 74, Copyright 1992, Laser Institute of America. All rights reserved.)

Preheating As indicated in Section 14.1.6, the absorptivity of a material increases with an increase in its temperature, for a given wavelength (Fig. 14.9). Thus, one effective way for enhancing absorption of the incident laser is to preheat the workpiece before the heat treatment process. The drawback with this approach, however, is that preheating introduces an additional step into the process. Furthermore, the preheating may affect the microstructure of the surrounding material. Preheating is further discussed in Section 11.2.2.3.

17.1.2.4 Initial Workpiece Microstructure The initial microstructure (e.g., size of ferrite grains and pearlite colonies) of the workpiece, prior to laser heat treatment, influences the depth of hardening that can be achieved. This is because the distance that the carbon atoms have to diffuse affects the homogenization of austenite, and thus the rate of structural transformation. When the initial or core microstructure is such that the carbon diffusion distance is very small (fine microstructure with a smaller grain size), deep surface-hardened casings are obtained. Thus, finer and more evenly distributed carbide dispersoids in a quenched and tempered structure (martensitic or bainitic microstructure prior to laser heat treatment) readily dissolve at lower temperatures within the available exposure time, and thus produce a better hardening effect. On the contrary, when the initial microstructure is coarse (e.g., spheroidized), then the carbon diffusion distance needed for the homogenization of austentite is long. This results in less uniform carbon distribution in the austentite, and thus a thin surface-hardened casing (Fig. 17.10). The transition from the full martensite hardness to the core hardness is also found to be more gradual in a coarse structure than in a fine structure.

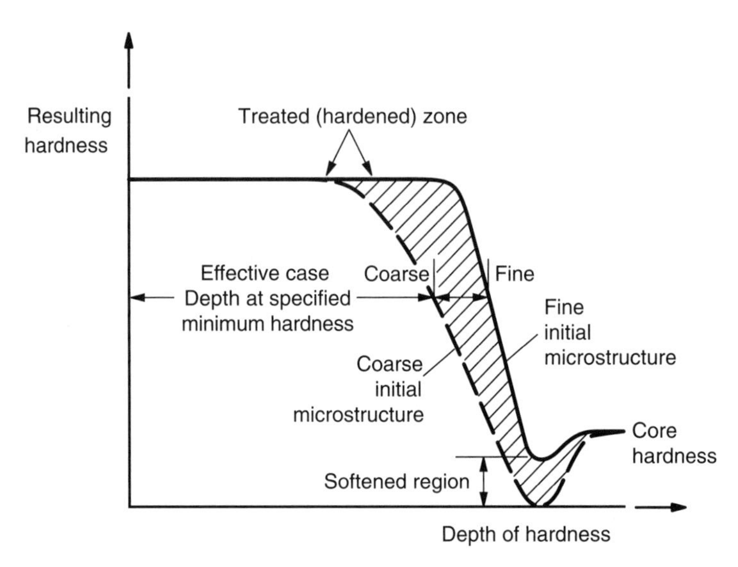

FIGURE 17.10 Effect of initial microstructure on laser surface treatment.

However, for steels with the same carbon content, both fine- and coarse-structured samples will have the same surface hardness since the extremely high surface temperature results in homogeneous austenitization in the surface region for both structures. Furthermore, since the thermal properties of steels do not change significantly with carbon and alloy content, the same cooling rates will be obtained for the same heat-treating conditions. Alloy content may increase the hardenability of the steel, enabling it to form more martensite for a given cooling rate. Hardenability is a measure of the amount of martensite that forms in the microstructure upon cooling. This is affected by both the alloying elements and carbon content. However, since the hardness of a steel is determined primarily by the carbon content, an alloy and a plain carbon steel of the same carbon content will have the same hardness of martensite that is formed, but not necessarily the same overall hardness.

17.1.3 Temperature Field

Analysis of heat flow during laser surface treatment can be made using methods that are discussed in Section 10.2. Thus only a brief reference will be made in this section. Since absorption of the laser beam in a solid metal is confined to a thin surface layer, the z-dependence of the absorption can be neglected, and the absorption considered to be two dimensional. Considering a Gaussian distributed heat source moving in a semi-infinite plate in the x-direction, the temperature distribution in the y–z plane (see Fig. 10.1) can be expressed as

$$T - T_0 = \frac{(1-R)q}{2\pi u_x k[t(t+t_0)]^{1/2}} \times \exp\left\{-\left[\frac{(z+z_0)^2}{4\kappa t} + \frac{y^2}{4\kappa(t+t_0)}\right]\right\} \tag{17.3}$$

where t is the interaction time, $t_0 = w^2/4\kappa$ is the time required for heat to diffuse a distance equal to the beam radius, and z_0 is the distance over which heat can diffuse during the beam interaction time, $\frac{w}{u_x}$. It is a characteristic length that limits the surface temperature to a finite value, and is obtained by matching solutions at $z = z_0$ to known solutions for the peak surface temperature.

17.1.4 Microstructural Changes in Steels

Steels are extensively discussed in this section because they are still among the most widely used metal alloys today. Some metals undergo polymorphic transformation (i.e., change their crystal structure) when heated above a certain temperature. The change in crystal structure also changes their properties. An example is iron which changes from α-iron with the BCC structure at room temperature to γ-iron with the FCC structure when heated above 910°C (see Appendix 17B). The solubility of carbon in γ-iron is greater than in α-iron, and this makes steel subject to significant changes in property when heat treated.

The overall transformation process during heat treatment of steels occurs in three stages:

1. Diffusion controlled transformation of pearlite to austenite (pearlite dissolution).
2. Diffusion controlled homogenization of carbon in austenite.
3. Diffusionless or displacive transformation of austenite to martensite.

These three stages are discussed in further detail in the following sections.

17.1.4.1 Pearlite Dissolution Pearlite dissolution involves the transformation of the pearlite colonies to austenite as the material is heated. For plain carbon steels, when the material is heated above the eutectoid (A_1) temperature, the pearlite colonies transform to austenite. For pre-eutectoid steels, as the temperature increases, the size of the austenite regions increase by outward diffusion of carbon from the high-carbon (0.8% C) pearlite colonies into the ferrite region. Let the spacing of the pearlite plates in a colony be l_p, (Fig. 17.11). Then lateral diffusion of carbon over a distance l_p might be considered sufficient to convert the colony to austenite. For an isothermal process, the time, t, required for such diffusion is given by the equation:

$$l_p{}^2 = 2D_a t \tag{17.4}$$

where D_a is the diffusion coefficient for carbon.

However, in the case of heat treatment where the temperature is time dependent, the term $D_a t$ is replaced by

$$D_a t = \int_0^\infty D_0 \exp\left(-\frac{Q_a}{R_g T(t)}\right) dt \approx D_0 c_d \tau_h \exp\left(-\frac{Q_a}{R_g T_p}\right) \tag{17.5}$$

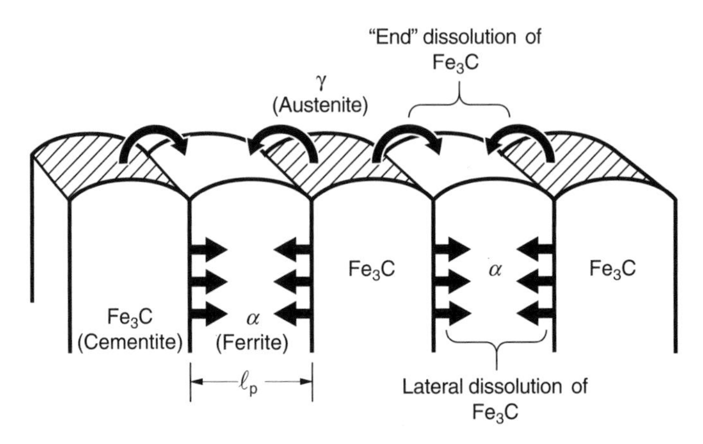

FIGURE 17.11 Transformation of pearlite to austenite. (From Ashby, M. F., and Easterling, K. E., 1984, *Acta Metallurgica*, Vol. 32, No. 11, pp. 1935–1948.)

where c_{d} is a constant, D_0 is a diffusion constant, Q_{a} is activation energy for the transformation, R_{g} is the gas constant, T_{p} is the peak temperature experienced during heating, and τ_{h} is the thermal time constant.

Thus,

$$l_{\mathrm{p}}^2 = 2D_0 c_{\mathrm{d}} \tau_{\mathrm{h}} \exp\left(-\frac{Q_{\mathrm{a}}}{R_{\mathrm{g}} T_{\mathrm{p}}}\right) \tag{17.6}$$

For a given pearlite plate spacing, l_{p}, equation (17.6) enables the peak temperature that is necessary for the transformation to take place to be estimated. The constant c_{d} is approximated by

$$c_{\mathrm{d}} = 3 \times \sqrt{\frac{R_{\mathrm{g}} T_{\mathrm{p}}}{Q_{\mathrm{a}}}} \tag{17.7}$$

To determine the processing conditions that will give the peak temperature, T_{p}, and thermal time constant τ_{h}, the following equations may be used:

$$T_{p1} = T_0 + \left(\frac{2}{e}\right)^{1/2} \frac{(1-R)q}{\pi \rho c_p w u_x (z + z_{01})} \qquad t << t_0 \tag{17.8}$$

$$\tau_{h1} = \frac{1}{\pi^2 k e \rho c_p} \left[\frac{(1-R)q}{u_x w (T_{\mathrm{p}} - T_0)}\right]^2 \qquad t << t_0 \tag{17.9}$$

$$T_{p2} = T_0 + \frac{2(1-R)q}{\pi e \rho c_p u_x (z + z_{02})^2} \qquad t >> t_0 \tag{17.10}$$

$$\tau_{h2} = \frac{(1-R)q}{2\pi k e u_x (T_p - T_0)} \qquad t >> t_0 \tag{17.11}$$

where e is the natural exponent, $z_{01}{}^2 = \frac{\pi \kappa w}{2eu_x}$, and $z_{02}{}^2 = \frac{w}{e}\left(\frac{\pi \kappa w}{u_x}\right)^{1/2}$.

Example 17.1 The surface of an AISI 1045 steel cam is to be heat treated using a 2.5 kW CO_2 laser to enhance the wear resistance. The defocused beam diameter is 5 mm, and the scan rate is set at 10 mm/s. Determine the spacing of the pearlite plates that will enable the pearlite colonies to be fully transformed to austenite under these conditions.

Assume

1. The temperature is time dependent.
2. The diffusion constant, D_0, for the diffusion of carbon in α-Fe (ferrite) is 0.020 $cm^2/s = 2 \times 10^{-6}$ $m^2/s = 2$ $mm^2/s = 2 \times 10^6$ $\mu m^2/s$ (Shewmon, 1983, Table 2-1, p. 64).
3. The activation energy for diffusion, Q_a, is 20.1 kcal/mol = 84.15 kJ/mol (Shewmon, 1983, Table 2-1, p. 64).
4. The reflection coefficient of the surface is 80%.
5. Ambient temperature is 25°C.

Solution:

The peak temperature and thermal time constant that will be obtained for the given conditions need to be determined. However, without additional information, we assume that t is indefinite, since we need to use equation (17.5). Therefore to calculate T_p and τ_h, we use equations (17.10) and (17.11) on the basis that $t >> t_0$:

$$T_{p2} = T_0 + \frac{2(1-R)q}{\pi e \rho c_p u_x (z + z_{02})^2} \qquad t >> t_0 \tag{17.10}$$

$$\tau_{h2} = \frac{(1-R)q}{2\pi k e u_x (T_p - T_0)} \qquad t >> t_0 \tag{17.11}$$

From Appendixes 10D and 10E, the properties of steel are approximated as

Average density, $\rho = 7870$ kg/m^3
Average specific heat, $c_p = 452$ J/kg K
Thermal conductivity, $k = 0.073$ W/mm K.

The thermal diffusivity, κ, of the material is

$$\kappa = \frac{k}{\rho c_p} = \frac{0.073}{7.87 \times 10^{-6} \times 452} = 20.52 \text{ mm}^2/\text{s}$$

also,

$$z_{02}{}^2 = \frac{w}{e}\left(\frac{\pi \kappa w}{u_x}\right)^{1/2} = \frac{2.5}{2.71828}\left(\frac{\pi \times 20.52 \times 2.5}{10}\right)^{1/2} = 3.69 \text{ mm}^2$$

$$\Rightarrow z_{02} = 1.92 \text{ mm}$$

Since the peak temperature will occur on the surface of the workpiece, $z = 0$. Thus, we have

$$T_{p2} = T_0 + \frac{2(1-R)q}{\pi e \rho c_p u_x (z + z_{02})^2}$$

$$= 25 + \frac{2 \times (1 - 0.8) \times 2500}{\pi \times 2.71828 \times 7.87 \times 10^{-6} \times 452 \times 10 \times (0 + 1.92)^2} = 918^\circ\text{C}$$

Therefore,

$$\tau_{h2} = \frac{(1-R)q}{2\pi k e u_x (T_p - T_0)}$$

$$= \frac{(1 - 0.8) \times 2500}{2 \times \pi \times 0.073 \times 2.71828 \times 10 \times (918 - 25)} = 0.045 \text{ s} = 45 \text{ ms}$$

The gas constant $R_g = 1.98$ cal/mol K. Thus,

$$c_d = 3 \times \sqrt{\frac{R_g T_p}{Q_a}} = 3 \times \sqrt{\frac{1.98 \times (918 + 273)}{20.1 \times 10^3}} = 1.027$$

Now the pearlite plate spacing is obtained from equation (17.6) as

$$l_p{}^2 = 2 D_0 c_d \tau_h \exp\left(-\frac{Q_a}{R_g T_p}\right)$$

$$\Rightarrow l_p{}^2 = 2 \times 2 \times 10^{-6} \times 1.027 \times 0.045 \times \exp\left(-\frac{20.1 \times 10^3}{1.98 \times (918 + 273)}\right)$$

$$= 0.367 \times 10^{-10} \text{ m}^2$$

or

$$l_p = 0.61 \times 10^{-5} \text{ m} = 6.1 \text{ μm}$$

17.1.4.2 Austenite Homogenization Austenite homogenization follows the transformation from pearlite to austenite, and is associated with the diffusion of carbon from the high-carbon austenite region (which transformed from pearlite) to the low-carbon austenite region (which transformed from ferrite; Fig. 17.12). The extent of

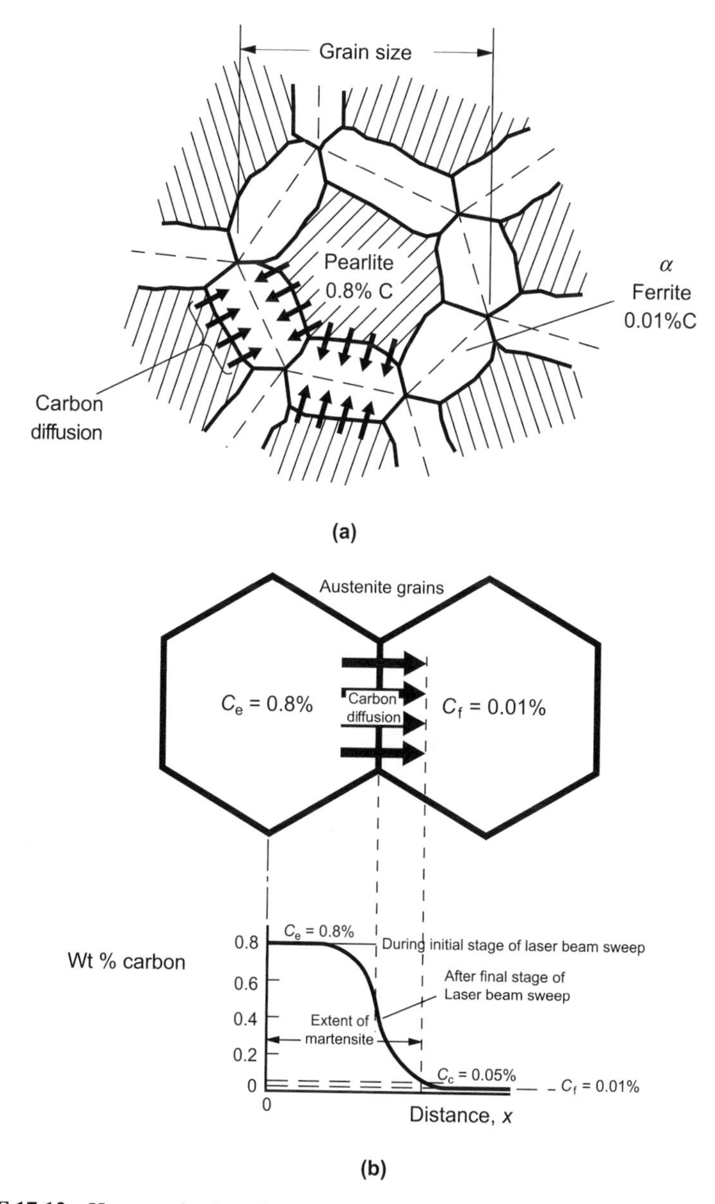

FIGURE 17.12 Homogenization of austenite. (a) Diffusion of carbon from pearlite to ferrite. (b) Variation of carbon content across the pearlite–ferrite interface. (From Ashby, M. F., and Easterling, K. E., 1984, *Acta Metallurgica*, Vol. 32, No. 11, pp. 1935–1948.)

homogenization depends on temperature and time. The variation of the carbon content with time as homogenization proceeds is obtained from the diffusion equation (see equation 12B.5) as

$$C_a(x, t) = 1/2(C_e + C_f) - 1/2(C_e - C_f)erf\left[\frac{x}{2\sqrt{D_a t}}\right] \tag{17.12}$$

where C_a is the carbon content in wt%, $C_e = 0.8\%$ is the pearlite carbon content, C_f is the carbon content of the ferrite, and x is the distance over which a carbon content of C_a is achieved in time t.

Now if C_c (see Fig. 17.12b) is the critical carbon fraction above which transformation to martensite occurs, then for $C_f << C_e$, the boundary of the region in which $C_a \geq C_c$ is approximated by

$$x = \frac{2}{\sqrt{\pi}} \ln\left(\frac{C_e}{2C_c}\right)\sqrt{D_a t} \tag{17.13}$$

Equation (17.13) defines the size of the zone over which martensite can be formed after the specimen is heated for a given time, t.

Example 17.2 Estimate the average size of the region where martensite can be formed, for the conditions in Example 17.1.

Solution:

Using equation (17.13) we have

$$x = \frac{2}{\sqrt{\pi}} \ln\left(\frac{C_e}{2C_c}\right)\sqrt{D_a t} \tag{17.13}$$

C_c is obtained from Fig. 17.12 as $C_c = 0.05\%$
thus,

$$x = \frac{2}{\sqrt{\pi}} \ln\left(\frac{C_e}{2C_c}\right)\sqrt{D_O c_d \tau_h \exp\left(-\frac{Q_a}{R_g T_p}\right)}$$

$$= \frac{2}{\sqrt{\pi}} \ln\left(\frac{0.8}{2 \times 0.05}\right)\sqrt{2 \times 10^{-6} \times 1.027 \times 0.045 \times \exp\left(-\frac{20.1 \times 10^3}{1.98 \times (918 + 273)}\right)}$$

$$= 1.01 \times 10^{-5}\ m = 10.1\ \mu\text{m}$$

17.1.4.3 Transformation to Martensite Martensite is normally formed when austenite is cooled at a high enough rate. It is extremely hard and brittle. The transformation from austenite to martensite is diffusionless, and takes place almost instantaneously.

The volume fraction f of martensite that is formed over a period t is given by

$$f = f_m - (f_m - f_i)\exp\left[-\frac{12 f_i^{2/3}}{g_s\sqrt{\pi}}\ln\left(\frac{C_e}{2C_c}\right)\sqrt{D_a t}\right] \tag{17.14}$$

where $f_i = \frac{C_a - C_f}{0.8 - C_f} \approx \frac{C_a}{0.8}$ is the volume fraction initially occupied by the pearlite colonies, which is also the minimum subsequent volume fraction of martensite

$$f_m = \begin{cases} 0 & \text{if} \quad T_p < A_1 \\ f_i + (1 - f_i)\frac{T_p - A_1}{A_3 - A_1} & \text{if} \quad A_1 < T_p < A_3 \\ 1 & \text{if} \quad T_p > A_3 \end{cases}$$

g_s is the average austenite grain size.

f_m is the maximum volume fraction permitted by the phase diagram; A_1 is the eutectoid temperature (Appendix 17B); and $A_3 = 1183 - 416C_a + 228C_a^2$ K is the lower temperature boundary of the austenite zone in the iron–carbon diagram. For the situation where the temperature is time dependent, the term $D_a t$ in equations (17.13) and (17.14) is replaced by $D_0 c_d \tau_h \exp\left(-\frac{Q_a}{R_g T_p}\right)$.

Example 17.3 Determine the volume fraction of martensite that can be formed, for the conditions in Example 17.1. Assume that the average austenite grain size is 200 μm.

Solution:

The volume fraction f can be obtained from equation (17.14) as

$$f = f_m - (f_m - f_i)\exp\left[-\frac{12 f_i^{2/3}}{g_s\sqrt{\pi}}\ln\left(\frac{C_e}{2C_c}\right)\sqrt{D_a t}\right] \tag{17.14}$$

Now

$$f_m = \begin{cases} 0 & \text{if} \quad T_p < A_1 \\ f_i + (1 - f_i)\frac{T_p - A_1}{A_3 - A_1} & \text{if} \quad A_1 < T_p < A_3 \\ 1 & \text{if} \quad T_p > A_3 \end{cases}$$

But
$f_i = \frac{C_a - C_f}{0.8 - C_f} \approx \frac{C_a}{0.8} = \frac{0.45}{0.8} = 0.5625$ for 1045 steel since C_a is then 0.45.
Also, $A_1 = 723°C$, and the A_3 temperature is estimated to be $A_3 = 769°C$.

Thus, since $T_p = 918°C$ from Example 17.1, $f_m = 1$. Therefore, we have

$$f = f_m - (f_m - f_i)\exp\left[-\frac{12 f_i^{2/3}}{g_s\sqrt{\pi}} \ln\left(\frac{C_e}{2C_c}\right)\sqrt{D_o c_d \tau_h \exp\left(-\frac{Q_a}{R_g T_p}\right)}\right]$$

$$= 1 - (1 - 0.5625)\exp\left[-\frac{12 \times 0.5625^{2/3}}{200 \times 10^{-6}\sqrt{\pi}} \ln\left(\frac{0.8}{2 \times 0.05}\right)\right.$$

$$\left.\times\sqrt{2 \times 10^{-6} \times 1.027 \times 0.045 \times \exp\left(-\frac{20.1 \times 10^3}{1.98 \times (918 + 273)}\right)}\right]$$

Thus, the volume fraction, f, of martensite formed is

$$f = 64.4\%$$

17.1.5 Nonferrous Alloys

Even though our discussion thus far has focused on phase transformation in steels, other materials can also be subjected to laser heat treatment to improve their mechanical properties. The hardness of some aluminum alloys, for example, can be increased by about 100% after laser treatment. The principal process for increasing the strength of nonferrous alloys is precipitation hardening. This requires the existence of a strong supersaturated solid solution at high cooling rates. Thus, it is only possible in alloys in which the solubility of one element in the solid solution increases with increasing temperature (Fig. 17.13). There are two steps involved in the precipitation hardening process:

1. Solution treatment.
2. Aging.

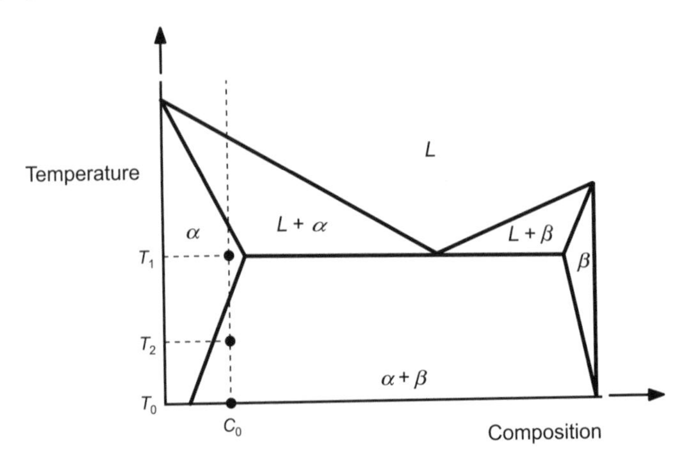

FIGURE 17.13 Solution treatment.

17.1.5.1 Solution Treatment The material is heated into the portion of the phase diagram where only a solid solution phase exists, say T_1 (Fig. 17.13) and allowed to stay until the solid solution phase, α, is formed. It is then quenched to T_2. The rapid cooling prevents the formation of the second phase, β, and the alloy formed at this stage is supersaturated, and initially relatively soft.

17.1.5.2 Aging If the alloy is maintained at T_2 for a while, nuclei of the second phase are formed and begin to grow throughout the material, resulting in second-phase particles. The formation of the second-phase particles impede the motion of dislocations. That means more stress is required to move the dislocations through them, and the strength of the material thus increases (Fig. 17.14a). The more particles are formed and grow, the greater the strength becomes (stage 1). Thus a finer distribution of second phase particles produces a stronger material. However, eventually the strength reaches a maximum (stage 2) and a further stay at T_2 only results in continued growth of the particles as smaller particles join together to form larger ones, and the spacing between them increases. The strength then begins to decrease (stage 3). This is overaging.

The lower the aging temperature, T_3, the longer it takes to reach the maximum strength, but the greater is the maximum strength since the second phase is more finely dispersed at the lower temperatures due to the increased nucleation rate (see Fig. 11.4; Fig. 17.14b and c). If T_3 is low enough, the second phase may not form at all.

Nonferrous alloys may also be hardened by laser treatment if at least one of the following conditions is satisfied:

1. Formation of a characteristic finely divided structure in the hardening zone.
2. Formation of metastable phases.
3. Dislocation density increase.

17.1.6 Hardness Variation

The variation of hardness and ductility for a surface-hardened workpiece is illustrated in Fig. 17.15. The top view (Fig. 17.15a) illustrates hardened regions where the hardness is higher and the ductility lower than the original material. As the cross-sectional profile of the surface-hardened region shows (Fig. 17.15b) hardness is a maximum at the surface and gradually decreases with depth from the surface. This is the trend observed for macrohardness measurements, which may be obtained using a Vickers indenter. However, microhardness measurement of individual martensitic regions for preeutectoid steels shows the martensite hardness to increase with distance below the surface. This is because the carbon content of the individual martensite formed is highest when the peak temperature experienced is just above the A_1 (eutectoid) temperature of the iron–carbon phase diagram (Appendix 17B). The A_1 temperature is achieved at a distance farther away from the heat source. At this temperature, the carbon content of the austenite that eventually transforms into martensite is about

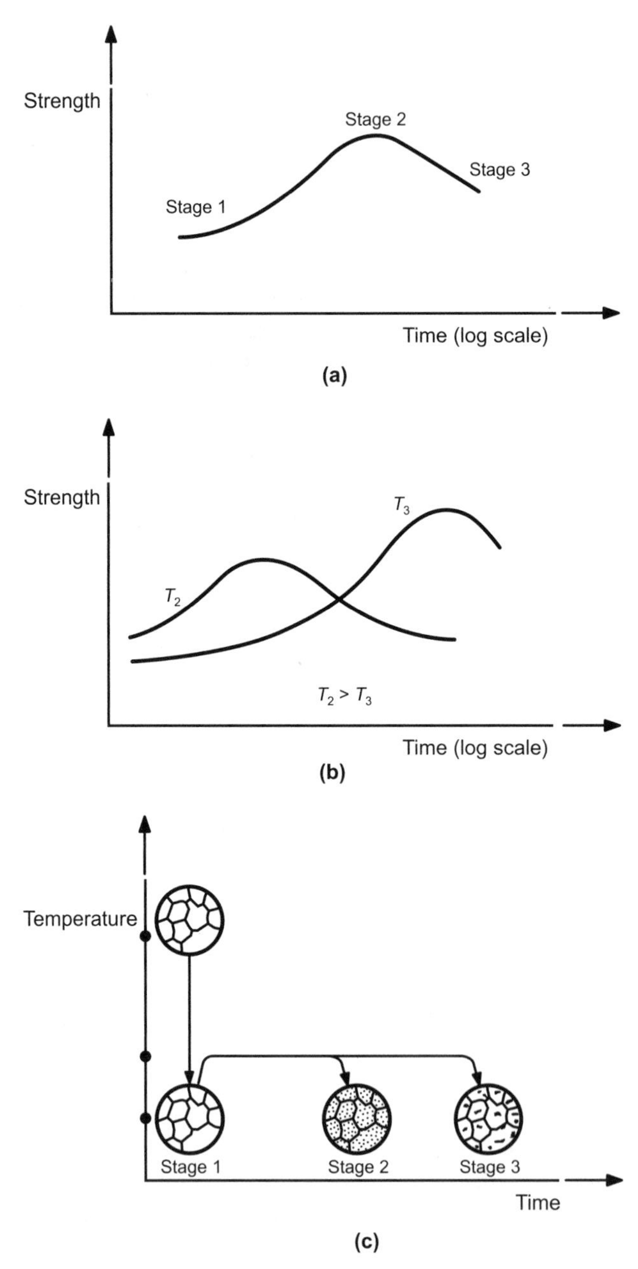

FIGURE 17.14 Illustration of the aging process. (a) Variation of hardness with time during aging. (b) Temperature dependence of the aging curve. (c) Variation of microstructure with time during aging. (From Barrett, C. R., Nix, W. D., and Tetelman, A. S., (1973) *The Principles of Engineering Materials*. Reprinted by permission of Pearson Education, Inc., Upper Saddle River, NJ.)

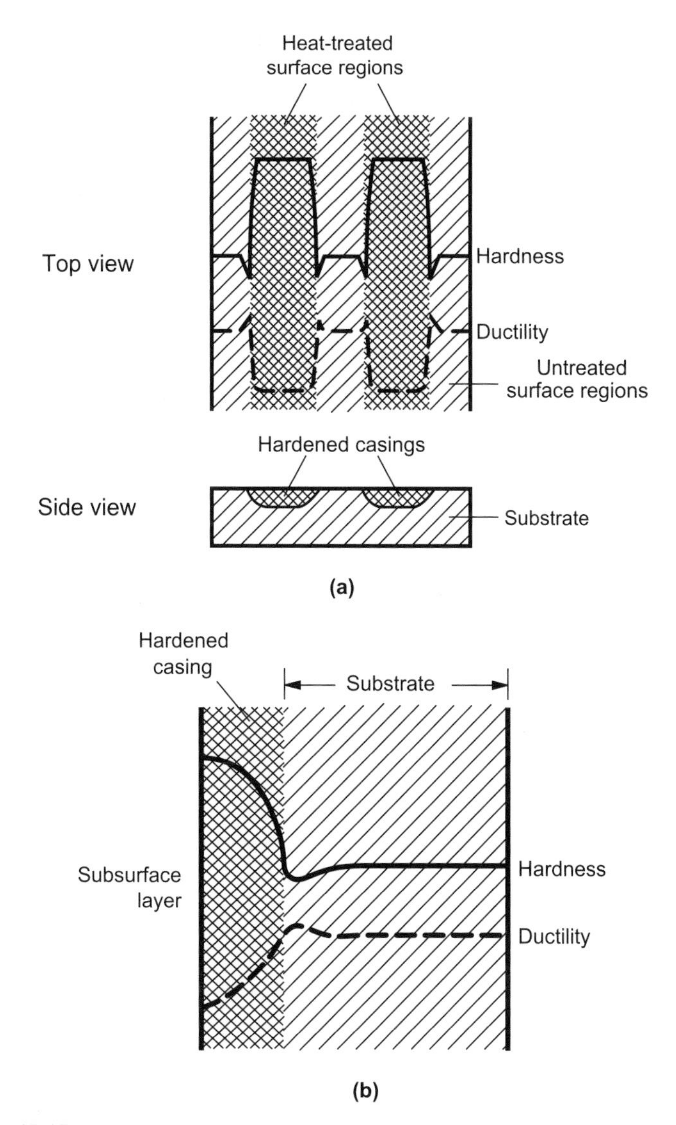

FIGURE 17.15 Hardness and ductility variation on a surface-hardened workpiece. (a) Hardness distribution on the top surface of the workpiece. (b) Hardness variation with depth. (From Gnanamuthu, D S., and Shankar, V S., 1985, *SPIE, Applications of High Powered Lasers.*)

0.8%, which is higher than that of the original steel. Also, at higher peak temperatures, carbon is able to diffuse faster outward into the surrounding low-carbon matrix. This reduces the local carbon concentration. Now since the hardness of martensite decreases with decreasing carbon content, the hardness of the individual martensitic regions is lower in the area where the peak temperatures are highest. On the contrary, more martensite is formed in the region where the peak temperatures are highest. Thus, the overall hardness in this region is higher than that in the lower temperature region, since the overall hardness of a material increases with the amount of the harder material.

The drop in hardness observed next to the hardened region is the part of the substrate that is heated to a high enough temperature for grain growth to occur, but where the cooling rate is not high enough to result in martensite formation.

17.1.7 Residual Stresses

Residual stresses are normally induced in the surface layers of a part that is subjected to surface modification. The key mechanisms responsible for the residual stresses are the following:

1. Thermal expansion and contraction associated with the process (see Chapter 13).
2. Phase transformation, as occurs during austenite to martensite transformation. The second mechanism is due to an increase in volume as austenite transforms to martensite, resulting in compressive stresses. These compressive residual stresses improve the fatigue life of the part.

The variation of transverse stresses across the laser track for a single scan laser heat treatment is shown in Fig. 17.16a, indicating tensile transverse stresses within the laser track, with balancing compressive stresses in the heat-affected zone outside the track. As Fig. 17.16b shows, with multiple overlapping scans, the residual stresses eventually become wholly compressive.

17.1.8 Advantages and Disadvantages of Laser Surface Treatment

17.1.8.1 Advantages The primary advantages of laser surface heat treatment include the following:

1. Rapid heating and cooling. This enables steels of lower hardenability to be heat treated. Thus it is often suitable for components that are difficult to harden using conventional methods such as flame and induction hardening.
2. The resulting heat-affected zone is minimal, compared to that associated with flame and induction hardening.
3. Minimal distortion due to low-heat input. This eliminates posthardening machining.
4. Localized heat input enables only desirable regions to be heat treated. Other regions are thus essentially unaffected. It also minimizes the total heat input to the part. Thus less energy is used.
5. Ease of processing complex shapes due to the ability to scan the beam over the part.

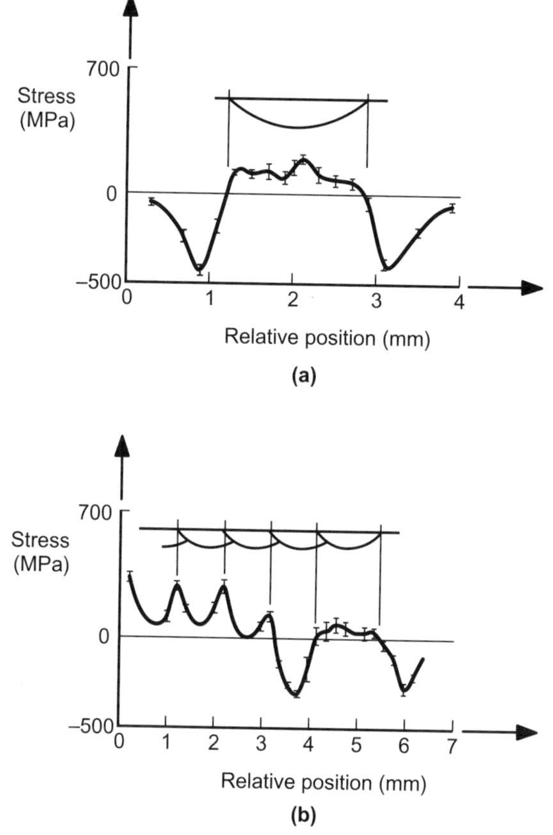

FIGURE 17.16 Transverse residual stress variation on a surface-hardened workpiece. (a) Single scan. (b) Multiple and overlapping scans. (From Van Brussel, B. A., and De Hosson, J. Th. M., 1993, *Materials Science and Engineering*, Vol. A161, pp. 83–89. Copyright Elsevier.)

6. More uniform case hardness.
7. No external quenching is necessary since the process often involves self-quenching.
8. Minimization of fumes and dirt that result from heating and quenching.
9. The short cycle time results in relatively fine-grained structure. This enables higher strength and good fatigue resistance to be achieved.

17.1.8.2 Disadvantages

1. Due to the short interaction times, coarse structures (or alloys that require long soaking times for heat treatment) tend to be difficult to laser heat treat. These include coarse pearlite, blocky ferrite, and steels that contain spheroidal carbides and cast irons that consist primarily of graphite without any pearlite. The

austenite may not be completely homogenized in such materials, and thus soft spots may occur in the hardened zone.

2. High-capital cost of the laser.
3. The need for sufficient mass for self-quenching.

17.2 LASER SURFACE MELTING

Steels with widely dispersed carbide or graphite (mainly low-carbon steels), are generally difficult to harden by solid-state transformation. These are more easily hardened by melting and solidification, since the diffusion rate of carbon is much higher in the molten state.

Laser surface melting (also known as skin melting or glazing) involves melting of a thin surface layer of material which subsequently undergoes rapid solidification as a result of self-quenching, resulting in alterations in the local microstructure. The process is similar to autogenous (no filler metal added) conduction mode laser welding. However, in this case, lower power densities are used at higher traverse speeds to ensure that only a thin layer of the substrate is melted, and that the resulting cooling rates are high enough to induce rapid solidification. Thus, the principles of rapid solidification discussed in Section 12.3 apply to this process.

The microstructural changes may be accompanied by changes in properties such as hardness, corrosion resistance, and wear resistance. The surface melting process results in four distinct zones in the material (Fig. 11.1):

1. Fusion zone.
2. Zone of partial melting.
3. Heat-affected zone.
4. Base material.

When the process involves multiple scanning, there also exists an overlapping zone when the individual beam scans are made to overlap.

The microstructure of the fusion zone depends on the alloy. For example, for AISI 4340 steel, it may consist of a fine cellular structure containing martensite, while for AISI 4140, it may consist primarily of fine-grained martensite with traces of austenite. The microhardness of the fusion zone is generally higher than that of the overlapping zone, whose hardness is in turn higher than that of the base material. Unlike the case of surface heat treatment (where no melting occurs), the microstructure of both the fusion and overlapping regions are independent of the thermal history of a steel, while that of the HAZ depends on the original matrix. The microstructure of the HAZ may consist of coarse martensite, since it is austenitized during heating, and then quenched.

In the fusion zone, increasing the scanning speed or decreasing the laser power results in a finer microstructure and thus higher microhardness. However, increasing the laser power increases the melt depth. For a single scan treatment, the microhardness is highest at the center of the path, and for overlapping treatments, reduces toward the overlap zone. Reducing the amount of overlap improves the overall microhardness.

17.3 LASER DIRECT METAL DEPOSITION

Laser deposition is a form of cladding, hardfacing, or coating technique for enhancing the corrosion and/or wear resistance of a part. This process involves using a laser beam to melt a very thin surface of a substrate. The molten substrate mixes with melted clad alloy, which is usually applied as powder (Fig. 17.17) to form a metallurgical bond. As in the other cases of laser surface modification, the rapid solidification and/or quench rates associated with the process enable either amorphous or nonequilibrium crystalline phases to be formed. The amorphous phases enable material properties to be obtained which cannot be obtained under equilibrium cooling. Laser cladding applications include engine valve components, boiler firewall, and turbine blades.

Potential problems that may arise during cladding include the following:

1. Porosity that may result from moisture and the entrapment of shielding gas.
2. Thermal stresses which may cause cracking, and are due to the following factors.

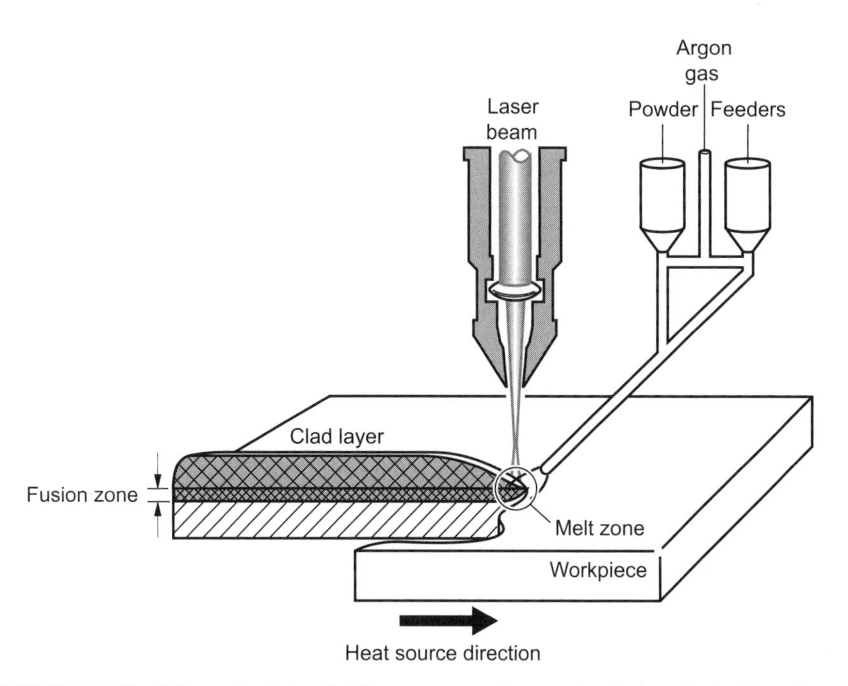

FIGURE 17.17 Schematic of the cladding process—the powder feed method. (From Jasim, K. M., and West, D. R. F., 1989, *SPIE, High Power Lasers and Laser Machining Technology*, Vol. 1132, pp. 237–245.)

(a) Difference between the melting points of the clad layer and substrate.
(b) Different coefficients of thermal expansion of the clad layer and substrate.
(c) Volume changes associated with phase changes that occur as the material cools down.

17.3.1 *Processing* Parameters

As an illustration, typical processing parameters for cladding of a Mg–Al alloy, for example, by powder feeding are listed in Table 17.3.

For Mg–Al alloys, the use of nitrogen shielding gas may result in the greatest depth of the alloyed region. Addition of helium or argon reduces the melt depth, but increases the hardness level by as much as 100%. Inert gas mixtures without any nitrogen result in the hardest cladding, with minimal porosity or cracking.

The clad tracks that are formed in laser cladding tend to be discontinuous with a high height-to-width (aspect) ratio at relatively low speeds, becoming shallower with a reducing aspect ratio as speed increases (Fig. 17.18). Thus, at very high speeds, about 30 mm/s, the cladding obtained is relatively thin, and the dilution (see Section 17.3.3) increases accordingly.

17.3.2 Methods for Applying the Coating Material

The coating material may be applied in a number of ways. These include the following:

1. Preplaced loose powder bed.
2. Presprayed surfacing powder.
3. Prepositioned chip or rod.
4. Wire or ribbon feed.
5. Powder feed.

TABLE 17.3 Typical Process Parameters Used in Laser Cladding

Parameters	Values	Units
Power	1–3	kW
Defocused beam diameter	3–5	mm
Power densities	1–100	kW/cm^2
Scanning velocities	1–60	mm/s
Powder feed rate	0.1–0.4	g/s
Clad layer thickness	0.1–2	mm
Feeder nozzle diameter	3	mm
Feeder nozzle orientation	30 with substrate	°
Shielding gas	Argon	–
Overlap between passes	50	%

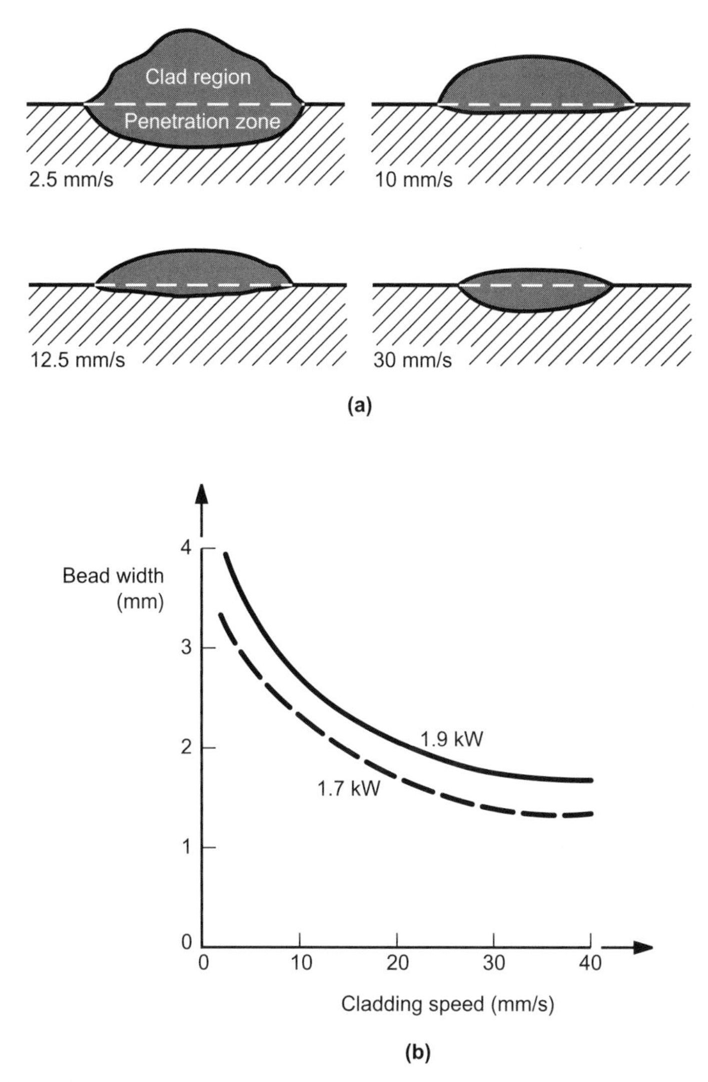

FIGURE 17.18 Effect of traverse speed on clad configuration. (a) Cross-sectional view of clad track. (b) Variation of bead width with cladding speed. (c) Variation of track height with cladding speed. (From Jasim, K. M., and West, D. R. F., 1989, *SPIE, High Power Lasers and Laser Machining Technology*, Vol. 1132, pp. 237–245.)

Other methods of coating include thermal or plasma spraying, screen printing, and electroplating. The listed methods will now be further discussed in the following sections.

Preplaced Powder—With this method (Fig. 17.19) the powder, either in loose form or in a slurry mixture, is placed on the substrate, and the laser beam is

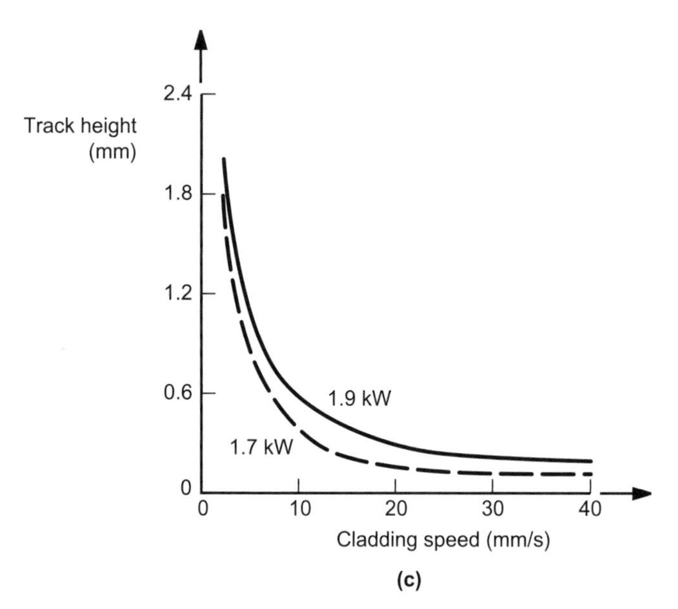

FIGURE 17.18 (*Continued*)

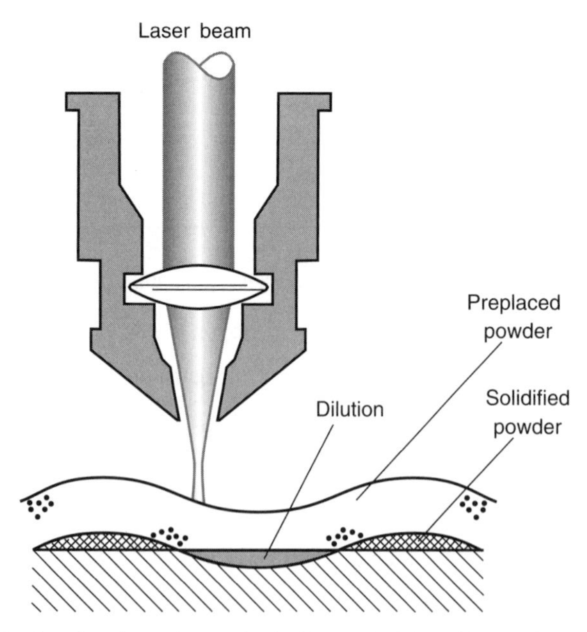

FIGURE 17.19 Preplaced powder method of laser cladding. (From Eboo, G. M., and Lindemanis, A. E., 1985, *SPIE, Applications of High Power Lasers*, Vol. 527, pp. 86–94.)

directed to melt through the powder and fuse it to the substrate. The major drawbacks of this method are listed here.

1. It is limited to flat beds. Generally, complex geometries are difficult to clad by this approach.
2. Shielding gas usage is difficult, as it tends to blow away the powder.
3. Evaporation or decomposition of the binder may lead to porosity.
4. Control of the molten zone between the clad layer and substrate is difficult. Variations in bed depth may cause either excessive dilution (small depth) or absence of metallurgical bonding (greater depth).
5. It requires a higher specific energy to melt through a specified bed depth than that for a powder feed system.

However, it has the advantage that it is very simple to use and is also relatively low cost.

Presprayed Surfacing Powder—This method involves first spraying surfacing powder on the substrate by thermal or plasma spraying, and subsequently consolidating by laser melting. The additional processing step involved is a drawback of the process.

Prepositioned Chip—A chip that is made to the shape of the component's surface is placed on the substrate, and consolidated by laser melting to form a metallurgical bond. The setback of this approach is also the extra processing step.

Wire Feed—This process is similar to arc or gas welding. In this case, the wire is fed into a laser generated melt zone at the substrate. It involves a single step, and complex shapes can be clad.

Powder Feeding—The setup for this method consists of the laser, powder delivery system, and shielding gas (Fig. 17.17). It involves continuously feeding the alloy powder through a nozzle that is positioned such that the powder feeds directly into the molten portion of the substrate. Different powder compositions can be fed in, enabling online variations in the alloy composition to be achieved.

The powder may be dispensed in one of two ways:

1. Gravity feeding, where the powder is delivered directly in front of the molten zone, and in that respect, resembles a preplaced powder bed. It may lead to segregation of widely different density particles.
2. Using a pneumatic screw feed system with a small amount of argon gas (about 1 L/min) flowing in the tube to aid powder flow and directional placement. Control of the powder flow rate into the molten zone is determined by the following factors:
   (a) Material flow characteristics.
   (b) Powder preparation, for example, preheat.
   (c) Powder injection angle.
   (d) Shielding gas characteristics.

To reduce oxidation and/or elemental loss, the process may be carried out in a slightly pressurized inert enclosure, with the laser beam being introduced through a NaCl window.

Since the cladding process involves addition of one or more materials to the substrate, this results in dilution or change in chemical composition of portions of the base metal.

17.3.3 Dilution

Dilution of the clad material can be defined either as the ratio of the area of substrate melted to the total melted zone, or in terms of the chemical composition difference between the starting powder and the solidified clad layer (Fig. 17.20). These may be expressed as

%Area dilution =

$$\frac{\text{Melted substrate cross sectional area}}{\text{Clad region cross-sectional area} + \text{Melted substrate cross-sectional area}}$$

$$= \frac{(2)}{(1) + (2)} \tag{17.15}$$

which is strictly applicable only to single track clads, and

$$\%\text{Chemical dilution} = \frac{C_{\text{clad}} - C_{\text{powder}}}{C_{\text{powder}}} \tag{17.16}$$

where C_{clad} is the elemental composition of the clad at a position near the substrate/clad interface and C_{powder} is the elemental composition of the starting powder.

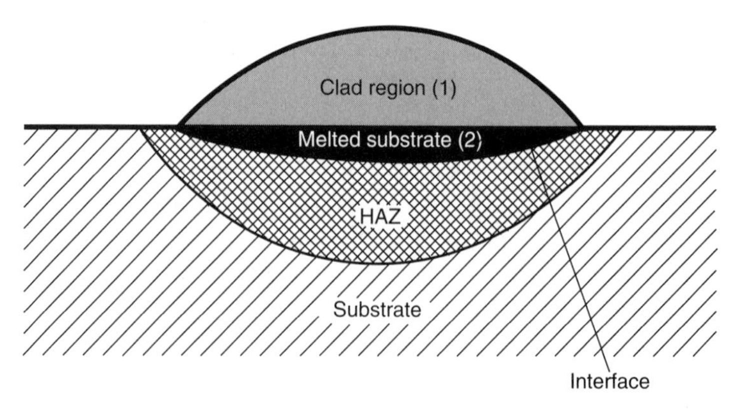

FIGURE 17.20 Dilution during the cladding process. (From Jasim, K. M., and West, D. R. F., 1989, *SPIE, High Power Lasers and Laser Machining Technology*, Vol. 1132, pp. 237–245.)

With increasing traverse speed, the area dilution tends to decrease to a minimum and then begin to increase, while the chemical dilution remains constant. The chemical dilution is a measure of the stoichiometry of the alloy formed by cladding, and thus is not expected to change significantly with processing conditions once melting occurs. However, the area dilution is a measure of the amount of substrate that melts, and thus is significantly influenced by the processing speed. A low value of area dilution may be appropriate for say, a hard coating application where it is desired to have a hard layer on the surface of the base material. On the contrary, where a different alloy composition is desired, then a high value of the chemical composition would be desirable.

17.3.4 Advantages and Disadvantages of Laser Cladding

17.3.4.1 Advantages Laser cladding has the following advantages over other more traditional methods:

1. Low-heat input, and thus narrow heat-affected zone and reduced distortion, with minimal machining being required.
2. Reduced alloy material loss.
3. Minimal clad dilution of base metal (less than 2%) enables the special properties of the clad material to be maintained.
4. Flexibility and ease of automation.
5. Complete metallurgical bonding to the substrate, resulting in high-integrity coating.

17.3.4.2 Disadvantages The primary disadvantages of laser cladding are as follows:

1. It is not the method of choice when the area to be hardened is relatively small.
2. It is more expensive than comparative processes that use plasma arc or gas tungsten arc heat sources.

In general, laser cladding is an effective means of changing the microstructure and thus mechanical properties of thin layers of a substrate where the surface area to be processed is large enough. Other methods of treatment exist that do not involve melting of the substrate, for example, laser physical vapor deposition and laser chemical vapor deposition. There are a number of other surface modification techniques that do not involve the use of lasers. However, in the following section, we further discuss the laser physical vapor deposition technique.

17.4 LASER PHYSICAL VAPOR DEPOSITION (LPVD)

The principle of LPVD is illustrated in Fig. 17.21. A CW laser beam is irradiated on a target in a vacuum chamber. The target is placed on a traveling holder and inclined to

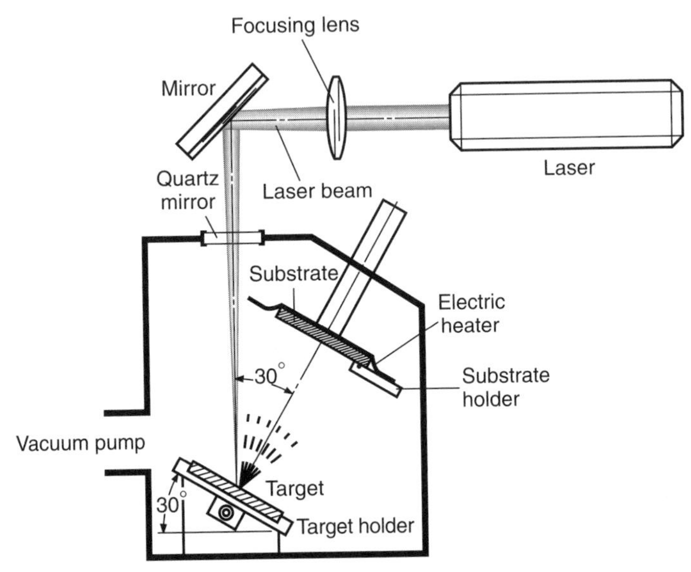

FIGURE 17.21 Schematic of the laser physical vapor deposition process. (From *ICALEO 1991 Proceedings*, Vol. 74, Copyright 1992, Laser Institute of America. All rights reserved.)

the beam axis, while the substrate is fixed and held parallel to the target. The target is moved at a preset speed, and the substrate may be preheated. The laser beam vaporizes material from the target, which is deposited on the substrate. As an illustration, typical key processing parameters would be as shown in Table 17.4.

The deposition rate increases with an increase in laser power, but decreases as the target–substrate distance increases (Fig. 17.22a). However, the resulting hardness decreases with an increase in laser power and increases as the target–substrate distance increases (Fig. 17.22b). The hardness also increases with an increase in substrate temperature, and this might be due to poor adhesivity to the substrate at lower temperatures.

TABLE 17.4 Typical Process Parameters Used in Laser Physical Vapor Deposition

Parameters	Values	Units
Power	100–200	W
Scanning velocities	3	mm/s
Chamber pressure	1×10^{-4}	Pa
Target orientation	30	°
Target–substrate distance	50	mm
Substrate temperature	800	K

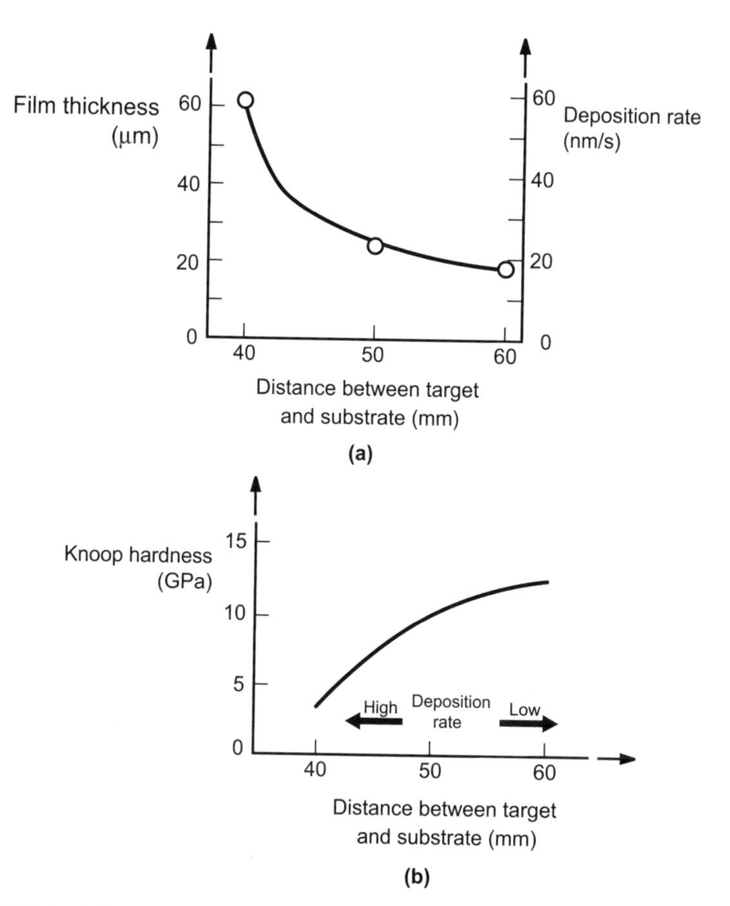

FIGURE 17.22 Effect of target–substrate distance on film thickness and hardness for a nickel substrate at a laser power of 100 W, substrate temperature of 673 K, and a pressure of 2.6×10^{-3} Pa. (a) Film thickness. (b) Film hardness. (From *ICALEO 1991 Proceedings*, Vol. 74, Copyright 1992, Laser Institute of America. All rights reserved.)

17.5 LASER SHOCK PEENING

Laser shock peening is a process that imparts compressive stresses on the surface of a material. It is used primarily for increasing fatigue life and improving resistance to cracking. One application of the process has been in making the leading edge of fan blades resistant to foreign object damage.

Laser shock peening is normally done using a thermoprotective coating or absorbing layer of black paint or tape applied to the surface of the part (Fig. 17.23). A layer of dielectric material that is transparent to the laser beam, usually glass or water, is placed on top of the absorbing layer. The laser beam instantly vaporizes the absorbing layer and produces plasma that expands rapidly, creating very high pressures at the interface, as a result of the recoil momentum of the ablated material. The pressure

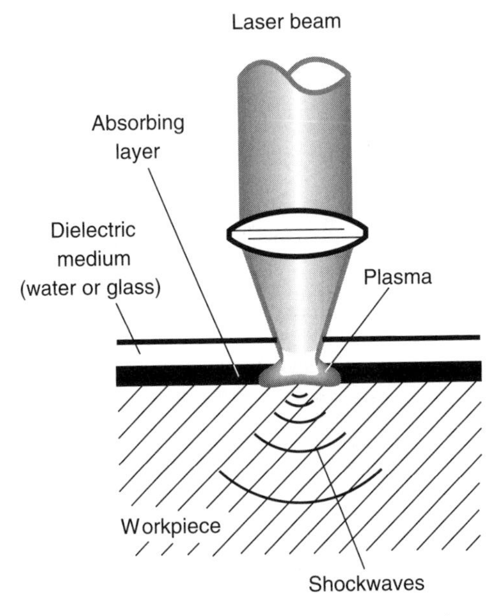

FIGURE 17.23 Schematic of the laser shock peening process.

results in shock waves that are transmitted throughout the material and induces compressive stresses in the material. With the plasma confined by the layer of dielectric material, pressures of the order of gigapascals are obtained. Pressures generated in the confined mode are typically an order of magnitude greater than in direct ablation, for similar laser irradiances. Also, using quartz as the overlay material results in much higher pressures than when water is used, since the acoustic impedance of water is about 10% that of quartz. Use of the thermoprotective coating avoids melting of the metallic surface during the irradiation. The process requires a laser system with high average power and large pulse energy for a high throughput. Typical process parameters are listed in Table 17.5.

Laser beams with a circular profile are often used for this process, but have the disadvantage that the resulting stress distribution is not uniform. Circular profiles produce a stress dip in the center, which can be a preferred location for crack initiation. For these reasons, a square profile is preferable. The percent of cold work and compressive residual stress increase as the number of pulses increases, with the

TABLE 17.5 Typical Process Parameters Used in Laser Shock Peening

Parameters	Values	Units
Power (average)	1	kW
Pulse power density	1–50	GW/cm^2
Pulse energy	100	J
Pulse duration	1–50	ns

magnitude of the compressive stress being proportional to the number of pulses per spot. The disadvantages of laser peening are that it is a relatively slow process; it can also be quite expensive, as layers of absorbing material must be applied before each peening application. However, it is becoming a more viable alternative to shot peening as new developments in laser systems allow faster processing.

17.5.1 Background Analysis

The pressure generated at the interface can be analyzed using the simplified setup shown in Fig. 17.24, considering glass to be the dielectric material. The pressure results in shock waves in both the workpiece and the glass overlay, and causes the interface walls to be displaced with velocities u_1 and u_2.

If the thickness of the interface at time t is $L_{\mathrm{p}}(t)$, then we have

$$L_{\mathrm{p}}(t) = \int_0^t [u_1(t) + u_2(t)]\mathrm{d}t \tag{17.17}$$

The velocities, $u_i (i = 1, 2)$, are related to the pressure P by

$$P = \rho_i U_i u_i = Z_i u_i \tag{17.18}$$

where $i = 1, 2$ is the material 1 or 2, U_i is the shock velocity (m/s), Z_i is the shock impedance (g/cm^2/s), and ρ_i is the density (kg/m^3).

For the situation where the two materials are solids, we can assume that their shock wave impedances, $Z_i = \rho_i U_i$ are constant. Then from equations (17.17) and (17.18), we get

$$\frac{\mathrm{d}L_{\mathrm{p}}(t)}{\mathrm{d}t} = \frac{2}{Z}P(t) \tag{17.19}$$

where $P(t)$ is the plasma pressure at time t, and $\frac{2}{Z}$ is $\frac{1}{Z_1} + \frac{1}{Z_2}$.

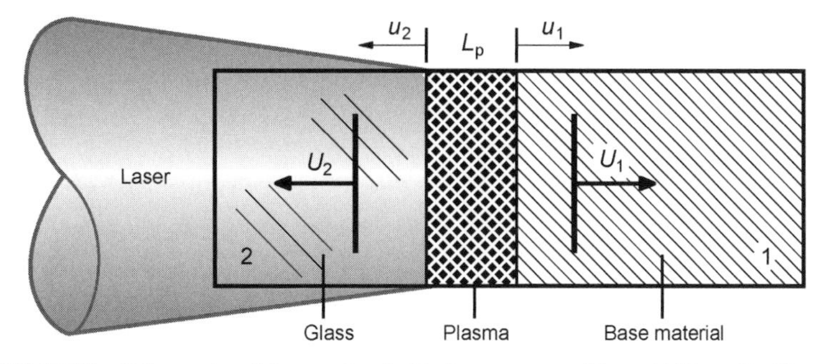

FIGURE 17.24 Schematic of the confined ablation process. (From Fabbro, R., Fournier, J., Ballard, P., Devaux, D., and Virmont, J., 1990, *Journal of Applied Physics,* Vol. 68, No. 2, pp. 775–784. Reprinted by permission of American Institute of Physics, Melville, NY.)

The laser energy that is absorbed increases the internal energy per unit volume, Q_e, of the interface plasma, and also does work to open up the interface. If the laser intensity is $I(t)$, then the energy per unit surface area deposited over the time interval dt is $I(t)dt$. Let the plasma thickness increase by dL_p during this time interval. The increase in internal energy is then given by $d[Q_e(t)L_p]$ (joules per unit area), and the work done per unit area by the forces resulting from the pressure is $P(t)dL_p$. Thus,

$$I(t) = P(t)\frac{dL_p(t)}{dt} + \frac{d[Q_e(t)L_p(t)]}{dt} \tag{17.20}$$

We shall now make the following assumptions:

1. The fraction, f_E, of the internal energy, $Q_e(t)$, which represents the thermal energy, $Q_T(t)$, is constant. The balance, $1 - f_E$, is used to ionize the gas. f_E is typically $\approx 0.1 - 0.2$.
2. The interface plasma obeys the ideal gas laws.

The relationship between the thermal energy, Q_T, of an ideal gas, and the resulting pressure, P, is given by

$$P(t) = \frac{2}{3}Q_T(t) = \frac{2}{3}f_E Q_e(t) \tag{17.21}$$

Equation (17.20) can then be rewritten as

$$I(t) = P(t)\frac{dL_p(t)}{dt} + \frac{3}{2f_E}\frac{d[P(t)L_p(t)]}{dt} \tag{17.22}$$

Further assume that the laser beam intensity, $I(t) = I_0 =$ constant, and that $L_p(0) = 0$. Then if the pulse duration is $\Delta\tau_p$, equations (17.19)–(17.22) can be solved to give the plasma pressure at the interface, P (which is then constant), as

$$P = 0.10 \times \sqrt{\left[\frac{f_E}{2f_E + 3}\right] Z I_0} \tag{17.23}$$

And the resulting plasma thickness is

$$L_p(\Delta\tau_p) = 2 \times 10^4 \frac{P\Delta\tau_p}{Z} \tag{17.24}$$

where the parameters have the following units:

P in kbars, Z in $g/cm^2\,s$, I_0 in GW/cm^2, L_p in μm, and $\Delta\tau_p$ in ns.

If, on the contrary, $L_p(0) = L_{p0} \neq 0$, then $P(t)$ is not constant. Under these circumstances, it can be shown, using equation (17.19) and neglecting the term $P(t)\frac{dL(t)}{dt}$

in equation (17.22), that

$$L_p(t)^2 = L_{p0}^2 + \left(\frac{2P_0 t}{Z}\right)^2$$

The pressure, which is then time dependent, becomes

$$P(t) = P_0\left(1 - \frac{L_{p0}^2}{L_p(t)^2}\right)^{1/2} \tag{17.25}$$

where $P_0 = 0.10 \times \sqrt{\left[\frac{f_E}{3}\right] Z I_0}$

Example 17.4 In shock peening the surface of an aluminum plate, the laser that is used generates a pulse energy of 80 J at a pulse duration of 10 ns, and a beam diameter of 3 mm. If the plasma thickness obtained at the end of the pulse is 40 μm, determine the interface pressure generated. Assume that the pressure is invariant with time; that 15% of the internal energy is expended as thermal energy; and that the pulse energy is uniform throughout the pulse duration.

Solution:

The plasma thickness is given by equation (17.24):

$$L_p(\Delta\tau_p) = 2 \times 10^4 \frac{P\Delta\tau_p}{Z} \tag{17.24}$$

$$\Rightarrow Z = 2 \times 10^4 \frac{P\Delta\tau_p}{L_p(\Delta\tau_p)} = 2 \times 10^4 \times \frac{P \times 10}{40} = 5 \times 10^3 \times P$$

thus from equation (17.23), we have

$$P = 0.10 \times \sqrt{\left[\frac{f_E}{2f_E + 3}\right] Z I_0} \tag{17.23}$$

$$= 0.10 \times \sqrt{\left[\frac{f_E}{2f_E + 3}\right] \times 5 \times 10^3 \times P \times I_0}$$

But the pulse power, q, is given by

$$q = \frac{80}{10 \times 10^{-9}} = 8 \times 10^9 \text{ W}$$

and

$$I_0 = \frac{8 \times 10^9}{\pi \times 0.3^2/4} = 113 \text{ GW/cm}^2$$

Thus,

$$P = 0.10 \times \sqrt{\left[\frac{0.15}{2 \times 0.15 + 3}\right] \times 5 \times 10^3 \times P \times 113}$$

$$\Rightarrow P = 256.8 \text{ kbars} = 25.7 \text{ GPa}$$

17.5.2 Advantages and Disadvantages of Laser Shock Peening

17.5.2.1 Advantages Laser shock peening has several advantages over traditional shot peening:

1. It produces compressive residual stresses that extend deeper into the material (over 2 mm) than can be obtained with shot peening.
2. The compressive residual stresses are also produced with less cold work than with shot peening, allowing for less thermal relaxation of these stresses when the part is subjected to high temperatures.
3. With conventional shot peening, the surface of the workpiece is dimpled, which can be undesirable aesthetically. Laser shock peening does not adversely affect the surface appearance.

17.5.2.2 Disadvantages The major disadvantage of laser shock peening is that it currently requires specially designed lasers for production runs, since it requires a laser system with high average power and large pulse energy for a high throughput. Thus, it is still primarily a developmental process.

17.6 SUMMARY

In this chapter, we have learnt that laser surface modification is used to change either the surface composition or microstructure of a material to give it certain desired properties, and the principal techniques include the following:

Laser Surface Heat Treatment—This involves exposing the surface of a material to a thermal cycle of rapid heating and cooling such that the surface layers are hardened, either through martensitic transformation in the case of steels, or through precipitation hardening, in the case of non-ferrous alloys. The process does not involve melting, and the transformations occur in the solid state.

Laser Surface Melting (Skin Melting or Glazing)—In this technique a thin surface layer of material is melted with the laser beam, and subsequently undergoes rapid solidification as a result of self-quenching, resulting in alterations in the local microstructure. The process is similar to autogenous (no filler metal added) conduction mode laser welding, but the resulting cooling rates are high enough to induce rapid solidification. This process is especially suited to materials which are generally difficult to harden by solid-state transformation, for example, steels with widely dispersed carbide or graphite (mainly low-carbon steels). The microstructural changes may be accompanied by changes in properties such as hardness, corrosion resistance, and wear resistance.

Laser Deposition (Cladding, Alloying, and Hard Coating)—This involves applying a second material to the surface and melting it with a laser beam either to alloy with a thin surface layer of the base material, or to bond to the surface. As in the other cases of laser surface modification, the rapid solidification and/or quench rates associated with the process enable either amorphous or nonequilibrium crystalline phases to be formed. The major methods for applying the coating include preplaced loose powder bed; presprayed surfacing powder; prepositioned chip or rod; wire or ribbon feed; and powder feed.

Laser Physical Vapor Deposition—In this method the second material is applied in vapor form and no melting takes place. A CW laser beam is irradiated on a target in a vacuum chamber, and vaporizes material from the target, which is deposited on the substrate, forming a thin layer. The deposition rate increases with an increase in laser power, but decreases as the target–substrate distance increases. The resulting hardness, on the other hand, decreases with an increase in laser power, and increases as the target–substrate distance increases.

Laser Shock Peening—This involves using the laser beam to vaporize a thin layer of thermoprotective coating or absorbing layer of black paint or tape. This is placed beneath a layer of dielectric material that is transparent to the laser beam, usually glass or water. The recoil momentum of the ablated material results in high pressure that generates shock waves, inducing compressive residual stresses on the surface of the material. It is used primarily for increasing fatigue life and improving resistance to cracking.

REFERENCES

Abbas, G., and West, D. R. F., 1989, Laser produced composite metal cladding, *SPIE, High Power Lasers and Laser Machining Technology*, Vol. 1132, pp. 232–236.

Ashby, M. F., and Easterling, K. E., 1984, The transformation hardening of steel surfaces by laser beams—I. Hypo-eutectoid steels, *Acta Metallurgica,* Vol. 32, No. 11, pp. 1935–1948.

Bataille, F., Kechemair, D., and Houdjal, R., 1991, Real time actuating of laser power and scanning velocity for thermal regulation during laser hardening, *SPIE, Industrial and Scientific Uses of High-Power Lasers*, Vol. 1502, pp. 135–139.

Bradley, J. R., 1986, Experimental determination of the coupling coefficient in laser surface hardening of steel, *SPIE, Laser Processing: Fundamentals, Applications, and Systems Engineering*, Vol. 668, pp. 23–30.

Bradley, J. R., 1988, A simplified correlation between laser processing parameters and hardened depth in steels, *Journal of Physics D: Applied Physics*, Vol. 21, pp. 834–837.

Braisted, W., and Brockman, R., 1999, Finite element simulation of laser shock peening, *International Journal of Fatigue*, Vol. 21, pp. 719–724.

Dane, C., Hackel, L., Halpin, J., Daly, J., Harrisson, J., and Harris, F., 2000, High-throughput laser peening of metals using a high-average-power Nd:glass laser system, *Proceedings of SPIE: The International Society for Optical Engineering*, Vol. 3887, pp. 211–221.

Ding, K., and Ye, L., 2003a, Three-dimensional dynamic finite element analysis of multiple laser shock peening processes, *Surface Engineering*, Vol. 19, No. 5, pp. 351–358.

Ding, K., and Ye, L., 2003b, FEM simulation of two sided laser shock peening of thin sections of Ti-6Al-4V alloy, *Surface Engineering*, Vol. 19, No. 2, pp. 127–133.

Eboo, G. M., and Lindemanis, A. E., 1985, Advances in laser cladding process technology, *SPIE, Applications of High Power Lasers*, Vol. 527, pp. 86–94.

Fabbro, R., Fournier, J., Ballard, P., Devaux, D., and Virmont, J., 1990, Physical study of laser-produced plasma in confined geometry, *Journal of Applied Physics,* Vol. 68, No. 2, pp. 775–784.

Fabbro, R., Peyre, P., Berthe, L., and Scherpereel, X., 1998, Physics and applications of laser-shock processing, *Journal of Laser Applications*, Vol. 10, No. 6, pp. 265–279.

Fairand, B. P., and Clauer, A. H., 1979, Laser generation of high-amplitude stress waves in materials, *Journal of Applied Physics*, Vol. 50, No. 3, pp. 1497–1502.

Fastow, M., Bamberger, M., Nir, N., and Landkof, M., 1990, Laser surface melting of AISI 4340 steel, *Materials Science and Technology,* Vol. 6, No. 9, pp. 900–904.

Festa, R., Nenci, F., Manca, O., and Naso, V., 1993, Thermal design and experimentation analysis of laser and electron beam hardening, *ASME Journal of Engineering for Industry*, Vol. 115, pp. 309–314.

Gnanamuthu, D. S., and Shankar, V. S., 1985, Laser heat treatement of iron-base alloys, *SPIE, Applications of High Powered Lasers*, Vol. 572, pp. 56–72.

Hill, M., DeWald, A., Demma, A., Hackel, L., Chen, H., Dane, C., Specht, R., and Harris, F., 2003, Laser peening technology, *Advanced Materials and Processes*, Vol. 161, No. 8, pp. 65–67.

Ivarson, A., Powell, J., Ohlsson, L., and Magnusson, C., 1991, Optimisation of the laser cutting process for thin section stainless steels, *ICALEO Proceedings 1991*, pp. 211–220.

Jasim, K. M., and West, D. R. F., 1989, Laser cladding of carbon steel with a ceramic/metallic composite, *SPIE, High Power Lasers and Laser Machining Technology*, Vol. 1132, pp. 237–245.

Jasnowski, K. S., Justes, S. P., Amatrian, J. F. Z., and Ibanez, F. G., 1989, Surface hardening: beam shaping and coating techniques, *SPIE, High Power Lasers and Laser Machining Technology*, Vol. 1132, pp. 246–256.

Koshy, P., 1989, Laser cladding techniques for application to wear and corrosion resistant coatings, *SPIE, Applications of High Power Lasers*, Vol. 527, pp. 80–85.

Kou, S., Sun, D. K., and Le, Y. P., 1983, A fundamental study of laser transformation hardening, *Metallurgical Transactions A,* Vol. 14A, pp. 643–653.

Kusinski, J., and Thomas, G., 1986, Effect of laser hardening on microstructure and wear resistance in medium carbon/chromium steels, *SPIE, Laser Processing: Fundamentals, Applications, and Systems Engineering*, Vol. 668, pp. 150–157.

Li, W.-B., Easterling, K. E., and Ashby M. F., 1986, Laser transformation hardening of steel—II. Hypereutectoid steels, *Acta Metallurgica,* Vol. 34, pp. 1533–1543.

Lin, J., and Steen, W. M., 1998, Design characteristics and development of a nozzle for coaxial laser cladding, *Journal of Laser Applications*, Vol. 10, No. 2, pp. 55–63.

Matsunawa, A., Katayama, S., Minonishi, M., Miyazawa, H., Hiramoto, S., and Oka, K., 1991, Formation of alumina film by laser PVD, *ICALEO Proceedings 1991*, pp. 363–370.

Mazumder, J., 1983, Laser heat treatment: the state of the art, *Journal of Metals,* Vol. 35, pp. 18–26.

Montross, C. S., Wei, T., Ye, L., Clark, G., and Mai, Y.-W., 2002, Laser shock processing and its effects on microstructure and properties of metal alloys: a review, *International Journal of Fatigue*, Vol. 24, pp. 1021–1036.

Navara, E., Bengtsson, B., Li, W.-B., and Easterling, K.E., 1984, Surface treatment of steel by laser transformation hardening, *Proceedings of the 3rd International Congress on Heat Treatment of Materials,* pp. 2.40–2.44.

Peyre, P., Sollier, A., Chaieb, I., Berthe, L., Bartnicki, E., Braham, C., and Fabbro, R., 2003, FEM simulation of residual stresses induced by laser peening, *The European Physical Journal Applied Physics*, Vol. 23, No. 2, pp. 83–88.

Rankin, J., Hill, M., Halpin, J., Chen, H., Hackel, L., and Harris, F., 2002, The effects of process variations on residual stress induced by laser peening, *Materials Science Forum*, Vols 404–407, pp. 95–100.

Seaman, F. D., 1986, Laser heat-treating, *Industrial Laser Handbook,* pp. 147–157.

See, D. W., Dulaney, J., Clauer, A., and Tenaglia, R., 2002, The Air Force manufacturing technology laser peening initiative, *Surface Engineering*, Vol. 18, No. 1, pp. 32–36.

Sepold, G., and Becker, R., 1986, Rapid solidification by laser beam techniques, *Proceedings of Science and Technology of the Undercooled Melt: Rapid Solidification Materials and Technologies Conference,* pp. 112–119.

Shewmon, P. G., 1983, Atomic theory of diffusion, *Diffusion in Solids*, J. Williams Book Company, Jenks, OK.

Shibata, K., Sakamoto, H., and Matsuyama, H., 1991, Absorptivity of polarized beam during laser hardening, *ICALEO Proceedings 1991,* pp. 409–413.

Shiue, R. K., and Chen, C., 1992, Laser transformation hardening of tempered 4340 steel, *Metallurgical Transactions A,* Vol. 23A, pp. 163–169.

Smith, P., Shepard, M., Prevey III, P., and Clauer, A., 2000, Effect of power density and pulse repetition on laser shock peening of Ti-6Al-4V, *Journal of Materials Engineering and Performance*, Vol. 9, No. 1, pp. 33–37.

Solina, A., De Sanctis, M. Paganini, L., Blarasin, A., and Quaranta, S., 1984, Origin and development of residual stresses induced by laser surface-hardening treatments, *Journal of Heat Treating,* Vol. 3, pp. 193–204.

Teresko, J., 2004, Peening with light: new laser, new benefits, *Industry Week*, Vol. 253, No. 3, p. 20.

Van Brussel, B. A., and De Hosson, J. Th. M., 1993, Residual stresses in the surface layer of laser-treated steels, *Materials Science and Engineering*, Vol. A161, pp. 83–89.

Wang, A. A., Sircar, S., and Mazumder, J., 1990, Laser cladding of Mg–Al alloys, *ICALEO Proceedings 1990*, pp. 502–512.

Wei, M.Y., and Chen, C., 1993, Predicting case depth in tempered steels hardened via laser processing, *Materials Science and Technology,* Vol. 9, pp. 69–73.

Yoshioka, Y., Akita, K., Suzuki, H., Sano, Y., and Ogawa, K., 2002, Residual stress measurements of laser peened steels by using synchrotron radiation, *Materials Science Forum*, Vols 404–407, pp. 83–88.

Zhang, W., Yao, L., and Noyan, I., 2004a, Microscale laser shock peening of thin films, part 1: experiment, modeling and simulation, *ASME Journal of Manufacturing Science and Engineering*, Vol. 126, pp. 10–17.

Zhang, W., Yao, L., and Noyan, I., 2004b, Microscale laser shock peening of thin films, part 2: high spatial resolution material characterization, *ASME Journal of Manufacturing Science and Engineering*, Vol. 126, pp. 18–24.

APPENDIX 17A

List of symbols used in the chapter.

Symbol	Parameter	Units
a	half length of resonator mirror	m
A_1	eutectoid temperature	°C
A_3	lower temperature boundary of austenite zone	°C
c_d	constant	–
C_a	carbon composition in % by weight	%
C_c	critical carbon fraction above which martensitic transformation occurs	%
C_{clad}	elemental composition of the clad at a position near the substrate/clad interface	%
C_e	carbon content of pearlite	%
C_f	carbon content of ferrite	%
C_{powder}	elemental composition of the starting powder	%
D_a	diffusion coefficient for carbon	mm/s
D_0	diffusion constant	mm/s
f	volume fraction of martensite formed at time t	–
f_E	fraction of internal energy that represents the thermal energy	–
f_i	volume fraction initially occupied by pearlite colonies	–
f_m	maximum volume fraction permitted by phase diagram	–
g_s	average austenite grain size	μm
l_p	spacing of pearlite plates in a colony	μm
$L_p(t)$	thickness of plasma interface at time t	μm

L_{p0}	initial thickness of plasma interface at time $t = 0$	μm
P_0	initial plasma pressure	kbar (100 MPa)
Q_a	activation energy for transformation	J
Q_e	internal energy of the interface plasma	J
Q_T	thermal energy of the interface plasma	J
t_0	time to diffuse a distance equal to beam radius	s
u_i	velocity of medium i	m/s
u_x	traverse velocity of the heat source in the x-direction	m/s
U_i	shock velocity of medium i	m/s
z_0	distance over which heat can diffuse during the beam interaction time $\frac{w}{u_x}$	mm
$Z = 2\frac{Z_1 Z_2}{Z_1+Z_2}$	overall shock impedance	g/cm^2 s
Z_i	shock impedance of medium i	g/cm^2 s
ρ_i	density of medium i	kg/mm^3
τ_h	thermal time constant	s
$\Delta\tau_p$	pulse duration	ns
θ_c	semiangle subtended by mirrors at each other's center	°
θ_d	semiangle of diffracted waves	°

APPENDIX 17B

The Iron–Carbon Diagram

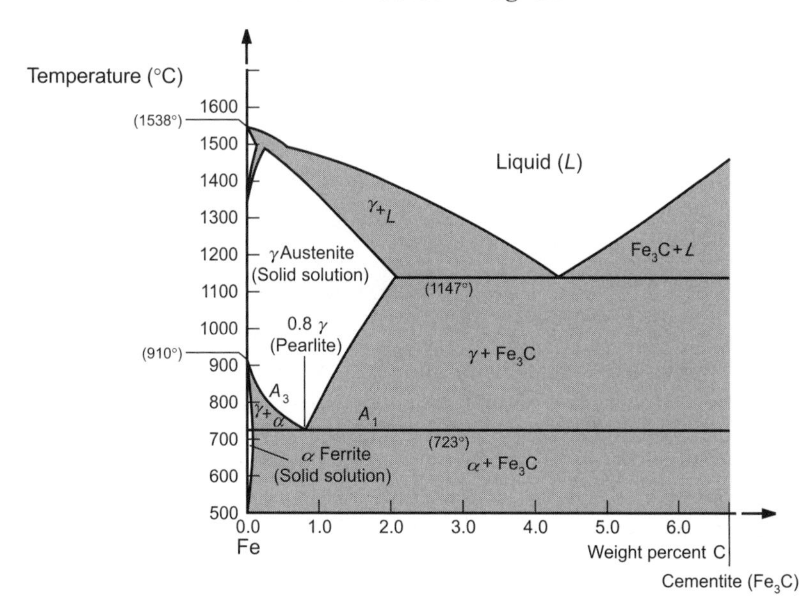

FIGURE B17.1 The iron-carbon diagram.

PROBLEMS

17.1. The surface of an AISI 1080 steel gear is to be heat treated using a 2.5-kW Nd:YAG laser. The purpose is to enhance the wear resistance. The defocused beam diameter is 5 mm, and the scan rate is set at 10 mm/s. Only a single pass of the laser beam will be used for the heat treatment. Determine the time, t, required for the diffusion controlled transformation of pearlite to austenite (pearlite dissolution).

Assume

(1) The temperature is time dependent
(2) The diffusion coefficient for carbon, $D_a = 9 \times 10^{-10}$ m²/s
(3) The diffusion constant, D_0, for the diffusion of carbon in α-Fe (ferrite) is 0.020 cm²/s $= 2 \times 10^{-6}$ m²/s (Shewmon, 1983, Table 2-1, p. 64)
(4) The activation energy for diffusion, Q_a, is 20.1 kcal/mol = 84.15 kJ/mol
(5) The reflection coefficient of the surface is 80%
(6) Ambient temperature is 25°C.

17.2. For the conditions in problem 17.1, determine the average size of the zone over which martensite can be formed.

17.3. For the conditions in problem 17.1, determine the volume fraction of martensite that can be formed. You may assume that the average austenite grain size is 200 μm.

17.4. A 7-kW CO_2 laser is used to heat treat the surface of an AISI 1045 steel cam at a beam diameter of 6 mm, and a scan rate of 25 mm/s. Determine the spacing of the pearlite plates that will enable the pearlite colonies to be fully transformed to austenite under these conditions.

Assume

(1) The temperature is time dependent
(2) The diffusion constant, D_0, for the diffusion of carbon in α-Fe (ferrite) is 0.020 cm²/s $= 2 \times 10^{-6}$ m²/s (Shewmon, 1983, Table 2-1, p. 64)
(3) The activation energy for diffusion, Q_a, is 20.1 kcal/mol = 84.15 kJ/mol (Shewmon, Table 2-1, p. 64)
(4) The reflection coefficient of the surface is 85%
(5) Ambient temperature is 25°C.

17.5. The leading edges of titanium turbine blades are to be shock peened using a laser that generates a pulse energy of 100 J with a pulse duration of 15 ns, and a beam diameter of 4 mm. If a plasma of thickness 50 μm develops by the time the pulse is over, determine the pressure that is generated at the interface between the substrate and the thermoprotective layer. Assume that the pressure is invariant with time; that 10% of the internal energy is expended as thermal energy; and that the pulse energy is uniform throughout the pulse duration.

17.6. The interface impedance associated with a shock peening process is determined to be $0.6 \times 10^6 \text{g/cm}^2\ \text{s}$. If the laser being used is capable of generating a pulse energy of 95 J, determine the pulse duration necessary for generating a plasma thickness of 60 μm at the end of the pulse. What is the resulting interface pressure? Assume that the beam diameter is 5 mm; that the pressure is invariant with time; that 12.5% of the internal energy is expended as thermal energy; and that the pulse energy is uniform throughout the pulse duration.

17.7. Explain how preheating affects the absorption of a material during laser surface heat treatment.

18 Laser Forming

Forming traditionally has been the manufacturing process by which the size or shape of a part is changed by the application of forces that produce stresses in the part that are greater than the yield strength and less than the fracture strength of the material. It requires the use of dies that determine the final shape of the part. Simple examples of parts produced by this method are beverage containers, angle brackets, or connecting rods. Thermomechanical forming, however, enables parts (sheet metal, rod, pipe, or shell) to be formed without external forces and does not require the use of dies. Laser forming is a type of thermomechanical forming and may be used to form an angle bracket, for example, without using dies. More complex parts, such as connecting rods that involve bulk forming, can only be made by traditional forming methods. However, where laser forming can be used, it also serves as a useful tool for rapid prototyping.

This chapter starts with a discussion on the basic principles of the laser forming process. Typical process parameters that may be used are then outlined. This is followed by a discussion on the various mechanisms that underlie the process. The mechanics of the process are then discussed, followed by the advantages and disadvantages of the process. Finally, possible applications are presented.

18.1 PRINCIPLE OF LASER FORMING

The technique for laser forming is very similar to that for laser surface heat treatment and involves scanning a defocused beam over the surface of the sheet metal to be formed. Moving the laser beam along a straight line without interruption causes the sheet to bend along the line of motion (Fig. 18.1).

The components of a laser forming system include

1. The laser source with beam delivery system.
2. Motion table unit on which the workpiece is mounted, or robot for holding a fiber-optic system.
3. Cooling system where necessary.
4. Temperature monitoring system.
5. Shape monitoring system.
6. Computer control system.

Principles of Laser Materials Processing, by Elijah Kannatey-Asibu, Jr.

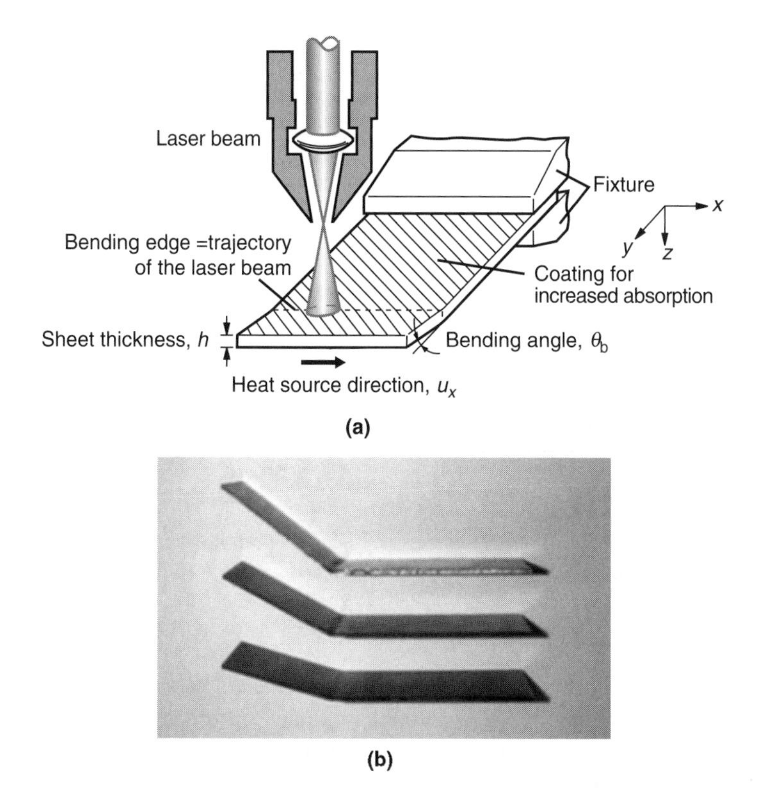

FIGURE 18.1 (a) Schematic of the laser beam bending process. (From Vollertsen, F., 1994b, *Lasers in Engineering,* Vol. 2, pp. 261–276.) (b) Photos of three sheet metals bent using a laser.

The typical bend angle that is achieved in a single step is about 2°, but may be as high as 10°. The total bend angle can be as high as 90° and higher by repetition of the process. The bend angle obtained after the first scan is greater than the bend angle obtained for each of the subsequent scans. However, after the first scan, the bend angle increases almost in proportion to the number of scans. More complex shapes can be obtained by offsetting each track by a small amount. In this case, the radius of the part produced depends, among other processing parameters, on the amount of offset of each track. The smaller the offset, the smaller the radius.

The bend angle that is achieved for each step increases with a decrease in sheet thickness (Fig. 18.2) due to a resulting decrease in bending restraint. The bend angle per step, however, decreases with a decrease in plate width. This is because with decreasing plate width, the volume of material that acts as a heat sink reduces and as a result, the temperature gradient associated with the process decreases, resulting in a reduced compressive strain and thus bend angle. On the contrary, for high width-to-thickness ratios greater than 10, the bend angle is almost independent of the plate width.

Processing parameters also affect the bend angle, with the angle increasing with increasing beam power and reducing scan rate. However, excessive power or very

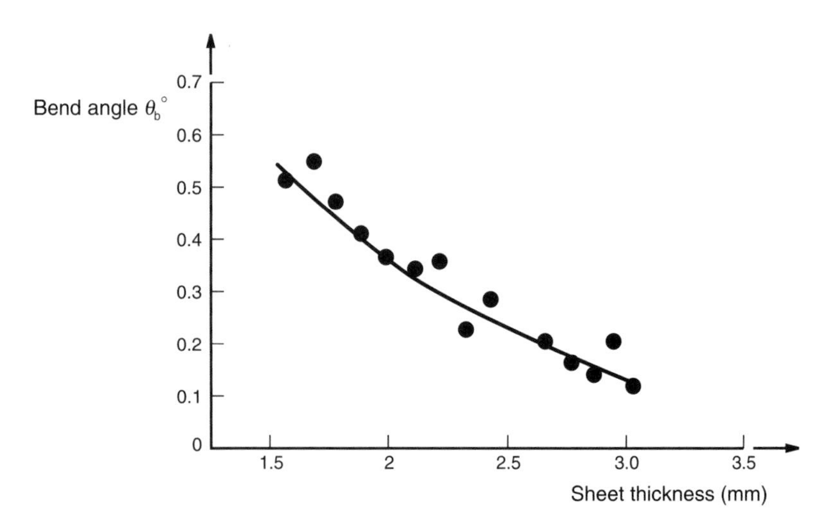

FIGURE 18.2 Influence of sheet thickness on bend angle for a plain carbon steel. (From Vollertsen, F., Geiger, M., and Li, W. M., 1993, *Proceeding of the Fourth International Conference on Technology of Plasticity*, pp. 1793–1798.)

low scan rates will form a molten pool. This is undesirable since it will result in the surface coating being absorbed into the workpiece. Surface coating is normally used to increase the absorption of the laser radiation by the workpiece (see Section 17.1.2.3). The method of cooling does not have very significant effect on the bend angle when laser forming.

18.2 PROCESS PARAMETERS

Typical processing conditions are listed in Table 18.1.

Thicker materials are more effectively bent by increasing the power input. However, the beam diameter needs to be increased accordingly to avoid melting the part. High scan rates are normally required for higher thermal conductivity materials such as aluminum alloys, while low thermal conductivity materials such as titanium alloys require lower scan rates. For the low thermal conductivity materials, if the scan rate is too high, only a thin surface layer will be heated. Efficiency of the process can be

TABLE 18.1 Typical Process Parameters Used in Laser Forming

Parameter	Value	Units
Power	1	kW
Defocused beam diameter	2–6	mm
Scanning velocities	20–600	mm/s
Sheet Thickness	Up to 10	mm

increased by proper application of absorptive coatings to increase the beam absorption, especially on metallic surfaces.

A major issue with the laser forming process is the influence of the irradiation on the material's microstructure. Martensitic transformation, for example, may occur during laser forming of high carbon or alloy steels, and intermetallic phases may be formed in some aluminum alloys, resulting in corrosion. The microstructural effects of laser processing in general are discussed in greater detail in Chapter 11.

18.3 LASER FORMING MECHANISMS

The basic mechanism for laser forming is one of local deformation of the material as a result of thermally induced residual stresses. This may result from one of the three process mechanisms, depending on the temperature field that is generated, the beam shape, and the dimensions of the part to be formed. The three mechanisms are

1. The temperature gradient mechanism.
2. The upsetting mechanism.
3. The buckling mechanism.

These are further discussed in the following sections.

18.3.1 Temperature Gradient Mechanism

The temperature gradient mechanism is illustrated in Fig. 18.3. It evolves from processing conditions such as rapid heating of the workpiece surface, coupled with a relatively low heat conduction that generate very high temperature gradients in the thickness or z-direction (Fig. 18.3a). Thermal expansion of the heated surface results in initial bending of the sheet away from the heat source, or toward the cold side of the workpiece during heating. This is often referred to as counterbending, and the bend angle at this stage is very small, $\approx 0.05°$. As the heated surface tries to expand, however, the constraint of the surrounding material, which is at a lower

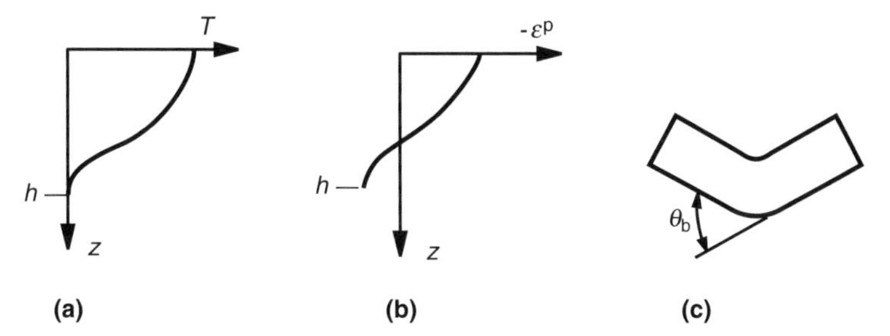

FIGURE 18.3 Temperature gradient mechanism. (a) Temperature variation in the thickness direction. (b) Strain variation in the thickness direction. (c) The bend angle. (From Geiger, M., 1994, *Annals of the CIRP*, Vol. 43, pp. 1–8.)

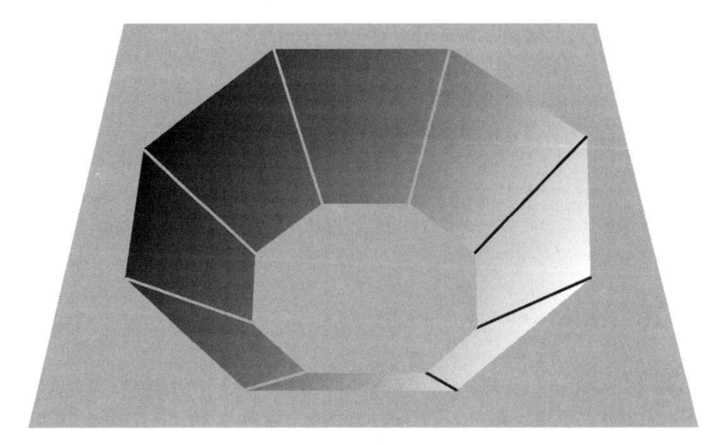

FIGURE 18.4 An illustration of a shape produced by the temperature gradient mechanism. (From Geiger, M., 1994, *Annals of the CIRP,* Vol. 43, pp. 1–8.)

temperature, restricts its free expansion. This results in compressive thermal stresses that induce plastic compressive strain in the heated layer (Fig. 18.3b). Plastic strains are induced since the yield strength of the material is lower at higher temperatures. With increasing ratio of plastic to elastic strain, the amount of final bend angle that results increases. Thus, ideally, it is desirable to minimize the elastic strain by inducing very high temperatures at the top surface layers, where the flow stress is very low.

On cooling, the heated surface layer contracts more than the lower layers ,which are not heated as much as the surface layers, resulting in bending of the sheet toward the heat source (Fig. 18.3c). Figure 18.4 illustrates a shape that is produced by the temperature gradient mechanism.

In the simplest case, a sheet of metal may be bent along a straight line. Repeating the process increases the bend angle, θ_b. During the bending process, the sheet thickness increases in the region along the bend as a result of the compressive strain. The temperature gradient mechanism is significant for relatively thick sheets. The ratio of the heated area diameter to the sheet thickness is relatively low, of the order of 1. It will not be effective for relatively thin materials of high thermal conductivity, since the resulting temperature gradient then will not be adequate to produce any effective bending.

18.3.2 Buckling Mechanism

The buckling mechanism occurs in relatively thin sheets where the ratio of the diameter of the heated area to the sheet thickness is relatively high, of the order of 10. The temperature gradient that arises in the thickness direction is then relatively small, so the material is heated almost uniformly through its thickness. The heated region tends to expand. However, this expansion is hindered by the unheated surrounding material. The restraint results in thermal compressive stresses being developed in the sheet, which for very thin sheets may lead to buckling when a critical stress value is reached (Fig. 18.5). Buckling is more likely to occur when the sheet is relatively thin. Otherwise the critical load for buckling will not be reached.

Bending of material close to the center of the heat source occurs plastically due to the high temperatures in that region, while farther away, close to the root of the buckle, the material is subjected to elastic bending as a result of the lower temperatures in that region. If the heat source traverses the entire length of the sheet, then no restraining forces exist to hold the elastically bent portion in place during cooling. Thus, elastic recovery occurs at the root, while the center remains bent, resulting in

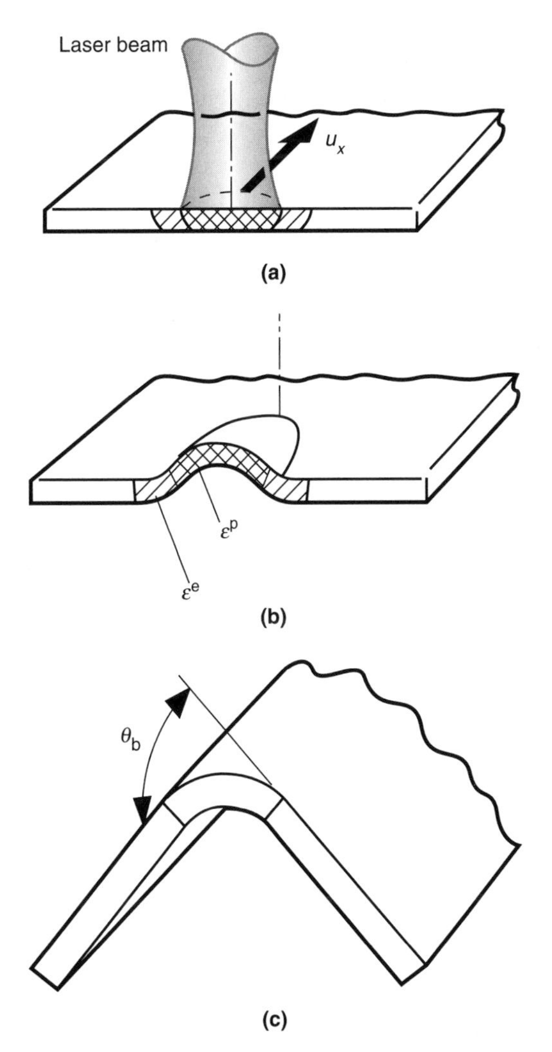

FIGURE 18.5 Buckling mechanism—sequence of steps leading to bending of a plate. (a) Beginning of heating. (b) Growth of buckle. (c) Development of bend angle. (From Vollertsen, F., 1994a, *Proceedings of LANE '94,* Vol. 1, pp. 345–360.)

the sheet being bent along the centerline. The direction of the bending that results from the buckling mechanism is unpredictable. It could be either toward or away from the heat source. This is because it is influenced by a number of factors such as the boundary conditions, the precurvature (or prebending) of the sheet, residual stresses, and external forces. However, it has been observed that at relatively high scan rates, say 15 mm/s (depending on the processing conditions), bending is always toward the laser beam. It only becomes unpredictable at lower speeds. Bending by the buckling mechanism does not increase the sheet thickness at the bend.

18.3.3 Upsetting Mechanism

The upsetting mechanism evolves when uniform heating of a localized zone is achieved through the thickness of the sheet (Fig. 18.6a and b). Thus, the process

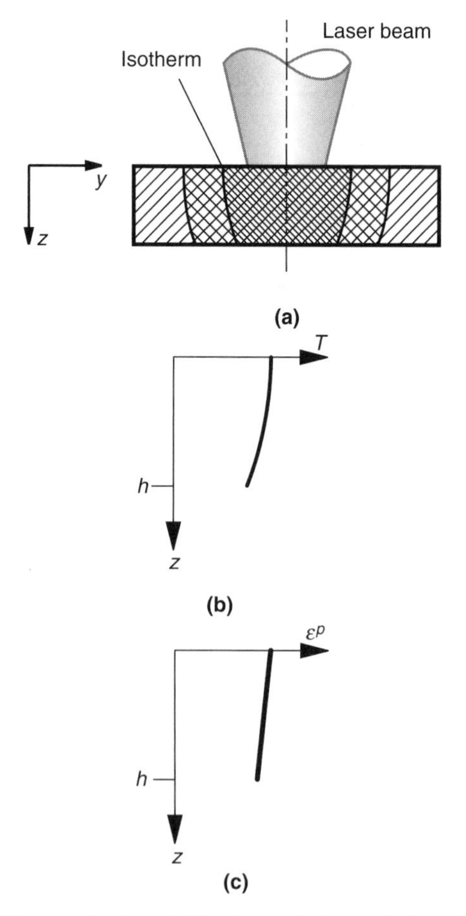

FIGURE 18.6 Upsetting mechanism. (a) Isotherms in the workpiece. (b) Temperature variation in the thickness direction. (c) Strain variation in the thickness direction. (From Vollertsen, F., 1994a, *Proceedings of LANE '94,* Vol. 1, pp. 345–360.)

FIGURE 18.7 Sample product resulting from the upsetting mechanism. (From Vollertsen, F., 1994a, *Proceedings of LANE '94,* Vol. 1, pp. 345–360.)

parameters may be similar to those of the buckling mechanism, except for the diameter of the heat source area that is relatively small. As a result of the near homogeneous heating of the sheet in the localized zone, and prevention of thermal expansion by the surrounding material, the sheet is subjected to near uniform compressive strain through its thickness (Fig. 18.6c). During cooling, the heated region contracts, resulting in corresponding deformation of the sheet. This mechanism enables shapes that are similar to deep drawn parts to be produced (Fig. 18.7).

18.3.4 Summary of the Forming Mechanisms

One common feature of all three mechanisms is the generation of strains by thermal stresses. For the temperature gradient and upsetting mechanisms, bending results primarily from plastic strains that evolve during heating. Since elastic strains are reversible, their presence tends to reduce the bend angle and are thus detrimental to the process. However, elastic strains are the driving force for forming in the case of the buckling mechanism.

The forming mechanisms are further summarized in Table 18.2.

TABLE 18.2 Comparison of the Three Mechanisms for Laser Forming

Parameter	Temperature Gradient Mechanism	Buckling Mechanism	Upsetting Mechanism
Temperature field	Steep temperature gradient	Nearly homogeneous with large diameter in x–y plane	Nearly homogeneous with small diameter in x–y plane
Sheet thickness	Increases	May remain constant	Increases
Forming direction	Toward laser beam; determined by process	Determined by precurvature, residual stresses, and so on.	Mostly toward laser beam; depends on original part geometry
Applications	Bending sheets along straight lines	Bending sheets along straight lines. Tube forming	Spatial sheet metal forming; profile forming

Source: Table 1, Vollerston, LANE 1994a

18.4 PROCESS ANALYSIS

The laser forming process is a coupled problem that involves four basic concepts:

1. Analysis of the temperature field. This is essentially a three-dimensional conduction heat transfer problem involving a moving heat source. Approximate use may be made of the quasistationary solution discussed in Chapter 10. For a more accurate analysis, however, the time-dependent problem incorporating temperature-dependent material properties has to be solved using numerical methods, with the appropriate boundary conditions.
2. Analysis of the plastic strains using the appropriate constitutive laws, incorporating temperature-dependent thermal expansion coefficient, elastic modulus, and flow stress. This would also involve a suitable yield criterion and restraint due to the geometry of the structure.
3. Analysis of the curvature during cooling due to the difference in plastic strains in the thickness direction. This would involve consideration of force and moment equilibrium, strain hardening, and flow stress.
4. Determination of the bend angle as an integral of the curvature that evolves with time.

In the remainder of this section, an approximate method is discussed. This provides some basic results that enable reasonable deductions to be made on the relationship between the process and product variables.

For the temperature gradient method, consider the plate shown in Fig. 18.8 and an associated Cartesian coordinate system, y–z, with origin O directly

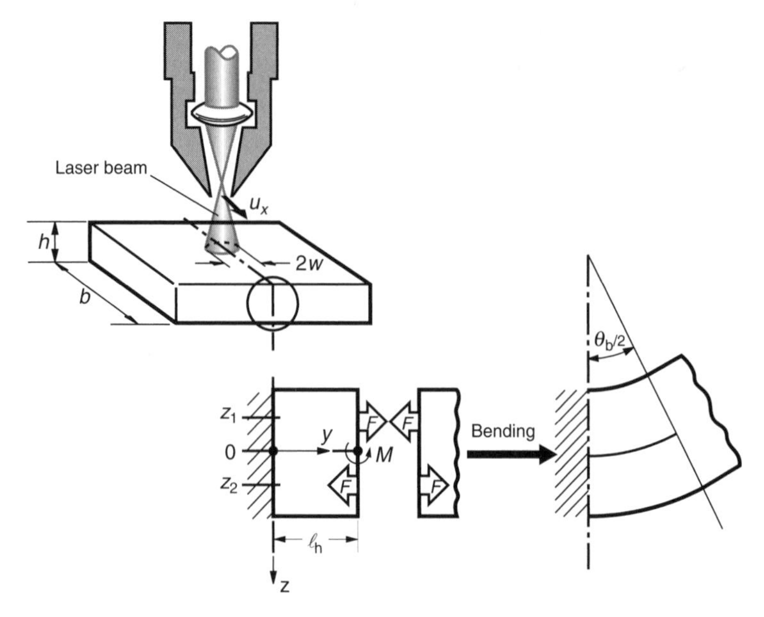

FIGURE 18.8 Two-layer model of laser beam bending. (From Vollertsen, F., 1994b, *Lasers in Engineering,* Vol. 2, pp. 261–276.)

below the centerline of the laser beam and halfway between the top and bottom surfaces. Assume that the sheet is made up of two layers in the heated zone of half-length, l_h, and a single layer for $y > l_h$. Further, assume that the plate width is relatively large compared to its thickness.

Let the centers of the two layers be located at z_1 and z_2. Then

$$z_1 - z_2 = \frac{h}{2} \tag{18.1}$$

where h is the thickness of the original sheet. For this simple case, the bend angle θ_b will be given by the geometry and difference in strains between the upper (ε_1) and lower (ε_2) layers:

$$\frac{\theta_b}{2} = \frac{l_h(\varepsilon_2 - \varepsilon_1)}{h/2} \tag{18.2}$$

where $\theta_b/2$ is the half-bend angle that is based on the plate symmetry.

Let the lower layer be heated to a homogeneous temperature of T_l and the upper layer to a homogeneous temperature of $T_u = T_l + \Delta T$. During the heating phase, the upper layer will be subjected to compressive forces, while the lower layer is subjected to tensile forces, The strain, ε_1, to which the upper layer is subjected will be a superposition of the strains due to the induced force, F, moment, M, and thermal expansion:

$$\varepsilon_1 = \frac{F}{E_{m1}A_1} - \frac{M_1}{(E_m I_m)_1} z_1 + \beta_l \Delta T \tag{18.3}$$

where A is the cross-sectional area (m^2), E_m is the elastic modulus (N/m^2), I_m is the second moment of area (m^4), M is the moment due to the forces (N m), F is the force acting on each layer (N), β_l is the linear coefficient of thermal expansion (K^{-1}), and ΔT is the temperature rise of the upper layer relative to the lower layer (K). The subscripts 1 and 2 indicate the upper and lower layers, respectively.

This analysis is based on the assumption that the thermal expansion of the upper layer is entirely converted into compressive plastic strain. The assumption neglects the temperature dependence of the yield stress, as well as the reversed straining during cooling. This is reasonable since at high temperatures, the yield stress of the material is relatively low.

As the sheet cools, conduction of heat into the lower layer results in a tendency for it to expand. This creates compressive stresses in this layer and that reduces the amount of tensile stress that develops in the upper layer. Thus, there is no cancellation of the compressive strain in the lower layer. The conversion of thermal expansion into plastic compression then results in

$$\varepsilon_1 = \frac{F}{E_{m1}A_1} - \frac{M_1}{(E_m I_m)_1} z_1 - \beta_l \Delta T_m \tag{18.4}$$

The corresponding strain in the lower layer is

$$\varepsilon_2 = -\frac{F}{E_{m2}A_2} - \frac{M_2}{(E_m I_m)_2} z_2 \tag{18.5}$$

where ΔT_m is the maximum temperature difference between the upper and lower layers.

The local curvature is given by the ratio of the moment M to the section modulus $E_m I_m$. For large curvatures, it can be assumed that

$$\frac{\theta_b}{2l_h} = \frac{M_1}{(E_m I_m)_1} = \frac{M_2}{(E_m I_m)_2} = \frac{M}{E_m I_m} \tag{18.6}$$

Substituting equations (18.4–18.6) into equation (18.2) gives

$$\theta_b = \frac{4l_h}{h}\left[-\frac{F}{E_{m2}A_2} - \frac{\theta_b}{2l_h} z_2 - \frac{F}{E_{m1}A_1} + \frac{\theta_b}{2l_h} z_1 + \beta_1 \Delta T_m\right] \tag{18.7}$$

Now, the bending moment is given by $M = F \times h/2$. Combining this with equation (18.6) gives

$$F = \frac{2M}{h} = \frac{\theta_b E_m I_m}{h l_h} \tag{18.8}$$

Also, substituting equations (18.1) and (18.8) into equation (18.7) gives

$$\theta_b = -\frac{4E_m I_m \theta_b (E_{m1}A_1 + E_{m2}A_2)}{h^2 E_{m1} A_1 E_{m2} A_2} + \theta_b + \frac{4 l_h \beta_l \Delta T_m}{h} \tag{18.9}$$

When $h_1 = h_2 = h/2$, and $I_m = bh^3/12$ for a rectangular plate that is heated through the upper half, equation (18.9) then becomes

$$\theta_b = \frac{3\beta_1 \Delta T_m l_h}{h} \tag{18.10}$$

And using an approximate solution for the temperature field, equation (18.10) can also be expressed as

$$\theta_b = \frac{3\beta_1 q}{\rho c_p u_x h^2} \tag{18.11}$$

where q is the power absorbed by the surface. The bend angle calculated is in radians. For the general case where the thicknesses of the upper and lower layers are different, we have

$$A_1 = bh_1 \tag{18.12a}$$

and

$$A_2 = b(h - h_1) \tag{18.12b}$$

The bend angle then becomes

$$\theta_b = \frac{12\beta_l \Delta T_m l_h h_1 (h - h_1)}{h^3} \tag{18.13}$$

An approximate expression for estimating the bend angle that results for each pass of the laser beam during laser forming has been derived in this section. The analysis is based on the temperature gradient mechanism. Similar expressions may be obtained for the other mechanisms. However, only one approach is presented since they are all approximate. The interested reader is referred to Vollertsen (1994a, 1994b). More exact determination of the deformed shape will require setting up the appropriate governing equations for both the heat flow and deformation and solving them numerically. This, however, is still an area of research and thus will not be discussed further.

Example A titanium plate is bent using a 1.5 kW Nd:YAG laser. The plate dimensions are thickness 4.5 mm, width 200 mm, and length 600 mm. The beam radius is 4 mm, and scanning is done at a rate of 40 mm/s along the width of the sheet metal. For a single scan, determine

1. The bend angle, θ_b.
2. The induced force, F.
3. The bending moment, M.
4. Strains, ε_1 and ε_2, in the upper and lower parts of the sheet, respectively.

Assume a surface reflection coefficient of 85%.

Solution:

From Appendices 10D and 10E, the properties of the titanium plate are approximated as average density, $\rho = 4507$ kg/m^3; average specific heat, $c_p = 99.3$ J/kg K; thermal conductivity, $k = 0.0114$ W/mm K; modulus of elasticity, $E = 116$ GPa (Appendix 13B); and linear coefficient of thermal expansion, $\beta_l = 10.8 \times 10^{-6}$/K.

1. From equation (18.11), the bend angle for a single pass is given by

$$\begin{aligned}\theta_b &= \frac{3\beta_l q}{\rho c_p u_x h^2} \\ &= \frac{3 \times 10.8 \times 10^{-6} \times (1 - 0.85) \times 1500}{4.507 \times 10^{-6} \times 99.3 \times 40 \times 4.5^2} = 0.02\ rad = 1.15^\circ \end{aligned} \tag{18.11}$$

2. The induced force is obtained from equation (18.8) as

$$F = \frac{2M}{h} = \frac{\theta_b E_m I_m}{h l_h} \tag{18.8}$$

The second moment of area is given by

$$I_m = \frac{bh^3}{12} = \frac{200 \times 4.5^3}{12} = 1518.75 \text{ mm}^4$$

Now if the half-length of the heated zone is estimated as being equal to the beam radius, then $l_h = 4$ mm. Thus,

$$F = \frac{0.02 \times 116 \times 10^3 \times 1518.75}{4.5 \times 4} = 196 \text{ kN}$$

3. The bending moment is obtained from equation (18.8) as

$$M = F \times \frac{h}{2} = 196{,}000 \times \frac{4.5}{2} = 440 \times 10^3 \text{ N mm} = 440 \text{ N m}$$

4. The strains are given by equations (18.4) and (18.5) as

$$\varepsilon_1 = \frac{F}{E_{m1} A_1} - \frac{M_1}{(E_m I_m)_1} z_1 - \beta_1 \Delta T_m \tag{18.4}$$

$$\varepsilon_2 = -\frac{F}{E_{m2} A_2} - \frac{M_2}{(E_m I_m)_2} z_2 \tag{18.5}$$

The maximum temperature difference between the upper and lower layers can be obtained from equation (18.10) as

$$\Delta T_m = \frac{h\theta_b}{3\beta_1 l_h} = \frac{4.5 \times 0.020}{3 \times 10.8 \times 10^{-6} \times 4} = 694°\text{C}$$

Therefore,

$$\varepsilon_1 = \frac{196 \times 10^3}{116 \times 10^3 \times 450} - \frac{440 \times 10^3}{(116 \times 10^3 \times 1518.75)} \times 1.125 - 10.8 \times 10^{-6} \times 694 = -0.00655$$

and

$$\varepsilon_2 = -\frac{196 \times 10^3}{116 \times 10^3 \times 450} - \frac{440 \times 10^3}{(116 \times 10^3 \times 1518.75)} \times 1.125 = -0.00657$$

18.5 ADVANTAGES AND DISADVANTAGES

18.5.1 Advantages

The primary advantages of laser forming include

1. The noncontact nature of the process that makes it independent of tool inaccuracies that might result from wear and deflection, since no external forces are involved. It also makes the process more flexible.
2. Ability to more accurately control the energy source and thus the forming process, compared to flame bending and mechanical forming in general.
3. Forming of parts in confined or inaccessible locations by remote application.
4. Minimal heat-affected zone size or material degradation compared to flame bending where the heat source is more diffuse.
5. Ability to form hard and brittle materials that are difficult to form by conventional methods.

18.5.2 Disadvantages

The major limitations are

1. The process is relatively slow compared to mechanical forming, by a factor of 5–20.
2. The high capital cost of the laser makes it more expensive than other forms of thermomechanical forming such as flame bending.
3. The low efficiency of the laser results in high energy consumption.
4. Due to the potential reflection of the laser beam in various directions, special safety precautions are essential.

18.6 APPLICATIONS

One area in which laser forming finds useful application is part straightening for mass production. This involves using the laser beam to deform and thus straighten parts that become distorted at one stage or another of their production. Due to the small size of the laser beam, and the ease with which it can be controlled, laser forming is highly suited for forming small components, that is, micrometal forming. Laser forming is also useful for forming of some automobile body components, even though it may not be suitable for the high production rates essential for automobile components. However, due to the low volume requirements in forming of ship planks and production of aerospace fuselage, laser forming is found to be useful for such applications. Finally, due to the high flexibility of the laser beam, it is well suited to the production of sheet metal components in space.

Since laser forming is slow relative to conventional forming it is most economical for maintenance work when there are no spare parts available. Also, since it does not require dies, laser forming is suited for complex components in small-lot production and fabrication of individual parts for rapid prototyping. This enables the performance of several variants of a design to be evaluated before committing to full production equipment.

18.7 SUMMARY

Laser forming is a thermomechanical forming process and thus enables sheet metal parts and pipes to be formed without external forces. Scanning the laser beam along a straight line causes the sheet metal to bend along the line of motion. More complex parts, however, can be formed by using a more complex trajectory. The three principal process mechanisms are

1. The temperature gradient mechanism—which results from processing conditions that generate very high temperature gradients in the sheet metal.
2. The buckling mechanism—which is more common in relatively thin sheets, where the ratio of the diameter of the heated area to the sheet thickness is relatively high, of the order of 10.
3. The upsetting mechanism—which evolves when uniform heating of a localized zone is achieved through the thickness of the sheet, using a heat source area that is relatively small.

Analysis shows that the bend angle achieved is directly proportional to the power input and inversely proportional to the traverse velocity, as well as the inverse of the plate thickness squared.

Compared to traditional sheet metal forming, laser forming is relatively slow. However, it is noncontact and is also suitable for applications such as rapid prototyping and micrometal forming. In addition, compared to other thermomechanical forming processes, it is appropriate for situations where very little microstructural damage to the base material can be sustained.

REFERENCES

1992, IMTS showcases laser forming, *Manufacturing Engineering,* Vol. 109, p. 42.

Bao, J. and Yao, L., 1999, Analysis and prediction of edge effects in laser bending, *ICALEO 99*, San Diego, CA, pp. 253–262.

Frackiewicz, H., 1993, Laser metal forming technology, *Fabtech, International '93,* The Society of Manufacturing Engineers, pp. 733–747.

Geiger, M., 1994, Synergy of laser material processing and metal forming, *Annals of the CIRP,* Vol. 43, pp. 1–8.

Geiger, M., Vollertsen, F., and Deinzer, G., 1993, Flexible straightening of car body shells by laser forming, *SAE International Congress and Exposition,* Detroit, MI, pp. 37–44.

Magee, J., Watkins, K. G., Steen, W. M., Calder, N. J., Sidhu, J., and Kirby, J., 1998, Laser bending of high strength alloys, *Journal of Laser Applications,* Volume 10, No. 4, pp. 149–155.

Masubuchi, K., 1992, Studies at M.I.T. related to applications of laser technologies to metal fabrication, *Proceedings of LAMP '92,* pp. 939–946.

Namba, Y., 1986, Laser forming in space, *International Conference on Lasers,* Wang, C. P., editor, pp. 403–407.

Namba, Y., 1987, Laser forming of metals and alloys, *Proceedings of LAMP '87,* pp. 601–606.

Pridham, M. S., and Thomson, G. A., 1994, Laser forming: a force for the future? *Materials World,* Vol. 2, pp. 574–575.

Scully, K., 1987, Laser line heating, *Journal of Ship Production,* Vol. 3, pp. 237–246.

Thomson, G. A., and Pridham, M. S., 1995, Laser forming, *Manufacturing Engineer,* Vol. 74, pp. 137–139.

Vaccari, J. A., 1993, The promise of laser forming, *American Machinist,* Vol. 137, pp. 36–38.

Veiko, Y. P., and Yakovlev, Y. B., 1994, Physical fundamentals of laser forming of micro-optical components, *Optical Engineering*, Vol. 33, pp. 3567–3571.

Vollertsen, F., Geiger, M., and Li, W. M., 1993, FDM and FEM simulation of laser forming: a comparative study, *Advanced Technology of Plasticity 1993—Proceeding of the Fourth International Conference on Technology of Plasticity,* pp. 1793–1798.

Vollertsen, F., 1994a, Mechanisms and models for laser forming, *Laser Assisted Net shape Engineering Proceedings of the LANE '94,* Vol. 1, pp. 345–360.

Vollertsen, F., 1994b, An analytical model for laser bending, *Lasers in Engineering,* Vol. 2, pp. 261–276.

Vollertsen, F., Komel, I., and Kals, R., 1995, The laser bending of steel foils for microparts by the buckling mechanism—a model, *Modelling and Simulations in Material Sciences and Engineering* Vol. 3, pp. 107–119.

Zhang, W., Yao, L., and Noyan, I., 2004a, Microscale laser shock peening of thin films, part 1: experiment, modeling and simulation, *ASME Journal of Manufacturing Science and Engineering*, Vol. 126, pp. 10–17.

Zhang, W., Yao, L., and Noyan, I., 2004b, Microscale laser shock peening of thin films, part 2: high spatial resolution material characterization, *ASME Journal of Manufacturing Science and Engineering*, Vol. 126, pp. 18–24.

Zhang, X. R., Chen, G., and Xu, X., 2002, Numerical simulation of pulsed laser bending, *ASME Journal of Applied Mechanics*, Vol. 69, pp. 254–260.

Zhang, X. R., and Xu, X., 2003, High precision microscale bending by pulsed and CW lasers, *ASME Journal of Manufacturing Science and Engineering*, Vol. 125, pp. 512–518.

APPENDIX 18A

List of symbols used in the chapter.

Symbol	Parameter	Units
I_m	second moment of area	m^4
l_h	half-length of heated layer	mm
T_l	temperature of lower layer	°C
T_u	temperature of upper layer	°C
ΔT	temperature difference between upper and lower layers	°C
ΔT_m	maximum temperature difference between upper and lower layers	°C
z_1, z_2	centers of two layers in the heated zone	mm
$\varepsilon_1, \varepsilon_2$	strains in upper and lower layers	–
εth	thermal strain	–
θ_b	bend angle	rad

PROBLEMS

18.1. A low carbon steel sheet metal is bent to form a shallow V-shape using a 2 kW CO_2 laser. The beam radius is 5 mm, and scanning is done at a rate of 25 mm/s along the width of the sheet metal. If the sheet thickness is 3 mm, width 250 mm, and length 400 mm, determine, after the first pass

1. The bend angle.
2. The induced force, F.

   Assume a surface reflection coefficient of 75%.

18.2. For the conditions in Problem 18.1, determine

1. The bending moment, M, that is induced during the process.
2. The strains, ε_1 and ε_2, in the upper and lower parts of the sheet, respectively.
3. How much time will be required to bend the sheet metal to an angle of 45°, if the beam rapid traverse is neglected?

18.3. A 6061-T6 aluminum alloy of thickness 4 mm, length 250 mm, and width 50 mm is to be bent along its width to form a 90° angle bracket for an automobile component. If the desired production rate for the part is 100 pieces per hour,

1. Calculate the power input to the workpiece that will be required for the process.
2. Select an appropriate laser for the process and estimate its desired power output.

18.4. A titanium plate is to be bent slightly to an angle of 2.5° using a laser. The plate width is 75 mm, and length 250 mm. How much power will be required to complete the process in a single scan if the beam radius is 5 mm, scanning is done at a rate of 50 mm/s along the width of the sheet metal, and the temperature rise between the top and bottom layers is $\Delta T_m = 775°C$? Assume a surface reflection coefficient of 75%. What force will be induced in the workpiece under these conditions?

19 Rapid Prototyping

Prototyping is a practice that has traditionally been used to facilitate the conceptualization of a design. Rapid prototyping (RP) is an advancement of this technology, and simply put, it involves the generation of physical objects from graphical computer data without the use of traditional tools. Rapid prototyping is referred to in the literature by several other names, for example, stereolithography (STL), solid freeform manufacturing (fabrication), layer manufacturing, desktop manufacturing, 3D printing, and in recent years, digital manufacturing. It is used in design conceptualization, to facilitate engineering stress analysis, and in making patterns and/or dies for manufacturing operations.

In this chapter, we shall start with a general discussion on the basic concepts of rapid prototyping, namely, computer-aided design (CAD), part building (processing), and post-processing. The major types of rapid prototyping systems in current use (including nonlaser-based systems) are outlined under part building. This is followed by a discussion on the main applications of rapid prototyping. A general overview of rapid prototyping is illustrated in Fig. 19.1, indicating some of the major processes currently in use.

Producing a part by rapid prototyping typically involves three main stages:

1. Computer-aided design.
2. Part building (processing).
3. Post-processing.

These basic steps are common amongst the different types of rapid prototyping techniques, and will be discussed in that order. The specific rapid prototyping techniques will be discussed under part building.

19.1 COMPUTER-AIDED DESIGN

The first stage of rapid prototyping involves either using an advanced 3D (three-dimensional) computer-aided design system to develop a computer model of the part to be produced or using reverse engineering to develop a computer model from a part. The reverse engineering approach normally involves using, for example, either

Principles of Laser Materials Processing, by Elijah Kannatey-Asibu, Jr.

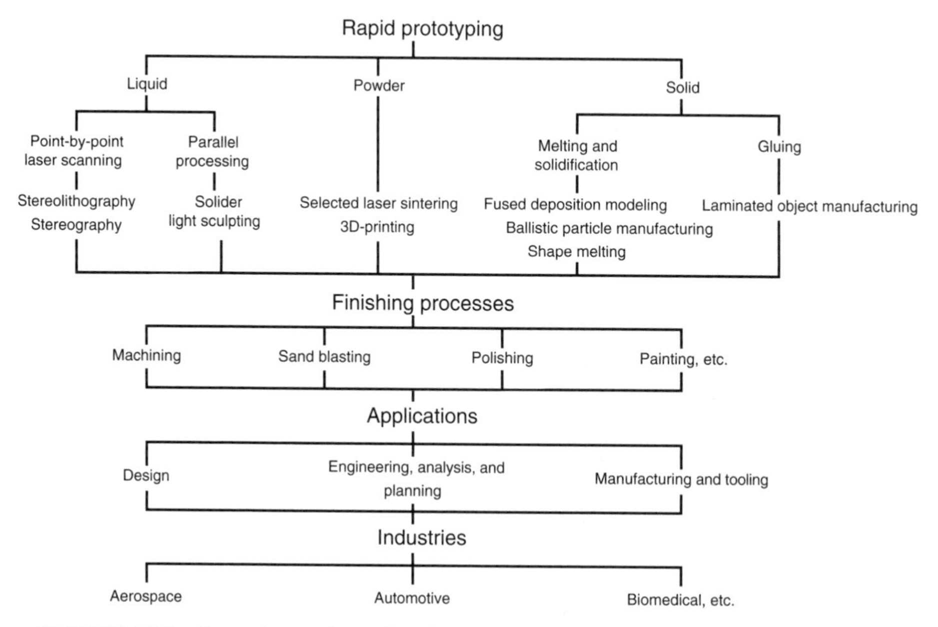

FIGURE 19.1 General overview of rapid prototyping systems. (From Kochan, D., 1993, *Solid Freeform Manufacturing*, Elsevier, Amsterdam, Holland. Copyright Elsevier.)

a coordinate measuring machine (CMM) or a laser digitizer to trace out the part. For the computer model to be most useful, it needs to be represented as closed surfaces that define an enclosed volume, thus ensuring that all horizontal cross sections are closed curves. A number of techniques are available for representing computer models of objects and they often use geometric transformations. These techniques normally involve the creation of curves, surfaces, and solids. The latter is generally referred to as solid modeling and is most suited for rapid prototyping applications. We start our discussion in this section with a brief background on geometric transformation, followed by representation of curves, surfaces, and solids.

19.1.1 Geometric Transformation

Geometric transformation is used to change (transform) a group or individual features of an object. The most common transformations are translation, scaling, and rotation. These operations can be represented in compact form using matrix notation as

$$\mathbf{Q} = \mathbf{C}\mathbf{T} \tag{19.1}$$

where **C** represents a set of data constituting the object to be transformed, **Q** represents the transformed object, and **T** is the transformation matrix.

The nature of the transformation matrix is determined by the operation to be performed, and in the following paragraphs, we present these for the operations listed. Due to space limitations, we focus only on three-dimensional transformations.

19.1.1.1 Translation Translation essentially moves the object from one location to another. Now let a line $\mathbf{C} = \begin{bmatrix} x_1 & y_1 & z_1 \\ x_2 & y_2 & z_2 \end{bmatrix}$ be translated to another location $\mathbf{Q} = \begin{bmatrix} X_1 & Y_1 & Z_1 \\ X_2 & Y_2 & Z_2 \end{bmatrix}$ such that

$$X_1 = x_1 + T_x, \quad Y_1 = y_1 + T_y, \quad Z_1 = z_1 + T_z$$
$$X_2 = x_2 + T_x, \quad Y_2 = y_2 + T_y, \quad Z_2 = z_2 + T_z$$

This may be achieved by the following operation:

$$\begin{bmatrix} X_1 & Y_1 & Z_1 & 1 \\ X_2 & Y_2 & Z_2 & 1 \end{bmatrix} = \begin{bmatrix} x_1 & y_1 & z_1 & 1 \\ x_2 & y_2 & z_2 & 1 \end{bmatrix} \begin{bmatrix} 1 & 0 & 0 & 0 \\ 0 & 1 & 0 & 0 \\ 0 & 0 & 1 & 0 \\ T_x & T_y & T_z & 1 \end{bmatrix} \tag{19.2}$$

where T_x, T_y, and T_z are the distances by which the line is translated in the x, y, and z directions, respectively.

19.1.1.2 Scaling Scaling results in the magnification or reduction in the size of an object. If a line $\mathbf{C} = \begin{bmatrix} x_1 & y_1 & z_1 \\ x_2 & y_2 & z_2 \end{bmatrix}$ is scaled to another line $\mathbf{Q} = \begin{bmatrix} X_1 & Y_1 & Z_1 \\ X_2 & Y_2 & Z_2 \end{bmatrix}$, we have

$$X_1 = x_1 T_x, \quad Y_1 = y_1 T_y, \quad Z_1 = z_1 T_z$$
$$X_2 = x_2 T_x, \quad Y_2 = y_2 T_y, \quad Z_2 = z_2 T_z$$

This may be achieved by the following operation:

$$\begin{bmatrix} X_1 & Y_1 & Z_1 & 1 \\ X_2 & Y_2 & Z_2 & 1 \end{bmatrix} = \begin{bmatrix} x_1 & y_1 & z_1 & 1 \\ x_2 & y_2 & z_2 & 1 \end{bmatrix} \begin{bmatrix} T_x & 0 & 0 & 0 \\ 0 & T_y & 0 & 0 \\ 0 & 0 & T_z & 0 \\ 0 & 0 & 0 & 1 \end{bmatrix} \tag{19.3}$$

where T_x, T_y, and T_z are the factors by which the line is scaled in the x, y, and z directions, respectively.

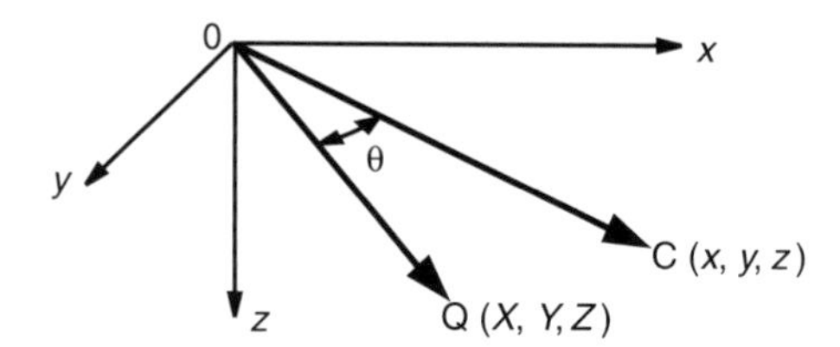

FIGURE 19.2 Rotation of a point through an angle θ about an axis.

19.1.1.3 Rotation Rotation of a point $\mathbf{C} = (x, y, z)$ about the z-axis, which passes through the origin, to another point $\mathbf{Q} = (X, Y, Z)$ (Fig. 19.2), can be achieved as follows:

$$\mathbf{Q} = \mathbf{C}\mathbf{T_R} \tag{19.4}$$

where

$$[\,\mathbf{T_R}\,] = [\,\mathbf{T_z}\,] = \begin{bmatrix} \cos\theta & \sin\theta & 0 & 0 \\ -\sin\theta & \cos\theta & 0 & 0 \\ 0 & 0 & 1 & 0 \\ 0 & 0 & 0 & 1 \end{bmatrix} \tag{19.4a}$$

and θ is the angle of rotation.

For rotation about the x-axis, we have

$$[\,\mathbf{T_R}\,] = [\,\mathbf{T_x}\,] = \begin{bmatrix} 1 & 0 & 0 & 0 \\ 0 & \cos\theta & \sin\theta & 0 \\ 0 & -\sin\theta & \cos\theta & 0 \\ 0 & 0 & 0 & 1 \end{bmatrix} \tag{19.4b}$$

And for rotation about the y-axis,

$$[\,\mathbf{T_R}\,] = [\,\mathbf{T_y}\,] = \begin{bmatrix} \cos\theta & 0 & -\sin\theta & 0 \\ 0 & 1 & 0 & 0 \\ \sin\theta & 0 & \cos\theta & 0 \\ 0 & 0 & 0 & 1 \end{bmatrix} \tag{19.4c}$$

Rotation about an axis that does not pass through the origin is accomplished through the following steps:

1. The object is first translated so that the axis of rotation passes through the origin, $\mathbf{T_{T1}}$.
2. It is then rotated about the axis, $\mathbf{T_R}$.
3. The object is finally translated back to its former position, $\mathbf{T_{T2}}$.

The overall transformation may then be expressed as

$$\mathbf{Q} = \mathbf{C}\mathbf{T}_{\mathbf{T1}}\mathbf{T}_{\mathbf{R}}\mathbf{T}_{\mathbf{T2}} \tag{19.5}$$

19.1.2 Curve and Surface Design

Curve and surface design is important since all objects are bounded by curves and surfaces. They are especially useful in the design of aircraft fuselage, automobile bodies, ship hulls, and sculptures. Two common curve forms that are used in geometric modeling are splines and Bezier curves.

19.1.2.1 Splines A spline is essentially a smooth curve that is made to go through a set of points. The simplest form is the cubic spline that can be used to fit a given set of data points, $x_0, x_1, \ldots, x_n$. This is done by first fitting each contiguous pair of points, $(x_j,\ x_{j+1})$, with a cubic polynomial such that the polynomial and its first and second derivatives are continuous on a set of intervals from x_0 to x_n.

A cubic polynomial can generally be expressed as:

$$y = f(x) = \sum_{i=0}^{3} a_i x^i \tag{19.6}$$

Thus, the cubic spline function can be expressed in the following form:

$$S_j(x) = a_{0j} + a_{1j}x + a_{2j}x^2 + a_{3j}x^3, \quad j = 0, 1, 2, \ldots, n-1 \tag{19.7}$$

where S_j represents span j of the spline, with the following conditions:

$$S(x_j) = f(x_j), \quad j = 0, 1, 2, \ldots, n \tag{19.7a}$$

$$S_{j+1}(x_{j+1}) = S_j(x_{j+1}), \quad j = 0, 1, 2, \ldots, n-2 \tag{19.7b}$$

$$S'_{j+1}(x_{j+1}) = S'_j(x_{j+1}), \quad j = 0, 1, 2, \ldots, n-2 \tag{19.7c}$$

$$S''_{j+1}(x_{j+1}) = S''_j(x_{j+1}), \quad j = 0, 1, 2, \ldots, n-2 \tag{19.7d}$$

The spline function is fully determined when the coefficients, a_{ij}, in equation (19.7) are determined. This is done by applying equations (19.7a)–(19.7d) to each span (S_j) of the spline, that is, between two contiguous points, $(x_j,\ x_{j+1})$. Unfortunately, this will result in a system of $n-1$ equations with $n+1$ unknowns. The other two equations are obtained by considering the boundary conditions at the end points of the spline, A, B (Fig. 19.3). If these are given by

$$S''(x_0) = 0$$

$$S''(x_n) = 0$$

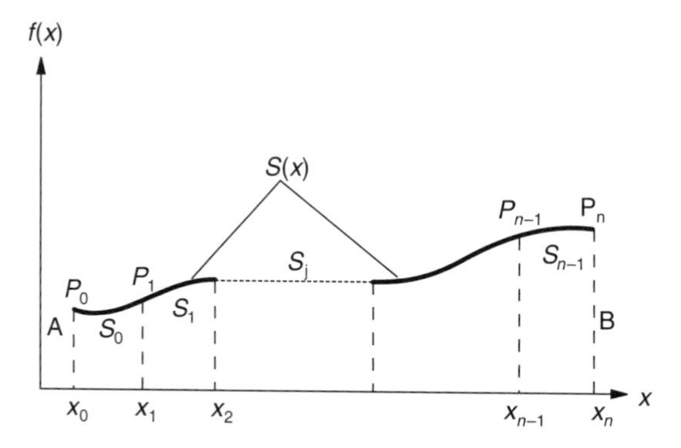

FIGURE 19.3 A spline function.

then the resulting spline is a natural spline. It is called a clamped spline if the boundary conditions are specified at the ends as follows:

$$S'(x_0) = f'(x_0)$$
$$S'(x_n) = f'(x_n) \tag{19.8}$$

where $f'(x_0)$ and $f'(x_n)$ are specified.

Equation (19.7) defines nonparametric cubic spline, and these have the property that they are single-valued functions. This makes them unsuitable for multivalued or closed curves, which are often encountered in geometric modeling. This problem may be overcome with the use of parametric splines. In parametric form, the variables x and y, equation (19.6), are expressed in terms of a third variable, t, such that

$$x = x(t) \tag{19.8a}$$

$$y = y(t) \tag{19.8b}$$

The cubic spline between two points is then expressed as

$$S(t) = \sum_{i=0}^{3} a_i t^i \quad t \in [0, 1] \tag{19.9}$$

And a point on the curve is given by

$$S(t) = [x(t) \;\; y(t)] \tag{19.10}$$

Again, the constants a_i in equation (19.9) are obtained by considering the boundary conditions for each spline and then also the overall boundary conditions for a natural spline or a clamped spline.

19.1.2.2 Bezier Curves The major drawback of cubic splines is that the curves are uniquely defined within a given interval and thus the shape cannot be controlled. Bezier curves permit control of the curve's shape. They are not constrained to pass through all the specified points. They only have to pass through the end points. The data points define the vertices of a polygon that defines the shape of the curve. The curve is tangent to the polygon at the end points and lies within the convex of the polygon (Fig. 19.4). The mathematical basis of the Bezier curve is derived from the Bernstein polynomial, which is given, for a degree n, by

$$B_{n,i}(t) = \begin{bmatrix} n \\ i \end{bmatrix} t^i(1-t)^{n-i}, \quad i = 0, 1, ..., n \tag{19.11}$$

where

$$\begin{bmatrix} n \\ i \end{bmatrix} = \frac{n!}{i!(n-i)!}$$

$n!$ is the n factorial, given by $\quad n! = n(n-1)(n-2)...2 \cdot 1$

The Bezier curve is given by

$$S(t) = \sum_{i=0}^{n} P_i B_{n,i}(t) \quad t \in [0, 1] \tag{19.12}$$

where $S(t)$ denotes points on the curve, P_i indicates specified data points, t is selected increment, n is degree of the Bezier polynomial, and $n + 1$ is number of data points, giving $n + 1$ vertices of the polygon.

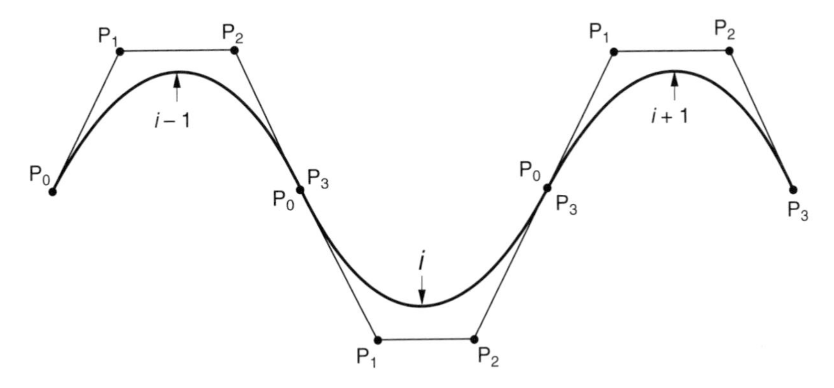

FIGURE 19.4 A Bezier curve.

For a three-dimensional curve, we have

$S(t) = [x(t) \quad y(t) \quad z(t)]$ = points on the curve
$P_i = [x_i \quad y_i \quad z_i]$ = coordinates of the data points.

19.1.2.3 Surface Representation A common approach used in constructing complex surfaces is to concatenate several small patches, (Fig. 19.5a). Another type is the ruled surface that contains two space parametric curves as opposite boundary curves. Linear interpolation is then used to connect the two curves at several points, thereby defining a surface (Fig. 19.5b). A curve may also be rotated about an axis to generate a surface of revolution, (Fig. 19.5c). A patch, for example, may be generated as a Bezier surface, which, for a region bounded by four curves, is expressed by

$$S(s, t) = \sum_{i=0}^{m} \sum_{j=0}^{n} P_{i,j} G_{i,m}(s) G_{j,n}(t) \tag{19.13}$$

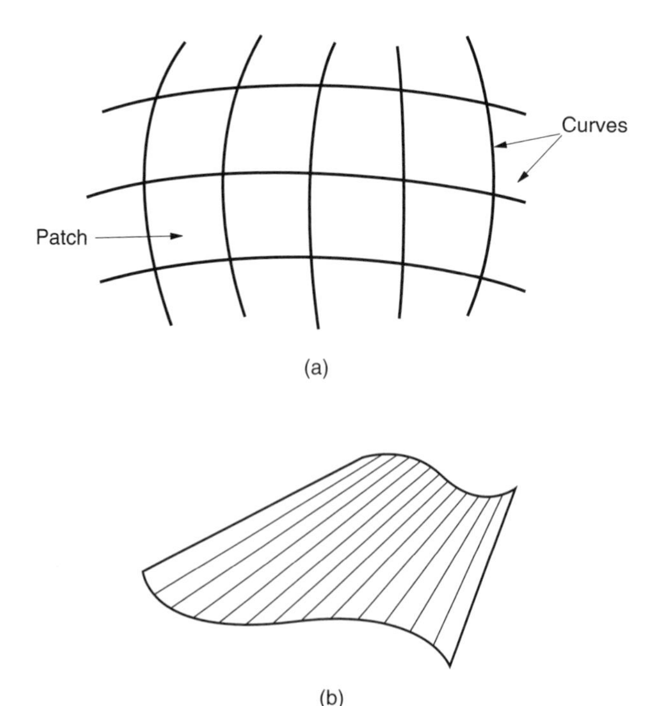

FIGURE 19.5 (a) A surface constructed from patches. (b) A ruled surface. (c) A surface of revolution.

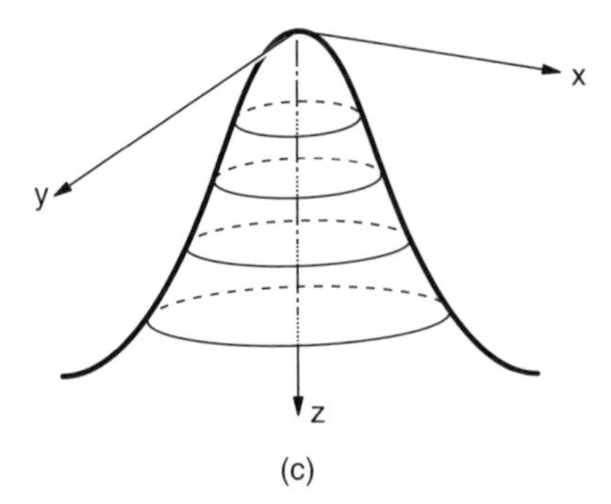

(c)

FIGURE 19.5 *(Continued)*

where $P_{i,j}$ indicates the specified data points $(m+1 \quad n+1)$, $G_{i,m}(s) = m!s^i(1-s)^{m-i}/i!(m-i)!$, and $G_{j,n}(t) = n!t^j(1-t)^{n-j}/j!(n-j)!$.

A point on the surface is given by $S(s,t) = [x(s,t)\ y(s,t)\ z(s,t)]$.

Example 19.1 Generate the Bezier curve for the data points given by P_0 [1, 2], P_1 [2, 3], P_2 [3, 0] and P_3 [4, 1]

Solution:

Since there are four points, $n = 3$. From equation (19.11), we obtain $B_{n,i}(t)$ for $n = 3$, $i = 0$ to 3 as follows:

$$B_{3,0}(t) = (1-t)^3$$
$$B_{3,1}(t) = t(1-t)^2$$
$$B_{3,2}(t) = t^2(1-t)$$
$$B_{3,3}(t) = t^3$$

With these functions, the Bezier curve is now given by

$$S(t) = P_0 B_{3,0} + P_1 B_{3,1} + P_2 B_{3,2} + P_3 B_{3,3}$$

For values of $t = 0.0,\ 0.25,\ 0.5,\ 0.75$, and 1.0, the corresponding Bezier coefficients are listed in Table 19.1.

TABLE 19.1 Bezier Function Values for Different *t* Values

t	$B_{3,0}(t)$	$B_{3,1}(t)$	$B_{3,2}(t)$	$B_{3,3}(t)$
0	1	0	0	0
0.25	0.421	0.421	0.14	0.015
0.5	0.125	0.375	0.375	0.125
0.75	0.015	0.14	0.421	0.421
1	0	0	0	1

Points on the Bezier curve are then given by

$$S(0) = [1, 2] \cdot 1 + [2, 3] \cdot 0 + [3, 0] \cdot 0 + [4, 1] \cdot 0 = [1, 2]$$

$$S(0.25) = [1, 2] \cdot 0.421 + [2, 3] \cdot 0.421 + [3, 0] \cdot 0.14 + [4, 1] \cdot 0.015 = [1.743, 2.12]$$

$$S(0.5) = [1, 2] \cdot 0.125 + [2, 3] \cdot 0.375 + [3, 0] \cdot 0.375 + [4, 1] \cdot 0.125 = [2.5, 1.5]$$

$$S(0.75) = [1, 2] \cdot 0.015 + [2, 3] \cdot 0.14 + [3, 0] \cdot 0.421 + [4, 1] \cdot 0.421 = [3.242, 0.871]$$

$$S(1) = [1, 2] \cdot 0 + [2, 3] \cdot 0 + [3, 0] \cdot 0 + [4, 1] \cdot 1 = [4, 1]$$

The corresponding Bezier curve is shown in the following graph:

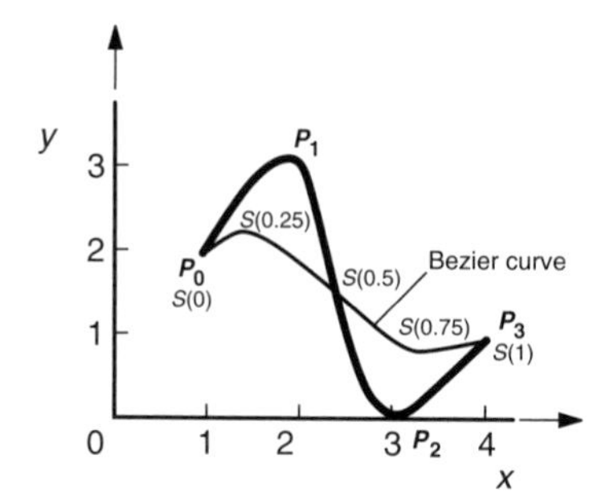

19.1.3 Solid Modeling

The curve generation methods discussed in the preceding sections are useful for generating wireframe models of objects. The wireframe method has the advantage of simplicity and speed of implementation. Even though quite useful for visualizing certain types of objects, it has a number of disadvantages. Since hidden lines cannot be removed, representation of complex objects using the wireframe method can often be ambiguous. Furthermore, it does not incorporate information on the physical characteristics of the object such as volume and mass. These drawbacks are eliminated with the use of solid models. Two of the most common solid modeling methods are constructive solid geometry (CSG) and boundary representation (BREP).

19.1.3.1 Constructive Solid Geometry The constructive solid geometry technique builds 3D objects using primitives, which are basic geometric shapes such as cubes, cylinders, cones, and spheres as building blocks. Boolean operators such as union ($\cup$), intersection ($\cap$), and difference ($-$) are used to combine them to create the final object. Figure 19.6 illustrates these operations using simple geometric figures. The process of union combines primitives into a new object, whereas the process of difference may be used to create a hole, for example. Intersection results in an object whose

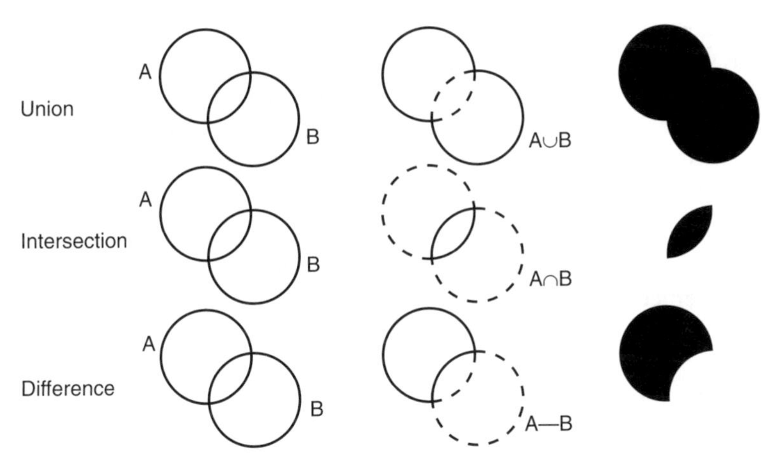

FIGURE 19.6 Boolean operations on simple geometric figures.

volume is common to the initial primitives and is useful for checking interference between objects.

The objects formed at intermediate stages of the CSG product development are organized in an orderly manner using a tree structure. A tree node represents an intermediate solid, which is obtained by combining primitive and intermediate solids lower down in the tree's hierarchy. This is illustrated for a simple object in Fig. 19.7.

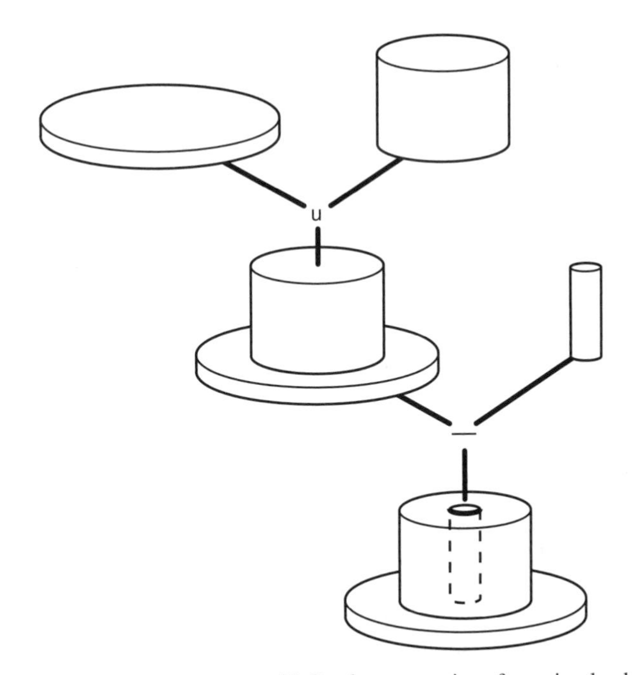

FIGURE 19.7 Tree structure with Boolean operations for a simple object.

One advantage of the CSG method is that it does not require as much memory storage as the other solid modeling methods, even though the computational time for constructing the object may be more. Also, the structure of the CSG tree generally parallels the sequence of operations that will be required to manufacture the object. Thus the designer is able to envisage the basic manufacturing plan for the object.

19.1.3.2 ***Boundary Representation*** With the boundary representation method, objects are defined by their enclosing surfaces or boundaries, in terms of their faces, edges, and vertices. Boolean operations may then be used to combine these entities to form the desired object. The end result is similar to a wireframe object, but can incorporate the object's physical properties. The BREP technique is useful for nonstandard shapes such as aircraft fuselage and automobile bodies and enables models to be constructed faster. However, it requires more storage, but less computational time in constructing the object.

A number of systems use a combination of these basic concepts. Common CAD systems that are based on solid modeling include I-DEAS, Unigraphics II, Pro/Engineer, CATIA, and Alias Designer. Generally, for convenience, and also to avoid potential errors downstream, all CAD values are positive numbers. That means the part geometry must reside in the positive x, y, and z octant.

19.1.4 Rapid Prototyping Software Formats

The CAD model is next converted into a format required by the rapid prototyping equipment. After the conversion, the file is then transmitted to the RP system's computer. Two common formats currently in use are the stereolithography (STL) and initial graphics exchange specifications (IGES) formats.

19.1.4.1 ***The STL (Stereolithography) Format*** This was developed by 3D Systems and is the most commonly used format. It approximates the outside surfaces of the model using polygons, usually, triangular facets (Fig. 19.8). For example, a rectangle is represented using two triangles, whereas a curved surface is approximated using a number of triangles. Of course, for curved surfaces, the larger the number of triangles used, the more accurate the representation. However, the file size or computation time then increases accordingly, even though the subsequent build (processing) time is not significantly affected. The desired accuracy is achieved by specifying a part resolution. This may be expressed either as a maximum absolute deviation from the true surface (Fig. 19.9a), or as a percentage of chord length (Fig. 19.9b), depending on the CAD system.

Each triangular facet is described by a set of X, Y, and Z coordinates for each of the three vertices and a unit normal vector to indicate which side of the facet is inside the object (Fig. 19.10). Since STL files do not contain topological data, the polygonal approximation models used tend to exhibit a number of errors including gaps (missing facets due to errors at the intersections between surfaces with large curvature, Fig. 19.8), and overlapping facets (resulting from numerical round-off errors).

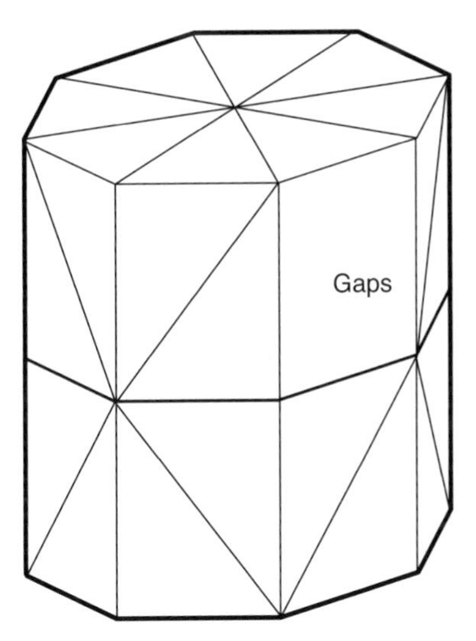

FIGURE 19.8 Tessellation of an object. (From Chua, C. K., and Leong, K. F., 1997, *Rapid Prototyping: Principles & Applications in Manufacturing*. Permission of World Scientific Publishing Co.)

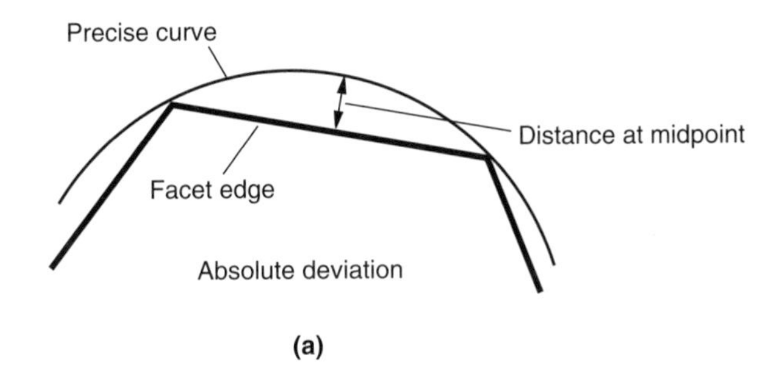

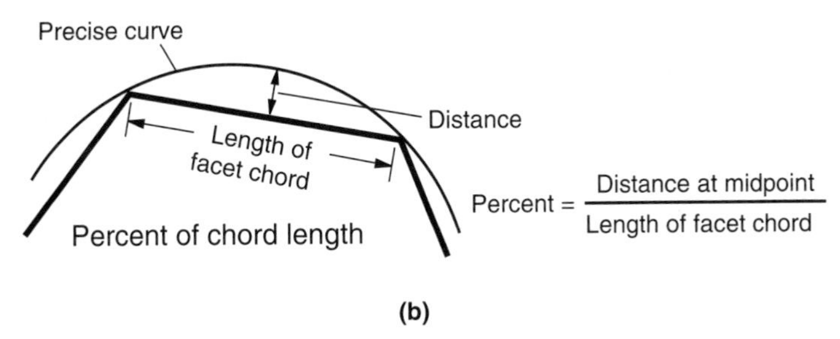

FIGURE 19.9 Accuracy specification for a CAD system. (a) Absolute deviation. (b) Percent of chord length. (From Jacobs, P. F., 1992, *Rapid Prototyping & Manufacturing: Fundamentals of Stereolithography*. Reprinted by permission of Society of Manufacturing Engineers, Dearborn, MI.)

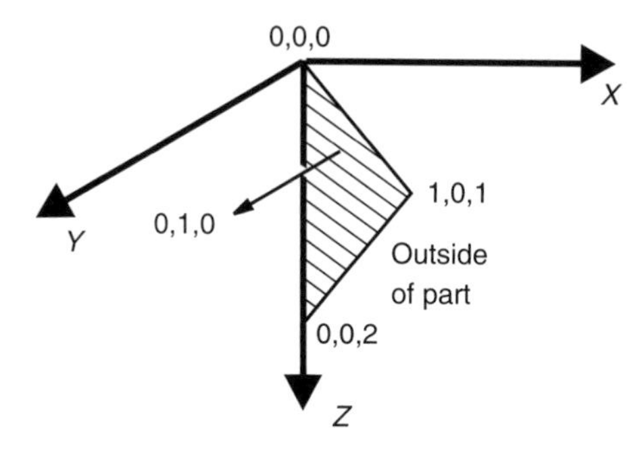

FIGURE 19.10 Identifying a triangular facet with a unit normal in an STL file. (From Chua, C. K., and Leong, K. F., 1997, *Rapid Prototyping: Principles & Applications in Manufacturing*. Permission of World Scientific Publishing Co.)

The STL format has the advantage that it is simple, providing simple files for data transfer. However, it has the disadvantage that its file is much larger than the original CAD file, since it carries a lot of redundant information. In addition, it is susceptible to the geometric errors mentioned in the preceding paragraph. Finally, slicing of the files takes a relatively long time due to the large file sizes.

19.1.4.2 The IGES Format This is a standard used to exchange graphics information between commercial CAD systems. It is capable of precisely representing CAD models and includes not only geometric but also topological information. The entities of points, lines, arcs, splines, surfaces, and solid elements are provided. Ways of representing the regularized operations for union, intersection, and difference are defined. IGES files, however, have the disadvantage that they often contain redundant information that is not needed for RP systems. Furthermore, the algorithms for slicing (see section 19.1.6) an IGES file are more complex than those for slicing an STL file. RP systems normally need support structures (see the next section). These cannot be created directly using the IGES format.

Other less commonly used formats include HPGL (Hewlett-Packard Graphics Language); CT (Computerized Tomography) that is more common in medical imaging; SLC (Stereolithography Contour); CLI (Common Layer Interface); RPI (Rapid Prototyping Interface); and LEAF (Layer Exchange ASCII Format).

19.1.5 Supports for Part Building

In rapid prototyping, supports are often needed to hold the part in place during the building phase (Fig. 19.11). They also allow safe removal of the part from the platform and prevent the recoater from colliding with the platform during the build stage. The supports are created in the CAD system, but, generally, the output files are separate from those of the part. Creating special supports is essential for most of the liquid-based systems (see Section 19.2.1), while for other systems such as laminated

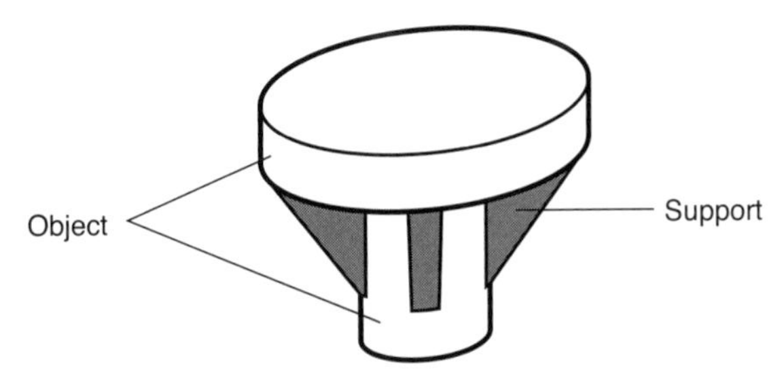

FIGURE 19.11 Sample support structure.

object manufacturing (LOM), selective laser sintering (SLS), and 3D Printing, no special supports are necessary since the supports are inherent in the part buildup. Where needed, supports may be generated by the engineer, CAD designer, or RP operator.

Sections for which supports are critical include areas that overhang preceding layers. If such a section is not supported, it will tend to pull on the preceding layer as it shrinks, resulting in curl, just as it occurs in a bimetallic strip. Supports normally consist of combinations of thin webs, and they need to overlap into the succeeding part layers by two to three layer thicknesses. An example of a support system is shown in Fig. 19.11.

19.1.6 Slicing

The rapid prototyping process generally involves building the part layer by layer, one on top of the other. To facilitate this process, the CAD model is first sliced into cross sections (Fig. 19.12). The slice thickness selected depends on the accuracy desired, and it typically ranges from 0.12 to 0.50 mm. Variation of the build time

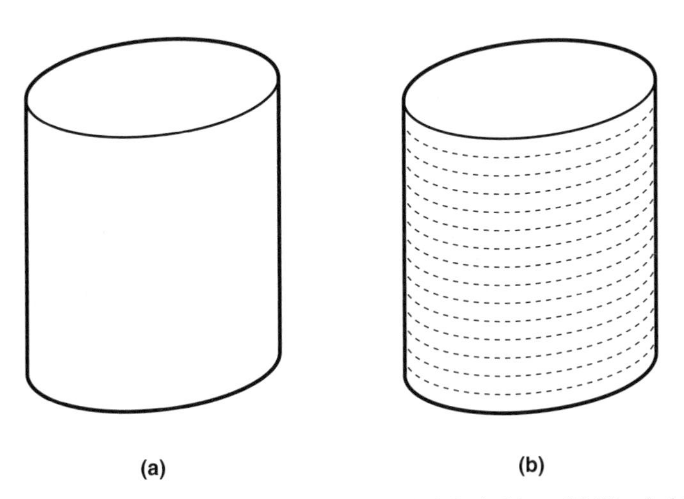

FIGURE 19.12 Slicing of a CAD model. (a) Original object. (b) Sliced object.

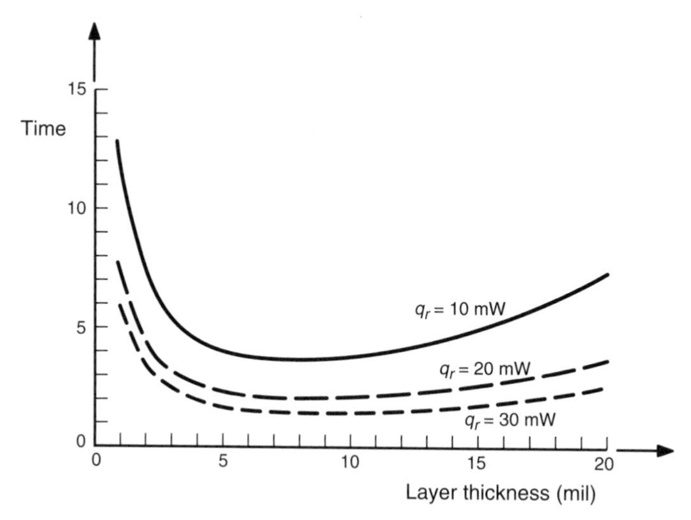

FIGURE 19.13 Variation of build time with slice thickness for liquid-based systems. (From Jacobs, P. F., 1992, *Rapid Prototyping & Manufacturing: Fundamentals of Stereolithography*. Reprinted by permission of Society of Manufacturing Engineers, Dearborn, MI.)

with slice thickness is shown in Fig. 19.13 for liquid-based systems. For very thin layers, the build time decreases as layer thickness increases. However, at relatively large layer thicknesses, the build time begins to increase with increasing layer thickness. This is because the scan velocity decreases exponentially with increasing cure thickness.

The CAD stage is normally the most time-consuming part of the rapid prototyping process. CAD systems that are used are normally those that have the capability for other design applications such as interference studies, stress analysis, FEM analysis, detail design and drafting, planning of manufacturing (including NC programming), and so on. This ensures that the same CAD package can be used for other design applications. In the next section, we discuss the basic concepts associated with part building.

19.2 PART BUILDING

There are several types of rapid prototyping methods, and these can be categorized by the configuration of the light source used, the materials, the energy source, or the type of application. Due to the significant changes in processing methodology associated with different materials, our discussion on part building will be based on the type of material to be processed. In this regard, we consider three main types of systems:

1. Liquid-based systems.
2. Powder-based systems.
3. Solid-based systems.

19.2.1 Liquid-Based Systems

Liquid-based rapid prototyping is normally referred to as stereolithography. The basic premise of the liquid-based systems involves building parts in a vat of photocurable organic liquid resin by exposing it to laser radiation to catalyze polymerization and thus solidify it. The wavelength of the laser radiation used is normally in the ultraviolet range. Thus, the most commonly used lasers are He-Cd and argon ion lasers. At any one time, only a thin layer (with depth equal to the slice thickness) of the resin is exposed to adequate enough radiation for polymerization to occur. The principle is illustrated in Fig. 19.14.

Before each layer is exposed to the laser beam, it needs to be leveled. This involves first checking the resin's z-level (height) using a sensor. If the resin level falls outside a tolerance band, a plunger is used to reposition the part in the z-direction until the desired level is achieved. Following leveling, the platform (on which the part rests) is lowered (deep dipping). A few seconds (z-wait), typically about 15–30 s, is then allowed for the resin to settle. The resin is now exposed to the laser beam. When the layer has solidified, the part is lowered to permit the next layer of resin to be formed over it. This is repeated until the part is completed. This is the general process, even though details may vary from one manufacturer to another.

A major advantage of stereolithography is that it is more accurate than any of the other rapid prototyping techniques listed. Even though almost all the processes enable internal hollow spaces to be produced, stereolithography has the added advantage that it enables the hollow interior to be completely drained. However, the process has the

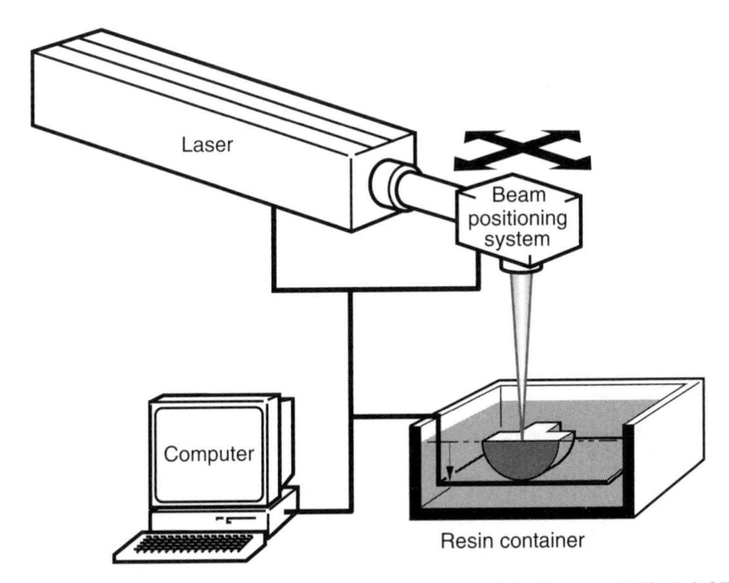

FIGURE 19.14 Basic principle of stereolithography. (From Kochan, D., 1993, *Solid Freeform Manufacturing*, Elsevier, Amsterdam, Holland. Copyright Elsevier.)

disadvantage that it is only useful for photosensitive materials. It is also a two-step process involving first the laser treatment, followed by curing in an oven.

There are two basic ways in which the resin is exposed to the laser beam:

1. Beam scanning.
2. Parallel processing.

19.2.1.1 ***Beam Scanning*** In beam scanning, the laser beam traces a path on the resin surface that is determined by the shape to be formed at that level (Fig. 19.14). First, the part borders for the given layer cross section are traced. The areas that will eventually become solid are then filled in or hatched. This essentially involves scanning the interior. There are different hatching styles available, (Fig. 19.15a and b). The scan lines of a given layer may be either perpendicular to each other or a combination of horizontal and 60°/120° lines that produce an internal pattern consisting of equilateral triangles. Scan lines of contiguous slices (layers) may either overlap (Fig. 19.15c) or just touch each other (Fig. 19.15d). They may also be in line (Fig. 19.15d) or be staggered (Fig. 19.15e). Hatch spacings are typically about 0.25 mm. Hatching results in vertical columns containing unsolidified resin, which is usually cured during post-processing. Other more sophisticated hatching techniques such as

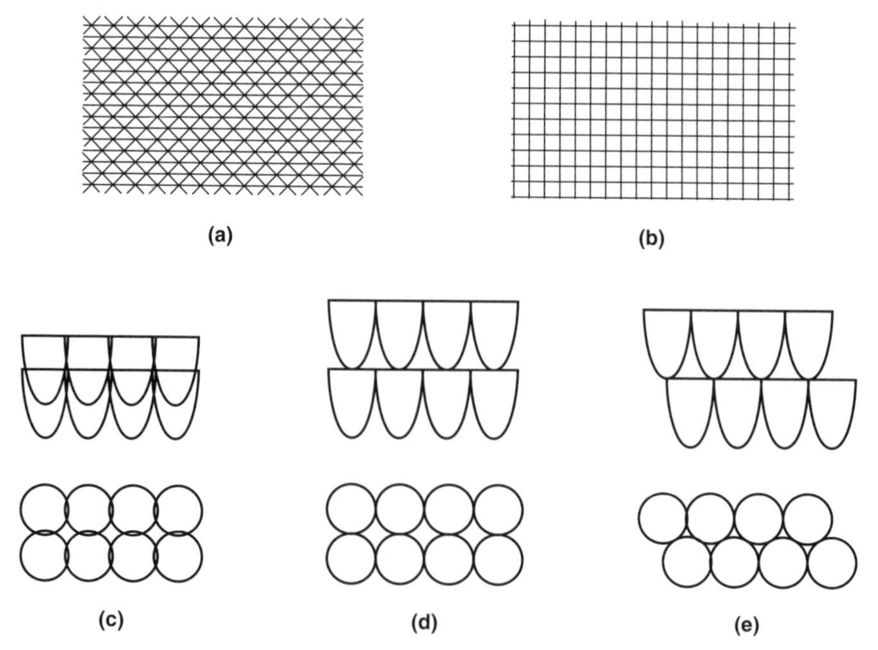

FIGURE 19.15 Hatch types. (a) Triangular hatch. (b) Rectangular hatch. (c) Overlapping layers. (d) Low overlapping layers. (e) Staggered hatch. (From Kochan, D., 1993, *Solid Freeform Manufacturing*, Elsevier, Amsterdam, Holland. Copyright Elsevier.)

TABLE 19.2 Sample Conditions for Rapid Prototyping by Beam Scanning.

Process Parameter	Range of Values	Units
Laser power	5–500	mW
Beam size	50–300	μm
Scan speed	0.75–24	m/s
Minimum layer thickness	0.01–0.3	mm

From Chua, C. K., and Leong, K. F., 1997, *Rapid Prototyping: Principles & Applications in Manufacturing*, John Wiley & Sons, Inc., New York.

the WEAVE style exist (see references). Sample conditions for rapid prototyping by beam scanning are shown in Table 19.2.

For effective scanning, the depth to which the laser beam penetrates the liquid has to at least equal the layer thickness. That defines the depth profile of the hatch lines shown in Fig. 19.15. To estimate the depth profile, let us consider the coordinate system shown in Fig. 19.16. Equation (1.33) can now be rewritten as

$$I(z) = I_0 e^{\frac{-z}{D_p}} \tag{19.14}$$

where I_0 is the intensity of the incident beam (W/cm^2) and $D_p = 1/\alpha =$ penetration depth (mm).

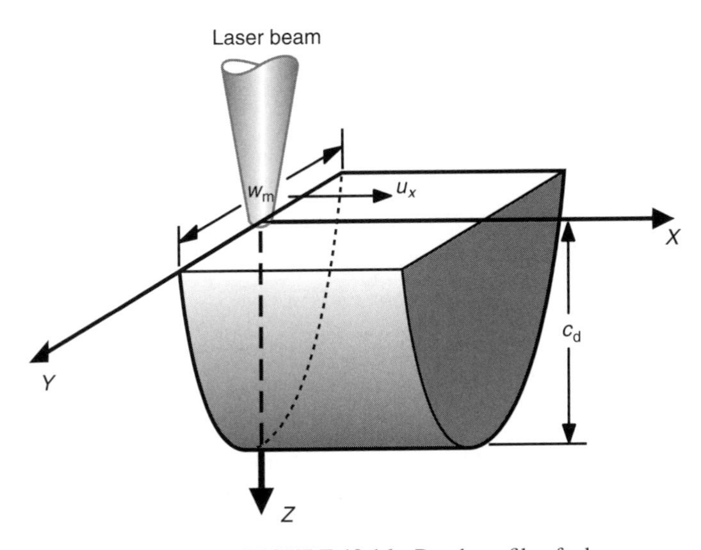

FIGURE 19.16 Depth profile of a laser scan.

For a Gaussian beam, the distribution of the incident beam, I_0, on the surface of the liquid layer is given by equation (2.40) as

$$I_0(r) = I_\mathrm{p} \mathrm{e}^{\left(-\frac{2r^2}{w^2}\right)} \tag{19.15}$$

where $I_0(r)$ = beam intensity at any radial position, r, on the surface ($\mathrm{W/cm^2}$), I_p = peak beam intensity ($\mathrm{W/cm^2}$), r = radial coordinate (μm), and w = radius at which beam intensity is $1/e^2 \times I_\mathrm{p}$ or $0.14 \times I_\mathrm{p}$. That means 86% of the total energy is contained in a spot of radius w (μm).

The peak intensity in equation (19.15), I_p, can be shown to be given (see Problem 19.4) by

$$I_\mathrm{p} = \frac{2q}{\pi w^2} \tag{19.16}$$

Thus, the surface intensity (power per unit area) becomes

$$I_0(r) = \frac{2q}{\pi w^2} \mathrm{e}^{\left(-\frac{2r^2}{w^2}\right)} \tag{19.17}$$

Since the laser beam is considered to scan linearly in the x-direction, by integrating equation (19.17) over the period of scanning (mathematically from $x = -\infty$ to $x = \infty$, but in reality the time it takes the laser beam to traverse its practical zone of influence) we get the exposure (energy per unit area) on the surface, Q_0, (see Problem 19.5) as

$$Q_0(y) = \sqrt{\frac{2}{\pi}} \frac{q}{w u_x} \mathrm{e}^{\left(-\frac{2y^2}{w^2}\right)} \tag{19.18}$$

Equation (19.18) is a function only of the y-coordinate since we are considering the exposure on the surface ($z = 0$), and quasistationary conditions are assumed, that is, there is no change in the x-direction. Now taking into account the propagation of the beam in the depth or z-direction using equation (19.14), we get the exposure through the thickness of the slice as

$$Q_\mathrm{p}(y, z) = \sqrt{\frac{2}{\pi}} \frac{q}{w u_x} \mathrm{e}^{\left(-\frac{2y^2}{w^2}\right)} \mathrm{e}^{\frac{-z}{D_\mathrm{p}}} \tag{19.19}$$

For a given level of exposure, the profile of the region that is affected by the laser beam is as shown in Fig. 19.16. Thus, if the resin that is being treated has a critical exposure value, Q_c, above which it polymerizes, then we can determine the cross section ($y - z$ or depth boundary) of the region that is polymerized by the laser beam for each scan by setting $Q_p = Q_c$, that is,

$$Q_c = \sqrt{\frac{2}{\pi}} \frac{q}{wu_x} e^{\left(-\frac{2y_c^2}{w^2}\right)} e^{\frac{-z_c}{D_p}} \tag{19.20}$$

where y_c, z_c = values of y, z when $Q_p = Q_c$.

Equation (19.20) can be shown (see Problem 19.6) to reduce to the form

$$z_c = ay_c^2 + b \tag{19.21}$$

where $a = -2D_p/w^2$, and $b = D_p \ln\left[\sqrt{2/\pi}\,(q/wu_x Q_c)\right]$.

This is the equation of a parabola. Thus the scanned line has a parabolic cross section (Fig. 19.16). From Fig. 19.16, it is evident that the maximum penetration that is cured (cure depth), C_d, can be obtained by setting y_c to zero. That gives

$$C_d = b = D_p \ln\left[\sqrt{\frac{2}{\pi}}\left(\frac{q}{wu_x Q_c}\right)\right] \tag{19.22}$$

Furthermore, the maximum width of the cured region (which occurs on the slice surface), W_m, is given by

$$W_m = w\sqrt{2\frac{C_d}{D_p}} \tag{19.23}$$

Example 19.2 An argon ion laser of output power 260 mW is used to process a photocurable resin at a scan speed of 2 m/s, with a beam radius of 200 μm. If the resin has a critical exposure of 12 mJ/cm^2 and the depth, z_c, at a distance of $y_c = 0.15$ mm from the scan line on the surface is 0.4 mm, determine the cure depth, C_d, for the process.

Solution:

From equation (19.20), we have

$$Q_c = \sqrt{\frac{2}{\pi}} \frac{q}{wu_x} e^{\left(-\frac{2y_c^2}{w^2}\right)} e^{\frac{-z_c}{D_p}} \tag{19.20}$$

$$\Rightarrow 12 \times 10^{-5} = \sqrt{\frac{2}{\pi}} \frac{260 \times 10^{-3}}{200 \times 10^{-3} \times 2000} e^{\left(-\frac{2\times 0.15^2}{200^2\times 10^{-6}}\right)} e^{\frac{-0.4}{D_p}}$$

Thus,

$$12 \times 10^{-5} = 16.84 \times 10^{-5} e^{\frac{-0.4}{D_p}}$$

$$\Rightarrow D_p = 1.18 \text{ mm}$$

Now from equation (19.22), the cure depth is obtained as

$$C_d = D_p \ln \left[\sqrt{\frac{2}{\pi}} \left(\frac{q}{w u_x Q_c} \right) \right] \tag{19.22}$$

$$\Rightarrow C_d = 1.18 \times \ln \left[\sqrt{\frac{2}{\pi}} \left(\frac{260 \times 10^{-3}}{200 \times 10^{-3} \times 2000 \times 12 \times 10^{-5}} \right) \right] = 1.72 \text{ mm}$$

The RP systems that are based on beam scanning include the following:

1. Stereolithography by 3D Systems.
2. Stereography by EOS.
3. SOUP by Mitsubishi.
4. SCS 3000 by Sony.
5. SOMOS 1000 by Du Pont.

19.2.1.2 Parallel Processing With parallel processing, a mask is used to expose the entire area of each layer to be solidified to the laser beam at the same time (Fig. 19.17). A different mask is used for each layer and the exposed area of the mask corresponds with the area to be solidified at that layer. This process is thus faster than the beam scanning process. It is also cheaper since the same glass plate is used to generate each of the masks. The masks are produced by transferring the information for each layer to a glass plate by electrostatically charging the areas to be masked, such that toner adheres to those areas. The regions corresponding to the part to be produced for that layer are thus left transparent, enabling the ultraviolet laser radiation to be transferred to the resin, polymerizing it in those regions. The unexposed regions remain liquid. After each layer is exposed to the beam, the resin that is not polymerized is sucked away and replaced by water-soluble wax, which provides support for the part. Thus, no supports are needed.

Systems that utilize this concept include

1. Solider by cubital.
2. Light sculpting.

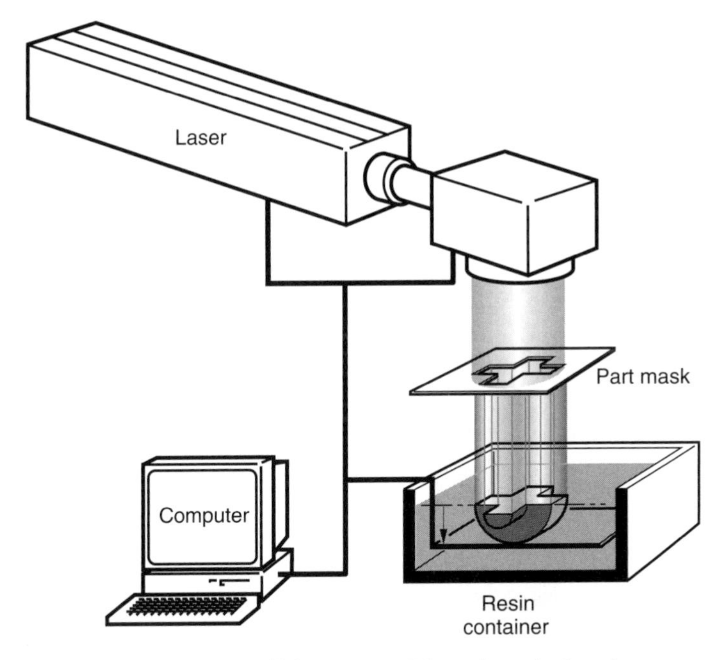

FIGURE 19.17 Parallel exposure of the resin to the laser beam.

19.2.2 Powder-Based Systems

There are different forms of the powder-based systems. Some of them have characteristics similar to the liquid-based systems described in the preceding section, while others are more similar to the solid-based systems to be discussed in the next section. The common feature among these systems is that they use powder (which may be plastic or metallic) as the starting material. The major types of powder-based systems include the following:

1. Selective laser sintering.
2. 3D printing.
3. Ballistic particle manufacturing (BPM).

Others include the direct shell production casting system and the multiphase jet solidification system. We shall only discuss the basic principles of the first three listed above. Of these, only selective laser sintering is laser based. The 3D printing and ballistic particle manufacturing do not involve the use of lasers. However, they are also discussed for the sake of completeness.

19.2.2.1 Selective Laser Sintering (SLS) This process is similar in concept to the stereolithography approach and involves first depositing a thin layer of the powder

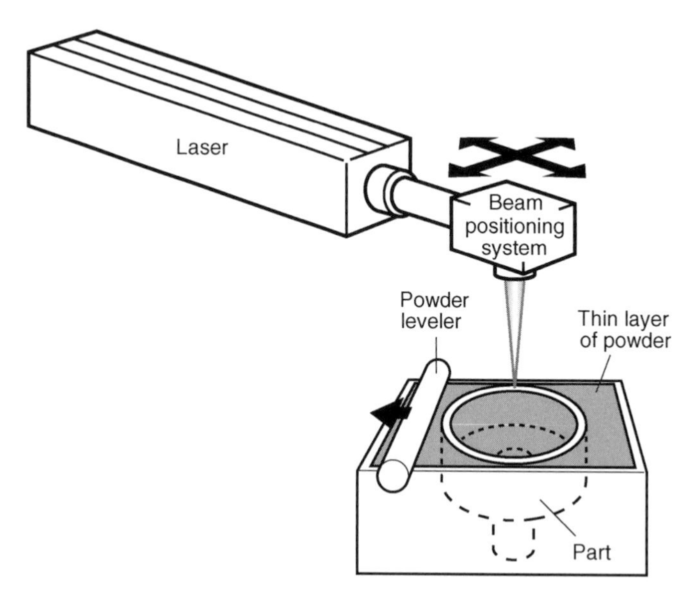

FIGURE 19.18 Schematic of the selective laser sintering process. (From Kochan, D., 1993, *Solid Freeform Manufacturing*, Elsevier, Amsterdam, Holland. Copyright Elsevier.)

material on the working platform within the process chamber. A CO_2 laser is then scanned on the powder layer to trace out the desired shape, heating up the scanned region to a temperature just below melting. The heat sinters the powder particles and thereby consolidates them, forming a solid mass. The beam is only directed to the regions defined by the part's designed shape and the rest of the powder material remains unchanged. On completion of this layer, another layer of powder is deposited on top of it, using a roller mechanism (Figure 19.18). The process of applying a powder layer and sintering is repeated until the part is completed, after which the unbound powder is emptied from the rest of the part.

During the process, the powder that is not sintered serves as a built-in support structure. Thus, no separate supports need to be incorporated in the design, simplifying both the design and the post-processing phases. However, the energy required may be about 300–500 times the energy required for photopolymerization of a layer of similar thickness. Also, an inert gas is used in the process chamber to prevent oxidation at the high temperatures. Various materials can be used, as long as they are in powder form; for example, thermoplastics, composites, ceramics, and metals. A typical machine used is the DTM Corporation's Sinterstation 2000. Sample conditions for rapid prototyping by selective laser sintering are shown in Table 19.3.

Selective laser sintering has the advantage of being useful for a wide range of materials—metals, ceramics, and plastics. Furthermore, different powders can be sintered at the same slice level, enabling a heterogeneous structure with specific desired characteristics to be fabricated. Any unsintered powder can be used again and is thus

TABLE 19.3 Sample Conditions for Selective Laser Sintering

Process Parameter	Range of Values	Units
Laser power	50–300	W
Beam size	400	μm
Scan speed	0.9–2	m/s
Minimum layer thickness	0.07–0.5	mm

(From Chua, C. K., and Leong, K. F., 1997, *Rapid Prototyping: Principles & Applications in Manufacturing*, John Wiley & Sons, Inc., New York.)

not wasted. The process is also one step and does not require any support. However, the resulting accuracy is less than that of stereolithography since it depends on the size of powder particles. Accuracy is further impaired by the tendency of neighboring particles to become attached to treated regions as a result of conduction. Also, internal hollow spaces that are created are not as easily cleaned. Thus, components made by the SLS method may not be suitable for surgical applications. In addition, an inert atmosphere is needed to avoid oxidation during sintering.

19.2.2.2 3D Printing The 3D printing process also starts with a layer of powder that is spread over a powder bed mounted on top of a piston and cylinder arrangement. However, this process is not laser-based, and in this case, the powder particles are consolidated by ejecting an adhesive bonding material onto the powder using technology similar to ink-jet printing, in the regions defined by the part's designed shape. The piston is then lowered, a new layer of powder is spread, and the binder applied (Fig. 19.19). The process is repeated until the part is completed. The part is then subjected to heat treatment to further strengthen it, after which the unbound powder is removed. This technology is still in the research and development phase, and there are no commercial systems available as yet.

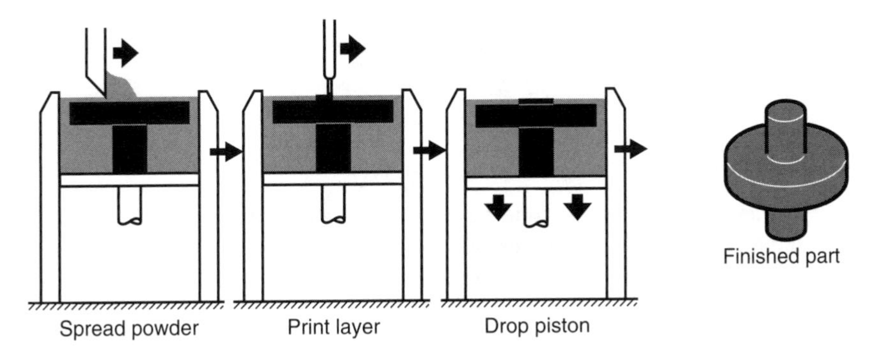

FIGURE 19.19 Principle of the 3D printing process. (From Kochan, D., 1993, *Solid Freeform Manufacturing*, Elsevier, Amsterdam, Holland. Copyright Elsevier.)

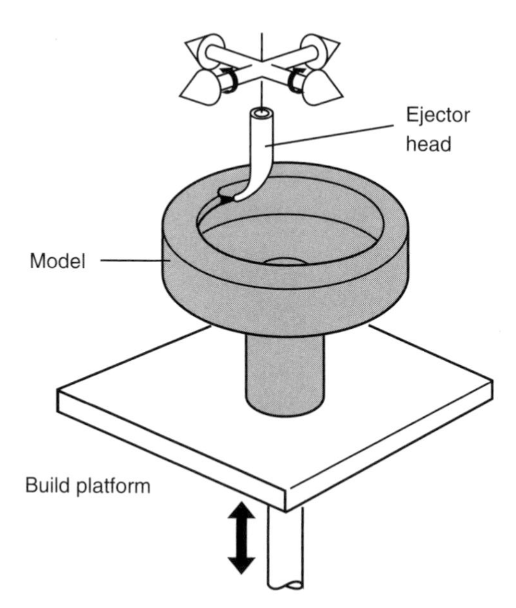

FIGURE 19.20 Principle of the ballistic particle manufacturing process. (From Chua, C. K., and Leong, K. F., 1997, *Rapid Prototyping: Principles & Applications in Manufacturing*. Permission of World Scientific Publishing Co.)

19.2.2.3 Ballistic Particle Manufacturing The BPM method is not laser based. With this process, each layer of material is applied by using a ceramic ejector head to shoot tiny uniform droplets at the rate of about 12,000/s at the target area where they solidify on contact (Fig. 19.20). The droplets are normally about 75 μm in diameter as they emerge, but then flatten to a height of about 50 μm on contact. The previous layer is softened by the impacting molten droplets, causing the two layers to bind together as they solidify. As the ejector head moves along, it is followed by a thermal iron that smoothens the droplets into an even layer. On completion of each layer, the platform is lowered for the next layer. Typical materials processed with this application are thermoplastics. One such system in commercial use is the BPM Technology's Personal Modeler 2100.

19.2.3 Solid-Based Systems

There are several forms of the solid-based systems, each differing in the manner in which the layer is applied. However, due to space constraints, we shall only discuss two of the most common systems. These are:

1. Fused deposition modeling (FDM)
2. Laminated object manufacturing (LOM).

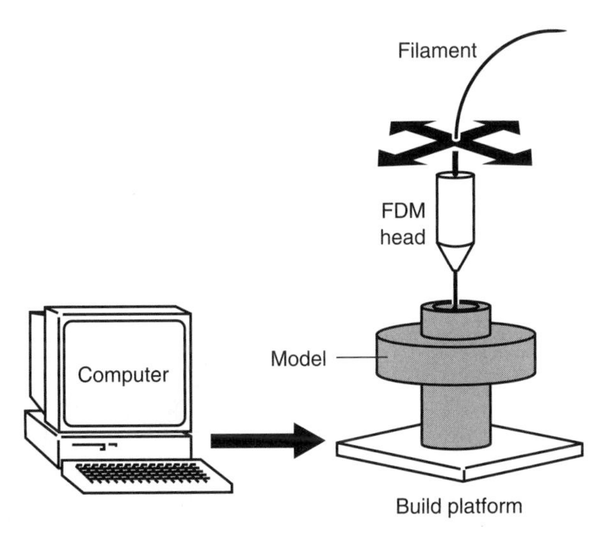

FIGURE 19.21 Principle of the fused deposition modeling process. (From Kochan, D., 1993, *Solid Freeform Manufacturing*, Elsevier, Amsterdam, Holland. Copyright Elsevier.)

LOM is laser based but FDM is not. Others are the selective adhesive and hot press (SAHP) system, the multijet modeling (MJM) system, the model maker MM-6B, and the hot plot.

19.2.3.1 Fused Deposition Modeling With this process, which is not laser based, the material is supplied as a filament wire (about 1.27 mm in diameter) on a spool. It is fed through a heated FDM head, where it becomes softened into a viscous state and then is extruded and deposited in ultrathin layers (Fig. 19.21). The extruded material quickly solidifies after deposition. The horizontal width of the extruded material can vary between 0.25 and 2.50 mm. As with the other RP processes, when one layer is completed, the next layer is laid on top of it until the part is done. Thermoplastics are typical materials that are processed by this method. An example of such a system is the Stratasys FDM 1650.

19.2.3.2 Laminated Object Manufacturing The principle of this process is illustrated in Fig. 19.22 and involves sequentially bonding together thin layers of adhesive-coated material. The sheet material is obtained from a supply roll. Once the sheet is in place, a laser beam is directed to cut the outline of the part in that sheet. Cross-hatches are then made by the beam in the excess material outside the part boundary, turning it into a support structure with a rectangular boundary that remains in place. After that, the platform is lowered with the formed stack, and a fresh section of the sheet material is moved on top of it. The platform is then raised and a heated roller rolls over it, bonding it to the preceding layer in one reciprocal motion. The part outline is then cut with the laser beam. This continues until the entire part is produced. The

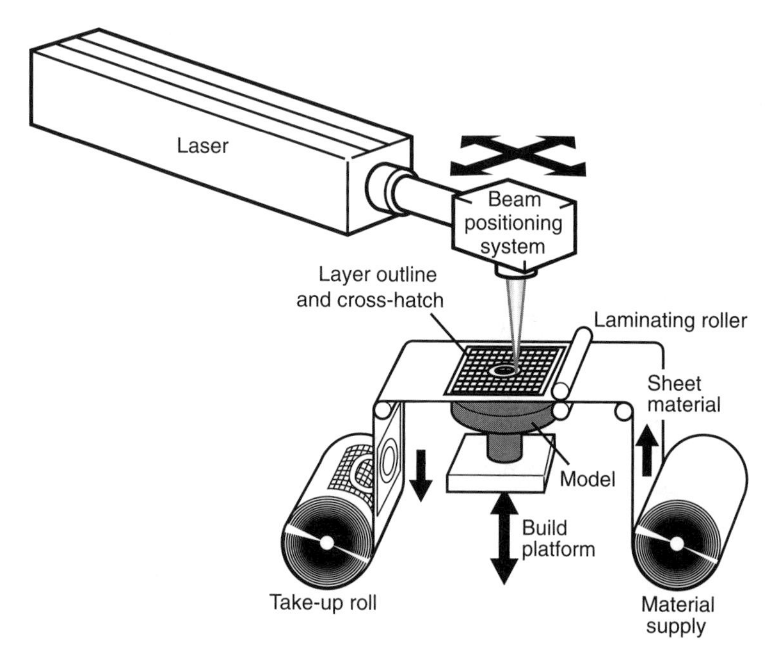

FIGURE 19.22 Principle of the laminated object manufacturing process. From Chua, C. K., and Leong, K. F., 1997, *Rapid Prototyping: Principles & Applications in Manufacturing*. Permission of World Scientific Publishing Co.

entire rectangular block then comes out, and the part is separated from the surrounding material that is easily removed as a result of the cross-hatching. Sample conditions for laminated object manufacturing are shown in Table 19.4.

Laminated object manufacturing is useful for a wide range of materials, including paper, plastics, metals, ceramics, and composites. However, it has the disadvantage that unused material resulting from internal hollow spaces can be removed only with difficulty. Also, finishing costs are generally higher than for the other laser-based processes.

An example of such a system is the Helisys LOM-2030.

TABLE 19.4 Sample Conditions for Laminated Object Manufacturing

Process Parameter	Range of Values	Units
Laser power	25–50	W
Beam size	250–400	μm
Scan speed	0.3–0.7	m/s
Material thickness	0.05–0.4	mm

From Chua, C. K., and Leong, K. F., 1997, *Rapid Prototyping: Principles & Applications in Manufacturing*, John Wiley & Sons, Inc., New York.

19.2.4 Qualitative Comparison of Some Major Systems

In this section, some of the major rapid prototyping techniques are compared qualitatively (Fig. 19.23). Quantitative comparison is difficult to achieve since a number of factors (such as system calibration, operator expertise, and method of measurement) do influence the outcome.

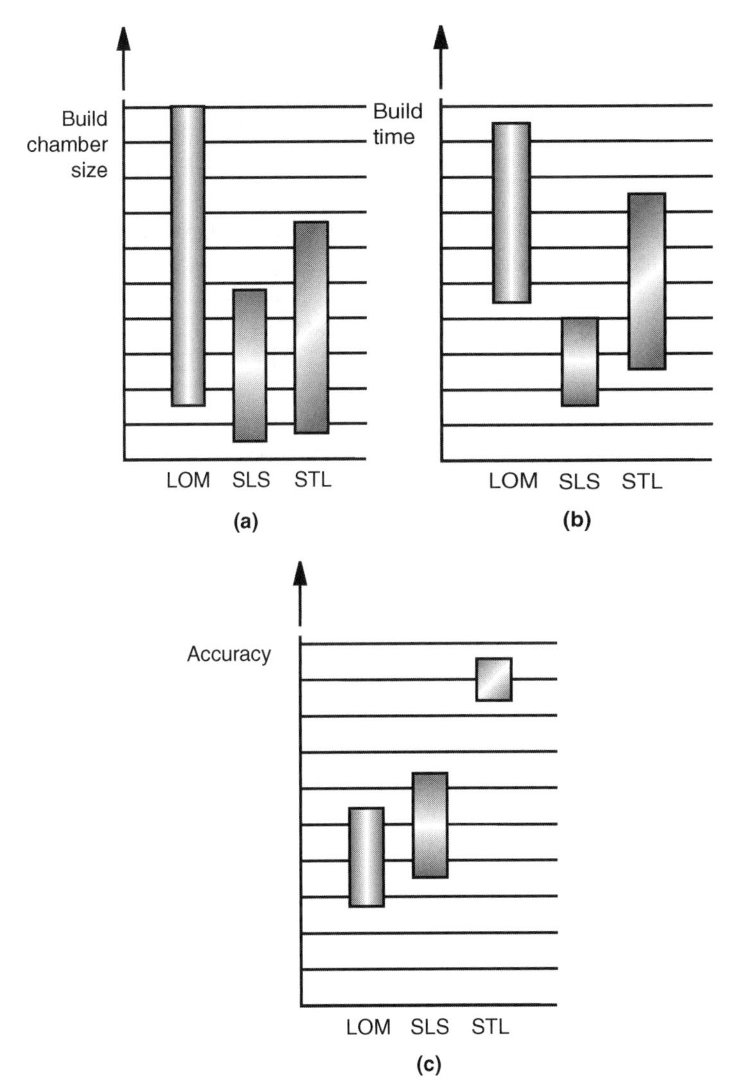

FIGURE 19.23 Qualitative comparison of laser-based rapid prototyping systems. (a) Build chamber size. (b) Build time. (c) Accuracy. (From Gebhart, A., 2003, *Rapid Prototyping*, Hanser Publishers, Munich, Germany.)

TABLE 19.5 Post-processing Operations

	Stereolithography	Selective Laser Sintering	Laminated Object Manufacturing
Postcuring	•		
Cleaning	•	•	•
Part finishing	•	•	•

19.3 POST-PROCESSING

Depending on the build methodology, post-processing may involve such operations as removal of the part, cleaning, postcuring, and part finishing. The major post-processing operations and the processes for which they are normally needed are summarized in Table 19.5.

For the liquid-based systems, once the part is completed, excess resin is allowed to drain back into the vat, and any resin adhering to the part is wiped off. A solvent is then used to clean it and low-pressure compressed air is subsequently used to dry it. Finally, the part is postcured using broadband ultraviolet radiation to complete the polymerization process and increase the strength of the part. This last step may take a few hours depending on the size of the part.

Finishing is often necessary because of factors that affect the process such as shrinkage, distortion, curling, and so on and depends on the intended application. Common finishing operations to which the part may be subjected include cutting, sandblasting, polishing, coating, and painting.

1. *Cutting* – Traditional cutting processes such as turning, grinding, milling, and drilling may be used to improve the surface finish of the part, remove unwanted supports, introduce additional form features, or improve dimensional accuracy.
2. *Sandblasting and Polishing* – Sandblasting may also be used to clean up the part, but for better appearance and higher accuracy, polishing would be necessary.
3. *Coating and Painting* – Coating is often used to provide a protective layer for the part, whereas painting is normally used for aesthetic purposes.

19.4 APPLICATIONS

Early RP systems were developed primarily for "touch-and-feel" applications that provided designers a means to visualize how the final product would look, without regard to their function. In recent years, as newer techniques have been developed, the range of applications has increased considerably. The major areas are

1. Design.
2. Engineering, analysis, and planning.
3. Manufacturing and tooling.

19.4.1 Design

Applications of RP systems in design include such issues as verification of CAD models, object visualization, proof of concept, and marketing. Rapid prototyping enables designers to have a physical part that confirms the design that they have created using CAD. This is especially important when aesthetic requirements are critical. A number of engineering components are quite complex, physically. Attempting to visualize the final form using CAD drawings can be quite challenging even for experienced designers, and more so for other engineers and management who may need to be involved in final design approval. The availability of a prototype significantly facilitates discussion among these groups. Finally, for marketing purposes, a physical model is always beneficial, enabling potential customers to get a feel for the final product. The prototype can also be used for marketing brochures.

19.4.2 Engineering, Analysis, and Planning

Aspects of engineering, analysis, and planning for which rapid prototyping is useful include scaling, form and fit, stress and flow analyses, mock-up parts, planning surgical operations, and custom prostheses and implants. RP techniques enable models to be scaled up or down by scaling the CAD design. When the appropriate size is selected, the CAD data are changed accordingly for creating the RP model. How a part or component fits in with the overall design is something that can be assessed using a prototype.

The flow characteristics associated with products that involve fluid flow are more easily evaluated with the aid of a prototype. Likewise, prototypes facilitate the evaluation of stresses that can develop in a part, using photo-optical methods. Furthermore, prototypes or mock-up parts can be assembled into the complete product and used for final testing. Finally, RP technology can also be used to produce custom prostheses and implants, as well as for planning surgical and reconstructive procedures.

19.4.3 Manufacturing and Tooling

Most applications of RP in manufacturing and tooling involve making patterns to be used in various processes, for example metal casting, metal spraying, and plastic vacuum casting with silicon molding. We shall illustrate the use of RP in these areas with metal casting.

Metal casting involves making a mold that has a cavity into which molten metal is poured and allowed to solidify. The mold is then opened and the casting taken out. The mold may be made of sand, plaster, or metal. For making a sand or plaster mold cavity, a pattern is often required. The pattern is made to the shape of the part to be produced, with only slight modifications, and is usually made out of wood, metal, or plastic material. Metal and plastic patterns can be made using rapid prototyping. Metal molds can also be made directly using rapid prototyping. We shall illustrate the application of RP in metal casting with the investment casting (lost wax) process (Fig. 19.24).

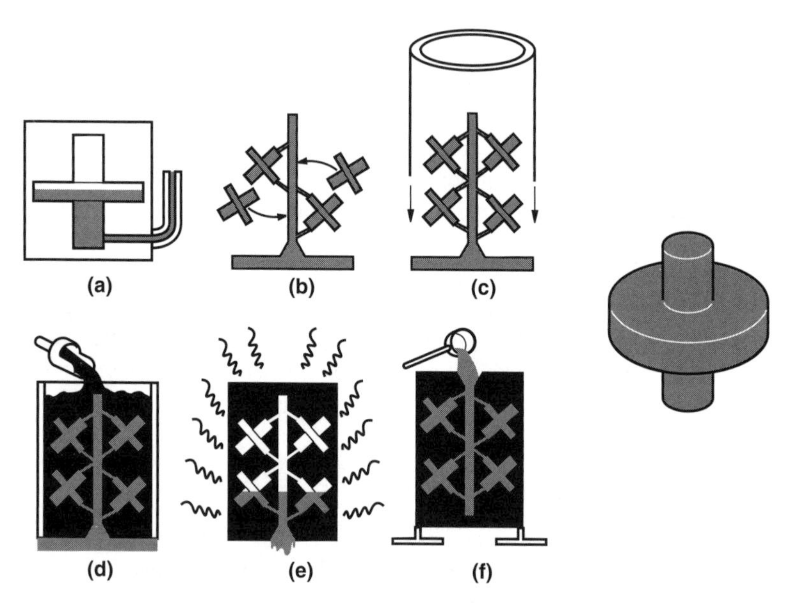

FIGURE 19.24 The investment casting process. (a) Making basic unit of pattern. (b) Joining pattern units into a more complex pattern. (c) Placing the flask. (d) Filling flask with mold material. (e) Melting off the pattern. (f) Pouring the molten material into the mold. (From Lindberg, R. A., 1977, *Processes and Materials of Manufacture*, Allyn and Bacon, Inc., Boston, MA.)

Traditionally, this process first involves preparing a pattern by injecting wax or thermoplastic resin into a die. A number of separate subpatterns may be joined together after they are removed from their respective dies, using an adhesive, to form a more complex pattern. A flask is then placed around the pattern and ceramic slurry poured into the flask to cover the pattern to form the mold. After the mold has set, it is heated to melt out the wax, leaving the mold cavity into which the molten metal can be poured for casting. Rapid prototyping can be used to make the pattern faster and more economically.

19.5 SUMMARY

Rapid prototyping involves the generation of physical objects from graphical computer data and is used in design conceptualization, to facilitate engineering stress analysis, and in making patterns and/or dies for manufacturing operations. To make a part by rapid prototyping, one has to go through the following three stages:

1. *Computer-Aided Design* – A CAD package is used to develop a computer model of the part to be produced. Where a physical object already exists, reverse engineering may be used to develop a computer model from the object. The CAD model is next converted into a format required by the rapid prototyping

equipment. A commonly used format is the STL format. For processing, the CAD model is sliced into cross sections to facilitate the layer-by-layer building of the part.

2. *Part Building (Processing)* – Different materials can be processed by rapid prototyping. The process can thus be categorized by the type of material, namely, liquid-based, powder-based, or solid-based system. In each case, the part is built one layer at a time. The liquid-based systems generally involve the use of lasers. One powder-based system that uses lasers is the selective laser sintering process. One solid-based system, laminated object manufacturing, also uses a laser.
3. *Post-processing* – After the part is made, it may be further finished by machining, sandblasting and polishing, and/or coating and painting.

REFERENCES

Amirouche, F. M. L., 1993, *Computer-Aided Design and Manufacturing*, Prentice-Hall, Englewood Cliffs, NJ.

Chua, C. K., and Leong, K. F., 1997, *Rapid Prototyping: Principles & Applications in Manufacturing*, John Wiley & Sons, Inc., New York.

Cooper, K. G., 2001, *Rapid Prototyping Technology*, Marcel Dekker, Inc., New York.

Davies, B. L., Robotham, A. J., and Yarwood, A., 1991, *Computer-Aided Drawing and Design*, Chapman & Hall, London.

Gebhart, A., 2003, *Rapid Prototyping*, Hanser Publishers, Munich, Germany.

Jacobs, P. F., 1992, *Rapid Prototyping & Manufacturing: Fundamentals of Stereolithography*, Society of Manufacturing Engineers, Dearborn, MI.

Kochan, D., 1993, *Solid Freeform Manufacturing*, Elsevier, Amsterdam, Holland.

Lindberg, R. A., 1977, *Processes and Materials of Manufacture*, Allyn and Bacon, Inc., Boston, MA.

Onwubiko, C., 1989, *Computer-Aided Design*, West Publishing Company, St. Paul, MN.

Pham, D. T., and Dimov, S. S., 2001, *Rapid Manufacturing*, Springer-Verlag, London.

APPENDIX 19A

List of symbols used in the chapter.

Symbol	**Parameter**	**Units**
a_i, a_{ij}	polynomial coefficients	–
$B_{n,i}$	Bezier polynomial	–
C_d	cure depth	m
$\mathbf{C}$	set of data constituting object to be transformed	–
D_p	penetration depth	m
n	degree of polynomial	–

P_i	points to be fitted with a polynomial	–
Q_c	critical exposure or energy per unit area	J/m^2
Q_0	incident energy per unit area	J/m^2
$\mathbf{Q}$	transformed object	–
S_j	span j of spline	–
t	dummy variable	–
$\mathbf{T}$	transformation matrix	–
$\mathbf{T_R}$	rotation transformation matrix	–
$\mathbf{T_T}$	translation transformation matrix	–
T_x, T_y, T_z	components of $\mathbf{T}$	–
W_m	maximum width of cured region	m
x_i	data points	–
y_c, z_c	values of y, z when $Q_p = Q_c$	m
x_i, y_i, z_i	components of $\mathbf{C}$	m
X_i, Y_i, Z_i	components of $\mathbf{Q}$	m

PROBLEMS

19.1. Determine the nonparametric natural cubic spline that fits the following data points:

$$(0, 0), \qquad (2, 3), \qquad (3.5, 2)$$

19.2. Obtain the Bezier curve that passes through the following data points:

$$(0, 2), \qquad (2, 3), \qquad (3.5, 6), \qquad (5, 1)$$

19.3. Obtain a natural cubic spline that fits the following data.

$$P_1[0, 5],\ P_2[1, 2],\ P_3[2, -3],\ P_4[3, -4],\ \text{and}\ P_5[4, 5]$$

19.4. For a Gaussian beam, the intensity of the incident beam, I_0, on the surface of the liquid layer is given by equation (19.15) as

$$I_0(r) = I_p e^{\left(-\frac{2r^2}{w^2}\right)} \tag{19.15}$$

Show that the peak intensity, I_p, is given by

$$I_p = \frac{2q}{\pi w^2} \tag{19.16}$$

19.5. Given the surface intensity of a laser beam on a slice:

$$I_0(r) = \frac{2q}{\pi w^2} e^{\left(-\frac{2r^2}{w^2}\right)} \tag{19.17}$$

Show that the corresponding surface exposure Q_0 becomes

$$Q_0(y) = \sqrt{\frac{2}{\pi}} \frac{q}{w u_x} e^{\left(-\frac{2y^2}{w^2}\right)} \tag{19.18}$$

19.6. The cross section of a scanned line is given by

$$Q_c = \sqrt{\frac{2}{\pi}} \frac{q}{w u_x} e^{\left(-\frac{2{y_c}^2}{w^2}\right)} e^{\frac{-z_c}{D_p}} \tag{19.20}$$

Show that the above equation reduces to the form

$$z_c = a{y_c}^2 + b \tag{19.21}$$

19.7. Given that the parabolic profile of the line-scan cross section is

$$z_c = a{y_c}^2 + b \tag{19.21}$$

Show that the maximum cure depth that results from a single line scan is

$$z_{max} = D_p \ln\left[\sqrt{\frac{2}{\pi}}\left(\frac{q}{w u_x Q_c}\right)\right] \tag{19.22}$$

19.8. Given that the parabolic profile of the line-scan cross section is

$$z_c = a{y_c}^2 + b \tag{19.21}$$

Show that the maximum cured width (on the surface of the slice) that results from a single line scan is

$$W_m = w\sqrt{\frac{2z_{max}}{D_p}} \tag{19.23}$$

19.9. An argon ion laser of output power 250 mW is used to process a photocurable resin at a scan speed of 2.5 m/s, with a beam radius of 250 μm. If the resin has a critical exposure of 13.5 mJ/cm^2 and the depth, z_c, at a distance of $y_c = 0.1$

mm from the scan line on the surface is 0.32 mm, determine the cure depth, C_d, for the process.

19.10. For the conditions given in Problem 19.9, determine

(a) The maximum cured linewidth.

(b) The equation of the cross section for the cured region.

(c) The peak intensity, I_p, of the incident beam.

20 Medical and Nanotechnology Applications of Lasers

Lasers have found extensive applications in the medical field over the years. With the field of nanotechnology being relatively recent, new developments are constantly evolving. Even though some of the nanoapplications are in the medical field, we separate the discussion into two sections. In the first section, we focus on medical applications of lasers, limiting the discussion to medical device manufacturing and therapeutic applications.

Nanotechnology applications are presented in the second section. We start by discussing the various ways in which nano features or structures can be produced. This is then followed by a discussion on a number of the different features that can be produced, specifically nanoholes and gratings, followed by nanobumps. The two-photon polymerization process is then presented, and finally the laser-assisted nanoimprint technology.

20.1 MEDICAL APPLICATIONS

Laser applications in the medical field cover a number of areas. These may be broadly categorized as

1. Medical device manufacturing.
2. Therapy.
3. Diagnostics.

There are several applications in each of these areas. However, due to space limitations, we shall discuss only the manufacturing and therapeutic applications, highlighting only one or two examples in each of those areas.

The use of the laser as a therapeutic tool and as a diagnostic tool in medicine is summarized in Tables 20.1 and 20.2, respectively.

In the next section, we shall discuss the application of lasers in the fabrication of medical devices, using stent manufacturing to illustrate the application.

Principles of Laser Materials Processing, by Elijah Kannatey-Asibu, Jr.

TABLE 20.1 Therapeutic Applications of the Laser

	Disruption	Coagulation	Cutting	Ablation
Surgery		•	•	
Gynecology		•	•	
Urology	•	•		
Ortholaringology		•	•	•
Ophthalmology	•	•		•
Dentistry		•	•	•
Orthopedics		•		•
Gastroscopy		•	•	
Dermatology	•	•		•

TABLE 20.2 Diagnostic Applications of the Laser

	Fluorescent Spectroscopy	Doppler Spectroscopy	Optical Tomography
Tumor recognition	•		•
Blood throughput		•	
Tissue differentiation	•		•
Tissue structures			•
Metabolic activity	•		

20.1.1 Medical Devices

A number of medical devices are fabricated using lasers. For metallic components, the basic principles discussed in preceding chapters do apply. We shall illustrate these with one specific example, stents (Fig. 20.1a). A stent is a small, lattice-shaped, metal tube that is inserted permanently into an artery. It provides a minimally invasive method of treating coronary artery diseases such as heart attacks and strokes. These diseases are normally a direct result of limited blood supply due to the constriction of blood vessels. Gradual build up of fats in the arteries is believed to be the main cause for arterial blockage. Earlier treatment of such blockage involved bypass surgery. The advent of stents in the 1980s began with balloon angioplasty and provided a less expensive and less traumatic alternative to bypass surgery for some patients. This evolved into the use of metallic stents, which are more durable. The stent helps hold open an artery so that blood can flow through it. The treatment simply involves placing the stent onto a balloon catheter assembly and inserting into the artery. The balloon is then inflated (Fig. 20.1b) forcing the artery to open up. When the balloon is subsequently deflated and removed, the stent is left in place, keeping the artery open.

Such a procedure is normally good for about 6 months, after which growth of muscle cells at stent–artery wall interface may lead to arterial reblockage, a phenomenon referred to as restenosis. Among other things, restenosis has been linked to the smoothness of finished stents, as investigations have shown that cells that cause

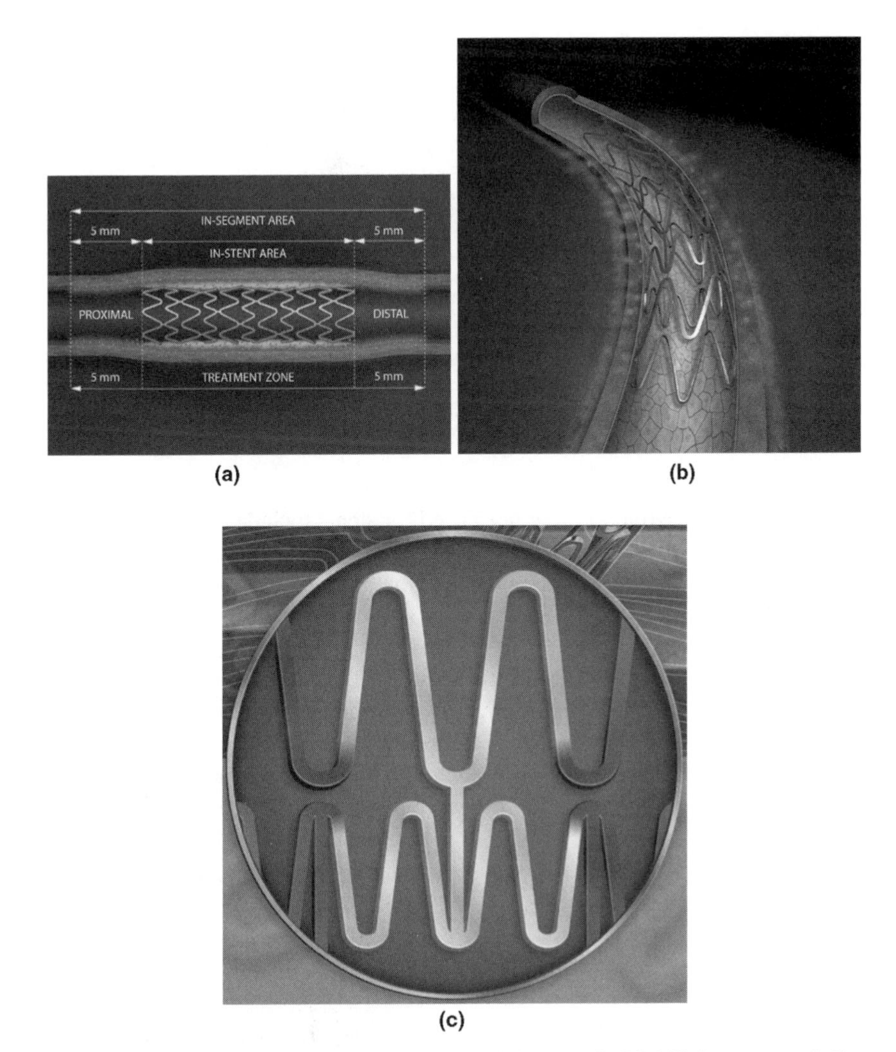

FIGURE 20.1 Coronary stent. (a) General multilink stent. (b) Multilink stent on a balloon. (c) Uniform scaffolding multilink stent. (By permission of Boston Scientific Corporation.)

restenosis preferentially grow at finished surface and arterial wall interfaces. In recent years, drug-eluting stents have been developed to minimize the onset of restenosis. These are very effective. However, they are also relatively expensive, compared to bare metal stents. Furthermore, recent studies have shown that drug-eluting stents may result in higher incidence of stroke or heart attack compared to bare metal stents.

Manufacturing of metallic stents involves the following steps:

1. Drawing of a tubing with dimensions that match the final stent diameter and wall thickness. Typical stent diameters range from 2 to 5mm, with wall thicknesses ranging from 50 to 200 μm and lengths from 6 to 38 mm.

2. Laser cutting: the appropriate design, as illustrated in Fig. 20.1c, is then cut out of the tubing, usually with a laser. Traditional laser cutting results in the formation of striations and dross. Typical surface roughness obtainable with conventional laser cutting is in the range 0.8–6.3 μm. Finishing operations are therefore necessary to ensure a desirable surface quality.
3. Finishing operations as commonly performed, in order, are
   a. Pickling, which involves chemical etching done in acid solution to remove most of the dross.
   b. Electropolishing, to remove any residual dross or sharp points that may be left. This involves passing electric current through an electrochemical solution.
4. The stent is finally placed onto a balloon catheter assembly, sterilized, and packaged.

Materials that have traditionally been used for stents include stainless steels (typically 316L series), cobalt-based alloys, tantalum alloys, nitinol (nickel and titanium) shape memory alloy, and biocompatible materials such as polyhydroxybutyrate (PHB).

A number of lasers such as CO_2, Nd:YAG, excimer, diode, copper vapor, and titanium sapphire lasers have been used for cutting 316L-series stainless steel material in stent manufacturing. With the traditional continuous wave or long pulse lasers, the thermal energy associated with the laser beam affects the microstructure of the material in the heat-affected zone (HAZ) and thereby the mechanical properties of the stent. This is undesirable for the following reasons:

1. Stents are permanently deformed by stretching during deployment in angioplasty.
2. Radial pressure from arterial walls and pressure from heart beats subject the stent to fatigue.

To mitigate these problems, ultrashort pulse lasers (pulse duration in the range 10^{-15}–10^{-11} s) are being investigated for stent fabrication. These are capable of significantly improving the surface quality, and thereby eliminating the need for additional finishing operations. Also, due to the reduced heat input and the short time available for the laser beam, the heat-affected zone size is minimized.

20.1.2 Therapeutic Applications

There are several therapeutic applications of the laser, and due to space limitations, we shall only discuss a few of these, specifically in surgery, ophthalmology, dermatology, and dentistry. In each of these areas, the procedure may be classified under photocoagulation, photodynamic therapy, photodisruption, photoevaporation, or photoablation. In a number of these applications, lasers have several advantages over traditional procedures. These include

1. Hemeostasis or the ability to minimize hemorrhaging through coagulation and thereby provide a clear surgical field.
2. Precision of the process.
3. Minimal damage to collateral tissue.
4. Rapid and relatively painless healing.
5. The ability to control the beam intensity, enabling control of depth and extent of procedure.
6. The ease with which the laser beam can be positioned and also used to access difficult-to-reach areas.

20.1.2.1 Surgical Procedures As Table 20.1 indicates, laser-based surgical procedures primarily involve coagulation and material removal, where they have significant advantages over mechanical incision.

One common surgical procedure that uses a laser is laser-assisted uvulopalatoplasty (LAUP). This is used to treat snoring that results from palatal flutter by reducing the tissues of velum and uvula, and stiffening them. This minimizes or eliminates obstruction of airway, as well as vibration of soft tissue at the soft palate level. It has the advantage of fewer long-term complications compared to the traditional uvulopalatopharyngoplasty. The procedure involves using the laser as either an incision tool to cut material or as a heat source to vaporize material. It is normally performed using local anesthesia.

Most laser systems can be used for this procedure. The CO_2 laser is quite commonly used, even though diode and Nd:YAG lasers are also extensively used. The latter two have the advantage of fiber transmission, with the diode laser having the added advantage of compactness. However, both the Nd:YAG and diode lasers tend to result in a deeper coagulation zone and usually take longer to heal, compared to the CO_2 laser. The postsurgical pain associated with the process is also more severe for the Nd:YAG and diode lasers.

20.1.2.2 Ophthalmology Lasers are ubiquitous in opthalmology and constitute a significant portion of ophthalmic therapy. Our focus here will be on photocoagulation and photodynamic applications.

As the name implies, photocoagulation involves the use of a laser (light source) to coagulate (clot) or destroy abnormal, leaking blood vessels and thereby stopping their growth. The process relies on the selective absorption of light around the green wavelength range by hemoglobin, the pigment in red blood cells. Various lasers are used for this procedure in ophthalmology, for example, the argon ion, krypton ion, diode, Nd:YAG, and dye lasers. They are usually used in the CW mode with maximum output power of 2 W, but only for short periods of about 0.1–1.0 s and with focal diameters of about 0.05–2 mm.

One common application of the procedure is in diabetic retinopathy where it is used to seal leaking blood vessels in the retina, slowing their growth, and thereby reducing the risk of vision loss. It is an outpatient procedure that uses local or topical anesthesia. Other applications of photocoagulation include treatment of detached retina, tumors of the retina, and age-related macular degeneration, which is a leading cause of blindness in the elderly.

In photodynamic therapy, a photosensitizer dye, which is a type of drug, is first injected into the patient. The dye has the tendency to accumulate in tumor tissue. When the tissue is now irradiated with light that is selectively absorbed by the photosensitizer, it produces a type of oxygen that induces cytotoxicity in the tissue cells, destroying them. It is usually an outpatient procedure and also has application in the treatment of age-related macular degeneration. Photodynamic therapy is also known by other names as phototherapy, photoradiation therapy, and photochemotherapy.

20.1.2.3 Dermatology Perhaps the most widely known use of lasers in medicine is in dermatology where they are used in skin resurfacing to improve skin appearance, in the treatment of skin tumors, in the treatment of scars and keloids, in the removal of tattoos, and so on.

The lasers that are most commonly used in laser skin resurfacing are the CO_2 and Er:YAG lasers. The coefficient of absorption in water of the Er:YAG laser beam is much higher than that of the CO_2 laser. Thus, for the Er:YAG laser, a significant amount of laser energy is absorbed in a thin layer of about 30 μm with a thermal damage zone of about 50 μm, compared to about a 100 μm depth and 150 μm damage zone for the CO_2 laser. Thus, there is less risk of skin damage with the Er:YAG laser. However, since it is necessary to penetrate depths of about 250–400 μm for effective laser skin resurfacing, about three passes are required with the CO_2 laser as opposed to the 12–15 passes required of the Er:YAG laser. Also, the outcome for deep wrinkles is better with the CO_2 laser than the Er:YAG laser. For facial resurfacing, general or local anesthesia may be used depending on the extent of the procedure.

A number of skin malignancies or tumors are treated using a variety of phototherapies, including photodynamic therapy, photovaporization, and photocoagulation. In photodynamic therapy, tunable dye lasers are commonly used, even though other forms of noncoherent light are becoming increasingly popular. CO_2 lasers are more extensively used in procedures requiring photovaporization. The beam may be defocused or focused, with the defocused beam being used to provide homogenous vaporization of superficial layers, while the focused beam is mainly used for tissue incision. A CW beam is normally used for large skin lesions, while a superpulse with a pulse duration ranging between 50 and 200 ms is preferred for small lesions. Power levels typically range from 5 to 20 W, with a beam diameter of about 2 mm. Photocoagulation for skin therapy is usually done with an Nd:YAG laser.

20.1.2.4 Dentistry The use of lasers in dentistry has largely been limited to soft tissue applications. Soft tissues include the tissue supporting the tongue, the gums, and the ligaments and fibers that bind tooth to the socket. Hard tissues refer to the tooth and the root. Major advantages of lasers in this field include the ability to induce hemeostasis during the procedure, and lower postoperative pain experienced, compared to the more traditional procedures. Dental applications for which lasers are used include gingivectomy (removing excess gum tissue), frenectomy (removal of frenula in the mouth), and removal of tumors. Lasers that are commonly used for dental procedures are the CO_2, Nd:YAG, argon ion, and holmium:YAG lasers. The collateral thermal damage that results when a CO_2 laser pulse of high peak power and short pulse duration is used on oral soft tissue is in the range 15–170 μm. The majority of the beam power is absorbed within the first 0.3 mm of soft tissue. The argon ion laser beam, however, is absorbed over a distance of 1–2 mm, with a collateral damage that is slightly greater than that of the CO_2 laser. The Nd:YAG laser tends to penetrate a lot deeper into soft tissue. On the contrary, in areas where bleeding can be significant, the Nd:YAG, argon ion, and holmium:YAG lasers are the ones to consider.

In using a laser for gingivectomy, an important issue is to prevent damage to underlying bone and tooth substrate. The CO_2 laser is effectively absorbed by watery tissue. Thus, it is the tool of choice for this procedure since it does not penetrate deep into the soft tissue and therefore has minimal impact on the underlying bone and tooth structures.

Different forms of oral lesion can be effectively removed using lasers. Oral leukoplakia is one common type of precancerous lesion that lends itself well to laser application. It forms as thick, white patches of the mucous membrane on the tongue, gums, or the inside of the cheeks and cannot be removed simply by scraping. It is normally caused by chronic irritations such as tobacco or alcohol abuse.

The procedure to remove the lesion may involve either excision or vaporization. In either case, no suturing is required. When the lesion is thick and white, then due to its paucity of water, it might be preferable to use excision. Also, since vaporization normally results in superficial layer removal, deeper layers may not be removed, enabling the lesion to recur. For the thin patches that lend themselves to vaporization, a defocused CO_2 laser beam may be used at a power level of 15–20 W. Vaporization normally leaves a thin carbonaceous deposit on the wound. It is not fully established whether or not this is beneficial to the healing process.

Compared to the medical field, the application of lasers in nanotechnology is a relatively new and rapidly growing field. In the next section, we outline some of the developments in nanotechnology that involve the use of lasers.

20.2 NANOTECHNOLOGY APPLICATIONS

Nanotechnology refers to the variety of techniques that are utilized in the creation of features and/or structures with minimum feature dimensions smaller than 100 nm. Several technologies have been developed for fabricating parts with nanosized components and/or features. Our focus in this section, however, will be on laser-based systems. This is an area that is mushrooming, and due to space limitations, we shall only discuss a few examples. Some of the features or structures that can be created using laser-based nanotechnology include nanoholes, grating, nanobumps, nanojets, and nanotubes.

The resolution of features created using laser technology is normally determined by the diffraction limit of the laser system, and this is of the order of half the wavelength of the radiation (see equation (7.20b)). However, with ultrashort or femtosecond pulse lasers, special techniques can be used to generate subwavelength features. These include selecting the peak laser fluence to be just above the threshold ablation fluence (see Fig. 15.28), using interferometry, using nanoparticles, and using the laser pulses in combination with an atomic force microscope or a scanning near-field optical microscope. In general, the feature resolution associated with femtosecond lasers can be approximated by

$$d_r = \frac{k_n \lambda}{\mathrm{NA} {q_e}^{1/2}} \tag{20.1}$$

where d_r is the femtosecond laser feature resolution, $k_n = 0.5 - 1$ is a constant, NA is the numerical aperture of the focusing optics, and q_e is the number of photons required to overcome the energy band gap.

20.2.1 Nanoholes and Grating

A matrix of nanoholes can be generated by ablating a thin film using four interfering femtosecond beams. Using a Ti:sapphire femtosecond laser (pulse duration, 90 fs; wavelength, 800 nm; pulse repetition rate, 10 Hz; beam diameter, 6 mm) with an average laser fluence of 120 $\mathrm{mJ/cm^2}$, a matrix of 800 nm diameter circular holes can be produced in a gold film of thickness 50 nm at intervals of 1.7 μm (Fig. 20.2a). Now with two interference beams instead of four, a grating structure is obtained (Fig. 20.2b).

Nanoholes can also be generated by directing the laser beam at the gap between the tip of an atomic force or scanning near-field optical microscope and the thin film. This phenomenon results from electromagnetic field enhancement below the tip and thermal expansion of the tip. When the beam intensity exceeds a threshold value (which is less than the ablation threshold without any particles), each pulse of the laser results in a nanohole being created in the film. The hole size and depth increase with an increase in the beam intensity. When the workpiece is moved linearly at a slow enough speed, a grating structure can be produced.

Finely dispersed nanoholes can also be created by first distributing nanoparticles that are transparent to ultraviolet radiation, for example, silica and polystyrene particles with diameters ranging between 150 and 1000 nm on, say, an aluminum film of thickness 35 nm on a silicon substrate. Illuminating the surface with a KrF excimer laser of wavelength 248 nm, pulse duration 23 ns, and fluence between 100 and 800 $\mathrm{mJ/cm^2}$ results in an array of nanoholes, with depths and diameters that increase with the laser fluence. Similar holes can also be produced on a silicon substrate using an infrared femtosecond laser after first depositing alumina particles (which are transparent to the infrared radiation) on the silicon substrate. This also results from optical enhancement effect between the particles and the substrate.

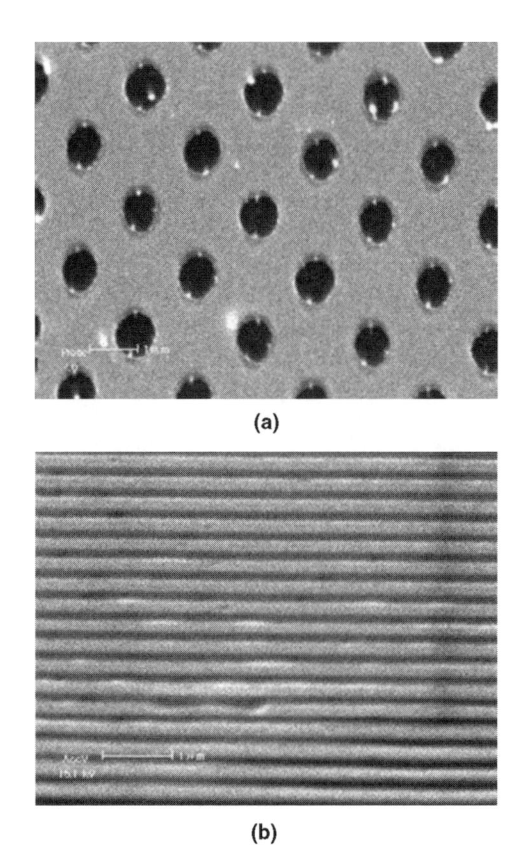

FIGURE 20.2 (a) A matrix of nanoholes generated using four femtosecond laser beams. (b) A grating generated using two femtosecond laser beams. (From Nakata, Y., Okada, T., and Maeda, M., 2004, *Proc. of SPIE*, Vol. 5339, pp. 9–19.)

20.2.2 Nanobumps

If the laser fluence is reduced to about 77 mJ/cm^2, a matrix of conical bumps are formed at locations of maximum interference of the four beams (Fig. 20.3a). The size of each bump grows with increasing fluence (Fig. 20.3b). At an even higher fluence, a bead or jet begins to form on top of each bump, Fig. 20.3c, and that grows with further increases in the fluence, until eventually the entire bump becomes a nanohole, (Fig. 20.3d). The hole diameter increases with the laser pulse energy. The minimum bump size obtainable is of the order of 8 nm height and 330 nm diameter. These are much smaller than the wavelength of the laser used. However, the typical bump aspect ratio (ratio of bump height to diameter) is about 0.44, which is about two orders of magnitude greater than that obtained when nanosecond lasers are used for laser texturing. This is because of the shorter thermal diffusion length (of the order of tens of nanometers) for a femtosecond laser compared to that of the nanosecond laser (of the order of tens of microns). Each of the nanofeatures, from the bump to

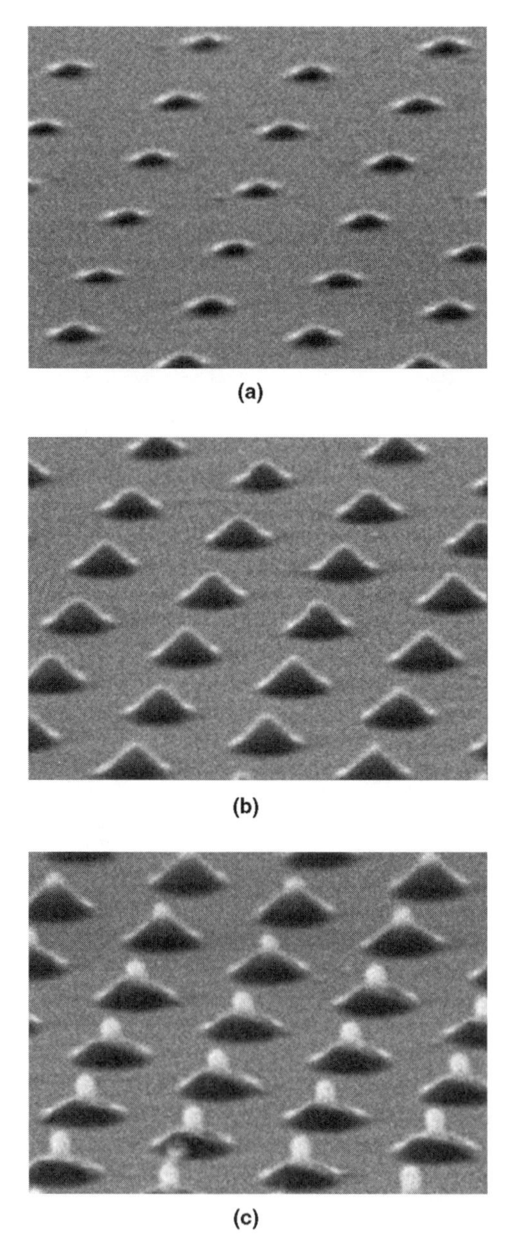

FIGURE 20.3 Nanofeatures on a gold film irradiated with four interfering femtosecond lasers. (a) A nanobump matrix at a fluence of 77 mJ/cm^2. (b) The nanobump matrix at a fluence of 89 mJ/cm^2. (c) Nanojets forming from the bumps at a fluence of 97 mJ/cm^2. (d) Nanoholes forming from the bumps at a fluence of 114 mJ/cm^2. (From Nakata, Y., Okada, T., and Maeda, M., 2004, *Proceedings of SPIE*, Vol. 5339, pp. 9–19.)

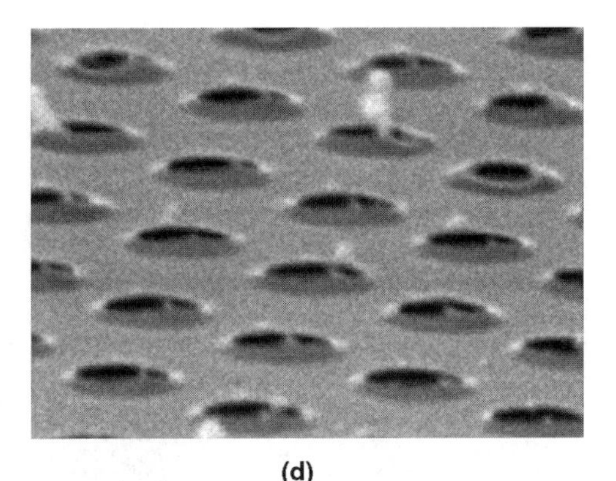

(d)

FIGURE 20.3 (*Continued*)

the hole formation, has a well-defined threshold. For some materials, for example, 100–200 nm chromium-coated layers, the hole may be surrounded by a molten ring along with some droplets in that region of the material where the beam intensity falls between the melting and ablation thresholds.

The bumps result from vapor pressure that is generated as part of the film at the locations of maximum interference. This peels and expands material that has been softened by heating at those locations. That the material is heated at those locations is supported by the conical shape of the bumps, indicating that the film is more elongated at the top of the bump than at the base. This is because the film is at a higher temperature, and therefore more ductile and thin at the top. The formation of the jet further supports this concepts and also indicates that the area where the jet forms may have been molten.

The bump structure that is obtained depends on the number of interfering laser beams. Ellipsoidal bumps are formed for three beams, Figs. 20.4a–c, and linear bumps are obtained for two beams (Figs. 20.5a and b).

One common way to generate the beam interference is to use a mirror beam splitter arrangement as shown in Fig. 20.6. The width of the interference region, b_n, is given by

$$b_n = \frac{c\tau_p}{\sin(\theta/2)} \tag{20.2}$$

where τ_p is the pulse duration and θ is the angle between two interfering beams.

For 100 fs laser beams interfering at a 10° angle, $b_n = 340$ μm.

In the next section, the concept of two-photon polymerization is discussed. This process enables three-dimensional polymer-based objects to be created within bulk resin material without resorting to layer-by-layer application of the resin, as is done in conventional rapid prototyping.

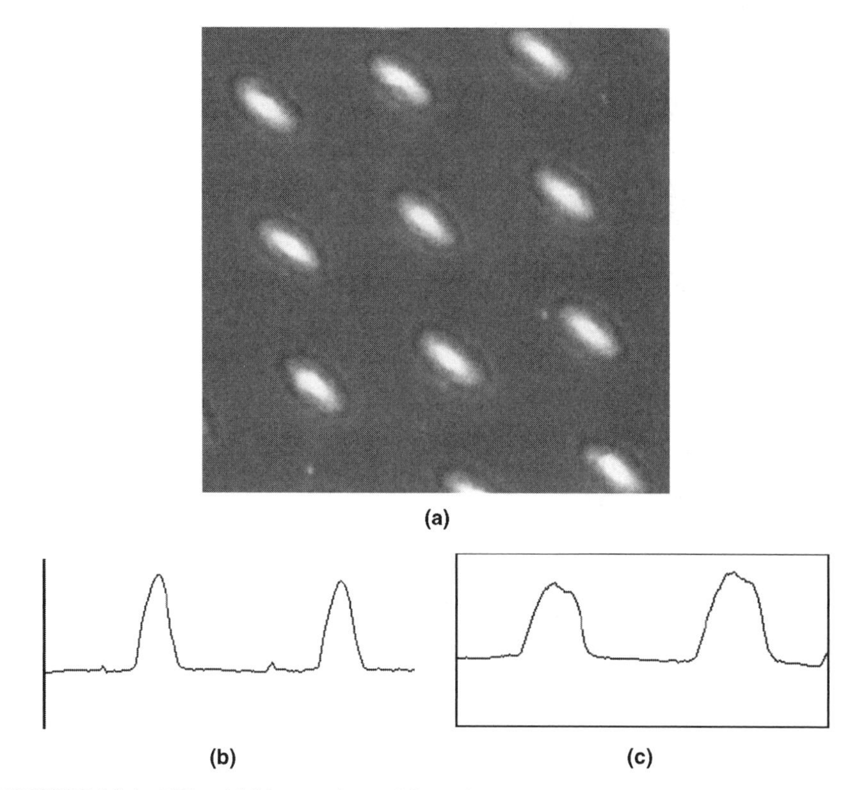

FIGURE 20.4 Ellipsoidal bumps formed from three beams. (a) Top view. (b) Atomic force microscope scan of a section through the minor axis. (c) Atomic force microscope scan of a section through the major axis. (From Nakata, Y., Okada, T., and Maeda, M., 2004, *Proceedings of SPIE*, Vol. 5339, pp. 9–19.)

20.2.3 Two-Photon Polymerization

Two-photon polymerization (2PP) is a process that is used to fabricate three-dimensional microcomponents that have nano features, using photosensitive resins, and is based on the two-photon absorption phenomenon (Section 1.7). The polymerization process is photoinitiated (as opposed to thermal initiation), and thus by controlling the intensity of the beam, the generation of radicals can be controlled. As a result, the process can be controlled with high precision. The resin used in photopolymerization is normally one that is catalyzed by ultraviolet radiation, together with a photochemical initiator that absorbs the radiation and becomes decomposed into free radicals. In two-photon polymerization, an infrared wavelength laser beam is used, so that instead of the single ultraviolet photon, two photons of near-infrared wavelength are simultaneously absorbed.

In addition to the photoinitiators, the resin also consists of monomers and oligomers, together with an appropriate concentration of reaction terminators. An

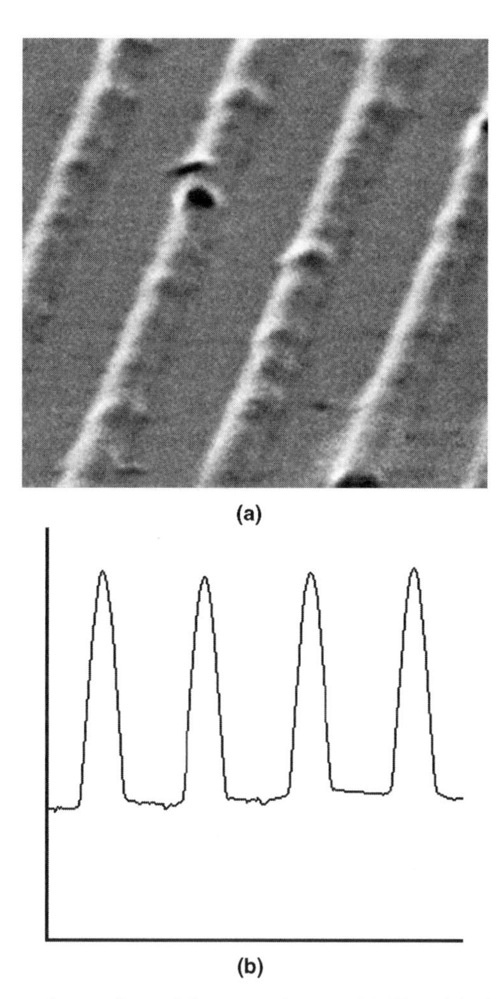

FIGURE 20.5 Linear bumps formed from two beams. (a) Pictorial view. (b) Atomic force microscope scan of a transverse section. (From Nakata, Y., Okada, T., and Maeda, M., 2004, *Proceedings of SPIE*, Vol. 5339, pp. 9–19.)

example would be urethane acrylate monomers/oligomers. The process of radical polymerization involves initiation, propagation, and termination and is illustrated below in equations (20.3)–(20.6). In photopolymerization, in general, one quantum of light decomposes a single initiator molecule into two free radicals. Thus, each initiator (I), which normally absorbs a single ultraviolet photon, absorbs two infrared photons simultaneously and is decomposed into two free radicals ($2\dot{R}$) (equation (20.3)). Each radical reacts with a monomer (M) or oligomer (M_n), generating a chain radical ($R\dot{M}$) at the ends of the monomers and oligomers (equation (20.4)). Each new radical combines with another monomer, generating a chain reaction, equation (20.5), which is terminated when the chained radical meets another chained radical, (equation (20.6)).

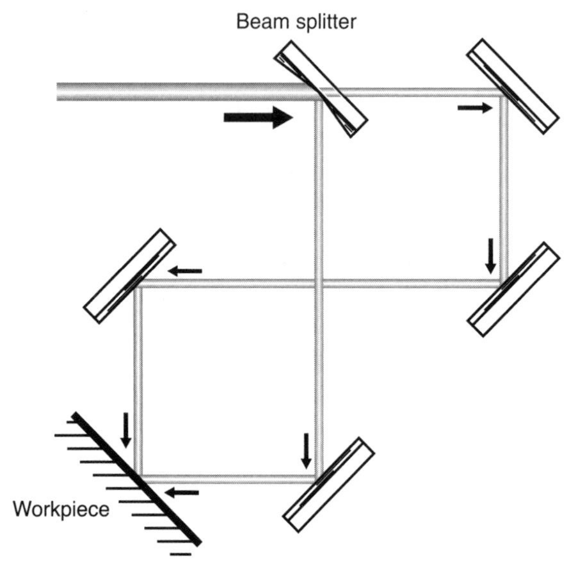

FIGURE 20.6 Mirror setup for generating interference beams. (From Nakata, Y., Okada, T., and Maeda, M., 2004, *Proceedings of SPIE*, Vol. 5339, pp. 9–19.)

Initiation

$$\mathrm{I} + h_{\mathrm{p}}\nu + h_{\mathrm{p}}\nu \rightarrow 2\dot{\mathrm{R}} \tag{20.3}$$

$$\dot{\mathrm{R}} + \mathrm{M} \rightarrow \mathrm{R}\dot{\mathrm{M}} \tag{20.4a}$$

For example,

$$\dot{\mathrm{R}} + \mathrm{CH_2} = \mathrm{CHX} \rightarrow \mathrm{R}\mathrm{C}\dot{\mathrm{H}}_2\,\underset{\substack{|\\ \mathrm{X}}}{\overset{\substack{\mathrm{H}\\ |}}{\mathrm{C}}} \tag{20.4b}$$

Propagation

$$\mathrm{R}\dot{\mathrm{M}} + \mathrm{M} \rightarrow \mathrm{R}\dot{\mathrm{M}}_1 \tag{20.5a}$$

$$\vdots$$

$$\mathrm{R}\dot{\mathrm{M}}_n + \mathrm{M} \rightarrow \mathrm{R}\dot{\mathrm{M}}_{n+1} \tag{20.5b}$$

For example,

$$
R-(CH_2CHX-)_n\dot{C}H_2\begin{array}{c}H\\|\\C\\|\\X\end{array} + CH_2 = CHX \rightarrow R--(CH_2CHX-)_{n+1}\dot{C}H_2\begin{array}{c}H\\|\\C\\|\\X\end{array} \tag{20.5c}
$$

Termination by combination

$$
R\dot{M}_n + R\dot{M}_n \rightarrow RM_{2n}R \tag{20.6a}
$$

For example,

$$
-\dot{C}H_2\begin{array}{c}H\\|\\C\\|\\X\end{array} + \begin{array}{c}H\\|\\C\\|\\X\end{array}\dot{C}H_2- \rightarrow -CH_2\begin{array}{c}H\\|\\C\\|\\X\end{array} - \begin{array}{c}H\\|\\C\\|\\X\end{array}CH_2- \tag{20.6b}
$$

Or termination by disproportionation

$$
R\dot{M}_n + R\dot{M}_n \rightarrow RM_n + RM_n \tag{20.6c}
$$

For example,

$$
-\dot{C}H_2\begin{array}{c}H\\|\\C\\|\\X\end{array} + \begin{array}{c}H\\|\\C\\|\\X\end{array}\dot{C}H_2- \rightarrow -CH_2\begin{array}{c}H\\|\\C\\|\\X\end{array} - H + \begin{array}{c}H\\|\\C\\|\\X\end{array} = CH- \tag{20.6d}
$$

Here, R is a photoinitiator and results from the removal of an electron from $\dot{R}$. Addition of a monomer to an oligomer M_n through the propagation process results in the new oligomer M_{n+1}. An oligomer consists of a finite number of monomers, as opposed to a polymer that theoretically has an infinite number of monomers.

The main reason for using an infrared wavelength laser is that since the resin is transparent to the infrared radiation, the beam is able to propagate, with little or no attenuation, to the focal zone. Material in the unfocused region (both solid and liquid) is not affected by the radiation. If a femtosecond laser of high enough intensity is

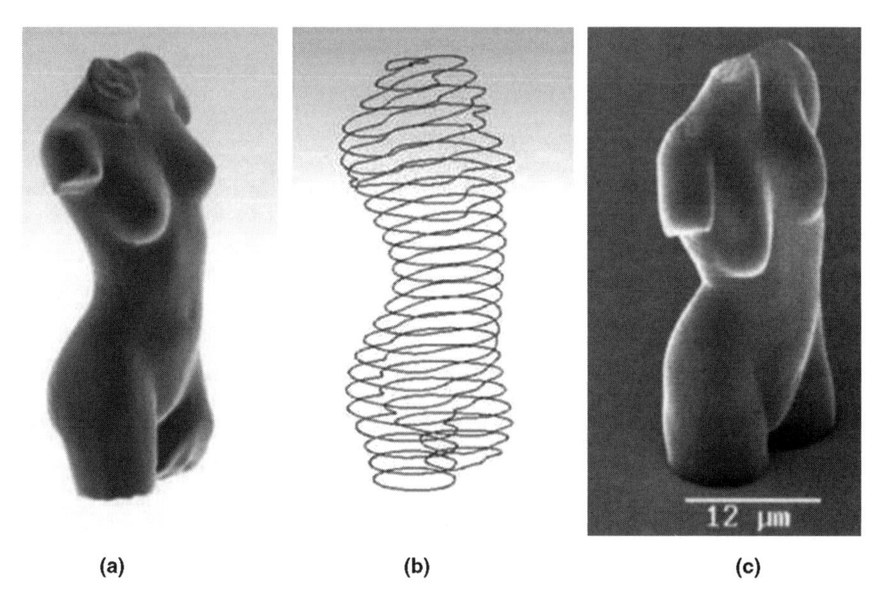

FIGURE 20.7 3D object produced by the two-photon polymerization process. (a) Original statue. (b) Computer scan of the statue. (c) Microscale statue produced by 2PP. (From Korte, F., Koch, J., Serbin, J., Ovsianikov, A., and Chichkov, B. N., 2004, *IEEE Transactions on Nanotechnology*, Vol. 3, No. 4, pp. 468–472.)

used with tight focusing, the power density in the focal zone can initiate two-photon absorption in that region. Since the rate of two-photon absorption is proportional to the square of the power density, the process is limited to a small region within the focal zone. Moving the focal point in 3D space through the resin enables any computer-generated 3D structure to be fabricated within the interior of the resin, Fig. 20.7, with a resolution better than 100 nm, and there is no need to fabricate it layer by layer, as is done with conventional rapid prototyping (Section 19.2.1). This improves the resolution of the process. The tight focusing also enables spatial resolution smaller than the laser diffraction limit to be achieved. If ultraviolet radiation was used (in which case the process would be single photon), the beam would polymerize material (while also being attenuated by it) through which it propagates.

As an example, the liquid resin ORMOCER, which is an organic–inorganic hybrid polymer, is sensitive to the 390 nm radiation and if it is exposed to a tightly focused femtosecond laser radiating at 780 nm (Ti:sapphire) with a pulse duration of 80 fs, and a pulse rate of 80 MHz, two-photon polymerization occurs. In the setup shown in Fig. 20.8, a microscope objective with a 100× magnification and a high numerical aperture of 1.4 is used, resulting in a strongly convergent and thus tightly focused beam. The threshold time for solidification of the polymer may be about 0.2 ms. After the process, the unpolymerized resin can be removed by dissolving in acetone or ethanol, depending on the polymer.

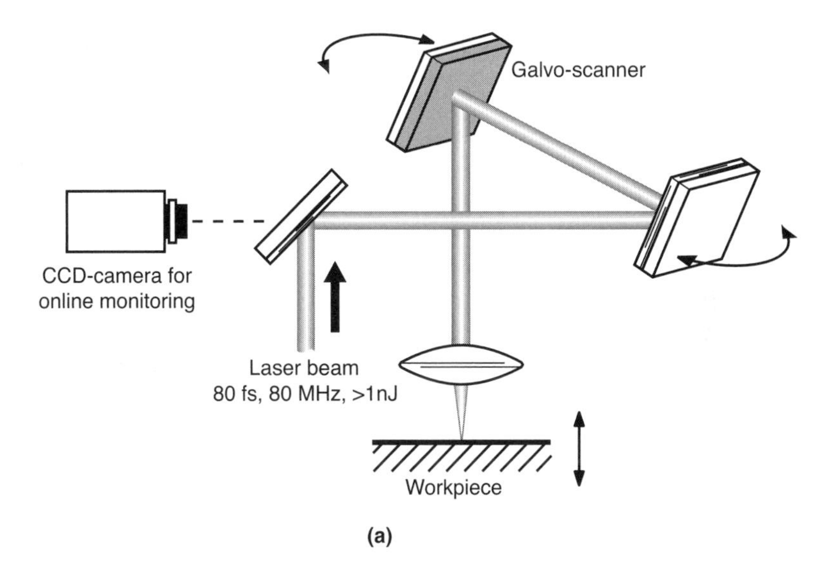

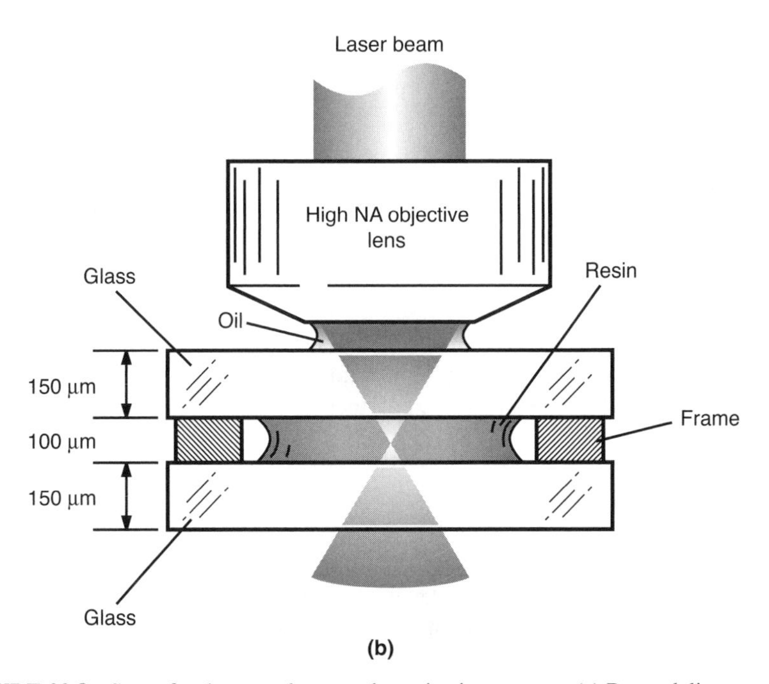

FIGURE 20.8 Setup for the two-photon polymerization process. (a) Beam delivery system. (b) Close-up of the polymerization setup. (From Korte, F., Koch, J., Serbin, J., Ovsianikov, A., and Chichkov, B. N., 2004, *IEEE Transactions on Nanotechnology*, Vol. 3, No. 4, pp. 468–472.)

The resolution of structures made by two-photon polymerization is determined by a number of factors:

1. Spherical aberration at the boundary between the sample and air due to their significantly different refractive indices. This causes deformation of the focused spot along the optical axis. This effect can be minimized by having oil between the lens and the sample to remove the refraction at the boundary between the sample and the air.
2. The diffusion of radicals and polymer chain growth beyond the radiation zone. However, since two-photon polymerization is only effective in the focused regions of high intensity, the resin does not polymerize much beyond the focal volume where the photon density is high. Thus, the resolution is much better than would be obtained for the conventional single-photon absorption process.

20.2.4 Laser-Assisted Nanoimprint Lithography

In traditional nanoimprint lithography (NIL), a mold (often a quartz template) is used to imprint features as small as 20 nm into a low-viscosity material, usually either a thermoplastic or UV-curable resist. Both two- and three-dimensional nanostructures can be produced with a single resist layer. The UV-based process does not require heating. However, multiple resist layers and multiple etching steps are often necessary, making it relatively expensive. The thermal-based process, on the contrary, requires heating of the polymer to soften it. This is a relatively slow process, and it creates a thermal expansion difference between the mold and substrate, which could result in misalignment.

Laser-assisted nanoimprint lithography (LAN) uses a laser beam that is transparent to the quartz mold (since the quartz material band gap is greater than the photon energy), for example, a 20 ns pulse of an XeCl excimer laser of wavelength 308 nm. The beam is directed at the polymer film on the substrate, and at the same time, the mold is imprinted into the polymer. The entire process can take less than 500 ns since only a single laser pulse is necessary. Lower imprint pressures are required since the laser beam reduces the viscosity of the polymer.

20.3 SUMMARY

In the medical field, lasers are used in the fabrication of medical devices such as stents, in therapeutic applications, and in medical diagnostics. In stent manufacturing, a laser is used to cut the stent geometry in a metal tubing, followed by finishing operations. The major therapeutic applications include surgery, where they have been used in laser-assisted uvulopalatoplasty; ophthalmology, with applications in diabetic retinopathy to seal leaking blood vessels in the retina; dermatology, for skin resurfacing to improve skin appearance, and also for treating skin tumors; and dentistry, where they are mainly used for soft tissue therapies such as gingivectomy (removing excess gum tissue) and tumor removal. In many of these applications, lasers can be used to minimize hemorrhaging through coagulation and have the advantage of process precision, minimal damage to collateral tissue, and rapid and relatively painless healing.

Laser-based nanotechnology applications include the production of features or structures such as nanoholes, grating, nanobumps, nanojets, and nanotubes. These can be produced in a number of different ways, such as selecting the peak laser fluence to be just above the threshold ablation fluence, using interferometry, using nanoparticles, and using the laser pulses in combination with an atomic force microscope or a scanning near-field optical microscope.

Three-dimensional microcomponents that have nano features can also be created using two-photon polymerization by tightly focusing a femtosecond infrared laser into a photosensitive resin. Finally, in laser-assisted nanoimprint lithography, a laser beam is used to heat and soften a polymer film before it is imprinted by a mold.

REFERENCES

Medical References

Berlien, H.-P., and Muller, G. J., 2003, *Applied Laser Medicine*, Springer-Verlag, Berlin.

Catone, G. A., 1994, Laser technology in oral and maxillofacial surgery. Part II: Applications. *Selected Readings in Oral and Maxillofacial Surgery*, Vol. 3, No. 5, pp. 1–35.

Constable, I. J., and Lim, A. S. M., 1990, *Laser — Its Clinical Uses in Eye Diseases*, Churchill Livingstone, Edinburgh, UK.

Peng, Q., Juzeniene, A., Chen, J., Svaasand, L. O., Warloe, T., Giercksky, K.-E., and Moan, J., 2008, Lasers in medicine, *Reports on Progress in Physics*, Vol. 71, pp. 1–28.

Singerman, L. J., and Coscas, G., 1999, *Ophthalmic Laser Surgery*, 3rd edition, Butterworth Heinemann, Boston, MA.

Visuri, S. R., 1996, *Laser Irradiation of Dental Hard Tissues*, Doctoral Dissertation, Northwestern University, Evanston, IL.

Nanotechnology References

Hong, M. H., Huang, S. M., Luk'yanchuk, B. S., Wang, Z. B., Lu, Y.F., and Chong, T. C., 2003, Laser assisted nanofabrication," *Proceedings of SPIE*, Vol. 4977, pp. 142–155.

Korte, F., Serbin, J., Koch, J., Egbert, A., Fallnich, C., Ostendorf, A., and Chichkov, B. N., 2003, Towards nanostructuring with femtosecond laser pulses, *Journal of Applied Physics A*, Vol. 77, pp. 229–235.

Korte, F., Koch, J., Serbin, J., Ovsianikov, A., and Chichkov, B. N., 2004, Three-dimensional nanostructuring with femtosecond laser pulses, *IEEE Transactions on Nanotechnology*, Vol. 3, No. 4, pp. 468–472.

Lu, Y.F., Zhang, L., Song, W. D., Zheng, Y. W., and Luk'yanchuk, B. S., 2002, Particle-enhanced near-field optical effect and laser writing for nanostructure fabrication, *Proceedings of the SPIE*, Vol. 4426, pp. 143–145.

Nakata, Y., Okada, T., and Maeda, M., 2004, Generation of uniformly spaced and nano-sized sructures by interfered femtosecond laser beams, *Proceedings of the SPIE*, Vol. 5339, pp. 9–19.

Takada, H., Kamata, M., Hagiwara, Y., and Obara, M., 2004, Nanostructure fabrication by femtosecond laser with near-field optical enhancement effect, *Proceedings of the SPIE*, Vol. 5448, pp. 765–772.

Xia, Q., Keimel, C., Ge, H., Yu, Z., Wu, W., and Chou, S. Y., 2003, Ultrafast patterning of nanostructures in polymers using laser assisted nanoimprint lithography, *Applied Physics Letters*, Vol. 83, No. 21, pp. 4417–4419.

Two-Photon Polymerization References

Belfield, K. D., 2001, Two-photon organic photochemistry, *The Spectrum*, Vol. 14, Issue 2, pp. 1–7.

Billmeyer Jr., F. W., 1962, Radical chain (addition) polymerization, *Textbook of Polymer Science*, Wiley Interscience, New York.

Galajda, P., and Ormos, P., 2001, Complex micromachines produced and driven by light, *Applied Physics Letters*, Vol. 78, No. 2, pp. 249–251.

Kawata, S., Sun H.-B., Tanaka, T., and Takada, K., 2001, Finer features for functional microdevices, *Nature*, Vol. 412, pp. 697–698.

Maruo, S., Nakamura, O., and Kawata, S., 1997, Three-dimensional microfabrication with two-photon-absorbed photopolymerization, *Optics Letters*, Vol. 22, No. 2, pp. 132–134.

Maruo, S., and Kawata, S., 1998, Two-photon-absorbed near-infrared photopolymerization for three-dimensional microfabrication, *Journal of Microelectromechanical Systems*, Vol. 7, No. 4, pp. 411–415.

Robinson, K., 2002, Two-photon absorption enables microfabrication, *Photonics TechnoWorld*, May 2002 edition.

Strickler, J. H., and Webb, W. W., 1990, Two-photon excitation in laser scanning fluorescence microscopy, *SPIE CAN-AM Eastern*, Vol. 1398, pp. 107–118.

Sun H.-B., Matsuo, S., and Misawa, H., 1999, Three-dimensional photonic crystal structures achieved with two-photon-absorption photopolymerization of resin, *Applied Physics Letters*, Vol. 74, No. 6, pp. 786–788.

Sun H.-B., Kawakami, T., Xu, Y., Ye, J.-Y., Matuso, S., and Misawa, H., Miwa, M., and Kaneko, R., 2000, Real three-dimensional microstructures fabricated by photopolymerization of resins through two-photon-absorption, *Optics Letters*, Vol. 25, No. 15, pp. 1110–1112.

Sun H.-B., Takada, K., and Kawata, S., 2001a, Elastic force analysis of functional polymer submicron oscillators, *Applied Physics Letters*, Vol. 79, No. 19, pp. 3173–3175.

Sun H.-B., Tanaka, T., Takada, K., and Kawata, S., 2001b, Two-photon photopolymerization and diagnosis of three-dimensional microstructures containing fluorescent dyes, *Applied Physics Letters*, Vol. 79, No. 10, pp. 1411–1413.

Sun H.-B., Tanaka, T., and Kawata, S., 2002, Three-dimensional focal spots related to two-photon excitation, *Applied Physics Letters*, Vol. 80, No. 20, pp. 3673–3675.

21 Sensors for Process Monitoring

Every process has inputs and outputs. The inputs are those parameters that can be changed and collectively used to control the process. The outputs are those parameters that give a measure of the state of the process and its quality. The inputs for laser processing include the beam power, beam size (focusing), and traverse velocity (welding or cutting speed). Others are gas flow rate, gas type (shielding gas, assist gas, suppressant gas), orientation, workpiece and filler material properties, workpiece geometry, depth of focus, beam mode structure, and joint configuration.

Primary outputs that provide a direct measure of the process state are usually difficult to measure in real time, that is, while the process is in progress and without destroying the part. Secondary outputs are more easily measured in real time, but not after the process. The primary outputs for laser welding, for example, include the penetration, bead width, reinforcement (collectively the bead geometry), heat-affected zone size, microstructure, mechanical properties (hardness and strength), residual stresses, and defects (cracks, inclusions, porosity, undercut, concavity, etc.). The secondary outputs include the temperatures (temperature distribution, peak temperatures, cooling rate), acoustic emission, and weld pool motion. In the case of laser cutting, the primary outputs include kerf width, cut edge squareness, microstructure, dross appearance, and surface roughness (striations). Secondary outputs of the laser cutting process include spark discharge beneath the cut (angle and intensity).

To ensure process quality, it is necessary to measure the outputs of interest and, sometimes where necessary, the inputs as well. The process outputs, however, often need special sensors. Ideally, the sensors have to be nonintrusive, rugged enough for the hostile environment, repeatable and reliable, and relatively inexpensive.

In the rest of the chapter, some of the sensors and sensing techniques that are commonly used or are under development for monitoring and diagnostics during laser processing are discussed. The discussion starts with monitoring of the laser beam itself. This is followed by monitoring of the process, with some of the systems that are discussed also being applicable to seam tracking.

21.1 LASER BEAM MONITORING

In this section, we discuss sensors that are used exclusively for monitoring and control of the laser beam. The principal beam characteristics that are often measured include

Principles of Laser Materials Processing, by Elijah Kannatey-Asibu, Jr.

the beam power, propagation (beam size and location), and mode structure. For a CW laser, the power level is of interest, while a pulsed laser is characterized by such parameters as beam energy, peak power, repetition rate, pulse duration, pulse shape, and average power. Beam propagation is characterized by the variation of diameter with distance (beam divergence), and beam center direction (pointing stability). The mode structure describes how power is distributed across a plane perpendicular to the propagation direction. Various techniques have been developed for monitoring each of these characteristics, and we shall briefly outline the principal ones in the following sections.

21.1.1 Beam Power

Drifts in the resonator and other disturbances often introduce fluctuations in the laser output power. Common techniques for measuring the output beam power are the pyroelectric detector and the standard beam dump.

21.1.1.1 Pyroelectric or Thermopile Detector This power measuring technique uses a beam splitter to divert a portion of the beam, which is sampled using a chopper. The resulting intermittent signal is then focused onto either a pyroelectric detector or a thermopile that measures the power level. Pyroelectric detectors are further discussed in Section 21.2.4.

21.1.1.2 Beam Dump High power lasers normally have a beam dump as a standard component, and its purpose is to dispose of the energy generated when the laser is idling. The beam dump may also be used as a calorimeter for measuring the output power of the laser. One such device consists of a conical calorimeter through which water is circulated (Fig. 21.1). With this device, the beam is intercepted by a water-cooled polished copper substrate and reflected into the cone. Thermocouples mounted at the inlet and outlet of the calorimeter enable the temperature rise in the water as it flows through the calorimeter to be determined. The temperature rise, ΔT, when

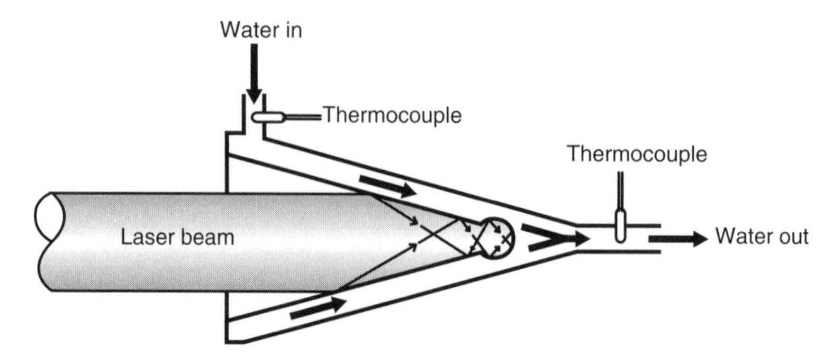

FIGURE 21.1 Schematic of a cone calorimeter beam dump. (From Steen, W. M., 2003, *Laser Material Processing*. By permission of Springer Science and Business Media).

combined with the flow rate, $\dot{m}$, as determined by a flow meter, enables the beam power q to be determined:

$$q = \dot{m} c_p \Delta T \tag{21.1}$$

One major disadvantage of this device is that the entire beam is absorbed by the beam dump, during which time the beam is not available for processing. It is thus not suitable for in-process monitoring. Furthermore, the response time is slow (with a time constant of the order of 10 s) and is also insensitive to small fluctuations in power.

Example 21.1 The inlet temperature of a cone calorimeter beam dump is measured to be 25°C, with the corresponding outlet temperature being 90°C. Determine the power of the laser beam incident on the calorimeter if the flow rate of water through it is 18 L/h and the specific heat of water is 4.186 J/g/K = 4186 J/Kg/K.

Solution:

From equation (21.1), the power is given by

$$q = \dot{m} c_p \Delta T \tag{21.1}$$

The density of water is 1 $\mathrm{g/cm^3}$ = 1 kg/L. The flow rate is therefore 18 kg/h = 0.005 kg/s. The power is thus

$$q = \dot{m} c_p \Delta T = 0.005 \times 4186 \times (90 - 25) = 1360 \text{ W} = 1.36 \text{ kW}$$

21.1.2 Beam Mode

The profile or mode of a laser beam can be measured using several techniques, including the laser beam analyzer and plastic burn analysis. These are discussed in the following sections. A third method, the array camera, is discussed in Section 21.2.5.1.

21.1.2.1 Laser Beam Analyzers Laser beam analyzers (LBA) are used to determine the beam mode (intensity distribution), overall power, and beam size (measured as the radius at which the beam power drops to $1/e^2$ of its peak (center) value). The device is normally mounted between the output window and shutter assembly, enabling power monitoring to continue even when the laser is idling. However, it can also be used to evaluate the beam profile at the focal point. A complete mapping of the beam profile may take about 5 s. The power loss associated with this technique is only about 0.1%, and thus it does not significantly interfere with the laser beam. The response time for the device can be as small as 0.01 s.

Two common techniques that have been developed for laser beam analysis are the rotating rod and the rotating hollow needle techniques.

Rotating Rod Technique This technique uses a rotating circular rod that is reflective. The rod is usually made of molybdenum and is rapidly swept through the laser beam in a plane perpendicular to the beam direction (Fig. 21.2a). Light is reflected from the rod surface in all directions as it sweeps through the beam. A small portion of the beam light is reflected from the rod onto two pyroelectric detectors that are located such that they measure light that is reflected in orthogonal directions. The detectors may have germanium-coated windows that will only transmit radiation of a specific wavelength. The output voltage of each pyroelectric detector is proportional to the laser power incident on it.

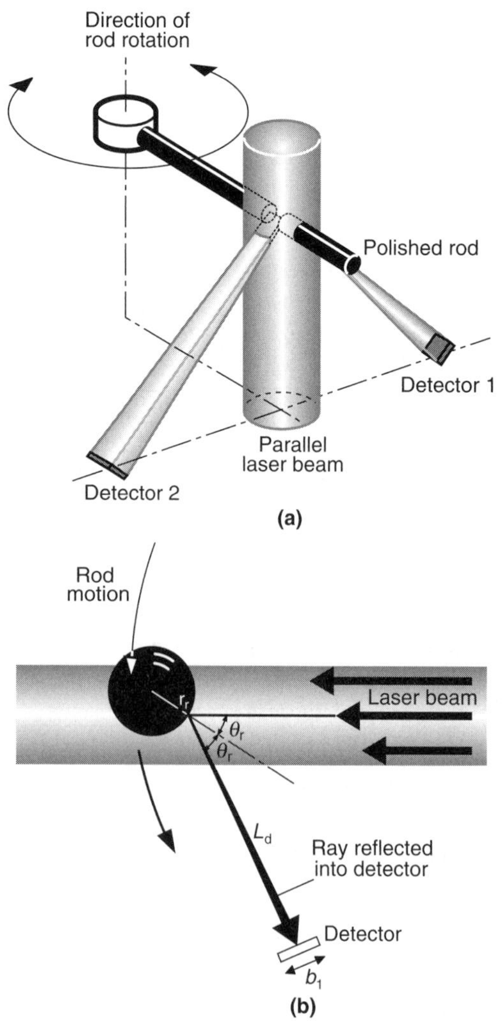

FIGURE 21.2 (a) Principle of the rotating rod laser beam analyzer. (b) Rotating rod laser beam analyzer view normal to the rod axis. (From Lim, G. C., and Steen, W. M., 1984, *Journal of Physics E: Scientific Instruments*, Vol. 17, IOP Publishing Ltd., pp. 999–1007.)

Let us now take a look at the basic principle of the device. First consider the case where the laser beam is parallel. A configuration of the system as obtained when looking along the length of the rod is illustrated in Fig. 21.2b, showing the laser beam perpendicular to the rod. Even though the rod reflects the beam in all directions, a detector positioned as shown receives a small portion of the reflected beam. Let the rod radius be r_{r}, and the angle between the incident beam and reflected beam received by the detector be $2\theta_{\mathrm{r}}$. Further, let the width of the detector be b_1, and its depth (into the paper), d_{d}. If the distance between the detector and point of reflection on the rod is L_{d}, then for $L_{\mathrm{d}} >> r_{\mathrm{r}}$ and $L_{\mathrm{d}} >> b_1$, the width, δr_{r}, of the beam that is just received by the detector is given by

$$\delta r_{\mathrm{r}} = \frac{b_1 r_{\mathrm{r}}}{2L_{\mathrm{d}}} \cos\theta_{\mathrm{r}} \tag{21.2}$$

δr_{r} is the resolution of the analyzer and represents the smallest width of the beam that can be detected. Thus at any instant, the area of beam whose power is received by the detector is given by $d_{\mathrm{d}} \times \delta r_{\mathrm{r}}$. It must be noted that equation (21.2) is based on the assumption that L_{d}, and for that matter, θ_{r} are essentially constant. However, in reality, both change as the rod sweeps across the beam. Typical values for the device parameters are $L_{\mathrm{d}} \approx 80$ mm, $b_1 = 3.5$ mm, $r_{\mathrm{r}} = 0.35$ mm, and $\theta_{\mathrm{r}} = 23°$.

Example 21.2 A rod of radius 0.35 mm is used to scan a laser beam. If the distance between the detector and point of reflection on the rod is 100 mm, and the angle between the incident beam and reflected beam received by the detector is 50°, determine the resolution of the beam analyzer. Assume the detector width is 3 mm.

Solution:

From equation (21.2), the resolution of the beam analyzer is given by

$$\delta r_{\mathrm{r}} = \frac{b_1 r_{\mathrm{r}}}{2L_{\mathrm{d}}} \cos\theta_{\mathrm{r}} \tag{21.2}$$

now $2\theta_{\mathrm{r}} = 50°$. Thus,

$$\delta r_{\mathrm{r}} = \frac{b_1 r_{\mathrm{r}}}{2L_{\mathrm{d}}} \cos\theta_r = \frac{3 \times 0.35}{2 \times 100} \cos 25 = 0.005 \text{ mm} = 5 \ \mu\text{m}.$$

In an actual system, two detectors (D_1 and D_2) are used and are placed directly opposite each other as shown in Fig. 21.3, which illustrates the system configuration in a plane normal to the beam axis. Since the rod rotates in a plane normal to the beam direction, the ray of light, say gD_1, that is reflected from the rod at g to detector D_1 is normal to the rod axis, Mg, where M is the rod pivot or center of the axis about which the rod rotates. It can be shown that during rotation of the rod, the detector D_1 sees the path through the laser beam in space, which is the arc of a circle with

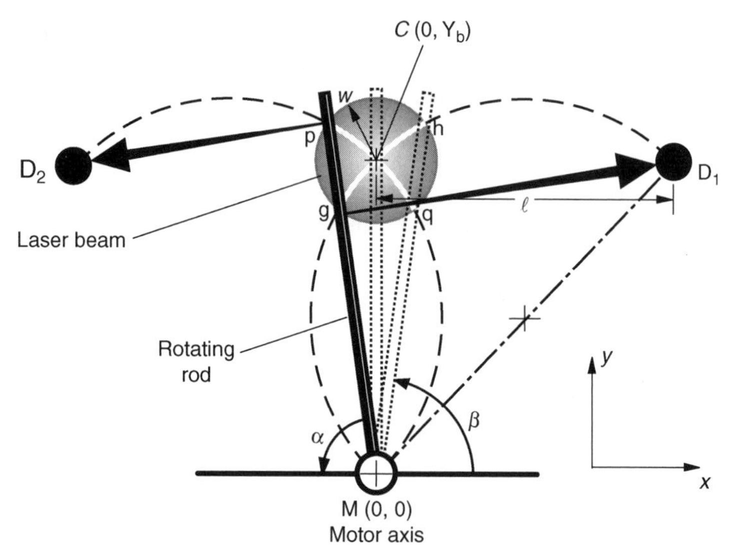

FIGURE 21.3 Rotating rod laser beam analyzer view normal to the beam axis. (From Lim, G. C., and Steen, W. M., 1984, *Journal of Physics E: Scientific Instruments*, Vol. 17, IOP Publishing Ltd., pp. 999–1007.)

MD_1 as diameter and passes through g, h, and C, where C is the center of the beam. Likewise, the detector D_2 sees the path through the laser beam in space, which is the arc of a circle with MD_2 as diameter and passes through C and say, q, p. Assuming the laser beam is circular in cross section, then it can be shown that the coordinates of the points g and h, where the rod intersects the boundary of the laser beam at any instant, are given by

$$x_g = \frac{w}{2l}\sqrt{2(l^2 - w\sqrt{2l^2 - w^2})} \tag{21.3a}$$

$$x_h = \frac{w}{2l}\sqrt{2(l^2 + w\sqrt{2l^2 - w^2})} \tag{21.3b}$$

$$y_g = \frac{1}{2l}(2l^2 - w^2 - w\sqrt{2l^2 - w^2}) \tag{21.3c}$$

$$y_h = \frac{1}{2l}(2l^2 - w^2 + w\sqrt{2l^2 - w^2}) \tag{21.3d}$$

where w is the beam radius and l is the distance from each detector to the beam center.

Equations (21.3a)–(21.3d) also give the locations of q and p since they are mirror images of g and h, respectively. It can therefore be shown that the angles α and β in Fig. 21.3 are equal. Points p, g, and M can thus be said to lie on a straight line, and likewise, h, q, and M. We can therefore deduce that when the rod is oriented at the angle α, detector D_1 sees light reflected from point g at the same time as detector D_2 sees light reflected from point p. Likewise for points h and q. If $l >> w$, then lines gh and pq are almost straight lines that are normal to each other. The outputs of the two detectors D_1 and D_2 thus correspond to the beam profile along two orthogonal axes. This analysis is strictly valid for an infinitesimally narrow detector window depth, that is, $d_d \to 0$.

Example 21.3 A rotating rod laser beam analyzer is used for monitoring a laser beam of diameter 10 mm. If the distance from each detector to the beam center is 75 mm, determine the x and y coordinates of the points of intersection of the rotating rod and the beam periphery.

Solution:

From Fig. 21.3 and equations (21.3a)–(21.3d), the x and y coordinates of the points of intersection of the rotating rod and the beam periphery are given by

$$x_g = \frac{w}{2l}\sqrt{2(l^2 - w\sqrt{2l^2 - w^2})} \tag{21.3a}$$

$$x_h = \frac{w}{2l}\sqrt{2(l^2 + w\sqrt{2l^2 - w^2})} \tag{21.3b}$$

$$y_g = \frac{1}{2l}(2l^2 - w^2 - w\sqrt{2l^2 - w^2}) \tag{21.3c}$$

$$y_h = \frac{1}{2l}(2l^2 - w^2 + w\sqrt{2l^2 - w^2}) \tag{21.3d}$$

Thus,

$$x_g = \frac{5}{2 \times 75}\sqrt{2(75^2 - 5\sqrt{2 \times 75^2 - 5^2})} = 3.5 \text{ mm}$$

$$x_h = \frac{5}{2 \times 75}\sqrt{2(75^2 + 5\sqrt{2 \times 75^2 - 5^2})} = 3.7 \text{ mm}$$

$$y_g = \frac{1}{2 \times 75}(2 \times 75^2 - 5^2 - 5\sqrt{2 \times 75^2 - 5^2}) = 71.3 \text{ mm}$$

$$y_h = \frac{1}{2 \times 75}(2 \times 75^2 - 5^2 + 5\sqrt{2 \times 75^2 - 5^2}) = 78.4 \text{ mm}$$

Reference to Fig. 21.4, the coordinates of the intersection points are therefore given by

$$p(x, y) = (-3.7, 78.4)$$

$$g(x, y) = (-3.5, 71.3)$$

$$q(x, y) = (3.5, 71.3)$$

$$h(x, y) = (3.7, 78.4).$$

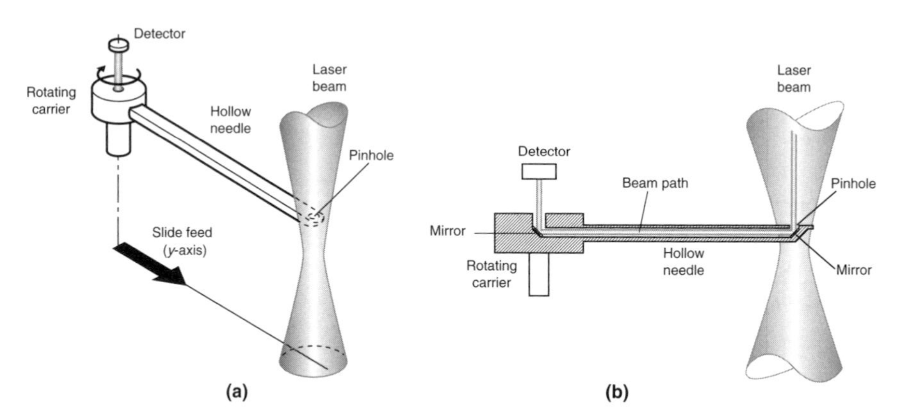

FIGURE 21.4 (a) Schematic of the rotating hollow needle-based laser beam analyzer. (b) Section view of the rotating hollow needle-based laser beam analyzer. (From *ICALEO 1992 Proceedings*, Vol. 75, Copyright 1993, Laser Institute of America. All rights reserved.)

Rotating Hollow Needle Technique With this technique, a long slender needle that is hollow inside has a pinhole drilled close to one end of it and is mounted such that the pinhole faces the laser beam (Fig. 21.4a). Photons that enter the pinhole are deflected to a pyroelectric detector using two mirrors (Fig. 21.4b). The needle rotation enables the laser beam to be scanned in one direction, say, the x-direction. Scanning in the y-direction to obtain a two-dimensional image is achieved by moving the axis of rotation in the y-direction. The size of pinhole used depends on the beam intensity. Pinhole diameters and measurement resolution are summarized in Table 21.1.

21.1.2.2 Plastic Burn Analysis Plastic burn analysis provides a qualitative measure of beam profiles, especially CO_2 laser beams. The material most commonly used is polymethylmethacrylate (PMMA or acrylic). It is a transparent polymer with a low thermal conductivity and high absorptivity at the CO_2 laser wavelength of 10.6 μm. These properties, in addition to its low sublimation temperature of 300°C, make it ideal for identification of beam mode. The process involves manually pulsing the laser at low power, about 50 W, until a deep enough cavity is formed in the PMMA material, and measuring the shape formed. The shape that is measured represents an imprint of the beam power distribution, since the depth to which the PMMA is sublimated at any location is determined by the beam power at that location.

TABLE 21.1 Rotating Hollow Needle Analyzer Parameters

Parameter	Unfocused Beam	Focused Beam
Pinhole diameter (μm)	100	20–50
Measurement resolution (μm)	75–1500	6–100

21.1.3 Beam Size

The beam size can be measured in a number of ways, but the easiest approach uses a Kapton film.

21.1.3.1 Kapton Film Kapton films are highly suited to measuring the sizes of pulsed beams. They are nonflammable, have low thermal conductivity, and have no melting point. The characteristics of the hole produced in the Kapton film depends on a number of factors, most important being the beam intensity. When the intensity exceeds a threshold value, the film undergoes pyrolytic decomposition to form a carbonaceous char. At yet higher intensities, the film is vaporized. For pulsed beams, the length of time affects the shape of hole produced. Well-defined holes are produced at intermediate pulse times. For very short or very long pulse times, an irregularly shaped hole is produced. Finally, the size of hole produced depends on the peak pulse power and film thickness.

The hole diameter d_h produced in a Kapton film by a pulsed multimode beam depends on the total pulse energy, Q_e, and is given empirically by

$$d_h = 2w_b - \frac{2\pi w_b{}^3 h \left[1 - \left(\frac{w_t}{w_b}\right)^3\right] \Delta H_v}{3(1-R)} \left(\frac{1}{Q_e}\right) \tag{21.4}$$

where h is the Kapton film thickness, R is the reflection coefficient, w_b is the base radius of the beam, whose intensity distribution is assumed to be of truncated shape, w_t is the top radius of the beam intensity distribution, and ΔH_v is the energy necessary to heat and vaporize a unit volume of the material.

From equation (21.4), it is evident that the hole diameter approaches the beam diameter as the pulse energy approaches infinity, that is,

$$d_h \to 2w_b \quad \text{as} \quad Q_e \to \infty$$

Thus, the beam diameter can be obtained by simply plotting the hole diameter versus the corresponding $1/Q_e$ and extrapolating the resulting line to the diameter axis.

For a Gaussian distributed beam, the hole diameter is given by

$$d_h{}^2 = 2w^2 \left(\ln Q_e - \ln\left[\frac{\pi w^2 h \Delta H_v}{2(1-R)}\right]\right) \tag{21.5}$$

where w is the $1/e^2$ beam radius.

When the beam radius is constant, the second term of equation (21.5) is constant. The square of the hole diameter is then directly proportional to the natural log of Q_e. The beam size can thus be obtained from the slope of the $d_h{}^2$ versus $\ln Q_e$ plot for a fixed mean power level. The beam size determined by either method is independent of the thickness of the Kapton film, which is typically in the range of 25–125 μm.

21.1.3.2 Other Methods A simple way to determine the beam size is to use a variable aperture to determine the area of the beam, which has 87.5 % of the total beam energy. The aperture is opened to first measure the total power, then reduced in size until the transmitted power is 87.5 % of the total. The output of the aperture is directed toward any of the power measuring devices outlined in Section 21.1.1. The radius of a circle that has the same area is defined as the beam radius. This method is most appropriate if the beam is circular.

Other methods for obtaining an approximate measure of the beam size include burning a hole in a metal foil or photographic paper and measuring the resulting diameter.

21.1.4 Beam Alignment

Beam wandering that occurs during processing can result from structural thermal expansions, mechanical vibration, and mirror mount instability. One technique for aligning the beam during the process uses a beam splitter placed in the beam path to divert a small portion of the beam to a beam position sensory unit (Fig. 21.5). The position sensory unit consists of a rotating disk with a small radial slit of width 0.5 mm, through which a slice of the split beam passes and is focused by a lens onto a detector. The disk, typically about 50 mm in diameter, rotates at a fixed angular speed

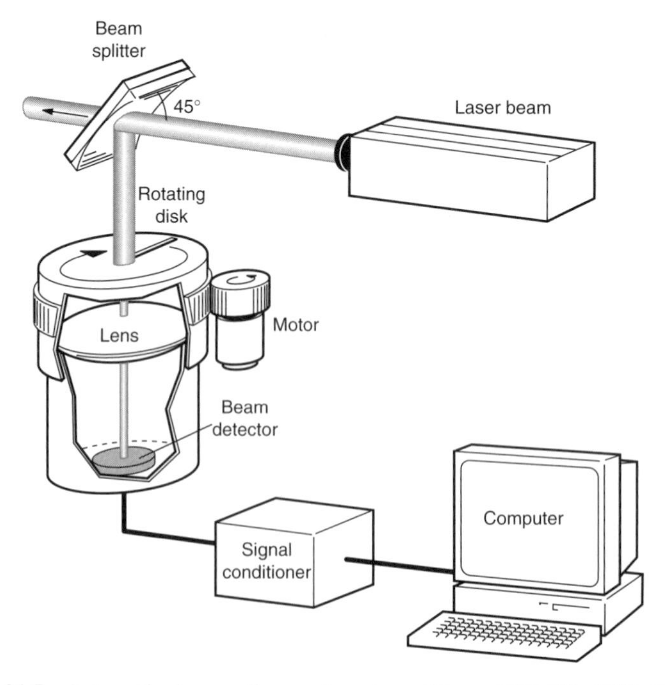

FIGURE 21.5 Automatic beam alignment system. (From *ICALEO 1990 Proceedings*, Copyright 1991, Laser Institute of America. All rights reserved.)

of about 125 RPM and has its center coinciding with the apex of the slit. The disk center and detector are arranged to be collinear with the lens optical axis. The disk diameter has to be much greater than the laser beam diameter. Let us now consider three cases:

(i) The beam energy is axisymmetrically distributed, and its center coincides with the disk center. Due to the symmetry, the amount of beam energy that reaches the detector through the slit is always the same, irrespective of the slit angular position. The detector output will thus appear constant on an oscilloscope and circular on a polar plot (Fig. 21.6a).

(ii) The beam energy is axisymmetrically distributed, but completely offset from the disk center. In this case, the detector output on an oscilloscope will increase to a peak and then decreases to zero. It will appear elliptical on a polar plot (Fig. 21.6b). If the beam radius w is known, then its center can be approximated by

$$r_{\mathrm{d}} = \frac{2w}{\Delta\phi} \tag{21.6}$$

where r_{d} is the distance between the disk and beam centers, t_1 and t_2 are the times of initial and final detection of the beam, $\Delta\phi = \omega_{\mathrm{v}}(t_2 - t_1)$ is the phase angle over which the beam is detected, and ω_{v} is the angular velocity of the disk.

(iii) The beam energy is axisymmetrically distributed and only partially offset from the disk center. The detector output will vary with time on both the oscilloscope and the polar plot (Fig. 21.6c).

This technique, even though not fully developed yet, has the potential to be used for automatic beam alignment or positioning the laser beam at the desired location on the workpiece.

Sensors that are used for monitoring the process quality are discussed in the next section.

21.2 PROCESS MONITORING

The principal parameters that need to be monitored during laser welding, for example, include the weld pool geometry (width and penetration), defects or discontinuities (cracking, porosity, etc.), microstructure (strength), residual stresses, and temperatures. For laser cutting, we have the kerf size, striations, dross, and microstructure. Among the most commonly used sensors for real-time monitoring are acoustic emission, audible sound (acoustic sensing), and infrared/ultraviolet detectors, all of which are secondary outputs.

21.2.1 Acoustic Emission

Acoustic emission (AE) refers to stress waves that are generated as a result of the rapid release of elastic strain energy within a material due to the rearrangement of its internal

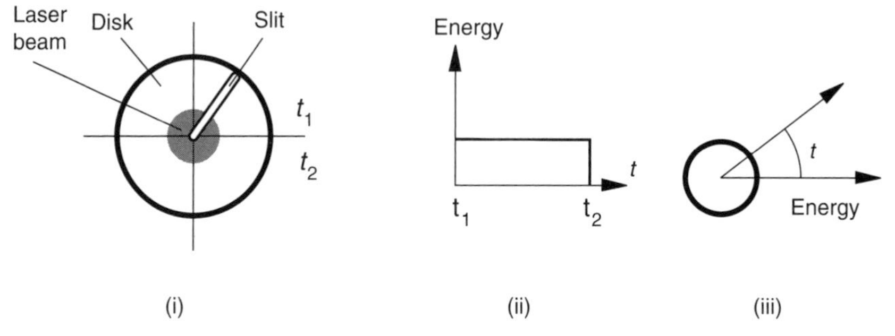

(a) Symmetrical beam, centrally located.

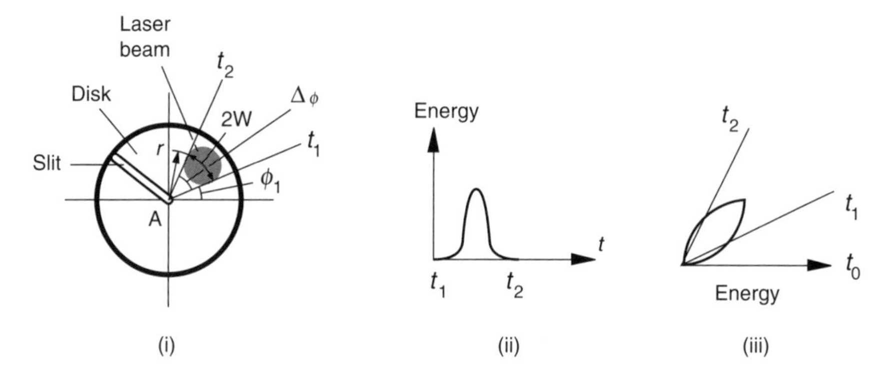

(b) Symmetrical beam, completely offset.

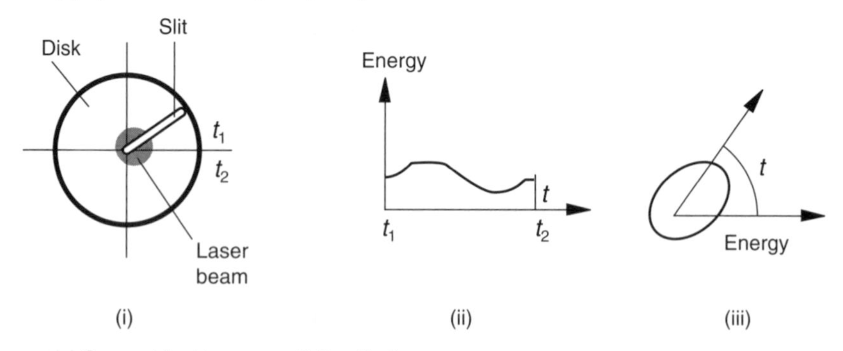

(c) Symmetrical beam, partially offset.

FIGURE 21.6 Automatic beam alignment system output showing (i) setup schematic, (ii) variation of detected energy with time, and (iii) polar plot of energy variation with time. (From *ICALEO 1990 Proceedings*, Copyright 1991, Laser Institute of America. All rights reserved.)

structure. It is also sometimes referred to as stress wave emission. The resulting stress waves propagate through the structure and produce small displacements on the surface of the structure. These are detected by sensors that convert the displacements into electrical signals. AE is an active phenomenon, since it is generated by the process under investigation. As a result, AE signals are well suited for real-time or continuous monitoring because they are generated while the phenomenon is undergoing change.

21.2.1.1 AE Detection Two types of transducers are normally used for AE signal detection, namely, piezoelectric and capacitive transducers. The piezoelectric transducer is more commonly used.

Piezoelectric Transducers The principle of piezoelectric transducers is based on the characteristics of a damped spring-mass system attached to a frame (Fig. 21.7). The signal to be detected subjects the base frame to an input displacement, y, assumed to be sinusoidal and given by

$$y = Y \sin \omega t \tag{21.7}$$

where Y is the maximum amplitude of the input base displacement (m) and ω is the angular frequency of the input base oscillation (rad/s).

Assuming that both the mass and base are displaced in the positive direction and that x is greater than y and $\dot{x}$ is greater than $\dot{y}$, then the equation of motion of the system shown in Fig. 21.7 is given by

$$m_a \ddot{x} = -k_s(x - y) - c_d(\dot{x} - \dot{y}) \tag{21.8}$$

where c_d is the damping coefficient (N s/m), k_s is the spring constant (N/m), m_a is the system mass (kg), and x is the displacement of the mass (m).

Now let $z = x - y$, the relative displacement between the base and mass. Then we have

$$m_a(\ddot{z} + \ddot{y}) + c_d \dot{z} + k_s z = 0 \tag{21.9}$$

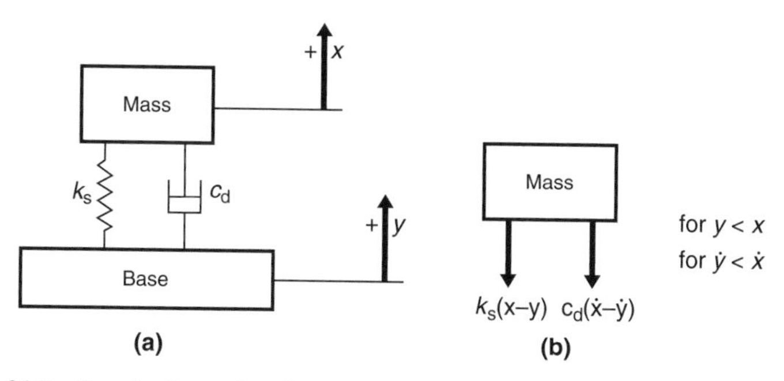

FIGURE 21.7 Simple damped spring–mass system. (a) Schematic. (b) Free-body diagram.

or

$$m_a \ddot{z} + c_d \dot{z} + k_s z = m Y \omega^2 \sin \omega t \tag{21.10}$$

The steady-state response obtained from the particular solution of equation (21.10) is given by

$$z = Z \sin(\omega t - \phi) \tag{21.11}$$

with

$$\frac{Z}{Y} = \frac{\omega^2}{{\omega_n}^2} \left[\left(1 - \frac{\omega^2}{{\omega_n}^2}\right)^2 + \left(2\zeta \frac{\omega}{\omega_n}\right)^2 \right]^{-1/2} \tag{21.12}$$

where Z is the maximum amplitude of the relative motion between the base and mass (m), ω_n is the natural frequency of vibration of the undamped system (rad/s), ϕ is the phase angle between the input motion y and the relative motion z (rad), and ζ is the damping ratio.

The phase angle is given by

$$\tan \phi = 2\zeta \frac{\omega}{\omega_n} \left(1 - \frac{\omega^2}{{\omega_n}^2}\right)^{-1} \tag{21.13}$$

From equation (21.12), we note that when $\omega_n << \omega$, that is, the natural frequency of the device is much lower than the lowest harmonic of the motion to be measured, then equation (21.12) approaches unity. Thus,

$$Y \simeq Z \quad \text{for} \quad \frac{\omega}{\omega_n} >> 1 \tag{21.14}$$

In other words, the relative motion between the mass and frame then approaches the absolute motion of the frame. In effect, the mass remains motionless in space. x is the displacement of the mass that becomes quasistatic at the high frequencies, that is $x \simeq 0$. The phase difference does not affect the magnitude of the reading. Thus by replacing the spring and damper system with a piezoelectric crystal (Fig. 21.8), the relative motion between the base and mass will strain the crystal, generating electrical signals that are a measure of the displacements due to the stress waves.

A material that is commonly used to induce the piezoelectric phenomenon in AE transducers is lead zirconate titanate (PZT), which is a ceramic element and is most commonly used because it has a relatively high dielectric constant. The relative dielectric constant of a material is the ratio of charge density arising from an electric field with and without the material present.

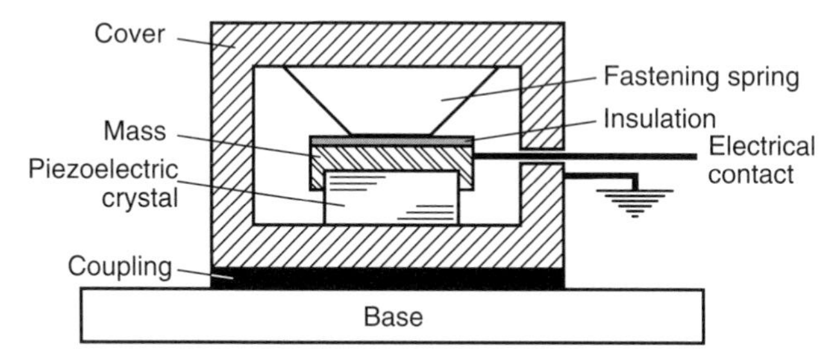

FIGURE 21.8 Schematic of a piezoelectric transducer.

Capacitive Transducers This type of transducer is based on the principle of the capacitor, where the capacitance between two parallel plates depends on the distance between the plates. The mass in this case is a metallic cylindrical piece placed with its flat surface parallel to the structure surface. The space between the mass and structure is statically charged, and changes in the gap result in a change in the capacitance of the system and consequently a change in the voltage across the capacitor.

21.2.1.2 Background This section provides a brief summary of some important factors associated with acoustic emission monitoring. Specifically addressed are the following:

Coupling—Since air is a poor transmitter of high-frequency waves, the transducer needs to be in very good acoustic contact with the workpiece under inspection. A couplant is normally applied to the transducer base for effective coupling. Typical couplants are oils, glycerin, and greases.

Preprocessing—The AE signals generated are normally of very low amplitude and high-frequency content and need to be amplified close to the signal source using a low-noise preamplifier, to minimize noise contamination as the signal is propagated over the transmission cable. Most preamplifiers have a gain of 40–60 db (100–1000 times).

Signal Forms—There are two principal types of AE signals:

1. *Burst signals*: These are typical of individual events, for example, cracking in weldments. An event is a single occurrence of an acoustic emission activity.
2. *Continuous signals*: These are typical of overlapping series of events, for example, continuous welding without defects.

Source Location—Acoustic emission waves travel at specific speeds in a given medium under given conditions, and this property can be used in locating the sources of emission signals, say cracking in structures. The method of triangulation is used for locating sources in plates, and involves the placement of a set of sensors (transducers), usually three or more on the structure under investigation. The transducers should not be collinear. The difference between

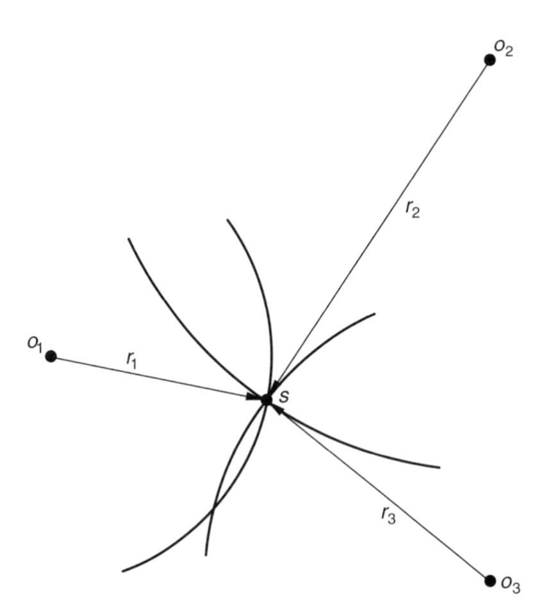

FIGURE 21.9 AE source location by triangulation.

the time it takes the waves to arrive at each transducer is used to compute the relative distance between the sensor and the source. The common point of intersection of circles drawn with the computed distances as radii, and the corresponding transducers as centers, gives the source location (Fig. 21.9). With appropriate modifications, the method can also be used in locating active defects in structures. For three-dimensional bodies, a minimum of six transducers are required. In applying this technique to laser processing, the effect of temperature and microstructure variation on wave speed will have to be considered.

Precautions—Some of the major precautions that need to be taken in applying conventional AE instrumentation include the following:

1. Protecting the transducer from the high temperatures of the processing environments.
2. Providing a highly reliable acoustic contact between the transducer and the structure.

To resolve the temperature problem, the transducer can be mounted on a waveguide. An appropriate shape of a waveguide has been determined to be a uniform-diameter cylindrical rod that terminates in a cone at the transducer end. The conical shape provides a gradual transition that prevents emission wave reflection that would be caused by sudden diameter changes. Suitable dimensions may be that of a rod of diameter 8 mm and total length 350 mm. Attenuation of the signal amplitude has been found to be more greatly influenced by a reduction in diameter than by an increase in length of the waveguide.

AE Sources and Major Uses—The common sources of AE include plastic deformation processes such as dislocation motion, twinning, and grain boundary sliding; phase transformations, such as martensitic transformation; decohesion of inclusions; fracture; and friction. The detectability of an acoustic emission signal depends on the energy release rate of the source. Thus slow processes such as creep and most phase transformations do not generate detectable AE. On the contrary, processes such as twinning, martensitic phase transformation, and a short pulse laser beam, which occur over a short period of time, tend to be prolific sources of AE.

Given the sources of acoustic emission, it would be most useful for detecting cracking in the fusion zone and martensitic transformation in both the fusion and heat-affected zones. It would also be sensitive to keyhole failure, and to some extent, porosity formation.

21.2.1.3 AE Transmission Stress waves normally propagate in a material in any of the four different wave modes, namely, shear (transverse) waves; longitudinal (pressure) waves; Rayleigh (surface) waves; and Lamb waves. The wave type that predominates in a detected signal depends on the locations of the transducer and source and the structure through which the signal propagates.

Longitudinal Waves—With longitudinal waves, particles of the propagation medium move in the same direction as the waves (Fig. 21.10a). Consequently, at positions equivalent to the crests and troughs of the shear waves, the medium is subjected to a series of compressions and rarefactions. This nature of the waves enables them to travel very easily through fluid media. Typical speeds of longitudinal waves is about 6000 m/s in steels, 1500 m/s in water, and 330 m/s in air.

Shear Waves—In this form of wave propagation, particles (atoms or molecules) of the propagating medium move in a direction perpendicular to the direction of wave travel (Fig. 21.10b). The propagating mode is similar to that observed

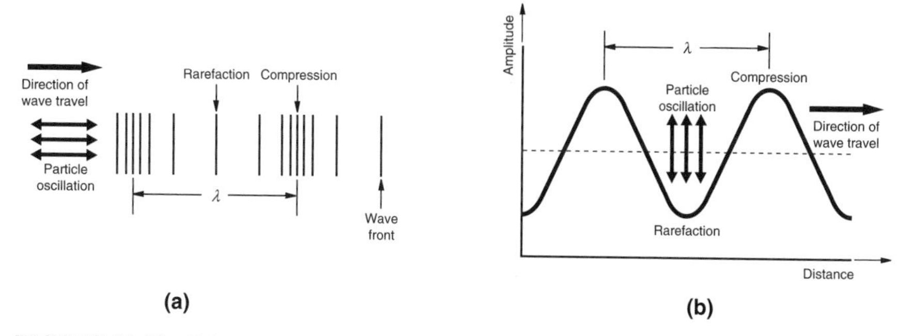

FIGURE 21.10 Schematic of (a) longitudinal wave and (b) shear wave. (From *Metals Handbook*, 8th edition, Vol. 11, 1976, *Nondestructive Inspection and Quality Control.* Reprinted with permission of ASM International. All rights reserved. www.asminternational.org.)

in a vibrating string. Typical speed of shear waves in metals is about half that of longitudinal waves.

Rayleigh Waves—Rayleigh waves generally propagate along the interface between media that have vastly different forces between their respective atoms or molecules, such as the interface between a solid and air (Fig. 21.11a). They are able to maintain their amplitudes in a given medium over longer distances than do shear and longitudinal waves. Particles in the solid medium oscillate in an elliptical fashion, and the waves are always reflected from a sharp corner but will continue if the corner is rounded.

Lamb Waves—Lamb waves occur principally in very thin plates of only a few wavelengths thick and involve slight distortions of the plate (Fig. 21.11b).

21.2.1.4 Traditional AE Signal Analysis In the course of propagation, waves often undergo considerable changes through scattering by structural defects; multiple reflections at interfaces; diffraction by crystal imperfections; and refraction when there is a medium change along the travel path. All these factors considerably modify the waveform by changes in phase, amplitude attenuation, and wave repetitions through reflections.

Statistical techniques are thus often used in analyzing AE data, and the following techniques have conventionally been used:

1. *Count and Count Rate*—This is a record of the signal ring-down pulses whose amplitude exceeds a preset threshold voltage.
2. *Amplitude Distribution Analysis*—An indication of the number of signals whose amplitude fall within a predefined range.

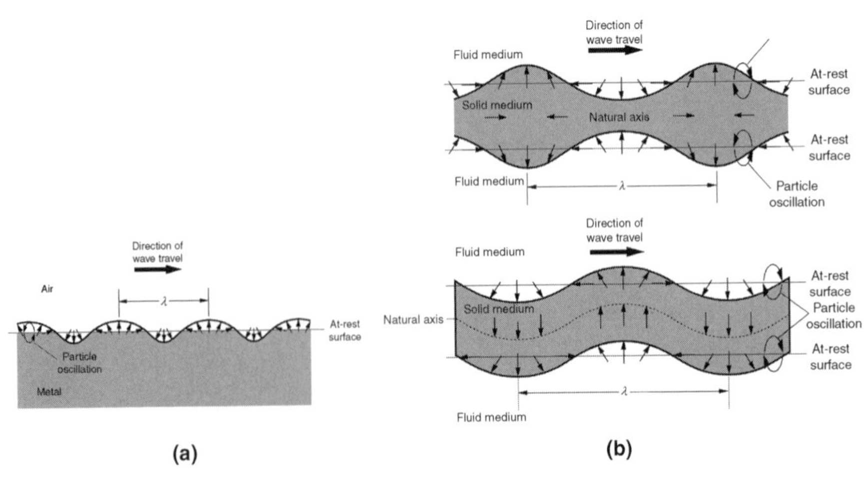

FIGURE 21.11 Schematic of (a) surface wave and (b) lamb wave. (From *Metals Handbook*, 8th edition, Vol. 11, 1976, *Nondestructive Inspection and Quality Control*. Reprinted with permission of ASM International. All rights reserved. www.asminternational.org.)

3. *Frequency Spectrum*—This shows the contribution of each frequency component to the total power.
4. *Autocorrelation Function*—This involves the comparison of a signal waveform $f(t)$ against a delayed version of the same waveform $f(t + \tau_d)$.
5. *Root Mean Square (RMS) Value of the Signal*—This is a measure of the signal intensity.

The original AE signal can be directly analyzed using any of these methods. However, due to the high sampling rate required to digitize the original signal, the RMS component of the signal is sometimes obtained first before further analysis is performed. On the contrary, with this approach, some of the higher frequency content of the signal may be lost.

The RMS value of the signal is a convenient measure of the signal energy content. The RMS value of an ac signal is that value of dc signal, which, if passed through the same circuit for the same period of time, would produce the same expenditure of energy as the ac signal and can be expressed as

$$\text{RMS} = \left[\frac{1}{\Delta t}\int_0^{\Delta t} V^2(t)\mathrm{d}t\right]^{\frac{1}{2}} \tag{21.15}$$

where $V(t)$ is the signal function (V) and Δt is the averaging time (s).

For a system with a significant amount of background noise that cannot be easily filtered from the measured signal, the RMS value of the actual signal can be obtained as

$$\text{RMS} = \left[\text{RMS}_\text{T}{}^2 - \text{RMS}_\text{n}{}^2\right]^{\frac{1}{2}} \tag{21.16}$$

where RMS is the RMS value of the actual signal (V), RMS_n is the RMS value of the background noise, and RMS_T is the total RMS of the measured signal.

There are a number of factors that influence the results of the AE signal analysis. These include the threshold setting (for count and count rate only); gain setting or amplification; and frequency bandwidth.

The threshold setting is a voltage level that is used for the signal counts. A signal count is registered each time the signal level exceeds the threshold setting. The latter is normally set higher than the background noise that could interfere with desired signals to eliminate their impact on the signal analysis. The frequency bandwidth cuts off unwanted higher and/or lower frequency signals, while the gain enables the initial low signal amplitude to be increased. Figure 21.12 illustrates some of the traditional AE signal processing terminology.

A variation of acoustic emission detection is what is termed the acoustic mirror. This is further discussed in the following section.

21.2.2 Acoustic Mirror

The acoustic mirror device arises from the observation that high frequency signals are generated when a high intensity laser beam is incident on a mirror. The signals

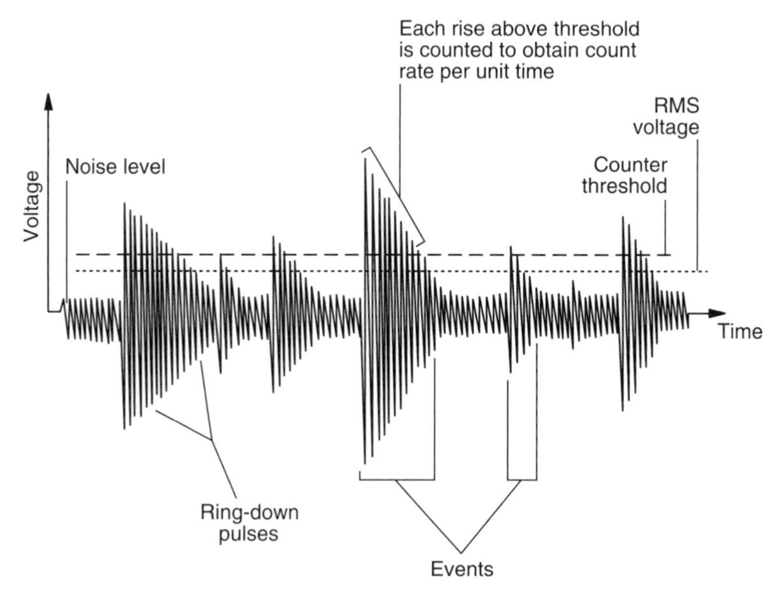

FIGURE 21.12 Traditional acoustic emission signal processing terminology. (From Kannatey-Asibu Jr., E., and Dornfeld, D. A., 1981, *ASME Journal of Engineering for Industry*, Vol. 103, pp. 330–340.)

are detected using a piezoelectric transducer mounted on the back-side of the mirror (Fig. 21.13a). The detected signals are sensitive to a number of variables, including the beam power, size, its location on the mirror, wavelength, laser cavity gas composition, and back reflections from the workpiece. The signal intensity increases with laser power and decreases with beam diameter on the detector mirror. The signals are stress waves that are generated by localized expansion of the mirror when it absorbs a certain percentage of the incident laser radiation that instantaneously heats the mirror surface. The recorded signal is due to power variation and not the absolute power. A typical signal output is shown in Fig. 21.13b, illustrating the long-term characteristics

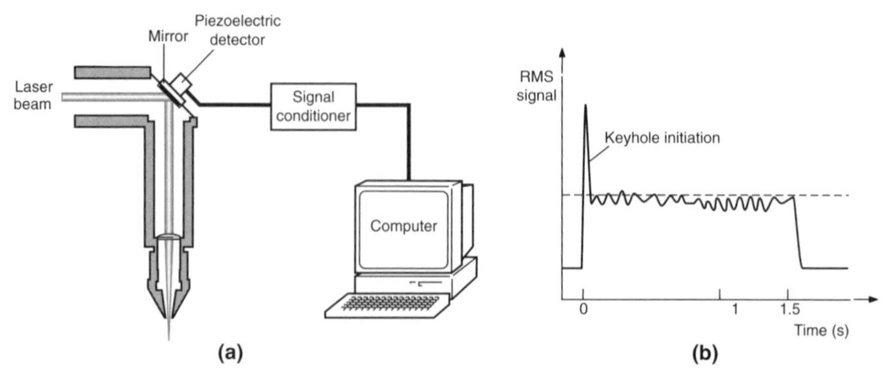

FIGURE 21.13 (a) An acoustic mirror detector. (b) An acoustic mirror output signal. (From Steen, W. M., and Weerasinghe, V. M., 1986, *SPIE Laser Processing*, Vol. 668, pp. 37–44.)

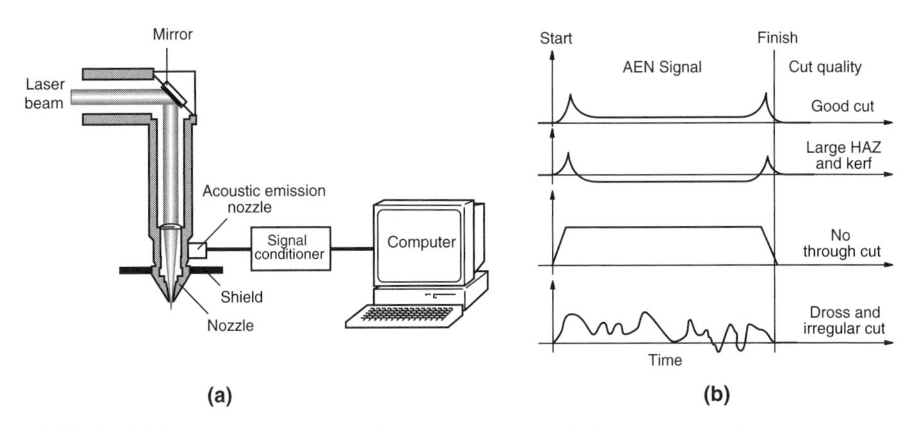

FIGURE 21.14 (a) Acoustic emission nozzle setup and (b) acoustic emission nozzle response during laser cutting. (From *ICALEO 1992 Proceedings*, Vol. 75. Copyright 1993, Laser Institute of America. All rights reserved.)

of the signal, as well as its dependence on keyhole initiation. Spectral analysis of the acoustic mirror signal normally shows two principal peaks, one around 100 kHz, and the other around 1 MHz. The low-frequency component is found to depend on the gas mixture, while the high-frequency component depends on the cavity design, and thus on the cavity oscillations. The dependence of the acoustic mirror signal on beam location on the mirror results from the variation in the distance between the beam location and the sensor location.

A major advantage of the acoustic mirror device is that it does not interfere with the incident laser beam and is also independent of the processing direction. The mirror can be located anywhere in the beam delivery path.

A variation of this monitoring technique is what is called the acoustic emission nozzle (AEN) and involves mounting a piezoelectric transducer on a beam delivery tube, which is thermally insulated from the nozzle by a washer. The detector is also protected from direct back reflections from the process by a shield (Fig. 21.14a). The characteristics of the signal detected under these conditions are found to be a good indication of the cut quality in laser cutting (Fig. 21.14b).

21.2.3 Audible Sound Emission

The audible sound (AS) technique uses a microphone to detect low frequency (15 Hz–20 kHz) acoustic signals generated during laser processing, in a sense, making use of the sound detected by an operator. The acoustic signals are the result of pressure variations at the process source, which propagate to the microphone. The intensity of the detected signal thus depends on the pressure and velocity of the fluid at the source. It also depends on the size of the source and the distance between the source and sensor. In laser cutting, the signal intensity depends on the density and velocity of the assist gas jet, and the resonant frequency of the detected signal is found to depend on the cutting front geometry, specifically the kerf width and depth of cut. The signals associated with gas flow rates up to 40 L/min, and velocities of the order of 50 m/s

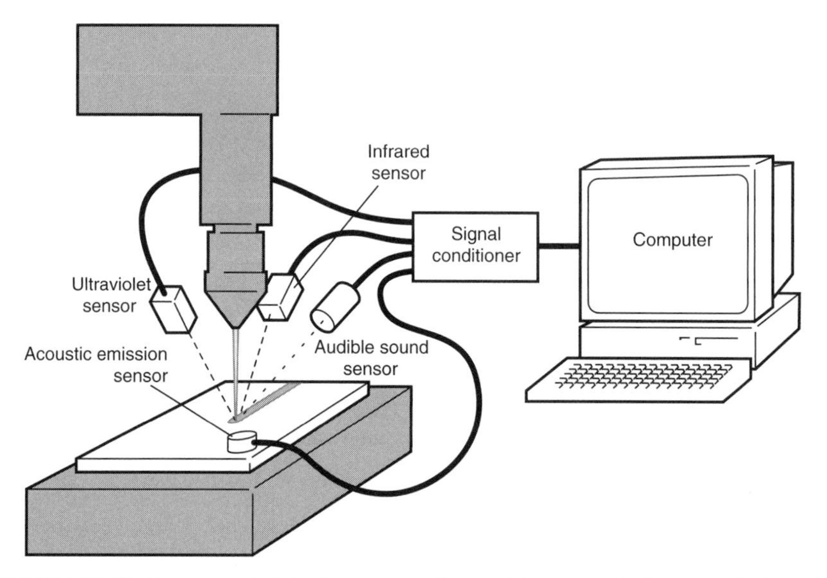

FIGURE 21.15 Noncontact detection system for audible sound, infrared, and ultraviolet signals. (From *Journal of Laser Applications*, Vol. 14, No. 2. Copyright 2002, Laser Institute of America. All rights reserved.)

in laser welding are barely measurable. The microphone is normally mounted in the immediate vicinity (approximately, 40 mm) and directed toward the process area (at an angle of about, 45°). It has the advantage of being noncontact. A typical setup is illustrated in Fig. 21.15.

Sources of sound emission during laser processing include the following:

1. Variations in gas momentum caused by heating and boiling processes (vapor formation).
2. Variations in gas pressure or momentum change associated with the impact of gas jets on the workpiece surface. This is especially important in laser cutting where the signals may result from vibrations caused by impingement of the assist gas on the cutting front.
3. Vapor expansion out of the keyhole opening in laser welding. The contribution of this signal source depends on:
   (a) Vapor pressure inside the keyhole
   (b) Size of the keyhole
   (c) Velocity of the expanding front.
4. Shock wave accompanying plasma ignition, which would result in a transient signal. However, when the plasma obstructs the laser beam and prevents it from reaching the workpiece, then the intensity of the subsequent signal is significantly reduced.
5. Defects such as cracking.

21.2.4 Infrared/Ultraviolet (IR/UV) Detection Techniques

In this section, general information is first presented on IR/UV detection and how this relates to laser processing. This is then followed by more detailed discussions of infrared detection and then ultraviolet detection.

IR/UV detection techniques analyze radiation emitted from the process zone in two wavelength bands of the electromagnetic spectrum. One is the infrared band in which most of the radiation from the hot material is concentrated, and the other is the ultraviolet band in which the plasma radiation is concentrated. Since a significant portion of the ultraviolet radiation is emitted by high-temperature plasma, changes in the temperature of the plasma emerging from the keyhole will result in a change in the ultraviolet signal, which can be used to provide a predictive indication of keyhole failure. This is effective because the response of liquid flow is much slower than that of the plasma. For example, should there be a sudden reduction in the plasma density, that could reduce the effectiveness of beam coupling to the keyhole, resulting in possible eventual failure of the keyhole and the welding process. Thus by detecting the change in the ultraviolet signal, the impending failure can be anticipated. The intensity of the ultraviolet signal is significantly reduced when the plasma is blown through the workpiece to form a cut, further confirming the dependence of the ultraviolet signal on the plasma. Changes in the infrared signal also provide information on variations in the melt pool.

A typical sensor used for the infrared radiation is a germanium photodiode fitted with a silicon filter having a spectral range from 1.0 to 1.9 μm. The ultraviolet radiation may be measured with a gallium phosphide (GaP) photodiode with a spectral range from 0.19 to 0.52 μm. Possible arrangements for the setup would be as shown in Fig. 21.15. Both the ultraviolet and infrared signal intensities increase with the laser power, while increasing shielding gas flow rate reduces the signal intensities, probably due to a reduction in plasma volume. Table 21.2 illustrates how the relative intensities of the infrared and ultraviolet signals reflect on weld quality.

21.2.4.1 Infrared Detection

Background At temperatures above absolute zero, all objects radiate infrared energy, which is part of the electromagnetic spectrum and has a wavelength range from 0.75 to 10^3 μm, (between the visible light and microwave bands). The bodies that emit

TABLE 21.2 Weld Quality and Its Dependence on Infrared/Ultraviolet Signals.

Ultraviolet Signal	Infrared Signal	Weld Quality
High	Low	Good
High	High	Poor
Low	High	Cut
Low	Low	Hole, low penetration
Oscillation	Oscillation	Hump and perforation

From Chen, H. B., Li, L., Brookfield, D. J., Williams, K., and Steen, W. M., 1991, *ICALEO*.

infrared radiation can be classified into three types based on the spectral characteristics of the radiation:

1. *Blackbody Radiation*—This type of radiation is emitted from a body with an emissivity of one. A blackbody is one that absorbs all incident radiation and reflects none.
2. *Graybody Radiation*—The emissivity distribution of this type of body is similar to that of a blackbody, but less than unity.
3. *Selective Radiation*—In this case, the emissivity varies with wavelength.

Emissivity is the ratio of the radiant energy emitted by a surface to that emitted by a blackbody at the same temperature. Also, the fraction of an incident beam of radiant power that is absorbed by a surface. It usually depends on the direction of the incident radiation and is greatest in the normal direction, and it declines in magnitude in proportion to the cosine of the angle with the normal. The intensity of the radiated power (Watts/mm^2) depends on both the temperature of the body and the wavelength. For a blackbody, the distribution is given by Planck's law, which can also be expressed as

$$I_\lambda = c_1 \lambda^{-5} / (e^{\frac{c_2}{\lambda T}} - 1) \tag{21.17}$$

where $c_1 = 2\pi c_0{}^2 h_\mathrm{p} = 3.74 \times 10^{-16}\ \mathrm{J\,m^2/s}$, $c_2 = c_0 h_\mathrm{p}/k_\mathrm{B} = 1.44 \times 10^{-2}\ \mathrm{m\,K}$, and I_λ is the amplitude of radiant energy at a given wavelength λ, at temperature T, per unit wavelength interval, per unit time, per unit solid angle, per unit area.

The energy distributions for a blackbody are shown in Fig. 21.16a and b on log–log and linear plots, respectively, and those for a graybody and nongraybody (where radiation is selective) are illustrated in Fig. 21.17. For a blackbody, each temperature results in a specific curve that increases rapidly from zero intensity at zero wavelength to a maximum, and then gradually decreases asymptotically toward zero at long wavelengths. The peak intensity increases with temperature and occurs at shorter wavelengths for higher temperatures. The wavelength λ_m at which the peak intensity occurs for a given temperature is given by Wien's law:

$$\lambda_m T = constant \tag{21.18}$$

For a blackbody at a given temperature, T, the total hemispherical radiation I per unit area and time is given by the area under the distribution curve for that temperature and is obtained by integrating Planck's equation. That results in the Stefan–Boltzmann law:

$$I = \int_0^\infty I_\lambda \mathrm{d}\lambda = \sigma_\mathrm{B}(T^4 - T_0{}^4) \tag{21.19}$$

where σ_B is the Stefan–Boltzmann constant $= 5.6704 \times 10^{-8}\ \mathrm{W/m^2K^4}$.

For a graybody, the total radiation is the product of that of the blackbody and the emissivity of the body in question. The emissivity generally depends on the material and its surface conditions. The temperature of a graybody whose emissivity and

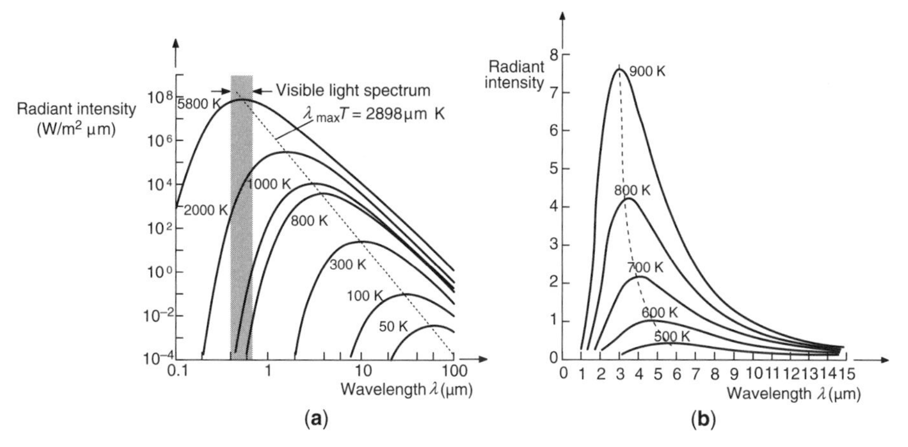

FIGURE 21.16 Energy distribution for (a) blackbody on a log scale and (b) blackbody on a linear scale. (From Benedict, R. P., 1984, *Fundamentals of Temperature, Pressure, and Flow Measurements*. Reprinted with permission of John Wiley & Sons, Inc.)

ambient temperature is known can be determined by measuring the radiance and using the Stefan–Boltzmann law.

Infrared Detectors The sensors that are used for detecting infrared radiation can be categorized into two broad groups:

1. *Thermal Detectors*—With this type of detector, the material absorbs radiation, resulting in the generation of phonons that cause lattice heating. The lattice

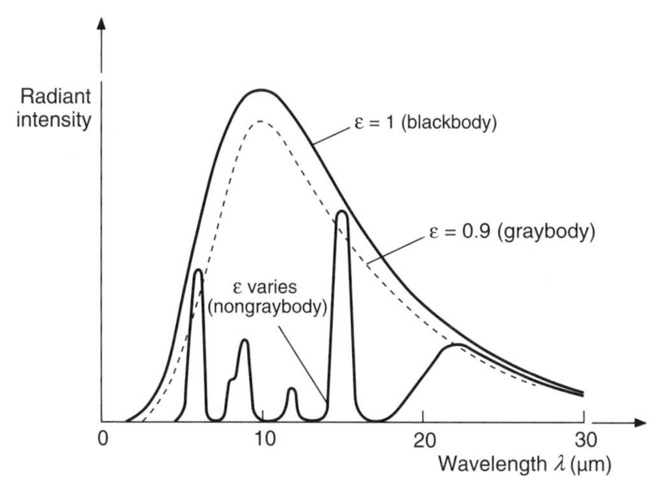

FIGURE 21.17 Energy distribution for a graybody and a nongraybody. (From Benedict, R. P., 1984, *Fundamentals of Temperature, Pressure, and Flow Measurements*. Reprinted with permission of John Wiley & Sons, Inc.)

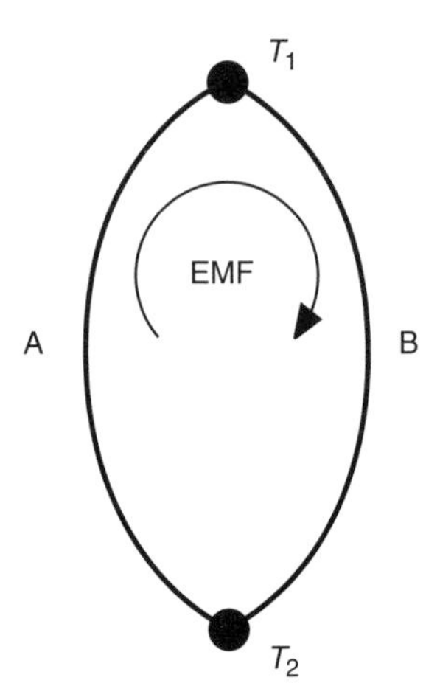

FIGURE 21.18 Basic schematic of a thermocouple junction.

heating produces a change in the electrical properties of the material that are detected. The transduction process thus involves two steps. Common types of thermal detectors include pyroelectric devices and thermocouples.

(a) *Pyroelectric Devices*—A change in temperature due to the incident radiation results in a change in dipole moment of the material, and thus charge difference between crystal faces, which is measured. They have fast response, but require chopped radiation source to define absolute levels.
(b) *Thermocouples*—The principle of thermocouples is based on the Seebeck effect, where thermal energy is converted into electric energy when two dissimilar metals are joined and one end is heated. If the joined wires form a closed circuit, in which case we also have two junctions, then electromotive force (EMF) is induced and results in current flow through the circuit. One of the junctions, the hot junction, is exposed to the incident radiation, while the cold junction remains at a relatively constant temperature. The cold junction is usually located near massive portions of the mounting or case.

For two dissimilar metals, A and B, joined as shown in Fig. 21.18, the net-induced EMF is given by

$$\begin{aligned} E_{\mathrm{v}} &= \int_{T_1}^{T_2} \alpha_{\mathrm{A}} \mathrm{d}T + \int_{T_2}^{T_1} \alpha_{\mathrm{B}} \mathrm{d}T \\ &= \int_{T_1}^{T_2} (\alpha_{\mathrm{A}} - \alpha_{\mathrm{B}}) \mathrm{d}T \end{aligned} \tag{21.20}$$

where $\alpha_{\mathrm{A}}, \alpha_{\mathrm{B}}$ are the absolute Seebeck coefficients of materials A and B, respectively (V/K).

The voltage, E_V, is thus used as a measure of the temperature difference between the hot and cold junctions.

2. *Photon Detectors*—With photon detectors, the radiation absorbed directly produces a change in the electrical properties of the detector and is thus one-step. Photon detectors thus have a faster response. Examples are photoconductive, photoresistive, photoemissive, and photovoltaic detectors.

   (a) *Photoconductive Detectors*—These consist of evaporated or chemically deposited multicrystalline films of cadmium sulfide, lead sulfide, lead telluride, lead selenide, etc. They become electrically conducting when exposed to radiation in wavelengths that they absorb. The resulting current flow is proportional in magnitude to the absorbed radiation.
   (b) *Photoresistive Detectors*—Photoresistive detecting elements are based on the same principle as photoconductive detectors, except that the current flow in this case is reverse to that of the photoconductive elements. Photodiodes involve two-electrode elements, while three-electrode elements are referred to as phototransistors.
   (c) *Photoemissive Detectors*—Metallic surfaces normally emit electrons when they absorb radiation. If such a surface is arranged to be a negative electrode, that is, a cathode, then the photoelectrons will move toward a positively charged surface, anode, thereby setting up current flow, which is a measure of the incident radiation.
   (d) *Photovoltaic Detectors*—These are also based on essentially the same principle as the photoconductive elements, with the radiation absorbed resulting in the release of photoelectrons, some of which accumulate at the cathode. A potential difference is thus built up between the anode and cathode, which is proportional to the intensity of the incident radiation.

Infrared Systems At the core of the infrared system is the detector, which may either be a single detector or an array detector.

1. *Single Detectors*—The single detector consists of a single element comprising one of the sensors discussed in the preceding section. The image of an object is obtained by using a scanner that sequentially focuses different sections of the object unto the detector. When it is only necessary to monitor the object along a line, a line scan is used. For area detection, the scan is done both horizontally and vertically. For spot detection, there is no need for scanning and the detector is focused at one spot.
2. *Linear Detectors*—The linear detector consists of a linear array of detectors, about 4096. Points along a line on the object are thus detected simultaneously, and thus for linear detection, no scanning is necessary. For area detection, the object is scanned vertically, if the array is horizontal.
3. *Array Detectors*—Two-dimensional array detectors are able to acquire an area image of an object without scanning and may have up to 4096 $\times$ 4096 elements.

Most commercial infrared systems use liquid nitrogen to cool the detectors, with the detector being mounted in the base of a Dewar flask. However, more recent developments may involve solid-state detectors that use the detector substrate as a heat sink, thereby making additional cooling unnecessary. During operation, the scanner detects heat from the source by capturing thermal radiation emitted as infrared electromagnetic waves. The use of a chopper within the scanner provides a radiance reference for the system. The detector output level may be set with this reference a number of times per second. This controls system drift that would otherwise be present when temperature difference measurements are made in the presence of a changing background temperature. The detected infrared energy is converted to an electrical signal and then quantified for each element using either color or different levels of gray to show discrete levels of thermal energy. Image processing may also be performed directly on the electrical signal. Most systems are either of the short wavelength or long wavelength type. The short wavelength systems have detectors that are primarily sensitive to the 1–6 μm range, while the long wavelength detectors are sensitive to the 8–14 μm range. Commercially available long wavelength systems are capable of measuring temperatures in the range of -20 to 2500°C with a resolution of 0.2°C.

Two major applications of infrared sensing are currently in use: (1) thermal imaging and (2) infrared signal analysis.

Thermal Imaging—In thermal imaging, thermal images are recorded on line and by image processing, process states such as weld geometry can be measured.

Infrared Signal Analysis—In infrared signal analysis, infrared signals are converted to electrical signals and analyzed using signal processing techniques.

Problem Sources

1. *Plasma Interference*—In addition to radiation from the object to be monitored such as the molten pool or joint, radiation may also be received from the plasma. This would result in erroneous interpretation of the data. For reliable monitoring, the plasma either needs to be screened from the detector or its radiation needs to be filtered from the incident radiation. The use of a detector that is most sensitive in the appropriate wavelength range would increase the signal-to-noise ratio.
2. *Emissivity*—The infrared radiation from a material is a function of both temperature and emissivity. Thus for a given detected radiation, the absolute temperature cannot be determined unless the emissivity is known. However, the emissivity can vary significantly for a given material, depending on both the surface condition and temperature. For a steel, for example, this may vary from about 0.07 for a polished surface to about 0.79 for an oxidized surface.

Application to Process Monitoring In this section, welding is used as a specific example. However, the concepts discussed are applicable to other processes that involve the formation of a molten pool. A regular welding process, for example, results in

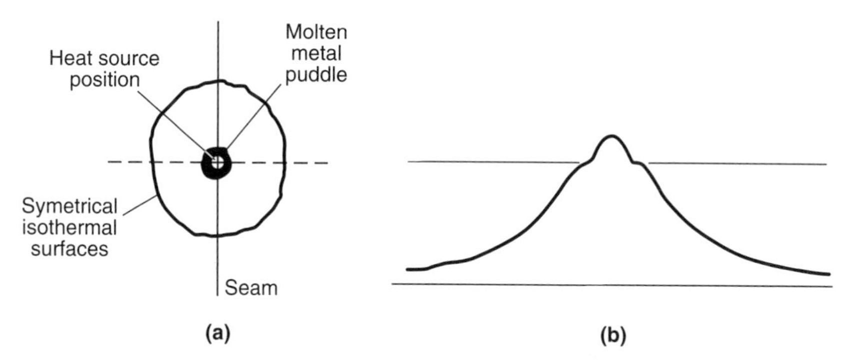

FIGURE 21.19 (a) Thermal image of a stationary weld pool with heat source directly above joint center. (b) Line scan of the thermal image. (From Chin, B. A., Madsen, N. H., and Goodling, J. S., 1983, *Welding Journal*, Vol. 62, pp. 227s–234s. By permission of American Welding Society.)

symmetric and repeatable patterns of the temperature gradients. Imperfections in the process, however, result in a discernible change in the thermal profiles.

The thermal image of a stationary welding process with the heat source positioned directly above the joint center is illustrated in Fig. 21.19a. The corresponding line scan across the center is shown in Fig. 21.19b, from which the distribution is found to be symmetrical. The average weld pool diameter can be obtained from the line scan and is given by the inflections around the peak temperature. When the heat source is shifted to one side of the joint center, the thermal image becomes distorted in shape, consisting then of half-moon shapes (Fig. 21.20a and b). This asymmetrical temperature distribution is caused by the excess energy, which is deposited on one side of the joint relative to the other, and the contact resistance at the joint, which

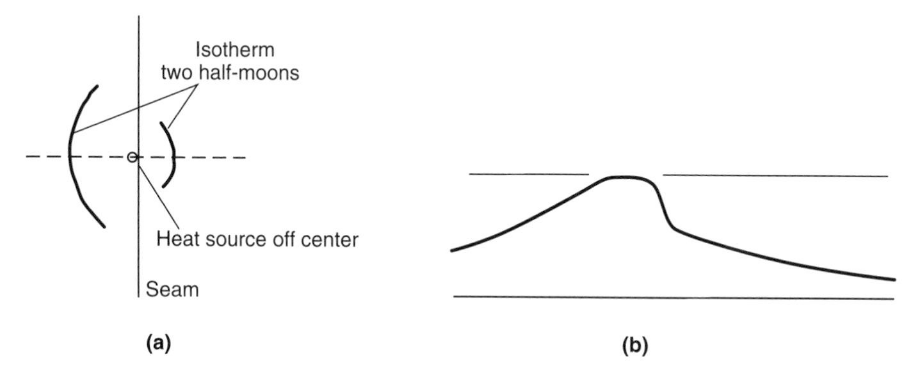

FIGURE 21.20 (a) Thermal image for a nonsymmetrically located heat source. (b) Corresponding line scan. (From Chin, B. A., Madsen, N. H., and Goodling, J. S., 1983, *Welding Journal*, Vol. 62, pp. 227s–234s. By permission of American Welding Society.)

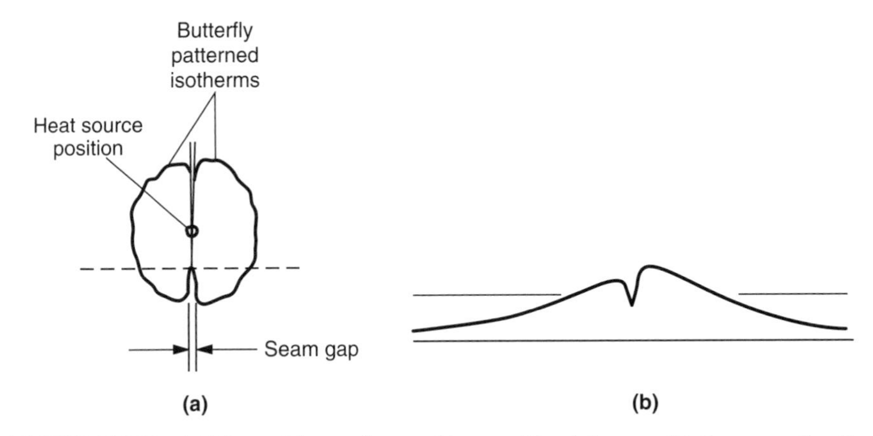

FIGURE 21.21 (a) Thermal image for a weldment with a joint gap. (b) Corresponding line scan. (From Chin, B. A., Madsen, N. H., and Goodling, J. S., 1983, *Welding Journal*, Vol. 62, pp. 227s–234s. By permission of American Welding Society.)

reduces heat flow across the joint, resulting in higher temperatures on the side with excess energy. The heat source could then be moved in the appropriate direction until the two radii are equal.

In addition to being used for joint tracking, the temperature isotherms could also be used to identify geometrical variations encountered in the welding process such as variations in the joint opening and mismatches. For example, a variation in the joint opening causes an indentation in the isothermal lines corresponding to a decrease from the peak temperatures of the metal surrounding the opening (Fig 21.21).

For the preceding monitoring schemes where the isotherms and temperature gradients are of interest, knowledge of the emissivity value is not essential. However, for evaluating the actual temperatures, it would be necessary to determine the emissivity value for the section of interest.

21.2.4.2 Ultraviolet Detection Radiation associated with the plasma is concentrated in the ultraviolet band of the electromagnetic spectrum (0.01–0.4 μm). Thus since welding plasma is primarily generated from the keyhole, ultraviolet sensing is usually correlated with keyhole formation and weld penetration.

There are two major applications of UV sensing: (1) spectroscopic analysis and (2) ultraviolet signal analysis.

Ultraviolet Spectroscopic Analysis Spectroscopic analysis is useful for determining the ion content in the plasma. This is important in determining the extent of plasma formation due to metal vapor and shielding gas. Especially important is plasma absorption of the laser power above the workpiece or coupling of laser energy to the material, a phenomenon that is highly influenced by the plasma size and its spectral characteristics. The spectrum provides information on the plasma electron temperature, T_{pl}, and also the electron density, n_e, which is normally obtained using techniques involving the relative line intensity and stark broadening of spectral lines. These techniques do not need calibration. The spectrum may be obtained using a spectograph coupled with a gated optical multichannel analyzer.

Plasma temperatures can be extracted from spectroscopic information by using relative emission intensity ratio of ion states of an element if the plasma is dense enough for local thermodynamic equilibrium (LTE) to exist.

Ultraviolet Signal Analysis While spectroscopic analysis is very important, recent developments in ultraviolet signal analysis have shown a lot of potential. In ultraviolet signal analysis, ultraviolet signals from the process are converted into electrical signals. Magnitude and frequency analyses of the output signals are then correlated with process states. In laser welding for example, when penetration occurs, a clear relationship between full penetration and the ultraviolet plasma signal is usually observed.

Fluctuations in the laser induced plasma do occur even when the laser power is constant. Such fluctuations can be monitored using a photodiode. Sudden changes in mass flow rate (and thus pressure) of the plasma out of a keyhole result in steep and narrow peaks in the photodiode signal. A key application of plasma monitoring is in detection of melt-through during laser welding, since this significantly reduces the pressure in the keyhole, and thereby the photodiode signal.

21.2.5 Optical (Vision) Sensing

Optical sensors are often used for monitoring molten pool geometry, kerf size, and observing flow on the free pool surface and chevron formation. In addition, they are also used for seam tracking. The basic components of an optical sensing system include the sensor or detector, illumination source, object, transmission elements, and the processor. The detector is described in the next section, followed by the measurement techniques in subsequent sections.

21.2.5.1 Optical Detectors The most common type of optical detector is the solid-state array camera, which usually has an array of charge injection or charge coupled device (CCD) light sensitive elements. The output of each element or pixel may be a 16-bit digitized data. The output from the camera is dumped into a memory buffer for analysis. An optical filter is usually used to separate the laser radiation from the heat radiation. The detector elements may require significant attenuation, especially when used for a focused beam.

21.2.5.2 Detector Setup Some of the problems associated with optical sensing include the extreme brightness of the plasma plume compared to that of the molten pool (high contrast) and dependence of the intensity on processing conditions. Spatter, fumes, and flux may also obscure the object to some extent. Separate illumination is thus normally used to enhance the system resolution. It may be in the form of either structured light or general illumination, that is nonstructured light. A structured light is a pattern of lines or a grid of light that is projected unto the object to help provide information on the three-dimensional shape of the object, based on the apparent distortion of the pattern (Fig. 21.22a).

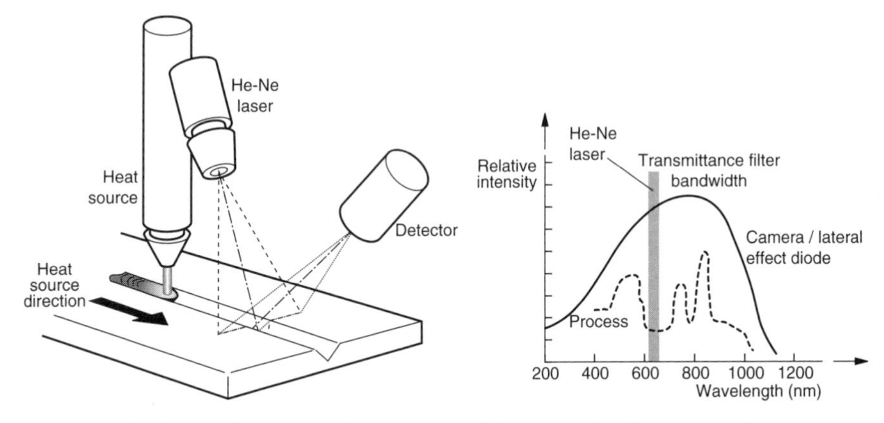

FIGURE 21.22 (a) Illustration of a structured light setup. (b) Illustration of detector and process spectral characteristics. (From Dornfeld, D. A., 1983, *Computer-Integrated Manufacturing*, ASME Bound Volume, Martinez, M., editor, pp. 89–97.)

The general illumination could come from an auxiliary high-intensity light source. An example of general illumination involves illuminating the object with a narrow bandwidth laser beam, with the beam bandwidth selected to be in the region where the sensitivity of the detector is high, based on the spectral characteristics of the detector (Fig. 21.22b). This is because the light spectrum from an object generally does not have equal intensity at all wavelengths (see Fig. 21.17). All light incident on the detector is then filtered except for the narrow bandwidth of the auxiliary beam, thereby subduing the effect of the bright light from the plume.

The detector arrangement for an optical sensing system is similar to that of an infrared array. (Section Infrared Systems). It may be a linear or two-dimensional array.

21.2.5.3 Edge Detection Methodology Edge detection is used to identify the molten pool boundary. This enables say the weld bead size to be determined, and hence, an estimate to be made of the load carrying capacity of the weldment. Using the setup shown in Fig. 21.23a, an example molten pool image data as obtained from the camera is shown in Fig. 21.23b. Figure 21.24 shows a three-dimensional view of the same image. The image could be contaminated by noise from a number of sources.

Noise Sources Potential sources of noise include the following:

1. *Dust on the Optical Components*—This results from condensation of metal vapor.
2. *Noise Due to Fiber Thread Boundary*—When a fiber optic cable is used, the image at the end of the cable is a combination of small images transmitted by individual fibers in the bundle. Very thin lines encircle each individual image cell, representing the boundaries of each fiber. These influence the image quality

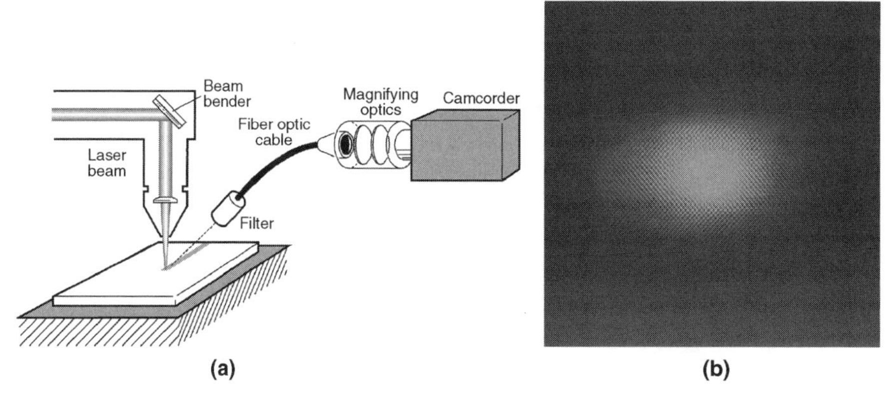

FIGURE 21.23 (a) Vision monitoring system. (b) Typical weld pool image before image processing. (From Tsai, F.-R., 1998, PhD Dissertation, University of Michigan, Ann Arbor, Michigan.)

because they result in uneven brightness, which is a noise source on the image and cannot be eliminated.

3. *Noise from the CCD Array*—Each individual photo diode cell of the CCD array converts a portion of the image into an electrical signal. Increasing the density of photo diodes in one CCD array results in a more vivid image. The image resolution is determined by the number of photo diodes used. The noise results when the CCD array converts the two-dimensional image into electrical signals. This is a minor noise source that cannot be corrected.
4. *Image Digitization Noise*—When the resolution of the frame grabber (say, 640×480 pixels) is different from that of the camcorder (say, 512×512 pixels), that can be a source of error.

As a result of these error sources, low pass filtering is necessary.

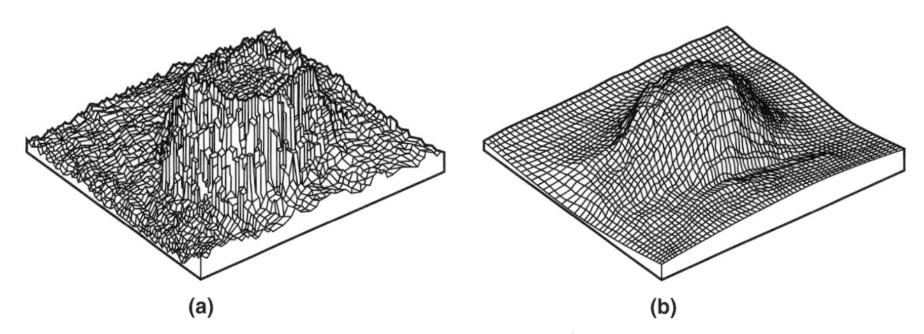

FIGURE 21.24 (a) Three-dimensional view of the weld pool image in Fig. 21.23 and (b) three-dimensional view of weld pool after nine-point averaging. (From Tsai, F.-R., 1998, PhD Dissertation, University of Michigan, Ann Arbor, Michigan.)

Filtering There are different types of filters available for analyzing optical signals. As an example, the nine-point averaging method, which is commonly used, is discussed. It involves replacing a given point $P(i, j)$ by the average of all eight neighboring points, plus the original value of the point:

$$P(i, j)_n = \frac{1}{9}\left[\sum_{j-1}^{j+1}\sum_{i-1}^{i+1} P(i, j)\right] \tag{21.21}$$

where $P(i, j)_n$ is the the new value of the point $P(i, j)$. Application of nine-point averaging to the original image of Fig. 21.24a is shown in Fig. 21.24b.

Edge Detection Following filtering, edge detection is then used to determine, for example, the molten pool boundaries. Two common techniques are directional and nondirectional edge detection.

Directional Edge Detection—The gradient value of a point $P(i, j)$ in the x-direction for directional edge detection is given by

$$\begin{aligned} x \text{ gradient of } P(i, j) = |&P(i-1, j-1) + P(i-1, j) + P(i-1, j+1) \\ &- P(i+1, j-1) - P(i+1, j) - P(i+1, j+1)| \end{aligned} \tag{21.22a}$$

Likewise, the gradient value in the y-direction is

$$\begin{aligned} y \text{ gradient of } P(i, j) = |&P(i-1, j-1) + P(i, j-1) + P(i+1, j-1) \\ &- P(i-1, j+1) - P(i, j+1) - P(i+1, j+1)| \end{aligned} \tag{21.22b}$$

The absolute value of the gradient is used to accommodate both positive and negative values. Boundary points are determined using a threshold, and the following equation may be used to identify candidate boundary points:

This is a candidate boundary point, if

$$(x - \textit{directional gradient} \geq \textit{threshold}) \text{ or } (y - \textit{directional gradient} \geq \textit{threshold}) \tag{21.23}$$

Nondirectional Edge Detection—The nondirectional edge detector is based on the gradient of an image $f(x, y)$ at location $P(x, y)$:

$$\nabla \mathbf{f} = \begin{pmatrix} \nabla_x f \\ \nabla_y f \end{pmatrix} = \begin{pmatrix} \frac{\partial f}{\partial x} \\ \frac{\partial f}{\partial y} \end{pmatrix} \tag{21.24}$$

where $\nabla_x f = \partial f/\partial x$ is the x-directional gradient and $\nabla_y f = \partial f/\partial y$ is the y-directional gradient.

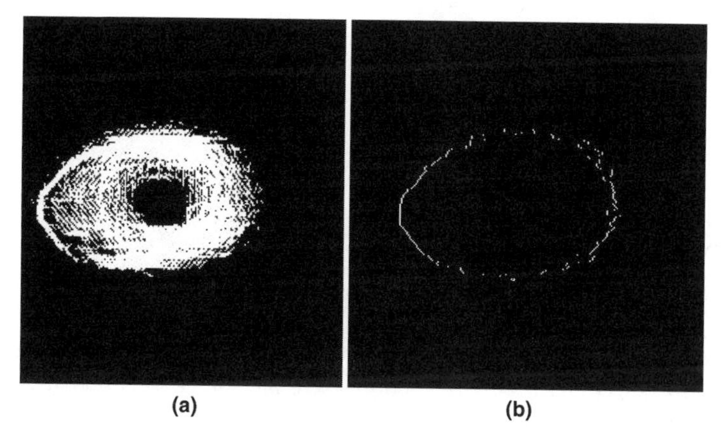

FIGURE 21.25 (a) Results of application of nondirectional edge detector. (b) Weld pool boundary points determined using the line-by-line searching method. (From Tsai, F.-R., 1998, PhD Dissertation, University of Michigan, Ann Arbor, Michigan.)

The gradient vector points in the direction of maximum change rate. The x- and y-directional gradient operators can be calculated using Sobel operators, which provide smoothing during the differentiation operation. Sobel operators in the x and y directions are represented as:

$$\begin{aligned}\nabla_x f = |&P(i-1, j-1) + 2P(i-1, j) + P(i-1, j+1) - P(i+1, j-1) \\ &-2P(i+1, j) - P(i+1, j+1)|\end{aligned} \tag{21.25}$$

$$\begin{aligned}\nabla_y f = |&P(i-1, j-1) + 2P(i, j-1) + P(i+1, j-1) - P(i-1, j+1) \\ &-2P(i, j+1) - P(i+1, j+1)|\end{aligned} \tag{21.26}$$

The values are always positive. Figure 21.25a shows the results (in two dimensions) of applying the nondirectional gradient detector to the image in Fig. 21.24b.

Finding the Boundary Points The actual boundary points need to be determined from the candidate points of Fig. 21.25a. A simple method is the line-by-line searching algorithm that searches the entire image row-by-row. One criterion for selecting outer boundary points might be

$$\text{outer boundary point} = \text{outer candidate point which has at least two other candidate points in its neighborhood.} \tag{21.27}$$

The resulting boundary for the image in Fig. 21.25a is shown in Fig. 21.25b.

Calibration Calibration is used to convert the initial unit of pixel to metric units and to determine the threshold values used in the boundary point search algorithm.

This may be done experimentally by first placing white circles of known diameter against a black background and capturing their images. The resulting images may be elliptical, depending on the camera orientation. The lengths of the major and minor axes of the image are then determined in pixels. From these measurements, the image resolution can be determined in millimeters per pixel.

21.3 SUMMARY

Sensors for monitoring laser processes can be categorized into three groups:

- *Beam Monitoring Sensors*—They are used for measuring beam characteristics such as power and beam distribution. Examples of such sensors are the beam dump (for power) and laser beam analyzers (rotating rod, hollow needle, or camera for beam distribution). The beam size can also be determined using a Kapton film.
- *Process Monitoring Sensors*—These are used to measure actual process features such as geometry of the molten pool, kerf size, and defects that are formed during the process. The top view of the pool geometry, for example, can be measured using vision, infrared, or ultraviolet sensors. Infrared and ultraviolet detectors can also provide information on the stability and therefore quality of the process. The individual detector elements can be arranged either in a linear array (for measurement along a line) or a two-dimensional array (for area images). Whereas the vision, infrared, and ultraviolet sensors can detect defects that are exposed to the surface, acoustic emission detectors can, in addition, detect defects that are embedded in the workpiece. In all cases, special precautions have to be taken to prevent contamination of the measured signal by noise.
- *Seam Tracking Sensors*—These are used to detect the joint along which the laser beam is expected to follow, enabling the beam to be guided automatically. Seam tracking can be done using vision and infrared sensors.

REFERENCES

Astic, D., Canosa, A., Vigiano, P., and Autric, M., 1991, Visible spectroscopy of laser produced aluminum plasma. Electron temperature and density determinations, *ICALEO*, pp. 94–103.

Autric, M., Vigiano, P., and Astic, D., 1988, Visible spectroscopy of laser produced plasma, *SPIE High Power* CO_2 *Laser Systems and Applications*, Vol. 1020, pp. 103–112.

Benedict, R. P., 1984, *Fundamentals of Temperature, Pressure, and Flow Measurements*, 3rd edition, John Wiley and Sons, New York.

Beyer, E., Abels, P., Drenker, A., Maischner, D., and Sokolowski, W., 1991, New devices for on-line process diagnostics during laser machining, *ICALEO*, pp. 133–139.

Carlson, N. M., and Johnson, J. A., 1988, Ultrasonic sensing of weld pool penetration, *Welding Journal*, Vol. 67, pp. 239s–246s.

Chen, H. B., Li, L., Brookfield, D. J., Williams, K., and Steen, W. M., 1991, Laser process monitoring with dual wavelength optical sensors, *ICALEO*, pp. 113–122.

Chen, S. L., Li, L., Modern, P., and Steen, W. M., 1991, In process laser beam position sensing, *SPIE Industrial and Scientific Uses of High-Power Lasers*, Vol. 1502, pp. 123–134.

Chin, B. A., Madsen, N. H., and Goodling, J. S., 1983, Infrared thermography for sensing the arc welding process, *Welding Journal*, Vol. 62, pp. 227s–234s.

Denney, P. E., and Metzbower, E. A., 1991, Synchronized laser-video camera system study of high power laser material interactions, *ICALEO*, pp. 84–93.

DiPietro, P., and Yao, Y. L., 1994, An investigation into characterizing and optimizing laser cutting quality: A review, *International Journal of Machine Tools & Manufacture*, Vol. 34, No. 3, pp. 225–243.

Dornfeld, D. A., 1983, Arc weld monitoring for process control—seam tracking, *Computer-Integrated Manufacturing*, ASME Bound Volume, Martinez, M., editor, pp. 89–97.

Dove, R. C., and Adams, P. H., 1964, *Experimental Stress Analysis and Motion Measurement*, Charles E. Merrill Publishing Co., Columbus, OH.

Farson, D. F., Fang, K. S., and Kern, J., 1991, Intelligent laser welding control, *ICALEO*, pp. 104–112.

Gatzweiler, W., Maischner, D., and Beyer, E., 1988, On-line plasma diagnostics for process-control in welding with CO_2 lasers, *SPIE High Power* CO_2 *Laser Systems and Applications*, Vol. 1020, pp. 142–148.

Gregersen, O., and Olsen, F. O., 1990, Beam analyzing system for CO_2 lasers, *ICALEO*, pp. 28–35.

Hamann, C., Rosen, H.-G., and LaBiger, B., 1989, Acoustic emission and its application to laser spot welding, *SPIE High Power Lasers and Laser Machining Technology*, Vol. 1132, pp. 275–281.

Holtgen, B., Treusch, H. G., and Aehling, H., 1988, Diagnostic system for lasers in the visible and near infrared region, *SPIE Laser Assisted Processing*, Vol. 1022, pp. 52–54.

Kannatey-Asibu Jr., E., and Dornfeld, D. A., 1981, Quantitative relationships for acoustic emission from orthogonal metal cutting, *ASME Journal of Engineering for Industry*, Vol. 103, pp. 330–340.

Kannatey-Asibu Jr., E., 1982, On the application of the pattern recognition method to manufacturing process monitoring, *Proceedings of the 10th North American Manufacturing Research Conference*, pp. 487–492.

Lewis, G. K., and Dixon, R. D., 1985, Plasma monitoring of laser beam welds, *Welding Journal*, Vol. 64, pp. 49s–54s.

Li, L., Qi, N., Brookfield, D. J., and Steen, W. M., 1990, On-line laser weld sensing for quality control, *ICALEO*, pp. 411–421.

Li, L., and Steen, W. M., 1992, Non-contact acoustic emission monitoring during laser processing, *ICALEO*, pp. 719–728.

Lim, G. C., and Steen, W. M., 1982, Measurement of the temporal and spatial power distribution of a high-power CO_2 laser beam, *Optics and Laser Technology*, pp. 149–153.

Lim, G. C., and Steen, W. M., 1984, Instrument for instantaneous *in situ* analysis of the mode structure of a high-power laser beam, *Journal of Physics E: Scientific Instruments*, Vol. 17, pp. 999–1007.

Liu, X., and Kannatey-Asibu Jr., E., 1990, Classification of AE signals for martensite formation from welding, *Welding Journal*, Vol. 69, No. 10, pp. 389s–394s.

Liu, Y., and Leong, K. H., 1992, Laser beam diagnostics for kilowatt power pulsed YAG laser, *ICALEO*, pp. 77–87.

Maischner, D., Drenker, A., Seidel, B., Abels, P., and Beyer, E., 1991, Process control during laser beam welding, *ICALEO*, pp. 150–155.

Maxfield, B. W., Kuramoto, A., and Hulbert, J. K., 1987, Evaluating EMAT designs for selected applications, *Materials Evaluation*, Vol. 45, pp. 1166–1183.

Metals Handbook, 8th edition, Vol. 11, 1976, *Nondestructive Inspection and Quality Control,* ASM, Metals Park, OH.

Mombo-Caristan, J.-C., Koch, M., and Prange, W., 1991, Seam geometry monitoring for tailored welded blanks, *ICALEO*, pp. 123–132.

Mombo-Caristan, J.-C., Koch, M., and Prange, W., 1993, Process controls for laser blank welding, *ALAW*, pp. 1–7.

Orlick, H., Morgenstern, H., and Meyendorf, N., 1991, Process monitoring in welding and solid state lasers by sound emission analysis, *Welding and Cutting*, Vol. 12, pp. 15–18.

Parthasarathi, S., Khan, P. A. A., and Paul, A. J., 1992, Intelligent laser processing of materials, *ICALEO*, pp. 708–718.

Richardson, R.W., Gutow, D.A., Anderson, R. A., and Farson, D. F., 1984, Coaxial arc weld pool viewing for process monitoring and control, *Welding Journal*, Vol. 63, pp. 43s–50s.

Sasnett, M. W., 1990, LIA Beam Characterization Report, *ICALEO*, pp. 1–12.

Smith, E. T., 1999, Monitoring laser weld quality using acoustic signals, *PhD Dissertation*, University of Michigan, Ann Arbor, MI.

Sokolowski, W., Herziger, G., and Beyer, E., 1989, Spectroscopic study of laser-induced plasma in the welding process of steel and aluminum, *SPIE High Power Lasers and Laser Machining Technology*, Vol. 1132, pp. 288–295.

Steen, W. M., and Weerasinghe, V. M., 1986, In process beam monitoring, *SPIE Laser Processing: Fundamentals, Applications, and Systems Engineering*, Vol. 668, pp. 37–44.

Steen, W. M., 2003, *Laser Material Processing*, Springer-Verlag, Berlin.

Sun, A., Kannatey-Asibu Jr., E., and Gartner, M., 1999, Sensor systems for real time monitoring of (laser) weld quality, *Journal of Laser Applications*, Vol. 11, No. 4, pp. 153–168.

Sun, A., Kannatey-Asibu Jr., E., and Gartner, M., 2002, Monitoring of laser weld penetration using sensor fusion, *Journal of Laser Applications*, Vol. 14, No. 2, pp. 114–121.

Tsai, F.-R., 1998, Vision sensing, modeling, and control of laser weld pool geometry, PhD Dissertation, University of Michigan, Ann Arbor, MI.

VanderWert, T. L., 1992, In-process workpiece and fixture sensing for flexible, repeatable laser processing of 3D parts, *ICALEO*, pp. 699–707.

Voelkel, D. D., and Mazumder, J., 1990, Visualization and dimensional measurement of the laser weld pool, *ICALEO*, pp. 422–429.

Wang, Z. Y., Liu, J. T., Hirak, D. M., Weckman, D. C., and Kerr, H. W., 1991, Measurement of pulsed laser beam dimensions using KAPTON films, *ICALEO*, pp. 74–83.

Weeter, L., and Albright, C., 1987, The effect of full penetration on laser-induced stress-wave emissions during laser spot welding, *Materials Evaluation*, Vol. 45, pp. 353–357.

Whitehouse, D. R., and Nilsen, C. J., 1990, Plastic burn analysis (PBA) for CO_2 laser beam diagnostics, *ICALEO*, pp. 13–27.

APPENDIX 21A

List of symbols used in the chapter.

Symbol	Parameter	Units
b_1	width of laser beam analyzer detector	mm
c_d	damping coefficient	N s/m
c_1	constant $= 2\pi c^2 h_p = 3.74 \times 10^{-16}$	J m^2/s
c_2	constant $= ch_p/k_B = 1.44 \times 10^{-2}$	m K
d_d	depth of laser beam analyzer detector	mm
d_h	diameter of hole in Kapton film	mm
ΔH_v	energy necessary to heat and vaporize a unit volume of the material	J/mm^3
I	total hemispherical radiation intensity	W/m^2
I_λ	amplitude of radiant energy	J/m^{3}s
k_s	spring constant	N/m
l	distance from each detector to the beam center	mm
L_d	distance between detector and point of reflection on rod	mm
$\dot{m}$	mass flow rate of fluid	kg/s
$P(i, j)$	location of a point representing a pixel	–
$P(i, j)_n$	new value of the point $P(i, j)$	–
Q_e	total pulse energy	J
r_r	radius of rotating rod	mm
r_d	distance between centers of disk and beam	mm
RMS	root mean square value of signal	V
Δr	resolution of laser beam analyzer	mm
δt	averaging time	s
T_{pl}	plasma temperature	K
$V(t)$	signal voltage function	V
w_b	base radius of laser beam intensity distribution	μm
w_t	top radius of laser beam intensity distribution, assumed to be of truncated shape	μm
x	displacement of mass	mm
y	input displacement of base frame	mm
Y	maximum amplitude of the input base displacement	mm
z	relative displacement between base and mass	mm
Z	maximum amplitude of the relative motion between the base and mass	mm
α_A	absolute Seebeck coefficient of material A	V/K
α_B	absolute Seebeck coefficient of material B	V/K
ω	angular frequency of input base oscillation	rad/s

ω_n	natural frequency of vibration of undamped system	rad/s
ω_v	angular velocity	rad/s
ϕ	phase angle between input motion y and relative motion z	rad
τ_d	time delay	s
θ_r	half angle between indident beam and reflected beam received by detector	°
λ_m	wavelength at which peak intensity occurs	μm
ζ	damping ratio	—

PROBLEMS

21.1. A laser beam of power 3 kW is dumped into a cone calorimeter beam dump. If the inlet temperature of the water flowing through the calorimeter is 30°C, determine the water flow rate through it if the outlet temperature cannot exceed 95°C. The specific heat of water is 4.186 J/g°C = 4186 J/kg°C.

21.2. Show that for $L_d >> r_r$ and $L_d >> b_1$, the width, δr_r, of the beam that is just received by the detector of a rotating rod laser beam analyzer is

$$\delta r_r = \frac{b_1 r}{2L_d} \cos\theta_r$$

21.3. Show that for a rotating rod laser beam analyzer, during rotation of the rod, the detector D_1 sees the path through the laser beam in space, which is the arc of a circle with MD_1 as diameter and passes through g and C where C is the center of the beam (see Fig. 21.3).

21.4. For a rotating rod laser beam analyzer, show that the angle $\alpha = \beta = w/\sqrt{2}l$ (Fig. 21.3).

21.5. Assuming a circular laser beam, show that the points g, q, h, and p, where the rod intersects the boundary of the laser beam at any instant, Fig. 21.3, are given by

$$y_g = \frac{1}{2l}(2l^2 - w^2 - w\sqrt{2l^2 - w^2})$$

$$y_h = \frac{1}{2l}(2l^2 - w^2 + w\sqrt{2l^2 - w^2})$$

$$x_g = \frac{b_1}{2l}\sqrt{2(l^2 - w\sqrt{2l^2 - w^2})}$$

$$x_h = \frac{b_1}{2l}\sqrt{2(l^2 + w\sqrt{2l^2 - w^2})}$$

where w is the beam radius and l is the distance from each detector to the beam center.

21.6. Show that the power loss due to reflection by a stationary rod in a rotating rod LBA is given by

$$\frac{4r_{\rm r}}{\pi b_1}$$

Further show that the power loss when the rod is not stationary is given by

$$\frac{4r}{\pi^2 R}$$

21.7. The rod used in a laser beam analyzer is of diameter 800 μm. The distance between the detector and point of reflection on the rod is 120 mm, and the angle between the incident beam and reflected beam received by the detector is 56°. Calculate the detector width if the resolution of the beam analyzer is 3 μm.

21.8. A rotating rod laser beam analyzer is used for monitoring a laser beam of diameter 15 mm. If the distance from each detector to the beam center is 80 mm, determine the x and y coordinates of the points of intersection of the rotating rod and the beam periphery.

21.9. Determine the amplitude of radiant energy at a wavelength of 2 μm by a blackbody at a temperature of 1500°C.

22 Processing of Sensor Outputs

In the preceding chapter, the major sensors that are used in monitoring laser processes were discussed. Process outputs, as reflected in the measured signals, are generally stochastic in nature. Thus, it is often necessary to further analyze the signals to extract pertinent information about the state of the process. There are several methods available for analyzing signals, but only a basic approach is considered here. The discussion in this chapter on processing of sensor outputs therefore focuses on the identification of faults that may occur during laser processing, for example, cracking, porosity, no weld, and so on and is limited to classification methods. Implementation of a classification scheme normally involves three steps (Fig. 22.1):

1. Sampling (digitizing) the input signal to produce the pattern space.
2. Feature extraction, which normally involves transformation of the signal from the pattern into the feature space, from which useful information can be obtained. This is usually accompanied by a reduction in the data size.
3. Classification of the feature space to identify the individual signal sources (classes).

Since sampling of an analogue signal influences the results of its transformation, the discussion on these two issues are combined into one, under signal transformation. The Fourier transform, the discrete Fourier transform, and their properties are discussed, along with advantages and pitfalls of digital analysis. This is followed by the sampling theorem, aliasing, leakage, windowing, picket fence effect, and segmental averaging. Next, data reduction is presented, focusing on optimum transform for data reduction, the variance criterion, and the class mean scatter criterion. Pattern recognition and neural network analysis are then discussed in terms of pattern classification. Finally, time–frequency methods such as the short-time Fourier transform, wavelet analysis, and time–frequency distributions are briefly outlined.

22.1 SIGNAL TRANSFORMATION

Measured signals are often contaminated by noise. To characterize the system from which the signal is being obtained, in the presence of the noise, the signal is first transformed into another domain to facilitate analysis and interpretation. There are various

---

Principles of Laser Materials Processing, by Elijah Kannatey-Asibu, Jr.

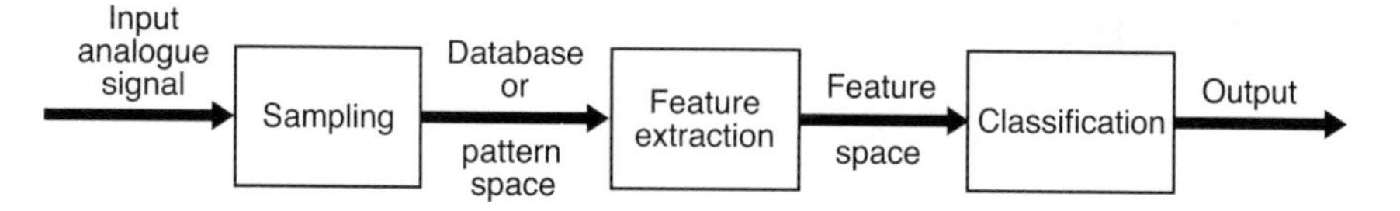

FIGURE 22.1 Basic classification scheme.

transforms that can be used, such as the Fourier transform, the Walsh–Hadamard transform, the Haar transform, and the Karhunen–Loeve transform. The Fourier transform is perhaps the most commonly used one because it permits a more useful engineering interpretation of the results and will therefore be the focus of our discussion.

22.1.1 The Fourier Transform

The Fourier transform involves the transformation of information from the time domain (time series) to the frequency domain (spectrum). The basis of spectral analysis is that a nonperiodic signal, $f(t)$, that is bounded and has a finite number of discontinuities can be represented in the frequency domain using the Fourier transform, which is given as

$$g(\omega) = \frac{1}{2\pi}\int_{-\infty}^{\infty} f(t)\mathrm{e}^{-i\omega t}\mathrm{d}t \tag{22.1}$$

and

$$f(t) = \int_{-\infty}^{\infty} g(\omega)\mathrm{e}^{i\omega t}\mathrm{d}\omega \tag{22.2}$$

where $i = \sqrt{-1}$, $\omega = 2\pi\nu$ is the angular frequency, and ν is the frequency.

Equation (22.1) is the Fourier transform of $f(t)$, and equation (22.2) is the inverse Fourier transform, and together they constitute the Fourier transform pair. The Fourier transform is a special form of the Fourier series and is applicable to nonperiodic functions. Note that there are several variants of the Fourier transform pair, and a different version is used in Chapter 5 for convenience.

$g(\omega)$ is a complex continuous function of ω and can be rewritten as

$$g(\omega) = A(\omega) + iB(\omega) \tag{22.3}$$

where $A(\omega)$ and $B(\omega)$ are the real and complex components, respectively, of $g(\omega)$.

The amplitude spectrum, $a(\omega)$, of the function $f(t)$ is then given by

$$a(\omega) = |g(\omega)| = \sqrt{[A(\omega)]^2 + [B(\omega)]^2} \tag{22.4a}$$

the power spectrum by

$$q(\omega) = |g(\omega)|^2 = [A(\omega)]^2 + [B(\omega)]^2 \tag{22.4b}$$

and the phase spectrum by

$$\phi(\omega) = \tan^{-1} \frac{B(\omega)}{A(\omega)} \tag{22.4c}$$

Direct use of the amplitude spectrum, $a(\omega)$, can sometimes present problems when there are pseudorandom components that can produce varying $a(\omega)$ that average out to zero over a long signal length, $\Delta\tau$. The use of the power spectrum eliminates this problem. To make the power spectrum independent of the duration, $\Delta\tau$, of the data, it can be reexpressed as

$$q_n(\omega) = \frac{1}{\Delta\tau} |g(w)|^2 \tag{22.5}$$

22.1.2 The Discrete Fourier Transform (DFT)

For digital analysis, the discrete equivalent of the Fourier transform is used. Let $\{x(j)\}$ denote a sequence $x(j)$, $j = 0, 1, ..., N_m - 1$ of N_m finite-valued real or complex numbers, representing a time series of data. The discrete Fourier transform of the sequence is given as

$$y(k) = \frac{1}{N_m} \sum_{j=0}^{N_m-1} x(j) e^{\frac{-2\pi i j k}{N_m}} \quad k = 0, 1, ..., N_m - 1 \tag{22.6}$$

and the inverse transform

$$x(j) = \sum_{k=0}^{N_m-1} y(k) e^{\frac{2\pi i j k}{N_m}} \quad j = 0, 1, ..., N_m - 1 \tag{22.7}$$

To illustrate some basic concepts of the DFT, a real-valued signal and its DFT are shown in Fig. 22.2. In this figure,

$\nu_0 = \frac{1}{\Delta\tau}$
$\Delta\tau = N_m \delta\tau$
ν_s (sampling rate) $= N_m \nu_0 = \frac{1}{\delta\tau}$
ν_f(folding frequency) $= \frac{\nu_s}{2}$.

The signal time series, $x(j)$, is considered to be periodic. Some points to note in relation to the DFT are as follows:

1. Each component of the transform is a harmonic. The actual frequency is $k\nu_0$, where ν_0 is the fundamental frequency.
2. For real $x(j)$, the real part of the digital transform, Re$[y(k)]$, is symmetric about the folding frequency ($\nu_f = \nu_s/2$). ν_s is the sampling rate. The imaginary component, Im $[y(k)]$, is antisymmetric.

A brief overview of some properties of the discrete Fourier transform is provided in the next section.

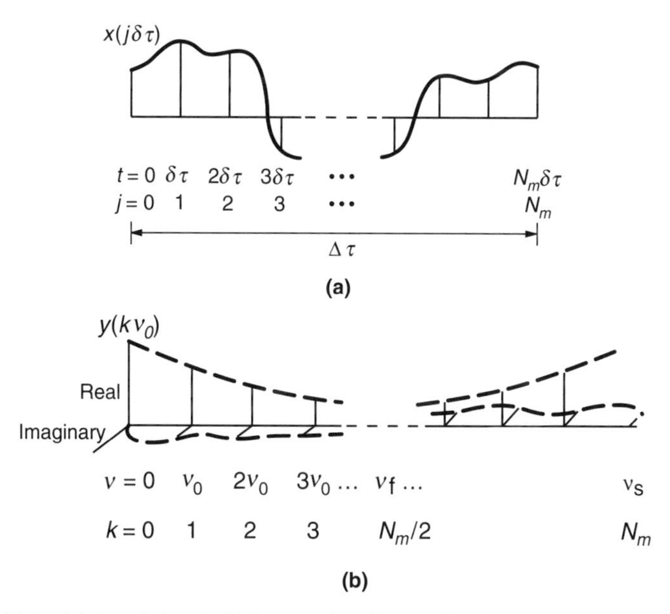

FIGURE 22.2 (a) A real signal. (b) Its complex discrete Fourier transform. (From Bergland, G. D., 1969, *IEEE Spectrum*, Vol. 6, pp. 41–51.)

22.1.3 Properties of the Discrete Fourier Transform

Some of the important properties associated with the discrete Fourier transform involve linearity, complex conjugage, shifting, convolution, and correlation characteristics. In the following discussion, it is assumed that the discrete Fourier transform of the time sequence represented by $x(j)$ is given by the sequence $y(k)$.

Linearity—The discrete Fourier transform is linear. In other words, if

$$x(j) = ax_1(j) + bx_2(j)$$

then

$$y(k) = ay_1(k) + by_2(k) \tag{22.8}$$

Complex Conjugate—If the time sequence $x(j) = [x(0), x(1), \ldots, x(N_m - 1)]$ is real valued, and $N_m/2$ is an integer, then its transform coefficients, $y(k)$, are such that

$$y(N_m/2 + l) = \tilde{y}(N_m/2 - l) \quad l = 0, 1, \ldots, N_m/2 \tag{22.9}$$

where $\tilde{y}$ is the complex conjugate of y.

This indicates that the power or amplitude spectrum is an even function about $N_m/2$, and thus only $k = 0, 1, \ldots, N_m/2$ components need to be computed.

Shifting—If $z(j) = x(j+h) \quad h = 0, 1, \ldots, N_m - 1$, and if we denote the DFT of $z(j)$ by $y_z(k)$ and that of $x(j)$ by $y_x(k)$, then

$$y_z(k) = \mathrm{e}^{\frac{2\pi ikh}{N_m}} y_x(k) \tag{22.10}$$

Convolution—If $\{z(j)\}$ and $\{x(j)\}$ are real-valued sequences, and their convolution is given by

$$R(j) = \frac{1}{N_\mathrm{m}} \sum_{h=0}^{N_\mathrm{m}-1} x(h)z(j-h) \quad j = 0, 1, \ldots, N_m - 1$$

then

$$y_R(k) = y_x(k)y_z(k) \tag{22.11}$$

Correlation—If $\{z(j)\}$ and $\{x(j)\}$ are real-valued sequences, and their correlation is given by

$$\hat{R}(j) = \frac{1}{N_\mathrm{m}} \sum_{h=0}^{N_\mathrm{m}-1} x(h)z(j+h) \quad j = 0, 1, \ldots, N_m - 1$$

then

$$y_{\hat{R}}(k) = \tilde{y}_x(k)y_z(k) \tag{22.12}$$

22.1.4 Advantages of Digital Analysis

Digital analysis has several advantages over analogue analysis:

1. A high dynamic range (signal-to-noise ratio) is attainable, since digital computations can be very precise.
2. The number of spectral components (resolution cells or data size) obtained is determined primarily by the memory capacity of the processor. There is no need for narrow band filters.
3. There is no drift of response characteristics.
4. There is greater flexibility in analysis.

22.1.5 Pitfalls of Digital Analysis

To understand some of the problems one might encounter in using the discrete Fourier transform, consider a simple cosine waveform. The continuous transform of this waveform consists of two impulse functions that are symmetric about zero frequency (Fig. 22.3a). For digital analysis, only a finite portion of the original cosine waveform

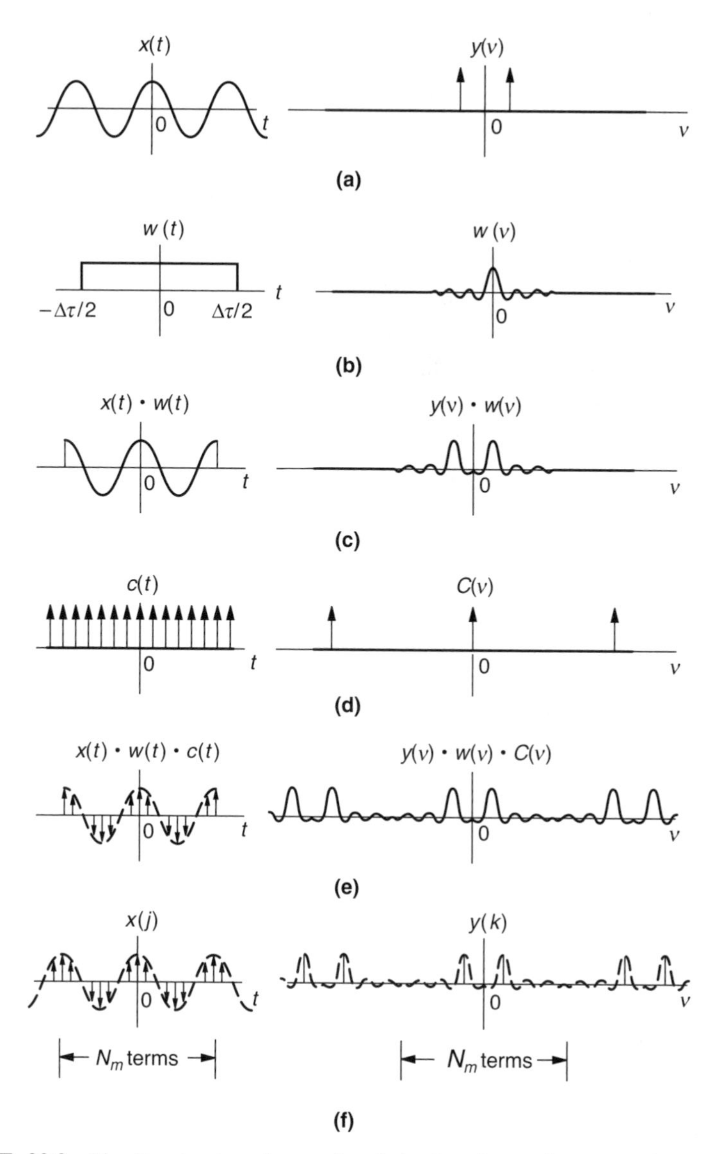

FIGURE 22.3 The Fourier transform of a finite length continuous cosine waveform. (a) Original analogue signal and its transform. (b) window function to be applied. (c) Windowed signal. (d) Sampling points. (e) Digitized or sampled windowed signal. (f) Sampled signal and its continuous frequency-domain function. (From Bergland, G. D., 1969, *IEEE Spectrum*, Vol. 6, pp. 41–51.)

signal is observed. This is equivalent to multiplying the signal by a unity rectangular window (see also Section 22.1.9.1). The transform of this rectangular window is a $(\sin x/x)$ function (Fig. 22.3b). This results in distortion of the transform of the finite waveform (Fig. 22.3c). Now sampling is equivalent to multiplying the signal by a train of impulses (Fig. 22.3d). The final transform of the finite sampled waveform is thus a repetition of the distorted transform in the frequency domain (Figs 22.3e and f). What this means is that a simple and straightforward sampling of a signal may introduce spurious frequency components where none existed.

22.1.6 The Sampling Theorem

The sampling theorem determines the lower limit of the sampling rate that needs to be used to avoid problems that can result from sampling. For the original continuous signal to be recoverable from its digital version, the digital form has to be obtained by sampling at a frequency that is at least twice the highest frequency in the original continuous signal, that is,

$$\nu_s \geq 2\nu_h \qquad (22.13)$$

where ν_h is the highest frequency in the signal, $\nu_s = 1/\delta\tau$ is the sampling rate (sampling frequency), and $\delta\tau$ is the sampling period.

The sampling theorem only establishes an ideal limit. In practice, the sampling rate has to be about 5–10 times the highest frequency. Some consequences of sampling, or more specifically, of using the discrete Fourier transform are aliasing, leakage, and the picket fence effect. These are briefly discussed in the following sections.

22.1.7 Aliasing

Aliasing is a phenomenon that happens when a signal is sampled at a rate that is lower than twice the highest frequency (ν_h in Fig. 22.4). When this occurs, frequencies above the folding frequency ($\nu_f = \nu_s/2$) are aliased or folded back into the frequency range below the folding frequency. In other words, a frequency component that is, say, 100 Hz higher than the folding frequency will be impersonated by a frequency component that is 100 Hz lower than the folding frequency. The low frequency component has the same amplitude as the high frequency component. In Fig. 22.5, the original high-frequency signal is impersonated by the low-frequency alias that also fits the sampled data. In general,

$$\nu_{\text{alias}} = \nu_{\text{sample}} - \nu_{\text{actual}} \qquad (22.14a)$$

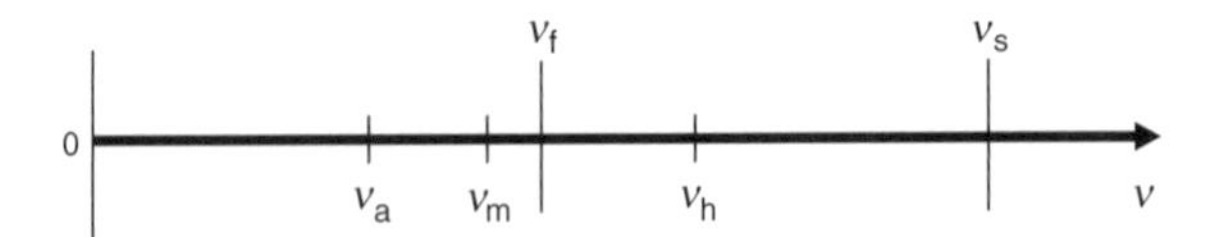

FIGURE 22.4 Illustration of aliasing on a frequency scale.

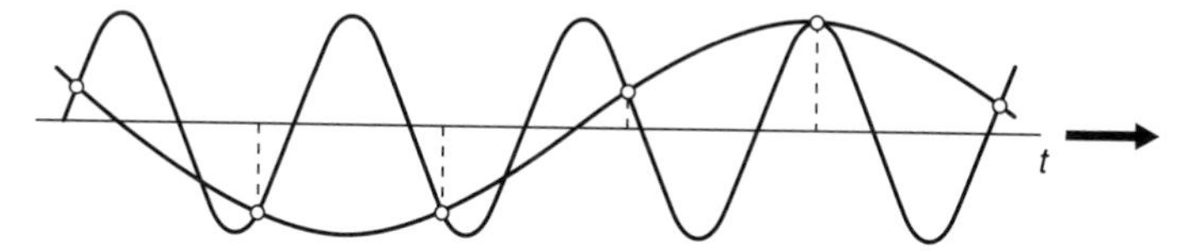

FIGURE 22.5 An example of low frequency "impersonating" a high frequency. (From Bergland, G. D., 1969, *IEEE Spectrum*, Vol. 6, pp. 41–51.)

or as illustrated in Fig. 22.4,

$$\nu_{\rm h} - \nu_{\rm f} = \nu_{\rm f} - \nu_{\rm a} \tag{22.14b}$$

where $\nu_{\rm a}$ is the aliased frequency and $\nu_{\rm f}$ is the folding frequency.

In other words, the aliasing signal frequency is always the difference between the actual signal and a harmonic (multiple) of the sampling frequency. In Fig. 22.4, ν_m is sampled without any problems since it is lower than half the sampling rate.

22.1.8 Leakage

Leakage results primarily from the fact that the signal being analyzed is finite in length. As indicated in Section 22.1.5, the finite length of data obtained by sampling over a period, $\Delta\tau$, is equivalent to multiplying the signal by a rectangular window (Fig. 22.6) (see also Section 22.1.9.1). This is the same as a convolution in the frequency domain. Since the rectangular window has a $\sin x/x$ function, the convolved signal is also of this form and centered at the spectrum of the original signal. The $\sin x/x$ function has a series of spurious peaks that decay very slowly (Fig. 22.7). These spurious peaks (side lobes) introduce high-frequency components into the original signal, and those may create aliasing problems where there was none (Fig. 22.8).

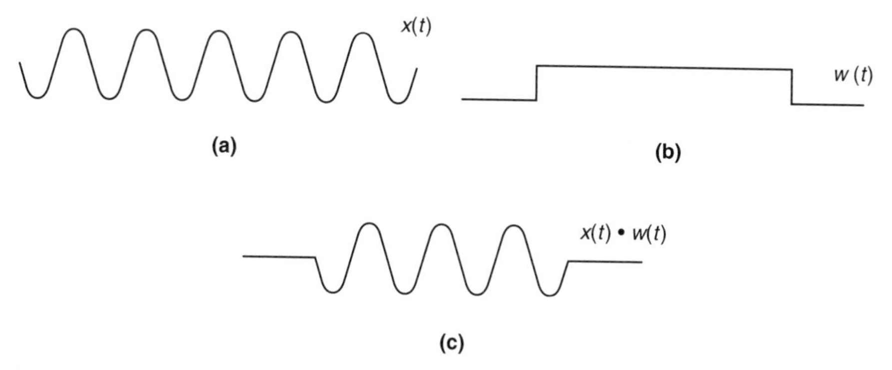

FIGURE 22.6 The rectangular data window implied when a finite record of data is analyzed. (a) Original signal. (b) Rectangular window function. (c) Windowed signal. (From Bergland, G. D., 1969, *IEEE Spectrum*, Vol. 6, pp. 41–51.)

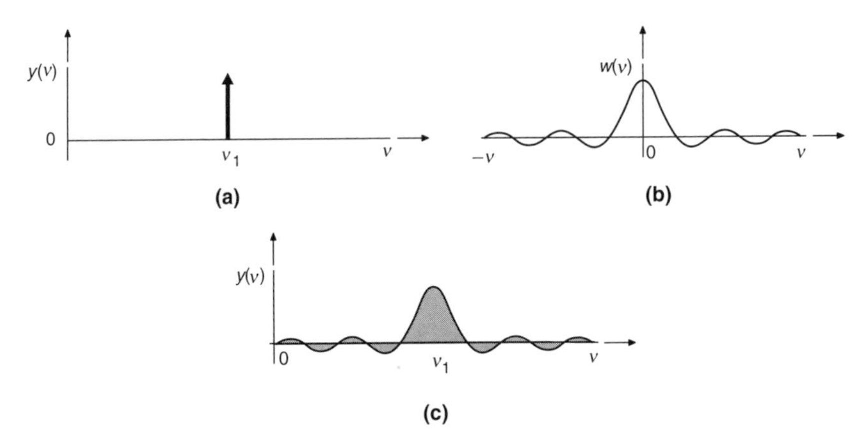

FIGURE 22.7 The leakage of energy from one discrete frequency into adjacent frequencies resulting from the analysis of a finite record. (a) Original signal frequency. (b) Frequency response of a rectangular window function. (c) Frequency response of the windowed signal. (From Bergland, G. D., 1969, *IEEE Spectrum*, Vol. 6, pp. 41–51.)

The effect of the leakage problem can be minimized by applying a window that has lower side lobes and is as close as possible to an impulse function. Ideally, this would require an infinite length of data. In lieu of this, windows that have the following characteristics may be used:

1. Main lobe of the window is as narrow as possible.
2. Maximum side lobe is relatively small.

In the following section, some common window functions are outlined.

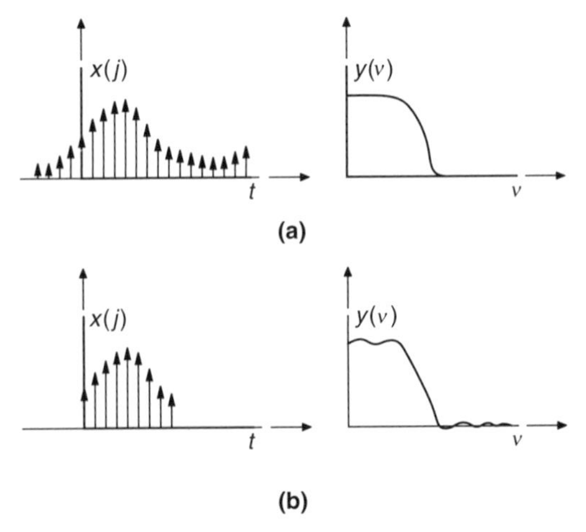

FIGURE 22.8 Illustration of leakage effect produced by abrupt termination of signal. (a) Infinite signal and its frequency response. (b) The signal and its frequency response after windowing.

22.1.9 Window Functions

For each of the window functions discussed in this section, the frequency response in the corresponding figure is expressed in decibel as

$$g_{\mathrm{db}}(\nu) = 20\log_{10}\frac{|g(\nu)|}{g(0)}$$

where $g(0)$ is the corresponding dc value.

22.1.9.1 Rectangular Window The rectangular window function is defined (see Fig. 22.9a) as

$$f(t) = \begin{cases} 1 & \text{for } |t| < \frac{\tau}{2} \\ 0 & \text{elsewhere} \end{cases} \tag{22.15a}$$

and the corresponding transform, Fig. 22.9b, is

$$g(\nu) = \frac{\tau\sin(\pi\nu\tau)}{(\pi\nu\tau)} \tag{22.15b}$$

22.1.9.2 Triangular Window This is defined, Fig. 22.10a, as

$$f(t) = \begin{cases} 1 - \frac{2|t|}{\tau} & |t| \leq \frac{\tau}{2} \\ 0 & \text{elsewhere} \end{cases} \tag{22.16a}$$

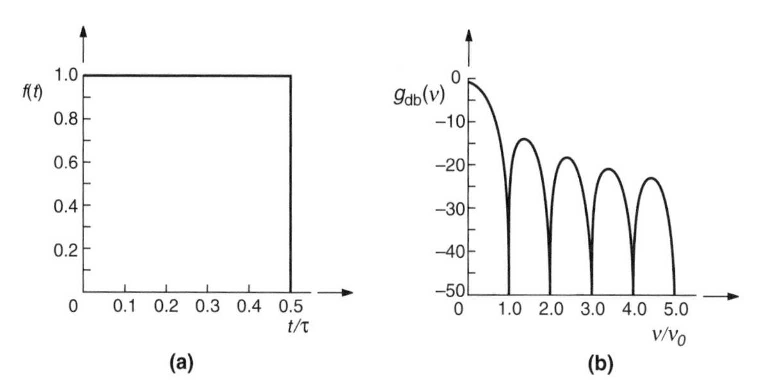

FIGURE 22.9 Rectangular window function. (a) Time domain signal. (b) Decibel amplitude response. (From Stanley, W. D., 1975, *Digital Signal Processing*. Reprinted by permission of Pearson Education, Inc., Upper Saddle River, NJ.)

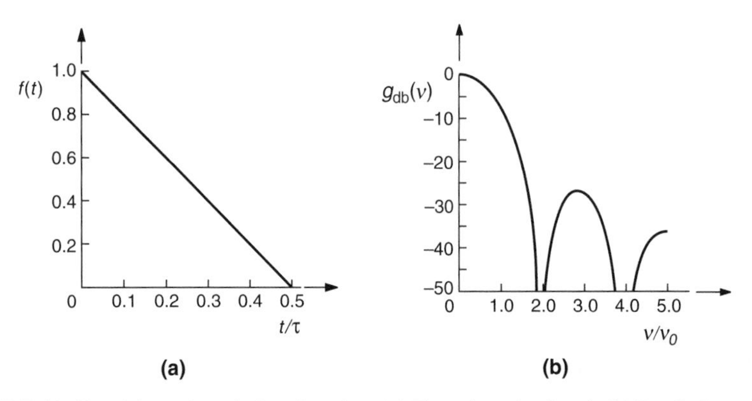

FIGURE 22.10 Triangular window function. (a) Time domain signal. (b) Decibel amplitude response. (From Stanley, W. D., 1975, *Digital Signal Processing*. Reprinted by permission of Pearson Education, Inc., Upper Saddle River, NJ.)

with the transform, Fig. 22.10b,

$$g(\nu) = \frac{\tau}{2}\left[\frac{\sin(\pi\nu\tau/2)}{(\pi\nu\tau/2)}\right]^2 \tag{22.16b}$$

22.1.9.3 Hanning Window The Hanning window is also called the cosine-squared function, Fig. 22.11a,

$$f(t) = \begin{cases} \cos^2\left(\frac{\pi t}{\tau}\right) = 1/2\left(1 + \cos\frac{2\pi t}{\tau}\right) & |t| \le \frac{\tau}{2} \\ 0 \ \text{elsewhere} \end{cases} \tag{22.17a}$$

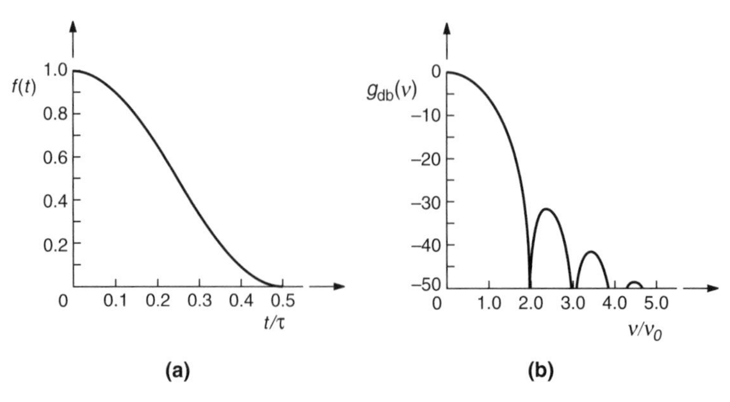

FIGURE 22.11 Hanning window function. (a) Time domain signal. (b) Decibel amplitude response. (From Stanley, W. D., 1975, *Digital Signal Processing*. Reprinted by permission of Pearson Education, Inc., Upper Saddle River, NJ.)

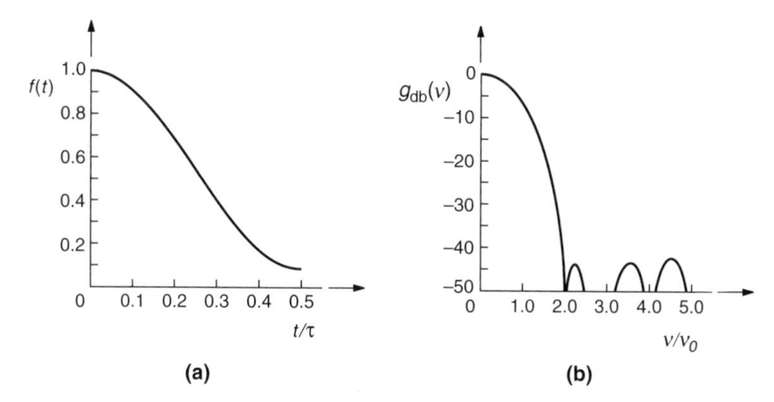

FIGURE 22.12 Hamming window function. (a) Time domain signal. (b) Decibel amplitude response. (From Stanley, W. D., 1975, *Digital Signal Processing*. Reprinted by permission of Pearson Education, Inc., Upper Saddle River, NJ.)

with the transform, Fig. 22.11b,

$$g(\nu) = \frac{\tau}{2}\frac{\sin \pi\nu\tau}{\pi\nu\tau}\left[\frac{1}{1-\nu^2\tau^2}\right] \tag{22.17b}$$

22.1.9.4 Hamming Window The Hamming function is given (Fig. 22.12a) by

$$f(t) = \begin{cases} 0.54 + 0.46\cos\frac{2\pi t}{\tau} & |t| < \frac{\tau}{2} \\ 0 \ \ \text{elsewhere} \end{cases} \tag{22.18a}$$

with the transform (Fig. 22.12b),

$$g(\nu) = \frac{\tau \sin \pi\nu\tau}{\pi\nu\tau}\left[\frac{0.54 - 0.08\nu^2\tau^2}{1-\nu^2\tau^2}\right] \tag{22.18b}$$

22.1.10 Picket Fence Effect

Since the discrete Fourier transform provides only the spectrum at discrete points of the fundamental frequency and its harmonics, it is possible to miss the peak of a particular component that might be between two consecutive spectral components.

This effect can be reduced by increasing the data period (length), in effect, improving the resolution of the spectrum. To do this without additional computations, the data length is increased by appending zeros to the end of the original data. This process does not change the continuous equivalent of the original signal. Zeros should be added only after windowing.

22.1.11 Segmental Averaging

Signals that are contaminated by a significant amount of noise can produce very erratic spectra that do not converge, irrespective of the data length or the sampling period. To minimize this effect, the signal is broken down into a number of segments, m, and the spectra of the individual segments averaged. However, such averaging reduces the resolution of the spectrum by smoothing the spectral components. The greater the number, m, of segments for the same length, $\Delta\tau$, of data, the worse the resolution.

The resolution, ν_0, of the averaged data is given by

$$\nu_o = \nu_s/N_m = \frac{1}{\Delta\tau} \tag{22.19}$$

where N_m is the number of sample points.

22.2 DATA REDUCTION

A prime consideration in the development of any real-time monitoring system is the response speed, since it determines how soon corrective action can be taken. The response speed in turn depends on the computational efficiency (how fast the appropriate calculations can be made). To improve the computational efficiency, the data size or dimensionality may be reduced by eliminating the feature components that convey the least information about the system. This stage of analysis is referred to as data reduction or feature selection. Data reduction is used to reduce the initial size of the data to make subsequent computations faster, with minimal loss of information.

The Fourier transform discussed in the preceding section presents the original data in a domain that facilitates physical interpretation. However, it is not necessarily the optimum transform with respect to data reduction. The optimum transform is one that would transform the data into a domain where the elimination of some components to reduce the data size would result in minimum error.

22.2.1 Optimum Transform for Data Reduction

To determine the optimum transform, let us consider the sampled data in the time domain to be represented by an N_m-vector whose transpose is

$$\mathbf{x}' = [x_1, x_2, x_3, \ldots, x_{N_m}] \tag{22.20}$$

and let it be transformed into the feature domain

$$\mathbf{y}' = [y_1, y_2, y_3, \ldots, y_{N_m}] \tag{22.21}$$

such that

$$\mathbf{y} = \mathbf{Tx} \tag{22.22}$$

where $\mathbf{T}$ is an orthogonal transform.

The objective of data reduction is to eliminate $N_m - n$ components of the data in the feature domain such that the original signal can be reconstructed from the resulting n-dimensional data without significant error. The error criterion to be used here is the mean square error criterion. Let the transform matrix $\mathbf{T}$ be represented by the column vectors

$$\mathbf{T}' = [\mathbf{T_1}, \mathbf{T_2}, ..., \mathbf{T_{N_m}}] \tag{22.23}$$

where $\mathbf{T_j}$ are N_m-vector columns, which are real valued and orthonormal, that is,

$$\mathbf{T_j}'\mathbf{T_k} = \begin{cases} 1 & j = k \\ 0 & j \neq k \end{cases} \tag{22.24a}$$

$$\mathbf{T}'\mathbf{T} = \mathbf{I} \tag{22.24b}$$

From equations (22.22)–(22.24b), we have

$$\mathbf{x} = \mathbf{T}'\mathbf{y} = \left[\mathbf{T_1}, \mathbf{T_2}, ..., \mathbf{T_{N_m}}\right]\mathbf{y} \tag{22.25a}$$

or

$$\begin{aligned}\mathbf{x} &= \left[\mathbf{T_1}, \mathbf{T_2}, ..., \mathbf{T_{N_m}}\right] \begin{bmatrix} y_1 \\ y_2 \\ \cdot \\ \cdot \\ \cdot \\ y_{N_m} \end{bmatrix} \\ &= y_1\mathbf{T_1} + y_2\mathbf{T_2} + \cdots + y_{N_m}\mathbf{T_{N_m}} \\ &= \sum_{j=1}^{N_m} y_j \mathbf{T_j} \end{aligned} \tag{22.25b}$$

After selecting n-components of $\mathbf{y}$, let us replace the $N_m - n$ components to be eliminated by preselected constants, c_j. The estimate $\hat{\mathbf{x}}$ of $\mathbf{x}$ is then

$$\hat{\mathbf{x}} = \sum_{j=1}^{n} y_j \mathbf{T_j} + \sum_{j=n+1}^{N_m} c_j \mathbf{T_j} \tag{22.26}$$

The error introduced by the data reduction is then given by

$$\Delta\mathbf{x} = \mathbf{x} - \hat{\mathbf{x}} \tag{22.27a}$$

or

$$\Delta \mathbf{x} = \sum_{j=1}^{N_m} y_j \mathbf{T_j} - \sum_{j=1}^{n} y_j \mathbf{T_j} - \sum_{j=n+1}^{N_m} c_j \mathbf{T_j}$$

$$= \sum_{j=n+1}^{N_m} (y_j - c_j) \mathbf{T_j} \tag{22.27b}$$

Now the mean square error, ε, is given by

$$\varepsilon = E\left[||\Delta \mathbf{x}||^{\mathbf{2}} \right] = E\left[\Delta \mathbf{x}' \cdot \Delta \mathbf{x} \right] \tag{22.28}$$

where E indicates the expected value. Thus,

$$\varepsilon = E\left[\sum_{j=n+1}^{N_m} \sum_{k=n+1}^{N_m} (y_j - c_j)(y_k - c_k) \mathbf{T_j}' \mathbf{T_k} \right] \tag{22.29}$$

From the orthogonality property, equation (22.29) reduces to

$$\varepsilon = \sum_{j=n+1}^{N_m} E\left[(y_j - c_j)^2 \right] \tag{22.30}$$

Now we want to select c_j and $\mathbf{T_j}$ to minimize ε.

Optimum c_j
The optimum value of c_j that minimizes ε is obtained by differentiating ε with respect to c_j:

$$\frac{\partial \varepsilon}{\partial c_j} = -2\left[E\{y_j\} - c_j \right] = 0$$

which gives

$$c_j = E\{y_j\} \tag{22.31}$$

This means that in estimating $\mathbf{x}$, the y_j's that are not measured should be replaced by their expected values. Now premultiplying equation (22.25b) by $\mathbf{T_j}'$ gives (again from the orthonormal condition)

$$y_j = \mathbf{T_j}' \mathbf{x} \tag{22.32}$$

which when substituted into equation (22.31) gives

$$c_j = E\{\mathbf{T_j}'\mathbf{x}\} = \mathbf{T_j}'E\{\mathbf{x}\}$$
$$= \mathbf{T_j}'\bar{\mathbf{x}} \tag{22.33}$$

Now let us rewrite equation (22.30) as

$$\varepsilon = \sum_{j=n+1}^{N_m} E\left[(y_j - c_j)(y_j - c_j)'\right] \tag{22.34}$$

If we now substitute from equations (22.32) and (22.33) for y_j and c_j, we get the minimum error as

$$\varepsilon_{\min} = \sum_{j=n+1}^{N_m} E\left[(\mathbf{T_j}'\mathbf{x} - \mathbf{T_j}'\bar{\mathbf{x}})(\mathbf{T_j}'\mathbf{x} - \mathbf{T_j}'\bar{\mathbf{x}})'\right]$$
$$= \sum_{j=n+1}^{N_m} \mathbf{T_j}'E\left[(\mathbf{x} - \bar{\mathbf{x}})(\mathbf{x} - \bar{\mathbf{x}})'\right]\mathbf{T_j}$$
$$= \sum_{j=n+1}^{N_m} \mathbf{T_j}'\Phi_{\mathbf{x}}\mathbf{T_j} \tag{22.35}$$

where

$$\Phi_{\mathbf{x}} = E\left[(\mathbf{x} - \bar{\mathbf{x}})(\mathbf{x} - \bar{\mathbf{x}})'\right] = \text{covariance matrix of } \mathbf{x}. \tag{22.36}$$

Optimum T_j

Due to the orthonormality condition, the process of minimizing ε with respect to $\mathbf{T_j}$ must be subject to the constraint

$$\mathbf{T_j}'\mathbf{T_j} = 1 \tag{22.37}$$

Lagrange multipliers must therefore be used, and the following function minimized:

$$\hat{\varepsilon} = \varepsilon - \sum_{j=n+1}^{N_m} (\mathbf{T_j}'\mathbf{T_j} - 1)\lambda_j \tag{22.38}$$

where λ_j is the Lagrange multiplier.

Now from equations (22.35) and (22.38), we get

$$\hat{\varepsilon} = \sum_{j=n+1}^{N_m} \left[\mathbf{T_j}'\Phi_{\mathbf{x}}\mathbf{T_j} - \lambda_j(\mathbf{T_j}'\mathbf{T_j} - 1)\right] \tag{22.39}$$

Differentiating $\hat{\varepsilon}$ with respect to $\mathbf{T_j}$ gives

$$\nabla_{\mathbf{T_j}}\hat{\varepsilon} = 2\Phi_{\mathbf{x}}\mathbf{T_j} - 2\lambda_j\mathbf{T_j} = 0 \tag{22.40}$$

$$\Rightarrow \Phi_{\mathbf{x}}\mathbf{T_j} = \lambda_j\mathbf{T_j} \tag{22.41}$$

Thus, $\mathbf{T_j}$ is an eigenvector of the covariance matrix $\Phi_{\mathbf{x}}$, and λ_j is the corresponding eigenvalue. The optimum transform with respect to the mean square error criterion is thus given by the eigenvectors of the covariance matrix of **x**. This is the Karhunen–Loeve transform.

When the system under consideration is a multiclass system where each class has a different *a priori* probability, the Karhunen–Loeve transform is then given by the overall covariance matrix

$$\Phi_{\mathbf{x}} = p_1\Phi_{\mathbf{x1}} + p_2\Phi_{\mathbf{x2}} + \ldots + p_s\Phi_{\mathbf{xs}} \tag{22.42}$$

where p_i is the *a priori* probability of class i. Here, we use the subscript i to indicate the class or state of a system. s is the number of classes.

One disadvantage of the Karhunen–Loeve transform is that there is no fast algorithm for it, since it is based on the covariance matrix of the data. Substituting equation (22.41) into (22.35) gives

$$\varepsilon_{\min} = \sum_{j=n+1}^{N_m} \lambda_j \tag{22.43}$$

Now the transform domain covariance matrix $\Phi_{\mathbf{y}}$ is obtained from the time domain covariance matrix as

$$\begin{aligned}\Phi_{\mathbf{y}} &= \mathbf{T}\Phi_{\mathbf{x}}\mathbf{T}^{-1} \\ &= \mathbf{T}\Phi_{\mathbf{x}}\mathbf{T}'\end{aligned} \tag{22.44}$$

And since $\mathbf{T}$ consists of the eigenvectors of $\Phi_{\mathbf{x}}$, $\Phi_{\mathbf{y}}$ has to be diagonal, with the diagonal elements comprising the eigenvalues of $\Phi_{\mathbf{x}}$, that is,

$$\Phi_{\mathbf{y}} = \text{diag}\,[\lambda_1, \lambda_2, \ldots, \lambda_{N_m}] \tag{22.45}$$

The fact that $\Phi_{\mathbf{y}}$ is diagonal indicates that the transform domain components, y_j, are uncorrelated.

Having determined the optimum transform that will enable the original data set to be reduced with respect to the mean square error, we need to have a criterion for selecting the reduced data set or features. One that is based on the preceding analysis is the variance criterion, which is briefly discussed in the following section.

22.2.2 Variance Criterion

From equation (22.43), it is evident that the mean square error will be minimum if the data components with the smallest eigenvalues are eliminated. But from equation (22.45), the eigenvalues are the diagonal elements of the transform domain covariance matrix, $\Phi_{\mathbf{y}}$. Thus, the eigenvalues are the variances of the transform components, y_j. The variance criterion then reduces to eliminating the $(N_m - n)$ feature components with the smallest variances.

There are a number of other data reduction criteria. A commonly used one is the class mean scatter criterion.

22.2.3 Class Mean Scatter Criterion

The class mean scatter criterion enables the data size to be reduced by selecting the features that minimize scatter of data within the individual classes, while at the same time increasing the scatter (or separation) between the classes. It involves first determining the feature mean for each class, $\bar{\mathbf{y}}_{\mathbf{i}}$:

$$\bar{\mathbf{y}}_{\mathbf{i}} = \frac{1}{M_i}\sum_{k=1}^{M_i} \mathbf{y}_{\mathbf{ik}} \qquad i = 1, \ldots, s \tag{22.46}$$

where M_i is the number of patterns in class s_i. In this and subsequent sections, the subscript i is used to indicate the class of the system. $\mathbf{y}_{\mathbf{ik}} \equiv \mathbf{y}$ is the kth pattern in class s_i.

The classes of interest in this case would normally include the signal sources, for example, good weld, cracking, porosity, no weld, and any other defects of interest. All other unwanted signals can then be grouped together and classified as noise.

The overall system mean, $\bar{\mathbf{y}}$, is then determined as

$$\bar{\mathbf{y}} = \sum_{i=1}^{s} p_i \bar{\mathbf{y}}_{\mathbf{i}} \tag{22.47}$$

where p_i is the *a priori* probability of class s_i.

The scatter within each class is obtained by calculating the covariance matrix $\Phi_{\mathbf{yi}}$ as

$$\Phi_{\mathbf{yi}} = \frac{1}{M_i}\sum_{k=1}^{M_i} (\mathbf{y}_{\mathbf{ik}} - \bar{\mathbf{y}}_{\mathbf{i}})(\mathbf{y}_{\mathbf{ik}} - \bar{\mathbf{y}}_{\mathbf{i}})' \tag{22.48}$$

which leads to an overall system covariance matrix $\Phi_{\mathbf{y}}$ of

$$\Phi_{\mathbf{y}} = \sum_{i=1}^{s} p_i \Phi_{\mathbf{yi}} \tag{22.49}$$

The scatter $\Phi_{\mathbf{ys}}$ between the individual classes is defined as

$$\Phi_{\mathbf{ys}} = \sum_{i=1}^{s} p_i(\bar{\mathbf{y}}_{\mathbf{i}} - \bar{\mathbf{y}})(\bar{\mathbf{y}}_{\mathbf{i}} - \bar{\mathbf{y}})' \tag{22.50}$$

and from this equation the feature selection criterion is defined as

$$\mathbf{Q} = \frac{\Phi_{\mathbf{ys}}(j, j)}{\Phi_{\mathbf{y}}(j, j)} \tag{22.51}$$

where $\Phi_{\mathbf{ys}}(j, j)$ and $\Phi_{\mathbf{y}}(j, j)$ are the jth diagonal elements of the covariance matrices $\Phi_{\mathbf{ys}}$ and $\Phi_{\mathbf{y}}$, respectively.

Since the objective is to minimize the scatter within the individual classes, while maximizing the scatter between the classes, the desired number of features with maximum $\mathbf{Q}$ values are selected as features. The number of features used is very important, since in addition to the computational efficiency, a high dimensionality requires a correspondingly large number of experimental data sets for training the system and developing the classifier. This is because the adequate number of training data sets has to be four or more times the number of features. However, an excessive number of training sets adds to the cost without necessarily improving the system performance. On the contrary, a small number of features may provide an insufficient description of the system.

Once the desired number of features is known, say n, the data sets can then be represented in an n-dimensional feature space to show their relationship to each other. For ease of graphical representation, however, two of the features may be plotted in a two-dimensional feature space (see, for example, Fig. 22.13).

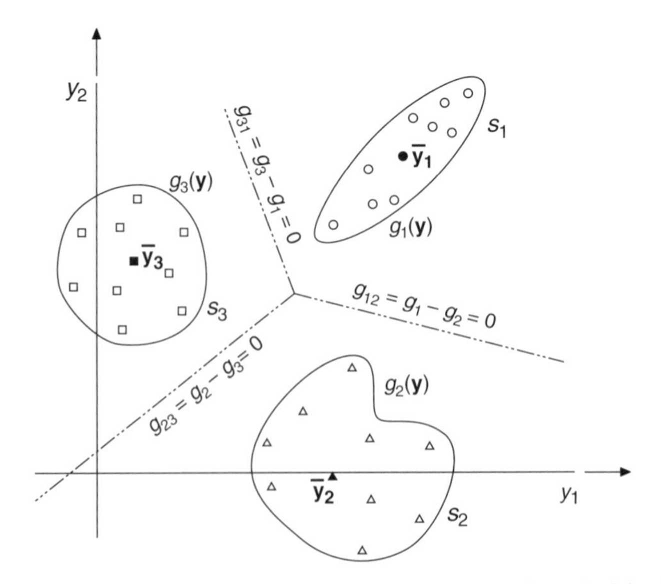

FIGURE 22.13 A two-dimensional, three-class feature space with linear decision boundaries.

Having reduced the data set to a manageable size without significant loss of information, the next step is to identify the source of the data or assign it to a specific class. This step is referred to as pattern classification and is discussed in the following section.

22.3 PATTERN CLASSIFICATION

There are several techniques available for classifying different patterns, but due to space limitations, our discussion focuses on two of the most common methods, pattern recognition and neural network analysis.

22.3.1 Pattern Recognition

The discussion on pattern recognition starts with the classical Bayes decision theory, which is a fundamental statistical approach to pattern classification.

22.3.1.1 Bayes Decision Theory The discussion on Bayes decision theory is approached using a specific example. Let us consider a process that can be in one of the s ($s \geq 2$) mutually exclusive states or classes and is monitored at some regular intervals. For $s = 2$, one of the states may be considered to be in control, while the other is out of control. For example, in laser welding, let s_1 indicate the state that the process is in good condition, while s_2 represents a defective weld. Further, let each class or state of the process have an *a priori* probability p_i (*a priori* probability of s_1 is p_1 and that of s_2 is p_2), such that

$$p_1 + p_2 = 1 \tag{22.52}$$

The *a priori* probability of a given state is obtained from previous knowledge of the likelihood that the next observation will belong to that particular state.

Now let the process be monitored by measurements of n-dimensional vectors of observations of an output, say the acoustic emission signal power spectrum, $\mathbf{y}$. Since we cannot at any time predict with certainty what the next observation will be, $\mathbf{y}$ is considered to be a random variable with a distribution that depends on the state. Furthermore, let the class-conditional probability density function for $\mathbf{y}$ be $p(\mathbf{y}|s_i)$. This is the probability density function for the observation $\mathbf{y}$ given that the state is s_i (Fig. 22.14).

The problem that we are concerned with is to determine, for any observation, $\mathbf{y}$, that is made, which one of the s mutually exclusive states has generated $\mathbf{y}$, that is, whether the observed $\mathbf{y}$ is from a good weld or a defective weld. For this, we first take a look at the Bayes Rule.

The Bayes rule expresses the *a posteriori* probability $p(s_i|\mathbf{y})$ (see Fig. 22.15) in terms of the *a priori* probability p_i and the class-conditional probability density,

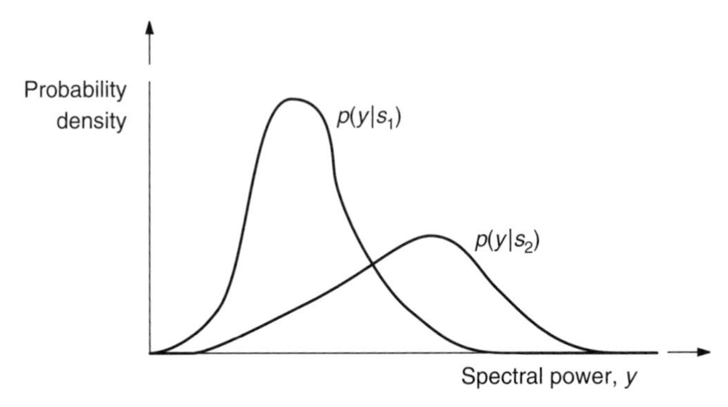

FIGURE 22.14 Class-conditional probability density functions for two hypothetical classes. (From Duda, R. O., and Hart, P. E., 1973, *Pattern Classification and Scene Analysis*. Reprinted with permission from John Wiley & Sons, Inc.)

$p(\mathbf{y}|s_i)$, that is,

$$p(s_i|\mathbf{y}) = \frac{p(\mathbf{y}|s_i)p_i}{p(\mathbf{y})} \tag{22.53}$$

where $p(\mathbf{y}) = \sum_{i=1}^{2} p(\mathbf{y}|s_i)p_i$ for the two-state example being considered.

22.3.1.2 Bayes Decision Rule for Minimum Error With the *a posteriori* probabilities determined using the Bayes rule, they can now be used to classify an observed signal in such a way as to minimize classification error. Now suppose that at some point in the welding process a single N_m-dimensional observation, $\mathbf{x}' = [x_1, x_2, x_3, ..., x_{N_m}]$, is made of say an acoustic emission signal that after transformation and data reduction becomes $\mathbf{y}' = [y_1, y_2, y_3, ..., y_n]$. Let us further suppose we are interested in determining whether the signal is from a good weld (1) or a defective weld (2), that is,

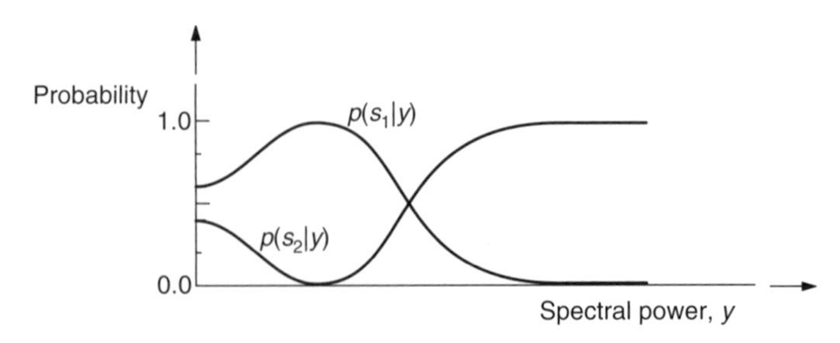

FIGURE 22.15 *A posteriori* probabilities for two hypothetical classes. (From Duda, R. O., and Hart, P. E., 1973, *Pattern Classification and Scene Analysis*. Reprinted with permission from John Wiley & Sons, Inc.)

whether the observation belongs to s_1 or s_2. If the observation $\mathbf{y}$ is such that

$$p(s_1|\mathbf{y}) > p(s_2|\mathbf{y}) \tag{22.54}$$

then we assign $\mathbf{y}$ to s_1. This is the Bayes decision rule for minimizing the probability of error. The probability of error associated with a decision is given by

$$p(\text{error}|\mathbf{y}) = \begin{cases} p(s_1|\mathbf{y}) & \text{if } \mathbf{y} \text{ is assigned to } s_2 \\ p(s_2|\mathbf{y}) & \text{if } \mathbf{y} \text{ is assigned to } s_1 \end{cases} \tag{22.55}$$

Using equation (22.53), the decision rule can also be expressed as

$$p(\mathbf{y}|s_1)p_1 > p(\mathbf{y}|s_2)p_2 \Rightarrow \mathbf{y} \in s_1 \tag{22.56a}$$

or

$$\text{if } \frac{p(\mathbf{y}|s_1)}{p(\mathbf{y}|s_2)} > \frac{p_2}{p_1}, \text{ then } \mathbf{y} \in s_1 \tag{22.56b}$$

$p(\mathbf{y}|s_i)$ is the likelihood of s_i with respect to $\mathbf{y}$. And

$$L_R = \frac{p(\mathbf{y}|s_1)}{p(\mathbf{y}|s_2)} \tag{22.56c}$$

is the likelihood ratio. Thus, the Bayes minimum error decision rule decides s_1 if the likelihood ratio exceeds a threshold value, which is independent of $\mathbf{y}$.

22.3.1.3 Discriminant Function Analysis The classification approach presented in the preceding section can be taken one step further by noting that for a multiclass system, the decision rules may be expressed as functions, discriminant functions, $g_i(\mathbf{y})$, with each class or state being associated with one function:

$$g_i(\mathbf{y}), \quad i = 1, 2, ..., s \tag{22.57}$$

The decision rule then becomes

$$\mathbf{y} \in s_i \text{ if } g_i(\mathbf{y}) > g_j(\mathbf{y}) \text{ for all } j \neq i \tag{22.58}$$

This may be represented schematically as in Fig. 22.13. For minimum error classification, the discriminant function may be the *a posteriori* probability, $p(s_i|\mathbf{y})$, that is,

$$g_i(\mathbf{y}) = p(s_i|\mathbf{y}) \tag{22.59}$$

Without loss of generality, we replace the function by the logarithm of the *a posteriori* probability, giving

$$g_i(\mathbf{y}) = \log p(s_i|\mathbf{y}) = \log[p(\mathbf{y}|s_i)p_i] \tag{22.60a}$$

or

$$g_i(\mathbf{y}) = \log p(\mathbf{y}|s_i) + \log p_i \tag{22.60b}$$

The decision rule divides the feature space into decision regions, Ω_i. Contiguous decision regions are bounded by decision boundaries such that

$$g_i(\mathbf{y}) = g_j(\mathbf{y}) \tag{22.61}$$

From equations (22.56) and (22.60), it is evident that the probability density function is necessary for classification. For simplicity, let us consider the case where the system being monitored has a Gaussian distribution. Then

$$p(\mathbf{y}|s_i) = G(\bar{\mathbf{y}}_\mathbf{i}, \Phi_{\mathbf{yi}}) = \frac{1}{(2\pi)^{n/2}|\Phi_{\mathbf{yi}}|^{1/2}} \exp\left[-\frac{1}{2}(\mathbf{y} - \bar{\mathbf{y}}_\mathbf{i})'\Phi_{\mathbf{yi}}{}^{-1}(\mathbf{y} - \bar{\mathbf{y}}_\mathbf{i})\right] \tag{22.62}$$

where $|\Phi_{\mathbf{yi}}| = |E\left[(\mathbf{y} - \bar{\mathbf{y}}_\mathbf{i})(\mathbf{y} - \bar{\mathbf{y}}_\mathbf{i})'\right]|$ is the determinant of the covariance matrix, $\bar{\mathbf{y}}_\mathbf{i} = E[\mathbf{y}_\mathbf{i}]$, and $\mathbf{y}$ is an n-dimensional vector.

Substituting equation (22.62) into (22.60) gives

$$g_i(\mathbf{y}) = -\frac{1}{2}(\mathbf{y} - \bar{\mathbf{y}}_\mathbf{i})'\Phi_{\mathbf{yi}}{}^{-1}(\mathbf{y} - \bar{\mathbf{y}}_\mathbf{i}) - \frac{n}{2}\log 2\pi - \frac{1}{2}\log|\Phi_{\mathbf{yi}}| + \log p_i \tag{22.63}$$

This is the general form of the discriminant function for a process with normal distribution.

22.3.1.4 Minimum-Distance Classifier The type of classifier that results from equation (22.63) when the features are uncorrelated, that is, statistically independent with the same variance, σ^2, for each feature and each class, that is,

$$\Phi_{\mathbf{yi}} = \sigma^2\mathbf{I} = \text{diagonal covariance matrix} \tag{22.64}$$

and also when the *a priori* probabilities are equal is the minimum-distance classifier.

Even for the slightly more general case where the *a priori* probabilities are not equal, the samples for the individual classes fall in hyperspherical clusters of equal size, with each cluster of the class being centered about its mean, $\bar{\mathbf{y}}_\mathbf{i}$ (Fig. 22.16). All the common terms in equation (22.63) can then be neglected, reducing the

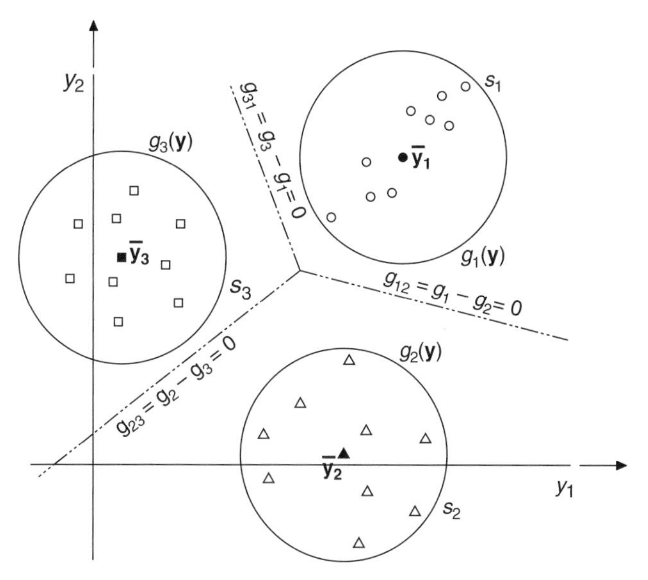

FIGURE 22.16 Decision boundaries for a minimum-distance classifier. (From Duda, R. O., and Hart, P. E., 1973, *Pattern Classification and Scene Analysis*. Reprinted with permission from John Wiley & Sons, Inc.)

discriminant function, $g_i(\mathbf{y})$, to the form

$$g_i(\mathbf{y}) = -\frac{||\mathbf{y} - \bar{\mathbf{y}}_\mathbf{i}||^2}{2\sigma^2} + \log p_i \tag{22.65}$$

with the Euclidean norm squared $||\mathbf{y} - \bar{\mathbf{y}}_\mathbf{i}||^2$ being

$$||\mathbf{y} - \bar{\mathbf{y}}_\mathbf{i}||^2 = (\mathbf{y} - \bar{\mathbf{y}}_\mathbf{i})'(\mathbf{y} - \bar{\mathbf{y}}_\mathbf{i}) \tag{22.66}$$

When equation (22.65) is expanded, it becomes

$$g_i(\mathbf{y}) = -\frac{1}{2\sigma^2}\left(\mathbf{y}'\mathbf{y} - 2\bar{\mathbf{y}}_\mathbf{i}{}'\mathbf{y} + \bar{\mathbf{y}}_\mathbf{i}{}'\bar{\mathbf{y}}_\mathbf{i}\right) + \log p_i \tag{22.67}$$

Since for any observation, that is, incoming signal, $\mathbf{y}$ is the same for all classes, the first term in equation (22.67) can be neglected, reducing it to

$$g_i(\mathbf{y}) = \mathbf{w}_\mathbf{i}{}'\mathbf{y} + w_{i0} \tag{22.68}$$

where $\mathbf{w}_\mathbf{i}{}' = \frac{1}{\sigma^2}\bar{\mathbf{y}}_\mathbf{i}{}'$ is the weighting coefficient vector and $w_{i0} =$ threshold $= -\frac{1}{2\sigma^2}\bar{\mathbf{y}}_\mathbf{i}{}'\bar{\mathbf{y}}_\mathbf{i} + \log p_i$.

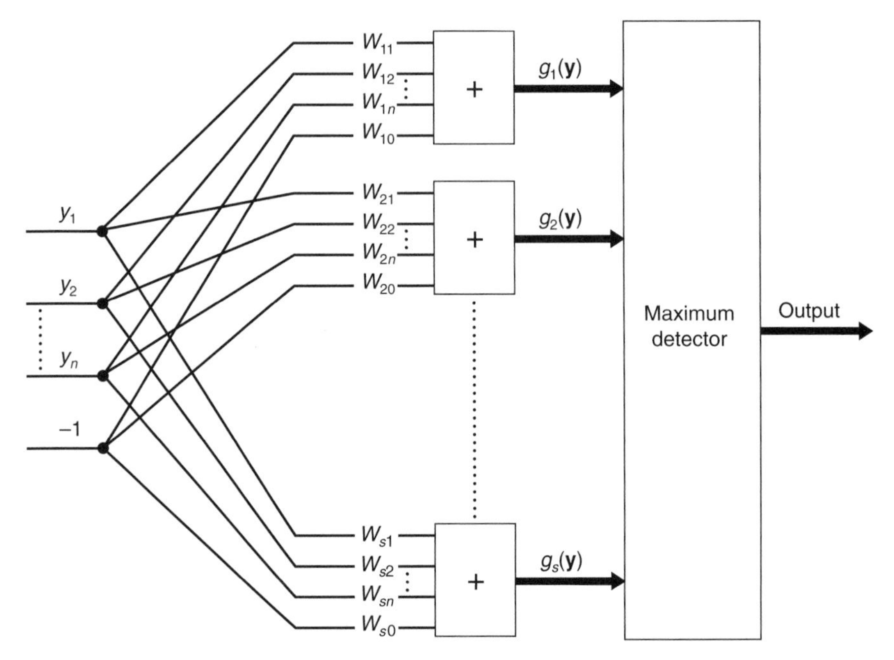

FIGURE 22.17 An n-dimensional s-class pattern classifier.

For each sampled signal, the classifier calculates the value of each discriminant function, $g_i(\mathbf{y})$, and assigns the signal to the class with the maximum discriminant function value (Fig. 22.17).

If the *a priori* probabilities, p_i, are equal for each of the classes, then equation (22.65) reduces to the norm squared. The classification process then involves computing the distance of the observation $\mathbf{y}$ from the mean $\bar{\mathbf{y}}_\mathbf{i}$ of each class and assigning $\mathbf{y}$ to the class mean to which it is closest, that is, the one with the minimum value of equation (22.66). This is the minimum-distance classifier. In this case, we need to only compute

$$g_i(\mathbf{y}) = \bar{\mathbf{y}}_\mathbf{i}{}'\mathbf{y} - \frac{1}{2}||\bar{\mathbf{y}}_\mathbf{i}||^2 \quad i = 1, 2, ..., s \tag{22.69}$$

Since the norm in equation (22.66) is minimum when equation (22.69) is maximum, equation (22.69) can also be used as a discriminant function. The classification process then involves assigning $\mathbf{y}$ to the class with the maximum discriminant function.

Equations (22.68) and (22.69) are linear and therefore termed linear discriminant functions. The decision boundary between two contiguous classes is then a hyperplane given by

$$g_i(\mathbf{y}) = g_j(\mathbf{y}) \tag{22.70}$$

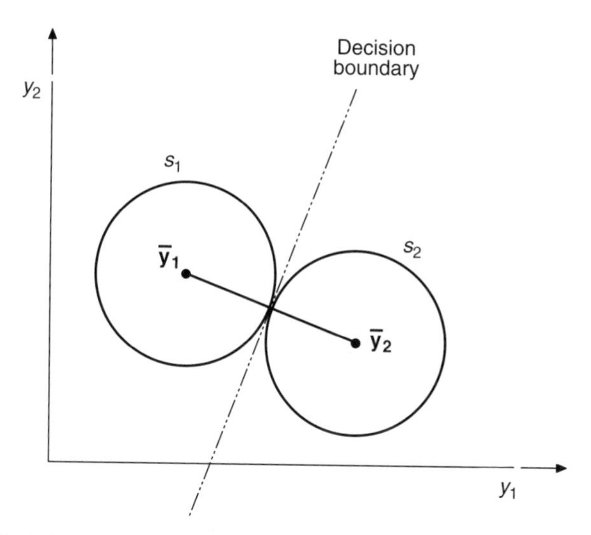

FIGURE 22.18 Minimum-distance classifier with equal *a priori* probabilities. (From Duda, R. O., and Hart, P. E., 1973, *Pattern Classification and Scene Analysis*. Reprinted with permission from John Wiley & Sons, Inc.)

and is orthogonal to the line joining the means. If the *a priori* probabilities are equal, then the hyperplane

$$g_i(\mathbf{y}) - g_j(\mathbf{y}) = 0 \tag{22.71}$$

intersects the line joining the means at the midpoint (Fig. 22.18).

22.3.1.5 General Linear Discriminant Function Let us now consider the situation where the features are not necessarily independent, but the covariance matrices for the individual classes are equal:

$$\Phi_{\mathbf{yi}} = \Phi_{\mathbf{y}} \tag{22.72}$$

The discriminant function, equation (22.63), then becomes

$$g_i(\mathbf{y}) = -\frac{1}{2}(\mathbf{y} - \bar{\mathbf{y}}_\mathbf{i})' {\Phi_\mathbf{y}}^{-1}(\mathbf{y} - \bar{\mathbf{y}}_\mathbf{i}) + \log p_i \tag{22.73}$$

Since the covariance matrices are equal, the quadratic term $\mathbf{y}' {\Phi_\mathbf{y}}^{-1}\mathbf{y}$ is the same for all classes. The discriminant function then reduces to

$$g_i(\mathbf{y}) = \mathbf{w_i}'\mathbf{y} + w_{i0} \tag{22.74}$$

where $\mathbf{w_i} = {\Phi_\mathbf{y}}^{-1}\bar{\mathbf{y}}_\mathbf{i}$ and $w_{i0} = -\frac{1}{2}\bar{\mathbf{y}}_\mathbf{i}' {\Phi_\mathbf{y}}^{-1}\bar{\mathbf{y}}_\mathbf{i} + \log p_i$.

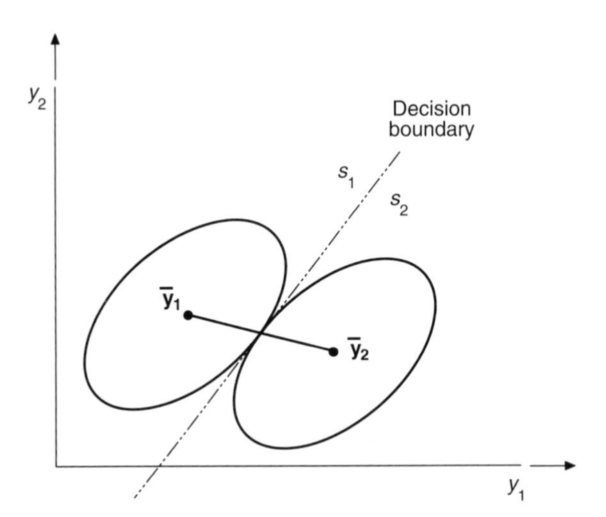

FIGURE 22.19 Decision boundary for a minimum Mahalanobis distance classifier. (From Duda, R. O., and Hart, P. E., 1973, *Pattern Classification and Scene Analysis*. Reprinted with permission from John Wiley & Sons, Inc.)

The resulting discriminant function is also linear. The individual classes in this case form hyperellipsoids of equal size and shape, with each cluster being centered about its mean, $\bar{\mathbf{y}}_\mathbf{i}$, Fig. 22.19. If the *a priori* probabilities are equal, then the classification process involves measuring, in essence, the squared Mahalanobis distance:

$$d^2 = (\mathbf{y} - \bar{\mathbf{y}}_\mathbf{i})' \Phi_\mathbf{y}{}^{-1} (\mathbf{y} - \bar{\mathbf{y}}_\mathbf{i}) \tag{22.75}$$

Again, the decision boundaries are hyperplanes but are generally not orthogonal to the line joining the means.

22.3.1.6 Quadratic Discriminant Function When the covariance matrix is different for each class, the quadratic term in equation (22.63) differs for each class and thus does not drop out. The resulting discriminant function is then quadratic and is given by

$$g_i(\mathbf{y}) = \mathbf{y}' \mathbf{R}_\mathbf{i} \mathbf{y} + \mathbf{w}_\mathbf{i}{}' \mathbf{y} + w_{io} \tag{22.76}$$

where $\mathbf{R}_\mathbf{i} = -\frac{1}{2} \Phi_\mathbf{yi}{}^{-1}$, $\mathbf{w}_\mathbf{i} = \Phi_\mathbf{yi}{}^{-1} \bar{\mathbf{y}}_\mathbf{i}$, and $w_{i0} = -\frac{1}{2} \bar{\mathbf{y}}_\mathbf{i}{}' \Phi_\mathbf{yi}{}^{-1} \bar{\mathbf{y}}_\mathbf{i} - \frac{1}{2} \log |\Phi_\mathbf{yi}| + \log p_i$.

22.3.1.7 Least-Squares Minimum Distance Classification Each data set (collectively, the patterns) that is analyzed can be plotted in a feature space using the individual feature components as coordinates. In the preceding classification schemes, it is assumed that the patterns in the feature space are clustered around their respective means, $\bar{\mathbf{y}}_{\mathbf{i}}$. However, this is not generally the case. Quite often, there is an overlap between the various classes. Thus, with a basic linear decision-making process, there can be a substantial amount of misclassification, where signals belonging to one class are erroneously assigned to another. To facilitate clustering of the patterns around their respective means, the patterns may be transformed from the feature space into a decision space. The objective is to map patterns of each class s_i into or to cluster around a predetermined point, $\mathbf{Q_i}$, $i = 1, 2, ..., s$ associated with that class.

Thus, for each class, s_i, a point, $\mathbf{Q_i}$, is defined in the decision space. An attempt is then made to transform all patterns belonging to s_i to cluster around $\mathbf{Q_i}$. Since there is still the likelihood of some misclassification even after this step, the transformation is made in such a way as to minimize the decision-making errors that will subsequently result. The transformation matrix, $\mathbf{T}$, is thus selected on the basis of the minimum mean square error criterion.

Let us indicate the n-dimensional data set for class s_i in the feature space as $\mathbf{y_{ij}}$, $j = 1, 2, ..., M_i$. M_i is the number of data sets in class s_i. Further, let this data set be mapped into the predetermined point $\mathbf{Q_i}' = [q_1, q_2, ..., q_s]$ in the decision space. Also, let the transformed point in the decision space be indicated by $\mathbf{z_{ij}}$:

$$\mathbf{z_{ij}} = \mathbf{T}\mathbf{y_{ij}} \tag{22.77a}$$

Since $\mathbf{z_{ij}}$ will not necessarily be located exactly at $\mathbf{Q_i}$, there will generally be an error ε_{ij}, associated with the transformation of each pattern or data $\mathbf{y_{ij}}$. This is given by

$$\varepsilon_{ij} = \mathbf{z_{ij}} - \mathbf{Q_i} = \mathbf{T}\mathbf{y_{ij}} - \mathbf{Q_i} \tag{22.77b}$$

The total mean square error, ε_i, associated with the transformation for class s_i is then

$$\varepsilon_i = \frac{1}{M_i}\sum_{j=1}^{M_i} ||\varepsilon_{ij}||^2 \tag{22.78}$$

Substituting from equation (22.77b), we get

$$\varepsilon_i = \frac{1}{M_i}\sum_{j=1}^{M_i}\left[\mathbf{y_{ij}}'\mathbf{T}'\mathbf{T}\mathbf{y_{ij}} - 2\mathbf{y_{ij}}'\mathbf{T}'\mathbf{Q_i} + ||\mathbf{Q_i}||^2\right] \tag{22.79}$$

To minimize ε_i with respect to $\mathbf{T}$, we find the gradient

$$\nabla_{\mathbf{T}}\varepsilon_i = 0$$

which gives

$$\frac{1}{M_i}\sum_{j=1}^{M_i}\left[\nabla_{\mathbf{T}}\left(\mathbf{y_{ij}}'\mathbf{T}'\mathbf{T}\mathbf{y_{ij}}\right) - \nabla_{\mathbf{T}}\left(2\mathbf{y_{ij}}'\mathbf{T}'\mathbf{Q_i}\right) + \nabla_{\mathbf{T}}\left(||\mathbf{Q_i}||^2\right)\right] = 0 \tag{22.80}$$

Equation (22.80) reduces to

$$\frac{1}{M_i}\sum_{j=1}^{M_i}\left[2\mathbf{T}\left(\mathbf{y_{ij}}\mathbf{y_{ij}}'\right) - \left(2\mathbf{Q_i}\mathbf{y_{ij}}'\right) + 0\right] = 0 \tag{22.81a}$$

or

$$\mathbf{T}\sum_{j=1}^{M_i}\left(\mathbf{y_{ij}}\mathbf{y_{ij}}'\right) = \sum_{j=1}^{M_i}\mathbf{Q_i}\mathbf{y_{ij}}' \tag{22.81b}$$

which gives the transformation matrix as

$$\mathbf{T} = \left[\sum_{j=1}^{M_i}\mathbf{Q_i}\mathbf{y_{ij}}'\right]\left[\sum_{j=1}^{M_i}\left(\mathbf{y_{ij}}\mathbf{y_{ij}}'\right)\right]^{-1} \tag{22.82}$$

In the general case, where each of the classes has a different *a priori* probability p_i, the transformation matrix is given by

$$\mathbf{T} = \mathbf{G_{Qy}}\mathbf{G_{yy}}^{-1} \tag{22.83}$$

where

$$\mathbf{G_{Qy}} = \sum_{i=1}^{s}\sum_{j=1}^{M_i}\frac{p_i}{M_i}\left(\mathbf{Q_i}\mathbf{y_{ij}}'\right) = E\left(\mathbf{Qy}'\right) = \text{cross-correlation matrix.}$$

$$\mathbf{G_{yy}} = \sum_{i=1}^{s}\sum_{j=1}^{M_i}\frac{p_i}{M_i}\left(\mathbf{y_{ij}}\mathbf{y_{ij}}'\right) = E\left(\mathbf{yy}'\right) = \text{autocorrelation matrix.}$$

The transformation matrix $\mathbf{T}$ obtained in this manner is an $s \times n$ matrix.

After transforming the signals into the decision space, appropriate discriminant functions can be developed for classifying the signals. To facilitate representation of the discriminant function in matrix form, the dimensionality of the pattern space is increased by 1 to $(n + 1)$-dimensions to account for the threshold in the discriminant function. Thus, $\mathbf{y}$ becomes $\hat{\mathbf{y}}$, where $\hat{\mathbf{y}}' = [y_1, y_2, ..., y_n, -1]$. In this form, the transformation matrix $\mathbf{T}$ is an $s \times (n + 1)$ matrix.

Applying minimum-distance classification to the patterns in the decision space, we determine the distance of the transformed point, $\mathbf{z_{ij}}$, from the preselected point, $\mathbf{Q_i}$, and assign $\mathbf{z_{ij}}$ to the class of the $\mathbf{Q_i}$ to which it is closest. Thus, we have for our discriminant function,

$$d_i^2 = ||\mathbf{z_{ij}} - \mathbf{Q_i}||^2 = ||\mathbf{z_{ij}}||^2 - 2\mathbf{Q_i}'\mathbf{z_{ij}} + ||\mathbf{Q_i}||^2 \quad i = 1, 2, ..., s \tag{22.84}$$

The resulting classifier is then known as the least-squares minimum distance classifier. To simplify the analysis, $\mathbf{Q_i}$ can be chosen, without loss of generality, as unit vectors. Thus,

$$||\mathbf{Q_i}||^2 = 1$$

and since $\mathbf{z_{ij}}$ is the same for all classes, the classification reduces to determining $\mathbf{Q_i}'\mathbf{z_{ij}}$, which is a maximum when d_i^2 is minimum. The discriminant function then becomes

$$g_i(\mathbf{y}) = \mathbf{Q_i}'\mathbf{z_{ij}} = \mathbf{Q_i}'\mathbf{T}\mathbf{y_{ij}} \tag{22.85}$$

and the class with the maximum discriminant function value is selected. As an illustration, for a three-class system, we have

$$\begin{bmatrix} g_1 \\ g_2 \\ g_3 \end{bmatrix} = \begin{bmatrix} 1 & 0 & 0 \\ 0 & 1 & 0 \\ 0 & 0 & 1 \end{bmatrix} \mathbf{Ty} = \mathbf{Ty} \tag{22.86}$$

where

$$\mathbf{T} = \begin{bmatrix} w_{11} & w_{12} & \cdots & w_{1n} & w_{10} \\ w_{21} & w_{22} & \cdots & w_{2n} & w_{20} \\ w_{31} & w_{32} & \cdots & w_{3n} & w_{30} \end{bmatrix}$$

This gives the discriminant function

$$g_i(\mathbf{y}) = w_{i1}y_1 + w_{i2}y_2 + \cdots + w_{in}y_n + w_{io} \quad i = 1, 2, \ldots, s \tag{22.87}$$

where w_{ik} is the weighting coefficient of the kth feature for class s_i and w_{i0} is the threshold for class s_i.

22.3.1.8 System Training To implement the classifier, it is necessary to determine the elements of the transformation matrix **T** or the weighting coefficients w_{ij} of the discriminant function, which for the linear case are given by equation (22.87). This requires training of the system. For that, data are generated under carefully controlled conditions that produce signals belonging to a particular class without any interference from signals of other classes. For example, if the application is such that cracks can be detrimental to the product, then one may be interested in distinguishing between, say, a sound weld and one with cracks.

In training such a system, then, it is essential to design experiments that will result in crack formation, without any other discontinuities being formed. The data obtained from such an experiment will then be used to define the discriminant function for the class associated with crack formation. A similar set of experiments is then conducted for a sound weld, and the data obtained are used to define the discriminant function for the class related to sound welds. It has to be emphasized that such training will have to be repeated for different conditions.

From this data, the mean, $\bar{\mathbf{y}}_{\mathbf{i}}$, and covariance matrix, $\Phi_{\mathbf{yi}}$, for each class is then obtained and substituted in equation (22.68), (22.74), (22.76), or (22.83) to give the corresponding discriminant function for each class.

The pattern recognition system is essentially a single-level network. Further improvements in classification may yet be obtained by using a multilayer network or a neural network. The neural network concept is briefly discussed in the next section.

22.3.2 Neural Network Analysis

Neural networks are based on a parallel processing architecture where the various processors or nodes are interconnected. The information available in the network is contained in the magnitude of the links between the processors. In the general case, each node may be connected to every other node. However, for computational efficiency, a simpler structure involving layers of nodes as illustrated in Fig. 22.20 is often used. In this case, each node of a hidden layer (layers other than the input and output layers) is connected to all nodes in the adjacent layers. It receives input from nodes of the preceding layer, and its output becomes part input to each node of the succeeding layer. The performance of the network depends on the nature of the process being modeled or monitored, the number of nodes used, and the number of layers, among other things. This simpler structure provides for efficient computation, and a more complex structure does not necessarily produce better results.

If we denote the input to the mth node of the kth layer by $I(m, k)$, and the output of that same node by $O(m, k)$, then the input $I(m, k)$ is given as

$$I(m, k) = \sum_{n} [q(m, n, k) \times O(n, k-1)] + r(m, k) \tag{22.88}$$

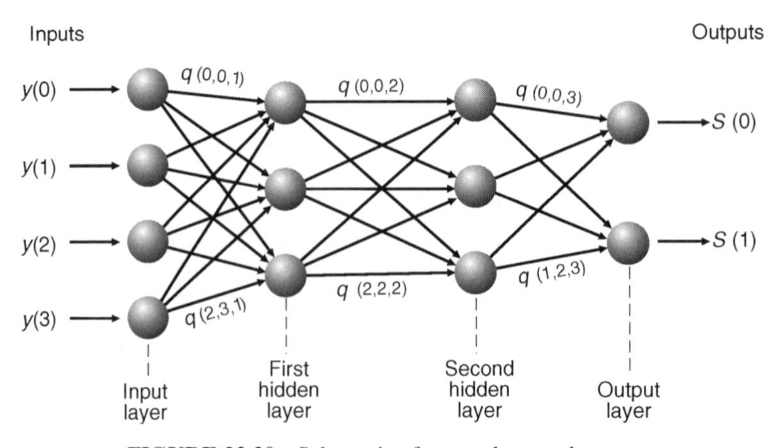

FIGURE 22.20 Schematic of a neural network system.

where $q(m, n, k)$ is the coefficient or magnitude of the link from node n in layer $k - 1$ to node m in layer k and $r(m, k)$ is the residue or threshold of node m in layer k.

The inputs to the overall network are denoted as $y(m)$ and the outputs as $s(m)$.

The network is linear when the input to each node is also taken as its output. For a useful network of this form, the input set has to be linearly independent, in which case the network is essentially equivalent to a two-layer network; that is, it can be shown to reduce to a two-layer network. Thus, a nonlinear relationship between the input and output of a node is essential for enhanced performance.

In the general case, the output $O(m, k)$ of each node is a specific function $f\{I(m, k)\}$ of the node input. A number of functions have been investigated that relate the input and output of a node. One that is commonly used is the monotonically increasing sigmoid function

$$O(m, k) = \frac{1}{1 + \exp[-I(m, k)]} \tag{22.89}$$

This output serves as part input to each node of the $(k + 1)$th layer, along with all other outputs of the kth layer. This structure is characterized as a feedforward neural network, and signals can propagate through the network only between adjacent layers, and only in the forward direction. From one layer to another, the signals are modified by the link magnitudes. The basic structure of the network thus consists of three stages, namely, the input, hidden layers, and the output, and the input signal propagates through the hidden layers in sequence, being modified at each stage, and finally coming out as the output. The hidden layers process the input information in a way that simulates the system of interest, thereby resulting in an output that is characteristic of the system input. The simplest structure would involve direct correspondence between the inputs and outputs. However, such simple systems are not as effective as those incorporating hidden layers, since the former are not capable of capturing the intricate details of the system. On the contrary, an unnecessarily large number of hidden layers may increase the number of iterations required to identify the network coefficients (magnitudes) without necessarily improving the performance.

Determination of the link coefficients or magnitudes to enable it to perform the mapping of the inputs to the outputs requires training of the network. This is accomplished by exposing the system to well-defined states of the process, say, known outputs corresponding to known inputs. This is repeated for various known input–output sets, and the total resulting error is determined. The link magnitudes are then modified, and the entire process repeated until an error criterion is satisfied. Once the system is subjected to the learning process and the information base established, it can be used for real-time application, even though the training itself is normally off-line. This is essentially similar to the training of the pattern recognition system.

Several methodologies have been developed for training multilayered neural networks. A common one is the back propagation method. Random values are initially assigned to the link coefficients, $q(m, n, k)$, and residues, $r(m, k)$, to define the network (typically in the range of $+1$ to -1). The weights must not start off with equal values, since it is generally not possible to reach a configuration of unequal weights under such circumstances. One set of inputs is then presented to the network and the outputs generated by the network compared with the corresponding actual or desired outputs. Due to the randomness of the initial network definition, the two sets of outputs will generally be different. The difference or error of each output is then determined. This is repeated for all the input sets available for training. The maximum sum of squares, ε, of the resulting errors is then determined, namely,

$$\varepsilon = \text{Max}_{i=1,\ldots,N_{io}} \sum_{m=1}^{M_o} [R(m,i) - s(m,i)]^2 \tag{22.90}$$

where $R(m,i)$ is the actual or desired mth output of data set i, $s(m,i)$ is the mth network output of data set i, N_{io} is the number of input–output training sets available, and M_o is the number of outputs available in a set.

This error measure is compared with a predetermined error threshold, ε_t, which serves as a convergence limit and if the measured error is less than the threshold, then the process is terminated. Otherwise, the network link coefficients are recalculated or updated iteratively by back propagation. In the back propagation method, the link coefficients preceding each node are determined as follows:

$$q(m,n,k)^{\{j+1\}} = q(m,n,k)^{\{j\}} + \alpha_q \beta(m,k) O(n,k-1) \tag{22.91a}$$

$$r(m,k)^{\{j+1\}} = r(m,k)^{\{j\}} - \alpha_r \beta(m,k) \tag{22.91b}$$

where α_i is the correction gain or learning rate that is positive and usually less than 1. $\beta(m,k)$ is the correction factor or error term for node m. The superscript $j+1$ indicates the next or updated link coefficient, while j represents the current one.

The learning rate α_i can significantly affect the performance of the network. Using a small value for α_i (typically between 0.05 and 0.25) ensures that a solution will be reached. However, that requires a large number of iterations. The normal approach is to start with a small value of α_i and increasing its value as learning proceeds. The higher value will increase the convergence speed. On the contrary, if the value is too high, that may cause the algorithm to oscillate around the minimum.

Convergence can be enhanced by adding a momentum term to smoothen weight changes:

$$\begin{aligned} q(m,n,k)^{\{j+1\}} &= q(m,n,k)^{\{j\}} + \alpha_q \beta(m,k) O(n,k-1) \\ &\quad + \gamma_q \left[q(m,n,k)^{\{j\}} - q(m,n,k)^{\{j-1\}} \right] \quad 0 < \gamma_q < 1 \end{aligned} \tag{22.92}$$

The goal of the momentum term is to keep the weight changes moving in the same direction. A typical value of the momentum parameter γ_q will be 0.9.

The correction factors are generally given by

$$\beta(m,k) = O(m,k)[1 - O(m,k)] \sum_k \sum_h q(h,m,k+1)\beta(h,k+1) \tag{22.93a}$$

where k is over all layers above the current layer, and h is over all nodes in those layers.

Those for the last or output layer are given by

$$\beta(m) = s(m)[R(m) - s(m)][1 - s(m)] \tag{22.93b}$$

Once the training is completed, that is, the maximum of the sum of squared errors converges to the threshold error, ε_t, the network is defined for the process in question such that for any process input that is presented to the network, the output predicted by the network should correspond to the actual process output.

22.3.3 Sensor Fusion

Whereas most sensors may be suitable for monitoring specific states or detecting specific defects associated with a process, no single sensor can reliably monitor all aspects of the process. Defects that may not be detectable to one sensor under a given set of processing conditions may be detectable to others. Thus, by combining multiple sensors, the advantages of each sensor are pooled to form a more reliable system. The drawback, however, is the cost and complexity of the multisensor system. Sensor fusion techniques can be categorized into three broad groups:

1. *Low Level or Data Fusion*: Where the original data from the different sensors is combined before features are extracted using any of the feature extraction methods described in Sections 22.1 and 22.2. This type of fusion is suitable for sensors that are identical or whose output data are comparable or can be merged, for example, infrared and ultraviolet output signals. The advantage with this approach is the minimum loss of information.
2. *Medium Level or Feature Fusion*: Where selected features may come from one or more sensors. The data from each sensor is first transformed into the feature space. The features from all the sensors are then concatenated into a single feature vector, before feature selection, and then classification. Since features that are orthogonal in one sensor set (due to the orthogonality of the FFT) may not be orthogonal to features in another sensor set, the resulting features may need to be processed by an orthogonal transform before pattern classification. The advantage with this approach is that it permits fusion of data from sensors that are not commensurate.
3. *High Level or Decision Fusion*: This invokes a voting system, with the data from each sensor being processed independently until a decision is made. The decisions from each sensor are then weighted and then combined to give the overall results through voting.

Whenever possible, it is generally recommended that the lowest level of fusion possible should be used, since it results in minimal loss of information. Hybrids or combinations of the different fusion methods may also be used.

22.3.4 Time–Frequency Analysis

The Fourier transform, as defined in Section 22.1.1, requires knowledge of the signal over the infinite range, that is, from $-\infty$ to $+\infty$. This mathematical expression represents the real physical situation only for specific cases, such as steady-state or stationary signals. For transients, this represents a distortion of the physical situation, even though it is mathematically correct since the results of the analysis are such that the signal vanishes outside its domain. Furthermore, it does not properly represent localized information, for example, spikes and high-frequency bursts. Such anomalies are addressed using time–frequency analysis, which is suitable for nonstationary signals since its frequency characteristics evolve with time. It enables the frequency content of a signal to be obtained locally in time (or at any "instant" in time).

Several techniques have been developed for time–frequency analysis. Perhaps the simplest one is a direct extension of the regular Fourier transform, the short-time Fourier transform (STFT), also known as windowed Fourier transform.

22.3.4.1 Short-Time Fourier Transform This technique involves taking the Fourier transform of short segments or windows of the signal one after another and concatenating the resulting spectrograms. Application of the window is illustrated in Fig. 22.21. In essence, well-localized slices of the function are taken one at a time by sliding the window function along the time axis and taking snapshots of the spectrograms. The result is a three-dimensional plot of the amplitude, frequency, and time (Fig. 22.22). Quite often a two-dimensional plot in the time–frequency domain (a phase space) is used, with amplitude being represented by a gray scale or color coding.

The short-time Fourier transform of a signal $f(t)$ can be expressed as

$$\mathrm{STFT}(\omega, \tau) = \int_{-\infty}^{\infty} f(t) w(t - \tau) \mathrm{e}^{-i\omega t} \mathrm{d}t \tag{22.94}$$

where $w(t)$ is the window function and $w(t - \tau)$ gives the region around the time of interest τ over which the spectrum is obtained.

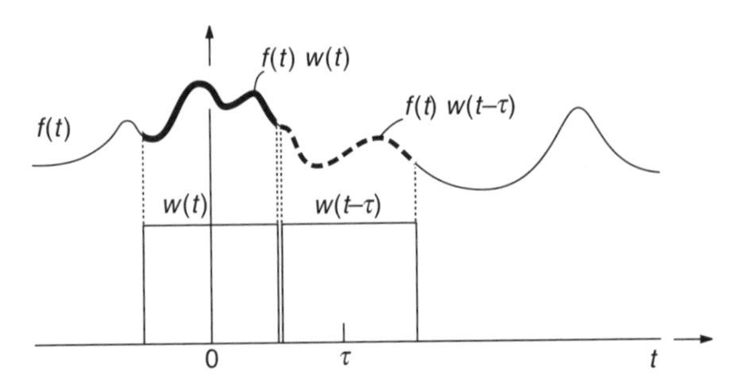

FIGURE 22.21 Windowed function to be used for the short-time Fourier transform.

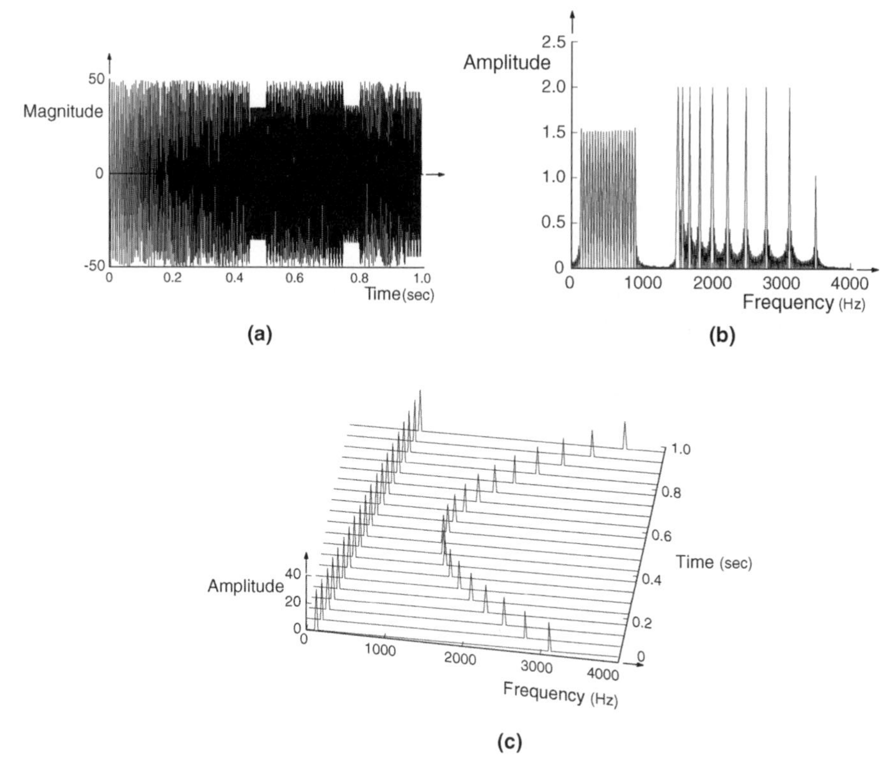

FIGURE 22.22 (a) Time domain signal of the function $f(t) = 30\sin(2\pi\nu_1 t) + 20\sin(2\pi\nu_2 t)$; $\nu_1 = 100 + 40(20t + 1)\nu_2 = 1500 + 20[10 - (20t + 1)]^2$ (b) Overall spectrum (c) Time-frequency plot.

The discrete form of the transform is given by

$$\text{STFT}(\omega, k) = \sum_{j=-\infty}^{\infty} x(j)w(j-k)\mathrm{e}^{-i\omega j} \quad k \text{ is discrete} \tag{22.95}$$

The signal $f(t)$ is suppressed by the window, $w(t-\tau)$, outside its domain, such that the Fourier transform results in a local spectrum. Several forms of the window function have been developed. The simplest case is that of the rectangular window function. However, as we know from Section 22.1.8, the rectangular window introduces side lobes into the frequency domain, and this can result in leakage. This can be mitigated by using any of the window functions discussed in Section 22.1.9. When a Gaussian window is used, then we have the Gabor transform

$$G_f(\omega, \tau) = \int_{-\infty}^{\infty} f(t)\mathrm{e}^{-\pi(t-\tau)^2}\mathrm{e}^{-i\omega t}\,\mathrm{d}t \tag{22.96}$$

From Section 22.1.2, we know that the resolution of the STFT is determined by the width of the window function. The wider the window, the better the frequency resolution. However, the time resolution then deteriorates. On the contrary, a narrower window results in better time resolution and poorer frequency resolution. This is one drawback of the STFT technique. To address this issue, other techniques have been developed for time–frequency analysis. The major ones include wavelet transforms and time–frequency distributions.

22.3.4.2 Wavelet Transforms The time and frequency resolution problem associated with the short-time Fourier transform is overcome using wavelets, since it is possible to adapt the window length of wavelets to the frequency. As a result, wavelets are able to track rapid variations in frequency. They are thus capable of giving good time resolution for appropriate regions of the signal (high-frequency segments) and likewise, good frequency resolution for the low-frequency segments. A wavelet, $\psi(t)$, is essentially a family of functions known as the mother wavelet that is characterized by position and frequency, and whose average value is zero:

$$\int_{-\infty}^{\infty} \psi(t)\mathrm{d}t = 0 \tag{22.97}$$

Associated with the mother wavelet are scaled and translated versions, often known as daughter wavelets. In essence, wavelets segment data into different frequency components and analyze each component with a resolution that is commensurate with its scale. The wavelet $\psi(t)$ is expressed as

$$\psi_{(a,b)}(t) = \frac{1}{\sqrt{|a|}}\psi\left(\frac{t-b}{a}\right) \tag{22.98}$$

where a is a dilation or scale parameter, known as the scalogram, that governs the frequency or width of the wavelet and b indicates the wavelet position. Thus, a variation in b implies a translation in time.

The factor $1/\sqrt{|a|}$ ensures that all scaled functions have the same energy. The smaller the absolute value of a, the higher the frequency content of the wavelet ψ.

A wavelet transform essentially represents a function by wavelets, and the transform of a function $f(t)$ with respect to wavelet $\psi(t)$ is given by

$$\mathrm{WF}(a, b) = \frac{1}{\sqrt{|a|}}\int_{-\infty}^{\infty} f(t)\widetilde{\psi}\left(\frac{t-b}{a}\right)\mathrm{d}t \tag{22.99}$$

where $\widetilde{\psi}(t)$ is the complex conjugate of $\psi(t)$.

The discrete wavelet transform (DWT) can be viewed simply as the wavelet transform of a sampled sequence, $x_j = x(j\delta\tau)$, which for a unit sampling period can be expressed as

$$\mathrm{DWT}[x_j; a, b] = \sum_j x(j)\widetilde{\psi}_{a,b}(j) \tag{22.100}$$

Several types of wavelet functions have been developed. Among the discrete wavelets are the Haar and Daubechies wavelets, and among the continuous ones are the Morlet and Mexican hat wavelets. The Haar wavelet is given by

$$\psi(t) = \begin{cases} 1 & 0 \le t < \frac{1}{2} \\ -1 & \frac{1}{2} \le t < 1 \\ 0 & \text{otherwise} \end{cases} \tag{22.101a}$$

with the scaling function

$$\phi(t) = \begin{cases} 1 & 0 \le t < 1 \\ 0 & \text{otherwise} \end{cases} \tag{22.101b}$$

Its main disadvantage is that it is not particularly good for approximating smooth functions since it is not continuous and therefore not differentiable. The Mexican hat wavelet is given by

$$\psi(t) = \frac{1}{\sqrt{2\pi\sigma^3}}\left[1 - \left(\frac{t}{\sigma}\right)^2\right] \mathrm{e}^{-\frac{1}{2}\left(\frac{t}{\sigma}\right)^2} \tag{22.102}$$

It is the second derivative of a Gaussian function that is normalized (Fig. 22.23).

22.3.4.3 *Time–Frequency Distributions* Time–frequency distributions are also useful for overcoming the drawbacks of spectrograms. They are particularly useful when a given output is composed of two or more signals such that each component requires a separate window for effective analysis. The first such distribution to be introduced was the Wigner–Ville distribution, which is given by

$$W(t, \omega) = \int_{-\infty}^{\infty} f\left(t + \frac{\tau}{2}\right) \widetilde{f}\left(t - \frac{\tau}{2}\right) \mathrm{e}^{-i\omega\tau} \mathrm{d}\tau \tag{22.103}$$

where $f(t)$ is the signal.

Even though it provides very good representation of the time–frequency structure for certain types of signals, the Wigner–Ville distribution can sometimes result in meaningless negative values. Furthermore, it sometimes shows energy concentration

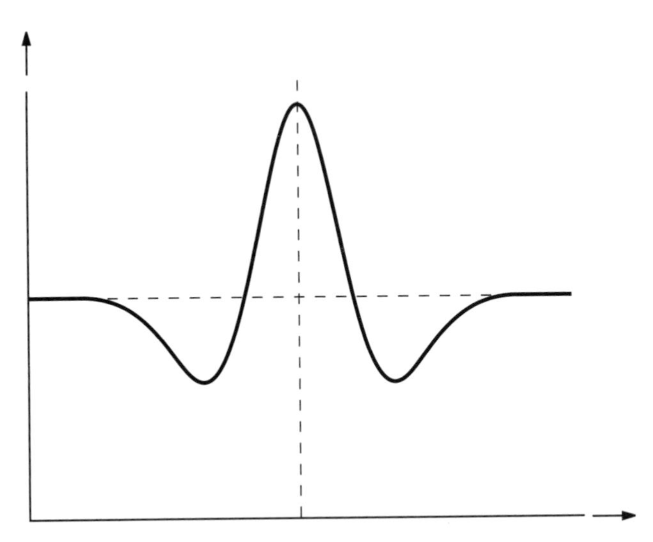

FIGURE 22.23 The Mexican hat wavelet.

in regions that are counterintuitive. As a result, a number of other distribution functions have been developed. Many of these are based on Cohen's class distribution function, which is defined as

$$C(t,\omega)=\int_{-\infty}^{\infty}\int_{-\infty}^{\infty}\int_{-\infty}^{\infty} f\left(u+\frac{\tau}{2}\right)\tilde{f}\left(u-\frac{\tau}{2}\right)\phi(\theta,\tau)\times \mathrm{e}^{i2\pi\theta u-i2\pi\theta t-i\omega\tau}\mathrm{d}u\mathrm{d}\theta\mathrm{d}\tau \tag{22.104}$$

where $\phi(\theta, \tau)$ is a two-dimensional function called a kernel, which determines the specific properties of the distribution.

The Cohen class distribution can also be written as

$$C(t,\omega)=\int_{-\infty}^{\infty}\int_{-\infty}^{\infty} M(\theta,\tau)\mathrm{e}^{-i2\pi\theta t-i\omega\tau}\mathrm{d}\theta\mathrm{d}\tau \tag{22.105}$$

where the generalized ambiguity function $M(\theta, \tau)$ is given by

$$M(\theta,\tau)=A(\theta,\tau)\phi(\theta,\tau) \tag{22.106}$$

and the characteristic function $A(\theta, \tau)$ is

$$A(\theta,\tau)=\int_{-\infty}^{\infty}\tilde{f}\left(u-\frac{\tau}{2}\right)f\left(u+\frac{\tau}{2}\right)\mathrm{e}^{i2\pi\theta u}\mathrm{d}u \tag{22.107}$$

To choose the proper kernel for time–frequency analysis, the signal may be plotted in the ambiguity domain using the ambiguity function. Once plotted in the ambiguity domain, a kernel could be selected based on the characteristics of the signal. This is possible because the kernel is multiplied with the signal in the ambiguity plane as shown by the characteristic function.

The ambiguity function is a bilinear function, and as such, it exhibits cross-components or cross-terms. These are nonzero values that are introduced into the time–frequency domain, but have no apparent physical meaning (cf. aliasing). These can make signal interpretation difficult by obscuring the true signal features and reducing autocomponent resolution. Kernels are therefore used to pass the autocomponents, while suppressing the cross-components. In essence, the kernel fits over the signal, acting as a filter to negate cross-terms that are undesirable. Thus, kernels have the advantage that it is possible to obtain distributions with certain properties by constraining the kernel. Furthermore, the properties of a new distribution can be deduced by inspection of the kernel. Selection of a kernel essentially characterizes the distribution.

Several kernels have been developed. The common ones include

- *Wigner–Ville Distribution Function*

$$\phi(\theta, \tau) = 1 \tag{22.108}$$

  In a sense, the Wigner–Ville distribution is a Cohen class distribution with a kernel of unity.

- *Choi–Williams Distribution Function*

$$\phi(\theta, \tau) = \mathrm{e}^{-a(\theta\tau)^2} \tag{22.109}$$

  where a is an adjustable parameter. The Choi–Williams distribution function uses an exponential function to suppress the cross-components. It is a reduced interference distribution kernel, and these have the ability to suppress artificial cross-terms.

22.3.4.4 Applications in Manufacturing The time–frequency technique that is used for a given manufacturing process depends on the process characteristics. For processes such as lathe turning or welding (including laser) of a long prismatic part, the short-time Fourier transform used with an appropriate window function should be adequate for process monitoring since the processes are essentially stationary. On the contrary, for processes such as milling, some sheet metal forming or bulk deformation operations, spot welding, or processes involving pulsed laser beams, it will be more appropriate to use wavelet analysis or time–frequency distributions due to the non–stationarity of the processes.

22.4 SUMMARY

Sensor outputs usually have to be analyzed to extract relevant information about the processes that they monitor. One approach is to classify the process output into different states, and this involves three steps: sampling (digitizing), feature extraction, and classification.

Sampling requires certain precautions to avoid contaminating the signal. According to the sampling theorem, the theoretical sampling rate has to be at least twice (in practice about 5–10 times) the highest frequency in the signal. Aliasing results if the sampling rate is less. Frequencies above the folding frequency are then folded back into the frequency range below the folding frequency. Using a finite length of sampled data is equivalent to multiplying the signal by a rectangular window, which introduces high-frequency side lobes into the signal, creating potential aliasing problems. This effect may be mitigated by using other window types such as the triangular, Hanning, and Hamming windows.

Feature extraction involves transformation of the signal, along with a reduction in the data size. Among the common transforms are the Fourier, Walsh–Hadamard, Haar, and Karhunen–Loeve transforms, with the Fourier transform being the most commonly used one. The discrete form of the Fourier transform has several advantages over the analogue transform: high signal-to-noise ratio, high resolution (no need for narrow band filters), no drift, and greater flexibility in analysis. However, again certain precautions need to be taken to get meaningful results.

Data reduction is used to reduce the initial data size to increase computation speed, with minimal loss of information. Among the most common criteria used for data reduction are the variance criterion (which eliminates the feature components with the smallest variances) and the class mean scatter criterion (which selects the features that minimize scatter of data within individual classes, while increasing the scatter between classes).

Two of the most common methods for distinguishing between different signal sources are pattern recognition and neural network analysis. Bayes' decision theory provides a fundamental basis to pattern recognition. It expresses the *a posteriori* probability in terms of the *a priori* probability and the class-conditional probability density. The decision can be based on minimum error or minimum risk. The resulting decision rules can be expressed as discriminant functions. This reduces to a minimum distance classifier if the source is assumed to be normally distributed with statistically independent features which have the same variance for each feature and each class. When the covariance matrix is different for the each class, the resulting discriminant function is quadratic. Neural networks represent the classifier by nodes that are interconnected, with information available in the network being contained in the magnitude of the links between the processors. Each classifier needs to be trained to determine the coefficients and thresholds of discriminant functions, or the link magnitudes of neural networks, before being used to monitor the process.

Since the traditional Fourier transform is unable to extract information on the time variation of a signal, it is mainly suitable for stationary processes. For nonstationary processes, it is necessary to use time–frequency methods. The simplest one is the short-time Fourier transform, which is essentially an extension of the traditional Fourier transform. Here, a series of transforms are obtained by sliding a window along the time domain. However, the frequency resolution tends to be poor when the time domain window is short. The use of wavelets and time–frequency distributions enable high-frequency resolutions to be obtained over short periods, and vice versa.

REFERENCES

Ahmed, R., and Rao, K. R., 1975, *Orthogonal Transforms and Digital Signal Processing*, Springer-Verlag, Berlin.

Akay, M., 1998, *Time Frequency and Wavelets in Biomedical Signal Processing*, IEEE Press, Piscataway, NJ.

Anderson, K., Cook, G. E., Karsai, G., and Ramaswamy, K., 1988, Artificial neural networks applied to arc welding process modeling and control, *IEEE Transactions on Industry Applications*, Vol. 26, pp. 824–830.

Bergland, G. D., 1969, A guided tour of the fast Fourier transform, *IEEE Spectrum*, Vol. 6, pp. 41–51.

Boashash, B., 1992, *Time-Frequency Signal Analysis—Methods and Applications*, Longman Cheshire, Melbourne, Australia.

Brooks, R. R., 1997, *Multi-Sensor Fusion: Fundamentals and Applications with Software*, Prentice Hall, Upper Saddle River, NJ.

Carpenter, G. A., Grossberg, S., and Reynolds, J. H., 1991, ARTMAP: supervised real-time learning and classification of nonstationary data by a self-organizing neural network, *Neural Networks*, Vol. 4, pp. 565–588.

Carpenter, G. A., Grossberg, S., Markuzon, N., Reynolds, J. H., and Rosen, D. B., 1992, Fuzzy ARTMAP: a neural network architecture for incremental supervised learning of analog multidimensional maps, *IEEE Transactions on Neural Networks*, Vol. 3, No. 5, pp. 698–713.

Choi, H. I., and Williams, W. J., 1989, Improved time–frequency representation of multicomponent signals using exponential kernels, *IEEE Transactions on Acoustics Speech Signal Processing*, Vol. 37, No. 6, pp. 862–871.

Cohen, L., 1966, Generalized phase–space distribution functions, *Journal of Mathematical Physics*, Vol. 7, pp. 781–786.

Daubechies, I., 1990, The wavelet transform, time–frequency localization and signal analysis, *IEEE Transactions on Information Theory*, Vol. 36, pp. 961–1005.

Duda, R. O., and Hart, P. E., 1973, *Pattern Classification and Scene Analysis*, John Wiley and Sons, New York.

Fausett, L., 1994, *Fundamentals of Neural Networks*, Prentice Hall, Englewood Cliffs, NJ.

Freeman, J. A., and Skapura, D. M., 1991, *Neural Networks—Algorithms, Applications, and Programming Techniques*, Addison-Wesley, Reading, MA.

Fukunaga, K., 1972, *Introduction to Statistical Pattern Recognition*, Academic Press, Inc., Orlando.

Graupe, D., 2007, *Principles of Artificial Neural Networks*, 2nd edition, World Scientific, Singapore.

Hall, D. L., 1992, *Mathematical Techniques in Multisensor Data Fusion*, Artech House, Boston, MA.

Haykin, S., 1994, *Neural Networks: A Comprehensive Foundation*, Macmillan, New York.

Hess-Nielson, N., and Wickerhauser, M. V., 1996, Wavelets and time–frequency analysis, *Proceedings of the IEEE*, Vol. 84, No. 4, pp. 523–540.

Hopfield, J. J., 1982, Neural networks and physical systems with emergent collective computational abilities, *Proceedings of the National Academy of Sciences USA*, Vol. 79, pp. 2554–2558.

Hush, D. R., and Horne, B. G., Jan. 1993, Progress in supervised neural networks, *IEEE Signal Processing Magazine*, pp. 8–39.

Kannatey-Asibu Jr., E., and Emel, E., 1987, Linear discriminant function analysis of acoustic emission signals for tool condition monitoring, *Journal of Mechanical Systems and Signal Processing*, Vol. 1, No. 4, pp. 333–347.

Klein, L. A., 2004, *Sensor and Data Fusion*, SPIE Press, Bellingham, WA.

Lippmann, R. P., April 1987, An Introduction to computing with neural nets, *IEEE ASSP Magazine*, pp. 4–22.

Liu, X., and Kannatey-Asibu Jr., E., 1990, Classification of AE signals for martensite formation from welding, *Welding Journal*, Vol. 69, No. 10, pp. 389s–394s.

Mallat, S., 1999, *A Wavelet Tour of Signal Processing*, 2nd edition, Academic Press, San Diego, CA.

Mertins, A., 1999, *Signal Analysis*, John Wiley & Sons, Inc., New York.

Pao, Y-H., 1989, *Adaptive Pattern Recognition and Neural Networks*, Addison-Wesley, Reading, MA.

Poggio, T., and Girosi, F., 1990, Networks for approximation and learning, *Proceedings of IEEE*, Vol. 78, pp. 1481–1497.

Rangwala, S., and Dornfeld, D. A., 1990, Sensor integration using neural networks for intelligent tool condition monitoring, *ASME Journal of Engineering for Industry*, Vol. 112, pp. 219–228.

Rumelhart, D., and McClelland, J., 1986, *Parallel Distributed Processing*, Vol. 1, MIT Press, Cambridge, MA.

Stanley, W. D., 1975, *Digital Signal Processing*, Reston Publishing, Reston, VA.

Sun, A., Kannatey-Asibu Jr., E., and Gartner, M., 2002, Monitoring of laser weld penetration using sensor fusion, *Journal of Laser Applications*, Vol. 14, No. 2, pp. 114–121.

Sun, A., Gartner, M., and Kannatey-Asibu Jr., E., 2000, Real-time monitoring of laser weld penetration using sensor fusion, *ICALEO 2000*, pp. E24–E34.

Sun, A., Kannatey-Asibu, E., Gartner, M., and Williams, W., 2001, Time–frequency analysis of laser weld signature, *Proceedings of SPIE Advanced Signal Processing Algorithms, Architectures, and Implementations XI*, Vol. 4474, pp. 103–114.

APPENDIX 22A

List of symbols used in the chapter.

Symbol	Parameter	Units
$a(\omega)$	amplitude spectrum	–
$A(\omega)$	real component of $g(\omega)$	–
$B(\omega)$	complex component of $g(\omega)$	–
c_i	constants	–
$f(t)$	signal function	–
$g_i(\mathbf{y})$	discriminant function	–
$g(\omega)$	Fourier transform of $f(t)$	–
$\mathbf{G_{Qy}}$	cross-correlation matrix	–
$\mathbf{G_{yy}}$	autocorrelation matrix	–
i, j, k, l, m	integers	–
$I(m, k)$	input to the mth node of the kth layer	–
L_R	likelihood ratio	–
M_j	number of patterns (data sets) in class s_j	–
M_o	number of outputs available in a set	–
N_m	number of sampled data points in the time series	–
N_{io}	number of input–output training sets available	–
$O(m, k)$	output of the mth node of the kth layer	–
p_i	*a priori* probability of class s_i	–
$p(\mathbf{y}\|s_i)$	class-conditional probability density function for $\mathbf{y}$	–
$p(s_i\|\mathbf{y})$	*a posteriori* probability	–
$q(\omega)$	power spectrum	–
$q(m, n, k)$	magnitude of link from node n in layer $k - 1$ to node m in layer k	–
$\mathbf{Q}$	feature selection criterion	–
$\mathbf{Q}_i$	cluster point i in decision space	–
$r(m, k)$	residue or threshold of node m in layer k	–
$R(m, i)$	the actual or desired mth output of data set i	–
s	number of classes	–
$s(m, i)$	the mth network output of data set i	–
s_i	ith class	–
$\mathbf{T}$	orthogonal transform	–
w_{ij}	weighting coefficient	–
$x(j)$	signal time series	–
$\mathbf{x}$	signal time series vector	–
$\hat{\mathbf{x}}$	estimate of $\mathbf{x}$	–
$y(k)$	discrete Fourier transform of $x(j)$	–
$\mathbf{y}$	Fourier transform of $\mathbf{x}$	–
$\mathbf{y_{ij}}$	jth pattern in class s_i	–
$\bar{\mathbf{y}}_{\mathbf{ij}}$	feature mean for class i	–

$\bar{\mathbf{y}}$	overall system mean	—
$\bar{\mathbf{y}}_{\mathbf{i}}$	mean of $\mathbf{y_i}$	—
$z(j)$	shifted time series	—
$\mathbf{z_{ij}}$	transformation of $\mathbf{y_{ij}}$ in decision space	—
α_i	correction gain or learning rate	—
$\beta(m, k)$	correction factor or error term for node m	—
$\delta\tau$	sampling period	s
$\Delta\tau$	signal duration	s
ε	mean square error	—
λ_i	Lagrange multiplier or eigenvalue	—
ν_{a}	aliasing frequency	Hz
ν_{f}	folding frequency	Hz
ν_{h}	highest frequency in a signal	Hz
ν_{s}	sampling frequency (sampling rate)	Hz
ω	angular frequency	$rad/$s
Ω_i	region in the domain of $\mathbf{y}$ for which $\mathbf{y} \in s_i$	—
$\Phi_{\mathbf{x}}$	covariance matrix of $\mathbf{x}$	—
$\Phi_{\mathbf{y}}$	covariance matrix of $\mathbf{y}$	—
$\Phi_{\mathbf{yi}}$	scatter within each class	—
$\Phi_{\mathbf{ys}}$	scatter between individual classes	—
σ^2	variance	—

PROBLEMS

22.1. Given a system with a normal conditional probability density function,

$$p(\mathbf{y}|s_i) = G(\bar{\mathbf{y}}_{\mathbf{i}}, \Phi_{\mathbf{yi}}) = \frac{1}{(2\pi)^{n/2}|\Phi_{\mathbf{yi}}|^{1/2}} \exp\left[-\frac{1}{2}(\mathbf{y} - \bar{\mathbf{y}}_{\mathbf{i}})' {\Phi_{\mathbf{yi}}}^{-1}(\mathbf{y} - \bar{\mathbf{y}}_{\mathbf{i}})\right]$$

with a different covariance matrix $\Phi_{\mathbf{yi}}$ for each class, obtain a general expression for the discriminant function, $g_i(\mathbf{Y})$, for classifying the system using the Bayes decision rule and the logarithm of the *a posteriori* probability if the *a priori* probability is p_i.

22.2. Derive the relationship

$$\mathbf{X} = [\mathbf{A}'\mathbf{A}]^{-1}\mathbf{A}'\mathbf{Y}$$

Given that $\mathbf{Y} = \mathbf{AX}$ and $\nabla_{\mathbf{X}}\|\mathbf{AX} - \mathbf{Y}\|^2 = 0$.

Also, $\mathbf{A}$, $\mathbf{X}$, and $\mathbf{Y}$ are $(n \times m)$, $(m \times 1)$, and $(n \times 1)$, respectively, with $[\mathbf{A}'\mathbf{A}]$ being nonsingular.

22.3. For the following two-dimensional training set

$$S_1 = \begin{bmatrix} 1 \\ 3 \end{bmatrix}, \begin{bmatrix} -1 \\ 4 \end{bmatrix}, \begin{bmatrix} -3 \\ 4 \end{bmatrix}, \begin{bmatrix} 0 \\ 2 \end{bmatrix}, \begin{bmatrix} -4 \\ 2 \end{bmatrix}$$

$$S_2 = \begin{bmatrix} 5 \\ 6 \end{bmatrix}, \begin{bmatrix} 4 \\ 5 \end{bmatrix}, \begin{bmatrix} 6 \\ 6 \end{bmatrix}, \begin{bmatrix} 8 \\ 8 \end{bmatrix}, \begin{bmatrix} 7 \\ 6 \end{bmatrix}$$

$$S_3 = \begin{bmatrix} 5 \\ 0 \end{bmatrix}, \begin{bmatrix} 6 \\ 1 \end{bmatrix}, \begin{bmatrix} 7 \\ -1 \end{bmatrix}, \begin{bmatrix} 8 \\ 0 \end{bmatrix}, \begin{bmatrix} 9 \\ 2 \end{bmatrix}$$

if the three classes have equal *a priori* probabilities, determine

(a) The discriminant functions for the classifier.

(b) The equations to the decision boundaries.

22.4. Consider $(n \times 1)$ vectors $\mathbf{X}$ and $\mathbf{Y}$ and an $(n \times n)$ matrix $\mathbf{A}$.

(a) Find F if

$$F = \nabla_{\mathbf{A}}[\mathbf{Y}'\mathbf{Y}]$$

(b) Determine

$$\nabla_{\mathbf{A}}[\mathbf{X}'\mathbf{A}'\mathbf{Y}]$$

22.5. Show that the following transformation matrix is orthonormal:

$$\mathbf{T} = \frac{1}{\sqrt{10}} \begin{bmatrix} 1 & 3 \\ -3 & 1 \end{bmatrix}$$

22.6. For the following two-dimensional training set

$$S_1 = \begin{bmatrix} 1 \\ 4 \end{bmatrix}, \begin{bmatrix} -1 \\ 5 \end{bmatrix}$$

$$S_2 = \begin{bmatrix} 4 \\ 7 \end{bmatrix}, \begin{bmatrix} 5 \\ 6 \end{bmatrix}$$

$$S_3 = \begin{bmatrix} 3 \\ 0 \end{bmatrix}, \begin{bmatrix} 5 \\ -1 \end{bmatrix}$$

if the three classes have equal *a priori* probabilities, determine

(a) The discriminant functions for the classifier.

(b) The equations to the decision boundaries.

22.7. Consider the one-dimensional form of the Cauchy distribution and let the class-conditional density be expressed as

$$p(y|s_i) = \frac{1}{\pi a}\left[\frac{1}{1+\left(\frac{y-b_i}{a}\right)^2}\right], \qquad i = 1, 2$$

Assuming equal *a priori* probabilities, show that the minimum probability of error is given by

$$p(\text{error}) = \frac{1}{2} - \frac{1}{\pi}\tan^{-1}\left|\frac{b_2 - b_1}{2a}\right|$$

22.8. Given $\mathbf{X}' = [x_1, x_2, \cdots, x_n]$, and

$$\mathbf{A} = \begin{bmatrix} a_{11} & a_{12} & \cdots & a_{1n} \\ a_{21} & a_{22} & \cdots & a_{2n} \\ \cdots & \cdots & \cdots & \cdots \\ a_{n1} & a_{n2} & \cdots & a_{nn} \end{bmatrix}$$

show that

$$\nabla_{\mathbf{A}}[\mathbf{X}'\mathbf{A}'\mathbf{A}\mathbf{X}] = 2\mathbf{A}(\mathbf{X}\mathbf{X}')$$

22.9. Obtain the covariance matrix $\Phi_{\mathbf{X}}$ of the following data set

$$\mathbf{X_1} = \begin{bmatrix} 0 \\ 2 \end{bmatrix}, \mathbf{X_2} = \begin{bmatrix} 1 \\ 2 \end{bmatrix}, \mathbf{X_3} = \begin{bmatrix} 1 \\ 1 \end{bmatrix}, \mathbf{X_4} = \begin{bmatrix} 1 \\ -1 \end{bmatrix}, \text{ and } \mathbf{X_5} = \begin{bmatrix} 2 \\ 3 \end{bmatrix}$$

22.10. Consider a two-dimensional training set to be as follows:

$$S_1 = \begin{bmatrix} -2 \\ 3 \end{bmatrix}, \begin{bmatrix} -3 \\ 3 \end{bmatrix}, \begin{bmatrix} -4 \\ 3 \end{bmatrix}$$

$$S_2 = \begin{bmatrix} 6 \\ 6 \end{bmatrix}, \begin{bmatrix} 6 \\ 7 \end{bmatrix}, \begin{bmatrix} 7 \\ 5 \end{bmatrix}$$

(a) Obtain the discriminant functions for the classifier, assuming equal *a priori* probabilities and using the least-squares minimum distance classifier.

(b) Determine the equation of the decision boundary.

22.11. A multivariate normal density has $\sigma_{ij} = 0$ and $\sigma_{ii} = \sigma_i{}^2$. Show that the normal density function for this case is given by

$$p(\mathbf{x}) = \frac{1}{\prod_{i=1}^{n} \sqrt{2\pi}\sigma_i} \cdot \exp\left[-\frac{1}{2}\sum_{i=1}^{n}\left(\frac{x_i - \mu_i}{\sigma_i}\right)^2\right]$$

where n is the rank or dimensionality of x, Σ indicates summation, and Π indicates product.

22.12. Show that for an orthonormal transformation matrix $\mathbf{T}$ and a real symmetric matrix $\mathbf{A}$,

$$\text{tr}(\mathbf{A}) = \text{tr}(\mathbf{TAT}') = \sum a_{jj}$$

where a_{jj} is the diagonal elements of $\mathbf{A}$.

23 Laser Safety

Lasers, like most other equipment used in processing materials, require caution for safe and effective utilization. In this chapter, the characteristics of the basic laser hazards that need to be understood for safe use of lasers are outlined, focusing on their application in materials processing. The discussion starts with the hazards associated with lasers. This is followed by the principal classification of lasers. Finally, the basic steps that need to be taken to prevent laser accidents are presented.

23.1 LASER HAZARDS

There are two principal forms of hazard associated with lasers:

1. Radiation-related hazards.
2. Nonbeam hazards.

23.1.1 Radiation-Related Hazards

Assessment of the radiation-related hazards associated with a laser or laser system is normally based on three primary issues:

1. The capacity of the laser and its ability to cause injury.
2. The environment in which the laser is to be used.
3. The personnel who will be associated with the laser (using or being exposed to it).

The damaging effects of a laser beam result primarily from

1. Its highly collimated nature that enables it to be focused to very small sizes.
2. Its monochromaticity that can result in selective interactions with human tissue.

Since different wavelengths interact differently with tissue, the type of injury that results normally depends on the type of laser being used. Injuries that result from radiation-related hazards are primarily to the human skin and eye, and given the same

---

Principles of Laser Materials Processing, by Elijah Kannatey-Asibu, Jr.

conditions, the seriousness of the injury depends on where it occurs. Whereas a burn on the hand may leave a permanent scar, an equivalent damage to the eye may cause blindness. The primary factors that determine the extent of damage sustained are

1. Beam power density.
2. Beam wavelength.
3. Irradiated area and its absorption characteristics.
4. Duration of exposure.

23.1.1.1 Mechanisms of Laser Damage To understand how damage may be caused, the principal mechanisms responsible for laser damage to human tissue are briefly outlined. These are

1. Photothermal effects.
2. Photochemical effects.
3. Photoacoustic effects.

Photothermal Effects With this mechanism, the radiation absorbed by the tissue increases the tissue's temperature until damage or a laser burn occurs. The process depends more on the rate at which incident energy is absorbed, that is, the power, rather than the total amount of energy absorbed. Thus, injuries that immediately follow exposure are more likely to be due to photothermal effects. Even though most cells can survive temperatures as high as 45°C, cells begin to die at temperatures of 60°C, and tissue evaporation would normally occur at about 350°C. Photothermal effects can thus result in immediate and permanent damage to body tissue (skin or eye).

Photochemical Effects Photochemical effects depend more on the total amount of energy absorbed by the tissue rather than on the rate at which it is absorbed. Its effect is thus cumulative. Thus, even though the power levels may not be sufficient to cause thermal damage, over long periods of exposure, the effect can be damaging. The mechanism involves changes in the chemical structure of the underlying tissue. For example, the creation of a skin tan by absorption of ultraviolet radiation from the sun is photochemical in nature. Even though such photochemical effects may be reversible once absorption of the radiation is stopped, cumulative effects may result in longer term damage.

As a result of this being an energy-dependent process, the photon energy plays a significant role. And since the photon energy increases with decreasing wavelength (see equation (1.2)), ultraviolet beams are more damaging than visible radiation, even though visible radiation can affect sensitive tissue such as the retina of the eye as a result of prolonged exposure. Infrared radiation generally does not cause damage by the photochemical mechanism because of the low photon energy.

Photoacoustic Effects Photoacoustic effects are generally associated with ultrashort duration pulses (typically less than 10 μs) of high energy that cause rapid expansion

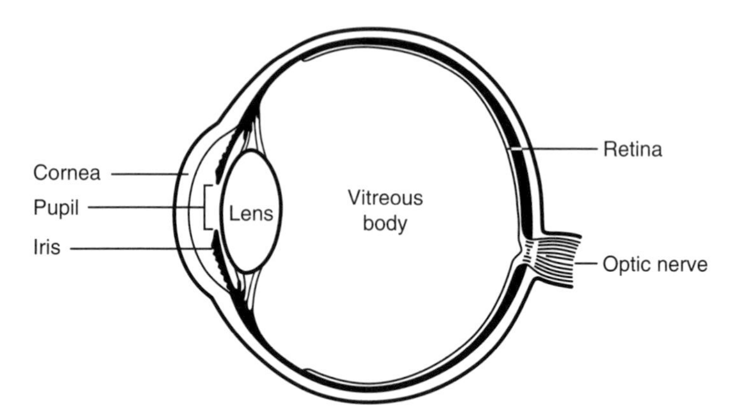

FIGURE 23.1 Schematic of the basic components of the human eye.

of the tissue on which they are incident. This generates an acoustic shock wave in the tissue that may cause mechanical disruption of cellular structures. This process is energy dependent.

The major forms of radiation-related hazards such as eye and skin hazards are now discussed.

23.1.1.2 Major Hazards

Eye Hazards For effective protection of the eye against laser hazards, it is necessary to have a basic understanding of the eye structure and how it interacts with incident radiation. Figure 23.1 shows the basic components of the eye. Light (radiation) enters the eye through the cornea, passes through the pupil, which is the opening within the iris, and is focused on the retina by the lens, which has an effective focal length of 17 mm. The diameter of the pupil can be as small as 2 mm and as large as 7 mm. Under normal circumstances, since light enters the eye from several directions, an image is formed over the entire retina (Fig. 23.2a). Thus, the power density of the light incident on the retina is relatively low. However, when a collimated beam is incident on the lens, it is focused onto a small spot (Fig. 23.2b) of diameter about

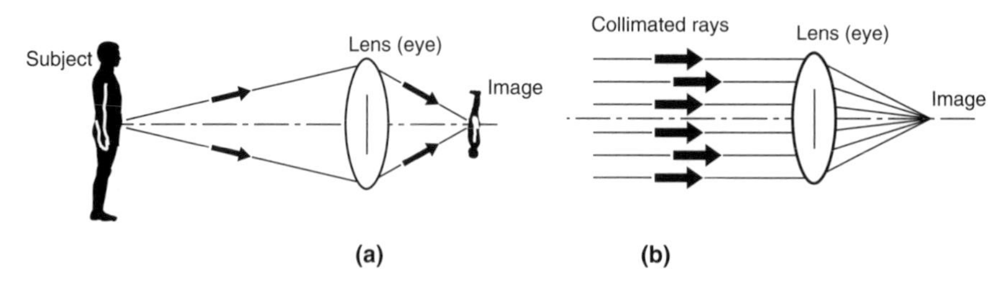

FIGURE 23.2 Focusing of (a) an object and (b) a collimated beam.

20 μm on the retina. This results in a magnification of the light intensity of about 10^5 on the retina. Thus, radiation that might appear to be of relatively low intensity on the outside could cause significant damage on the retina.

Figure 23.3 summarizes the absorption characteristics of the different components of the eye. All radiation in the wavelength range below 0.3 μm (ultraviolet) and above 3 μm (far infrared) is absorbed by the cornea. Radiation that falls between 0.3 and 0.4 μm (near ultraviolet) and also between 1.4 and 3 μm (midinfrared) is absorbed by the lens. The remaining wavelength range between 0.4 and 1.4 μm is focused on the retina. This wavelength range is known as the ocular focus or retinal hazard region. It must be noted that a part of the radiation in this range (0.4–0.7 μm) is visible, while that in the infrared region (0.7–1.4 μm) is not visible.

Any damage that is restricted to the surface layer (up to 50 μm depth) of the cornea tends to heal in about 48 h since the tissue in this region regenerates itself rather easily. However, if the deeper layers of the cornea are damaged, these are likely to result in scarring and the cornea becoming permanently opaque. Damage to the lens may cause cataract (the lens becoming opaque). Photochemical damage may result in changes

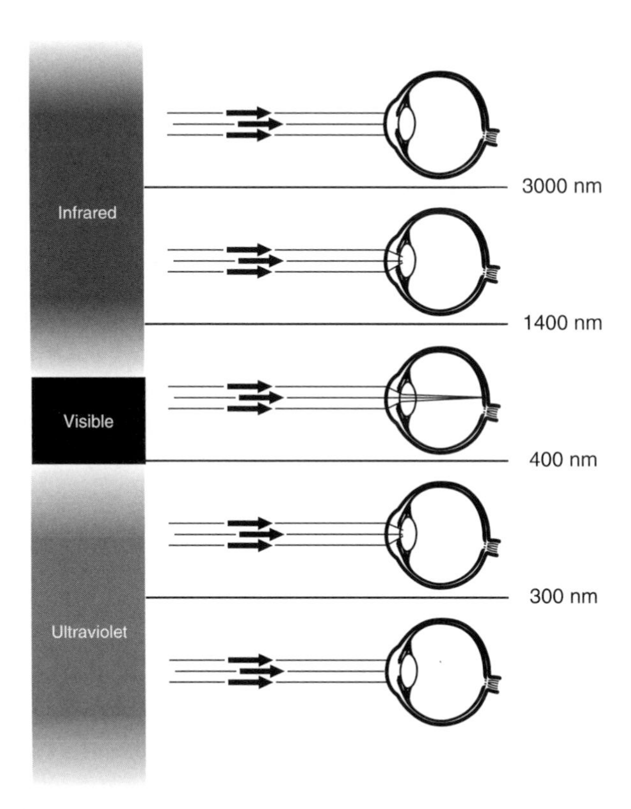

FIGURE 23.3 Absorption characteristics of the different components of the eye. (From Henderson, A. R., 1997, *A Guide to Laser Safety*. By permission of Springer Science and Business Media.)

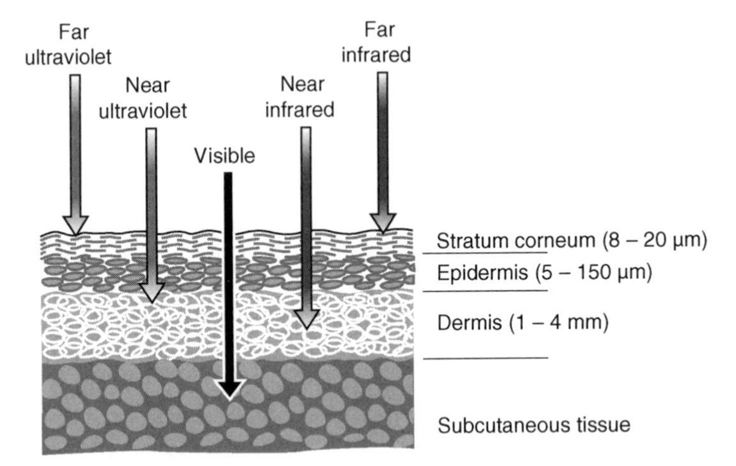

FIGURE 23.4 Extent to which different wavelengths of radiation penetrate the human skin. (From Henderson, A. R., 1997, *A Guide to Laser Safety*. By permission of Springer Science and Business Media.)

to night and color vision if the retina is subjected to long-term exposure at radiation levels that may not cause immediate injury. More severe damage to the eye includes the formation of a blind spot, distortion of vision, loss of high-resolution capability in the eye, and partial or total blindness.

Skin Hazards The extent to which different wavelengths of radiation penetrate the human skin is illustrated schematically in Fig. 23.4. It is evident from this figure that the radiation in the visible wavelength regime results in the highest penetration. Ultraviolet and far-infrared radiation are mainly absorbed in the outer protective layer of dead cells. Damage to the skin includes sunburn and tanning (due to photochemical reaction resulting from ultraviolet radiation), inflammation due to dilation of blood capillaries and a localized laser burn (due to a photothermal action resulting from a high intense beam), premature aging and skin cancer (from prolonged exposure), and tissue cutting due to ablation.

23.1.2 Nonbeam Hazards

In addition to the hazards associated with the beam itself as outlined above, other forms of hazard are associated with the overall laser system and environment, for which appropriate precautions are necessary. Such hazards can be life threatening and include

1. Electrical hazards.
2. Chemical hazards.
3. Environmental hazards.

4. Fire hazards.
5. Explosion hazards and compressed gases.
6. Other hazards.

We shall elaborate on some of these hazards, and the reader is referred to the American National Standards Institute (ANSI) Standards Z136.1-2007 or the International Electrotechnical Commission (IEC) Standards for further details.

23.1.2.1 Electrical Hazards A number of lasers need very high voltages for exciting the active medium, and this is especially the case for the high-power lasers. Electric shock can result from contact with voltages exceeding 50 V. Such contacts usually occur when equipment protective covers are removed, for example, during laser setup, installation, maintenance, service, and so on. Capacitor banks that can store significant electric charge are often used in pulsed lasers. Deaths have occurred in connection with laser systems as a result of electric shock. The major sources of problems are

1. Uncovered electrical terminals.
2. Improperly insulated electrical terminals.
3. Improperly displayed power on warning lights.
4. Improperly grounded equipment.

Other problem sources are inadequate first aid training, nonadherence to standards, and wires or cables on floor, creating slip hazards.

23.1.2.2 Chemical Hazards Some materials used in connection with lasers may be toxic and carcinogenic. These include some laser gases, especially those used with some excimer lasers, and some laser dyes.

23.1.2.3 Environmental Hazards Interaction of some lasers (especially the high-power ones) with matter may generate air contaminants such as fumes, dust, and vapors, which may be toxic and/or noxious. The materials may be plastics, composites, metals, or tissue. Certain materials used for laser windows may also generate hazardous substances when a threshold beam intensity is exceeded. The impact of such contaminants can be significantly reduced by using

1. Proper exhaust ventilation.
2. Respiratory protection.
3. Process isolation.

For exhaust ventilation, it is preferable to use hoods whenever possible since they are more effective. Respiratory protection is normally used as a temporary measure. Process isolation involves isolating the system using physical barriers or using remote control.

23.1.2.4 Fire Hazards This is especially important for Class 4 lasers (see Section 23.2). Accidental exposure of combustible material to high-power beams (exceeding 0.5 W) can cause ignition (author's personal experience in the laboratory), with the risk of starting a fire. Care should therefore be taken to avoid placing combustible materials in the beam path. For enclosure of the work area, flame-retardant materials should be used, whenever possible.

23.1.2.5 Explosion Hazards and Compressed Gases Some components of the laser system can be explosive if they eventually disintegrate. These include high-pressure arc lamps, capacitor banks, and some chemicals that may be used. In addition, proper precautions must be taken to ensure that the compressed gases used in a number of laser systems do not pose a health risk. Gas tanks must be properly secured and not left freestanding. Hazardous gases must be appropriately labeled.

23.1.2.6 Other Hazards Other hazards associated with laser systems include mechanical hazards involving rotating machinery, and the use of robots; noise; ergonomics issues associated with handling workpieces and other laser components; and generation of other forms of radiation by system components.

In the next section, we outline the various classes of lasers. This, to a large extent, determines the precautions that need to be taken to avoid injury in using lasers.

23.2 LASER CLASSIFICATION

The classification of a laser is based primarily on the ability of the main or reflected beam to cause injury to the eye or skin, and this depends on the level of accessible emission for that laser. For a laser in a given class, the output of that laser must not exceed the *accessible emission limit (AEL)* for that class. The AEL is the maximum accessible emission level permitted within a particular class. In this section, we provide a brief outline of the major classes of lasers. The interested reader is referred to the ANSI Z136.1-2007 document on the "Safe Use of Lasers," for further details.

There are four major laser classes:

Class 1: The radiation levels produced by Class 1 lasers are not considered high enough to cause any form of injury either during operation or during maintenance. Thus, no special precautions are necessary. An example is the laser in an audio compact disc player.

Class 2: These are low-power lasers and are further subdivided into 2 and 2M. The radiation from Class 2 lasers in general fall within the visible spectrum. Since the radiation is visible and of low power, natural reflexes such as blinking (with a response time of about 0.2 s) are considered adequate in protecting the eye against injury. However, damage may occur if one deliberately stares into the beam.

Class 3: These are medium-power lasers and are also further subdivided into 3R and 3B. This type of laser is not a skin hazard, but definitely an eye hazard.

Direct and specularly reflected Class 3 beams can cause eye injury in a time period shorter than what it takes for the eye to blink, even though the diffusely reflected beam is not normally hazardous. It also does not pose a fire hazard.

Class 4: With the high-power inherent in this class of lasers, they can be a hazard to the eye and skin, as well as pose a fire hazard. There is no upper power limit to this class of lasers.

23.3 PREVENTING LASER ACCIDENTS

In this section, we outline some of the major steps that need to be taken to prevent, or at least minimize, the occurrence of laser accidents. A more detailed description of control measures and precautions that need to be taken are provided by the ANSI standard. The major issues that we shall touch on are

1. Laser safety officer (LSO).
2. Engineering controls.
3. Administrative and procedural controls.
4. Protective equipment.
5. Warning signs and labels.

23.3.1 Laser Safety Officer

A laser safety officer is someone who is given the responsibility and authority for ensuring safety in connection with the laser system. The appointment of an LSO is especially important for Class 3 and Class 4 laser installations. The LSO's responsibilities include all the control measures that are necessary for ensuring safe use of the laser as specified in the ANSI Standard Z136.1-2007 and summarized in the next four subsections.

23.3.2 Engineering Controls

Engineering controls that are necessary for ensuring the safe use of lasers include the provision of protective housings, interlocks, key control, beam path enclosure, remote interlock connector, emergency stop button, laser activation warning systems, area control, equipment labels, and area sign posting.

Protective housing is normally provided by the equipment manufacturer. However, in situations where it may be necessary to use the laser without protective housing, such as in research and development, it will be necessary to provide such controls as access restriction and other safety measures. It is also necessary to provide protective housings with interlock systems that are activated to interrupt the beam, for example, when the protective housing is opened during operation or maintenance. Enclosure of the laser equipment and beam path is one of the most important steps for minimizing hazards.

The provision of a master switch with a key or coded access is necessary for Class 3B or 4 lasers, and in addition, a permanently attached emergency stop button should be provided. Laser activation warning systems such as an alarm or warning light that indicate activation of high voltage and radiation emission are also necessary. It is preferable to have such warnings activated before the emission of laser radiation.

It is also necessary to enclose the beam path. Where this is not completely possible, the LSO will establish a nominal hazard zone (NHZ), unless this is already provided by the laser manufacturer. A *nominal hazard zone* describes the space within which the level of the direct, reflected, or scattered radiation during operation exceeds the applicable maximum permissible exposure (MPE). A *maximum permissible exposure* (Tables 23.1 and 23.2) is the level of radiation to which a person may be exposed without hazardous effect or adverse biological changes in the eye or skin. Under such circumstances, the work area should be designated as a laser controlled area with restricted access that is limited only to necessary personnel. A remote interlock connector that terminates laser emission when opened is then necessary for Class 3B and 4 lasers. Such a system is illustrated in Fig. 23.5. It may also be designed with an interlock override capability, so that appropriate personnel can enter and leave the facility without necessarily interrupting the laser operation.

23.3.3 Administrative and Procedural Controls

Administrative and procedural controls specify rules and/or work practices that implement engineering controls. These include standard operating procedures that are maintained with the laser equipment for reference by the operator, maintenance, or service personnel. They also include specifying limitations on output radiation emission; education and training; specification of authorized personnel (operators, maintenance, and service personnel); alignment procedures; specification of protective equipment; and control of spectators.

23.3.4 Protective Equipment

The most important piece of protective equipment necessary for minimizing laser hazards is protective eyewear. Others include windows, barriers, curtains, and skin protective devices.

23.3.4.1 Protective Eyewear Eye protection is essential when personnel are either working with or are in the NHZ of Class 3B or 4 lasers that are operating. Its main purpose is to reduce any possible exposure to safe levels (not more than the MPE), while at the same time permitting enough radiation to be transmitted across the visible spectrum such that visual performance is not unduly degraded. Thus, the transmittance of the filter at the laser wavelength should not exceed $\tau_{\lambda_{\max}}$, which is given by

$$\tau_{\lambda_{\max}} = \frac{\text{MPE}}{\text{maximum possible exposure}} \tag{23.1}$$

TABLE 23.1 Maximum Permissible Exposure for the Eye

Wavelength (μm)	Exposure duration, t (s)	MPE J/cm^2	MPE W/cm^2	Notes
Ultraviolet				In the dual limit wavelength region (0.180–0.400 μm), the lower MPE considering photochemical and thermal effects must be chosen.
Dual limits for λ between 0.180 and 0.400 μm				
Thermal				
0.180–0.400	10^{-9} to 10	$0.56\,t^{0.25}$		
Photochemical				
0.180–0.302	10^{-9} to 3×10^4	3×10^{-3}		
0.302–0.315	10^{-9} to 3×10^4	$10^{200(\lambda-0.295)} \times 10^{-4}$		
0.315–0.400	10 to 3×10^4	1.0		
Visible				
0.400–0.700	10^{-13} to 10^{-11}	1.5×10^{-8}		In the wavelength region (0.400–0.500 μm), T_1 determines whether the photochemical or thermal MPE is lower.
0.400–0.700	10^{-11} to 10^{-9}	$2.7t^{0.75}$		
0.400–0.700	10^{-9} to 18×10^{-6}	5.0×10^{-7}		
0.400–0.700	18×10^{-6} to 10	$1.8\,t^{0.75} \times 10^{-3}$		
0.500–0.700	10 to 3×10^4		1×10^{-3}	
Thermal				
0.450–0.500	10 to T_1		1×10^{-3}	
Photochemical				
0.400–0.450	10 to 100	1×10^{-2}		
0.450–0.500	T_1 to 100	$C_B \times 10^{-2}$		
0.400–0.500	100 to 3×10^4		$C_B \times 10^{-4}$	
Near infrared				
0.700–1.050	10^{-13} to 10^{-11}	$1.5C_A \times 10^{-8}$		
0.700–1.050	10^{-11} to 10^{-9}	$2.7C_A\,t^{0.75}$		
0.700–1.050	10^{-9} to 18×10^{-6}	$5.0\,C_A \times 10^{-7}$		
0.700–1.050	18×10^{-6} to 10	$1.8\,C_A t^{0.75} \times 10^{-3}$		
0.700–1.050	10 to 3×10^4		$C_A \times 10^{-3}$	
1.050–1.400	10^{-13} to 10^{-11}	$1.5\,C_C \times 10^{-7}$		The wavelength region λ_1 to λ_2 means $\lambda_1 \leq \lambda < \lambda_2$, for example, 0.180–0.302 μm means $0.180 \leq \lambda < 0.302$ μm.
1.050–1.400	10^{-11} to 10^{-9}	$27.0\,C_C t^{0.75}$		
1.050–1.400	10^{-9} to 50×10^{-6}	$5.0\,C_C \times 10^{-6}$		
1.050–1.400	50×10^{-6} to 10	$9.0\,C_C t^{0.75} \times 10^{-3}$		
1.050–1.400	10 to 3×10^4		$5.0\,C_C \times 10^{-3}$	
Far infrared				
1.400–1.500	10^{-9} to 10^{-3}	0.1		*Note*: The MPEs must be in the same units.
1.400–1.500	10^{-3} to 10	$0.56\,t^{0.25}$		
1.400–1.500	10 to 3×10^4		0.1	
1.500–1.800	10^{-9} to 10	1.0		
1.500–1.800	10 to 3×10^4		0.1	
1.800–2.600	10^{-9} to 10^{-3}	0.1		
1.800–2.600	10^{-3} to 10	$0.56\,t^{0.25}$		
1.800–2.600	10 to 3×10^4		0.1	
2.600–1000	10^{-9} to 10^{-7}	1×10^{-2}		
2.600–1000	10^{-7} to 10	$0.56\,t^{0.25}$		
2.600–1000	10 to 3×10^4		0.1	

TABLE 23.2 Maximum Permissible Exposure for the Skin

		MPE	
Wavelength (μm)	Exposure Duration, t (s)	(J/cm^2) Except as Noted	(W/cm^2) Except as Noted
Ultraviolet			
Dual limits for λ between 0.180 and 0.400 μm			
Thermal			
0.180–0.400	10^{-9} to 10	$0.56\,t^{0.25}$	
Photochemical			
0.180–0.302	10^{-9} to 3×10^4	3×10^{-3}	
0.302–0.315	10^{-9} to 3×10^4	$10^{200(\lambda-0.295)} \times 10^{-4}$	
0.315–0.400	10 to 10^3	1.0	
0.315–0.400	10^3 to 3×10^4		1×10^{-3}
Visible and near infrared			
0.400–1.400	10^{-9} to 10^{-7}	$2C_A \times 10^{-2}$	
0.400–1.400	10^{-7} to 10	$1.1C_A\,t^{0.25}$	
0.400–1.400	10 to 3×10^4		$0.2C_A$
Far infrared			
1.400–1.500	10^{-9} to 10^{-3}	0.1	
1.400–1.500	10^{-3} to 10	$0.56\,t^{0.25}$	
1.400–1.500	10 to 3×10^4		0.1
1.500–1.800	10^{-9} to 10	1.0	
1.500–1.800	10 to 3×10^4		0.1
1.800–2.600	10^{-9} to 10^{-3}	0.1	
1.800–2.600	10^{-3} to 10	$0.56\,t^{0.25}$	
1.800–2.600	10 to 3×10^4		0.1
2.600–1000	10^{-9} to 10^{-7}	1×10^{-2}	
2.600–1000	10^{-7} to 10	$0.56\,t^{0.25}$	
2.600–1000	10 to 3×10^4		0.1

Protective devices that can be used include goggles, face shields, and spectacles or prescription eyewear using special filter materials or reflective coatings. Such protective eyewear should withstand either direct or diffusely scattered beams. Important factors that need to be considered in selecting protective eyewear are

1. Wavelength(s) of laser output.
2. Potential for multiwavelength operation.
3. Radiant exposure or intensity level.
4. Potential exposure time.
5. Maximum permissible exposure.
6. Optical density (OD) requirement of eyewear filter at laser output wavelength.

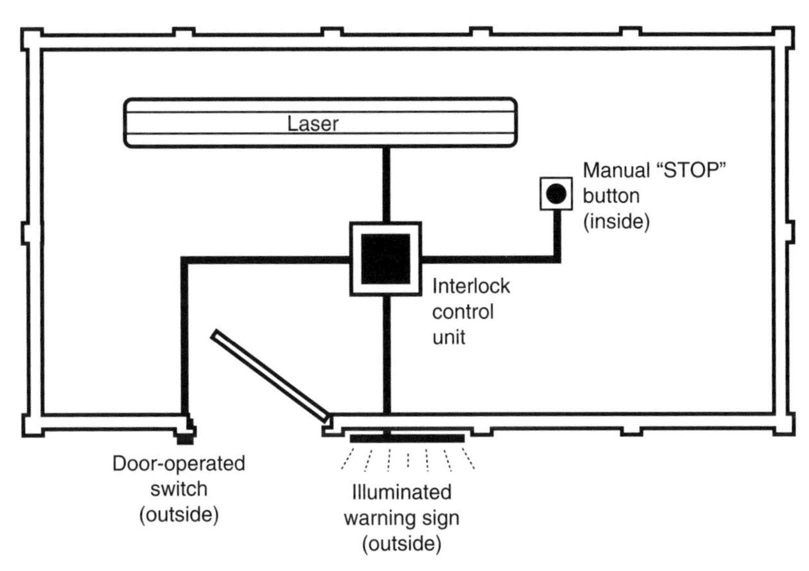

FIGURE 23.5 Illustration of interlocked access to a laser hazard area. (From Henderson, A. R., 1997, *A Guide to Laser Safety*. By permission of Springer Science and Business Media.)

The optical density, OD_λ, is the factor by which incident radiation at a specific wavelength would be attenuated and is given by

$$OD_\lambda = \log_{10} \frac{I_p}{MPE} = -\log_{10} \tau_\lambda \tag{23.2}$$

where I_p is the highest level of intensity (power per unit area) to which the eye could be exposed and is given by the beam power over the limiting aperture (which is 7 mm for the 0.4–1.4 μm region) (W/cm^2), *MPE* is the maximum permissible exposure (W/cm^2), and τ_λ is the transmittance of the filter at a given wavelength.

I_p and MPE have to be in the same units. The value of OD_λ that is calculated from equation (23.2) is rounded up to the next whole number. For a given potential laser output, the MPE for that wavelength has to be determined. Then, from equation (23.2), the optical density that will indicate the type of protective eyewear to be used can be calculated. For example, if the laser exposure is 10^5 times greater than the MPE, then an eyewear with an optical density of at least 5 would be needed.

Table 23.3 shows the relationship between the optical density, transmittance, and so on at a given wavelength and indicates, for a calculated optical density for that wavelength, the percentage of the beam that will be transmitted, or attenuated, or the level of protection to be expected.

For ease of reference, optical densities necessary for various laser outputs are summarized in Table 23.4. For completeness, the MPE for both eye and skin exposure are provided in Tables 23.1 and 23.2, respectively. It must be noted that the MPE value used depends on the duration of exposure to the beam.

TABLE 23.3 Relationship Between the Optical Density, Transmittance, and Level of Protection

Transmittance	Level of Protection (MPE)	OD_λ
1	1	0
10^{-1}	10	1
10^{-2}	10^2	2
10^{-3}	10^3	3
10^{-4}	10^4	4
10^{-5}	10^5	5
10^{-6}	10^6	6
10^{-7}	10^7	7

Example 23.1 As a Laser Systems Engineer with XYZ Company, you are in charge of installing a new 3 kW Nd:YAG laser system in the plant. Calculate an OD for the eyewear to be used for protection against long-term exposure.

Solution:

Wavelength of the Nd:YAG laser is 1.06 μm.

From Table 23.1, the MPE for long-term exposure for this wavelength is 5.0 $C_c \times 10^{-3}$ W/cm^2.

From Table 23.5 (Table 6 of ANSI Standards Z136.1-2007), $C_c = 1.0$ for the 1.06 μm wavelength.

Thus, MPE $= 5.0 \times 10^{-3}$ W/cm^2.

Using a limiting aperture of 7 mm = 0.7 cm, the highest level of intensity to which the eye could be exposed, I_p, is given by

$$I_p = \frac{3000}{\pi 0.7^2/4} = 7794 \text{ W/cm}^2$$

Therefore,

$$OD_\lambda = \log_{10}\left(\frac{I_p}{\text{MPE}}\right) = \log_{10}\left(\frac{7794}{5.0 \times 10^{-3}}\right) = \log_{10}(1.56 \times 10^6) = 6.19$$

which gives a final rounded-up optical density of 7. This compares with a value of 8 to be read directly from Table 23.4. Correction factors used in various tables are listed in Table 23.5.

23.3.4.2 Other Protective Equipment Other protective equipment include laser protective windows, barriers and curtains, and skin protection clothing such as gloves or even "sunscreen" creams.

TABLE 23.4 Optical Densities for Various Laser Outputs

Q-Switched Laser (10^{-9}–10^{-2} s)		Non-Q-Switched Lasers (0.4×10^{-3}–10^{-2} s)		Continuous-Wave Lasers Momentary (0.25–10 s)		Continuous-Wave Lasers Long-Term Staring (<1 h)		Attenuation	
Maximum Output Energy (J)	Max Beam Radiant Exposure (J/cm^2)	Max Laser Output Energy (J)	Max Beam Radiant Exposure (J/cm^2)	Max Power Output (W)	Max Beam Irradiance (W/cm^2)	Max Power Output (W)	Max Beam Irradiance (W/cm^2)	Attenuation Factor	OD
10	20	100	200	10^{5a}	2×10^{5a}	100^a	200^a	10^8	8
1	2	10	20	10^{4a}	2×10^{4a}	10^a	20^a	10^7	7
10^{-1}	2×10^{-1}	1	2	10^{3a}	2×10^{3a}	1	2	10^6	6
10^{-2}	2×10^{-2}	10^{-1}	2×10^{-1}	100^a	200^a	10^{-1}	2×10^{-1}	10^5	5
10^{-3}	2×10^{-3}	10^{-2}	2×10^{-2}	10	20	10^{-2}	2×10^{-2}	10^4	4
10^{-4}	2×10^{-4}	10^{-3}	2×10^{-3}	1	2	10^{-3}	2×10^{-3}	10^3	3
10^{-5}	2×10^{-5}	10^{-4}	2×10^{-4}	10^{-1}	2×10^{-1}	10^{-4}	2×10^{-4}	10^2	2
10^{-6}	2×10^{-6}	10^{-5}	2×10^{-5}	10^{-2}	2×10^{-2}	10^{-5}	2×10^{-5}	10	1

[a] Not recommended as a control procedure at these levels. These levels of power could damage or destroy the attenuating material used in the eye protection. The skin also needs protection at these levels.

Use of this table may result in optical densities (OD) greater than necessary.

TABLE 23.5 Correction Factors.

Parameters/Correction Factors	Wavelength (μm)
$C_A = 1.0$	0.400 to 0.700
$C_A = 10^{2(\lambda-0.700)}$	0.700 to 1.050
$C_A = 5.0$	1.050 to 1.400
$C_B = 1.0$	0.400 to 0.450
$C_B = 10^{20(\lambda-0.450)}$	0.450 to 0.600
$C_C = 1.0$	1.050 to 1.150
$C_C = 10^{18(\lambda-1.150)}$	1.150 to 1.200
$C_C = 8$	1.200 to 1.400
$T_1 = 10 \times 10^{20(\lambda-0.450)}$***	0.450 to 0.500

***$T_1 = 10$ s for $\lambda = 0.450\ \mu$m, and $T_1 = 100$ s for $\lambda = 0.500\ \mu$m.

Note 1: Wavelengths must be expressed in micrometers for calculations.

Note 2: The wavelength region λ_1 to λ_2 means $\lambda_1 \leq \lambda < \lambda_2$, 0.550 to 0.700 μm means $0.550 \leq \lambda < 0.700\ \mu$m.

23.3.5 Warning Signs and Labels

Another effective means of preventing accidents involves the use of warning signs and labels on the lasers. These inform anyone who uses the laser about the type of hazard associated with it. Such signs may also be placed at the entrance to the NHZ of the laser. The design of signs in terms of dimensions, color, letter size, and so on are specified in the American National Standard Specification for Accident Prevention Signs, ANSI Z535. The ANSI standards for signs used to denote Class 2, 3, and 4 lasers are shown in Fig. 23.6. For Class 2 lasers, the word "Caution" is used with all signs and labels, while for 3B and 4 lasers, the word "Danger" is used. The International Standard is based on the symbol shown in Fig. 23.7 (IEC Publication 825). Additional wording that is required, depending on the type of laser, is as follows:

Class 2: "Laser Radiation—Do Not Stare into Beam."

Class 2M and some 3R: "Laser Radiation—Do Not Stare into Beam or View Directly with Optical Instruments."

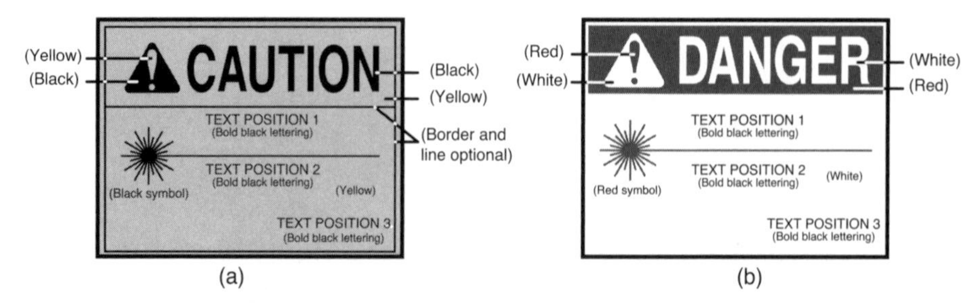

FIGURE 23.6 ANSI warning signs for lasers. (a) Class 2 and 2M lasers. (b) Class 3R, 3B, and 4 lasers. (From American National Standard for Safe Use of Lasers ANSI Z136.1-2007. Copyright 2007, Laser Institute of America. www.laserinstitute.org. All rights reserved.)

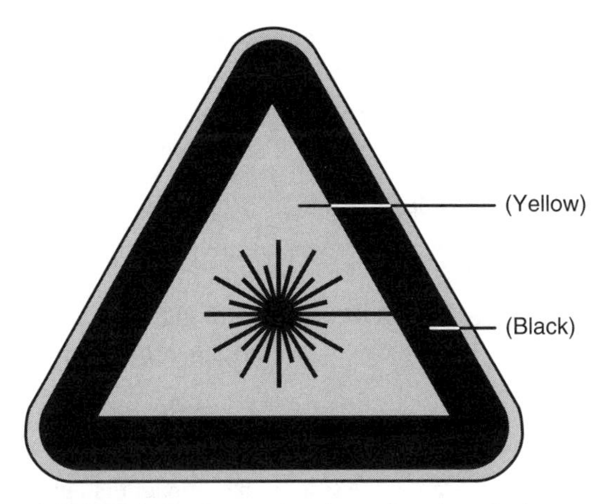

FIGURE 23.7 International warning signs for lasers.

Class 3R (others): "Laser Radiation—Avoid Direct Eye Exposure."

Class 3B: "Laser Radiation—Avoid Direct Exposure to Beam."

Class 4: "Laser Radiation—Avoid Eye or Skin Exposure to Direct or Scattered Radiation."

Other pertinent information to be provided on the sign includes

Position 1: Invisible Laser Radiation; Knock Before Entering; Do Not Enter When Light is On; Restricted Area; etc.

Position 2: Type of laser; emitted wavelength; pulse duration; maximum output

Position 3: Class of the laser.

23.4 SUMMARY

Lasers need to be used with caution for safety, since improper usage can cause serious injury. Injuries associated directly with the laser beam generally occur with respect to human skin and eye. The damage that is sustained depends on the power density, wavelength, the irradiated area and its absorption characteristics, and the exposure duration.

Damage to human tissue may result from photothermal, photochemical, or photoacoustic effects. With regard to the eye, focusing of a collimated beam on the retina results in magnification of the light intensity by a factor of about 10^5. Thus, relatively low-intensity radiation can cause significant damage to the retina. This is especially so for radiation in the visible and near-infrared regions. Damage to the surface layer of the cornea tends to heal in about 48 h. However, damage to deeper layers of the cornea can result in scarring and the cornea becoming permanently opaque. Damage to the lens may cause cataract. More severe damage to the eye includes the formation of a blind spot, distortion of vision, loss of high-resolution capability in the eye, and blindness. With regard to the skin, radiation in the visible wavelength regime results in the highest penetration. Ultraviolet and far-infrared radiation are mainly absorbed in the outer protective layer of dead cells. Damage to the skin includes sun burn and tanning, inflammation due to dilation of blood capillaries, premature aging, skin cancer, and deep burn (incision).

There are other hazards that do not directly involve the beam, and these are associated with electrical, chemical, environmental, fire, explosion, and compressed gas sources.

There are four major laser classes. For each class, the laser output must not exceed the accessible emission limit for that class. Details are provided in the ANSI Standard Z136.1-2007. The four classes are

Class 1: Radiation levels for this class of lasers are not high enough to cause any form of injury, for example, the laser in a compact disc player.

Class 2: These are low-power lasers, fall within the visible spectrum, and thus natural reflexes such as blinking are adequate in protecting the eye against injury. However, damage may occur if one deliberately stares into the beam.

Class 3: These are medium-power lasers, and even though not a skin hazard, are definitely an eye hazard. They also do not pose a fire hazard.

Class 4: With the high-power inherent in this class of lasers, they can be a hazard to the eye and skin, as well as pose a fire hazard. There is no upper power limit to this class of lasers.

To ensure that appropriate steps are taken to minimize laser accidents, it is necessary to have

1. A laser safety officer who ensures safety in connection with the laser system.
2. Engineering controls that include the provision of protective housings, interlocks, beam path enclosure, emergency stop button, laser activation warning systems, area sign posting, and so on.
3. Administrative and procedural controls that specify rules and/or work practices that implement engineering controls.
4. Protective equipment that include protective eyewear.
5. Warning signs and labels that inform anyone who uses the laser about the type of hazard associated with it.

REFERENCES

Henderson, A. R., 1997, *A Guide to Laser Safety*, Chapman and Hall, London.

Laser Institute of America, 2007, *American National Standard for Safe Use of Lasers*, ANSI Z136.1-2007, Orlando, FL.

Matthews, L, and Garcia, G., 1995, *Laser and Eye Safety in the Laboratory*, IEEE Press, The Institute of Electrical and Electronic Engineers, Inc., New York.

Winburn, D. C., 1990, *Practical Laser Safety*, 2nd edition, Marcel Dekker, Inc., New York.

APPENDIX 23A

List of symbols used in the chapter.

Symbol	Parameter	Units
AEL	accessible emission limit	W
C_A, C_B, C_C	correction factors	–
I_p	highest irradiance level	W/cm^2
MPE	maximum permissible exposure	W/cm^2
NHZ	nominal hazard zone	–
OD_λ	optical density	–
τ_λ	filter transmittance at laser wavelength	–

PROBLEM

23.1. Show that the eye magnifies the intensity of a collimated beam by about 10^5 times.

Overall List of Symbols

Symbol	Parameter	Units
A	cross-sectional area	m^2
b	width of plate	mm
c	velocity of light in free space = 3×10^8 (exactly 299,792,458)	m/s
c_m	velocity of light in a medium = c/n	m/s
c_p	specific heat at constant pressure	J/kg K
c_v	specific heat at constant volume	J/kg K
$c_{vol} = \rho c_p$	heat capacity (specific heat per unit volume)	J/m^3 K
C	alloy composition or solute concentration	%
C_0	initial liquid composition	%
CE	carbon equivalent	–
CR	cooling rate	°C/s
D	average grain size	m
D_l	diffusion coefficient of solute in liquid	m^2/s
D_0	initial average grain size	m
e	natural exponent = 2.718281828	—
$e(\nu)$	energy density (energy per unit volume) at the frequency ν	J/m^3 Hz
erfc(x)	$= \frac{2}{\sqrt{\pi}} \int_x^\infty e^{-\zeta^2} d\zeta$ = complementary error function	–
E	total energy of electron	J
ΔE_{ij}	energy difference between two energy levels i and j	J
E	electric field vector	V/m
E_l	magnitude of electric field vector (electric field strength)	V/m
E_m	Young's modulus	Pa (N/m^2)
E_0	amplitude of electric field vector	V/m
E_x, E_y, E_z	x, y, z components of **E**	V/m
f	volume fraction	–

Principles of Laser Materials Processing, by Elijah Kannatey-Asibu, Jr.

f_l	focal length	m
F	force	N
F_e	energy of the Fermi level	J
F_q	laser fluence or energy per unit area	J/m^2
F_r	Fresnel number	–
F_x, F_y, F_z	external body forces in the x, y and z directions, respectively	N/m^3
FWHM	full width at half maximum	–
g	gravitational acceleration	m/s^2
$\mathbf{g}$	gravitational acceleration vector	m/s^2
$g(\omega, \omega_o)$	line-shape function	–
G	shear modulus	Pa (N/m^2)
G_{Tl}	temperature gradient of liquid near solid–liquid interface	$^\circ$C/m
G_{Ts}	temperature gradient of solid near solid–liquid interface	$^\circ$C/m
ΔG	total free energy change	J
h	thickness	m
HAZ	heat affected zone	–
h_p	Planck's constant $= 6.625 \times 10^{-34}$	J s
$i = \sqrt{-1}$	complex variable	–
I	intensity or power per unit area (irradiance)	W/m^2
I_0	intensity of incident beam	W/m^2
I_p	peak intensity	W/m^2
I_r	intensity of reflected beam	W/m^2
J	flux of atoms	/m^2 s
k	thermal conductivity	W/m K
k_B	Boltzmann's constant $= 1.38 \times 10^{-23}$	J/K
$k_w = 2\pi/\lambda$	wavenumber	/m
K	K factor	–
$K_0(\chi)$	modified Bessel function of the second kind of order zero	–
l	optical path difference between rays	m
L	length of active medium (distance between active mirrors)	m
L_m	latent heat of melting (fusion) per unit mass	J/kg
L_{mv}	latent heat of melting (fusion) per unit volume	J/m^3
L_v	latent heat of vaporization per unit mass	J/kg
L_{vol}	latent heat of vaporization per unit volume	J/m^3
m	integer	–

$\dot{m}$	mass flow rate	kg/s
$\mathrm{m_a}$	mass	kg
$\mathrm{m_e}$	electron mass $= 9.11 \times 10^{-31}$	kg
M	moment due to forces	N m
M^2	M^2 factor	—
$M_c(\nu)$	number of modes in a closed cavity	/m^3
M_0	momentum	N m/s
n	index of refraction	–
$\underline{\mathbf{n}}$	unit vector normal to surface or curve	—
n_e	electron number density	/m^3
n_i	components of a unit vector normal to a surface ($i = 1, 2, 3$)	—
N	total population (number of atoms per unit volume) of all the energy levels	/m^3
ΔN	difference in population between two energy levels	/m^3
N_i	population (number of atoms per unit volume) of energy level i	/m^3
NA	numerical aperture	-
p	probability	–
p_t	probability per unit time	/s
P	pressure or surface traction per unit area	Pa (N/m^2)
P_x, P_y, P_z	pressure components in the x, y, z directions, respectively	Pa (N/m^2)
q	power or energy rate	W
q_a	net energy absorption rate	W
Q	energy	J
Q_a	energy absorbed by the workpiece	J
Q_f	quality factor	—
Q_l	energy lost by conduction, convection, and radiation	J
r	$\sqrt{\xi^2 + y^2 + z^2}$ or $\sqrt{\xi^2 + y^2}$ = radius or radial distance from a point	m
r_{mi}	radius of mirror i	m
r_s	radius of curvature of surface	m
R	reflection coefficient	–
R_g	gas constant	8.314 J/mol K or 1.987 cal/mol K
R_i	reflection coefficient of mirror i	–
S_y	yield strength	Pa (N/m^2)
SR	solidification rate	m/s
t	time	s
T	temperature	K
T_a	ambient temperature	K

T_l	liquidus temperature	K
T_s	solidus temperature	K
ΔT_e	degree of supercooling (undercooling)	K
$\Delta T_f = T_l - T_s$	equilibrium freezing range	K
$T_m = T_e$	melting (equilibrium) temperature	K
ΔT_m	temperature change from ambient to melting temperature	K
T_0	initial workpiece temperature	K
T_p	peak temperature	K
T_s	solidus temperature	K
T_v	vaporization (evaporation) temperature	K
ΔT_v	temperature change from melting to vaporization temperature	K
TEM_{mn}	transverse electromagnetic mode, m, n	–
$\mathbf{u}$	velocity vector	m/s
u_i	velocity components (u_x, u_y, u_z)	m/s
u_o	object distance	m
u_s	speed of sound	m/s
u_x	velocity component in the x-direction or heat source velocity	m/s
v_i	image distance	m
V	volume	m^3
V_c	cavity volume	m^3
w	beam radius	m
w_f	focused beam radius	m
w_m	beam radius at cavity mirrors	m
w_0	incident beam radius	m
w_w	beam radius at cavity center or beam waist	m
W_p	pumping rate	/s
x, y, z	coordinate system	–
ξ, y', z'	coordinate system attached to the moving heat source	–
u, v, w	displacements in the x, y, z directions	m
x_R	Rayleigh range	m
α	absorption coefficient	/m
β_C	solutal volume expansion coefficient	/%
β_l	linear thermal coefficient of expansion	/K
β_T	volumetric thermal coefficient of expansion	/K
ϵ_p	electric permittivity	F/m
$\epsilon_{p0} = 8.9 \times 10^{-12}$	permittivity of free space	F/m
ε	strain	–
$\varepsilon_1, \varepsilon_2, \varepsilon_3$	principal strains	–

$\varepsilon_x, \varepsilon_y, \varepsilon_z$	normal strains in the x, y, z directions, respectively	–
ε^e	elastic strain	–
ε^p	plastic strain	–
η	overall efficiency	%
γ	surface tension	J/m^2 or N/m
γ_e	interface surface energy	J/m^2
γ_{ij}	shear strain in a plane normal to i, in direction j	–
κ	thermal diffusivity	m^2/s
λ	wavelength	m
μ	dynamic or absolute viscosity	Pa s
μ_m	magnetic permeability	(H/m) or N/A^2
ν	frequency	Hz
$\Delta\nu$	frequency difference	Hz
ν_o	oscillating frequency	Hz
$\delta\nu$	minimum oscillating linewidth	Hz
ν_p	Poisson's ratio	–
ω	angular frequency	rad/s
$\Delta\omega$	oscillating bandwidth	rad/s
ω_0	central or resonant (transition or oscillation) frequency	rad/s
$\Delta\omega_0$	full width at half maximum	rad/s
ϕ	phase angle (shift)	$^\circ$
ρ	density	kg/m^3
σ	stress	N/m^2
$\sigma_1, \sigma_2, \sigma_3$	principal stresses	Pa (N/m^2)
$\sigma_x, \sigma_y, \sigma_z$	normal stresses in the x, y, z directions, respectively	Pa (N/m^2)
σ_B	Stefan–Botzmann constant $= 5.6704 \times 10^{-8}$	$W/m^2\ K^4$
σ_e	electric conductivity	(S/m) or A/V m
$\Delta\tau_p$	pulse width or pulse duration	s
τ_c	cavity lifetime	s
τ_{ij}	shear stress in a plane normal to i, in direction j	Pa (N/m^2)
τ_p	pulse duration	s
τ_{sp}	lifetime due to spontaneous emission	s
θ	angle	$^\circ$
θ_B	Brewster angle	$^\circ$
subscript g	refers to a gaseous medium	–
subscript l	refers to the molten metal (liquid)	–
subscript ls	refers to saturated property at liquid temperature, T_{ls}	–

subscript s	refers to a solid medium	–
subscript v	refers to vapor	–
subscripts x, y, z	refer to the x, y, z directions, respectively	–
superscripts e and p	refer to the elastic and plastic states, respectively	–

INDEX

Principles of Laser Materials Processing, by Elijah Kannatey-Asibu, Jr.